Instrumentation
For Scientists Series

ELECTRONIC MEASUREMENTS
FOR SCIENTISTS

HOWARD V. MALMSTADT
University of Illinois

CHRISTIE G. ENKE
Michigan State University

STANLEY R. CROUCH
Michigan State University

with

GARY HORLICK
University of Alberta

as coauthor of Module 4

W. A. BENJAMIN, INC.
Menlo Park, California · Reading, Massachusetts
London · Amsterdam · Don Mills, Ontario · Sydney

**THE MALMSTADT-ENKE
INSTRUMENTATION FOR SCIENTISTS SERIES**

The material contained in this book appeared originally in four separate publications, each of which was published in a **Text with Experiments** version and a **Text with Lab Summaries** version:

Electronic Analog Measurements and Transducers, by Howard V. Malmstadt, Christie G. Enke, and Stanley R. Crouch (Module 1)

Control of Electrical Quantities in Instrumentation, by Howard V. Malmstadt, Christie G. Enke, and Stanley R. Crouch (Module 2)

Digital and Analog Data Conversions, by Howard V. Malmstadt, Christie G. Enke, and Stanley R. Crouch (Module 3)

Optimization of Electronic Measurements, by Howard V. Malmstadt, Christie G. Enke, Stanley R. Crouch, and Gary Horlick (Module 4)

ISBN 0-8053-6902-3
ABCDEFGHIJ-HA-7987654

Semiconductor diode

Zener or breakdown diode

npn transistor

pnp transistor

n-channel JFET

p-channel JFET

Unijunction transistor

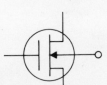

n-channel MOSFET

p-channel MOSFET

SCR and GTO

SCS

Four-layer diode

PUT

Diac

Silicon bilateral switch (SBS)

Triac

Light-emitting diode

Phototransistor

Light-activated SCR

LASCS

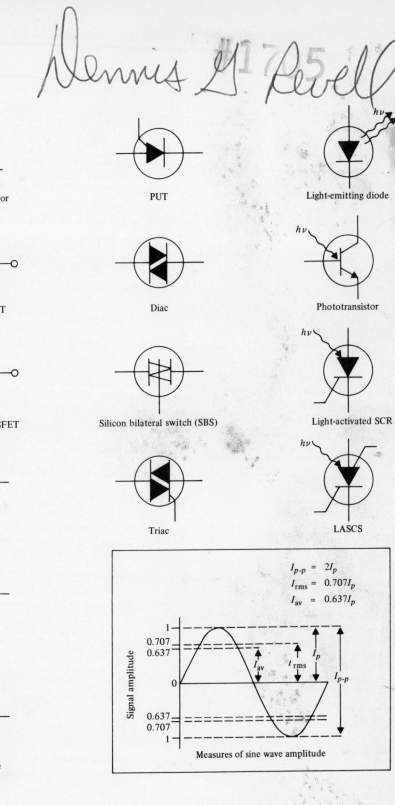

$$I_{p\text{-}p} = 2I_p$$
$$I_{\text{rms}} = 0.707I_p$$
$$I_{\text{av}} = 0.637I_p$$

Measures of sine wave amplitude

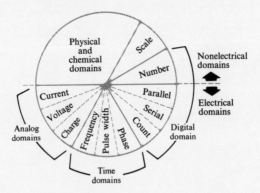

Data domain map

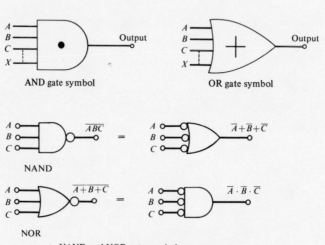

AND gate symbol

OR gate symbol

NAND

NOR

NAND and NOR gate symbols

Exclusive-OR gate symbol

Instrumentation
For Scientists Series

ELECTRONIC MEASUREMENTS
FOR SCIENTISTS

Contents

MODULE 3   DIGITAL AND ANALOG DATA CONVERSIONS

MODULE 4 OPTIMIZATION OF ELECTRONIC MEASUREMENTS

Preface

The standard laboratory tools in the modern scientific world include increasing numbers of complex measurement and control instruments. However, scientists frequently have difficulty keeping up with the rapid development of new devices and instruments. Scientists learn how to use the instruments which are commonly applied in their particular field, but their general understanding of instrumentation is often insufficient for them to adapt easily to new measurement concepts and unfamiliar equipment. This is partially caused by the fact that important electrical principles, concepts, and circuits, which are basic to the operation and application of modern instrumentation, are classically studied as topics isolated from their applications in medical science, chemistry, physics, and other areas. Consequently, most students have difficulty relating their knowledge of physical and electrical principles to the reliable setup, modification, and application of electronic measurement systems. The time delay between the study of the principles and the scientific measurement application, and the lack of text materials and experiments that coordinate the principles and applications, make it difficult for the student to put them both in perspective. This is not only unfortunate but must be considered unacceptable, especially for the non-electronics majors and scientists who have only a short time to learn "useful electronics."

After many years of teaching applied electronics to scientists and engineers and testing ideas of presentation, the authors are convinced that the basic electrical principles and experiments can be integrated in a meaningful way with major measurements and control concepts and instruments that are used daily in industrial and academic laboratories. This integration of materials makes the study of the basic principles both interesting and exciting because they can be put to work immediately on important projects. We have tested this approach with many classes and found it highly rewarding in terms of the conceptual understanding and direct applicability of the material. This book is an outgrowth of our experiences, and we have found it very beneficial to divide the material into modules. Four text modules are included in this book. In each text module the student starts with those topics that are required as a basis for intelligent experimentation and then proceeds to build up an increasingly sophisticated background—always integrating concepts, principles, and laboratory "know-how." The order of presentation and specific material for each of the text modules have been repeatedly refined by the help of many discussions with undergraduate and graduate students and practicing scientists. The text and experiment summaries have been evaluated on the basis of how well the students could understand, utilize, and extrapolate the information in practical situations at the conclusion of each topic.

For Module 1 it has been assumed that the reader has little or no background in electronic instrumentation. Therefore, the concepts and principles concerning electrical quantities are developed and at the same time related to important devices or instrumentation systems. With this procedure it has been possible to start at essentially "ground zero" and move quickly but systematically to important applications. Module 1 ends with the development and application of the modern automatic null comparison measurement systems that are found in the most elegant scientific instruments. The presentation of analog measurements and transducers first in Module 1 provides an overall perspective in which the study of circuit elements, semiconductor devices, and control systems in Module 2 becomes both meaningful and exciting. We have found that this approach benefits not only the undergraduate students but also the chemists, physicists, engineers, and medical scientists who want to gain new insights into their measurements and new ideas about the instruments they need or use in their work. The integration of the fundamentals and the applications leads to unifying concepts that often result in new ideas on how to solve measurement and control problems.

Module 2 begins with a study of the basic electrical properties of resistors, capacitors, and inductors, which provide control of electrical quantities, and then moves systematically to the semiconductor control elements that have revolutionized modern scientific instrumentation. The concepts and principles of electrical switching are developed in relation to specific solid state switching devices. In this manner it is possible to present first the basic principles and characteristics of semiconductor devices with their relatively simple and intuitive ON-OFF control applications. Then after the basic device characteristics have been introduced, the more advanced concepts of continuous control are presented through a study of basic amplifiers, including voltage, current, operational, and instrumentation amplifiers. This module ends with the development of the elegant automatic feedback control concepts and systems which are widespread in modern instrumentation systems.

Throughout Module 2 the important applications of the devices and circuits are integrated with concepts and principles. The intimate interplay of fundamentals and applications not only unifies the study of electrical control, but also stimulates the student in his study of the principles because he can see the immediate "payoff" in important applications. By studying both Modules 1 and 2 the student can gain a thorough background in dc analog electronics, including dc circuits, transducers, solid state devices, measurements, operational amplifier circuits, semiconductor principles, and servo control concepts and systems. The material in Modules 1 and 2 is presented so that it can be used as a first course in engineering electronics, electronics for physicists, instrumental methods, laboratory practice, measurement instrumentation, or as the first part of a more advanced course in Electronic Measurements for Scientists.

It has been our goal in Module 3, as in the other Modules, to present the new electronics within the context of its present or potential application in scientific measurement and control instrumentation. The presentation is based on the simple but elegant concept of *data domains,* the ways in which data are encoded at particular instants or points in the measurement process. Module 3 begins with an introduction to the encoding of measurement data in the electrical analog, time, and digital domains. The sequences of data domain conversions that are used to convert the information from the quantity being measured to the numerical result are illustrated for a variety of modern measurement systems. With this data domain view of the measurement process as a foundation, a detailed study of the circuits and techniques employed with analog, time, and digital signals is presented. This study includes signal processing within each domain (such as multipliers, phase-lock loops, and serial-to-parallel converters) and conversions between domains (such as digital frequency measurement, and analog-to-digital and digital-to-analog converters). The presentation of these domain conversion concepts and circuits includes the material for a modern introductory course on operational amplifier applications and digital electronics.

As scientists strive to make more difficult measurements or to improve the accuracy and precision of routine measurements, the question of how to optimize and upgrade the electronic measurement system is being asked frequently. In Module 4 of this book, the basic principles and techniques which are used for optimizing electronic measurements are presented and illustrated wtih many scientific measurement examples. Module 4 begins with a consideration of the characteristics of electrical signals and electrical noise. In this discussion it is shown how signal information can differ from noise in its amplitude spectrum, in its phase spectrum, and in other characteristics. After this introductory discussion, the principles and methods which are used to control the frequency response of measurement systems are presented to develop this important step in the optimization process. Then modulation and demodulation are introduced and illustrated with many examples. The various modulation-demodulation schemes are treated with an emphasis on their utility in scientific measurements. The implementation of many sophisticated electronic measurement techniques requires sampling an electronic signal. Hence a discussion of the considerations involved in the choice of sampling parameters and the errors which can result from an improper choice is included. This module ends with the development of the elegant and sophisticated signal-to-noise ratio enhancement techniques which are becoming increasingly widespread in scientific measurements. The principles, implementation, and instrumentation of lock-in amplification, boxcar integration, multichannel averaging, and correlation techniques are presented along with a brief discussion of related computer-based techniques.

The combination of Modules 1 through 4 in this book provides a firm foundation in analog and digital electronics and instrumentation concepts and their implementation. With this background it is possible to understand the design and application of digital and analog instruments, computer interfacing, and automated systems.

Through the division of the book materials into modules, we have attempted to provide the individual with the means to select the level and the depth of study most appropriate for his needs or curriculum. In addition to this hardbound text combining Modules 1 through 4, the individual modules are available as paperbacks in two forms: **Text with Experiments** modules and **Text with Lab Summaries** modules. Each **Text with Experiments** module contains detailed circuits and instructions for the experiments in workbook form and assumes implementation with specific types of instruments. The text material in each **Text with Lab Summaries** module is identical to that in the corresponding **Text with Experiments** module, but the experiments are presented as summaries that state the experimental objectives and general approach; all detailed instructions and circuits for the experiments are omitted. The different methods of packaging the text materials are intended to fulfill the requirements of different groups as closely as possible. Some want to take advantage of the many thousands of hours that have gone into writing and testing specific experiments that are uniquely designed to illustrate the important measurement principles and provide firsthand experience with state-of-the-art instrumentation. Others want to use the text materials contained in one or more modules as supplementary material for many types of courses in physics, engineering, chemistry, and the life sciences and have no desire to perform the specific experiments. Others want a basic text for a lecture course or a series of courses that includes all of the principles and concepts covered as in this text *Electronic Measurements for Scientists,* and either do not have an interest in laboratory experiments or have a separate lab course that uses experiments and systems quite different from those in the **Text with Experiments** modules.

We are grateful for the assistance of many others in the preparation of this book. The art work was done by E. C. Smith Associates under the direction of Stan Parnell. Peri Warstler, Parna Westfall, Louise Ziola, and Donna Abernathy all participated in typing the manuscript in its several stages of development. Scott Goode, Edward Codding, Brian Hahn, Al Scheeline, and Terry Woodruff were very helpful with proofreading, and along with Phil Notz, Kelly O'Keefe, Keith Caserta, and Dave Rothman assisted with the problems and problem solutions. We also greatly appreciate the patience and support of our families and our students who have borne with us through this extensive project.

For anyone who picks up this book there is probably no need to indicate the importance of measurement instruments, but we find it reassuring to recall the words of Lord Kelvin, "I often say that when you can measure what you are speaking about, and express it in numbers, you know something about it; but when you cannot measure it, when you cannot express it in numbers, your knowledge is of a meagre and unsatisfactory kind; it may be the beginning of knowledge, but you have scarcely, in your thoughts, advanced to the stage of *Science,* whatever the matter may be."

January 1974

H. V. M.
C. G. E.
S. R. C.
G. H.

ELECTRONIC MEASUREMENTS
FOR SCIENTISTS

ELECTRONIC ANALOG MEASUREMENTS AND TRANSDUCERS

Module I

Introduction

The specific scientific information that each scientist or engineer seeks to obtain, or process that he needs to control, depends of course on the area of his specialty. But regardless of the field of study, the final data (numbers) obtained or processes controlled generally require several transformations of the way in which the measurement or control data are represented or encoded. A suitable sequence of data encoding transformations can provide the desired measurement or control function. Because of the elegance, economy, speed, and convenience of modern electronic devices, most measurement and control systems utilize electronic instruments at some stage in obtaining, manipulating, processing, storing, displaying and/or interpreting the sought-for data.

To the uninitiated, there seems to be a never-ending variety of electronic devices which are used in the process of transforming physical or chemical information into numbers. In this text, all measurements are shown to be a simple combination of comparison and transformation of data. In electronic instruments, the data are encoded as electrical signal quantities. Since the number of ways in which data can be encoded electrically is limited, the data encoding techniques used in an instrument are an excellent means of explaining the electronic measurement process and classifying the devices used in electronic instruments.

In an electronic analog measurement device or system, the data are encoded as the magnitude of one of four basic electrical quantities of charge, current, voltage, or power. In this module, the application of the quantities of charge, current, and voltage in electronic analog measurements is described. The development of the inherent nature of each quantity and its relation to the others is followed by a description of the principal devices and techniques which are used to transform information to the analog quantity and then from the electrical quantity to the numerical measurement result. In this "measurement concept" oriented presentation, the basic applications, functions, operations, limitations, and relative merits of many common electronic analog instruments and transducers are seen in a new perspective. The instruments presented in this module include laboratory meters, oscilloscopes, recording potentiometers, operational amplifiers, and digital voltmeters. Also included are transducers which are used in the measurement of temperature, light, radioactivity, magnetic field, and chemical concentration.

Electronics-Aided Measurement

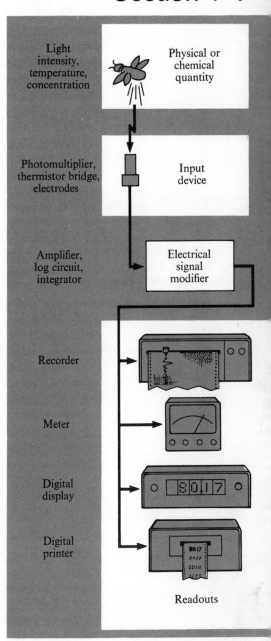

Fig. 1–1 Electronics-aided measurement.

In an electronics-aided measurement, the quantity to be measured is converted into an electrical signal and then amplified or otherwise modified to operate a device which visually displays the numerical value of the measured quantity. This process is illustrated for a typical case in the pictorial block diagram of Fig. 1–1. An input device such as a photodetector, thermistor bridge, glass pH electrode, or strain gage circuit is used to convert the quantity to be measured into an **electrical signal.**[1] Some characteristic of this signal is related in a known way to the quantity to be measured. The measured quantity is now encoded as an electrical signal. The electrical signal from the input device is then modified by an electronic circuit to make it suitable to operate a readout device. The electronic circuit is frequently an amplifier with the appropriate adjustable parameters (zero, standardization, position, etc.) and sometimes with automatic compensation for nonlinearities, temperature variation, etc. of the input device. It is a truism that the end result of any measurement is a number. The output number is obtained from a readout device such as a meter or chart recorder where the position of a marker against a numbered scale is observed, or from lights in the shape of numbers which are turned on. In the example of Fig. 1–1, the measured quantity was encoded in at least three different ways: first as the physical or chemical quantity or property, second as some characteristic of an electrical signal, and finally as a number. Each different way in which data can be encoded is called a **data domain.**

1–1.1 DATA DOMAINS

The many different electronic devices, circuits, and instruments used in science are rather overwhelming, and at first it seems impossible to understand even a small fraction of them. However, all electronic measurement

---

Note 1. Electrical Signal.

*An **electrical signal** is an electrical quantity or a variation in an electrical quantity that represents information.*

or control systems must use some means of representing or encoding data electrically. That is to say, the data must exist in the system in one or more electrical data domains. Since the number of electrical data domains is relatively small, the electronic part of the system can be simply described and analyzed in terms of the data domains employed and the methods of transforming or converting the data from one domain to another. A consideration of data domains is also very useful in modifying and designing electronic measurement systems and devices and in assessing and minimizing the sources of measurement errors. In addition, a much better understanding of the instrumental data handling process is gained as a result of the study and application of the data domain concepts. The first three concepts of data domain analysis are given in Table 1–1.

Table 1–1 Data domain concepts

1. Measurement data are represented in an instrument at any instant by a physical quantity, a chemical quantity, or some characteristic of an electrical signal. Each different characteristic or property used to represent data is called a **data domain.**

2. As the data proceed through the instrument, a change in the characteristic or property used to represent the measured data is called a **data domain conversion.**

3. Each measurement system can be described as a data domain converter or as a sequence of data domain converters, each of which can be analyzed separately.

Using the data domain concepts, the basic electronic measurement of Fig. 1–1 is described as follows: The measurement data exist first as the physical or chemical quantity to be measured. The input device encodes the original data into an electrical signal and thus into one of the **electrical domains.** The measurement data remain in an electrical domain (though not necessarily the same one) through the electrical signal modifier. Then the output device converts information contained in the electrical signal into some readable form such as the relative positions of a marker on a scale, i.e., a nonelectrical domain. Thus the entire measurement can be described in terms of conversions between data domains. The characteristics of each interdomain converter and each signal modifier affect the quality of the measurement. To take advantage of special input devices, particular readouts, and available signal processing techniques, an instrument may involve many data domain conversions and signal modifiers. The data domain concepts allow each step to be blocked out and analyzed separately.

1-1.2 MAPPING DOMAIN CONVERSIONS

It is increasingly common for modern laboratory instruments to use three or more domain conversions to perform the desired measurement. Knowing the sequence and characteristics for the data domain conversions in a particular instrument can help in understanding its operation, applications, limitations, and adaptability as part of a larger measurement system. When analyzing an instrument it is helpful to use a "map" of the data domains as shown in Fig. 1–2.

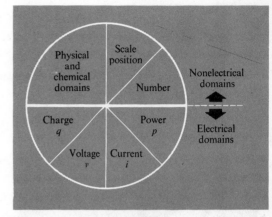

Fig. 1–2 Data domain map.

In this map some of the ways in which information can be represented are shown as separate areas. The nonelectrical domains, such as the physical, chemical, scale-position, and number domains are shown in the top half, while the electrical quantities are shown as domains in the bottom half. These four electrical domains in which data can exist are electrical charge q, electrical current i, which is the rate of flow of charge, electrical voltage v, which is the electrical pressure which causes the motion of charge, and electrical power p, which is equal to the rate of performing work and is the product of current and voltage. The application of these basic electrical quantities in electronic instruments and devices will be elaborated on throughout this module. The path of the data can be traced out on the map as it is followed through the instrument. In a complete measurement system, the data begin in a physical or chemical domain as the quantity being measured. The measurement path may then go through several other domains as the data are converted from one form to another. The path always ends in the number domain, since the end result of a measurement is a number. When the data path includes any of the electrical domains (those below the horizontal median in Fig. 1–2) the measurement is at least partly electronic. In an electronic measurement, then, the data path crosses the median at least twice: once when the information about the physical or chemical quantity is converted to an electrical domain, and once when the electrical quantity is converted to a scale position or number. From the point of view of the electrical domains, devices which convert information in a nonelectrical domain are called **input transducers,** and those which convert data in an electrical domain to a nonelectrical domain are called **output transducers.** The term "transducer" is used to describe a great variety of devices. However, there are so many electrical and non-electrical domains that it is much more descriptive to include the specific data domains the transducer converts from and to when a specific transducer is being discussed, e.g., a "temperature-to-voltage" transducer or converter. The domain mapping process is illustrated here for a few instruments, and the concepts are developed in subsequent discussions.

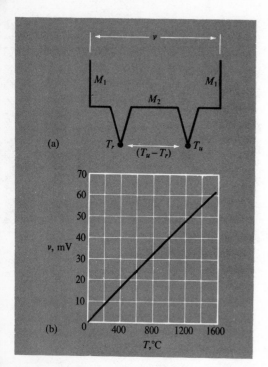

Fig. 1–3 (a) Thermocouple schematic. (b) Graph of transfer function.

Thermocouple Thermometer

The thermocouple is a temperature-to-voltage (physical-to-electrical domain) transducer that is applicable for a wide range of temperature measurements. When two dissimilar metals are joined together as shown in Fig. 1–3a, a voltage v is developed between the open ends which is a function of the temperature difference between the junctions. For some metal pairs this **thermoelectric effect** provides a reproducible relationship between the voltage v and the temperature difference between the two junctions $(T_u - T_r)$. This relationship is called the **transfer function.** The transfer function for a chromel/alumel thermocouple is given graphically in Fig. 1–3b and can be expressed by the equation

$$v = \mathcal{A}T_u + \tfrac{1}{2}\mathcal{B}T_u^2 + \tfrac{1}{3}\mathcal{C}T_u^3$$

when T_r is $0°\text{C}$. Since the coefficients $\mathcal{B}$ and $\mathcal{C}$ are quite small in most cases, the transfer function for the thermocouple can be approximated by the simple linear equation $v = \mathcal{A}(T_u - T_r)$. The coefficient $\mathcal{A}$ is approximately 4×10^{-6} V/°C for the chromel/alumel thermocouple.

The voltage output of the thermocouple is amplified and converted to a proportional current of sufficient magnitude to operate a meter, as illustrated in the block diagram Fig. 1–4. The current amplitude causes a proportional deflection of the meter pointer. A number is then visually observed by "reading" the position of the pointer against the calibrated scale. The signal path on the data domain map for the temperature measurement using the thermocouple is shown in Fig. 1–5. Note that there are

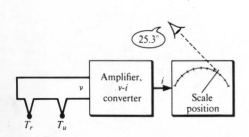

Fig. 1–4 Thermocouple temperature measurement.

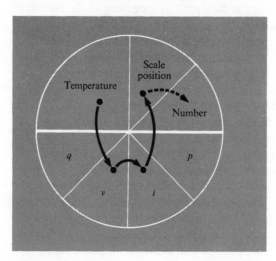

Fig. 1–5 Domain map for temperature measurement.

three instrumental interdomain conversions and one "human" interdomain conversion.

pH Meter

The measurement of the hydrogen ion activity (pH) in a solution can be accomplished with an information signal path similar to that illustrated in Fig. 1–5. However, the transducer is now an electrode pair where one electrode responds to hydrogen ion activity and the other remains constant so as to develop a potential difference proportional to pH. That is, v is a function of pH ($v = f(\text{pH})$), where the relationship is again called the transfer function.

If it is desired to record the pH as a function of time, a strip chart recorder can be used as illustrated in Fig. 1–6. The deflection of the pen is now proportional to the pH, and the chart paper is moved under the pen at a constant rate (e.g., 2 in./min). The signal path is still the same as in Fig. 1–5 but now a motor or meter coil can cause a pen to move instead of a pointer, and the pH value at any time can be read from the chart at any later time.

Whereas the temperature and pH examples presented here follow the same signal path, there are many other signal paths utilized in other measurement instruments. The overall measurement accuracy of any instrument is dependent on the reliability of the information transfer from one domain to another. The final significance of the numbers read at the output depends on the transfer functions of all the interdomain converters.

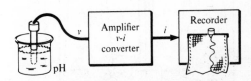

Fig. 1–6 Recording pH measurement.

1–1.3 MEASUREMENT ERRORS

To understand the sources of error in electronic measurements, it is helpful to review briefly the basic measurement process. Measurement can be defined as *the determination of a particular characteristic of a sample in terms of a number of standard units for that characteristic.* The comparison of the quantity or property to be measured with standard units of that quantity or property is implicit in this definition. The comparison concept in measurement is illustrated by Fig. 1–7. The quantity to be measured Q_u is compared with a reference standard quantity Q_r. The difference ($Q_u - Q_r$) is converted to another form (domain) such as scale position. The quantity measured Q_o is then the sum of the standard units in the reference quantity Q_r and the difference detector output ($Q_u - Q_r$) calibrated in the same standard units.

All measurement devices involve both a difference detector and a reference standard, although there is wide variation in the degree to which one or the other is relied upon in the measurement. As an example, let

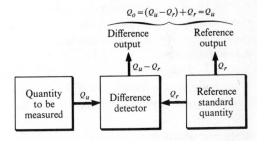

Fig. 1–7 Elements of the basic measurement system.

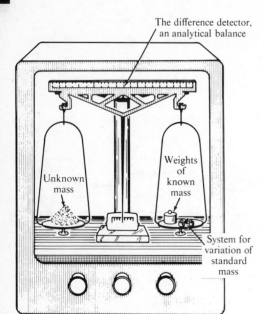

The difference detector, an analytical balance

Unknown mass

Weights of known mass

System for variation of standard mass

Fig. 1–8 Double pan balance.

Note 2.   Sensitivity.
*The **sensitivity** of a measurement is the smallest change in the quantity to be measured which produces a detectable change in the output. In this case it is synonymous with the term **minimum detectability.***
Note 3.   Precision.
*The **precision** of a measurement is related to its degree of repeatability and resolution.*
Note 4.   Accuracy.
*The **accuracy** of a measurement is the degree of agreement between the true and measured values.*
Note 5.   Buoyancy Assumption.
It is assumed that the difference in the buoyancy effects of air on the standard and unknown masses is negligible or corrected for.

us compare three mass measuring devices. With a double pan balance, the unknown mass is compared with standard weights, whole units and fractional, until the difference detector points to zero (null), as illustrated in Fig. 1–8. This kind of measurement is called a **null comparison** because the reference standard is adjusted until the difference detector output is null. Used in this way the difference detector is often called a **null detector.** The analytical balance is the null detector. This assumes, of course, that the reference standard can be adjusted in sufficiently small increments to establish a true null indication. In this case the measurement **sensitivity**[2] depends on the design and construction of the null detector: the greater the sensitivity of the balance to any mass difference, the better the **precision**[3] of the measurement. For example, if a difference in mass of 0.0001 g can just be detected by a given balance, then the reproducibility (precision) of determining a mass of 0.1000 g will be 1 part/1000 or 0.1%. However, the **accuracy**[4,5] of determining the unknown mass not only depends on the sensitivity of the balance but also on the accuracy of the standard masses. It is possible for the standard weights to have a large error because of rough use or contamination. If the standards are inaccurate by 1%, the unknown mass will be measured with an error of 1% because of an erroneous standard, even though the null detector is operating perfectly within its design capability to show differences within 0.1% for the 0.1 g mass.

In the double pan balance, the accuracy and resolution of the standard weights determine the accuracy of the measurement if the difference detector is sufficiently sensitive. There is *no* accuracy requirement for the off-null calibration of the difference detector because *the standard is adjusted to be equal in value to the unknown.* It is, however, essential that the null detector indicate the true null point within the required accuracy and precision. For a true null measurement, then, the null detector need only indicate when a difference exists between the unknown and standard quantities. The location of the null comparison point is greatly facilitated if the null detector indicates the *sign* of the difference when off null. However, the magnitude of the difference never contributes to the final measurement in a null comparison measurement. Therefore the off-null indicator of a null detector need not give *any* indication of the magnitude of the difference. Such a device is called a **comparator** since it indicates only whether the unknown quantity is greater or less than the reference standard. The value of the standard for which the smallest change in value causes the comparator output to indicate a change in the sign of the difference is the null point. The comparator will later be shown to be an important and often used type of null detector.

The other extreme of making a mass measurement is with a spring-loaded scale such as the bathroom scale. In this case, the manufacturer uses standard reference weights to calibrate the off-null scale markings for the prototype instrument. The accuracy of each particular scale depends on how similar the components and construction details of the production scales are to the prototype scale. The reference standard is remote, and the instrument is subject to nonlinearities, calibration drifts, and consequently larger errors. It is not surprising, therefore, that precision scientific measurements often use the null comparison method where the unknown quantity, property, or parameter is directly compared against a standard reference of the same quantity, property, or parameter.

In making certain comparison measurements it is often inconvenient, too slow, or unnecessary to establish a perfect null point. An example of this is the "single pan" balance where a close approximation to the null point is established against internal standard weights and then the off-null deflection of an optical beam is calibrated in fractional mass units. If the off-balance amount is only a small fraction of the total amount, then relatively large errors in the off-balance detector are insignificant. For instance, if the reference standard is set to a precisely known value within 1% of the null point and the difference detector has an error of 3% of the difference, the measurement error will be no more than 0.03%. This kind of analysis isolates the sources of errors in measurement devices and helps in choosing among available devices for a particular application. Similar comparisons and analyses will be made for electronic measurement devices as they are described in this book. After a variety of measurement systems have been considered in this way, the statements in Table 1–2 concerning measurement devices appear to have general validity.

Table 1–2   General characteristics of measurement devices

---

1. All measurement devices employ both a difference detector and a reference standard quantity.

2. Either the difference detector or the reference standard can affect the accuracy of the measurement.

3. The reference standard quantity has the same property or characteristic as that which is being measured.

Domain Converters

The measurement process just reviewed is a kind of data domain conversion where the quantity to be measured is converted into a *number* of standard units of that quantity. A data domain conversion does not have to result in a number but it can. In general, then, *a data domain conversion is the transformation of a number of units of some physical, chemical, or electrical characteristic into a related number of units of a different characteristic.* For example, the conversion of units of temperature to voltage is a data domain conversion performed by a thermocouple. A test of whether or not a data domain conversion takes place is whether or not there is a change in the units of the characteristic used to encode the data. Devices for converting data from one domain to another encode units of one characteristic in terms of another. Therefore, *all interdomain converters have the general characteristics of the measurement devices listed in Table 1–2.*

The thermocouple can be used as an example to illustrate the applicability of the statements in Table 1–2 to interdomain converters.

Statement 1:   The thermocouple pair is the difference detector; the reference standard is typically the ice point (0°C) or the steam point (100°C). Other known temperatures (such as ambient room temperature) could also be used as the reference.

Statement 2:   The conversion error (difference between the determined and actual temperatures) depends upon the accuracy of the thermocouple transfer function from temperature to voltage, and the difference between the standard and the unknown temperatures. The greater the temperature difference between the standard and unknown, the more the conversion accuracy depends upon the accuracy with which the characteristics of the thermocouple (difference detector) are known.

Statement 3:   The reference standard is temperature, the units of which are being converted to electrical potential difference (voltage). Note that another data domain conversion would be required to convert the thermocouple voltage to a number. This combination of data domain converters would then comprise a complete temperature measurement system.

If one becomes familiar with the identification of interdomain converters in measurement systems, the basis of the measurement and the sources of error are more easily analyzed. The measurement accuracy depends directly upon the standard source and difference detector characteristics in each domain converter.

Current Measurement and Transducers

Whenever there is a net motion of electrical charge in a material, an **electrical current** exists in that material. The magnitude of the current is equal to the quantity of charge passing a given point in the material per unit time. Electrical current is used to communicate information in many ways. Thus an understanding of electrical current is basic to the study of electronic measurements in the scientific laboratory. Many transducers produce a current which is related to a physical or chemical phenomenon. In addition, the many materials used in modern electronic devices are characterized primarily by the mechanism and degree of their conduction of electrical current. In this section the nature of electrical current is described. The measurement of current, the electrical effects of current in various combinations of conductors, and some of the current sources used in scientific measurements are introduced.

1–2.1 MOTION OF CHARGE

The atoms which make up matter are composed of negatively charged electrons, equally but positively charged protons, and electrically neutral neutrons. The number of electrons and protons in a neutral atom are essentially equal so that there is no net charge.

When a neutral atom loses an electron, it becomes a positively charged **ion.** Its positive electrical charge is equal to the negative charge Q_e on the electron it lost, namely 1.603×10^{-19} **coulombs** (C).[6] If the electron becomes associated with a neutral atom, the atom becomes a negative ion, again with a charge Q_e or 1.603×10^{-19} C. Energy is required to separate positive and negative charges because they are attracted to each other. When two charges are separated, a potential energy difference exists be-

Note 6. Coulomb.
The coulomb (C) is the amount of charge required to convert 0.00111800 g of silver ions (Ag^+) to silver metal.

tween them. **Electrical potential difference,** or **electromotive force (emf),** is the potential energy difference in joules per coulomb of charge separation and has the units **volts** (V). In the space between separated unlike charges, the emf changes continuously from the point where one charge is located to the point where the second charge is located. A region in which the emf changes from point to point is said to contain an **electric field.** The **electric field strength** is the change in emf per unit change in position in volts per centimeter. If a third charged particle is placed between the first two it will be attracted toward the charge of opposite sign and repelled by the charge of similar sign. This is due to the electric field which exists between the separated charges. As the particle approaches the charge of opposite sign, its potential energy decreases. If the charged particle is free to move in that direction, it will. Thus a separation of charge requires energy and creates an electrical potential difference which establishes an electric field, and movable charged particles in an electric field will move in the direction of the opposite charge. In virtually every kind of matter, movable electrons or ions exist or can be produced.

The force by which neutral atoms combine to form molecules or crystalline solids is primarily due to the interaction (sharing or transfer) of the outermost **(valence)** electrons of each atom. In solid (or liquid) metals, the valence electrons of each atom are shared by so many neighboring atoms that a substantial fraction of the valence electrons are essentially free to migrate from atom to atom in the material. In some other materials, some of the electrons require little energy to leave their valence orbit. The result is two charged particles: an electron and a positive ion. Some crystalline solids are formed entirely of positive and negative ions, held together partly by electrostatic charge. Defects in the crystal structure allow the motion of ions in solids. The application of an electric field to a material results in the net motion of the mobile charged particles in that material.

Conductivity

Consider a 1 cm section of a cylinder of material with a cross section of a square centimeters as shown in Fig. 1–9. A uniform electric field strength $\mathcal{E}$ of 1 V/cm is applied between opposite faces of the cylinder. The separation of charge which created the electric field used one joule of energy for each coulomb of charge required. The number of coulombs required to establish the given field strength depends upon the geometry of the charged materials and upon the nature of the intervening material. All negatively charged mobile particles will be accelerated in the upward direction but in liquid or solid materials will quickly reach a limiting average velocity due to collisions with other particles. Similarly, all posi-

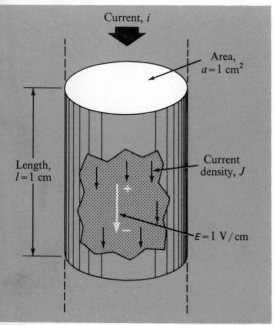

Fig. 1–9 Cylindrical section of conducting material.

tively charged mobile particles will move downward. Both positive and negative mobile particles contribute to the flow of charge through the cylinder and thus are termed **charge carriers.**

The rate of charge motion per unit area, J, is called the **current density** and has the units C/sec-cm². The current density J_1 due to species-1 charge carriers crossing either face of the cylinder is

$$J_1 = n_1 Q_1 u_1, \tag{1-1}$$

where n_1 is the number of species-1 charge carriers per cm³, Q_1 is the charge on each species-1 carrier in coulombs, and u_1 is the average limiting velocity of species-1 carriers in cm/sec. The average velocity of a charge carrier is proportional to its charge, the electric field strength, and its ease of motion m through the material. Since the ease of motion is generally not dependent upon the electric field strength $\mathcal{E}$,

$$u_1 = Q_1 \mathcal{E} m_1. \tag{1-2}$$

Combining Eqs. (1-1) and (1-2), we have

$$J_1 = \mathcal{E} Q_1{}^2 n_1 m_1. \tag{1-3}$$

Whatever the sign of Q_1 in Eq. (1-3) (+ or − charge carrier), $Q_1{}^2$ is positive. Thus the sign of J_1 is always the same as the sign of $\mathcal{E}$. This means that the direction of the current is the same as the direction of the electric field regardless of the sign of the charge carrier. Note that positive charges move in the direction of the electric field while negative charges move in the opposite direction. Both contribute to the *current* which is in the direction of the field.

For conduction in solids, where the charge on a charge carrier is almost always the single electron charge, Q_e, it is customary to define a term **mobility** μ_1, which is $m_1 Q_e$ and has the units cm²/V-sec. Thus Eq. (1-3) becomes

$$J_1 = \mathcal{E} Q_e n_1 \mu_1. \tag{1-4}$$

All the charge carriers crossing each square centimeter of either end of the cylindrical section each second contribute to the total charge flux density J in coulombs/cm²-sec. Thus the total J is the sum of the J's for the p different charge carrier species in the material:

$$J = \mathcal{E} Q_e (n_1 \mu_1 + n_2 \mu_2 + \cdots + n_p \mu_p). \tag{1-5}$$

The unit C/sec is given the name ampere, A, and is the standard unit for the electric current i or I.[7] Thus J has the units A/cm². Since each material will differ in the concentration and mobility of charge carriers, the summation in Eq. (1-5) is a constant which is characteristic of the material

Note 7. Symbols for Variables.
It is general practice to use a lower case letter, for example, i,v,q, as the symbol for the value of a quantity when the variation in that quantity is of interest. Capital letters, for example, I,V,Q, are used as symbols for quantities which are considered to be constant in value. For a table of quantities, symbols, and units see the rear endpapers of this text.

under given conditions of temperature, humidity, etc. This constant is the **conductivity** σ. Therefore

$$\sigma = Q_e(n_1\mu_1 + n_2\mu_2 + \cdots + n_p\mu_p) \qquad (1\text{–}6)$$

and

$$J = \mathcal{E}\sigma, \qquad (1\text{–}7)$$

where σ has the units A/V-cm.

Thus we see that the electric current density for most materials is proportional to the electric field strength and to the conductivity of that material. It is also common to write Eq. (1–7) as

$$J = \mathcal{E}/\rho \qquad (1\text{–}8)$$

where ρ is the reciprocal of σ and is called the **resistivity** of the material.

Ohm's Law

When an electric field of $\mathcal{E}$ volts/cm is applied to a piece of material of nonunit dimensions, the electric current I in the direction of the field which results is

$$I = Ja = \mathcal{E}\sigma a, \qquad (1\text{–}9)$$

where a is the cross-section area in cm^2 taken at right angles to the electric field. The total potential difference V applied to the material is equal to $\mathcal{E}l$, where l is the length of the material in centimeters in the direction of the electric field. Thus

$$I = V\sigma a/l. \qquad (1\text{–}10)$$

The quantity $\sigma a/l$ is called the **conductance.** The conductance G and its reciprocal, the **resistance** R, relate the current through a conductor to the potential difference across it according to Ohm's Law, which is written

$$I = VG = V/R \qquad \text{or} \qquad i = v/R. \qquad (1\text{–}11)$$

R has the units V/A called **ohms** (Ω). The conductance G has the units A/V, which is given the name mho (Ω^{-1}). A conducting device which is used specifically for its resistance and which obeys Ohm's Law in that application is called a **resistor.**

According to Ohm's Law, when a resistor of resistance R ohms is connected between two points of potential difference V volts, charge will flow through the resistor at a rate of V/R amperes or C/sec (1 A of current is $1/Q_e = 6.23 \times 10^{18}$ electron charges/sec). Charge will continue to be transported between the two points at the rate V/R amperes as long as there is a finite potential difference between the points. Thus there can never be an uneven charge distribution which creates an electric field in a conductor without an immediate flow of charge (current) acting to reduce the electric field to zero. However, if a source of energy is present which can maintain a potential difference between the two points (such as at the terminals of a battery), the electrical current in the conductor will be continuous.

A Complete Circuit

A diagram of a conducting path connected between the terminals of a battery is shown in Fig. 1–10. The conducting path includes the switch contacts, a light bulb, and the connecting wires. The connecting wires offer very little resistance to the flow of charge. The filament wire in the light bulb is a less perfect conductor. When the switch is closed, completing the conducting path between the battery terminals, there will be a steady current through the conductor. This is indicated by the current arrow. In accordance with the prevalent convention, the arrow is drawn to indicate the direction of the flow of positive charges. It might be more "graphic" in some cases to indicate the direction of negative charge or electron flow, but the positive-current convention adopted in this book is that used in all college-level texts and scientific journals.

The current will be the same through every segment of the conducting path at any instant. This must be so because unequal currents could occur only if there were an accumulation of charge in some segments of the conducting path and a corresponding loss of charge in others. As shown above, the potential difference resulting from an uneven charge distribution in a conductor would cause a charge-equalizing current.

If the current in a conducting path is the same through every segment there must be a source of charge at one end of the path and an acceptor of charge at the other end. When this condition is met, a **complete circuit** for the flow of charge through the path is said to exist.

If the battery voltage and the conductor resistance are known, the current in the circuit can be calculated from Ohm's Law. For instance, for the circuit of Fig. 1–10, assume that the battery voltage is 1.5 V ($V = 1.5$ V), the resistance of the light bulb is 2.5 Ω ($R = 2.5$ Ω), and the resistance of the wires and closed switch are 0 Ω. In this case, the current I will be

$$V/R = 1.5 \text{ V}/2.5 \text{ Ω} = 0.60 \text{ A}$$

($I = 0.60$ A).

When the switch contacts are open, charge cannot flow through the switch. The circuit is now incomplete and is often referred to as an **open circuit.** Since the current is the same through every **element**[8] of the circuit,

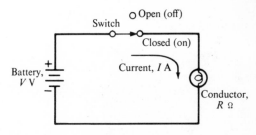

Fig. 1–10   Simple dc circuit.

Note 8.   Circuit Element.
*An **element** is any component or device such as a resistor, light bulb, or battery that is used to make up an electrical circuit.*

the current through every element is zero. Since there is no current, there must be no electric field in the circuit. The connecting wire from the + terminal of the battery to the switch must be at the + terminal potential everywhere. Similarly, the circuit elements connecting the − terminal of the battery to the other switch contact are at the battery's − terminal potential at every point. The full battery voltage therefore appears between the open switch contacts.

1–2.2 CURRENT METERS

There are several methods for measuring the *rate of charge flow* (current) in an electrical circuit. In this section the classical moving coil current meter and the related galvanometer and recording galvanometer are described. The moving-coil meter remains popular as a readout device because of its simplicity, ruggedness, and low cost. Some other methods of current measurement involve the measurement of the proportional voltage drop across a known resistance ($v = iR$ whence $i = v/R$), and they are discussed in Section 1–3.3 on Voltage Measurement.

D'Arsonval Meter

In order to measure current in an electrical circuit, the force exerted by a magnetic field on moving charges is often used to develop a mechanical motion which deflects a pointer in proportion to the current magnitude. The existence of a magnetic field can be demonstrated by sprinkling iron filings on a sheet of paper under which a horseshoe magnet is placed as shown in Fig. 1–11a. The lines of force suggested by the arrangement of the filings are known as magnetic field lines. The basic measure of magnetic field strength is called the **magnetic field** and given the symbol B.

The fundamental property of a magnetic field is that it exerts a force F on a moving charge which is perpendicular to both the direction of the magnetic field and the direction of motion. This fundamental property is responsible for such familiar phenomena as the deflection of an ion beam in a mass spectrometer and the curvature of a charged particle track in a bubble chamber immersed in a magnetic field. Since an electric current is a net flow or motion of charge, a current-carrying wire in a magnetic field also experiences a force at right angles to the magnetic field and the current direction. If the wire is at right angles to the magnetic field as shown in Fig. 1–11b, the upward force on the wire is given by $F = IlB,$ where l is the length of wire.[9] The force on a wire parallel to the magnetic field is zero.

Consider a rectangular loop of current-carrying wire placed in a magnetic field $B,$ so that the plane of the loop is parallel to the magnetic field, as shown in Fig. 1–11c. There will be a net force on the two sides of

Note 9. *Force and Magnetic Field Units.* *Force F is in newtons when I is in amperes, l in meters, and B in weber/(meter)²* *(Wb/m^2). Another common unit for B is* *the gauss, which is $10^{-4}\ Wb/m^2$.*

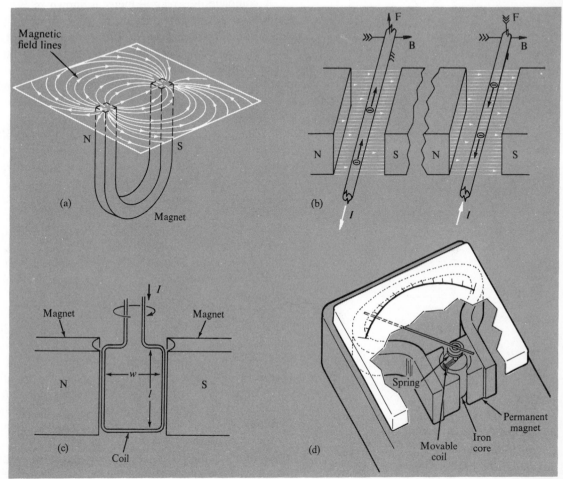

(a)

Magnetic field lines

N   S

Magnet

(b)

F

B

N   S   N   S

I   I

(c)

Magnet   Magnet

N   w   S

l

Coil

I

(d)

Spring

Movable coil

Iron core

Permanent magnet

the loop which are perpendicular to the magnetic field. If the loop is of length l, the force exerted on the conductor at the right is IlB in a clockwise direction as shown. A similar force is exerted on the left-hand conductor in a clockwise direction. The clockwise forces impart a mechanical rotation force or **torque** (force $\times$ moment arm) to the loop. Since the torque is equal to the product of the force times one-half the width w of the loop, the torque on each conductor is $\frac{1}{2}IlwB$, and the total torque is $IlwB$. Because lw is the area, a, of the loop, the total torque can also be expressed as IaB. For a multiturn coil of n turns suspended in a magnetic field the total torque is $nIaB$. The addition of a cylinder of soft iron in the space between the magnet poles increases the magnetic field and improves its uniformity.

Fig. 1–11 Moving coil meter. (a) Magnetic field around a magnet. (b) Force on current-carrying wire in a magnetic field. (c) Force on a current-carrying loop in a magnetic field. (d) Moving coil meter movement.

A complete moving coil meter is shown in Fig. 1–11d. A coil of fine wire wound on a frame is mounted in the air space between the poles of a permanent horseshoe magnet. Hardened-steel pivots attached to the coil frame fit into jeweled bearings so that the coil rotates with a minimum of friction. Another method of mounting the coil is to use flexible bands of metal in place of the pivot axles. These bands, firmly fixed to the frame and the coil along the axis of rotation and stretched tight, serve as both axle, pivot, and restoring spring. These "taut-band" meter movements are more rugged than quality jeweled movements and offer better accuracy at moderate cost. An indicating pointer is attached to the coil assembly, and springs or the taut-band arrangement are used to return the needle (and coil) to a fixed reference point. The springs provide a counter-torque that cancels out the magnetic torque on the coil and produces a steady angular deflection, ϕ. The restoring torque of the springs can be expressed as $k\phi$, where k is a spring constant. If the restoring torque is equated to the magnetic torque on the coil, $nIaB$, the angular deflection of the coil ϕ is equal to $nBaI/k$ or $\phi = KI$, where K is a constant. Thus the coil deflection which is indicated by a pointer is directly proportional to the magnitude of the current.

The moving coil meter is a current-to-scale-position interdomain converter. It is used as a readout device or output transducer because the scale behind the pointer can be calibrated in the numbers of the units being measured. The transfer function for the ideal current meter is $\phi = KI$. The proportionality constant K is a measure of the **sensitivity** of the meter and can be expressed as milliamperes full scale, microamperes per centimeter, microamperes per degree, etc. (Note that for domain converters, the term "sensitivity" refers to the slope of the transfer function.) Two other characteristics of current meters that influence their effectiveness in a given application are the internal resistance (the dc resistance of the coil in ohms), and the accuracy (normally $\pm0.5\%$ to $\pm3\%$ of full scale deflection for general-purpose meters, and $\pm0.1\%$ to $\pm1\%$ for laboratory types).

Galvanometer

The usual galvanometer is a sensitive moving coil meter similar in principle to the D'Arsonval current meter. In the galvanometer, however, the pivots are replaced by vertical filamentary suspensions and the pointer by a mirror and light-beam system as shown in Fig. 1–12. A small rotation of the very sensitive coil is amplified into a large horizontal displacement of the light beam reflected from a small mirror mounted on the coil. The light beam is projected on a scale usually divided into millimeter segments that can be estimated to about 0.2 mm. Sensitivity is often given as the current, in

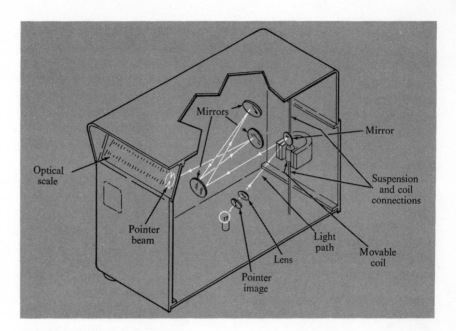

Fig. 1–12 Moving coil galvanometer.

microamperes, necessary to give a 1 mm deflection (usually on a scale at a distance of 1 m).

Unless the action of a galvanometer is damped, a change in the current will cause the coil to oscillate back and forth across the new rest point several times before coming to rest. This is due to the large coil mass and low spring torsion required for high sensitivity. It is usually desirable, therefore, to damp the meter critically by adjusting the resistance in the circuit external to the galvanometer. The motion of the coil in the stationary magnetic field creates a potential difference across the coil. The external circuit resistance puts a load on this "generator," which uses energy from the momentum of the coil and provides a kind of electromagnetic damping. Thus, if the external resistance is too small, the meter is **underdamped;** if it is too large, the meter approaches a new rest point slowly **(overdamped);** when **critically damped,** the meter reaches each new reading without oscillation and in the shortest possible time. The value of the external circuit resistance which provides critical damping is called the **critical damping resistor—external** (CDRX). A comparison of important characteristics of a few types is given in Table 1–3. In general, the more sensitive the galvanometer, the more fragile it is, and the longer the period of oscillation.

Table 1–3   Characteristics of a few dc moving coil galvanometers

Type	Sensitivity, $\mu A/mm$	CDRX*, Ω	Coil resistance, Ω	Period, sec
Box type	0.005	400	25	2.5
	0.005	25,000	550	3
Pointer type	2	20	12	2.5
with taut	0.25	1,800	250	3
suspension				
Laboratory	0.0005	10,000	650	6
reflecting type	0.0001	22,000	500	14

* Critical damping resistance, external.

Recording Meters

A recording galvanometer is shown in Fig. 1–13. A writing stylus is used instead of a pointer. The stylus can have an ink tip, or it can have a tip that is the contact for an electrosensitive, heat-sensitive, or pressure-sensitive paper. The roll of chart paper is usually driven by a constant-speed motor. If a writing arm of fixed length is used, the ordinate will be curved. In order to convert the curvilinear motion of the writing tip into rectilinear motion, various writing mechanisms have been devised to change the effective length of the writing arm as it moves across the chart. Some recording galvanometers use a mirror and light beam like an ordinary galvanometer. A high intensity ultraviolet light source and light-sensitive paper are used. These devices are sometimes called **oscillographs.**

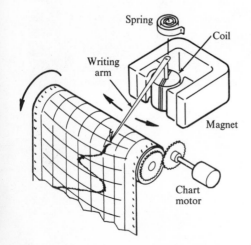

Fig. 1–13   Recording galvanometer.

1–2.3 CONDUCTORS IN SERIES

It has been shown that when two points of different electrical potential are connected by an electrical conductor a current will exist through the conductor. From Ohm's Law, the magnitude of the current is $I = V/R$, where V is the potential difference across the conductor and R is the resistance of the conductor. Stated in another way, a current will exist whenever a conducting path exists between two points of different potential. This is true regardless of how circuitous or complex the conducting path may be

The conducting path of Fig. 1–14 is composed of two resistors and a battery. The two resistors are said to be **in series** because as the conducting path is traced from one battery terminal to the other the resistors are traversed in sequence, or in series. Note that in a series circuit, connecting

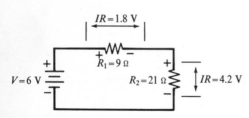

Fig. 1–14   *IR* drops in a series circuit.

elements have only one terminal of each element in common. The rate of charge movement in the circuit depends on both resistors. The total resistance to the flow of charge in the circuit is the sum of the resistance of the two resistors, that is,

$$R_s = R_1 + R_2, \qquad (1\text{--}12)$$

and in general for a series of N resistors

$$R_s = R_1 + R_2 + R_3 + \cdots + R_N, \qquad (1\text{--}13)$$

where N is any number.

In other words, a series combination of resistors, $R_1, R_2, R_3, \ldots, R_N$, *is equivalent to* a single resistor of value R_s. The current through the series of resistors will be

$$I = V/R_s, \qquad (1\text{--}14)$$

or in the case of Fig. 1–14, $I = 6 \text{ V}/(9 \ \Omega + 21 \ \Omega) = 0.2$ A. Of course, the current is the same anywhere in the series circuit at any instant. Clearly the current in the series circuit is affected by the values of all the series resistors.

Current Measurement

Suppose the current in the battery and light bulb circuit of Fig. 1–10 were to be measured with a current meter which has a resistance of 0.1 Ω. In the resulting circuit, shown in Fig. 1–15, the current meter is placed *in series* with the light bulb so that the current in the light bulb and the meter will be the same. The value of the current measured will be $I_m = V/(R_l + R_m) = 1.5 \text{ V}/2.6 \ \Omega = 0.577$ A. When this is compared with the value of 0.6 A obtained by calculation without the current meter in the circuit, it is clear that *the introduction of the current meter affects the value of the current in the circuit.*

The relative magnitude of the error introduced by inserting a current meter into the circuit of Fig. 1–15 can be described quantitatively by calculating the **relative perturbation error** (RPE) which is given by

$$\text{RPE} = \frac{I - I_m}{I}, \qquad (1\text{--}15)$$

where I is the current in the circuit without the meter inserted, and I_m is the current with the meter present.

If the relationships $I = V/R_l$ and $I_m = V/(R_l + R_m)$ are substituted into Eq. (1–15), the relative perturbation error is given by

$$\text{RPE} = \frac{R_m}{R_l + R_m}. \qquad (1\text{--}16)$$

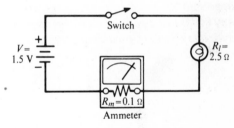

Fig. 1–15 Current measurement.

From Eq. (1–16) it can be seen that the error is negligible when R_m is negligible compared to R_l. For a maximum error of 1%, the meter resistance R_m should be 0.01 R_l. For 0.1% error, R_m should be 0.001 R_l, etc.

Since a conventional current meter must be placed in series in the current path and since all current meters have some resistance, this error will occur to some extent in all current measurements. However, it should be pointed out that while the measured current is not the same as it was without the meter, *the meter is measuring the actual circuit current which exists while the meter is in the circuit.*

Ohm's Law can also be used to calculate the potential difference across each series resistor. A rearrangement of $I = V/R$ gives

$$V = IR, \qquad (1\text{–}17)$$

which indicates that whenever a current I exists in a conductor of resistance R, a potential difference V appears across that conductor. This potential difference is sometimes spoken of as the IR **potential drop** or just IR **drop.** For the resistors of Fig. 1–14,

$$IR_1 = 0.2 \text{ A} \times 9 \text{ } \Omega = 1.8 \text{ V} \qquad \text{and} \qquad IR_2 = 0.2 \text{ A} \times 21 \text{ } \Omega = 4.2 \text{ V.}$$

The sum of the IR drops in the series circuit must be equal to the potential difference at the battery terminals, for there are no other elements in the circuit across which a potential difference can exist.[10] The potential across the current meter in Fig. 1–15 is

$$V_m = 0.577 \text{ A} \times 0.1 \text{ } \Omega = 0.0577 \text{ V.}$$

The potential across the lamp V_l can be obtained by subtracting V_m from the battery voltage:

$$V_l = 1.5 \text{ V} - 0.0577 \text{ V} = 1.4432 \text{ V.}$$

Note 10.   Kirchhoff's Voltage Law. Kirchhoff's Voltage Law states that the algebraic sum of the voltages from all voltage sources and IR drops encountered at any instant as a path is traced out around a complete circuit (closed loop path) is zero.

Experiment 1–1   *Measurement and Control of Current in a Series Circuit*

This experiment illustrates an application of Ohm's Law and the measurement of current with a meter while providing some interesting information about a lamp. Experience is gained wiring a series circuit.

A variable resistor is used to control the current through a lamp in a simple series circuit. Current meter and light intensity observations are made for several resistance values. From the measured data, the resistance of the lamp is calculated for each current value.

Experiment 1–2   *The Equivalent Resistance of Series Resistors*

This experiment verifies that the effect of several series resistors on a circuit is the same as that of a single resistor which has a value equal to the sum of the values of the several resistors. This is done by measuring the current in a circuit containing first the series resistors and second a resistor of equivalent value.

Measurement of Potential Difference with the Current Meter

The moving coil meter has a constant resistance, so that the current through the meter is proportional to the voltage across it. In this sense, the current meter can be used to measure voltage. The full-scale deflection sensitivity V_{fs}, in volts, is the full-scale deflection current I_{fs} times the resistance of the meter R_m. To extend the voltage range of the meter, it is necessary only to add resistance in series with the meter circuit as in Fig. 1–16. Now $V_{fs} = I_{fs}(R_s + R_m)$. If a 200 μA meter movement is to be used for a 0–20 V voltmeter, the total meter resistance should be

$$20\,\text{V}/(0.2 \times 10^{-3}\,\text{A}) = 100\,\text{k}\Omega.$$

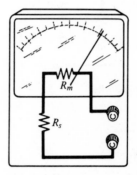

Fig.   1–16   A dc voltmeter.

The resistance of the meter movement would be about 300 Ω, i.e. sufficiently small compared with the total resistance that in this case its contribution could be neglected.

Note that $(R_s + R_m)/V_{fs}$ is a constant. This constant is the **ohms-per-volt** rating of the voltmeter. The resistance of the voltmeter is the ohms-per-volt rating times the full-scale deflection voltage. A 20,000 Ω/V voltmeter will have a resistance of 30,000 Ω on the 1.5 V scale and 600 kΩ on the 30 V scale. The current sensitivity of a meter with a 20,000 Ω/V rating is 50 μA full-scale deflection. It is important to have a voltmeter resistance which is high compared with the resistance of the circuit being measured. The ohms-per-volt rating of the meter will be higher if a moving coil meter of greater current sensitivity is used.

Experiment 1–3   *Measurement of Voltage with the Current Meter*

This experiment demonstrates that an ordinary sensitive current meter can be made into a voltmeter by using a large series resistor. The resulting voltmeter is used to measure several voltages, and its limitations and characteristics are determined.

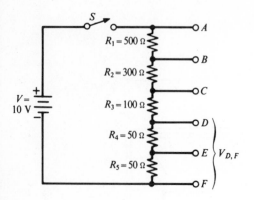

Fig. 1–17   Voltage divider.

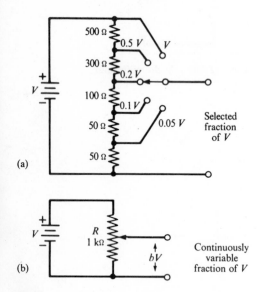

(a)

(b)

Fig. 1–18   Voltage dividers. (a) Selector switch. (b) Potentiometer.

Note 11.   Potentiometers.
The word "potentiometer" has two meanings.
When used to describe a continuously variable
voltage divider, it is frequently shortened to
"pot." A potentiometer can also mean a
complete comparison system for voltage
measurements as discussed in Section 1–3.

Voltage Divider

The resistors of Fig. 1–17 are connected in series. As explained earlier, there is only one current path in the complete circuit so that the current in each component is the same. When switch S is closed, the 10 V source is connected to the total resistance between points A and F. Recall that the total resistance R_s in a series circuit is the sum of the separate resistances:

$$R_s = R_1 + R_2 + R_3 + R_4 + R_5. \tag{1–18}$$

The current $I = V/R_s = 10 \text{ V}/1000 \ \Omega = 0.01$ A. The total potential difference V is divided among the separate resistors as in Fig. 1–14. The potential difference (voltage) between any two points can be determined by calculating the IR drop.

$$V_{D,F} = (R_4 + R_5)I = 100 \ \Omega \times 0.01 \text{ A} = 1.0 \text{ V} \tag{1–19}$$

The fraction of the total applied voltage which appears between points D and F is

$$\frac{V_{D,F}}{V} = \frac{(R_4 + R_5)I}{V} = \frac{R_4 + R_5}{R_s}, \tag{1–20}$$

and this ratio is independent of the voltage V. In other words, the output voltage is a certain constant fraction of the applied voltage. For this reason this circuit is called a **voltage divider.**

It can be seen from Eq. (1–20) that the voltage between two points in a series circuit will be negligible compared to the total applied voltage if the resistance between the two points is very much smaller than the total series resistance. Similarly, the potential between two points separated by a very large fraction of the total series resistance will be very nearly equal to the total applied voltage.

A selector switch may be used to select various fractions of the input voltage, as in Fig. 1–18a. The output voltage increments may be made as small as desired. Selector switches commonly have up to 11 positions, but switches with more positions are also available. For finer adjustment of the voltage-divider fraction *b,* a resistor with a continuously variable slider contact called a **potentiometer**[11] is used, as shown in Fig. 1–18b.

Since the current in every element of the series circuit is equal, a single switch placed anywhere in the current path is sufficient to turn off the current in the entire circuit. Similarly, a single fuse is sufficient to protect all components against excessive current. *The order of elements in a series circuit is immaterial.* If any element of the circuit burns out or becomes disconnected, the entire circuit is disabled. This is a common occurrence with "series-string" Christmas-tree lights and with series-connected vacuum-tube filaments in inexpensive television sets.

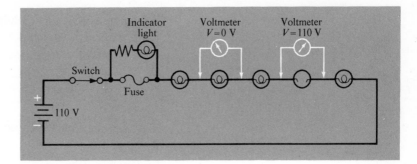

Fig. 1–19   Voltages in open series circuit.

Because there is no current in the incomplete series circuit, there will be no *IR* drop in any of the circuit elements. The entire source voltage must then be across the "open" element terminals, as shown in Fig. 1–19 for the case of the series-wired lamps. This is a convenient method for finding the open element, because the potential drop across every other element will be zero. Another interesting application of this principle is the indicator light across the fuse in Fig. 1–19. Assuming that the burned-out lamp has been replaced, the full 110 V will appear across the fuse terminals should the fuse blow, and the "blown-fuse" indicator light would then light. The resistor is used to limit the current to a safe value for the indicator light bulb.

Experiment 1–4   *Voltage Dividers*

The fact that the source voltage is divided among the elements in a series circuit is demonstrated in this experiment by measuring *IR* drops. This concept is then applied to make and test a precision voltage divider.

1–2.4   CONDUCTORS IN PARALLEL

Often several conducting paths exist between two points in a circuit, providing independent alternate routes for the transport of charge between the two points. The separate conducting paths between two points are said to be connected in **parallel.** Resistors connected in parallel to a voltage source are shown in Fig. 1–20. Note that a conducting path which is in parallel with other conducting paths has one end of the path connected to one end of all the other parallel paths and the opposite end connected to the opposite ends of the other parallel paths. Therefore the same voltage is applied to each conducting path in parallel. Since each resistor in Fig. 1–20 provides

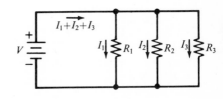

Fig. 1–20   Parallel circuit.

a separate conducting path between the points of potential difference V, the total parallel conductance G_p is the sum of the conductances of the separate paths:

$$G_p = G_1 + G_2 + G_3 = \frac{1}{R_1} + \frac{1}{R_2} + \frac{1}{R_3}. \qquad (1\text{--}21)$$

The resistance equivalent to a number of parallel conductors is $R_p = 1/G_p$. Thus

$$R_p = \cfrac{1}{\cfrac{1}{R_1} + \cfrac{1}{R_2} + \cfrac{1}{R_3} + \cdots + \cfrac{1}{R_N}}, \qquad (1\text{--}22)$$

where N is the number of parallel conductors. For two resistors in parallel,

$$R_p = \frac{1}{1/R_1 + 1/R_2} = \frac{R_1 R_2}{R_1 + R_2}. \qquad (1\text{--}23)$$

The current through each resistor in Fig. 1–20 can be calculated from Ohm's Law:

$$I_1 = V/R_1, \qquad I_2 = V/R_2, \qquad I_3 = V/R_3. \qquad (1\text{--}24)$$

The sum of the currents in the parallel conductors must be equal to the current supplied by the battery,[12] *as shown,* in order to avoid a net accumulation or depletion of charge at some point in the conducting circuit.

Current Splitting

When two conductors R_1 and R_2 are connected in parallel, the circuit current I_t is split between them. The fraction of I_t which exists in R_1 is

$$\frac{I_1}{I_t} = \frac{V/R_1}{V/R_p}.$$

Substituting for R_p from Eq. (1–23), we have

$$\frac{I_1}{I_t} = \frac{R_2}{R_1 + R_2}. \qquad (1\text{--}25)$$

Thus the fraction of the total current through either resistor is the resistance of the *other* resistor divided by the sum of the resistance of both resistors. This "current splitting" relationship is frequently useful in calculating the desired values for parallel resistors.

Note 12. Kirchhoff's Current Law. Kirchhoff's Current Law states that the algebraic sum of all currents encountered at any instant at a junction (point where three or more conductors are joined) must be zero.

Experiment 1–5   *Resistors in Parallel*

Kirchhoff's Current Law and the current-splitting relationship are studied and demonstrated in this experiment. Two resistors are connected in parallel across a voltage source. The current through each resistor and the total circuit current are measured and compared with the expected values.

Shunting the Current Meter

The moving coil meter responds to current, and the current range of any specific meter is determined by the design of the coil, springs, and magnet. To avoid having to use a separate meter for each current measurement range required, **shunting** is applied to obtain multiple ranges with a single sensitive meter. A meter is **shunted** by providing an alternate (parallel) conducting path around the current meter so that only a fraction of the circuit current exists in the meter coil, as shown in Fig. 1–21a. From the current-splitting equation (1–25),

$$\frac{i_m}{i_t} = \frac{R_{\text{sh}}}{R_m + R_{\text{sh}}}. \qquad (1\text{–}26)$$

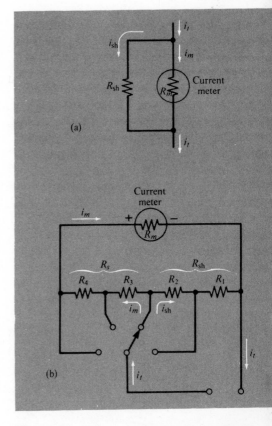

If the range of the meter is to be increased tenfold, only one-tenth of the total current in the circuit should pass through the meter and nine-tenths through the shunt resistor. From Eq. (1–26), $i_m/i_t = 1/10$ and $R_{\text{sh}} = R_m/9$. If a 1 mA meter movement which is to read 100 mA full scale has a resistance of 46 Ω, the shunt resistor R_{sh} should be $R_m/99 = 0.455$ Ω. Usually this resistance needs to be accurate to only 1% because the accuracy of most panel meters is only 1% to 3%. Accurate low-resistance shunts must be wired so that the resistances of the connections to the meter and shunt are not part of R_{sh}. If the calculated R_{sh} is impractically low, one usually puts a resistance in series with the meter coil so as to increase the effective value of R_m.

A common type of shunt for high current measurements which uses resistors in series with the meter is the Ayrton shunt shown in Fig. 1–21b. The measured current i_t divides between two conducting paths, the shunt path R_{sh}, which is $R_1 + R_2$ when the switch is in position, as shown in Fig. 1–21b, and the meter path, which is $R_s + R_m$ where $R_s + R_{\text{sh}}$ is the sum of $R_1 + R_2 + R_3 + R_4$. Since the voltage drop is the same across both resistance branches,

$$i_m(R_m + R_s) = (i_t - i_m)R_{\text{sh}}.$$

This reduces to

$$\frac{i_m}{i_t} = \frac{R_{\text{sh}}}{R_s + R_m + R_{\text{sh}}}. \qquad (1\text{–}27)$$

Fig. 1–21 Meter shunts: (a) Simple shunt. (b) Ayrton shunt.

Therefore, if a 50 μA meter is used and it is desired to have current ranges of 15 mA, 150 mA, 0.5 A, and 15 A, then the ratios $(R_s + R_m + R_{sh})/R_{sh}$ should be 300, 3000, 10,000 and 300,000 respectively. If $R_m = 5000 \; \Omega$, the correct ratios are obtained if

$$R_{sh} = R_1 + R_2 + R_3 + R_4 = 16.7 \; \Omega$$

for the 15 mA range; and

$$R_{sh} = R_1 + R_2 + R_3 = 1.67 \; \Omega, \qquad R_{sh} = R_1 + R_2 = 0.5 \; \Omega, \qquad \text{and}$$

$$R_{sh} = R_1 = 0.0167 \; \Omega$$

for the 150 mA, 0.5 A, and 15 A ranges, respectively. It follows that $R_2 = 0.483 \; \Omega$, $R_3 = 1.17 \; \Omega$, and $R_4 = 15 \; \Omega$.

Experiment 1–6   *Current Meter Resistance*

A method for measuring the resistance of a meter movement is illustrated in this experiment, and the resistance of the particular meter in the experimental station is obtained. The technique used is to measure a circuit current with the meter and then adjust an added shunt resistor to obtain exactly half the original current reading.

Experiment 1–7   *Shunts for a Current Meter*

A sensitive meter movement is made suitable for measurement of higher currents by placing precision resistors in parallel with the meter. A shunt resistance is calculated for a specific desired current range. Measurements are made with the resulting current meter, and its effective resistance is calculated. Multimeter shunt resistor values are calculated.

1–2.5 TRANSDUCERS WITH CURRENT OR CHARGE OUTPUT

Various transducers such as radiation and ionization detectors produce an electrical charge or current at the output which is a known function of the phenomenon being measured. From a data domain point of view, information in the physical domain, such as light intensity, radioactivity, and molecular concentration, is converted or encoded by the transducer into a related magnitude of electrical charge or current at the output of the transducer. Several types of transducers convert the capture of an ion or a photon of light into a related quantity of electric charge. Thus a measurement of the total charge produced is directly related to the total number of ions or photons captured. Many charge-output transducers are closely related to, and are often the same as, current-output transducers since the current is simply the rate of charge produced. Thus the current output from many charge conversion transducers is related to the *rate* of capture of ions or photons. Whether the quantity of charge or the current from the transducer is measured depends upon whether we wish to measure the number of charge-producing phenomena or the rate of their occurrence. The use of a single transducer in both charge and current output modes is illustrated by the first transducer discussed in this section, the photomultiplier tube.

From a practical viewpoint a **current source** is a signal source where the rate of charge flow through a load connected to its output is controlled according to some desired relationship—whether it be a constant fixed current, a series of current pulses, or a current that varies according to some complex signal pattern.

In this section the very important photomultiplier (PM) tube, phototube, photodiode, ionization detectors, and oxygen electrode are introduced. All of these transducers produce a charge or current which is related to the transduced phenomena by convenient transfer functions.

Photomultiplier (PM) Tube

The quality of tens of thousands of research and routine spectrophotometers depends directly on the operating characteristics of photomultiplier (PM) tubes and their associated electronic readout and power supply circuits. The great variety of instruments in which PM tubes are utilized is indicative of their importance. The PM tube is now found in essentially all of the commonly used molecular UV-visible absorption, fluorescence, phosphorescence, reflectance, Laser-Raman, atomic absorption, atomic emission, and atomic fluorescence spectrometers; and also in specialized rapid-scan, submicrosecond time-resolved, T-jump, and chromatogram scanning spectrometers, multichannel spark source direct readers, densitometers, and other types of instruments.

The electrode arrangement of a typical PM tube is shown in Fig. 1–22a. The input radiation is incident on the **photocathode** from which photoelectrons are ejected with an efficiency (b_λ) which depends upon the energy ($h\nu$) of the photon and the type of cathode surface. A large fraction of the photoelectrons ejected by the photocathode are attracted to the first **dynode** because the dynode is maintained at a positive voltage relative to the photocathode. The fraction of photoelectrons collected by the first dynode is the collection efficiency, b_c, typically about 75%. Depending upon the voltage and dynode characteristics, a certain probable number of **sec-**

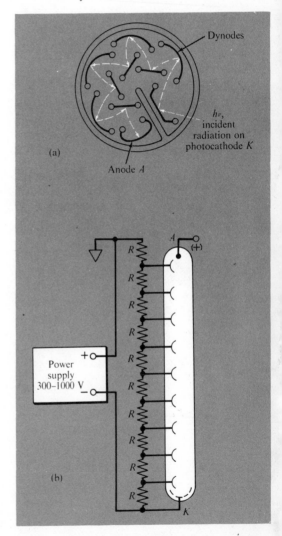

Fig. 1–22 Photomultiplier transducer. (a) Cross-sectional view of side-illuminated PM tube. (b) Schematic showing connection to power supply.

ondary electrons (e.g., 4 or 5) will be ejected from the dynode for each photoelectron striking it. The secondary electrons from the first dynode are attracted to the second dynode, where each electron ejects more secondary electrons—and likewise around the successive dynodes until the electrons are collected at the anode. The transfer efficiency, b_t, of electron bursts between dynodes is almost unity so that the number of anode pulses is nearly equal to the number of photoelectrons reaching the first dynode. Each successive electrode must be at a higher potential than the previous one so that electrons are attracted to the next dynode. A voltage divider that provides the required potential differences between successive dynodes is shown in Fig. 1–22b.

Thus each photoelectron collected by the first dynode is greatly augmented within the PM tube by the process of secondary electron emission to form a relatively large packet or **pulse** of electrons at the anode. This pulse is typically only a few nanoseconds long.

The number of anode pulses, N_a, over certain boundary conditions can be written

$$N_a = b_t b_c b_\lambda P_k = b P_k, \qquad (1\text{--}28)$$

where P_k is either the number of photons that strike the photocathode because of some isolated event such as a light burst from a firefly, a spark discharge, or a shutter opening, or the number of photons that strike the cathode during a unit time interval (seconds), which is called the **radiant flux.**

The effective gain, G, of the PM tube, that is, the number of electrons in each anode pulse, depends greatly on the PM voltage, but is typically 10^5 to 10^7. Even at constant PM voltage, the statistical nature of secondary emission from the dynodes introduces a rather large amplitude fluctuation of the anode pulses. Because the charge on the electron (Q_e) is 1.6×10^{-19} C and assuming an *average* internal electron gain, $\overline{G}$, of 10^6, we find that the *average* charge delivered to the anode for each photoelectron reaching the first dynode is $Q_e\overline{G}$, or 1.6×10^{-13} C per anode pulse.

In general, then, the average charge $\overline{Q}$ delivered to the anode over the prescribed boundary conditions D (event or time interval) can be expressed as

$$\overline{Q}\left(\frac{\text{coulombs}}{D}\right) = N_a\left(\frac{\text{anode pulses}}{D}\right)\overline{G}\left(\frac{\text{electrons}}{\text{anode pulse}}\right)Q_e\left(\frac{\text{coulombs}}{\text{electron}}\right).$$
$$(1\text{--}29)$$

Under a given set of operating parameters (PM voltage, area of cathode exposed, wavelength, etc.), the product $b\overline{G}Q_e$ can be considered to have a constant average value k_λ in coulombs/photon, so that combining Eqs. (1–28) and (1–29), and substituting k_λ, we get

$$\overline{Q} = k_\lambda P_k \ \text{(coulombs/event)}. \qquad (1\text{–}30)$$

In this case, P_k is the total number of photons during a specific event and k_λ is the sensitivity. The linear transfer function, Eq. (1–30), relates the information from the physical domain (photons) to the transduced information now in the electrical domain (charge) so that it can be processed and measured. The PM tube typically has a wide linear dynamic range of about 10^6 or more.

If the total charge caused by some isolated event is the measured quantity, then the PM is a transducer with *charge* output. If the flow of charge per unit time (current) caused by a continuous light source is the measured quantity, then the PM is a transducer with *current* output. Thus, although the output at the anode of the PM tube is inherently a series of charge packets or current pulses, the output signal resulting from a given intensity of light can be expressed as an average rate of flow of electrons/sec, or current $\overline{I}$, as summarized in Eq. (1–31):

$$\overline{I} = b Q_e \overline{G} P_k \ \text{(coulombs/second)}. \qquad (1\text{–}31)$$

For example, if the number of incident photons/sec, P_k, equals 10^7, the average electrons/anode pulse, $\overline{G}$, equals 10^6, and the probability of an incident photon producing an anode pulse is $b = 0.1$, then

$$\overline{I} = 0.1 \times 1.6 \times 10^{-19} \times 10^6 \times 10^7 = 1.6 \times 10^{-7} \, \text{A}.$$

Thus Eq. (1–31) is the transfer function relating PM tube current to the incident radiant flux P_k.

Frequently the transfer function is expressed in terms of output current as a function of input radiant power, or the sensitivity is expressed in amperes/watt (A/W). Since the radiant power of an incident beam is $P_k h\nu$, where $h\nu$ is the energy of the incident photon, the sensitivity k'_λ is given by

$$k'_\lambda = \frac{b Q_e G}{h\nu} = \left(\frac{b_\lambda Q_e}{h\nu} \right) b_c b_i G.$$

The quantity in parentheses in the above expression is characteristic of a given photocathode and is called the **radiant cathodic sensitivity** in amperes/watt.

It is apparent from Eq. (1–31) that the output current of a PM tube fluctuates as P_k and G fluctuate. That is, the random time behavior of the anode pulses as a result of the incident radiation, and the amplitude distribution from random secondary electron emission cause the so-called "shot noise" on the PM tube analog signal. By using suitable electronic filters, the fluctuations are averaged or integrated so that the average current, $\bar{I}$, can be observed. As a "rule of thumb," for many PM tubes a current of 10^{-7} A is equivalent to about 10^6 photons/sec. Bceause an average current can be related to the number of pulses in unit time, it is frequently convenient to determine the noise component of an analog PM tube signal by utilizing count statistics. This is illustrated in Module Four, *Optimization of Electronic Measurements*.

If it is assumed that there is negligible pileup of anode pulses, i.e., nearly all anode pulses are separated in time, then the number of anode pulses per event, N_a, as expressed by Eq. (1–28) can be counted by digital techniques over the prescribed boundary conditions.

It is the counting of the anode pulses, which are directly related to the number of photons incident on the cathode, that provides the measurement technique that is appropriately called "photon counting." This technique is described in greater detail in Module Three, *Digital and Analog Data Conversions*.

Radioactive Scintillation Detector

There are various crystalline materials called **scintillators** that emit fluorescent (scintillation) radiation pulses in the ultraviolet or visible spectral regions when they are bombarded by radioactive particles or gamma radiation. Therefore, if a scintillator is deposited on the window of a PM tube so that it is close to the photocathode, the many emitted photons caused by each radioactive particle will strike the photocathode and cause ejection of many photoelectrons. These photoelectrons will be internally amplified by the PM tube in the manner previously described.

For high counting efficiency the radioactive material can be put into a small cavity in the scintillator, as shown in Fig. 1–23. PM tubes with transparent photocathodes and end-on windows are generally used because of the optimum geometry for collecting the emitted fluorescent photons. The number of emitted photons per radioactive particle is dependent on the energy dissipated by the particle in the scintillation crystal. Thus the detector current or charge output is related to the energy of the radioactive particle(s).

The major difference between "scintillation counting" and "photon counting" is that in the former, the anode pulses are larger and easier to detect. In photon counting, each anode current pulse is produced by a single photoelectron ejected by a photon. In contrast, each radioactive particle causes the emission of a large number of photons that strike the photocathode and eject many photoelectrons in a very short time, and these cause much larger anode current pulses than result from a single photoelectron.

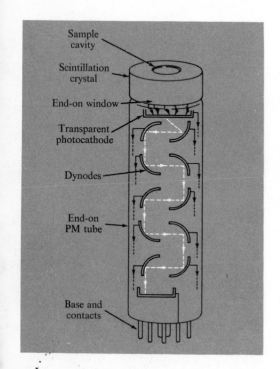

Sample cavity

Scintillation crystal

End-on window

Transparent photocathode

Dynodes

End-on PM tube

Base and contacts

Fig. 1–23 PM tube with end-on illumination and scintillator crystal for radioactivity measurements.

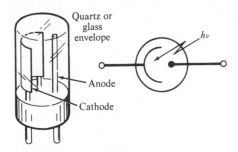

Fig. 1–24 Vacuum phototube pictorial and schematic symbol.

Fig. 1–25 Photocathode spectral response curve (the five numbers designate specific types of photocathodes).

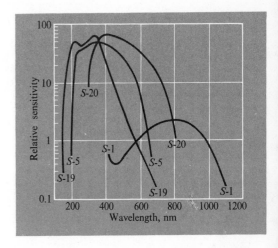

Vacuum Phototube

The ordinary two-electrode phototube contains a photocathode and anode in an evacuated glass envelope, but does not contain the dynodes which provide internal amplification as for the PM tube. A typical phototube and its schematic symbol are shown in Fig. 1–24. Because there are no dynodes, the photoelectrons ejected from the cathode are attracted directly to the anode, which is held at a positive potential with respect to the cathode.

The materials used for the photocathodes of phototubes and PM tubes are similar. The photoemission sensitivity of each material is different so that the spectral range over which a specific type of tube can be used depends on the photocathode, as illustrated by the spectral sensitivity curves in Fig. 1–25. It is important to select the tube that has high sensitivity (quantum efficiency, b_λ) in the spectral range of interest. The characteristic current-voltage curves (*i-v* curves) for a phototube and a photocurrent measurement circuit are shown in Fig. 1–26. To obtain these curves, the photocurrent i is measured as the supply voltage V is varied while the phototube is exposed to a preset light intensity. The arrow through the voltage source symbol indicates that it can provide a variable voltage. These curves are typical of devices which become **current limited.** That is, there are many devices in which the charge carriers arrive at the electrode terminals at a maximum rate which is independent of the applied voltage over a wide range. The maximum rate is determined by some limiting process such as the average rate of ejection of photoelectrons,

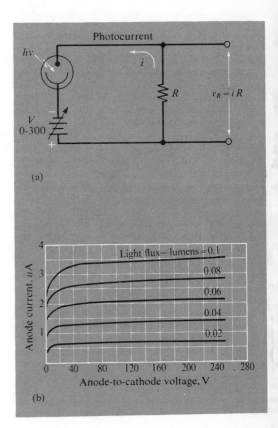

Fig. 1–26 (a) Phototube circuit. (b) Current-voltage curves for various light intensity levels.

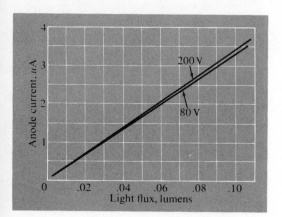

Fig. 1–27 Phototube anode current vs. light intensity for several voltages.

which is dependent on the rate of arrival of photons at the photocathode; or, in another example, the limiting rate of diffusion of reducible ions to a cathode is determined by the concentration of that ion in solution. Because the rate of movement of charge carriers is limited by fundamental limiting processes under the experimental conditions, any increase in applied voltage above a certain critical voltage cannot increase the current unless some other process sets in so as to produce charge carriers at a higher rate. The heights of the current plateaus in Fig. 1–26 are thus determined by the rate of arrival of photons at the photocathode and are relatively independent of the applied voltage over a wide range of values.

When the phototube is operated in the current-limiting region, the transfer function plot of output current versus light intensity (photons/sec incident on the photocathode) is a straight line and the slope is almost unaffected by the applied voltage, as shown in Fig. 1–27.

Photodiode and Phototransistor

The photodiode consists of a semiconductor *pn* junction that produces charge carriers at a rate dependent on the rate of arrival of incident photons at the junction. The specific mechanism by which charge carriers are produced by promoting electrons from the valence band to the conduction band will be better understood after a study of the *pn* junction as described in Module Two, *Control of Electrical Quantities in Instrumentation.* However, the operation of the device is similar in principle to the phototube. That is, the number of charge carriers produced is proportional to the incident light intensity so that the output current is a measure of light intensity. The current-voltage curves for the photodiode are similar to those for the phototube. The phototransistor is another transducer that provides a current output that is related to the radiation incident on a semiconductor *pn* junction. It combines the current amplification of a transistor with the light detection of a semiconductor photodiode.

Experiment 1–8 *Light Intensity Measurements*

Experience with a transducer that converts radiant power into a related electrical current is obtained in this experiment. The photon source is a small tungsten lamp which illuminates the photocathode of a small photodiode. A semiconductor photodiode or phototransistor may also be used. A voltage source appropriate for the photodetector is used to allow the collection of the photoelectrons. The most sensitive current scale of a multimeter is used to measure the photocurrent.

Flame Ionization Detector

If two metal plates are placed opposite to each other and just above a small hydrogen-air flame, as shown in Fig. 1–28, and if molecules enter the flame which are readily ionizable by the flame energy, then there will be a flow of charge carriers (both electrons and positive ions) between the electrodes. The rate of flow of electrons that occurs in the external circuit depends on the rate of formation of charge carriers, which is determined by the concentration of ionizable molecules in the flame at any given time. Therefore, the device can provide an output current related to the molecular concentration. Indeed, this type of flame ionization detector is one of the most important transducers in gas chromatography. The various constituents in a complex sample are selectively separated as they pass through the column and as each exits from the end of the column it passes into the flame ionization detector.

The current i is equal to the number of charge carriers (N_i) produced per second by ionization that reach each detector electrode plate times the unit charge per charge carrier (Q_e). For example, if the number of electrons produced equals 10^{10}/sec, then, since $Q_e = 1.6 \times 10^{-19}$ C,

$$i = 1.6 \times 10^{-9} \text{ A.}$$

The typical currents measured with the flame ionization detector are from about 10^{-6} to 10^{-11} A.

The peak heights of the current pulses caused by ionization of each molecular species in the flame are a quantitative measure of the concentration of each separated species. However, a better quantitative measure of each constituent can be obtained by integrating the current during the time it takes each species to pass through the flame. The integration provides a measure of the total number of charge carriers produced while a specific species is passing through the detector. Thus, any variation in the peak shape due to a change in column parameters would not have so much effect on the integrated value as on the peak height.

The Oxygen (O₂) Electrode

The determination of oxygen concentration levels in biological systems, in lakes, rivers, oceans, and many laboratory solutions is often based on an electrode system that is a "current source" transducer. The electrode and experimental systems are shown in the schematic diagram, Fig. 1–29. A platinum wire is sealed in glass with the wire tip just protruding out of the sealed end so as to form a microelectrode that makes contact with the test solution. A silver–silver chloride electrode provides the reference potential. A variable voltage source and a current meter are connected in series, and

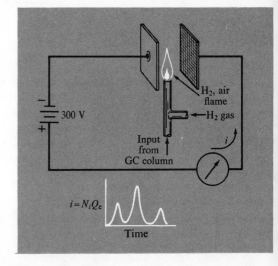

Fig. 1–28 Flame ionization detector for gas chromatography.

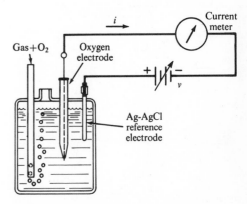

Fig. 1–29 Schematic for O₂ electrode.

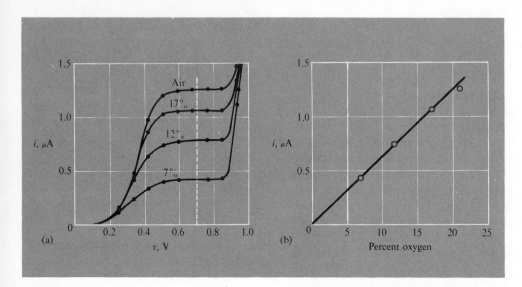

(a)

(b)

Fig. 1–30 (a) Current-voltage curves for O_2 electrode. (b) Output current vs. percent oxygen for $v = 0.7$ V.

the combination is attached between the test (O_2) and reference electrodes as shown in Fig. 1–29.

If the voltage from the source is varied between 0 and 1.0 V, current-voltage (i-v) curves are obtained (Fig. 1–30a) that have characteristics similar to those shown for the phototube (Fig. 1–26b). That is, there is again the characteristic plateau where a change of applied voltage has little effect on the measured current. The height of the current plateau is a function of the oxygen concentration as illustrated by the i-v curves. Therefore, the current-limiting process must be related to the oxygen concentration.

When the platinum microelectrode is made about 0.6 to 0.9 V negative with respect to the reference electrode, the oxygen molecules that reach the platinum surface immediately accept electrons and become reduced. Therefore, electrons move through the external circuit to the oxygen electrode at a rate determined by the rate of arrival of oxygen molecules at the platinum surface. It can be shown that the rate of arrival of O_2 molecules at the electrode surface is governed by fundamental mass transport processes and is directly dependent on the oxygen concentration in the bulk solution. Therefore, the magnitude of the limiting current is a measure of the O_2 concentration, as shown by the transfer function for $v = 0.7$ V in Fig. 1–30b.

PROBLEMS

1. The current in the circuit of Fig. 1–10 was found to be 300 mA when $V = 10$ V and a lamp was used as the conductor. What was the lamp resistance?

2. A series circuit is composed of a 5 V source, a 1 kΩ resistor, and a 500 Ω resistor. What is the current in the circuit and the *IR* drop across each resistor?

3. What is the basic current sensitivity of the meter movement in a 100 kΩ/V voltmeter? What series resistance is required to make this a 1 V full-scale voltmeter if the meter resistance is 10 kΩ?

4. Design a voltage divider to provide 1, 3, 9, 15 and 30 V from a 50 V dc source.

5. What voltage must be applied to a resistance of a 5 kΩ in order to have 3.5 mA of current through the resistor?

6. Design a switch-selectable precision voltage divider with taps for attenuations of $\times 10$, $\times 100$, and $\times 1000$ and a total resistance of 10 kΩ.

7. A current meter has a resistance of 5 kΩ. It is inserted into a circuit containing a 10 V source and a 12 kΩ resistor. (a) What is the current in the absence of the meter? (b) What is the current when the meter is present. (c) What is the relative perturbation error?

8. What is the maximum resistance a current meter can have to measure the current in a circuit consisting of a 6 V battery and a 10 kΩ resistor, given that the error is not to exceed 1%?

9. What size resistor must be placed in parallel with a 3 kΩ resistor to obtain a total resistance of 1.2 kΩ?

10. A current meter of 1 mA full-scale sensitivity and 50 Ω resistance is to be used in measurements. A 6 mA full-scale meter is desired. What shunt resistance should be used?

11. Resistors of 1 kΩ, 3 kΩ, and 10 kΩ are in parallel across a 5 V source. What is the total current from the source and the current through each of the three resistors?

12. Resistors of 10 kΩ and 20 kΩ are in parallel across a source of *V* volts. What fraction of the total current will pass through each resistor?

13. For an Ayrton shunt similar to that shown in Fig. 1–21b, what values of R_1, R_2, R_3, and R_4 must be used with a 50 μA, 1 kΩ meter movement to provide scales of 1 mA, 10 mA, 50 mA and 100 mA?

14. A photomultiplier with an average gain of 5×10^5 and a probability of 0.05 that incident photons will produce an anode pulse has an average output current of 10^{-8} A. What is the radiant flux at the photocathode in photons/sec?

15. A photomultiplier with an average gain of 10^6 has a collection efficiency of 0.8, a quantum efficiency of 0.1, and a transfer efficiency of 0.99 at the wavelength of interest. If 5000 photons/sec strike the photocathode, what is the anode pulse rate? What is the average anode current and the transfer function?

Voltage Sources, Transducers, and Measurement

As described in Section 1–2, **voltage** is a measure of the electric pressure or **potential difference** between two points. It is an essential quantity in all electronic circuits. Therefore, an investigation of the generation, application, and measurement of voltages is important for an understanding of electronic instruments.

In this section, after some general concepts related to the generation of voltages and voltage standards have been introduced, several types of transducers are presented which provide a reproducible and useful relationship between various physical or chemical quantities and the output voltage. The principles and characteristics of various voltage measurement systems are then investigated, including the oscilloscope, field effect transistor voltmeter (FET VM), and potentiometer. The major concepts of null measurements are developed and illustrated by potentiometric voltage measurement. Several null detectors and their application to potentiometry are described. The null measurement concepts introduced in this section lead directly to the servo voltage and current measurement systems discussed in Section 1–4. Some of the most important measurement concepts and techniques utilized in nearly all areas of scientific instrumentation are presented in this section.

1–3.1 VOLTAGE GENERATION

As outlined in Section 1–2.1, the establishment of a potential difference between two terminals involves the separation of charged particles. This requires **work** (energy). The unit of potential difference is the **volt,** and the number of volts or voltage V times the total units of charge Q (coulombs) equals the work W (joules) on the Q units of charge, that is,

$$W \text{ (joules)} = Q \text{ (coulombs)} \, V \text{ (volts)}. \qquad (1\text{–}32)$$

Stated in another way, the voltage is equal to the work (joules) performed per unit of charge (coulomb) ($V = W/Q$). The common practice of using the terms **potential difference, voltage,** and **electromotive force** interchangeably will be followed in this book. These terms are symbolized by V, v, or emf. (In some texts, the symbols e and E are also used for voltage.)

The term "voltage" implies a separation of charge carriers so as to establish an electric field. Thus, if two terminals of a voltage source are connected across a load resistor, there will be a current through the resistor in the direction of the electric field. Some source of energy is required to separate the charge carriers and thus produce a voltage between two terminals. Some classic methods of producing a voltage, such as the mechanical separation of charges in an electrostatic generator, the chemical separation in a cell or battery, and the motion of a conductor in a magnetic field as in an electromagnetic generator are presented in this section, and some basic relationships are introduced.

Separation of Charge

First, let us consider two metallic conductors, 1 and 2, separated by an insulator as in Fig. 1–31. Assume that, to begin with, there is no potential difference between them. This can be assured by connecting them momentarily with a wire. Now, if an electron is removed from conductor 1 and placed in conductor 2, a charge separation is performed. The charge on 2 will be $-Q_e$ and on 1, it will be $+Q_e$. This will result in a potential difference between the two conductors of V volts. According to the laws of electrostatics, the magnitude of the potential difference produced by the separation of one electron charge depends upon the geometry of the separated conductors and the nature of the insulator between them. The energy required to transfer another electron from 1 to 2 is greater than for the first because the electron must be removed from a region of increased positive charge and put in a region of increased negative charge. This produces a higher potential difference between the conductors. Each increment of charge transferred increases the potential difference by the same increment as the first. Thus, the potential difference between the conductors is proportional to the amount of charge Q transferred, so that

$$V = kQ. \qquad (1\text{--}33)$$

Equation (1–33) indicates that to establish a potential difference of 1 V between two conductors, it will be necessary to separate $1/k$ coulombs of charge; that is, $Q = 1$ (volt)$/k$ (volts/coulomb). To produce the potential difference it is, of course, necessary to transport charge carriers in a direction opposite to the direction in which the resulting electric field is trying to move them.

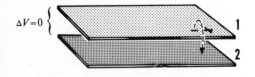

$\Delta V = 0$

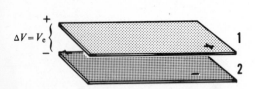

$\Delta V = V_e$

Fig. 1–31 Separation of charge.

Charge Storage.   Any two conductors of any shape or arrangement that are separated from each other can have a potential difference between them. From Eq. (1–33), the amount of charge Q (coulombs) that must be transferred from one conductor to the other to cause a 1 V difference in potential between the two conductors is $1/k$. This constant is called the capacitance, C, so that

$$C = Q/V. \qquad\qquad (1\text{–}34)$$

From Eq. (1–34) it can be seen that C has the units coulombs/volt; 1 C/V has been designated the **farad** (F). If the conductors in Fig. 1–31 have a capacitance of 1 F, then from Eq. (1–34) we can see that a transfer of 1 C of charge will result in a potential difference of 1 V. Conversely, if this same device has a potential difference of 1 V between its terminals, then it must contain (by storing) 1 C of separated charge. Devices which are made and used for their property of capacitance are called **capacitors.** They are symbolized in diagrams as shown in Fig. 1–32. However, it is clear from the above discussion that any pair of conductors has the property of capacitance. The amount of capacitance exhibited by a pair of conductors increases with an increase in the area of the conductors, decreases with an increase in distance between the conductors, and is affected by the nature of the insulator between the conductors. Practical capacitors are described in Module Two, *Control of Electrical Quantities in Instrumentation.*

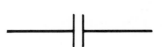

Fig. 1–32 Schematic symbol for capacitor.

The separation of charge required during the charging of a capacitor requires energy. A charged capacitor (one with a finite potential difference between the plates) stores electrical potential energy. This stored energy can be converted to some other form of energy by "discharging" the capacitor, i.e., connecting a conductor or **load** which can make heat or light or do mechanical work. However, the current through the load occurs at the expense of the charge on the capacitor. As the charge decreases, the capacitor voltage decreases since

$$Q = CV,$$

so that eventually the charge separation, the voltage, and the current are all zero.

When a load is first connected to a capacitor charged to a voltage V, the first electron charge through the load loses an energy W of Q_eV joules as given by Eq. (1–32), where Q_e is the electron charge. However, the voltage across the capacitor when the last electron charge is passed is essentially zero, so the energy lost for this charge is also essentially zero. Since the capacitor voltage is proportional to the charge at any time, the average energy lost per electron is the same as the energy lost by the electron which passed at half charge. Thus the energy or work obtained in discharging a capacitor with charge Q completely is

$$W_d = \frac{QV}{2} = \frac{Q^2}{2C} = \frac{CV^2}{2}, \qquad (1\text{--}35)$$

which is also the amount of potential energy (in joules) stored by the capacitor.

Many sources of steady voltage can be described as capacitors with a means of maintaining or replenishing their charge as needed. When a load of resistance R is connected to a steady voltage source of V volts, the current supplied by the voltage source, or its rate of discharge, is $I = V/R$ coulombs/sec. Clearly, the voltage source will not maintain its no-load output voltage if the rate of discharge exceeds the rate by which the charge separation can be maintained within the source. There is, therefore, a limit to the current which a voltage source can supply before the value of its output voltage is significantly decreased. If the load is connected for t seconds, the total charge discharged, Q, is It coulombs. The work performed on the load during discharge is

$$W = QV = ItV$$

as given by Eq. (1–32). Power is defined as the rate of doing work, or W/t. A work rate of 1 J/sec is equal to 1 watt (W) of power. Therefore,

$$P = W/t = QV/t = ItV/t,$$

from which we obtain

$$P = IV. \qquad (1\text{--}36)$$

Thus any voltage source under load supplies power to that load. However, all measurement transducers and signal sources can supply only limited power to a load without a significant change in their output voltage and/or transfer function.

Electrostatic Voltage Generator

Probably everyone has experienced an unexpected spark discharge in reaching for a doorknob or light switch. These sudden shocking discharges are, of course, caused by a person becoming charged to a considerably different potential than conductors in the room. By the mechanical friction of clothing rubbing on objects and especially shoes moving across carpets, there are many charge carriers mechanically separated and then stored so as to build up a potential difference (often several thousand volts) between the body and its surroundings. The high voltage that is built up can cause a spark to develop between the body and some object. Fortunately, the

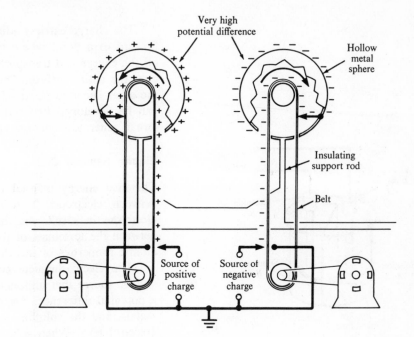

Fig. 1–33 Van de Graaff generator.

accumulated charge is relatively low so that, even though the voltage is very high, the number of charge carriers that must move to discharge the body is relatively small. Thus the discharge is harmless to the individual except for the scare. However, these high voltage electrostatic discharges can be harmful to certain semiconductor devices used in electronic equipment.

The mechanical transport of charge carriers to insulated conductors to create very high voltages is used in the well-known Van de Graaff voltage generator. This voltage generator can establish voltages of a few million volts, and it has been used effectively in the production of high-energy particles for basic studies. The steady current output capability is generally in the 10^{-6} to 10^{-3}A range. One type of Van de Graaff voltage generator is illustrated in Fig. 1–33. The principle of operation is quite simple. Each of two hollow metal spheres (often several meters in diameter) is mounted on top of a long insulating column. A pulley and belt system driven by a motor provides the means of mechanically transporting the charge carriers to the metal spheres. The belts are made of rubber, silk, paper, or some other flexible insulating material. Each belt runs between electrodes which effectively spray charge carriers into the belt. The electrodes are arranged so that positive charges are sprayed on one belt and negative charges on the other belt.

The charge carriers adhere so tightly to the belt that they are carried upward to a point where a sharp electrode acts as a brush and sweeps off the charge and transports it to the metal sphere, where it is distributed over the entire surface. Thus one sphere becomes positively charged and the other becomes negatively charged, and a very high potential difference can be developed between the two spheres. In some designs of Van de Graaff generators only one sphere is used.

Electrochemical Cell

Chemical energy is used in batteries to produce a potential difference between electrodes. That is, chemical reactions can occur at each of two electrodes in a cell, and these reactions cause a net difference of charge between the terminals of the electrodes. The cell voltage depends on the atomic properties of the chemical species that make up the cell.

The electrochemical cell, shown in Fig. 1–34, is an example of a voltage source that utilizes chemical reactions at two electrodes to develop a potential difference. The two electrodes can be of the same or different metals, and the solutions can contain the same or different ionic species (electrolytes). When solutions A and B are different, then a diaphragm or membrane of porous material, called a **separator,** is used to prevent the solutions from mixing. The separator provides contact between solutions A and B but the transport of solution through it is slow enough to avoid any significant mixing of the two solutions.

The actual chemical process that occurs at each electrode is called a half-cell reaction. Consider, for example, a cell where one electrode is cadmium in a cadmium sulfate solution (Cd^{++} and $SO_4^{=}$ ions), and the other electrode is silver in a silver nitrate solution (Ag^+ and NO_3^- ions), as illustrated in Fig. 1–35. The half-reactions are

$$Cd^{++} + 2e^- \rightleftharpoons Cd(metal) \tag{1–37}$$

and

$$Ag^+ + e^- \rightleftharpoons Ag(metal). \tag{1–38}$$

When either half-reaction proceeds to the right, metal ions combine with electrons in the metal to become metal atoms, adding positive charge to the metal. Conversely, when either half-reaction proceeds to the left, metal atoms become metal ions, leaving the associated electron in the metal and thus making it more negative. If the concentrations of cadmium ions (Cd^{++}) at the cadmium electrode and of silver ions (Ag^+) at the silver electrode are each 1 mole/liter, the chemical energy of the half-reaction proceeding to the right is much greater for silver than for cadmium. Thus

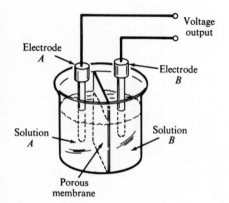

Fig. 1–34   Electrochemical cell.

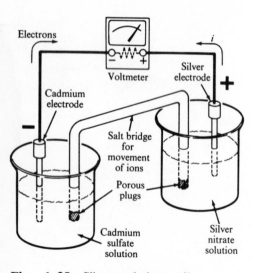

Fig. 1–35   Silver cadmium cell.

the silver electrode is observed to be more positive than the cadmium electrode by approximately 1.2 V ($V = 1.2$ V). Therefore, if a resistive load is connected between the electrodes, there will be a **flow of electrons** through the connecting wires and the load from the cadmium to the silver electrode, as illustrated in Fig. 1–35. The electrons at the surface of the silver electrode combine with silver ions at the metal-solution interface forming more silver metal, which deposits on the silver electrode. That is, reaction (1–38) proceeds to the right **reducing** the positive ions to metallic silver. The electrode at which reduction occurs is called the **cathode.** To produce the electrons that move away from the cadmium electrode through the external circuit in Fig. 1–35, the cadmium metal must go into solution to form cadmium ions. That is, reaction (1–37) proceeds to the left, providing the electrons and **oxidizing** the cadmium metal to ions. The electrode at which oxidation occurs is called the **anode.** Note that although the electrons are the charge carriers in the external circuit, it is the ions that move through the solution.

Electrochemical cells can be used as chemical concentration transducers since the cell voltage is a function of the chemical composition of the cell. The use of a very stable electrochemical cell as a voltage reference source is described in the next section. Chemical cells (batteries) that have been designed to be used as power supplies rather than voltage reference sources are discussed in Module Two, *Control of Electrical Quantities in Instrumentation.*

Voltage Standard. The so-called "working standard" of electromotive force is the cadmium Weston cell shown in Fig. 1–36. The positive electrode is mercury in contact with mercurous sulfate, the negative electrode is cadmium amalgam, and the solution is saturated in cadmium sulfate (excess crystals). This "normal" cell has a value of 1.0183 international volts.[13]

The saturated normal cell has a high temperature coefficient of voltage and is not portable. A standard cell for ordinary measurements is made by introducing a few modifications to the standard cell of Fig. 1–36. The concentration of the cadmium sulfate solution is adjusted so that it will reach saturation at 4°C. Therefore, it is unsaturated at ordinary room temperatures. Also some type of retaining member (cork or rubber septum) is placed over each electrode to prevent displacement of the material around the electrode. This "unsaturated" standard cell is usually surrounded with copper and encased in plastic for portability.

The emf of the unsaturated cell is somewhat higher than that of the saturated cell, and is usually in the range 1.0190 to 1.0194 absolute volt. Each cell is individually calibrated. The temperature coefficient of voltages of the unsaturated cell is much less than that of the saturated cell (only

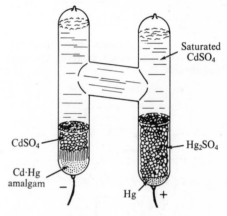

Fig. 1–36 Weston standard cell.

Note 13.   The International Volt.
The international volt as a working standard is defined as: 1 V = 1/1.0183 *of the emf of the Weston Normal (saturated) cell at 20°C.*

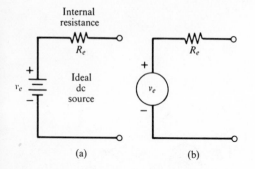

Fig. 1–37 Voltage source equivalent circuits. Both voltage source symbols are used interchangeably.

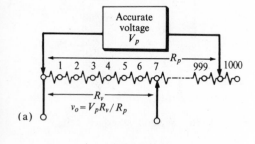

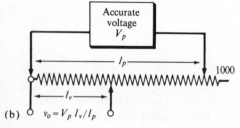

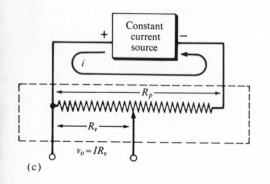

Fig. 1–38 Variable standard voltage. (a) Resistance varied in equal steps. (b) Resistance slidewire. (c) Voltage proportional to preset constant current.

about 10 µV/°C). This low temperature coefficient is also desirable for general laboratory applications. Currents in excess of 0.0001 A should never pass through the cell, and even currents of 10 µA should not pass for more than a few seconds at a time. In other words, the standard cell should only be used as a voltage reference and never in circuits that require significant current to pass through the cell.

Voltage Source Equivalent Circuit. All dc voltage sources may be represented as an ideal battery in series with an internal resistance as shown in Fig. 1–37. The battery symbol is often used to represent a steady voltage source even when an actual battery is not supplying the voltage. Thus the battery symbol in Fig. 1–37a really represents any voltage source. The generator symbol in Fig. 1–37b is also frequently used to represent a voltage source. In this module and other works, these symbols are used essentially interchangeably.

For voltage transducers and reference sources the internal resistances are usually quite high (about 10^2–10^{10} Ω). For power supplies the internal resistance is generally quite low (less than a few ohms). A dc voltage source obtained by rectification of an ac voltage and also utilizing filtering and regulation (as described in Module Two) may also be represented by the simple schematic diagram of Fig. 1–37.

Laboratory Voltage Source

It is frequently important to utilize a voltage reference source (VRS) for scientific measurements or control functions. Sometimes only one fixed voltage reference such as 10.000 V is required. In other cases accurate reference voltages are required over a wide range—for example, continuously adjustable from 0–2.0000 V. To make a voltage source that is continuously variable over a desired range, a resistance network can be used that acts as a **voltage divider.**

In principle, a variable dc standard voltage is obtained simply by connecting an accurate constant voltage across a variable-resistance voltage divider. In Fig. 1–38a the resistance is varied in discrete equal resistance steps so that R_v is a fraction of the total resistance R_p and $R_v/R_p = v_o/V_p$. If the applied voltage $V_p = 1.000$ V and $R_v/R_p = 7.00/1000$ (limit of error of each resistance step is 0.1%), then $v_o = 0.00700$ V. In Fig. 1–38b a linear slidewire is used to vary the ratio R_v/R_p. The length l_v from one end of the wire to the sliding tap (slider or wiper) is a fraction of the total length l_p across which the accurate voltage V_p is applied. Therefore, $v_o = V_p l_v / l_p$. It is apparent from Fig. 1–38a and b that the accuracy of v_o is dependent on the accuracy of V_p, and on the linearity of R_p, but not on the absolute value of R_p. In practice, however, R_p is usually made as an accurately known resistance. This is done so that accurate resistances can be added in series, and in parallel, to change the voltage across the slide-wire by known amounts without changing the voltage supply.

It is useful to consider that in effect an accurate constant current I is provided through the resistance R_p to provide an accurate total applied voltage $IR_p = V_p$, as shown in Fig. 1–38c. Since the resistance R_v varies from 0 to R_p, the voltage varies from 0 to IR_p. If the current $I = 10.00$ mA and $R_p = 10.00$ Ω, then v_o will vary from 0 to 100.0 mV and at any position will equal $100.0 \, R_v/R_p$ mV.

One type of VRS that is widely used is obtained by conversion of the ac line voltage to a dc voltage, regulation of the dc voltage, and voltage division with a Kelvin-Varley divider.

The Kelvin-Varley Divider. Precision resistors can be connected as shown in Fig. 1–39 to form an accurate voltage divider known as the Kelvin-Varley divider. It is especially useful because the effective resistance of the **network**[14] remains constant as the slider and switch positions are changed. The switch contacts in Fig. 1–39 move in pairs so that two series resistances in one branch (for example, KV-1) are always in parallel with the effective resistance of the next branch. By suitable selection of resistance values each parallel resistor combination has an effective resistance equal to one of the 11 series resistors. Therefore, the resistance combination in each branch is equal to 10 equal resistors in series; one-tenth of the voltage in each branch appears at the switch contacts and one-tenth of the voltage appears across each of the other resistors in the branch.

For the divider of Fig. 1–39, the 100 Ω variable potentiometer KV-3 is in parallel with two 50 Ω resistors in series, so that the parallel combination is 50 Ω. Thus the resistance of the KV-2 branch is $10 \times 50 = 500$ Ω in parallel with a precision 125 Ω resistor so that the parallel combination is

$$125 \times 500/(125 + 500) = 100 \ \Omega.$$

This 100 Ω resistance of the KV-2 branch is in parallel with two 50 Ω series resistors in the KV-1 branch. It can thus be seen that the effective resistance of the network is 500 Ω regardless of the switch or slider positions.

With 10 V across the input to the divider, the change in successive switch positions in branch KV-1 provides 1 V increments across the input to branch KV-2. Likewise the change in successive switch positions in

Note 14. Electrical Network.
A network is any combination of resistors and/or other circuit elements that are connected to perform a specific function.

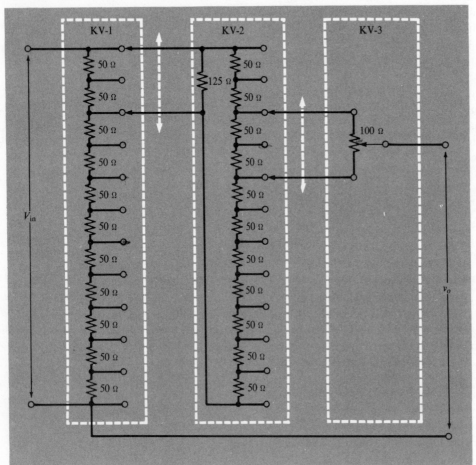

Fig. **1–39** Kelvin-Varley voltage dividers.

KV-2 provides 0.1 V increments. The variation of the slider from one end to the other provides continuous variation of the output voltage over a 0–0.1 V span. With the switch positions set as in Fig. 1–39, the output voltage would be 9.750 V.

Experiment 1–9 *Voltage Reference Source*

A laboratory voltage reference source is characterized in this experiment. The experiment illustrates the manner in which a voltage divider is used to provide different output voltages from a constant source, the dependence of the equivalent resistance of the source on the position of the voltage divider, and the accuracy of the source. The decade voltage dividers in a laboratory source are used to provide a wide range of output voltages for calibration and other purposes. Voltmeter measurements are made to determine the accuracy of the source, and a load is placed on the source in order to measure its equivalent resistance. The equivalent resistance is seen to be dependent on the voltage selected.

Electromagnetic Voltage Generator

If a conductor is moved in a magnetic field, an emf is induced in the conductor which is proportional to the rate with which the conductor traverses the magnetic field lines. If the conductor is part of a closed circuit, the induced emf causes an induced current, and the direction of the current depends on the direction of conductor motion across the magnetic field. This is the basis for many electrical transducers and power generators.

Figure 1–40a shows a metal wire moving downward at a velocity u through a uniform magnetic field B which is constant in time. Since there is a force exerted upon charged particles moving in a magnetic field, the mobile electrons in the conductor will have a net motion in the direction shown. The force on the electrons is proportional to the magnetic field, B, and the velocity of the conductor, u. The resulting excess of electrons at one end of the wire creates an electric field or emf in the wire. The electrons will continue to move until the force on the electrons due to the resulting electric field is equal and opposite to the force on the electrons due to the movement of the conductor in the magnetic field. Thus the emf produced is also proportional to B and u, and emf $= kuB$. The quantity uB is proportional to the rate with which the wire traverses the magnetic field lines, indicating that an equivalent voltage can be produced by moving a wire quickly through a weak magnetic field or more slowly through a stronger magnetic field.

If the conductor is in a closed circuit, the magnetically induced emf causes a current i in the direction indicated by the arrow coming out of the end of the conductor. If the direction of conductor movement is reversed as shown in Fig. 1–40b, then the directions of the electric field and the current are reversed.

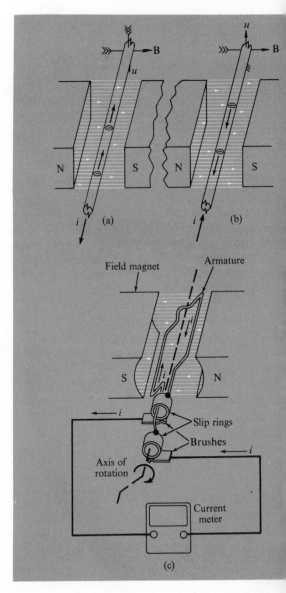

Fig. 1–40 Induced voltage and current. (a) By movement of a conductor downward through a magnetic field. (b) By movement of a conductor upward through a magnetic field. (c) By rotation of a conducting loop in a magnetic field.

The rotation of a loop or coil of wire in a magnetic field is illustrated in Fig. 1–40c. What happens in this case is the basis for the existence of electrical power generators and important types of transducers. The conductor loop is called the armature. The armature is connected to an external closed circuit through graphite brushes which ride against each slip ring attached to the rotating armature.

It can readily be seen in Fig. 1–41 that as the armature rotates, the rate of traversal of the magnetic field lines by the armature wires changes. In Fig. 1–41a the conductor loop is first shown moving parallel to the magnetic field so that the rate of traversal of magnetic field lines is zero and the induced emf is zero. When the conductor loop is moving at right angles to the magnetic field, as in the second position, the induced emf is at a maximum. Therefore the induced voltage changes from zero to a maximum value as the armature rotates from 0° to 90°. As the armature loop rotates, there is a change in the **magnetic flux,** ϕ, of the loop, that is, the number of magnetic field lines which are enclosed by the loop. The rate of change of ϕ is, of course, exactly equal to the rate with which the armature wires are traversing the magnetic field lines. Therefore, the emf induced in the armature is proportional to the rate of change of flux, $d\phi/dt$, which is called the **flux linkage.** Thus emf $= k'd\phi/dt$. Then, as the armature continues to rotate from 90° to 180°, the induced voltage returns to zero and then reverses polarity as the armature continues through one rotation from 180° to 270°. The polarity reversal is due to the fact that the armature wire which was moving *down* through the field is now moving *up* through the field and vice versa. At 270° the induced voltage is a maximum of equal magnitude but opposite polarity to the maximum at 90°. As the rotation of the armature continues back to the starting point, the induced voltage decreases from the maximum to zero. Thus one rotation of the armature produces the familiar sinusoidal wave shape descriptive of induced alternating (ac) voltage and ac current shown in Fig. 1–41a.

The magnitude of the sine wave voltage at any instant is called the **instantaneous voltage** v; it attains a maximum value V_{max} twice (once $+$ and once $-$) during each cycle. The rate of traversal of the magnetic field lines is proportional to the sine of the angle θ for the conductor loop relative to the starting position shown in Fig. 1–41. At $\theta = 0°$ the plane of the loop is perpendicular to the flux lines. The instantaneous voltage $v = V_{max}$ sin θ, and the instantaneous current $i = I_{max}$ sin θ. The important ac relationships are discussed in detail in Module Four, *Optimization of Electronic Measurements.*

If the slip rings of the ac generator are replaced with two semi-cylindrical segments (commutator), as illustrated in Fig. 1–41b, a dc generator is obtained. The effect of the commutator contact is to reverse the external connections to the rotating loop every half-rotation, as shown.

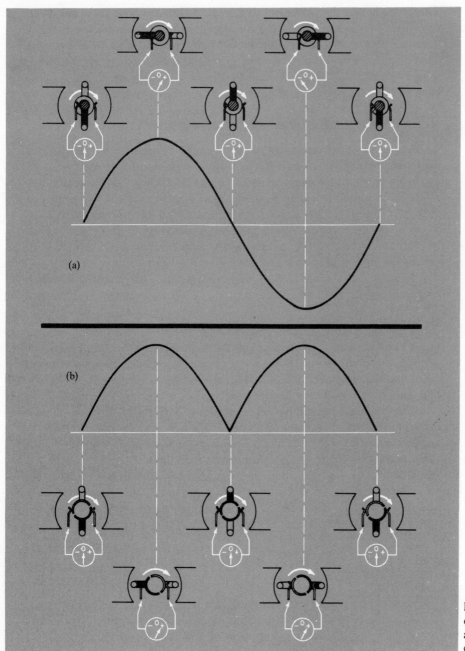

(a)

(b)

Fig. 1–41 Induced voltage and current caused by rotation of a conducting loop in a magnetic field. (a) Ac connection. (b) Dc connection.

The result is that the current in the external circuit is unidirectional. When the current variations due to the rotation of the armature are too rapid for a meter pointer to follow, the meter pointer will point to the average current value. For the ac generator, Fig. 1–41a, the average current output is zero. However, the dc generator, Fig. 1–41b, has an average output current which is proportional to the rate of rotation of the armature.

1–3.2 TRANSDUCERS WITH VOLTAGE OUTPUT

A thermocouple produces an output voltage related to the temperature difference between two metallic junctions; a glass/calomel electrode pair produces an output voltage related to solution acidity (pH); specific ion/ reference electrode pairs can be obtained that respond selectively to provide voltages related to the activity or concentration of calcium, potassium, fluoride, or other specific ions in solution; a piezoelectric device can provide an output voltage related to changes in pressure. In these examples the sought-for information is transduced into a related voltage which is then measured. Characteristics of several transducers that produce a voltage output are described in this section.

Faraday Cage

A transducer that can be used to monitor electron or ion beams and convert the number (and charge) of incident particles to a proportional output voltage is shown in Fig. 1–42. It is known as a Faraday cage or cup, and consists of a well insulated metallic cup.

The beam of charged particles enters the cage through a small hole as illustrated in Fig. 1–42. An aperture in front of the cup opening prevents incident particles from reaching the outside of the cage, where they could cause secondary electrons to be knocked out of the walls and collected. The beam of ions or electrons that enter the cage can be reflected and produce secondary electrons, but the cage design prevents any significant loss of charged particles. Therefore, if the incoming beam consists of electrons, and the capacitance between the cage and common is C, the output voltage is

$$v = \frac{q}{C} = \frac{n_e Q_e}{C} = k n_e, \tag{1–39}$$

where n_e is the number of incoming electrons over the collection period, and $k = Q_e$ (charge of electron in coulombs)$/C$ (capacitance in farads).
If the incoming beam is composed of positive ions, then

$$v = \frac{q}{C} = \frac{n_1 Q_1 + \ldots + n_n Q_n}{C}, \tag{1–40}$$

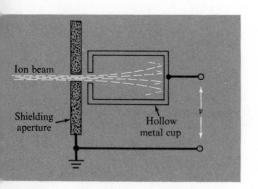

Fig. 1–42   Faraday cage.

Ion beam

Shielding aperture

Hollow metal cup

v

where n_1 and Q_1 are the number and charge (coulombs) per ion of ion 1, and n_n and Q_n are the same for other ion species in the incoming beam.

As few as 10^3 charged particles can be measured with a Faraday cage.

Charge Excess Transducer

In studies of atmospheric electricity, the excess charge in air is a significant quantity. This can be determined by a charge-collector transducer as illustrated in Fig. 1–43.

A metal cylinder is insulated from ground and shielded with a grounded metal enclosure as shown. The cylinder is packed with fine steel wool. After the cylinder has initially been discharged, the air that contains ions of either polarity is pumped through the steel wool. The incoming charged particles transfer their charge to the cylinder, and if the positive and negative charges are equal in number, the cylinder will remain uncharged. If there are more positive charges than negative ones in the air, then the cylinder will become positively charged with respect to ground, as indicated by the polarity of the output voltage. The polarity of the output voltage will be reversed for an excess of negative charges.

The net difference in the positive and negative charges that are drawn into the transducer produces an output voltage proportional to the difference, so that

$$v = \frac{q}{C} = \frac{n_d M t Q_e}{C}, \qquad (1-41)$$

where n_d is the excess of positively charged particles $(n_+ - n_-)$ per cubic meter (m^3) of air, M is the gas-volume flow velocity in m^3/sec, and t is the collection time.

Piezoelectric Transducer

When certain crystalline substances are mechanically deformed (by twisting, bending, or shearing forces), charge carriers are displaced and separated along certain crystalline axes. If electrodes are placed on specific faces of the crystal, then the mechanical deforming force in a suitable direction will be transduced into a linearly related output voltage across the electrodes. This characteristic of crystalline material is called the **piezoelectric** effect. The phenomenon is reversible in that the application of a potential difference across the electrodes causes a distortion of the crystal. The two effects are responsible for the use of piezoelectric devices in many applications.

The basic ways of deforming piezoelectric crystals are shown in Fig. 1–44, and some typical mountings are shown in Fig. 1–45.

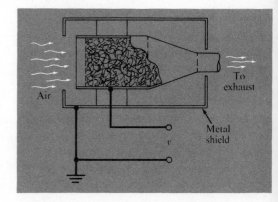

Fig. 1–43 Charge-excess transducer.

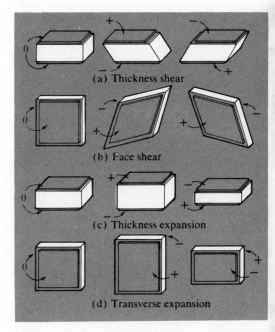

(a) Thickness shear

(b) Face shear

(c) Thickness expansion

(d) Transverse expansion

Fig. 1–44 Piezoelectric crystal deformations.

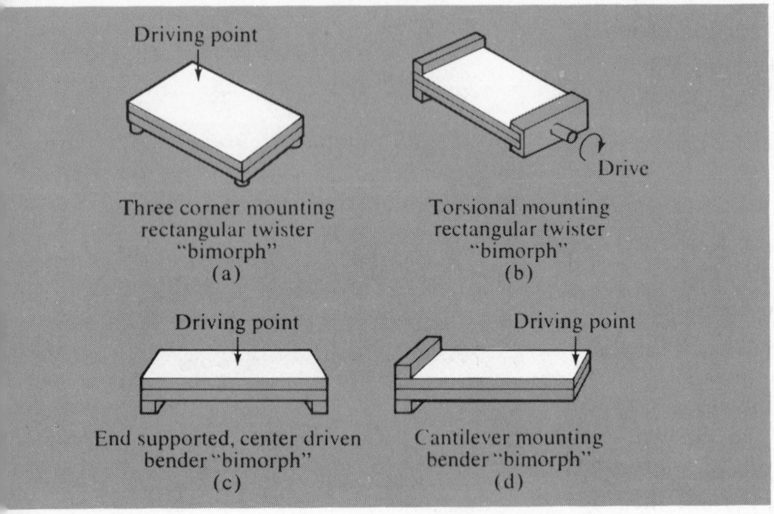

Driving point

Three corner mounting
rectangular twister
"bimorph"
(a)

Drive

Torsional mounting
rectangular twister
"bimorph"
(b)

Driving point

End supported, center driven
bender "bimorph"
(c)

Driving point

Cantilever mounting
bender "bimorph"
(d)

Fig. 1–45 Mountings for piezoelectric crystals.

Piezo crystals are relatively stiff and the tolerable distortions are small. For example, the deformation of crystals used in phonograph pickups is about 10 μm/g of weight. Small crystals of this type develop outputs of a fraction of one volt, whereas some very large crystals, under high forces, can develop outputs of several hundred volts.

Because of the leakage resistance of piezo crystals the output voltage is not maintained when a constant force is applied. The device acts like a capacitor with a resistive load which slowly discharges after the charging source has been removed. Consequently piezo crystals can be used only where the mechanical forces change periodically, usually at rates of a few cycles to megacycles per second.

In the biomedical field piezoelectric transducers have been used for converting heart sounds, muscle pull, respiration, arterial and venous pulses to related output voltages. The use of piezoelectric crystals in many types of phonograph pickups, crystal microphones, and industrial pressure transducers is widespread.

Photovoltaic Cell

When input photons are incident on certain types of dissimilar material boundaries or junctions, the energy in the radiation can displace charge carriers and produce voltages across the junction. These devices are known as barrier-layer or photovoltaic cells, and a typical unit is illustrated in Fig. 1–46. This unit consists of a thin coating of selenium on an iron base plate; a thin transparent film of cadmium metal is adjacent to, but insulated

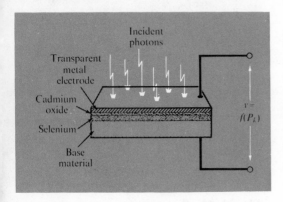

Incident
photons

Transparent
metal
electrode

Cadmium
oxide

Selenium

Base
material

$V = f(P_\lambda)$

Fig. 1–46 Photovoltaic cell.

from the selenium by a layer of cadmium oxide. The insulation region constitutes a barrier layer across which a potential difference develops when photons are incident on the device. The metal film becomes negative and the selenium positive.

Selenium cells are sensitive to radiation in the spectral range 300 to 700 nm, with the sensitivity peaking at about 560 nm, as shown in Fig. 1–47. It is therefore easy to combine the selenium cell with a filter such that the spectral response of the combination is similar to the human eye. Selenium cells are especially useful in camera exposure meters and simple colorimeters.

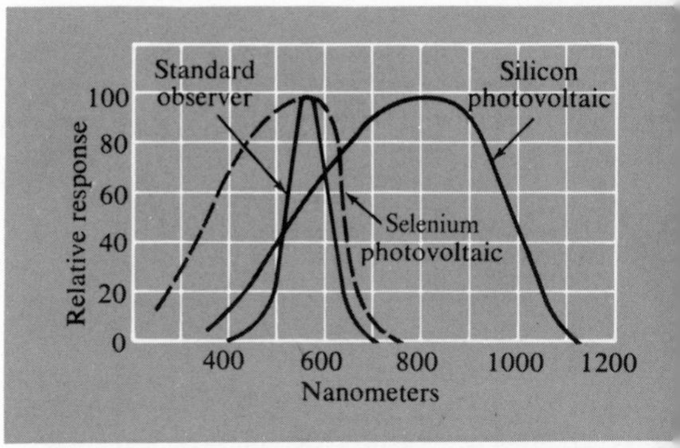

Another prominent type of photovoltaic cell is made with a silicon semiconductor. Since it can provide relatively large currents, it can be used as a power source. Because of the relatively high efficiency of the silicon cell in producing power from sunlight, the unit is often called a **solar battery.** The use of the solar battery as a power source is discussed in Module Two, *Control of Electrical Quantities in Instrumentation.*

Arrays of silicon cells have been built for the purpose of reading hole patterns in punched-card and punched-tape systems. They are also part of many other light detection systems.

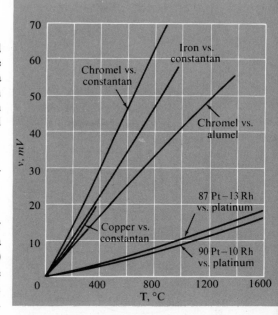

Fig. 1–47 Spectral response curves for selenium and silicon photovoltaic cells and the human eye.

Experiment 1–10   *Photovoltaic Cell Light Measurements*

Light measurements are made in this experiment with a photovoltaic cell to show that the cell is a transducer which provides an output voltage related to the light intensity and that the cell can be characterized by a voltage source equivalent circuit. The voltage output of the cell in room light and in darkness and the maximum output voltage are measured with a voltmeter. The equivalent resistance is determined by shunting the cell with a load resistor and an equivalent circuit is drawn.

Temperature-to-Voltage Transducer

The thermocouple was introduced in Section 1–1.2 as an illustration for mapping domain conversions. This very important transducer produces a relatively low output voltage which is related to temperature (about 1–10 $\mu V/°C$). Typical temperature-voltage curves for several thermocouple combinations are given in Fig. 1–48. The sensitivity, or change in voltage for a given change in temperature, dv/dT, depends on the chemical composition and the physical treatment of the materials used to make the thermocouple.

Fig. 1–48 Graph of transfer functions for several common thermocouples.

If several thermocouples are combined as shown in Fig. 1–49, the assembly is called a **thermopile.** The greater the number of thermocouples connected together, the greater the sensitivity of the pile; the sensitivity increase is in direct proportion to the number of thermocouples in the thermopile.

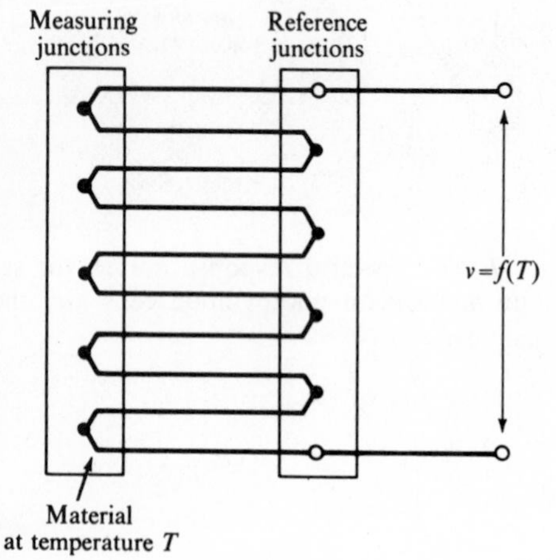

Ion-Selective Electrodes

The possibility of inserting a transducer into a solution which then responds selectively to a specific chemical constituent is indeed attractive. For many years it has been possible to insert an electrode pair into a solution and obtain an output voltage related to pH (solution acidity). In recent years it has also become possible to use other ion-selective electrodes that respond to specific cations and anions as shown in Table 1–4. These electrodes utilize different types of membranes to obtain their specificity.

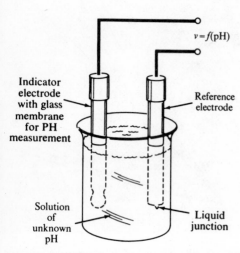

Measuring junctions / Reference junctions

$v = f(T)$

Material at temperature T

Fig.   1–49   Thermopile.

Table 1–4   Common ion-selective electrodes

Cation electrodes		Anion electrodes	
Ion	Membrane type	Ion	Membrane type
Cadmium	Solid state	Bromide	Solid state
Calcium	Liquid ion exchange	Chloride	Solid state
Copper	Solid state	Cyanide	Solid state
Hydrogen	Glass	Fluoride	Solid state
Lead	Solid state	Nitrate	Liquid ion exchange
Mercury	Solid state	Perchlorate	Liquid ion exchange
Sodium	Glass	Sulfide	Solid state
Univalent	Glass	Thiocyanate	Solid state
Divalent	Solid state, liquid ion exchange		

$v = f(\text{pH})$

Indicator electrode with glass membrane for PH measurement

Reference electrode

Solution of unknown pH

Liquid junction

Fig.   1–50   pH/reference electrode pair.

The potential of an indicator electrode, v_{ind}, is measured with respect to a reference electrode as illustrated in Fig. 1–50. The potential of the reference electrode, V_{ref}, is independent of the test solution composition, but there is generally a small contact potential difference or **liquid junction potential,** V_j. The emf of a complete electrochemical cell is

$$v_{\text{cell}} = v_{\text{ind}} + V_{\text{ref}} + V_j. \tag{1–42}$$

Since the reference electrode potential is independent of sample composition and V_j is usually either negligible or constant, the two can be considered a constant voltage V_{rj} so that

$$v_{\text{cell}} = v_{\text{ind}} + V_{rj}. \qquad (1\text{-}43)$$

From the Nernst equation the potential of an indicator electrode can be written as

$$v_{\text{ind}} = V^0 + \frac{RT}{n\mathcal{F}} \ln a_{\text{ind}}, \qquad (1\text{-}44)$$

where R, T, n, and $\mathcal{F}$ are the gas constant, absolute temperature, ionic charge, and Faraday constant, respectively. The constant standard potential for the indicator electrode is given by V^0, and a_{ind} is the activity of the ionic species to which the electrode responds. By combining V^0 and V_{rj} into one term V', the observed cell potential is given by

$$v_{\text{cell}} = V' + \frac{RT}{n\mathcal{F}} \ln a_{\text{ind}}. \qquad (1\text{-}45)$$

Since V' is constant, the measured voltage is seen to vary with the logarithm of the ionic activity of a specific ion as determined by the selected indicator electrode.

The indicator electrode shown in Fig. 1–51a has a glass membrane which is sensitive to changes in pH. The glass electrode has a high resistance, typically 50–500 MΩ. The indicator electrode shown in Fig. 1–51b has a thin layer of ion exchanger which provides a membrane potential that changes as a function of calcium ion activity.

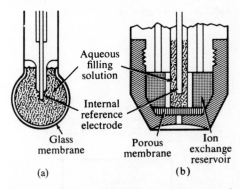

Aqueous filling solution

Internal reference electrode

Glass membrane

Porous membrane

Ion exchange reservoir

(a)

(b)

Fig. 1–51 Specific ion membrane electrodes. (a) Glass. (b) Liquid ion exchange.

Electromagnetic Transducers

In Section 1–3.1 an electromagnetic voltage generator was described in which the induced voltage is proportional to the change in the magnetic flux ϕ in a coil of wire per unit time t, $d\phi/dt$. Also, the induced voltage v is proportional to the number of turns in the coil so that

$$v = -N \frac{d\phi}{dt}.$$

Since the movement of a conductor in a magnetic field can induce an emf, it is not surprising that the electromagnetic effect is the basis for angular and linear velocity transducers.

Linear Velocity Transducers.   A simple form of an electromagnetic linear velocity transducer is illustrated in Fig. 1–52. The induced voltage is

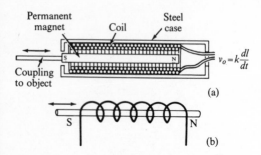

Fig. 1–52 An electromagnetic velocity transducer: (a) Pictorial. (b) Schematic.

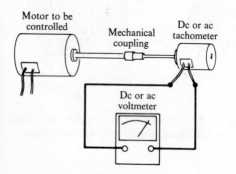

Fig. 1–53 Tachometer generator application for motor speed control.

proportional to the velocity at which the magnetic core moves in or out of the coil. The object whose velocity is to be measured is attached to the permanent magnet. Alternatively, the magnet may be held stationary, and the coil attached to the object.

Angular Velocity Electromagnetic Transducers.   Devices that are designed to determine the rotation rate of mechanical systems by measuring angular velocity are called **tachometers.** In a dc tachometer generator a permanent magnet stator can be used with a standard generator winding and a commutator-type rotor. The output voltage is typically a few volts per 1000 revolutions per minute (rpm). The polarity of the output voltage is dependent on the direction of rotation.

An ac permanent magnet tachometer is similar in principle to the simple ac electromagnetic voltage generator shown in Fig. 1–40c. An application of a dc or ac tachometer is illustrated in Fig. 1–53. A small tachometer generator is attached to a rotating shaft so as to develop a voltage output proportional to the shaft speed. A change of shaft speed will produce a change of tachometer output voltage. This change in voltage produces an error signal that can be used to restore the speed to the desired value.

Hall Effect Transducers

An important type of interaction, known as the *Hall effect,* occurs if a current-carrying conductor is introduced into a magnetic field. When a metal strip is fixed in position with its plane perpendicular to a magnetic field, and the control current i_c is in a direction through the strip, as shown in Fig. 1–54a, then a potential difference is developed across the strip at right angles to the current flow and to the magnetic field. This potential difference is called the *Hall voltage, v_H,* and is expressed as

$$v_H = ki_c B/d,$$

where B is the field strength (gauss), d is the thickness of the strip in centimeters, and k is the Hall coefficient which depends on the type of material and the temperature.

The Hall coefficient, k, is very large for n-type germanium and indium arsenide or antimonide and quite small for most metals except for silicon, bismuth, and tellurium. Therefore, most devices are made with n-type semiconductors. The output Hall voltage is typically in the millivolt range per kilogauss at the rated control current. Although it might seem from the above equation that the Hall voltage can be increased by increasing the control current i_c and decreasing the thickness d, both changes would cause the strip to become excessively hot with concomitant changes in its characteristics. The internal resistances of typical Hall devices (generators) vary from a few ohms to several hundred ohms.

Since a change in magnetic field strength B provides a proportional change in the Hall voltage v_H, Hall effect devices can be used in many ways to transduce position into an electrical voltage. Several Hall effect transducer configurations are illustrated in Figs. 1–54b through d.

1–3.3 DIRECT VOLTAGE MEASUREMENT

The measurement of the voltage output from a transducer or other source can be accomplished in several ways. One method is illustrated in Section 1–2.3, where a sensitive current meter (current-to-position transducer) is put in series with a resistor and the series combination is connected across the voltage source terminals. The resistance of the combination can be considered as a voltage-to-current transducer. In this type of voltmeter significant power is required to move the meter needle, so that the relatively large current drawn from the source can significantly change (load) the voltage output. This can cause rather large measurement errors. To decrease the **loading** effect of voltmeters, it is shown in this section that the indicating device can be preceded by field-effect transistors or electrometer tubes which draw only very small currents from the voltage source. One of the most important instruments for display of voltage information, the oscilloscope, is also presented.

An ideal voltage measurement device requires no current or charge flow from the voltage source. Hence the measurement will produce no change in the source voltage due to a change in the electrical charge that provides the voltage, and no IR voltage drop caused by measurement current in the internal resistance of the voltage source. The ideal of zero current or charge flow in a voltage measurement is not perfectly attainable, but using the comparison technique (next section) and modern devices, one can make the charge flow negligibly small for most measurements.

Loading

One of the important considerations in making voltage measurements is **loading.** Any device that will draw current from a voltage source is called a **load.** Measurement devices such as oscilloscopes, vacuum tube voltmeters (VTVM), transistor voltmeters (TVM), and digital voltmeters (DVM) are often unsuspected significant loads on the sources from which they obtain data.

If a voltmeter draws too much current, the data obtained from the meter can be greatly in error. The question is: How much is *too* much current? To obtain an answer, it is probably easiest to represent a typical

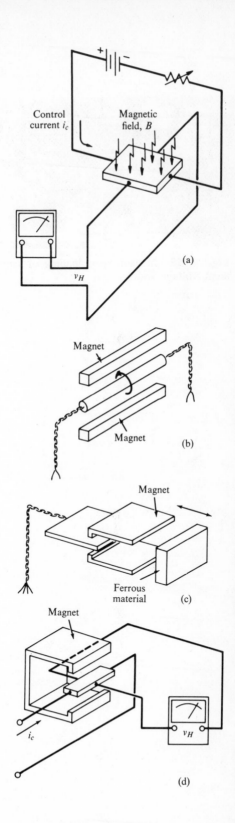

Fig. 1–54 Hall effect device. (a) Pictorial representation. (b, c, d) Some transducer configurations and applications.

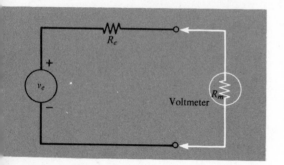

Fig. 1–55 Measurement loading of equivalent voltage source.

voltage source by an equivalent circuit as shown in Fig. 1–55. Even complex voltage sources can be represented by an equivalent voltage source, v_e, the open-circuit or "no-load" voltage, and a series resistance, R_e, the internal source resistance. All sources have some internal resistance —some have only a fraction of an ohm, but others have many megohms.

If a meter with an effective resistance R_m across its input terminals is connected to the voltage source, it can be seen that a current i will flow through the series resistances $R_e + R_m$. From Ohm's Law,

$$i = \frac{v_e}{R_e + R_m}.$$

The voltage across the meter input v_m is

$$v_m = iR_m.$$

If the above two equations are combined, v_m is given by

$$v_m = v_e - iR_e = \frac{v_e R_m}{R_m + R_e}. \tag{1–46}$$

In other words, the voltage as "seen" by the meter is essentially equal to the actual source voltage only if the voltage drop IR_e is negligible compared to v_e, and this qualification depends on the accuracy desired for the measurement. If R_e is very small and R_m is very large, the voltage drop iR_e will not introduce a "significant" measurement error. However, if $R_e = R_m$, the measured voltage is only one-half of the *actual* source voltage.

The relative perturbation error (RPE) caused by the introduction of the voltage measurement device is given by

$$\text{RPE} = \frac{v_e - v_m}{v_e}. \tag{1–47}$$

If the value for v_m from Eq. (1–46) is substituted into Eq. (1–47), the RPE is

$$\text{RPE} = -\frac{R_e}{R_m + R_e}. \tag{1–48}$$

Thus for a maximum error of 1%, the meter resistance R_m should be at least 100 R_e; for a 0.1% RPE, R_m must be 1000 R_e, etc.

Loading a Divider. Another typical loading problem is associated with voltage dividers. Often a voltage divider such as the one in Fig. 1–18 is made from relatively high resistance values to avoid loading the voltage source. When the series resistor network is connected across the voltage source terminals, five known fractions of the source voltage are switch-selectable. However, when the divider is "loaded" (a resistive load placed across the divider output terminals), the voltage across the load can be considerably less than that of the "unloaded" divider. This would be expected from consideration of the resistance values for the divider and the load. If it is assumed that the internal resistance of the voltage source is small compared to the divider resistance, one can readily understand the effective or equivalent resistance of the divider as the divider fraction changes. Consider the continuously variable voltage divider of Fig. 1–18b. When the potentiometer is at its bottom extreme, and the fraction b is zero, the equivalent resistance R_e of the divider must be 0 Ω since the output is essentially a short circuit. When the potentiometer is at its top extreme and b is one, the equivalent divider resistance must also be nearly 0 Ω since the output terminals are connected directly across the voltage source V, which has been assumed to have negligible internal resistance. When the potentiometer is in the middle of its range, the equivalent resistance must, therefore, reach a maximum value. As derived in Module Two, the equivalent divider resistance can be expressed by the equation $R_e = Rb(1 - b)$, where R is the total resistance of the divider. Thus, in Fig. 1–18a, where $R = 1$ kΩ, the equivalent divider resistance at the 0.05V setting is 1 kΩ $\times$ 0.05 $\times$ 0.95 = 47 Ω. At the 0.5V setting, however, $R_e = 1$ kΩ $\times$ 0.5 $\times$ 0.5 = 250 Ω, showing the divider to be more susceptible to loading at midsetting.

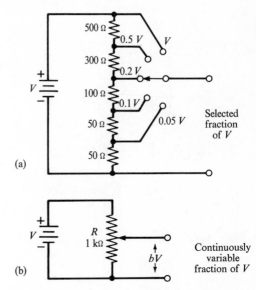

Fig. 1–18 (repeated) Voltage dividers. (a) Selector switch. (b) Potentiometer.

Experiment 1–11   *Loading Errors in Voltage Measurements*

Voltage measurements are made from a source of high equivalent resistance and from a source-voltage divider combination to illustrate that the introduction of the measurement devices can cause a significant loading or perturbation error. The voltage at the terminals of a battery and a large series resistor is measured with a relatively low resistance voltmeter. A precision voltage divider is connected to a battery, and the output voltage of the divider is measured with the low resistance meter. Equivalent circuits are drawn to explain the perturbation error.

Field-Effect Transistor Voltmeter

The rather large loading errors that can be caused by connecting a moving-coil voltmeter across a voltage source can be greatly decreased by using a junction field-effect transistor as a voltage-to-current domain converter.

The basis of operation of a field-effect transistor (FET) is easy to understand. A semiconductor provides a current path whose resistance to the flow of the charge carriers (electrons) is varied by applying a transverse

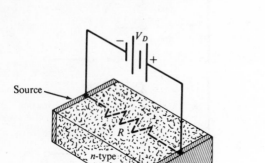

Fig. 1–56   An *n*-type semiconductor.

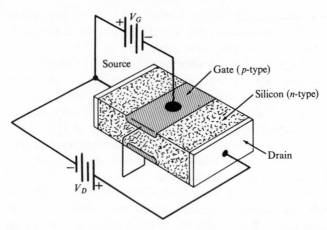

Fig. 1–57   Field-effect transistor showing applied voltages.

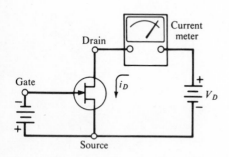

Fig. 1–58   Simple FET voltmeter.

electric field. Figure 1–56 shows the semiconductor material with the two contacts called the **source** and the **drain.** The material between the source and drain is the **channel.** A third contact is made to the channel through a region called the **gate.** When the gate voltage is negative with respect to the source (for an electron-conducting semiconductor), the contact between gate and the channel is nonconducting. As shown in Fig. 1–57, the gate-channel potential difference creates an electric field in the channel which tends to reduce the number of charge carriers in that region. The stronger the electric field, the fewer charge carriers and the higher the channel resistance. When the FET is connected in the circuit of Fig. 1–58, the source-drain current i_D will depend upon the gate-source potential difference. Since the gate-channel contact is essentially nonconducting, the voltage source for the gate circuit needs to provide only a very small current (10^{-7} to 10^{-12} A) to the FET.

The drain current can be used to operate a current meter. The use of an FET with an indicating meter to form a simple FET voltmeter is illustrated in Fig. 1–58. The increment of change in drain current for a given change in gate potential ($\Delta i_D / \Delta v_G$) is essentially constant, is called the **transconductance,** and has the units of reciprocal resistance.

Multiple ranges in a voltmeter are obtained by using a voltage divider connected across the input as shown in Fig. 1–59. For inexpensive FET voltmeters a divider with a total resistance of 100 MΩ is commonly used. For many applications a 100 MΩ divider input resistance causes negligible current from the voltage sources measured with the FET voltmeter. However, for some applications even a 100 MΩ input resistance represents a significant load.

Experiment 1–12 *Field-Effect Transistor Voltage-to-Current Converter*

The FET is studied in this experiment to show that it is a voltage-actuated device which can supply current to a load while drawing only a very small current from a source and, used as the input stage to a meter, it is an excellent device for reducing loading errors. The drain current of the FET is measured as a function of gate (input) voltage, and the slope of the transfer function plot (transconductance) is determined. An upper limit for the input current to the FET is determined. The FET or a similar high input resistance device such as a vacuum tube is used for a voltmeter input in Experiment 1–13.

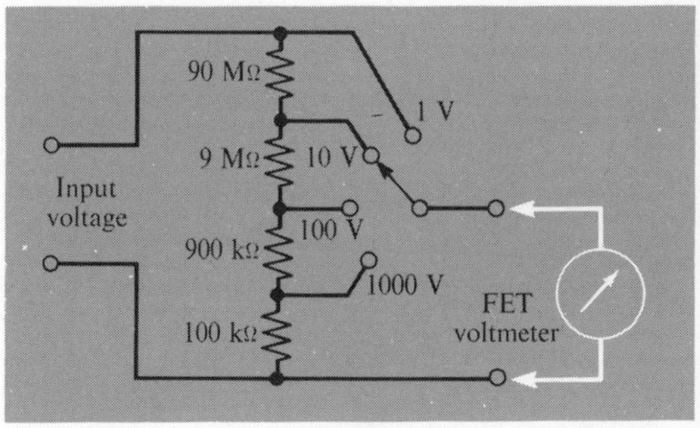

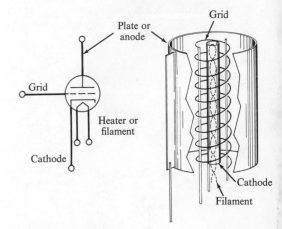

Fig. 1–59 Input voltage divider for FET voltmeters.

The Vacuum Tube Voltmeter (VTVM)

A vacuum tube can be used in the same way as a FET to make a voltmeter with a high input resistance. The **electrometer** tube is a special type of vacuum triode. The **cathode** of the vacuum triode, Fig. 1–60, is heated to the temperature required for the thermionic emission of electrons. During normal operation, the **anode** (plate) is positive with respect to the cathode, and electrons flow from the cathode to the anode. A control grid is placed relatively close to the cathode. Because of its proximity, a small negative potential on the grid with respect to the cathode can counteract the attractive force of a relatively large positive potential on the anode with respect to the cathode. As the grid is made less negative (more positive), more electrons will be attracted to the anode; thus, the grid-cathode potential controls the current from a power supply connected between cathode and anode.

A simple dc vacuum-tube electrometer is shown in Fig. 1–61. The plate current as a function of the grid potential for a typical vacuum triode is also shown in Fig. 1–61. It can be seen from this transfer characteristic that the potential applied between the grid and the cathode will control the current through a meter that is inserted in the plate circuit. The slope of this curve is the **transconductance** of the electrometer tube.

Fig. 1–60 Vacuum triode. (a) Schematic symbol. (b) Pictorial.

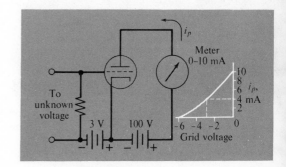

Fig. 1–61 Simple vacuum-tube voltmeter. Circuit and plot of i_p vs. grid voltage.

The circuit described has a limited range of input voltages, is nonlinear at the ends of the transfer characteristic, and, once calibrated, requires that the tube maintain constant characteristics. These limitations can be overcome to a great extent by more complex circuits. However, the principle of the operation of all vacuum tube electrometer amplifiers is the same as that of Fig. 1–61.

Both FET and vacuum tube electrometer amplifiers can greatly reduce loading effects on the circuit to be measured. With sufficient care some electrometer circuits load the measured voltage source with a current as low as 10^{-17} A, representing only about 64 electrons/sec. One of the most familiar applications of electrometer tube amplifiers was in pH meters, but it is now more common to use FET amplifiers.

Experiment 1–13   *Field-Effect Transistor Voltmeter or Vacuum Tube Voltmeter*

A field-effect transistor (FET) voltmeter or a vacuum tube voltmeter (VTVM) is used in this experiment in the measurement of source voltages. The input resistance of the voltmeter is determined in order to compare this type of voltmeter with the series resistance moving coil type voltmeter. The input resistance is also determined as a function of the full-scale voltage setting, and a suitable input voltage divider for an FET voltmeter or VTVM is designed. The accuracy of the voltage measurement is compared with the ± 2%–3% of full scale which is typical for these meters.

The Oscilloscope

The oscilloscope has become an almost indispensable laboratory instrument because of its ability to display voltage information as a function of time t, or to plot one voltage x as a function of another voltage y. The x vs. y or x vs. t plots are displayed on the face of a cathode-ray tube (CRT). The CRT is the indicating device for an oscilloscope in the same sense that the moving-coil meter is the indicating device for a voltmeter. The tube consists of the so-called "electron gun," shown in Fig. 1–62a, and the deflection plates (Fig. 1–62c) combined in a vacuum tube with a fluorescent screen on the enlarged end, as shown in Fig. 1–62b. The purpose of the electron gun is to provide a beam of electrons that is focused to a sharp point at the fluorescent screen. The end of the tube is coated with various phosphors which emit visible phosphorescent radiation at the point of bombardment with electrons. The intensity of the emitted visible light depends on the

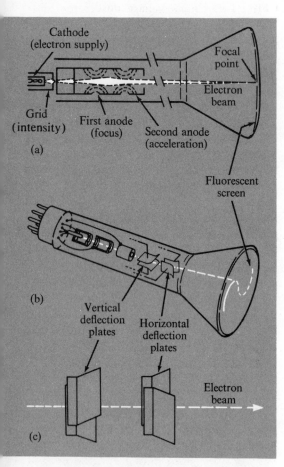

Fig. 1–62 Cathode-ray tube. (a) Electron gun. (b) Cathode-ray tube. (c) Deflection plates.

number of electrons per unit time striking a given area of the screen; the larger the number, the more intense the light. By partially surrounding the cathode with a metal grid, and applying a voltage such that the grid is negative with respect to the cathode, the rate of electron flow through the grid can be regulated. The more negative the grid potential, the greater the repulsion of electrons, the fewer the electrons reaching the fluorescent screen, and the lower the intensity of the visible light spot. In other words, a variable grid-cathode voltage supply acts as the **intensity control.**

To obtain the point source of electrons on the screen, it is necessary to focus and accelerate the electrons by two anodes that are at high positive potentials with respect to the cathode. For a typical 5 in. tube, the second anode is often 2000 to 10,000 V more positive than the cathode, and the first anode is about 350 to 750 V more positive. The electrostatic field that exists between the two hollow cylindrical anodes provides the necessary focusing of the electron beam. The diverging electrons entering the first hollow anode are forced to converge because of the field that exists between the first and second anode. The desired focus point is, of course, the fluorescent screen, and the correct focus can be obtained by varying the voltage on one anode with respect to the other. The **focus control** is another front-panel adjustment on the oscilloscope.

Electron Beam Deflection.   The electron beam then passes through the first set of deflection plates that are mounted in the horizontal plane, as shown in Fig. 1–62. These plates are referred to as the **vertical deflection plates.** Next, the beam passes between two plates mounted perpendicular to the first set called the **horizontal deflection plates.** The horizontal and vertical deflection plates of the CRT serve to deflect the electron beam. The CRT deflection plates are shown schematically in Fig. 1–63. The point in the center represents the spot on the scope face caused by the undeflected electron beam.

If one vertical deflection plate is made positive with respect to the other vertical plate, the electron beam is deflected toward the positive plate. This may be accomplished by applying a voltage as shown in Fig. 1–64. The larger the magnitude of the applied voltage, the greater the deflection of the electron beam. A horizontal deflection of the electron beam may be accomplished in an analogous manner using the horizontal deflection plates. If voltages are applied to both sets of plates at the same time, the position of the electron beam in the *xy*-plane will depend on the sign (positive or negative) and the magnitude of the two deflection voltages. In this manner the electron beam can plot out in the *xy*-plane one deflection voltage as a function of the other.

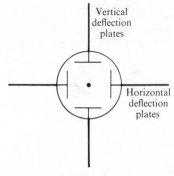

Fig. 1–63 Schematic representation of the CRT deflection plates.

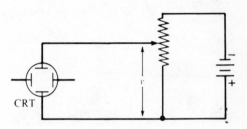

Fig. 1–64 Application of a deflection voltage to the vertical CRT plates.

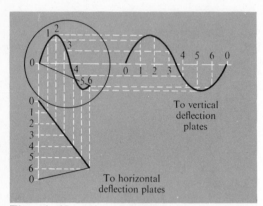

Fig. 1–65 Linear horizontal sweep for display of vertical deflection voltage vs. time curve.

Linear Sweep for Horizontal Deflection Plates. For most applications of the oscilloscope, the voltage applied between the horizontal deflection plates produces a **linear trace.** That is, the light spot is moved horizontally across the face of the tube at a uniform rate. The linear trace is important because it provides a uniform time scale against which another voltage can be plotted by applying it across the vertical deflection plates. This is illustrated in Fig. 1–65 where a **sawtooth** voltage provides the linear horizontal movement of the electron beam, the voltage applied to the vertical deflection plates is plotted vs. time, and the curve is displayed on the screen.

Since the electron beam is "swept" across the screen at a uniform rate, it is customary to refer to the sawtooth waveform as the **sweep.** The oscillator circuit that produces the sweep is called the **sweep generator.** The rate of change of voltage must be very uniform in order to obtain a reliable time base. Good linearity of the sweep is a major requirement for a reliable oscilloscope. In most modern scopes, the time base is calibrated in sec/cm and is accurate to 3%. At the end of the sweep the electron beam is moved back across the screen to the starting point by lowering the sweep voltage back to zero. The finite time required to change the sweep voltage back to the starting value can cause a visible trace of low intensity, known as the **retrace** or **flyback.** By applying a negative grid voltage during the retrace time interval, it is possible to blank out the retrace so that it is not visible.

Triggered Sweep Oscilloscope. To provide a continuous, stable display of a repetitive waveform signal on the oscilloscope screen, the starting time of the sweep must coincide with a single point on the waveform of the signal to be observed. For a triggered sweep oscilloscope, this involves using the input signal to start the sweep at the same point on the waveform for each sweep. The triggering feature allows considerable flexibility in choosing the display mode. For instance, it is possible to choose whether the time base will trigger on a positive or a negative slope of the input waveform, and it is possible to select the triggering level, that is, the sign and magnitude of signal required to trigger the sweep.

Other advantages of a triggered sweep are that the signal input need not be repetitive to be displayed (i.e., single events can be shown since the event itself starts the sweep), and a calibrated time base (usually sec/cm) is provided, which makes it possible to perform time interval and frequency measurements. The scope displays observed for various triggering conditions are illustrated on the right side of Fig. 1–66.

Some inexpensive oscilloscopes use a continuous sawtooth sweep generator called a **synchronized sweep.** For a stable display, the sweep generator must oscillate at exactly the same frequency as the repetition

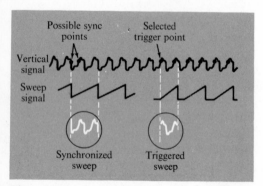

Fig. 1–66 Synchronized and triggered sweeps.

rate of the signal or some submultiple thereof. The sweep generator is synchronized to the signal frequency by using the input signal to control the sweep generator frequency over a limited range. A manual adjustment of the coarse sweep generator frequency is required. The relationship of the sweep signal and the input waveform is shown in Fig. 1–66. Since the sweep rate depends on the frequency of the signal, the exact time relationship between various parts of the trace is not known.

Basic Components of an Oscilloscope. In addition to the CRT and the sweep generator, other circuits are required to make up an oscilloscope, as shown in the block diagram of Fig. 1–67. The voltage required to deflect the electron beam is usually much higher than the voltage levels that are measured. For this reason, amplifiers are used to increase the signal voltage to a level required by the deflection plates. The quality and type of oscilloscope are greatly dependent on its amplifiers. Very high frequency and dc amplifiers are available in some oscilloscopes, covering the range from dc to 10 MHz for a good scope, or beyond 500 MHz for certain wideband scopes. The input resistance of most oscilloscopes is typically 1–10 MΩ.

A probe with an internal resistance, as shown in Fig. 1–68, is frequently used to increase the input resistance of an oscilloscope tenfold. The combination of R_p and R_m forms a voltage divider or **attenuator** which decreases the maximum sensitivity of the oscilloscope tenfold.

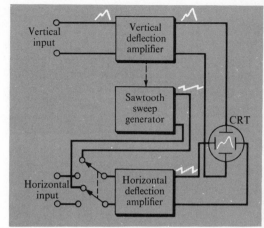

Fig. 1–67 Block diagram of basic oscilloscope.

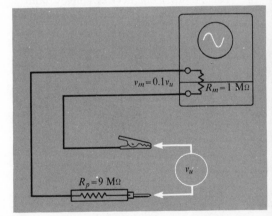

Fig. 1–68 A "×10" attenuator probe.

Experiment 1–14 *Voltage Measurements with the Oscilloscope*

The oscilloscope is used and studied here as a voltage measurement instrument in order to demonstrate its functions and versatility in measurements. Voltages are applied to the vertical and horizontal inputs, and the deflections of the trace are measured. Input voltages are also applied simultaneously to both inputs in order to obtain *x-y* deflections. The oscilloscope time base and triggering controls are introduced and used in the measurement of a time-dependent signal. A typical oscilloscope probe used for attenuation and increased input resistance is characterized, and a circuit diagram of the probe is drawn.

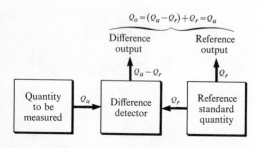

$$Q_o = (Q_u - Q_r) + Q_r = Q_u$$

Difference output   Reference output

$Q_u - Q_r$ Q_r

| Quantity to be measured | Q_u → | Difference detector | ← Q_r | Reference standard quantity |

Fig. 1–69 Elements of the basic measurement system.

Table 1–2

1. All measurement devices employ both a difference detector and a reference standard quantity.

2. Either the difference detector or the reference standard can affect the accuracy of the measurement.

3. The reference standard quantity has the same property or characteristic as that which is being measured.

1–3.4 VOLTAGE MEASUREMENT BY NULL COMPARISON TECHNIQUES

In Section 1–1.3 of this module the basic measurement system shown in Fig. 1–69 is discussed and the three measurement concepts in Table 1–2 were outlined. This table is repeated here for convenience as these concepts will be referred to often in this section.

The difference detector is used to measure the difference between the unknown and standard quantities. The measured value is the sum of the difference and reference standard outputs. The ordinary voltmeter made from the moving coil current meter has a zero adjustment only. That is, the reference standard used in conjunction with the voltage detector output is a zero voltage standard. Once zeroed, the entire measurement information comes from the voltmeter (difference detector) and depends upon the accuracy and stability of the voltmeter's transfer function. When a reference standard voltage is used in conjunction with a voltmeter, the range and accuracy of the resulting measurements can be increased by as much as two orders of magnitude. This technique, called **offsetting,** is described in this section and is used to introduce the technique of null comparison voltage measurement.

If the standard quantity is adjustable in sufficiently fine increments, the difference output can be adjusted to zero. This process is called a **null comparison.** Since, at null, the measured value is exactly equal to the reference standard value, the off-null output of the difference detector need not be calibrated. The only requirement of the difference detector is that it be able to indicate, with sufficient sensitivity, that a difference between the standard and unknown quantities exists.

Because the null comparison measurement technique depends on a local reference standard quantity and not on the transfer function of the difference detector, it is the most direct and accurate measurement method for many quantities. In order to be applied, a variable and accurately calibrated reference standard of the measured quantity and a suitable null detector must be available. The fact that these requirements are readily met in the case of voltage measurements has resulted in the development and wide application of a variety of null comparison voltage measurement methods. In this section, the application of the null comparison concept to voltage measurements is introduced, common null detectors and their characteristics are described, and the laboratory potentiometer and the technique of "potentiometry" are reviewed in this context.

Voltage Measurements with Offset

The measurement of voltage using offset is represented by the circuit of Fig. 1–70. A source of stable and adjustable voltage, V_{os}, is placed in series

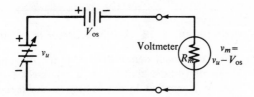

Fig. 1–70 Voltage measurement with offset.

with the unknown voltage source, v_u. The polarity of V_{os} is such that it is opposite to that of v_u in the series circuit. Thus the voltage applied to the voltmeter is $v_u - V_{os}$. The value of V_{os} is generally set to a large fraction (0.5 to 0.99) of v_u. Thus the voltage applied to the voltmeter is only 1% to 50% of v_u. This technique is thus useful in extending the range of the voltmeter or allowing it to be set on a more sensitive range. The value of v_u is the sum of V_{os} and the voltmeter reading, v_m. Both the accuracy and resolution of a measurement can be improved by offsetting. For instance, if V_{os} is 1.000 V and the voltmeter reads 0.047 V on the 0.1 V scale, $v_u = 1.047$ V. This many significant figures would not be available from a direct voltmeter measurement on the 1 V or 1.5 V scale. Furthermore, if the voltmeter accuracy is $\pm$ 2% of full scale, the measured value is 1.047 $\pm$ 0.002 V, whereas in a direct measurement on the 1.5 V scale, the reading would be 1.05 $\pm$ 0.030 V. These examples assume that the accuracy and stability of the offset source are within the accuracy and readability limits of the voltmeter.

Another application of the offset measurement technique is the measurement of small changes in a relatively large source voltage. In this case V_{os} is set to or near the initial, average, or smallest value of v_u, and the difference between v_u and V_{os} is indicated by the voltmeter. The voltmeter range can be set at the magnitude of the expected change which can then be read with much greater accuracy than if the entire magnitude of v_u were connected to the voltmeter. This technique is frequently used for such measurements as the variation of source voltage due to loading or the change in a voltaic cell voltage due to temperature variation or change in composition.

An additional advantage that occurs with offset measurements is reduced loading of the unknown voltage source. The current through the voltmeter is $(v_u - V_{os})/R_m$. If offset were not used the current would be v_u/R_m, which is larger by the factor $v_u/(v_u - V_{os})$ assuming a constant value for R_m. This advantage is not, however, realized with the moving coil voltmeter if the voltmeter sensitivity is increased to give the same deflection (meter-current value) as would have existed without offset on the less sensitive range.

Experiment 1–15   *Voltage Offset in Measurements*

This experiment demonstrates the use of voltage offset in the measurement of small voltage changes in a relatively large average voltage. An offset voltage is applied in opposition to the measured voltage, which allows a high sensitivity scale to be used on the measurement device. The effect of a load on the voltage supplied by a 9 V battery is measured on the oscilloscope. Changes due to loading are in the millivolt range, and the 9 V static battery voltage is offset with a constant reference voltage. Measurement of the change in battery voltage due to loading is used to calculate the equivalent internal resistance, R_e, of the battery.

Voltage Comparison Techniques

The measurement of voltage by a null comparison procedure is represented schematically in Fig. 1–71. As in the measurement with offset, an adjustable reference standard voltage v_r is in series with the unknown voltage v_u. The only basic difference is that the reference standard can be adjusted in sufficiently small increments that v_r can be made almost exactly equal to v_u ($v_r = v_u$). Since the reference standard has calibrated settings, the magnitude of v_u can then be read from the dial settings which determine the magnitude of v_r. A null detector is used to detect an off-null condition and to aid in adjusting v_r to be exactly equal to v_u. The null detector is a kind of difference detector except that it only needs to indicate the existence of a potential difference at its terminals. When a null measurement is complete, the magnitude of the difference is always zero (within the sensitivity of the null detector). A null detector which indicates the sign of the

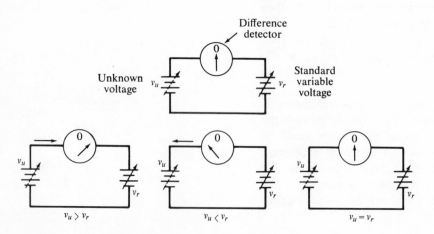

Difference detector

Unknown voltage v_u

Standard variable voltage v_r

$v_u > v_r$ $v_u < v_r$ $v_u = v_r$

Fig. 1–71 Null comparison voltage measurement.

potential difference is illustrated in Fig. 1–71. The information about the sign of the difference between v_r and v_u is very helpful in achieving a null since it indicates whether v_r should be increased or decreased to approach the null point.

The difference in voltage between v_r and v_u can be detected by either a sensitive voltmeter with a high resistance or a sensitive current meter with a low resistance. Each has its advantages in certain cases. This is illustrated in Fig. 1–72. The internal resistances of the unknown (R_u) and standard (R_r) voltage sources are shown, as well as the resistance of the null detector (R_{nd}). When the circuit is off-null, there will be a current i_{nd} which depends upon the off-balance voltage and the circuit resistance[15]:

$$i_{nd} = (v_r - v_u)/(R_u + R_r + R_{nd}). \qquad (1-49)$$

From Eq. (1–49) it is clear that i_{nd} will be greatest when R_{nd} is negligible compared to $R_u + R_r$. Thus, if a null detector measures the current resulting from the voltage imbalance between v_u and v_r, it should have a low resistance. It is also clear that the sensitivity of a current-type null detector will increase as the source resistance decreases and vice versa. A low-resistance reference standard should be used with such detectors.

On the other hand, a voltage-sensitive null detector senses a fraction of the voltage difference between v_u and v_r. From Fig. 1–72, the voltage across the null detector with input resistance R_{nd} is $i_{nd}R_{nd}$ so that

$$v_{nd} = i_{nd}R_{nd} = (v_r - v_u) \frac{R_{nd}}{R_u + R_r + R_{nd}}. \qquad (1-50)$$

The maximum value for v_{nd} is thus $v_r - v_u$, which is obtained under the condition that R_{nd} is very much larger than $R_u + R_r$. Thus voltage null detectors are useful with high resistance voltage sources as long as the resistance of the null detector is higher than the source resistance.

It should be emphasized that there is no measurement error due to loading the unknown or reference voltage sources with either detector type. When v_r is adjusted so that i_{nd} (or v_{nd}) is zero, there is no current in the circuit and consequently no IR drop across R_u or R_r. The advantage of zero loading error is inherent in all null comparison voltage measurements, and thus voltage sources with high internal resistances can be measured without error. However, to locate the null point with a low-resistance current-sensitive null detector, the voltage sources must supply current when v_r is off-null. According to Eq. (1–49), a very high value of R_u might reduce the sensitivity of the null detector to the point of uselessness. Also some voltage sources such as the glass pH electrode do not soon regain a stable output voltage after they have been loaded. High-input-resistance voltage-sensitive null detectors are required for such high-resistance or load-sensitive sources.

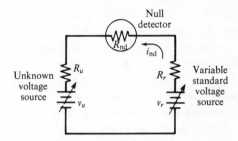

Fig. 1–72 Circuit to illustrate the effect of null detector resistance.

Note 15. Direction of Current.
When the direction of the current in a circuit is not known, the equation for the current can be derived, assuming a particular direction (for example, the direction of current shown by the arrow in Fig. 1–72, which was used in deriving Eq. 1–49). If the numerical solution of the equation indicates a negative current (for example, if $v_u > v_r$ in Eq. 1–49), it indicates that the actual current direction is opposite to that assumed.

Note 16.   Commons and Grounds.
Electronic circuits generally have a very
low resistance conductor (a rather large
copper wire or foil) to which one
terminal of power supplies, signal
sources, and many other circuit
components are connected. This
conductor, which provides a point of
common potential, is frequently called
circuit common. *It is indicated in a circuit*
diagram by the triangle symbol (▽).
Because it is a common point, the
voltages at all other points in the circuit
are generally referenced to (measured
wtih respect to) the circuit common.

When a circuit is linked to other
circuits in a system, the commons of
those circuits are often connected
together to provide the same common
potential for the system. The circuit or
system common may also be connected
to the universal common, **earth ground,**
by connection to a ground rod, water
pipe, or power-line common. When a
circuit common is not connected to a
system common or when a circuit or
system common is not connected to earth
ground, it is said to be **floating.** *The*
circuit common of an instrument is
sometimes connected, either internally or
externally, to its case. For safety, the
metal cases of all power-line operated
instruments are, or should be, connected
to the power-line common. Therefore,
any circuit that has its common
connected to its case is connected to
earth ground whenever the instrument is
plugged into the power line.

Null Comparison Circuit Variations.   The basic null comparison circuit of v_u, v_r and the null detector in series is very simple and is, in general, subject only to a variation of the devices used in the three major circuit elements. An important consideration in the choice of devices for practical measurement systems is the placement of the system common.[16] The three possible locations for the system common connection are shown in Fig. 1–73. Note that all three circuits in Fig. 1–73 are exactly the same as the circuit of Fig. 1–71. The position of the common point does not affect the basic series connection of v_u, v_r and null detector from which the basic null comparison voltage measurement characteristics were derived. In the first location (Fig. 1–73a), both v_u and v_r are connected directly to the system common. The null detector connections are at voltages v_u and v_r with respect to the common. In this case the null detector must have

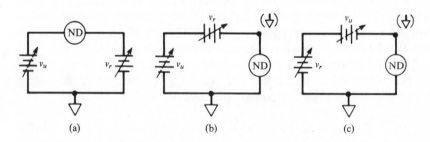

Fig. 1–73 Locations of system common in the null comparison circuit. (a) Floating null detector. (b) Floating reference source. (c) Floating unknown source.

true **difference inputs,** since a difference between v_u and v_r must be detected regardless of the magnitude of either. To accomplish this, the null detector inputs must either be completely independent of the system common (floating) or be perfectly balanced with respect to the common point. When the null detector can meet the input balance requirement this common point is often used.

The variation shown in Fig. 1–73b places the ground between the unknown source v_u and the null detector. Now one connection to the null detector is always at the common voltage. When the system is at null ($v_u = v_r$) the other terminal of the null detector is also at the common voltage. Thus when null is detected, both null detector connections are at the common voltage. This eliminates the stringent balance requirements of the detector input connections with respect to ground. On the other hand, the reference voltage source is now *floating,* that is, not connected to common. Special care is required in the design of a reference source if its output voltage is to remain stable and accurate as the voltage between it and the system common changes. We would appear to have simply traded one problem for another except that at null, the point between the v_r source and the null detector is at the common potential. This point is called a **virtual common**[17] because when the null condition is fulfilled, even though it is not connected directly to common, it is at the common voltage. Thus one connection of the v_r source is at the common voltage when the null measurement is complete. This relieves some of the problems of maintaining the stability of the floating reference source since one terminal is at virtual common when the reference setting is taken for the measurement value.

A third location for the system common is shown in Fig. 1–73c. This circuit is the same as that of Fig. 1–73b except that the positions of v_u and v_r have been interchanged. Now one terminal of v_r is connected to common and the v_u source is floating. The best position for the common connection depends upon which source can best tolerate being disconnected from the common. Neither circuit (Fig. 1–73b or c) requires a floating null detector and both establish a virtual common at the null detector.

In the discussions of null detectors and practical null comparison systems that follow, the factors of difference input balance, source floating, and virtual common are considered.

Experiment 1–16   *Voltage Comparison Measurements with the Current Meter*

A null comparison technique is used for voltage measurements in order to demonstrate the high precision, accuracy, and other advantages of the method. In Experiment 1–15, one arrangement of the null comparison method was used to measure the no-load voltage of the battery. In this experiment an alternative type of potentiometric measurement is made in which neither lead of the null detector is grounded. A stable voltage reference source is used as the variable reference standard, while a multimeter is used as the null detector. Unknown voltages from a battery and a voltage divider are measured by the null comparison method.

Null Detectors

For several decades the standard null detector for voltage comparison measurements was the laboratory galvanometer. This is the very sensitive version of the moving coil meter which uses a light beam for a pointer as

Note 17.   Virtual Ground or Common. A **virtual common point** *in a circuit is a point which is virtually at the circuit common potential, but is not connected to the circuit common point. Since the term "ground" is often loosely used for circuit common even when the common is not connected to earth ground, the term* **virtual ground** *is frequently used to identify a point at the circuit common potential. The term virtual ground should be used only to describe a point at ground potential which is not connected to ground.*

discussed in Section 1–2.4. The chief characteristic of the galvanometer is that the off-null signal must supply the total energy required to turn the coil and mirror against the restoring force of the suspension springs. As scientists strove to make increasingly accurate and sensitive measurements, the sensitivity of the galvanometer was increased. This was achieved by increasing the light beam (pointer) length and decreasing the spring force. In order to keep reasonable damping, the mass of the coil and mirror must be reduced with the spring force. The result of increasing the sensitivity beyond a certain point is a substantial increase in size and delicacy. The galvanometer, being energized entirely by the input signal, requires no connections to the system common. Being completely independent of the common allows it to be used in the floating position of Fig. 1–73a as well as in the two variations shown in Fig. 1–73b and c.

Recent advances in the stability and sensitivity of electronic amplifiers have made it possible to use them to amplify the difference signal in a comparison measurement to a voltage or current large enough to operate a variety of indicators. The principle of amplification of a measurement signal to reduce loading and drive readout displays was introduced in Section 1–3.3 with the FET and vacuum-tube voltmeters, and the oscilloscope. These devices can be, and often are, used as null detectors in null comparison measurements. However, they are not ideally suited to such service for three reasons: First, the amplifiers used in voltmeters and oscilloscopes often drift with time and temperature changes. This requires fairly frequent "zero" adjustment in order to ensure that a null output indication is obtained when the difference input is truly zero. In the null comparison application, the accuracy of the null indication is the most important consideration and it should not require frequent checking or adjustment. Second, the linear transfer function, calibrated output, and multiple ranges of most voltmeters and oscilloscopes are superfluous in the null detector application. Finally, most voltmeters and oscilloscopes have a connection to the system common. If they have difference inputs (many don't), the inputs are often imperfectly balanced with respect to the common. This generally limits the application of such devices to the grounded null detector configurations shown in Fig. 1–73b and c.

The Comparator.   An amplifier designed specifically for null comparison measurement is the **comparator.** The comparator is a very high gain amplifier with well-balanced difference inputs and controlled output limits. The symbol for the comparator and a plot of its transfer function are shown in Fig. 1–74. Ideally, the comparator has just two output states: +limit and —limit. The state of the output indicates whether the condition $v_u < v_r$ or $v_u > v_r$ exists. When a comparator is used as a null detector the reference voltage is adjusted to the point where the smallest increment

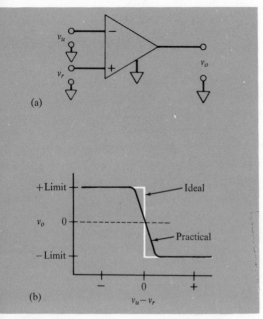

(a)

(b)

Fig. 1–74 (a) Comparator symbol. (b) Transfer function.

of change in v_r causes the comparator output to change state. In a practical comparator, the region of the transfer function at zero difference has a finite slope. This is the region where the magnitude of the difference times the gain of the amplifier is less than the +limit or −limit. This slope approaches the vertical ideal as the comparator gain or the smallest difference of interest increases.

The comparator input circuits are designed for high stability of the null point and good balance with respect to common. However, the device is perfect in neither respect. It is important to check the specifications of a comparator with respect to its zero drift to be sure it is within the allowable null detection error. Common values for drifts due to time and temperature are in tens of microvolts.

Inputs which are perfectly balanced with respect to the common voltage will respond only to a difference in voltage at the two inputs. When the comparator is in the configuration shown in Fig. 1–73a the voltage with respect to common at one input is v_u, and at the other input, it is v_r. This is shown in Fig. 1–75a. To evaluate the balance of the comparator inputs, it is useful to consider that the sources v_u and v_r provide a **difference voltage** v_d at the comparator inputs, which is $v_u - v_r$, and a **common mode voltage** v_{cm}. The common mode voltage is the average of the two input voltages with respect to the common. However, when the system is at or near null, the two input voltages v_u and v_r are approximately equal so that their average is essentially equal to v_r. In Fig. 1–75b, the common mode voltage v_{cm} is shown approximately equal to v_r. The **difference response** is the output voltage change Δv_o for a given input difference voltage change Δv_d or $\Delta v_o/\Delta v_d$. The **common mode response** is similarly $\Delta v_o/\Delta v_{cm}$. Ideally, the common mode response should be zero. The commonly used measure of the perfection of input balance is the **common mode rejection ratio** (CMRR) which is simply the ratio of the difference response to the common mode response. Therefore

$$\text{CMRR} = \frac{\Delta v_o/\Delta v_d}{\Delta v_o/\Delta v_{cm}}. \qquad (1\text{--}51)$$

Of course, the better the balance, the higher the CMRR. A typical comparator CMRR is 10^4 or more. If a comparator is used in the floating null detector configuration (Fig. 1–73a), the null indication for $v_u = v_r = 1$ V could be different from that for $v_u = v_r = 0$ V by 1 V/CMRR ≈ 100 μV. This error could be reduced by choosing a comparator with a higher CMRR and avoided entirely by choosing not to float the null detector. Another reason for not floating the comparator null detector is that in the comparator amplifier each input has a finite resistance to the system common. These resistances can load the sources v_u and v_r in the configuration

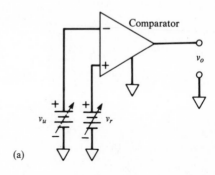

(a)

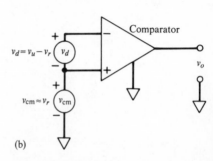

(b)

Fig. 1–75 Equivalent input voltage sources that show the difference and common mode voltages.

of Fig. 1–73a even when $v_u = v_r$. Even though the input resistances to common can be 10 MΩ or more, the zero-loading advantage of the null comparison measurement is partially lost. This loading does not occur in the null configurations in which the null detector is connected to common, since, at null, the voltage across the input resistances is zero.

Experiment 1–17   *Operational Amplifier Voltage Comparator*

In this experiment an operational amplifier (OA) voltage comparator in conjunction with a readout device is used as a null detector for a null comparison voltage measurement in order to compare and contrast this null detector with the others previously studied. The OA comparator is used with neither input grounded (differential mode) and with one input grounded (single-ended mode). An unknown battery voltage is compared with a variable reference voltage, and the measurement precision is estimated. The comparator output is used to drive an indicator light to signal when the sign of the input difference voltage changes.

Chopper-Input Amplifier.   As we can see from the foregoing discussion, the ideal null detector would have no zero drift, perfectly balanced difference inputs (infinite CMRR), and very high sensitivity. The chopper-input amplifier shown in Fig. 1–76 comes very near to achieving this ideal. The characteristics of this amplifier are best illustrated when the system common is in the floating null detector configuration shown in Fig. 1–76. The "chopper" is a continuously alternating switch connected so that the amplifier input is alternately connected to v_u and v_r. The waveform at point Ⓐ is the chopper position control signal. The signal at point Ⓑ is alternately v_u and v_r. This signal is connected to the amplifier input through a capacitor C. The amplifier input resistance R keeps the *average* voltage with respect to the common at point Ⓒ equal to zero. The average voltage with respect to the common at point Ⓑ is the average of v_u and v_r or v_{cm}. Thus, in operation, the capacitor C charges up to the common mode voltage v_{cm}. The resistance R is generally so large that the charge on capacitor C does not change appreciably during each alternation of the chopper. Thus the voltage across capacitor C is essentially constant at v_{cm}. The capacitor voltage serves as a kind of offset voltage of v_{cm} V between points Ⓑ and Ⓒ. The changes in voltage at Ⓑ due to the difference between v_u and v_r thus appear as variations about the common voltage at point Ⓒ. These variations are then amplified by the amplifier. Since the signal at point Ⓒ is affected only by the difference between v_u and v_r, the common mode response of the system is very low, that is, the CMRR is very high. Zero drift is also eliminated by the chopper which provides a

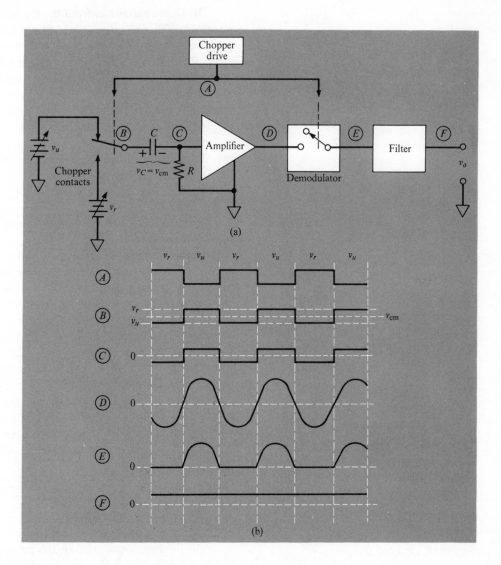

Fig. 1–76 (a) Chopper input amplifier. (b) Waveforms.

signal variation at point Ⓑ only when there is a real difference between v_u and v_r.

Note that with the chopper-input amplifier, only the chopper is floating. The introduction of the capacitor which always charges to the voltage v_{cm} allows the null detector amplifier to be connected to the system common so that all three active null comparison circuit elements can be connected directly to the system common.

The signal variations from the chopper are greatly amplified to produce waveform Ⓓ. An off-null condition can be detected by observing the amplified variation at this point, but the sign of the input difference is not clear. Restoration of the sign of the input difference is accomplished by a "demodulator" circuit. This is also a switch which connects the amplifier to the output only when the chopper is in one particular position. (The v_r position is used in Fig. 1–76.) The result is a signal with positive pulses if $v_r > v_u$ and negative pulses if $v_r < v_u$. A filter between the demodulator and output is often used to smooth the pulses into a steady voltage which may be millions of times greater than the difference between v_u and v_r at the input. Of course, the output voltage is limited in range so that the transfer function for the chopper-input amplifier is similar to that of the comparator in Fig. 1–74.

The chopper-input amplifier would appear to have almost infinite resistance between the inputs since the v_u and v_r chopper contacts are never connected together. In practice, the resistance between inputs is limited only by the resistance of the open chopper switch. However, some loading of the voltage sources does occur in two circumstances; one is when v_u or v_r changes and the other is when the system is off-null. The charge required to bring capacitor C to the voltage v_{cm} is supplied by the input voltage sources. Whenever v_{cm} changes (due to a change in v_u or v_r) the charge on capacitor C must change. When v_u and v_r are not equal, there is a non-zero voltage at Ⓒ and a small current through resistor R. This current is supplied by the voltage source connected at that instant. By careful design, the amount of charge required can be made negligibly small for most sources. On the other hand, at null, when $v_u = v_r$, v_{cm} is constant and the voltage across resistor R is zero so that no charge or current is required from either source and a zero-load null measurement is achieved.

Experiment 1–18   *Chopper-Amplifier Null Detector*

This experiment illustrates the use of a chopper amplifier as the null detector in null comparison measurements. A comparison of this null detector with the multimeter used in the previous experiment is made in order to demonstrate the high sensitivity which can be obtained with the chopper amplifier. An unknown battery voltage and a variable reference standard voltage are the inputs to the chopper-amplifier stages of a servo recorder. The amplified difference voltage is seen to drive a servo motor and move a pen which is mechanically linked to the motor. The motor is observed to move when the unknown voltage and the reference voltage are sufficiently different. When the voltages are very nearly equal, the motor, and thus the indicator pen, are observed to come to rest.

The Laboratory Potentiometer

The standard laboratory potentiometer, the classic precision voltage measurement device, is simply a combination of a precision adjustable standard voltage source (literally, the potentiometer) and a sensitive null detector in a single instrument. It measures voltage by the null comparison technique described in this section. It also includes a provision for a null comparison standardization of the variable reference voltage source against a standard Weston cell. The circuit of a basic potentiometer with a voltage span of 100.0 mV is shown in Fig. 1–77.

To provide a standard constant current of 10.00 mA through the precision divider resistance R_p, the switch S_1 is thrown to the left to close contacts a–b; the resistance R_A is adjusted until the null detector reads zero, indicating that $IR_s = 1.018$ V, and since $R_s = 101.8$ Ω, the current I must be 10.00 mA. Now, in order to measure an unknown voltage v_u, switch S_1 is thrown to the right to close contact b–c; the precision divider is then varied until $IR_v = v_u$, as indicated again by the fact that there is no current through the difference detector. The voltage $IR_v = IR_pb = 100.0b$ mV for the example of Fig. 1–77, where b is the fraction R_v/R_p as shown on the potentiometer indicator dials. The location of the system common in a laboratory potentiometer can take any of the forms indicated in Fig. 1–73, depending upon the devices used for the null detector and reference source. In the classical potentiometer, the reference source is battery-operated, and the null detector is a galvanometer or a battery-operated amplifier. Thus none of the potentiometer circuit elements have a connection to a laboratory common. This allows the user to connect the common and the potentiometer instrument case to those points most convenient for the circuit being measured.

Because for many years the potentiometer was the only device used for making null comparison voltage measurements, null comparison voltage measurements are called **potentiometry.** These terms are still applied in reference to other null comparison voltage measurement devices and techniques. The development of potentiometric systems in which the null balancing procedure is automated was an important development. Such "automatic potentiometers" have extended the advantages of null comparison voltage measurements into applications where the time or tedium of manual null balancing make manual potentiometric measurements impractical. Several automatic null comparison voltage measurement systems are introduced in the next section.

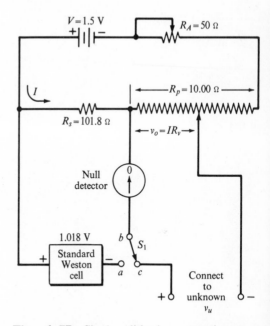

Fig. 1–77 Single slidewire potentiometer with span of 0 to 100.0 mV.

Experiment 1–19   *Characterization of a Laboratory Potentiometer*

In this experiment a conventional laboratory potentiometer is used and characterized in order to compare it with the null comparison systems used previously. The null detector of the potentiometer is first characterized and compared to other null detectors. The complete potentiometer is then used in the measurement of unknown voltages. The potentiometer working voltage standard is compared in turns of resolution, stability and accuracy to that of a laboratory voltage reference source, and the potentiometer is used to calibrate the laboratory reference source.

PROBLEMS

1.   A parallel plate capacitor has a capacitance of 0.02 μF. (a) How much charge must be separated to produce a potential difference of 1.3 V between the plates of the capacitor? (b) How much work would be obtained by discharging the above capacitor?

2.   A 1.5 V alkaline flashlight battery has a maximum output current of 0.9 A. What is the internal resistance of the cell? Compare the internal resistance of the alkaline battery with that of a 2 V lead-acid storage cell which can deliver 500 A.

3.   The input voltage to the Kelvin-Varley divider of Fig. 1–39 is 5.00 V. Draw the switch positions for an output voltage of 1.83 V.

4.   Some ion-selective electrodes may have an equivalent resistance as high as 10^{10} Ω. Assume that such an electrode with a reference electrode produces a cell voltage of 1.0 V. What is the necessary input resistance for a voltmeter used to measure the cell voltage if the error is not to exceed 1 mV?

5.   A voltage divider is constructed across a 3 V source which consists of a 10 kΩ resistor and a 20 kΩ resistor. The voltage across the 20 kΩ resistor is to be measured with a voltmeter of 100 kΩ resistance. (a) What voltage appears across the 20 kΩ resistor when the meter is absent? (b) What is the voltage when the meter is present? (c) What is the relative perturbation error?

6.   The voltage divider shown in Fig. P–6 is used to produce 1 V and 5 V from a 10 V source. A voltmeter with an input resistance of 1 MΩ is used to measure the divider output. (a) What is the relative perturbation error when the 5 V output is being measured? (b) What is the relative perturbation error when the 1 V output is being measured?

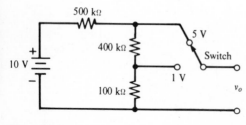

Fig.   P–6

7. A voltage of 5.00 V is applied to the 1 kΩ voltage divider in Fig. 1–18b. A voltmeter of 10 kΩ input resistance is used to measure the output. (a) What is the relative error when 5.00 V is selected? (b) What is the error when 1.00 V is selected?

8. The voltage from a source is measured as 6.1 V on the 15 V full-scale range of a 20 kΩ/V meter. When measured on the 5 V full-scale range of the same meter, 4.37 V is obtained. (a) What is the true voltage? (b) What is the equivalent source resistance?

9. A comparator is to be used as a null detector in a voltage measurement. The output voltage changes 10 V for an input difference voltage of 500 μV, while the output voltage changes by 1 V for a common mode input voltage of 500 mV. What is the common mode rejection ratio (CMRR) of the comparator?

10. For the potentiometer of Fig. 1–77, what value of R_s should be substituted if the span is to be calibrated for 500 mV full-scale?

11. A variable reference standard is made from a 1 kΩ center-tapped Helipot and a 1.34 V Hg cell as shown in Fig. P–11. What is the value of R_s such that v_r may be varied from +0.5 V to −0.5 V?

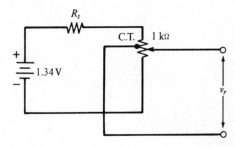

Fig. P–11

12. The galvanometer in the null comparison circuit of Fig. P–12 has a sensitivity of 3.0×10^{-7} A/mm and a resistance of 150 Ω. The slide-wire used to vary v_r is 1 m long and has a resistance of 100 Ω. The voltage drop across the slidewire is adjusted to 2.0000 V. Assume that a 1 mm deflection of the galvanometer can just be seen and that the wiper can be positioned to the nearest 0.1 mm. With what accuracy can v_u be measured if R_u is (a) 100 Ω, (b) 10 kΩ, (c) 1 MΩ?

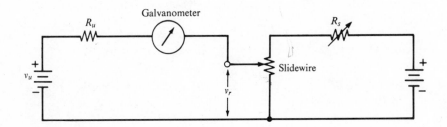

Fig. P–12

Servo Measurement Systems Section 1–4

The null comparison concept that is presented in the previous section can provide the nearly perfect measurement: it is direct because the unknown quantity is compared directly with standard amounts of the same quantity; it is potentially accurate since the accuracy is not dependent upon the calibrations of the null detector output; it causes a minimum perturbation of the measured quantity because that quantity is perfectly balanced by the standard quantity at null.

In this section, it will be shown how the automation of the null comparison measurement results in the powerful concept of feedback control. Feedback control systems, called **servo systems,** are the basis of virtually all automatic measurement or control systems. The application of the automated feedback control to measurements has resulted in a family of automated null comparison measurement systems including the potentiometric recorder, the potentiometric amplifier, the operational amplifier voltage follower, digital voltmeters and the summing and integrating circuits in an analog computer. These important measurement tools will be discussed in this section as examples of the application of the servo system to null comparison measurements of voltage and current.

1–4.1 SERVO VOLTAGE FOLLOWERS

The process of a manual null comparison voltage measurement is shown in Fig. 1–78a. The experimenter adjusts the continuously variable reference v_r manually while he visually observes the null detector. The null detector indicates the direction in which v_r should be adjusted to achieve the null condition of $v_r = v_u$. The value of v_r is obtained by observation of the indicator scale which is mechanically coupled to the v_r adjustment. If, after balance is complete, v_u changes, a new null point must be set at which v_r is once again equal to v_u. When the system is continuously nulled

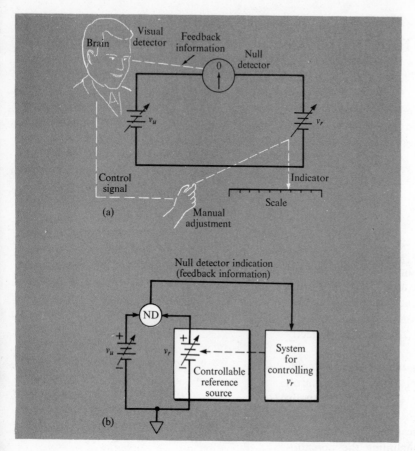

Fig. 1–78 Null comparison measurements of unknown voltage. (a) Manual potentiometric measurement. (b) Automatic null-balancing potentiometer.

in this way, the reference voltage v_r follows the measured voltage v_u. If the unknown voltage changes frequently or rapidly, the manual adjustment of v_r to follow v_u becomes impossible. For this reason, automatic nulling systems have been devised following the general pattern of Fig. 1–78b. In this case the reference source is adjustable by mechanical motion or by electrical signals. An electrical signal is obtained from the null detector indicating $v_u > v_r$, $v_u = v_r$, or $v_u < v_r$. The system for controlling v_r must interpret the null detector indication signal and provide the appropriate motion or signals to the controllable reference source so that v_r approaches the value of v_u. This process continues until the null is achieved and is repeated at regular intervals or when v_u changes so that v_r automatically follows v_u.

The concept which makes it possible to control v_r so that it follows v_u even when it is not known how or when v_u will change is **feedback control.** In the manual comparison measurement or voltage follower of

Fig. 1–78a, the eye monitors the null detector indicator to obtain information about the difference between the actual state of v_r and the desired state of $v_r = v_u$. If a difference is detected, the information concerning the sign of this difference is fed back through the eye to the brain, which interprets the feedback information and issues appropriate commands to the hand adjusting v_r. The only information required in order to control v_r to follow v_u exactly is the difference between v_u and v_r which remains after each adjustment of v_r. The null detector provides the difference information in the form of the sign of the difference between v_u and v_r. This information is fed back from the difference detector to the controller. The information feedback path is illustrated by the automatic voltage follower of Fig. 1–78b. The system for controlling v_r must translate the feedback information into control signals appropriate to adjust the output of the controllable reference source (v_r) in the proper direction.

All systems which detect a difference between the actual and the desired states of a controllable quantity (e.g. voltage, temperature, position) and then feed the difference information back to a controlling device which causes that difference to become essentially zero fall under the general classification of **servo systems.** The word "servo" comes from the Latin word *servus,* meaning a slave. A servo system, then, could be considered as a slave that follows a command. The servo system can compare the result of its effort with the order and make corrections to follow the command exactly. If the quantity to be controlled is the position of an object and the controlling device is an electric motor or an electrically controlled hydraulic system, the **servo system** is an **electromechanical servo system** or a **servomechanism.** The position controlling devices in a servomechanism can be very powerful so that they can position heavy loads in response to command signals obtained with relatively little effort. The "power amplification" capability of servomechanisms are used to advantage to move the control surfaces (such as rudders, elevators, and stabilizers) of large ships and jet planes in response to the position of finger-tip controls on the pilot's console. Because of their inherent precision, servomechanisms are often used to position critical parts of delicate instruments in the precise location required for optimum instrument performance.

Several of the feedback control or servo systems which have been used to make an automatic **voltage follower** are introduced in this section. When a servomechanism is used to provide a mechanical adjustment of v_r, the automatic voltage follower is called a **servomechanical potentiometer.** When an incremental signal is used to set both an incrementally adjustable reference source and a numerical display, the system is called a **digital voltmeter.** If v_r is the output voltage of an amplifier, the voltage follower

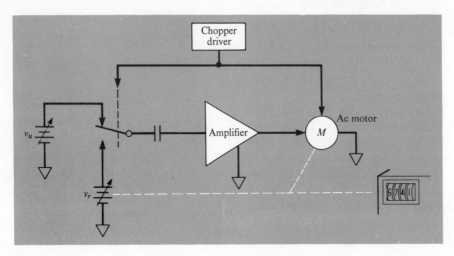

Fig. 1–79 Servomechanical potentiometer with ac servo motor and turns-counting readout.

system is called a **potentiometric amplifier** or an **operational amplifier voltage follower.** As this section demonstrates, all these systems are simply potentiometers which use an automatic technique for adjusting v_r to follow v_u.

The Servomechanical Potentiometer and Servo Recorder

The servomechanical potentiometer is an electromechanical servo system acting as a voltage follower. Since the value of v_r is kept equal to that of v_u by a mechanical motion, this same motion is used to position an indicating pointer or rotate the shaft of a turns-counting dial from which the value of v_r can be read. The basic servomechanical voltage follower is shown in Fig. 1–79. A chopper input amplifier is used as a null detector to compare the measured voltage v_u and the variable reference voltage v_r. The latter is varied by turning a shaft which is mechanically connected to servo motor M. The action of the servo motor can best be understood by referring to the chopper-input amplifier waveforms, Fig. 1–76b. When v_r is not equal to v_u, the difference appears as a signal variation at the amplifier input (waveform $Ⓒ$). This variation is amplified by the amplifier to a signal strength sufficient to turn the motor (waveform $Ⓓ$). Note that the condition illustrated, $v_r > v_u$, results in an amplifier output (waveform $Ⓓ$) which is positive when the chopper drive signal (waveform $Ⓐ$) is negative. This combination of polarities applied to an ac servo motor causes the motor to turn in the direction that will decrease v_r. For the condition $v_r < v_u$, the amplifier output polarity is the same as the chopper drive

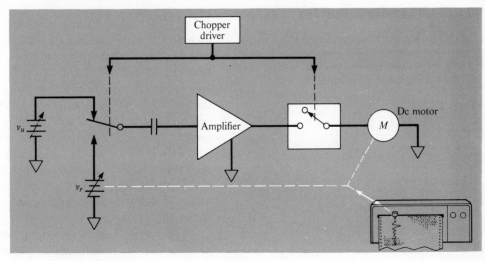

Fig. 1–80 Servo recorder with dc servo motor and recorder readout.

signal. In this case the motor will turn in the opposite direction, and v_r will increase.

If the servo system were ideal, even the slightest difference voltage between v_u and v_r would provide a sufficient servo motor signal to correct the difference. However, there is always a minimum null detector input signal which will result in servo motor rotation and a small zero drift or input imbalance in the null detector. The combination of these effects results in a small uncertainty or error v_s, so that $v_u = v_r \pm v_s$. Many servo systems used in potentiometric measurements maintain the error to within a few microvolts for full scale spans of 1 mV or more. For certain specialized instruments the error is reduced to as low as a few nV. For practical purposes the error signal is often negligible, and v_r effectively follows v_u.

When the servo system is at balance the accuracy of the measurement of v_u depends on the accuracy of the reference voltage v_r and the magnitude of v_s. The amplification (gain) of the servo amplifier should, of course, be made very high so that v_s is negligible. Note that the amplifier gain does not have to be constant since it affects only the magnitude of the error term v_s. As long as it remains high the gain can vary over rather wide limits because of component aging, temperature, etc., without affecting the measurement accuracy. The relative independence of amplifier gain and linearity exhibited by feedback control systems is a natural consequence of the earlier assertion that the exact transfer function of null detectors in null comparison measurements is unimportant.

In some servomechanisms advantage is taken of the mechanical motion to provide a recording function. A servo recorder block diagram is shown in Fig. 1–80. The recorder is similar to the servomechanical potentiometer except that the value of v_r is indicated by the position of a pen along one axis of a sheet of graph paper. The pen-positioning mechanism, the servo motor, and the v_r adjustment are all mechanically coupled. Like the chopper-input amplifier of Fig. 1–76, the servo system amplifier shown in the recorder of Fig. 1–80 uses a demodulator circuit to obtain a steady output voltage of a polarity equal to the sign of $v_r - v_u$. This signal is used to drive a dc servo motor for which the direction of rotation reverses when the polarity of the drive signal reverses. The motor is connected such that a positive amplifier output signal will cause the motor to decrease v_r. Recently the dc servo motor has become favored over the ac servo motor in many applications because of its smaller size and lower cost for equivalent torque.

The servo systems shown in Figs. 1–79 and 1–80 show the system common in the floating null detector position, Fig. 1–73a. Because the chopper input provides excellent characteristics for the floating null detector application, this is the most often used configuration for the electromechanical servo. However, the floating v_r configuration, Fig. 1–73b, is also found. A floating reference source is relatively easy to obtain in the electromechanical servo because the v_r adjustment is made by mechanical rather than electrical connection to the null detector output.

In a strip-chart recorder, a long strip of graph paper is moved at right angles to the pen motion at a constant rate. Since the pen position is linearly related to v_r and v_r follows v_u exactly, the variation of v_u with time is recorded on the chart paper. An x-y recorder uses two complete servomechanisms: one for positioning the pen along the x-axis of a graph paper in response to v_x, and another for positioning the pen along the y-axis in response to v_y. As v_x and v_y vary, a plot of v_x vs. v_y is produced.

For all servo systems, a certain amount of time is required for the motor to adjust v_r in response to a change in the value of v_u. For servo recorders the response time will typically allow full-scale travel of the pen in 0.5 to 2 sec. Large changes in v_u occurring within a time span of 1 sec or less will not be recorded accurately. One is not assured of an accurate measurement until the servo motor (and hence the pen) motion has stopped, since the motion of the servo motor indicates the possibility of a significant difference signal at the null detector input.

The Recorder Variable Reference Source.   For convenience and versatility, the variable reference standard voltage used in a recorder servo system has an output with an adjustable zero position. A typical circuit is shown in Fig. 1–81. A stable voltage source V is placed across the potentiometer R_{sw}, the **slidewire potentiometer** which is operated by the servo motor. A manually adjusted zero potentiometer, R_z, is also placed across the voltage source V. A variable calibration resistor R_c is used to provide a precise

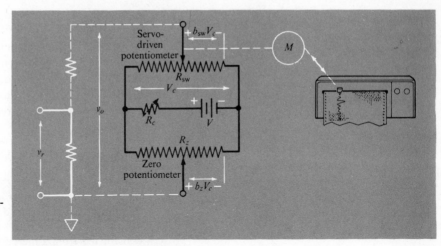

Fig. 1–81 A recorder variable reference source.

calibrated voltage V_c across R_{sw}. The reference voltage v_o is the voltage between the R_{sw} tap and the R_z tap. Since V_c is divided by the linear potentiometer R_{sw}, the voltage from the tap of R_{sw} to its right side is $b_{sw}V_c$, where b_{sw} is the fraction of R_{sw} from the adjustable tap to the right side. Similarly, the voltage from the tap of R_z to its right side is b_zV_c. The total output voltage is the difference of the fractions of V_c to the right of the taps of the two potentiometers. Thus

$$v_o = b_{sw}V_c - b_zV_c = V_c(b_{sw} - b_z). \qquad (1\text{--}52)$$

It is clear from Eq. (1–52) that a change in b_{sw} due to the movement of the motor and pen affects the output voltage v_o linearly as expected. The effect of setting the tap of R_z will be illustrated by a few examples. If $b_z = 0$, that is, if the zero tap is at the extreme right position, v_o will vary from 0 to $+V_c$ V as the slidewire tap moves from left to right. This places the pen position corresponding to 0 V at the extreme right. If $b_z = 1$, v_o will vary from $-V_c$ to 0 V as the pen moves from right to left. Now 0 is at the left with increasing *negative* voltage to the right. The zero voltage pen position can be placed in mid-chart if $b_z = 0.5$. In this case, v_o will change from $-V_c/2$ to $+V_c/2$ as the pen is moved from right to left. Note in these examples that the span of v_o is always V_c V and that the right side is the more negative. The polarity can be changed to positive right by reversing the polarity of the source V.

The span can be changed by changing V_c or by putting a voltage divider across v_o. If $V_c = 1.00$ V and a 1000 : 1 voltage divider is placed between the variable reference output v_o, the variable reference voltage v_r connected to the chopper input is just 1/1000 of v_o and the full-scale span of the recorder would be 1.00 mV. The reference divider technique is often used for sensitive recorders. Several taps could be placed in the reference divider to provide full-scale spans from 1 mV to 1 V. Some recorders do this. Others have a fixed span of 1 mV or 10 mV and divide the input voltage (v_u) to obtain other ranges. Unfortunately putting a voltage divider across v_u places the divider resistance (often 10 MΩ) across the measured voltage and destroys the no-load advantage of the null comparison measurement. Such recorders are not "potentiometric" on all ranges.

Experiment 1–20 *Servo Slidewire Reference Voltage Source*

The slidewire reference voltage source which is employed in typical servo recorders is built in order to help the student understand how a fixed range reference can be obtained, how variable ranges are achieved, and how a variable zero output can be added. A fixed 0–1 V voltage reference is first constructed from a mercury (Hg) battery, a linear potentiometer similar to the recorder slidewire, and a voltage divider. The divider resistors are then changed to vary the full-scale span of the reference voltage. Next a zero adjust potentiometer is added to make it possible to vary the zero position of the reference voltage while keeping the full-scale span constant.

Experiment 1–21 *Automatic Balancing Potentiometer*

Previous experiments have demonstrated the use of a chopper amplifier as a null detector in voltage comparison measurements and the construction of a servo slidewire reference voltage. This experiment demonstrates that an automatic balancing potentiometer can be obtained by mechanically coupling a servo motor to the slidewire potentiometer and connecting the output of the slidewire reference voltage back to one of the chopper-amplifier inputs. The necessary connections for the automatic potentiometer are introduced and the output of the servo slidewire reference voltage, as well as the position of a pen, is observed as a function of input voltage. The zero output control of the recorder is studied, and a voltage divider is constructed at the output of the slidewire to change the full-scale span. Unknown voltages are measured and the recorder dead zone is determined.

Potentiometric Amplifier

Whereas the servomechanical potentiometer described in the previous section utilized an electromechanical system to maintain the reference voltage equal to the unknown voltage, the same voltage follower (automatic null comparison) operation can be accomplished with an all-electronic system. This system uses a chopper input amplifier; the principle of operation is illustrated in Fig. 1–82.

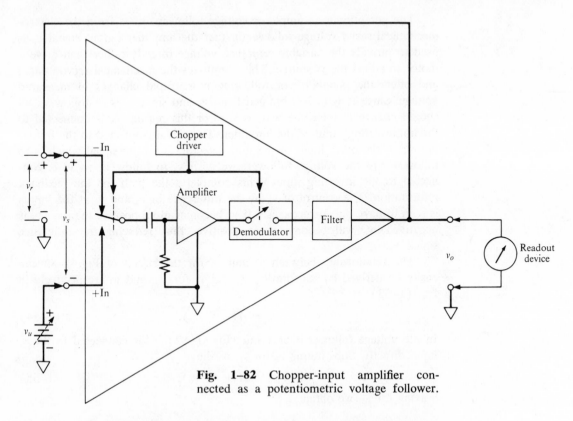

Fig. 1–82 Chopper-input amplifier connected as a potentiometric voltage follower.

The chopper-input amplifier of Fig. 1–76a is shown in the larger triangle. The input chopper alternately connects v_u and v_r at the amplifier input and any difference voltage ($v_s = v_u - v_r$) is amplified by the amplifier which has very high amplification (A). The amplified signal is demodulated (converted back to dc) and filtered (to remove the variation caused by chopping) so that the output dc voltage is

$$v_o = A(v_+ - v_-) = -Av_s, \qquad (1\text{--}53)$$

where v_+ and v_- are the signal voltages connected to the +in and −in inputs of the amplifier, respectively. These inputs are labeled +in and −in to indicate the direction of output voltage change for a positive change in signal voltage at that input. That is, a positive voltage change at +in will produce a positive change in v_o while a positive voltage change at −in will produce a negative change in v_o. Because the voltage changes at the −in input appear at the output with reversed polarity, the −in terminal is often called the **inverting input.** This leads to calling the +in terminal the **noninverting input.**

The potentiometric amplifier voltage follower differs from the electromechanical servo voltage follower in that the amplifier output signal v_o is used to provide the variable reference voltage directly rather than drive a motor to adjust the v_r source. This eliminates the mechanical servo system and offers the possibility of following more rapid changes in measured voltage. Since v_r is to be obtained from v_o and since v_r is to follow v_u, v_o should change in the same way as v_u. For this reason, v_u is connected to the noninverting input of the amplifier. The signal connected to the inverting input, the other input of the null detector, is v_r, the signal which is to follow v_u. In the voltage follower, we wish v_o to follow v_u so v_o is connected to the inverting input. This completes the path for the feedback information. The control of v_o by the difference in v_u and v_r at the inputs is seen to have the correct sense. If v_u becomes more positive than v_r (v_s is negative), v_o will become more positive, thus reducing the difference signal v_s.

The relationship between v_o and v_u for the circuit of Fig. 1–82 can easily be derived by substituting v_u and v_r for v_+ and v_- respectively in Eq. (1–53):

$$v_o = A(v_u - v_r). \qquad (1\text{–}54)$$

In the voltage follower connection (Fig. 1–82), v_o is connected to the v_r input directly. Substituting v_o for v_r, we have

$$v_o = A(v_u - v_o). \qquad (1\text{–}55)$$

Solving for v_o, we obtain

$$v_o = v_u \frac{A}{A+1} \quad \text{or} \quad v_o \approx v_u \quad \text{for} \quad A \gg 1. \qquad (1\text{–}56)$$

Since A is very much larger than 1, often 10^4 to 10^7, v_o is equal to v_u with an accuracy of 1 part in A. The error is due to the fact that, according to Eq. (1–53), there must be some difference input v_s in order to maintain a finite output voltage. However, as Eq. (1–56) shows, the resulting error is small as long as A is high. If the error is due to maintaining a nonzero output voltage, one would expect the error to be proportional to the magnitude of v_o. This is shown by Eq. (1–57),

$$v_o = v_u - v_o/A, \qquad (1\text{–}57)$$

which is a rearrangement of the exact part of Eq. (1–56) in which the term $-v_o/A$ is the difference between v_o and v_u in volts. For an amplifier with a maximum v_o of 10 V and an amplification A of 10^6, the maximum error is 10 μV. Note again that the absolute value of A is not important as long as it is large.

The potentiometric amplifier voltage follower is not a complete voltage measurement system in that its output is a voltage which must still be measured to obtain the value of v_u. In the domain sense, the amplifier output information is in the voltage domain which must be somehow converted to the number domain to complete the measurement. The great merits of the potentiometric amplifier are that the inputs of the amplifier have the very high resistance and balance (CMRR) that are characteristic of the chopper-input amplifier, and that the output is a low-resistance grounded source of voltage exactly equal to v_u. This allows almost any convenient voltage measuring device to be connected to v_o to convert from the voltage to the scale position or number domains.

One terminal of the output signal v_o of the potentiometric amplifier is the circuit common connection. This precludes the possibility of the connection of the amplifier in the common point configuration of Fig. 1–73b which requires a floating reference source. Thus only the floating null detector configuration shown in Fig. 1–82 and the floating unknown source configuration are possible. Because it is frequently inconvenient to have a floating signal source and because the chopper input amplifier performs well in the floating null detector configuration, the latter is most often used. For ease in connection to a variety of signal sources, the circuit common of the potentiometric amplifier is frequently floating with respect to the system commons and the earth ground.

Potentiometric Amplifier with Gain.   When more measurement sensitivity is needed than is inherent in the readout device, the potentiometric amplifier can be connected to provide an output voltage v_o that is a constant multiple of the measured voltage v_u. This is done in the same manner used to increase the servo recorder sensitivity; i.e., only a fraction of the output voltage v_o is used for v_r. The resulting circuit is shown in Fig. 1–83, where the triangular symbol represents the entire chopper-input amplifier. A fraction, β, of the output voltage is obtained by the voltage divider made up of R_1 and R_2. The resulting voltage, βv_o, provides the v_r signal at the inverting ($-$) input of the amplifier. Substituting v_u and βv_o for v_+ and v_- in Eq. (1–53),

$$v_o = A(v_u - v_r) = A(v_u - \beta v_o), \qquad (1\text{–}58)$$

and solving for v_o, we have

$$v_o = v_u \left(\frac{A}{\beta A + 1} \right) \quad \text{or} \quad v_o \approx v_u \left(\frac{1}{\beta} \right) \qquad \text{for} \quad \beta A \gg 1. \qquad (1\text{–}59)$$

As Eq. (1–59) indicates, when βA is very much larger than 1, the signal v_u is amplified by the constant factor $1/\beta$. That is, if R_1 and R_2 are 99 kΩ and 1 kΩ respectively so that $\beta = 1/100$, then $v_o = 100 v_u$ and the amplifier has a gain of 100. The precision and constancy of this gain depend upon the values of resistors R_1 and R_2 and not upon the amplifica-

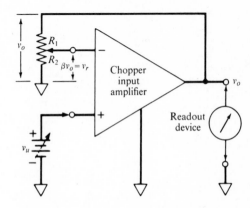

Fig. 1–83 Chopper-input amplifier connected as a potentiometric amplifier with gain.

tion A, as long as βA is very large. It is also clear that the larger the desired gain ($1/\beta$), the larger the amplification A must be for an equivalent error. This is substantiated by the rearrangement of the exact portion of Eq. (1–59) given in Eq. (1–60),

$$v_o = \frac{v_u}{\beta} - \frac{v_o}{\beta A}, \tag{1–60}$$

where the second term on the right is the error in the output voltage in volts. If, as before, an amplifier with a maximum v_o of 10 V and an amplification of $A = 10^6$ is used with $1/\beta = 100$, the maximum error voltage is 1 mV in 10 V.

Because of its near-ideal input characteristics and its precise gain, the potentiometric amplifier is used as an input signal amplifier with a variety of readout devices such as recorders, pH meters, and digital voltmeters. Frequently the readout device has a single sensitivity or full-scale deflection. In these cases, a range of measurement sensitivities is obtained by selecting values of β from a precision multitap voltage divider. While the potentiometric amplifier is inherently capable of following faster signal changes than the electromechanical servo system, its response rate is still limited to signal changes which are slow compared to the time of one chopper alternation. Typical chopper rates are 60, 400, and 1000 alternations/sec. For following signal changes in the millisecond or faster time range, chopper-input amplifiers are generally not practical. More often the operational amplifier described in the next subsection is employed.

Experiment 1–22   *Potentiometric Amplifier*

This experiment demonstrates that a null comparison technique can be used to provide precision voltage amplification under potentiometric conditions. A servo amplifier card is used to construct a potentiometric amplifier with a gain selected from two precision resistors. The experimental gain is compared to that expected. Then a commercial potentiometric amplifier is used as a precision voltage amplifier.

Operational Amplifier Voltage Follower

The operational amplifier (OA or op amp) is designed to perform, as well as possible, the functions of the chopper input amplifier but with increased response speed. To accomplish this, the internal design of the operational amplifier is quite different from the chopper input amplifier, as shown in Fig. 1–84. The chopper input is replaced with a high-quality difference

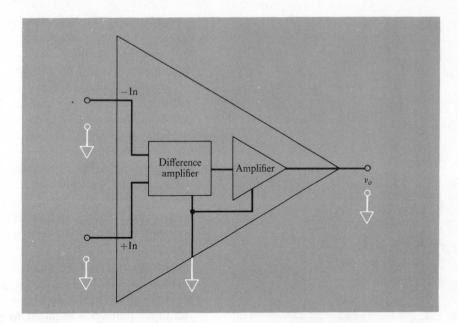

Fig. 1–84 Basic operational amplifier.

amplifier similar to the input of the comparator discussed earlier. The difference signal is amplified further to provide a high amplification within the operational amplifier, typically, $A = 10^4$ to 10^8. Since the input signals are not chopped, the sign of the difference is preserved at the difference amplifier output, thus eliminating the need for a demodulator or filter. All OA input and output voltages are applied or measured with respect to the system common. Because this is understood to be the case with OA's, the common connections shown in white in Fig. 1–84 are often omitted. The elimination of the slowly responding elements of the chopper-input amplifier provides much greater simplicity and response speed. Many OA's can follow signal changes in microseconds or less. These advantages are obtained at some sacrifice in input characteristics. These are: a finite input resistance at each input, imperfect balance in the inputs, and zero drift as described with respect to the comparator.

Continuous improvement in OA design has led to input characteristics which cause negligible error over an extremely wide range of applications. Typical input resistance values are from 10^{12} to 10^{14} Ω for FET input operational amplifiers. The common mode rejection ratio (CMRR), a measure of the input balance, is usually about 10^5, with higher values available. The actual input difference signal which results in a zero output voltage is called the **input offset.** Ideally this offset is zero, and a provision is usually made for adjusting the offset to zero. However, the offset of a difference amplifier can drift with time, temperature and circuit supply

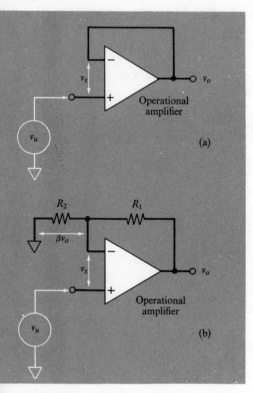

Fig. 1–85 (a) Operational amplifier voltage follower. (b) Follower with gain.

voltage causing an equivalent error in the comparison of the two input signals. In ordinary amplifiers and circuits, zero drift is less than 1 mV. With special amplifiers and care in application, the offset error can be held to within a few microvolts.

In most applications, the connections for the chopper-input amplifier and the operational amplifier are identical. The circuit for the OA floating null detector voltage follower is shown in Fig. 1–85a. The external connections, basic operation, and signal relationships are identical to those of the potentiometric amplifier voltage follower, Fig. 1–82 and Eqs. (1–53) through (1–57). The output voltage, v_o, is controlled to be equal to the input voltage, v_u, by the feedback of the v_o signal, the comparison of v_o with v_u at the OA difference input, and the application of a correcting change in level to the amplifier output. It is also analogous to the servomechanical voltage follower of Figs. 1–79 and 1–80 and is one of the family of servo systems.

Ideally, the voltage follower output voltage v_o is exactly equal to the input voltage v_u. As indicated in Fig. 1–85a, the error, or difference, $v_o - v_u$, is equal to v_s. The maximum value for the error, v_s, is the sum of (1) the input difference required to produce v_o volts of output (v_o/A from Eq. 1–57), (2) the common mode error ($v_{cm}/\text{CMRR} = v_o/\text{CMRR}$ from Eq. 1–51), and (3) the offset voltage. Assuming a maximum output voltage of 10 V, an amplification of 10^6, a CMRR of 10^5, and an offset voltage v_{off} of 100 µV, the maximum difference between v_o and v_u would be $10/10^6 + 10/10^5 + 10^{-4} = 2.1 \times 10^{-4}$ V. Since the common mode voltage is equal to v_u, the CMRR is a critical characteristic for the voltage follower application. Furthermore, the CMRR rating of some OA's decreases as the common mode voltage increases.

Amplification of the input signal with the OA voltage follower is obtained the same way as with the recorder and potentiometric amplifier, that is, by comparison of a fraction of v_o with v_u. The circuit is shown in Fig. 1–85b; it is essentially identical to that of Fig. 1–83. Since β in Fig. 1–85b is equal to $R_2/(R_1 + R_2)$, the resulting amplifier gain ($1/\beta$ from Eq. 1–59) is equal to $(R_1 + R_2)/R_2$. This circuit is so similar to that of the voltage follower that it is often called the **follower with gain.** Again, the difference between βv_o and v_u, or the **input error,** is equal to the sum of the amplification error, the common mode error, and the offset. These are v_o/A, $\beta v_o/\text{CMRR}$, and v_{off} respectively. The **output error** is equal to the input error times the amplifier gain, $1/\beta$, since the total input signal is amplified by this factor. Therefore the output error is the sum of $v_o/\beta A$ (as given in Eq. 1–60), v_o/CMRR, and v_{off}/β. For the same amplifier used in the follower error illustration, and with a $1/\beta$ gain of 100, the maximum input error is $10/10^6 + 10^{-1}/10^5 + 10^{-4} = 1.1 \times 10^{-4}$ V relative to a maximum usable input voltage of $\beta v_{o\ max} = 10^{-1}$ V. Note that v_{off} now contributes the major source of error and that the common mode error is relatively less important. The maximum output error is $10/10^{-4} + 10/10^5 + 10^{-2} = 1.1 \times 10^{-2}$ V relative to the full-scale value of 10 V. Both input and output errors in this example are about 1 part per thousand. Choosing an amplifier with better characteristics will reduce the error but, as we have seen, not necessarily to the same degree for all applications.

The common connection of the output voltage v_o of the OA makes it impossible to use the OA in the floating reference voltage null comparison configuration. Just as with the chopper-input amplifier, the floating null detector voltage follower (Fig. 1–85) was just described and is most often used. However, the input balance and resistance characteristics of the OA are generally not as good as those of the chopper-input amplifier. Therefore, with OA null voltage comparisons, the advantages of the only other possibility, the floating unknown source configuration, are worth considering. This configuration is repeated from Fig. 1–73c in Fig. 1–86a. Notice that one input of the null detector is connected to the common and that v_u is connected between v_r and the other input. With a voltage follower, $v_o = v_r$, so the resulting floating input signal voltage follower circuit is that shown in Fig. 1–86b. Such new circuits should always be tested to verify that the v_r correction is in the direction to reduce v_s. Assume that v_u becomes more positive. This causes v_s to become more negative which, in turn, causes $v_o = v_r$ to become more positive. Thus v_o follows v_u and acts to minimize v_s. (If the test indicates an inverse direction of v_o, the connections to the OA inputs should be interchanged.)

Since $v_o = -Av_s$, and $v_s = v_o - v_u$,

$$v_o = A(v_u - v_o),$$

which is identical to Eq. (1–55) for the floating null detector voltage follower. Therefore the input vs. output equations previously derived apply to both circuit configurations. Zero drift, or v_{off}, also affects both configurations identically. However, an inspection of Fig. 1–86b shows that the common mode signal at the amplifier inputs is always zero. Since the common mode input voltage doesn't change, there can be no error due to a finite CMRR. Because both OA inputs are essentially at the circuit common, there is negligible voltage applied to the amplifier input resistance, and the loading of the signal source is greatly reduced. The conclusion to be reached, then, is that the floating signal source configuration is superior to the more common voltage follower circuit when the signal source is not already connected to the OA circuit common. The same considerations would apply to the floating signal source follower with gain circuit shown in Fig. 1–86c.

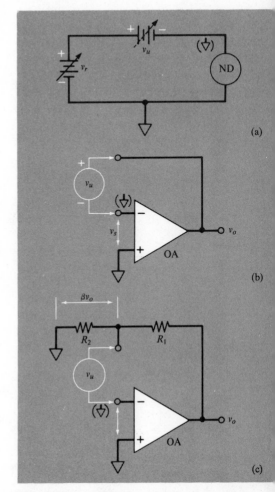

Fig. 1–86 (a) Floating signal source comparison measurement. (b) Voltage follower. (c) Follower with gain.

Experiment 1–23   *Operational Amplifier Voltage Follower*

This experiment illustrates that a voltage follower is highly useful in isolating voltage sources from loads because it can supply fairly large amounts of current to a load while drawing only small amounts of current from a source. The input-output relationship of a unity gain follower is determined, and the output current capabilities of the follower are observed. The very low input current of the follower is estimated from the voltage drop across a large input resistor. A follower with gain is then constructed and the experimental gain compared to the gain expected.

The Digital Voltmeter

Most digital voltmeters (DVM's) are another form of servo system in which the measurement information is displayed in numerical form. The general block diagram for a servo digital voltmeter is shown in Fig. 1–87. The number which is the measurement output is stored in a **register** and displayed. The number is converted to a proportional voltage which is compared with the measured voltage. If the comparator output indicates a difference, the numerical driver causes the number in the register to increase or decrease so as to reduce the magnitude of the difference. Since the digital voltmeter presents the measured value in numerical form, it is a complete measurement device. Its numerical output is read more easily and accurately than scale position indicators, and it is often possible to connect the numerical output to a printer or a computer for recording or performing calculations on the measured voltage values.

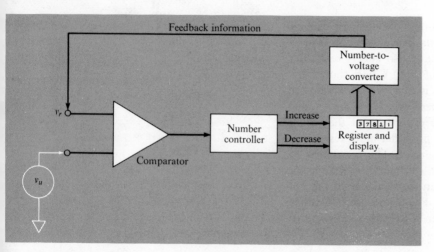

Fig. 1–87 A digital voltmeter.

The type of DVM most like the servo systems previously discussed uses an electronic counter for the register. The output from the number controller causes the counter to **count up** or **count down** (increase or decrease by 1 count) until the number in the register, converted to voltage units, matches the measured voltage. However, other types of registers and a variety of techniques are used to adjust the stored number to match the input voltage. One popular type of DVM called the **dual-slope DVM** is also a null comparison, or servo measurement device. It matches the charge transported by the unknown voltage with an equal charge transported by a standard voltage. Digital measurement systems, including digital voltmeters, are discussed in much greater detail in Module Three, *Digital and Analog Data Conversions*.

Most DVM's use an input resistance divider to change sensitivity ranges. Therefore, unless the DVM input is potentiometric, the measurement loading error will be that due to the input divider resistance just as with a FET VM. If the input is potentiometric, the input loading error will be determined by the comparator type and characteristics. A measurement loading error acceptable for a FET VM might not be acceptable for a DVM because of the greater potential accuracy and resolution of the DVM. For instance, if a voltmeter with a 10 MΩ input resistance is used to measure a 7.342 V source with an internal resistance of 50 kΩ, a value of 7.305 V is measured. If the meter is a FET VM, the meter movement probably has an accuracy no better than $\pm 1\%$ of full-scale. If full-scale is 10 V, the measured voltage is 7.31 $\pm$ 0.1 V, and the meter reading error exceeds the loading error. However, if the voltmeter is a four-digit DVM, it is seen that a significant error appears in the second-to-the-least significant digit. In order to measure the above voltage source with an error of ± 1 in the least significant digit, so that the full digital display is accurate, the input resistance of the four-digit DVM would have to be 360 MΩ!

Experiment 1–24 *Digital Voltmeter Measurements*

A digital voltmeter (DVM) is used as a voltage measurement device to demonstrate the high accuracy and resolution which can be obtained with typical DVM's. A calibrated laboratory reference source is used to check the accuracy and precision of the DVM, and experience is gained in adjusting the DVM to obtain optimum resolution without overranging. The accuracy of the DVM is compared to its specified values on several different full-scale ranges.

1–4.2 SERVO CURRENT FOLLOWERS

The concept of null comparison can be applied to the measurement of current as well as voltage. A variable reference current source is used to offset the unknown current source. The reference current is then adjusted to bring the null detector to a zero difference indication. As this section will show, the same null detectors used for null comparison voltage measurements can also be used for null comparison current measurements. Also, the many different servo systems which are used to automate potentiometric measurements can also be used to automate the null comparison current measurement. The result of automation is a current source which is continuously adjusted to follow the measured current. The automatic current follower systems described in this section are widely used in the precision measurements of small currents and are the basic circuits used in modern analog computers.

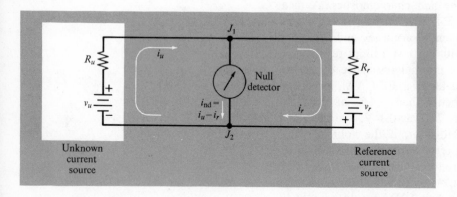

Fig. 1–88 Basic current comparison circuit, series equivalent current source.

Current Null Comparison Measurement

The basic current null comparison circuit is shown in Fig. 1–88. The null detector is connected so as to complete the current path from the unknown current source on the left. This is the same position a current meter would be in for measuring the magnitude of i_u. A reference current source is then connected to the same null detector in such a way that its current through the null detector is opposite that of the unknown current. The current path for i_u is split at junction J_1 and rejoins at J_2. Part of the current goes through the null detector (i_{nd}) and the other part (i_r) goes through the reference current source. Therefore i_u must be the sum of the currents in the other two branches:

$$i_u = i_{nd} + i_r \qquad \text{and} \qquad i_{nd} = i_u - i_r. \qquad (1\text{–}61)$$

Therefore the null detector current, i_{nd}, is equal to the difference between the unknown and reference source currents. The sign of the null detector current, i_{nd}, indicates whether i_r is larger or smaller than i_u. Clearly, then, when i_r is adjusted to be equal to i_u, the current through the null detector will be zero.

If the current through the null detector is zero, the voltage across the null detector must be zero also. Therefore, at null, the potential difference between J_1 and J_2 is zero.

The condition at null and the effects of the source and detector resistances can be most easily seen if parallel equivalent current source circuits are used for the unknown and reference currents. Just as a voltage source can be represented by an ideal voltage source in series with an internal resistance, a current source can be represented by an ideal current generator of i_e amperes in parallel with an internal resistance of R_e ohms as shown in Fig. 1–89. A current generator will deliver its maximum theoretical current when its output terminals are connected directly together (shorted). If $R_L = 0 \ \Omega$ in Fig. 1–89, all the current i_e will go through R_L as long as R_e is greater than zero. Note that in this case, the output voltage, $i_e R_L$, is zero. When the current source is connected to a nonzero load, the generator current, i_e, will split between R_e and R_L so that the load current i_L is no longer equal to i_e. If R_e is very much larger than R_L, the current through R_e can be neglected and the current source behavior is essentially ideal. Therefore, the *higher* the parallel equivalent resistance, R_e, of a current source, the *more ideal* the source. It can also be said that *all current sources are ideal when connected so that their output voltage is zero,* for this means that the voltage across R_e is also zero and none of the equivalent current, i_e, will be lost internally.

The basic current null comparison circuit is redrawn in Fig. 1–90, this time using the parallel equivalent current source circuits. Note that all the components in the circuit are in parallel. Part of the current i_u goes

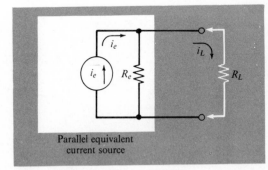

Fig. 1–89 Parallel equivalent current source.

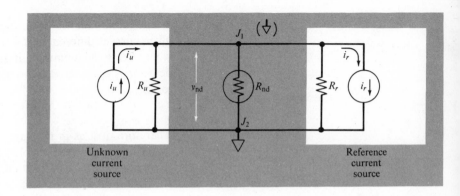

Fig. 1–90 Current comparison circuit, parallel equivalent current source.

through the reference generator, i_r. The remainder must go through the three parallel resistors R_u, R_{nd}, and R_r. The voltage across the parallel circuit is thus the current through the three resistors, $i_u - i_r$, times the equivalent resistance of the three parallel resistors. Thus

$$v_{nd} = (i_u - i_r) \frac{R_{nd}R_uR_r}{R_{nd}R_u + R_{nd}R_r + R_uR_r}. \tag{1-62}$$

Several important facts can be obtained from Eq. (1–62). The first is that when $i_u = i_r$, $v_{nd} = 0$. That is, at balance, the output voltage of both current sources is zero. Both sources are therefore under ideal conditions, delivering the full values of i_u and i_r to the external circuit. Thus the current comparison at null idealizes the current sources just as the null voltage comparison unloaded the voltage sources. If point J_2 is connected to the circuit common, point J_1 is also at the common potential at null. Another important observation from Eq. (1–62) is the relationship between the current-difference sensitivity and the null detector resistance R_{nd}. Since v_{nd} is proportional to R_{nd}, the voltage sensitivity for the null detector increases as R_{nd} increases. If R_{nd} is very much larger than R_u or R_r, Eq. (1–62) reduces to

$$v_{nd} \approx (i_u - i_r) \frac{R_uR_r}{R_u + R_r} \qquad \text{for} \quad R_{nd} \gg R_u, R_r, \tag{1-63}$$

which is the difference current times the parallel resistance of R_u and R_r. Therefore the voltage sensitivity of the null detector increases with the ideality of the current sources and is limited by the less ideal of the two.

A current-sensitive null detector can also be used in the circuit of Fig. 1–90. The current through the null detector is obtained by dividing v_{nd} from Eq. (1–62) by R_{nd}. The result is

$$i_{nd} = (i_u - i_r) \frac{R_uR_r}{R_{nd}R_u + R_{nd}R_r + R_uR_r}. \tag{1-64}$$

This equation shows again that $i_{nd} = 0$ if $i_u = i_r$. It also shows that for a given current difference, $i_u - i_r$, the null detector current i_{nd} will be largest when R_{nd} is a minimum. If the resistance R_{nd} of a current-sensitive null detector is considered to be much less than R_u or R_r, Eq. (1–64) reduces to

$$i_{nd} \approx (i_u - i_r) \qquad \text{for} \quad R_{nd} \ll R_u, R_r. \tag{1-65}$$

Therefore, just as in the case of the null comparison voltage measurement, either a high-resistance voltage-sensitive or a low-resistance current-sensitive null detector can be used. Both provide, at null, ideal load conditions for both unknown and reference current sources.

You will recall that for the basic null voltage comparison circuit of

Fig. 1–73, there were three possible positions for the connection of the circuit common due to the series arrangement of the two sources and the null detector. Note from Fig. 1–90 that in the basic current comparison circuit, all components of the circuit are connected in parallel. There are, then, only two positions for the connection of the circuit common: J_1 and J_2. Brief inspection shows that these two positions are equivalent. Therefore, *there is only one common point configuration for the current comparison circuit, and all components are connected to the circuit common.* Note also that the other connection to each of the components is at the common potential when the system is at null. This is indicated by the virtual common symbol ($\overset{+}{\triangledown}$) next to J_1. This fact eliminates the need for floating sources and the concern about common mode error at the input to the null detector.

Two null comparison measurements of current are shown in Fig. 1–91. The more common technique is the first, in which a precision load resistor, R_L, is connected across the current source. The resulting *IR* potential drop across R_L is measured potentiometrically by adjusting v_r so that the null detector indicates that $v_r = i_L R_L$. This null voltage technique eliminates loading of the variable voltage reference source since there is zero current through the null detector or the equivalent voltage source resistance R_r at null. However, the voltage across the unknown current source at null is not zero. The unknown current source is thus not connected to an ideal load at null. This technique is accurate only if the equivalent current source resistance, R_u, is so much larger than R_L that $i_L = i_u$ within the measurement accuracy limitations. Therefore, at balance, $i_u = v_r/R_L$. For unknown currents in the ampere and milliampere ranges, for which R_L can be 1 kΩ or less, this is often a reasonable assumption.

The current null comparison circuit shown in Fig. 1–91b is similar to the null voltage *IR* measurement except the positions of the precision resistor and null detector have been reversed and so has the polarity of the voltage source. The precision resistor, R_L, is now connected as a load across the variable voltage reference source to provide a variable standard current source of $i_r = v_r/R_L$. At null, i_r is exactly equal to i_u and the voltage across the null detector is zero, providing an ideal load for the unknown current source. However, the current i_r represents a load on the voltage source v_r. This load will be negligible if the precision resistor R_L is so much larger than the equivalent voltage source resistance R_r that $i_r R_r$ is a negligible fraction of v_r. In this case, $i_r = i_u = v_r/R_L$ at null. For unknown currents in the microampere and less ranges for which R_L can be 100 kΩ or more, the loading of v_r is generally negligible. Many modern voltage reference sources have equivalent output resistances of a fraction of an ohm, which would allow their use in the current comparison circuit for measurement of unknown currents up to the milliampere range. Clearly,

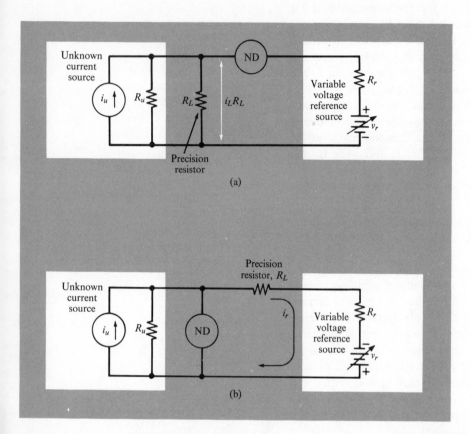

Fig. 1–91 Null comparison current measurement. (a) Potentiometric measurement of *IR* drop. (b) Current null comparison.

the null current comparison technique is far superior to the *IR* drop measurement technique for small currents since it requires no assumption about the ideality of the unknown current source.

Despite the obviously parallel advantages of null comparison measurements for voltage and current, the technique of null current measurement has been largely neglected by standard treatments on laboratory measurements while potentiometry is always included. Since this neglect can hardly be due to a lack of applications, it must be due to a general failure to recognize the possibility of extending the null measurement concept to current measurements. There are doubtless many accepted laboratory measurements which could be substantially improved by carefully considering the application of the null comparison measurement concept.

Even though manual null current comparison measurements are rarely employed, a number of circuits which are very commonly used with servomechanical systems and operational amplifiers are in fact automatic null current comparison systems. Several of these circuits are described in the

Experiment 1–25 *Current Comparison Measurements with the Current Meter*

The principle of current comparison measurements is illustrated in this experiment and several advantages of the technique are demonstrated. An ordinary moving coil meter is first used to measure current in a simple circuit by conventional techniques and is shown to introduce a perturbation error because of its internal resistance. The current is measured by a null comparison method with the same current meter as the null detector. The comparison method is shown to be ideal for current measurements because of the low perturbation error of the meter at null.

remainder of this section. The current comparison concept greatly aids in understanding the operation and characteristics of this family of circuits.

Servomechanical Current Follower

The versatile servo system is easily applied to the automation of the null adjustment in a null current comparison measurement. A servomechanical recording null current measurement system is shown in Fig. 1–92. The null detector inputs (chopper contacts) are connected directly across the unknown current source. The usual recorder variable reference voltage source (Fig. 1–81) applies the voltage v_o across the load resistor R_f resulting in a reference source current, i_r.

Fig. 1–92 Recording null comparison servomechanism.

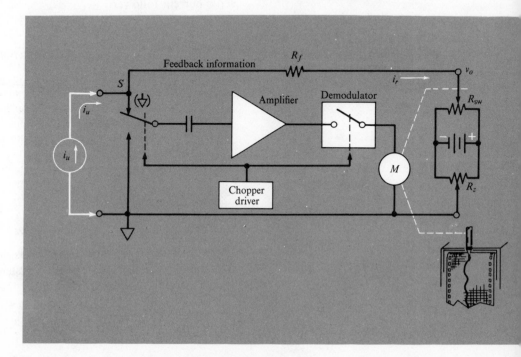

The magnitude of i_r is the feedback information that is compared with i_u to obtain information on the desired adjustment of v_o. Because the load resistor is in the feedback path, it is often called the feedback resistor and given the symbol R_f. When i_u and i_r are unequal, the voltage at point S is not 0 V. This off-null signal is detected by the chopper input and amplified, causing the motor to turn and move the recorder pen and the slidewire potentiometer to reduce the off-null signal and keep point S at the common potential. Thus the current, i_r, and the position of the pen follow the unknown current, i_u. This system is the servomechanical equivalent of Fig. 1–91b. As such the loading of the unknown current source is ideal while that on the servo-adjusted voltage reference source is not. For v_o to be the expected linear function of the position of R_{sw} and the pen, the loading of the slidewire circuit must be negligible. This system has many advantages for the measurement of small currents when R_f can be large. If the range of v_o is 1 V and R_f is 10 MΩ, the full-scale recorder deflection sensitivity would be 10^{-7} A. To record relatively high currents, it is preferred to connect a load resistor across the current source, and use the resulting iR drop for v_u in the servo potentiometer of Fig. 1–80.

Since the null detector in Fig. 1–92 is connected to the system common, the chopper input is not needed for its excellent common mode rejection characteristics. However, its other characteristics of high input resistance and freedom from zero drift are advantageous in this application where very low signal levels are involved. These same advantages apply to the chopper-input amplifier when it is used in an all-electronic servo current follower.

Experiment 1–26 *Servo Current Follower*

This experiment demonstrates that a servo system can be used to make current comparison measurements automatically. The servo recorder is wired as a current follower by inserting a resistor between the slidewire reference voltage output and the input of the chopper amplifier. The servo current follower is used to measure currents from a standard current source and the photocurrent from a phototransistor radiation transducer. Results of the light measurements are compared to those obtained in previous experiments.

Operational Amplifier Current Follower

When the current recording function is not needed, or high-speed balance is required, an all-electronic servo system can be used as a current follower. For this purpose, the chopper-input or operational amplifiers are

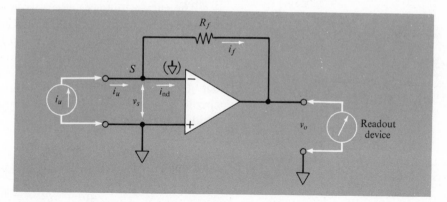

Fig. 1–93 OA current follower.

used. As we have seen, they are essentially equivalent except for their response rate and input characteristics. The basic electronic servo current follower is shown in Fig. 1–93. The unknown current source is connected directly to the null detector input terminals. The OA output voltage v_o is connected through the precision load resistor, R_f, to one of the null detector inputs. The negative (inverting) input is used for the feedback current information (generally symbolized i_f for OA circuits) because it provides the proper corrective direction at the output. If i_u increases, point S becomes more positive, making v_o more negative and thus increasing i_f as desired. The other OA input is then connected to the circuit common to complete the current path for i_f, the reference source current.

From the currents at point S in Fig. 1–93, it can be seen that

$$i_u = i_f + i_{nd}, \tag{1–66}$$

and by summing the potentials in the loop including v_s, the voltage drop $i_f R_f$ across the feedback resistor, and v_o, we get

$$v_o = v_s - i_f R_f. \tag{1–67}$$

Substituting for i_f from Eq. (1–66) and v_s from $v_o = -Av_s$ (Eq. 1–53) into Eq. (1–67), we obtain

$$v_o = -i_u R_f + i_{nd} R_f - \frac{v_o}{A} \quad \text{or} \quad v_o = -R_f(i_u - i_{nd})\left(\frac{A}{1+A}\right). \tag{1–68}$$

From either form of Eq. (1–68) it can be seen that when A is very large and i_{nd} is very much less than i_u, v_o is proportional to i_u:

$$v_o \approx -i_u R_f \quad (\text{for } A \gg 1 \quad \text{and} \quad i_{nd} \ll i_u). \tag{1–69}$$

Thus i_f follows i_u and produces an output voltage proportional to i_u.

The current follower circuit is in many ways analogous to the voltage follower. The output voltage produced is the same as that which would occur if R_f were simply a load resistor across the current source i_u. However, the current follower presents a near ideal load to the current source and also is an output voltage source suitable for driving a variety of readout devices. The effective input resistance R_{cf} of the current follower circuit as seen by the unknown current source is the input voltage, v_s, divided by the input current, i_u:

$$R_{cf} = v_s/i_u. \tag{1--70}$$

Substituting for v_s from $v_o = -Av_s$ and for i_u from Eq. (1–69), we have

$$R_{cf} = R_f/A, \tag{1--71}$$

which shows that the loading effect of R_f on the current source i_u is improved A times by using the current follower. Considering that A is generally 10^5 to 10^7, the use of the current follower can and does provide a dramatic improvement in current measurements.

The range of currents that can be measured by a current follower and readout is limited on the low end by the input current, i_{nd}, of the OA and on the high end by the OA's output current capability. The input current characteristic of OA's varies from 10^{-8} to 10^{-15} A, making the current follower useful into the picoampere current range with careful choice of the OA. However, it should be noted from Eq. (1–68) that the OA input current is multiplied by the same factor as i_u. The output current capability of OA's is generally 2 to 100 mA. The OA output must supply both the feedback current i_f and the readout device current. Another possible source of error is the input voltage drift of the OA. This appears as a variation in v_s. From Eq. (1–67), we see that a drift in v_s affects v_o by an amount equivalent to the drift. Since the OA output is 1–10 V for most measurements and input drifts are generally much less than 1 mV, input voltage drift is not a serious problem with the current follower.

In addition to its obvious applications in current measurement systems, the current follower, in several variations, is the basic active circuit in modern analog computers. Since simple analog computation circuits are also very useful in instruments, two of these variations, the inverting and summing amplifiers, are described below.

Experiment 1–27 *Operational Amplifier Current Follower*

The operational amplifier (OA) current follower is shown in this experiment to perform current comparison measurements in a manner analogous to the servo current follower. The advantages of fast response time and no mechanical couplings are illustrated. An OA current follower or current-to-voltage converter is constructed, and the output voltage–input current relationship is studied. Deviations from ideality are noted. The OA current follower is used to measure the current from a phototransistor transducer.

OA Inverting Amplifier

The inverting amplifier is a simple variation of the current follower circuit in which the unknown current source is a voltage source and resistor in series. The resulting circuit is shown in Fig. 1–94. Summing the voltages in the input circuit gives

$$v_u = i_{in}R_{in} + v_s. \tag{1-72}$$

Summing the voltages in the feedback loop, we have

$$v_o = v_s - i_f R_f. \tag{1-73}$$

Recalling that $v_o = -Av_s$ and assuming that the amplifier input current, i_{nd}, is negligible with respect to i_{in} so that $i_f = i_{in}$, Eqs. (1–72) and (1–73) can be combined and solved for v_o:

$$v_o = -v_u \frac{R_f}{R_{in}} \left(\frac{A}{1 + A + R_f/R_{in}} \right). \tag{1-74}$$

From the expression in parentheses in Eq. (1–74), it can be seen that if A is very much larger than 1 and R_f/R_{in}, then

$$v_o \approx -v_u \frac{R_f}{R_{in}} \qquad (\text{for } A \gg R_f/R_{in}, 1). \tag{1-75}$$

Thus the output voltage is a constant times the input voltage and the constant $(-R_f/R_{in})$ is dependent only upon the values of the resistors R_f and R_{in}. This provides the possibility for precision amplification of a voltage signal. The voltage follower with gain of Fig. 1–85b also provides precision gain with a different circuit. In contrast to the voltage follower, the inverting amplifier inverts the input signal as indicated by the minus sign in Eq. (1–75). Also the inverting amplifier has one of its inputs connected to the circuit common which eliminates the common-mode rejection error, and it has a simpler relationship between gain and resistance which allows whole decade values of resistors to produce whole decade values of system gain, for example, $R_f = 1$ MΩ, $R_{in} = 10$ kΩ, for $G = 100$. On the other hand, the input source, v_u, is loaded by the input resistor, R_{in}, in the inverting amplifier and is potentiometrically measured in the follower with gain. Each amplifier circuit has its merits depending upon the desirability of inversion and the problems of source loading.

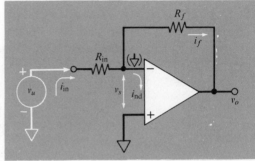

Fig. 1–94 OA inverting amplifier.

Experiment 1–28   *Operational Amplifier Inverting Amplifier*

This experiment demonstrates the use of an operational amplifier (OA), in the current comparison mode, as an inverting voltage amplifier. An input resistor is inserted between the voltage source and the OA summing point to provide the input current. The output voltage of the inverting amplifier is measured as a function of input voltage, of feedback resistance, and of input resistance. Results are compared to those expected from theoretical equations, and any deviations are noted.

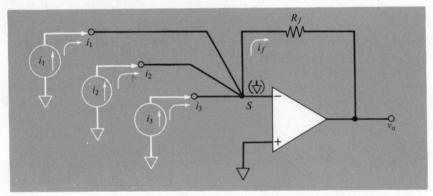

Fig. 1–95 Summing current follower.

Summing Amplifier

Since the current follower amplifier of Fig. 1–93 provides an input that is maintained at the common potential, a current source connected to point S has the same output that it would have if it were connected to the common. Therefore, several current sources can be connected to point S without interfering with each other. The result is shown in Fig. 1–95. Each current source applies its current to the virtual common point S. Since the sum of all currents to point S must be equal to i_f,

$$i_f = i_1 + i_2 + i_3. \tag{1–76}$$

For this reason, point S is often called the summing point of the OA in current follower applications. Now, from Eq. (1–68), neglecting i_{nd}, we see that

$$v_o = -R_f(i_1 + i_2 + i_3)\,\frac{A}{1 + A}, \tag{1–77}$$

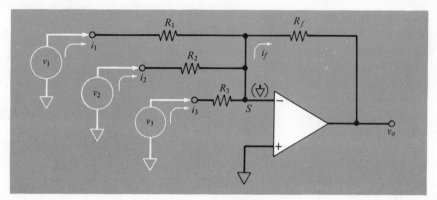

Fig. 1–96 Summing amplifier.

which is

$$v_o \approx -R_f(i_1 + i_2 + i_3) \qquad \text{(for } A \gg 1\text{)}. \qquad (1\text{–}78)$$

Therefore the output voltage is proportional to the sum of the input currents. This circuit has many uses in instrumentation including the summing of currents from several transducers or the offsetting of a portion of the current measured from one transducer by a constant value.

Again, the current sources can be voltages with series resistors as shown in Fig. 1–96. Assuming point S to be at the common potential, $i_1 = v_1/R_1$, $i_2 = v_2/R_2$, and $i_3 = v_3/R_3$. Substituting these relations for the current in Eq. (1–78), we obtain

$$v_o \approx -\left(v_1 \frac{R_f}{R_1} + v_2 \frac{R_f}{R_2} + v_3 \frac{R_f}{R_3}\right) \text{ (for } A \gg R_f/R_{\text{in}}\text{)}, \quad (1\text{–}79)$$

which shows that the output voltage is the sum of the input voltages, each multiplied by its own R_f/R_{in} ratio. This circuit is called the **weighted summing amplifier** because the contribution of each input voltage to the sum is weighted by its individual gain factor. Frequently, a simple sum is desired so that R_1, R_2, and R_3 are all equal to R_{in}. Then

$$v_o \approx \frac{-R_f}{R_{\text{in}}} (v_1 + v_2 + v_3) \text{ (for } A \gg R_f/R_{\text{in}}\text{)}. \qquad (1\text{–}80)$$

This circuit obviously has the same loading characteristics as the inverting amplifier. It serves the same type of applications in instruments as the summing current follower and is the summing amplifier used in analog computers.

Experiment 1–29   *OA Summing Amplifier*

This experiment demonstrates the use of an operational amplifier as a summing amplifier to provide an output voltage related to the sum or difference of two or more input currents. Voltage sources are connected with different input resistances to the summing point of an operational amplifier current follower. The output voltage is measured as a function of the input voltages. Addition and subtraction of currents are demonstrated by changing the polarity of the input sources. Results are compared to those expected from theory.

PROBLEMS

1. A slidewire reference source similar to that of Fig. 1–81 is to be constructed. The zero adjust potentiometer R_z and the slidewire potentiometer R_{sw} are both 1 kΩ. What value of calibrated source voltage V_c and what fraction of R_z are needed to make the output voltage vary from -0.3 V to $+0.7$ V?

2. A variable range servo recorder is shown in Fig. 1–81. If R_{sw} and R_z are both 500 Ω potentiometers and $V = 1.50$ V, what values of R_c and the divider resistors are necessary to give 1.00 V across the slidewire and a 10 mV full-scale range?

3. A potentiometric amplifier has an amplification A of 5×10^4. The amplifier has a maximum output v_o of 5 V. What is the maximum error in V between v_o and the input voltage v_u?

4. A potentiometric amplifier with gain is constructed similar to that in Fig. 1–83. Resistors R_1 and R_2 are made 999 kΩ and 1 kΩ respectively. The amplifier has an amplification of 10^4. What is the maximum error voltage if the maximum output voltage is 10 V?

5. A Weston cell is to be used to make a precision voltage source. However, any current drawn from the Weston cell tends to reduce its stability and reproducibility. An operational amplifier voltage follower and a unity gain inverting operational amplifier circuit were suggested as isolation devices. Discuss the merits of both circuits for this application.

6. The operational amplifier circuit of Fig. 1–85b is constructed. What is the gain if (a) $R_1 = 100$ kΩ and $R_2 = 10$ kΩ, (b) $R_1 = R_2 = 10$ kΩ, and (c) $R_1 = 100$ kΩ and $R_2 = 1$ kΩ. Assume that the amplification A is very large.

7. The servomechanical current follower of Fig. 1–92 is to provide full-scale current ranges of 0.1 μA, 0.5 μA, 1.0 μA, and 5.0 μA. If the slidewire output voltage v_o is 1.20 V, what values of feedback resistance R_f are necessary for the above spans?

8. The circuit of Fig. 1–94 is used to measure a voltage source of 1.0 V having an equivalent resistance of 1 kΩ. If $R_{in} = 10$ kΩ and $R_f = 100$ kΩ what would v_o be?

9. An inverting operational amplifier circuit of the type shown in Fig. 1–94 is to be wired. What is the highest voltage gain, accurate to 1%, obtainable with an amplifier whose amplification A is (a) 1000, (b) 10^4, (c) 10^6.

10. Design operational amplifier circuits which will produce (a) a 1 V output for a 10 μA input current, (b) an output voltage which is $-(5v_1 + 3v_2)$, where v_1 and v_2 are input voltage sources, and (c) an output voltage which is $v_1 - v_2$. Hint: Use more than one OA.

Bibliography

"Electricity and Magnetism," in *Berkeley Physics Course,* Volume 2, McGraw-Hill, New York, N.Y., 1965.

A basic text which includes a discussion of the principles of electrical and magnetic quantities, properties, and interactions. Many excellent diagrams help the reader to understand these topics.

Enke, C. G., "Data Domains—An Analysis of Digital and Analog Instrumentation Systems and Components," *Analytical Chemistry,* **43**, Jan., 1971, p. 69A.

A formal development of the data domain concepts as they are related to the basic measurement process and applied in scientific instrumentation.

Geddes, L. A., and L. E. Baker, "Principles of Applied Biomedical Instrumentation," Wiley, New York, N.Y., 1968.

A transducer-oriented treatment of modern biomedical measurement devices and systems. Written for the life scientist.

Halliday, D., and R. Resnick, *Physics, Parts I and II,* Wiley, New York, N.Y., 1966.

A standard college physics text which includes basic discussions of electrical quantities and magnetism.

Lion, K. S., *Instrumentation in Scientific Research—Electrical Input Transducers,* McGraw-Hill, New York, N.Y., 1959.

An older "classic" which covers mechanical, temperature, magnetic, electrical, and radiation transducers. Contains principles and applications.

Malmstadt, H. V., and C. G. Enke, *Digital Electronics for Scientists,* W. A. Benjamin, Menlo Park, Calif., 1969.

The first exposition of the use of data domains for the analysis of measurement systems. An analog section includes null-comparison voltage measurements and an introduction to servo systems and operational amplifiers and their applications.

Malmstadt, H. V., C. G. Enke, and E. C. Toren, Jr., *Electronics for Scientists,* W. A. Benjamin, Menlo Park, Calif., 1962.

An earlier work containing many still-relevant sections including electrical measurements, comparison measurements, servo systems, and operational amplifiers.

Malmstadt, H. V., M. L. Franklin and G. Horlick, "Photon Counting for Spectrophotometry," *Analytical Chemistry,* **44**, July, 1972, p. 63A.

An introduction to the photomultiplier tube as a transducer with both current and charge domain output. A summary of the important characteristics of photon counting systems, especially as applied to spectrophotometry.

Oliver, F. J., *Practical Instrumentation Transducers,* Hayden, New York, N.Y., 1971.

A thorough survey of physical-to-electrical domain transducers and their associated circuits used in measurement and control systems.

Stout, M. B., *Basic Electrical Measurements,* 2nd ed., Prentice-Hall, Englewood Cliffs, N.J., 1960.

Detailed descriptions of various types of potentiometers and bridges and their components. Many references and problems.

Solutions to Problems

SOLUTIONS TO PROBLEMS IN SECTION 1–2

1. Since $I = V/R$ from Eq. (1–11), $R = V/I = 10 \text{ V}/0.300 \text{ A} = 33.3 \text{ }\Omega$.

2. First the resistance equivalent to the 500 Ω and 1 kΩ resistors in series must be calculated. From Eq. (1–12), $R_s = R_1 + R_2 = 500 + 1000 = 1500 \text{ }\Omega$. The current in each element of the series circuit is now obtained from Ohm's Law, $I = V/R_s = 5 \text{ V}/1.5 \times 10^3 \text{ }\Omega = 3.33 \times 10^{-3} \text{ A}$. The IR drop across the 500 Ω resistor is $IR_1 = 3.33 \times 10^{-3} \times 500 = 1.67 \text{ V}$ and across the 1 kΩ resistor is $IR_2 = 3.33 \times 10^{-3} \times 1.00 \times 10^3 = 3.33 \text{ V}$.

3. Since the voltmeter resistance is 100 kΩ per volt of full-scale deflection, the resistance of the voltmeter $(R_m + R_s$ in Fig. 1–16) on the 1 volt scale would be 100 kΩ. The current through the meter when the full-scale voltage is applied to it is thus $I = V/R = 1.00/1.00 \times 10^5 = 10^{-5} \text{ A} = 10 \text{ }\mu\text{A}$. Thus the full-scale deflection sensitivity of the meter movement is 10 μA. Since the total voltmeter resistance should be 100 kΩ for a 1 V full-scale deflection, and $R_m = 10 \text{ k}\Omega$, $R_s = 100 \text{ k}\Omega - 10 \text{ k}\Omega = 90 \text{ k}\Omega$.

4. A divider to produce 5 output voltages of less than the supply voltage is shown in Fig. S–4. The total divider resistance $R_s = R_1 + R_2 + R_3 + R_4 + R_5 + R_6$. The voltage at V_5, which is to be 1 V, will be 50 V $\times R_6/R_s = 1$ V. Therefore $R_6 = (1/50)R_s$. Similarly, for $V_4 = 3$ V, $R_5 + R_6 = (3/50)R_s$. For $V_3 = 9$ V, $R_4 + R_5 + R_6 = (9/50)R_s$ and for $V_2 = 15$ V, $R_3 + R_4 + R_5 + R_6 = (15/50)R_s$, and for $V_1 = 30$ V, $R_2 + R_3 + R_4 + R_5 + R_6 = (30/50)R_s$. The optimum value for R_s depends on the nature of the voltage source and

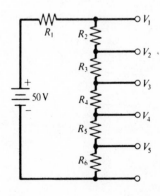

Fig. S–4

117

the loads that are to be connected to the divider outputs. If, for example, $R_s = 1$ kΩ,

$$R_6 = (1/50)R_s = 20 \text{ Ω},$$
$$R_5 + R_6 = (3/50)R_s = 60 \text{ Ω}, \qquad R_5 = 60 - 20 = 40 \text{ Ω},$$
$$R_4 + R_5 + R_6 = (9/50)R_s = 180 \text{ Ω}, \qquad R_4 = 180 - 60 = 120 \text{ Ω},$$
$$R_3 + R_4 + R_5 + R_6 = (15/50)R_s = 300 \text{ Ω}, \qquad R_3 = 300 - 180 = 120 \text{ Ω},$$
$$R_2 + R_3 + R_4 + R_5 + R_6 = (30/50)R_s = 600 \text{ Ω}, \qquad R_2 = 600 - 300 = 300 \text{ Ω},$$
$$R_1 = R_s - 600 = 1000 - 600 = 400 \text{ Ω}.$$

5. From Ohm's Law, $V = IR = 3.5 \times 10^{-3} \text{ A} \times 5 \times 10^3 \text{ Ω} = 17.5$ V.

6. The divider circuit is shown in Fig. S-6. $R_s = R_1 + R_2 + R_3 + R_4 = 10$ kΩ. In the $\times 1000$ position, $V_{\text{out}} = V_{\text{in}}/1000$ and $R_4 = R_s/1000 = 10$ Ω. In the $\times 100$ position, $V_{\text{out}} = V_{\text{in}}/100$ and $R_3 + R_4 = R_s/100 = 100$ Ω, so $R_3 = 100 - 10 = 90$ Ω. In the $\times 10$ position, $V_{\text{out}} = V_{\text{in}}/10$ and $R_2 + R_3 + R_4 = R_s/10 = 1000$ Ω, so $R_2 = 1000 - 100 = 900$ Ω. $R_1 = R_s - 1000 = 9000$ Ω.

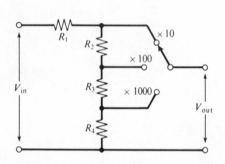

Fig. S–6

7. a) $I = V/R_l = 10$ V$/12$ kΩ $= 8.33 \times 10^{-4}$ A.
   b) $I_m = V/(R_l + R_m) = 10$ V$/(12$ kΩ $+ 5$ kΩ$) = 5.88 \times 10^{-4}$ A.
   c) From Eq. (1–15), RPE $= (I - I_m)/I = 2.45 \times 10^{-4}/8.35 \times 10^{-4} = 0.294$ or 29.4%, or from Eq. (1–16), RPE $= R_m/(R_l + R_m) = 5$ kΩ$/17$kΩ $= 0.294$.

8. From Eq. (1–16), RPE $\leqslant 0.01 \leqslant R_m/(10{,}000 + R_m)$, $0.99R_m \leqslant 100$, $R_m \leqslant 101$ Ω. From the circuit voltage, the minimum full scale current sensitivity can be calculated. $I_{fs} \geqslant V/R = 6$ V$/10$ kΩ $= 6 \times 10^{-4}$ A.

9. From Eq. (1–23), $R_p = R_1 R_2/(R_1 + R_2)$. Since $R_p = 1.2$ kΩ and $R_2 = 3$ kΩ, $3000R_1/(R_1 + 3000) = 1200$, $R_1 = 2$ kΩ.

10. Substitute into Eq. (1–26), $I_m = 10^{-3}$ A, $I_t = 6 \times 10^{-3}$ A, $R_m = 50$ Ω. $R_{\text{sh}}/(R_m + R_{\text{sh}}) = I_m/I_t$, $R_{\text{sh}}/(50 + R_{\text{sh}}) = 1/6$, $R_{\text{sh}} = 10$ Ω.

11. The equivalent resistance of the 3 parallel resistors is given by Eq. (1–22):

$$R_p = \cfrac{1}{\cfrac{1}{1 \text{ kΩ}} + \cfrac{1}{3 \text{ kΩ}} + \cfrac{1}{10 \text{ kΩ}}} = 698 \text{ Ω}.$$

Now $I_t = V/R_p = 5/698 = 7.17 \times 10^{-3}$ A. The current through each resistor is $I = V/R$:

$$I_1 = 5/1 \text{ kΩ} = 5 \times 10^{-3} \text{ A},$$

$$I_2 = 5/3 \text{ k}\Omega = 1.67 \times 10^{-3} \text{ A},$$
$$I_3 = 5/10 \text{ k}\Omega = 5 \times 10^{-4} \text{ A},$$

from which we can confirm $I_t = I_1 + I_2 + I_3 = 7.17 \times 10^{-3} \text{ A}$.

12. Substituting in Eq. (1–25), where $R_1 = 10 \text{ k}\Omega$ and $R_2 = 20 \text{ k}\Omega$, we find the current through R_1 to be

$$I_1 = \frac{R_2}{R_1 + R_2} I_t = \frac{20 \text{ k}\Omega}{10 \text{ k}\Omega + 20 \text{ k}\Omega} I_t = \frac{2}{3} I_t$$

and the current through R_2 to be

$$I_2 = \frac{R_1}{R_1 + R_2} I_t = \frac{10 \text{ k}\Omega}{10 \text{ k}\Omega + 20 \text{ k}\Omega} I_t = \frac{1}{3} I_t.$$

13. Substitute in Eq. (1–27), $i_m = 50 \times 10^{-6} \text{ A}$, $R_m = 10^3 \ \Omega$. For the 1 mA position, $i_t = 10^{-3} \text{ A}$, $R_{\text{sh}} = R_1 + R_2 + R_3 + R_4 = R_t$, and $R_s = 0$.

$$\frac{I_m}{I_t} = \frac{R_{\text{sh}}}{R_s + R_m + R_{\text{sh}}} = \frac{50 \times 10^{-6}}{10^{-3}} = \frac{R_t}{1 \text{ k}\Omega + R_t}, \qquad R_t = 52.6 \ \Omega.$$

For the 100 mA position, $I_t = 10^{-1} \text{ A}$, $R_s = R_2 + R_3 + R_4$, and $R_{\text{sh}} = R_1$.

$$\frac{50 \times 10^{-6}}{10^{-1}} = \frac{R_1}{R_t + 1 \text{ k}\Omega} = \frac{R_1}{1052.6}, \qquad R_1 = 0.526 \ \Omega.$$

For the 50 mA position, $I_t = 5 \times 10^{-2} \text{ A}$, $R_s = R_3 + R_4$, and $R_{\text{sh}} = R_1 + R_2$.

$$\frac{50 \times 10^{-6}}{5 \times 10^{-2}} = \frac{R_1 + R_2}{R_t + 1 \text{ k}\Omega} = \frac{R_1 + R_2}{1052.6}, \qquad R_1 + R_2 = 1.0526 \ \Omega, \qquad R_2 = 0.526 \ \Omega.$$

For the 10 mA position, $I_t = 10^{-2} \text{ A}$, $R_s = R_4$, and $R_{\text{sh}} = R_1 + R_2 + R_3$.

$$\frac{50 \times 10^{-6}}{10^{-2}} = \frac{R_1 + R_2 + R_3}{1052.6},$$

$$R_1 + R_2 + R_3 = 5.263 \ \Omega, \qquad R_3 = 4.210 \ \Omega,$$

$$R_4 = R_t - (R_1 + R_2 + R_3) = 52.6 - 5.26 = 47.3 \ \Omega.$$

14. Substitute in Eq. (1–31), $\bar{I} = 10^{-8} \text{ A}$, $\bar{G} = 5 \times 10^5$, $Q_e = 1.6 \times 10^{-19} \text{ C}$, and $b = 0.05$. Then

$$P_k = \bar{I}/(bQ_e\bar{G}) = 10^{-8}/(0.05 \times 1.6 \times 10^{-19} \times 5 \times 10^5)$$

$$= 2.5 \times 10^6 \text{ photons/sec.}$$

15. Substitute in Eq. (1–28), $b_t = 0.99$, $b_c = 0.8$, $b_\lambda = 0.1$, $P_k = 5000$. Then $N_a = b_t b_c b_\lambda P_k = 396$ pulses/sec. The average anode current from Eqs. (1–31) and (1–28) is

$$\bar{I} = N_a Q_e \bar{G}$$

$$= 396 \times 1.6 \times 10^{-19} \times 10^6 = 6.34 \times 10^{-11} \text{ A.}$$

The transfer function for current is $\bar{I} = kP_k$, where

$$k = \frac{6.34 \times 10^{-11}}{5 \times 10^3} = 1.27 \times 10^{-14} \text{ A-sec/photon.}$$

SOLUTIONS TO PROBLEMS IN SECTION 1-3

1. Substituting in Eq. (1–34), we get $Q = CV = 2 \times 10^{-8} \times 1.3 = 2.6 \times 10^{-8}$ C. From Eq. (1–35), $W_d = QV/2 = 1.69 \times 10^{-8}$ J.

2. The maximum current is obtained from a voltage source when its output terminals are shorted together. From Fig. 1–37, the resulting current would be $I_{max} = v_e/R_e$. Solving for R_e in each case, we get

$$R_e = 1.5/0.9 = 1.67 \ \Omega \qquad \text{for the alkaline battery,}$$

$$R_e = 2.0/500 = 4 \times 10^{-3} \ \Omega \text{ for the lead-acid cell.}$$

3. With a 5 V supply, KV-1 adjusts the output in 0.5 V steps, KV-2 in 0.05 V steps, and KV-3 has a 0.05 V total range. To obtain 1.83 V, KV-1 should be in the 1.5 to 2.0 V range; that is, the contacts should bridge the fourth and fifth resistors from the bottom. There is 0.5 V across KV-2, of which 0.30 to 0.35 V is needed. Therefore, the divider should be in position 6, one position below that pictured in Fig. 1–39. There is 0.05 V across KV-3, of which 0.03 V is needed, so KV-3 should be 6/10 of the way from the bottom to the top.

4. Substitute in Eq. (1–46), $V_e = 1.0$ V, $IR_e \leqslant 10^{-3}$ V, $R_e = 10^{10} \ \Omega$. Then

$$0.999 \leqslant \frac{1.0 \, R_m}{R_m + 10^{10}},$$

$$R_m \geqslant 0.999 \times 10^{13} \approx 10^{13} \ \Omega.$$

5. a) For a voltage divider, $V_o = V_{in}R_1/(R_1 + R_2)$, so $V_o = 3 \times 20$ $\text{k}\Omega/30 \text{ k}\Omega = 2$ V.

b) With meter connected, there is a 100 kΩ resistance in parallel with the 20 kΩ divider resistor. The equivalent parallel resistance $R_p = 20$ kΩ $\times$ 100 kΩ/120 kΩ $= 16.7$ kΩ. Now recalculating the output voltage from the above relationship, we get

$$v_o = 3 \times 16.7 \text{ k}\Omega/26.7 \text{ k}\Omega = 1.88 \text{ V}.$$

c) From Eq. (1–47),

$$\text{RPE} = (2 - 1.88)/2 = 0.06 \quad \text{or} \quad 6\%.$$

6. a) By the reasoning in problem 5 above, when the meter is connected, $v_o = 10 \times 333$ kΩ/(500 kΩ $+ 333$ kΩ) $= 4$ V. Therefore, RPE $= (5–4)/5 = 0.2$ or 20%.

b) Similarly, v_o (measured) $= 10 \times 91$ kΩ/(900 kΩ $+ 91$ kΩ) $= 0.92$ V and RPE $= (1–0.92)/1 = 0.08$ or 8%.

7. a) When 5.00 V is selected, the full 1 kΩ divider appears in parallel with the meter, and the total 5.00 V must appear across the parallel combination. Hence RPE $= 0$.

b) When 1 V is selected, 200 Ω from the divider appears in parallel with the meter. The equivalent parallel resistance is $R_p = 200$ Ω $\times$ 10,000 Ω/(200 $+$ 10,000) $= 196$ Ω. Then

$$v_o = 5 \text{ V} \times 196 \text{ } \Omega/(196 + 800) \text{ } \Omega = 0.984 \text{ V},$$

$$\text{RPE} = (1.0 - .984)/1.0 = 0.016 = 1.6\%.$$

8. From Fig. 1–55 and the equations leading to Eq. (1–46), the current through the voltmeter in each case can be calculated. For the 15 V scale,

$$I_{15} = 6.1/300 \text{ k}\Omega = 20.3 \times 10^{-6} \text{ A},$$

and for the 5 V scale,

$$I_5 = 4.37/100 \text{ k}\Omega = 43.7 \times 10^{-6} \text{ A}.$$

From Eq. (1–46),

$$V_m = 6.1 = V_e - 20.3 \times 10^{-6} R_e$$

and

$$V_m = 4.37 = V_e - 43.7 \times 10^{-6} R_e.$$

Subtracting the lower equation from the upper to eliminate V_e, we get $1.73 = 23.4 \times 10^{-6} R_e$, from which $R_e = 73.9$ kΩ. We can obtain V_e from either equation,

$$V_e = 6.1 + 20.3 \times 10^{-6} \times 73.9 \times 10^3 = 7.60 \text{ V},$$

$$V_e = 4.37 + 43.7 \times 10^{-6} \times 73.9 \times 10^3 = 7.60 \text{ V}.$$

9. Substitute in Eq. (1–51), $\Delta v_o/\Delta v_d = 10/5 \times 10^{-4}$ and $\Delta v_o/\Delta v_{\text{cm}} = 1/0.5$. Then

$$\text{CMRR} = \frac{\Delta v_o/\Delta v_d}{\Delta v_o/\Delta v_{\text{cm}}} = \frac{2 \times 10^4}{2} = 10^4.$$

10. First calculate I for 0.500 V across R_p:

$$I = 0.500/10.00 = 0.05000 \text{ A}.$$

To standardize, IR_s must be equal to the 1.018 V from the Weston cell. Therefore, $R_s = 1.018/I = 20.36 \ \Omega$.

11. The total desired span of adjustment is 0.5 V $-$ (-0.5 V) $= 1.0$ V. Therefore, there should be 1.00 V across the 1 kΩ resistor and the remainder of the 1.34 V across R_s. From the equation for a voltage divider,

$$v_o = v_{\text{in}} \frac{R_1}{R_1 + R_2}, \qquad 1.00 = 1.34 \cdot \frac{1 \text{ k}\Omega}{R_s + 1 \text{ k}\Omega},$$

from which $R_s = 340 \ \Omega$. Or the current I through the divider must be 1.00 V/1 k$\Omega = 1.00$ mA. For $1.34 - 1.00 = 0.34$ V IR drop across R_s, $R_s = 0.34/1 \times 10^{-3} = 340 \ \Omega$.

12. The maximum accuracy for the measurement of v_u will be determined by the resolution of the slidewire or the minimum observable deflection of the null detector, whichever is larger. The resolution of the slidewire is 2.0 V $\times 10^{-4}$ m/1 m $= 2 \times 10^{-4}$ V. The minimum observable deflection of the galvanometer is

$$3 \times 10^{-7} \text{ A/mm} \times 1 \text{ mm} = 3 \times 10^{-7} \text{ A}.$$

The voltage sensitivity of the measurement ($v_r - v_u$) can be calculated from the current sensitivity with Eq. (1–49), where $i_{\text{nd}} = 3 \times 10^{-7}$ A, $R_{\text{nd}} = 150 \ \Omega$.

$$(v_r - v_u)_{\text{(min)}} = i_{\text{nd(min)}}(R_u + R_r + R_{\text{nd}}).$$

R_r can be neglected or estimated to be 25 Ω from the equation for the equivalent output resistance of a voltage divider, given in the paragraph below Eq. (1–48) as $R_e = Rb \ (1 - b)$, where b is the output voltage fraction. R_e is a maximum when $b = 1/2$, so $R_e = 100 \ (1/2)(1/2) = 25 \ \Omega$.

a) $(v_r - v_u)_{min} = 3 \times 10^{-7} (100 + 25 + 150) = 8.25 \times 10^{-5}$ V.
   Therefore, the slidewire setting limits the accuracy to 2×10^{-4} V.

b) $(v_r - v_u)_{min} = 3 \times 10^{-7} (10^4 + 25 + 150) = 3.05 \times 10^{-3}$ V.
   Now the null detector limits the accuracy to about 3 mV.

c) $(v_r - v_u)_{min} = 3 \times 10^{-7} \times 10^6 = 0.3$ V, showing that the potentiometer is nearly useless for very high values of R_u.

SOLUTIONS TO PROBLEMS IN SECTION 1-4

1. Substituting in Eq. (1–52) for the condition at each extreme value of b_{sw}, we get

$$- 0.3 = V_c(0 - b_z) = - V_c b_z,$$

$$+ 0.7 = V_c(1 - b_z) = V_c - V_c b_z.$$

Subtracting the lower equation from the upper, we get $+ 1.0 = + V_c$, and substituting 1.0 V for V_c in either equation, we find $b_z = 0.3$.

2. The parallel equivalent resistance of R_{sw} and R_z is $R_p = 500 \times 500/(500 + 500) = 250$ Ω. When $V_c = 1.00$ V and $V = 1.50$ V, there must be 1.00 V across R_p and $1.50 - 1.00 = 0.50$ V across R_c. Then

$$1.00/250 = 0.50/R_c,$$

$$R_c = 125 \text{ Ω}.$$

For a maximum loading error of 0.1%, the total divider resistance must be 1000 times the maximum output resistance of the slidewire bridge. The maximum output resistance of each divider is $500b (1 - b)$ which, when $b = \frac{1}{2}$, equals 125 Ω. Since there are two dividers, $R_{o(max)} = 250$ Ω and the minimum divider resistance is 250 Ω. To obtain $V_r = 10$ mV, the divider relationship must be

$$0.010 = 1.0 \frac{R_1}{R_1 + R_2},$$

where R_1 is the lower resistor. Thus $R_2 = 99R_1$. Since $R_1 + R_2 = 250$ kΩ, $R_1 = 2.5$ kΩ and $R_2 = 247.5$ kΩ.

3. From Eq. (1–57), the error $(v_o - v_u)$ equals $-v_o/A$, which is a maximum when v_o has the maximum value of 5 V. Thus $(v_o - v_u)_{max} = -5/5 \times 10^4 = -10^{-4}$ V.

4. From Eq. (1–60) the error term is $-v_o/\beta A$, which is a maximum when $v_o = 10$ V. The fraction $\beta = 1/(999 + 1) = 10^{-3}$. Thus the maximum error is $-10/10^{-3} \times 10^4 = 1$ V or 10%!

5. The voltage follower is preferred because of its much greater input impedance. In addition, the accuracy of its gain is not dependent on the accuracy or stability of gain-determining resistors.

6. a) For $R_1 = 100$ kΩ and $R_2 = 10$ kΩ, $\beta = 10$ k$\Omega/110$ k$\Omega = 0.091$ and from Eq. (1–59) for very large A,

$$v_o/v_u = 1/\beta = 11.00.$$

b) $R_1 = R_2 = 10$ kΩ, $\beta = 10$ k$\Omega/20$ k$\Omega = 0.500$. Then

$$v_o/v_u = 1/\beta = 2.00.$$

c) $R_1 = 100$ kΩ, $R_2 = 1$ kΩ, $\beta = 1$ k$\Omega/101$ k$\Omega = 0.0099$. Then

$$v_o/v_u = 1/\beta = 101.0.$$

7. $R_f = v_{o(\text{fs})}/i_{r(\text{fs})}$. In all cases, $v_{o(\text{fs})} = 1.20$ V.

For 0.1 μA, $R_f = 1.20/10^{-7} = 1.20 \times 10^7$ Ω.

For 0.5 μA, $R_f = 1.20/5 \times 10^{-7} = 2.4 \times 10^6$ Ω.

For 1.0 μA, $R_f = 1.20/10^{-6} = 1.20 \times 10^6$ Ω.

For 5.0 μA, $R_f = 1.20/5 \times 10^{-6} = 2.4 \times 10^5$ Ω.

8. For $v_u = 1.0$ V, $R_e = 1$ kΩ, and $R_{\text{in}} = 10$ kΩ, $i_{\text{in}} = 1.0/(1$ k$\Omega + 10$ k$\Omega) = 9.091 \times 10^{-5}$ A. For large A, $i_f = i_{\text{in}}$ and $v_o = -i_f R_f$, so

$$v_o = -10^5 \times 9.091 \times 10^{-5} = -9.091 \text{ V}.$$

If R_e were zero, the output voltage would be

$$v_o = -v_u \frac{R_f}{R_{\text{in}}} = -1.0 \frac{100 \text{ k}\Omega}{10 \text{ k}\Omega} = -10.00 \text{ V}.$$

Thus the error due to R_e is almost 10%. To calculate v_o for various values of R_e directly, R_e can simply be considered as part of R_{in}.

9. Rearranging Eq. (1–74), we get

$$v_o(1 + \frac{1}{A} + \frac{R_f}{AR_{\text{in}}}) = -\frac{R_f}{R_{\text{in}}} v_u,$$

where the difference between the term in parentheses and unity is the gain error. For 1% error,

$$1 + \frac{1}{A} + \frac{R_f}{AR_{\text{in}}} \leqslant 1.01.$$

a) For $A = 10^3$, $1 + 10^{-3} + 10^{-3}R_f/R_{in} \leqslant 1.01$, $10^{-3}R_f/R_{in} \leqslant 0.009$, and $R_f/R_{in} \leqslant 9$.

b) For $A = 10^4$, $1 + 10^{-4} + 10^{-4}R_f/R_{in} \leqslant 1.01$, and $R_f/R_{in} \leqslant 99$.

c) For $A = 10^6$, $1 + 10^{-6} + 10^{-6}R_f/R_{in} \leqslant 1.01$. and $R_f/R_{in} \leqslant 10^4$ (within 1%).

10. a) Use the circuit of Fig. 1–93 with $R_f = 1$ V/10^{-5} A $= 10^5$ Ω.

b) Use the circuit of Fig. 1–96 with $R_f/R_1 = 5$ and $R_f/R_2 = 3$. For instance, $R_f = 30$ kΩ, $R_1 = 6$ kΩ, and $R_2 = 10$ kΩ.

c) Use the inverting amplifier of Fig. 1–94 with unity gain ($R_f = R_{in}$) to obtain $-v_2$. Use the two-input summing amplifier of Fig. 1–96 to add v_1 and $-v_2$ with unity gain ($R_f = R_1 = R_2$).

CONTROL OF ELECTRICAL QUANTITIES IN INSTRUMENTATION

Module 2

Introduction

Many types of components are used in electronic circuits, but none more frequently than resistive and conductive elements. The movement of electrical charge in circuits is directed by conductors, and the rate of flow of charge can be controlled by resistors. In the first section of this module the characteristics of conducting materials used in electronic instrumentation are introduced through a basic examination of their structures, and practical resistors and conductors and their limitations are discussed. Transducers that convert temperature, light intensity, and strain into related resistance values, and the techniques for resistance measurement and network analysis, are then presented.

Capacitors and inductors play a major role in many instruments for the control of electrical charge and voltage. When they are combined in various ways with other elements, the resulting circuits can provide a variety of important functions. The characteristics of practical capacitors and inductors and their applications in basic RC and RL circuits are described, and the types of transducers that convert various physical quantities into related values of capacitance and inductance are considered. The important operational amplifier integrators and differentiators that utilize capacitors for charge storage, and their use in several applications, are described in the second section.

The development of inexpensive high speed switches is basic to the modern instrumentation/automation revolution. Also, new types of power switches have greatly facilitated the design of reliable and compact instruments. In the third section of this module the important types of diodes, transistors, FET's, optoelectronic devices, SCR's, and other types of switching elements that control electrical quantities are introduced. The application of program-controlled switches for waveform generation is illustrated.

All measurement and control instruments require various types of power supplies in order to operate. Most modern instruments operate from the ac power line, but generally require regulated dc voltages such as $+5$ V and ± 15 V. The conversion from ac to the required dc voltages is described in the fourth section. There is also a discussion of electrochemical and solar batteries that are used in portable equipment.

It is often necessary to amplify without distortion the signals from input transducers in order to provide reliable measurements or control. Various elegant amplification concepts were introduced in Module 1, *Electronic Analog Measurements and Transducers,* and several practical linear amplifiers were described on a functional block basis. In the final two sections of this module the internal operation of the amplifier functional block is considered in more detail and some of the major concepts and basic relationships for linear amplification and control are discussed. Very accurate types of current and voltage regulators with linear feedback control are also presented.

Resistive Devices for Measurement and Control

The electrical properties of a material or device determine the way it responds to the electrical quantities of charge, potential difference, current, and power. All electrical and electronic devices possess in some degree the properties of resistance, capacitance, and inductance, and all device characteristics can be explained in terms of these three properties.

When a device is designed to have a single property (e.g., a resistor to have only the property of resistance), the presence of the other electrical properties is considered an imperfection in that device. The magnitude of the electrical property of an ideal "linear" device is independent of the magnitude of the electrical quantities applied. Thus the resistance of a linear resistor is independent of the applied voltage or current; the capacitance of a linear capacitor is independent of the quantity or sign of stored charge; and so on. The conditions under which deviations from the expected linear behavior of devices are encountered constitute possible limitations on the device's applicability. For example, the maximum power dissipation rating of a resistor limits the constant current-voltage product that can be applied to it without risking overheating or destruction. On the other hand, many devices such as diodes are intentionally made to be very nonlinear. The properties of nonlinear devices can be varied in response to applied electrical quantities.

The magnitudes of the electrical properties of devices are often affected by environmental conditions such as temperature, humidity, pressure, radiation, and magnetic field strength. In some cases the sensitivity of a device to these factors limits the scope of its applications. In other cases, devices for which some electrical property is intentionally sensitive to a particular physical parameter are used to convert the magnitude of the physical parameter into an electrical property that can be measured electrically. Thus

131

the measured resistance of a temperature-dependent resistor is related to the temperature of the resistor.

No electrical device has complete freedom from all unwanted electrical characteristics, or complete independence from age or ambient conditions. Various materials and methods of construction result in various combinations of the desired qualities and limitations. This explains why there are so many different types of electrical components even though there are only three basic electrical properties. In this section, practical conductors and resistors will be described; the measurement of resistance will be discussed; and the characteristics of circuits containing combinations of resistors will be investigated.

2–1.1 CONDUCTANCE AND RESISTANCE

When an electrical voltage is applied across a piece of material, a net motion of the mobile charge carriers in that material is generated in the direction of the electric field created by the applied voltage. This is illustrated in Fig. 2–1 for a material with cross-sectional area a cm² and length l cm.

The net motion of charge carriers, or electrical current, is measured by the net charge crossing a plane (which is perpendicular to the motion) each second. The current, in coulombs per second, is given by

$$I = \frac{aVQ_e}{l} \, (n_1\mu_1 + n_2\mu_2 + \ldots + n_p\mu_p), \qquad (2\text{–}1)$$

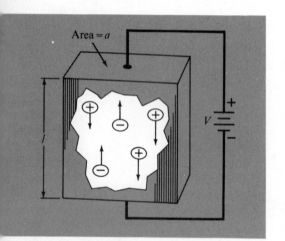

Fig. 2–1 Motion of charge carriers in an applied electric field.

where there are p different kinds of charge carriers designated 1, 2, 3, and so on up to p. For the kind of charge carrier designated 1, n_1 is the density in carriers/cm³, Q_e is the electron charge in coulombs, and μ_1 is the mobility of the carrier in cm²/volt-sec. It is assumed in Eq. (2–1) that the charge carriers each have a charge equal to Q_e.

The conductance G in mhos of the piece of material as connected in Fig. 2–1 is thus

$$G = \frac{I}{V} = \frac{aQ_e}{l} \, (n_1\mu_1 + n_2\mu_2 + \ldots + n_p\mu_p). \qquad (2\text{–}2)$$

The conductivity σ, characteristic of the material, is the conductance between opposite faces of a 1 cm tube with an applied voltage of 1 volt (or the current per cm² with an applied electric field strength of 1 volt per cm). Therefore

$$\sigma = Q_e(n_1\mu_1 + n_2\mu_1 + \ldots + n_p\mu_p), \qquad (2\text{–}3)$$

where σ has the units amperes/volt-cm. Resistance R in ohms is defined as the reciprocal of conductance. Therefore

$$R = \frac{V}{I} = \frac{l}{a\sigma} \tag{2-4}$$

and the resistivity ρ in volt-cm/ampere is $1/\sigma$. Thus the resistivity of any material depends on the population density, charge, and mobility of the mobile charge carriers in the material; the resistance of a device made of that material depends on the dimensions of the material and its resistivity.

The work required to move charge through a conducting material is equal to the product of the total charge moved and the potential difference across the conductor:

$$W = QV. \tag{2-5}$$

When the units of Q and V are coulombs (C) and volts (V) respectively, the work W is in joules (J). This is work done in the conductor which produces heat, light, or force. If a current of I amperes exists in a conductor for t sec, the total charge moved is $Q = It$ coulombs, and the work performed on (and dissipated by) the conductor is

$$W = ItV. \tag{2-6}$$

Power (P) is defined as the rate of doing work, or W/t. A work rate of 1 J/sec is equal to 1 watt (W) of power. Therefore,

$$P = W/t = ItV/t = IV. \tag{2-7}$$

By combining Ohm's Law (Eq. 2–4) with Eq. (2–7), the power dissipated in a conductor can be related to the resistance of the conductor and the current through it (or the voltage across it). Thus

$$P = I^2R = V^2/R, \tag{2-8}$$

which shows that the power dissipated in a resistor increases as the square of the applied current or voltage.

2–1.2 MECHANISMS OF CONDUCTION

It was shown in the previous section that the conductivity of any material depends on the density of mobile charge carriers, n, and their mobility, μ. If n and μ were constant quantities like Q_e, that is all we would need to know to characterize resistive devices. However, n and μ (and thus σ and R) can change with changes in temperature, humidity, light intensity, voltage, and other factors. Furthermore, the way in which the population density and mobility of the charge carriers vary with these environmental parameters depends on the mechanisms by which the charge carriers are created and transported. Materials used for electronic devices are classed according to their mechanism of charge carrier generation and motion. They generally fall into three main categories: metallic conductors, semiconductors, and insulators. In the next section the basic mechanisms of conduction will be described, and it is from these that the characteristics of the common types of conducting materials will be developed.

Since matter is generally composed of electrically neutral atoms, a mechanism must exist for the separation of electrons from some of the atoms in conducting materials to form free negatively charged electrons and positively charged ions. The differences in the mechanisms of **ionization** among each of the types of conducting material contribute greatly to the differences in their conductive properties. Free electrons are the principal charge carriers for many conducting materials but the positive ions also often play a significant, and sometimes dominant, role in the conduction process.

The electrons in an isolated atom are confined to specific orbitals or energy levels. Each energy level can be populated by two electrons of opposite spin. There are many more energy levels than are needed to accommodate the atom's normal (neutral) complement of electrons. In its lowest energy state the electrons in the atom populate the lowest possible (innermost) energy levels. An electron may be promoted to a higher energy level by absorbing energy from a photon or by a collision with another particle. The atom is now said to be in an **excited state.** Excited state atoms generally return to the ground state or lowest energy state by losing energy through collisions and/or by emission of radiation. If the energy source which excites an atom is sufficiently energetic, it can provide enough energy for an electron to escape the atom completely. The atom is then said to be **ionized,** and, in the process, two charged particles are created. In the presence of an electric field, the electrons will be accelerated in the direction of the positive charge and the positive ions will be accelerated in the direction of the negative charge.

Metals

The outermost and most easily removed electrons of an atom are frequently called the **valence** electrons. It is the interaction of the valence electrons of atoms that is primarily responsible for the formation of crystals and molecules. When similar atoms combine to form a pure crystalline solid, the discrete energy levels of the valence electrons broaden into bands of allowed energy levels. The maximum electron population of each energy band is twice the number of atoms in the solid, but the electrons may have any energy within the band limits. One such band is the **valence** or **bonding**

band. In this band the valence electron(s) of each atom share orbitals with the electrons of the nearest neighbor atoms to form **covalent bonds** between the atoms. If there is an odd number of valence electrons in each atom and there are two available energy states per atom in each band, one of the energy bands will be only half filled. The unoccupied energy states in the energy band allow the electrons in that band to be accelerated by an electric field. Thus electrons in a partially filled band are **conductive electrons** and a partially filled band is a **conduction band.** Monovalent metals such as lithium, sodium, and potassium are observed to have one free conducting electron per atom as expected. The metal atoms in this case are positively charged ions and would also contribute to the metal's conductivity if they were reasonably free to move.

In materials with an even number of valence electrons in the lowest energy state, the valence bands are filled and therefore the electrons in them are not free to move. An allowed energy band higher than the highest valence band will be empty. This is illustrated in Fig. 2–2. If some of the valence electrons can be energized (promoted) into the next higher band, they will be free to move in this partially filled band. This next higher band is thus the conduction band in this case. The magnitude of the energy gap between the filled valence band and the conduction band determines the conductance type of a given material. If the energy gap is zero, it is a **metallic conductor;** if it is small, it is a **semiconductor;** and if it is large, it is an **insulator.**

In any solid material, there is an equilibrium distance between the atoms. This is the distance for which the energy of combination is a minimum. Variations in this equilibrium distance with temperature are responsible for variations in bulk dimensions with temperature. The closer atoms are to each other, the more their electron orbitals interact. This causes the energy range of the allowed bands to increase and that of the forbidden bands to decrease. This is shown in Fig. 2–3. In many metals the interatomic distance is so small (as evidenced by the high density of such materials) that the allowed valence and conduction bands overlap. When this occurs, the combined allowed energy band is unfilled and the valence electrons can conduct. This explains the high conductivity of bivalent metals such as magnesium, calcium, and strontium even at temperatures near absolute zero (minimum energy state). If the atoms have an even number of valence electrons *and the valence and conduction bands do not overlap,* the material must be a semiconductor or an insulator.

The electrical properties of most metals are consistent with a model in which the metal atoms are positive ions in a "sea" of mobile electrons. The resistance of the metal results from the collision of the mobile electrons with the metal atoms. At very low temperatures (near absolute zero) the metal structure is almost perfectly periodic so that the electron motion,

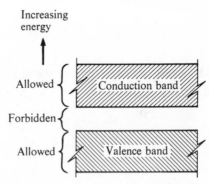

Fig. 2–2   Electron energy bands in a solid.

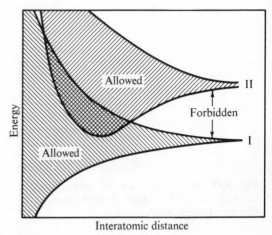

Fig. 2–3   Energy bands vs. distance.

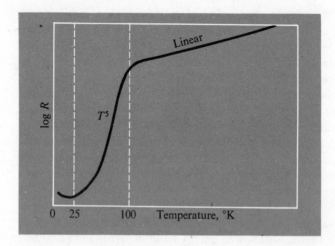

Fig. 2–4 Effect of temperature on the resistance of a metal.

described as a wave function, meets little (in a superconductor, zero) resistance. As the temperature increases from 0°K, random vibrations of the metal atoms cause a rapid increase in the collision rate. The resistivity in the low temperature region actually increases as the fifth power of T for many metals, as shown in Fig. 2–4. At higher temperatures (still much lower than room temperature) the resistivity increases less rapidly and eventually increases approximately linearly with temperature. The resistivity of most metals increases from 3 to 6 parts per thousand per degree centigrade in the linear region. Such changes can cause difficulty in precision circuits but they are turned to advantage in the platinum resistance thermometer.

Semiconductors

A material with an even number of valence electrons and a forbidden energy gap between the valence and conduction bands cannot conduct electricity at 0°K. At absolute zero temperature the valence band is filled and the conduction band is empty. A source of energy is required to promote valence electrons to the conduction band. This energy might be provided by quanta of radiation (photons), resulting in photoconduction, or by thermal excitation at higher temperatures. An energy source at least equal in energy to the forbidden gap is obviously required for excitation to conduction. The result of promoting an occasional valence electron to the conduction band is shown in Fig. 2–5. For each electron in the conduction band, there is a lack of one electron in the valence band. The point where a valence or bonding electron is missing is called a **hole.** It must be kept in mind that while the conduction band electrons are free to move throughout the material (and do), the valence electrons are limited to interactions

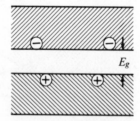

Fig. 2–5 Band structure of semiconductor or insulator with valence electrons excited to the conduction band.

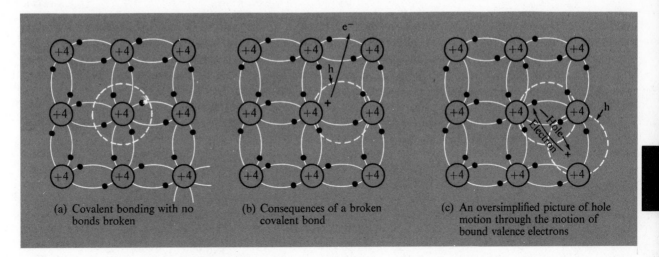

(a) Covalent bonding with no bonds broken

(b) Consequences of a broken covalent bond

(c) An oversimplified picture of hole motion through the motion of bound valence electrons

Fig. 2–6 Motion of conduction electron and hole in semiconductor or insulator.

with near neighbor atoms. Thus the area of a missing valence electron (the hole) is a localized region of positive charge. When an electric field is applied to a material with excited electrons, the conduction electrons will drift in the direction of increasingly positive potential. Valence electrons can only move in response to the electric field if they are adjacent to a hole. The net result of the valence electron motion is a drift of the positive hole in the direction of increasingly negative potential. This action is illustrated in Fig. 2–6.

The population density of electrons n_e in the conduction band at any given temperature is greater for materials with lower forbidden gap energies, E_g, and increases with increasing temperature for any material. The relationship between n_e, T, and E_g is given approximately by the following equation:

$$n_e = 2(2\pi m_e kT/h^2)^{3/2} e^{-E_g/2kT}, \qquad (2-9)$$

where m_e is the mass of the electron, k is Boltzmann's constant, and h is Planck's constant. For every conduction electron there is also a hole which acts as a positive charge carrier. Thus $n_h = n_e$ and from Eq. (2–3) the total conductivity is

$$\sigma = Q_e(n_e\mu_e + n_h\mu_h) = n_e Q_e(\mu_e + \mu_h). \qquad (2-10)$$

The energy gap and mobilities for several pure metals and semiconductor materials are given in Table 2–1.

From Eqs. (2–9) and (2–10) it can be seen that if μ_e and μ_h are not strongly dependent on temperature, then the conductivity will increase approximately exponentially with increasing temperature, as is illustrated in

Table 2–1 Conducting characteristics of several pure materials at 25°C

Material	Energy gap, eV	Mobility at 25°C, cm²/V-sec		Carrier density, n	Resistivity ρ, Ω-cm
		Electrons	Holes		
Elements					
Copper	—	35	—	$\cong 10^{23}$	$\cong 1.5 \times 10^{-6}$
Silver	—	56	—	$\cong 10^{23}$	$\cong 1.5 \times 10^{-6}$
Diamond	5.47	1,800	1600	—	—
Germanium	0.80	3,900	1900	2.5×10^{12}	8.9×10^{4}
Silicon	1.12	1,500	600	1.6×10^{10}	2.5×10^{5}
III-V compounds					
Ga Sb	0.67	4,000	1400		
Ga As	1.43	8,500	400	1.1×10^{7}	
In Sb	0.16	78,000	750		
In·As	0.33	33,000	460		
II-VI compounds					
CdS	2.42	300	50		
CdSe	1.7	800			
ZnO	3.2	200			

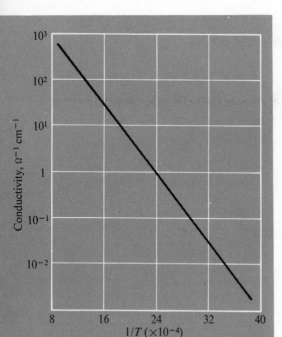

Fig. 2–7 The electrical conductivity vs. reciprocal absolute temperature for the intrinsic semiconduction region in germanium.

Fig. 2–7. This thermal dependence is the basis of the temperature-dependent resistor (thermistor).

A semiconductor which conducts by equal numbers of holes and electrons created by the excitation of valence electrons is called an **intrinsic semiconductor.** All *pure* semiconductors and insulators are intrinsic semiconductors. Whether a material is a semiconductor or an insulator depends on the magnitude of the energy gap E_g compared with the operating temperature or energy of excitation. At low enough exciting energies, all such materials are insulators, and at high enough exciting energies all such materials are conductors. Table 2–1 shows how markedly different the resistivity can be for such materials at room temperature.

Extrinsic Semiconductors. The majority of semiconductors used in electronic materials are not pure, or intrinsic, semiconductors as described above, but are impure, **doped** or **extrinsic semiconductors.** The intentional impurity (or dopant) is an extremely small amount of an element with 3 valence electrons per atom, such as boron or indium, or 5 valence electrons per atom, such as arsenic or antimony. The impurity level used is generally of the order of 1 part in 10^8 so that the structure, dimensions, and chemical characteristics of the semiconductor material are essentially unaffected by doping. The impurity atoms simply occupy scattered sites in the normal semiconductor crystal lattice.

An atom with 5 valence electrons in a crystal structure which uses 4 valence electrons per atom for bonding, has one valence electron more than it needs. The energy level of this additional electron is only a few hundredths of an electron-volt from the conduction band for the semiconductor. Thus at normal temperatures, essentially all the impurity atoms are ionized and have contributed an electron to the conduction band. No holes are produced in this process since each atom still has 4 bonding electrons. The valence band is still filled and thus holes are only produced by intrinsic type excitation. In germanium there are about 10^{23} atoms per cm³. A concentration of arsenic dopant of only 1 part in 10^8 will produce 10^{15} conducting electrons per cm³. This charge carrier density is roughly 100 times larger than the intrinsic charge carrier density of germanium at room temperature. Thus a trace impurity can have an enormous effect on the electrical properties.

The 5-valent impurity is called a **donor** impurity because it donates electrons to the conduction band. The resulting material is called an **n-type semiconductor** because the majority of charge carriers are negatively charged. In any semiconductor, the product of n_e, the conducting electron density, and n_h, the hole density, is a constant at any given temperature. Thus

$$n_e \cdot n_h = n_i^2(T), \tag{2–11}$$

where $n_i^2(T)$, the charge density product, is a function of temperature *but not of impurity concentration*. For intrinsic germanium at room temperature, $n_i^2(T) \approx (10^{13})^2 \approx 10^{26}$. Since this constant is independent of n_e or n_h, for the example above where $n_e = 10^{15}$ electrons/cm³, $n_h = 10^{26}/10^{15} = 10^{11}$ holes/cm³. Thus there are 10^4 more electron charge carriers than hole charge carriers in the doped germanium in this example. The electrons are truly the majority of charge carriers and are aptly called **majority carriers.**

The addition of 3-valent impurity atoms to a semiconductor material has a complementary effect to that just described. When 3-valent atoms take their place in the lattice, one valence electron is missing per atom. This creates, in effect, a hole in the valence band which greatly reduces the energy required to ionize a nearby semiconductor atom. Just as was shown with the 5-valent impurity, the required ionization energy is small compared to the thermal energy (kT) at room temperature. An impurity level of 10^{15} atoms of indium per cm³ of germanium would thus produce essentially 10^{15} holes per cm³ at room temperature. Since by Eq. (2–7), n_e would be 10^{11} electrons per cm³, the holes are the majority carriers. The 3-valent dopant is called an **acceptor** and the acceptor-doped semiconductor is called a **p-type semiconductor.**

If both donor and acceptor impurities are present, they tend to neutralize each other. The effective impurity material is the one present in excess and its effective concentration is equal to its excess concentration.

The temperature dependence of the carrier concentration of a doped semiconductor is illustrated in Fig. 2–8. At low temperatures the number of charge carriers n_i contributed by the dopant increases as kT becomes significant compared to E_i, the impurity ionization energy. At temperatures between $100°$ and $300°K$, essentially all the impurity atoms are ionized and the charge density is quite constant. Essentially all the charge carriers are being created by the extrinsic or impurity mechanism. Above room temperature, kT becomes significant with respect to E_g, the intrinsic ionization energy gap, and the charge carrier density increases by the intrinsic mechanism. The operation of extrinsic semiconductors in the intrinsic temperature range poses a threat to the component and is generally avoided. In the intrinsic range, an increase in temperature can cause an increase in current through the device which can cause further heating of the material, and so on. For germanium devices, the maximum safe operating temperature is about $85°C$, while silicon devices, with a higher E_g, can be safely operated up to $200°C$.

2–1.3 RESISTORS

A large variety of electronic components are made for the sole purpose of introducing resistance in an electric circuit. Such components are called **resistors.** They are used to limit the current in a circuit, to convert current to voltage, and to divide voltages and currents. Resistors are made from a variety of materials and come in many shapes, sizes, and values. They are characterized primarily by the nominal resistance value, the accuracy with which the actual resistance agrees with the nominal value, the power dissipation, the stability of the resistance with respect to time, temperature, humidity, etc. The characteristics of several of the most common resistor types will be discussed in this section.

Wire

Metallic wire is an essential part of all electronic systems. Its low resistance can provide a nearly ideal electrical connection between components. It is also used in the fabrication of many devices such as inductors, transformers, and wire-wound resistors. The usual electrical hookup wire is made of copper, a metal which provides very high conductivity and good flexibility at moderate cost. The copper wire is often plated with a thin layer of another

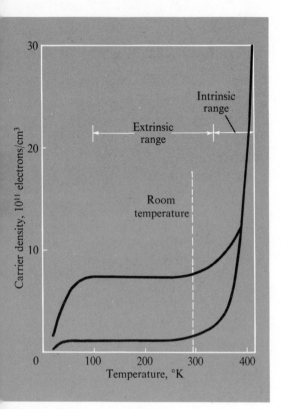

Fig. 2–8 Carrier density vs. temperature in two n-doped germanium crystals.

Table 2–2 Copper wire characteristics

A.W.G.*	Number of strands	Diameter per strand, mm	Resistance, Ω/m	Current capacity, A
30	1	0.255	0.346	0.144
24	1	0.511	0.0804	
24	7	0.022	0.0804	0.577
22	1	0.644	0.0501	
22	7	0.025	0.053	0.918 (5)
20	1	0.812	0.0316	
20	7	0.032	0.0332	1.46 (7.5)
18	1	1.024	0.0198	
18	7	0.040	0.0203	2.32 (10)
16	1	1.291	0.0125	
16	19	0.029	0.0130	3.69 (13)
14	1	1.628	0.0078	5.87 (17)
12	1	2.053	0.0049	9.33 (23)

* American Wire Gauge (A.W.G.) is a means of specifying relative wire diameter. The lower the A.W.G. number, the larger the diameter and the lower the resistance per unit length.

metal such as silver or tin to make it easier to solder other components to it,[1] and is then covered with an insulator, usually plastic. Copper wire is available in many diameters, called **gauges.** From Eq. (2–4), it is clear that the larger the diameter, the lower the resistance per unit length. Table 2–2 gives the resistance of several sizes of copper wire. A conductor made of several collected small wires is called **stranded wire.** It offers improved flexibility and is less likely to break under repeated flexing.

To provide some perspective, 22 gauge wire is the normal hookup wire size, 18 gauge is used for household lamp cords, and 12 and 14 gauge are used for house wiring. The current values given in parentheses in Table 2–2 would bring the temperature of the wire to 100°C if the wire were bundled or enclosed and the ambient temperature were 57°C (135°F).

Carbon Resistors

The most common type of resistor is made of hot-pressed carbon granules **(composition resistor)** or a thin film of finely divided carbon granules deposited on an insulating substrate **(carbon film resistor).** These are the fa-

Note 1.   Tinned Wire.

A copper wire that has been coated with a layer of tin or solder to prevent corrosion and to simplify soldering.

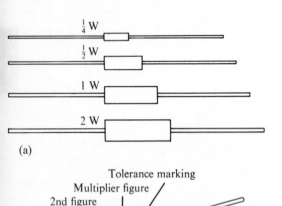

(a)

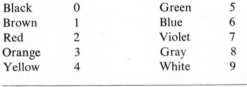

(b)

Fig. 2–9   Carbon resistors: (a) physical size for given power dissipation; (b) color code.

miliar resistors shown in Fig. 2–9a. They have the advantages of being inexpensive, remarkably free of capacitance and inductance, and available over a very wide range of values (from 2.7 Ω to 2.2 $\times$ 10⁷ Ω and from $\frac{1}{8}$ to 2 W). On the other hand, they are only moderately accurate (tolerances are $\pm 5\%$ and $\pm 10\%$ from the nominal value for carbon composition resistors and $\pm 1\%$ and $\pm 2\%$ for carbon film resistors), they introduce more electrical noise into circuits than do other resistor types, and they have a relatively high **temperature coefficient of resistance** (variation of resistance with temperature).

Color Code. The nominal resistance value of most carbon resistors is marked on them by three colored bands painted around the body of the resistor. These three colors can be translated into a number with two significant figures by using the standard resistor color code. This is illustrated in Fig. 2–9b. Each number from 0 to 9 has been assigned to a color as shown in Table 2–3. Note that from 2 to 7 the colors follow the spectrum.

Table 2–3   Resistor color code

Black	0	Green	5
Brown	1	Blue	6
Red	2	Violet	7
Orange	3	Gray	8
Yellow	4	White	9

The color of the band closest to the end of the resistor represents the first figure of the resistance; the second band, the second figure; and the third band, the number of zeros to add to the first two figures to get the total resistance. Thus a resistor which is coded with yellow, violet, and red is 4700 ohms or 4.7 kΩ. Blue, gray, green is 6,800,000 Ω, or 6.8 MΩ; and green, blue, black is 56 Ω. For resistance between 1 and 10 Ω, gold is used for the third band. Thus orange, white, gold is 3.9 Ω.

The Accuracy of the Nominal Resistance. The limit of accuracy assured by the manufacturer of the resistor is called the **tolerance.** This is expressed in percent of the nominal value. For instance, a 100 Ω resistor with a tolerance of 10% may have any value from 90 to 110 Ω. In color-coded resistors the tolerance is indicated by the color of the fourth band, silver being 10%, gold being 5%. Closer tolerances are printed on the resistor body.

The 10% tolerance requirement of most common resistor applications has resulted in the apparently strange values of the resistors manufactured. The standard values of resistances available in 10% tolerance between 1 and 10 kΩ are 1, 1.2, 1.5, 1.8, 2.2, 2.7, 3.3, 3.9, 4.7, 5.6, 6.8, 8.2, and 10 kΩ. Note that there is a value within 10% of any required resistance.

The use of "round" numbers for resistance values would result in an unnecessarily large number of values required to cover the same resistance range. Where 5% tolerance is specified, 5% resistors are supplied in the values 1.1, 1.3, 1.6, 2.0, 2.4, 3.0, 3.6, 4.3, 5.1, 6.2, 7.5, and 9.1 kΩ, in addition to all the standard 10% values. In all other decades, the 5 and 10% values are the same as those listed above.

Many factors affect the accuracy of a resistor. For resistors of very low values (less than 10 Ω) the resistance of the connection to the resistor can become a significant part of the total resistance of the circuit. For resistors of very high values (greater than 10 MΩ) leakage of current between the contacts on the surface of the resistor can significantly decrease the resistance of the circuit. This effect, which will depend on the ambient humidity, can be reduced through the use of special insulating materials. In applications where stability is important it is desirable to use a resistor with a much higher power rating than the circuit requires. This will reduce temperature changes in the resistor.

Wire-Wound Resistors

Resistors are made of wire for three different applications: low resistance, high precision, and high power dissipation. All are made by winding a coil of a relatively high resistivity metal alloy wire onto a spool (often ceramic) and making contact to each end of the wire as shown in Fig. 2–10a. Low resistance wire-wound resistors require little wire, are inexpensive, and are more stable than composition resistors. Resistors of low resistance are often called upon to dissipate relatively large amounts of power. For dissipation greater than 2 W, **power resistors** are used. For this application, rather low gauge resistance wire is wound on a ceramic form which will withstand high temperatures. Power resistors should be mounted so that they can readily dissipate heat to the air or the instrument case.

When a low temperature coefficient alloy wire is used, very precise and stable resistors can be made. Accuracy and stability of the order of $\pm 1\%$ to $\pm 0.001\%$ are obtainable in this way. Wire-wound resistors become less desirable as the nominal resistance increases beyond 1 kΩ. The longer wire lengths required increase the cost and the inductance of the resistor.

Metal Film Resistors

The improvement in the techniques of metal film deposition and the development of better substrate materials have provided a valuable addition to resistor types. Metal film resistors are made by vacuum-depositing a thin metal layer on a low thermal expansion substrate. The resistance of the film is increased and adjusted by etching or grinding a pattern through the

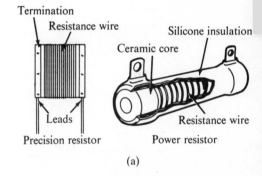

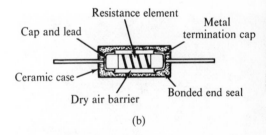

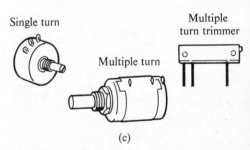

Fig. 2–10 Practical resistors: (a) wire wound; (b) metal film; (c) variable resistors (potentiometers).

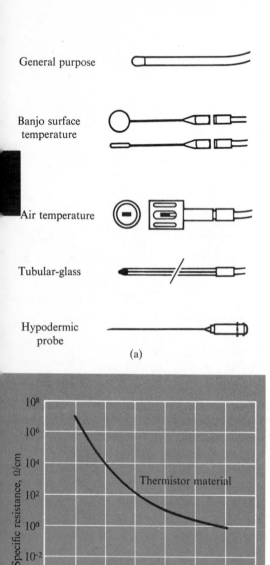

General purpose

Banjo surface
temperature

Air temperature

Tubular-glass

Hypodermic
probe

(a)

Thermistor material

Platinum

Specific resistance, Ω/cm

Temperature, °C

(b)

deposited film as illustrated in Fig. 2–10b. Accuracy and stability approaching the better wire-wound resistors are now available. Metal film resistors are much easier to fabricate in high resistance values than are wire-wound resistors, are essentially noninductive, and introduce a minimum of *electrical noise*. Noise characteristics of resistors are discussed in Module 4.

Variable Resistors

Some resistors are made to have an adjustable or variable resistance. This is accomplished by exposing the resistive element so that contact can be made at any point along the resistor by a movable contact, or "wiper." Such a device is called a **rheostat** or a **potentiometer.** The value of the total or maximum resistance is usually stamped on the case. The resistive element is usually either a coil of resistance wire or a strip of resistive film. Several types of potentiometers are shown in Fig. 2–10c.

2–1.4 RESISTIVE TRANSDUCERS

If the resistance of a device is a known and reproducible function of temperature, light, position, strain, voltage, magnetic field, or another physical or chemical parameter, the device can be used as an interdomain converter which converts from the domain of the functional dependence to the domain of resistance. The functional relationship constitutes the transfer function for the conversion. The resistance is then measured by one of the techniques to be described in Section 2–1.5, and the magnitude of the temperature, light, or strain is deduced from the transfer function. In this section, a number of common resistive transducers and their characteristics are described.

Temperature-Dependent Resistors

As illustrated in Section 2–1.2, the conductance of all conducting materials varies with temperature. If the resistance-temperature relationship for a particular conducting device is characterized and reproducible, the measurement of that device's resistance can be used to determine the temperature of the device. In this application both metallic and semiconductor

Fig. 2–11 Resistance-temperature response of typical thermistor material, compared with platinum and several practical thermistor probes.

materials are used. The most common metallic thermal resistance sensor is the platinum resistance thermometer. Temperature-dependent semi-conductor resistors are called **thermistors.**

Platinum Resistance Thermometer. The sensing element of a platinum resistance thermometer is a high-purity platinum wire (99.999% pure) wound on a ceramic core and placed in the tip of a probe. Copper contacts are made to the ends of the sensing element wire and brought out through the other end of the probe shaft. The resistance of a platinum resistance thermometer is generally 100 to 500 Ω at 0°C. As in the case of all common metallic conductors, the resistance increases with increasing temperature. The resistance-temperature relationship for pure platinum is essentially linear over the range from $-270°C$ to $1100°C$ with a resistance change of 0.392% per °C. Platinum resistance thermometers are extremely accurate, sensitive, and stable.

Thermistor. The thermistor is made from an intrinsic semiconductor. As we have seen, the number of charge carriers in intrinsic semiconducting material increases linearly with $1/T$. Thus the resistance of the thermistor decreases sharply with increasing temperature. Typical resistance-temperature curves are shown in Fig. 2–11 together with some typical configurations of thermistor probes. The relatively large resistance change per degree makes the thermistor a useful device in temperature measurement or control devices requiring high accuracy or resolution. Temperature changes as small as 5×10^{-4} °C can be detected. Thermistors are used in the temperature range of $-100°C$ to $+300°C$. Any current in a resistance does work which appears as heat. Care must be taken that the amount of heat produced by the current in the thermistor or platinum resistance thermometer does not affect the temperature of the measured system.

Light-Dependent Resistors

When light falls on a semiconductor, the photon energy may be sufficient to move valence electrons into the conduction band. The resulting increased density of charge carriers due to the light makes the material more conducting. Thus the resistivity decreases as the light intensity increases. The semiconducting sulfide, selenide, and telluride salts of cadmium are most often used for this purpose. Some practical photoconductive cells and the spectral response curves of such devices are shown in Fig. 2–12. For each material to photoconduct, the incident light must have a short enough wavelength (high enough energy) for the photons to have enough energy to move electrons from the valence to conduction states. As the energy of the light increases beyond the minimum required for promotion of valence electrons to the conduction band, the absorption of the

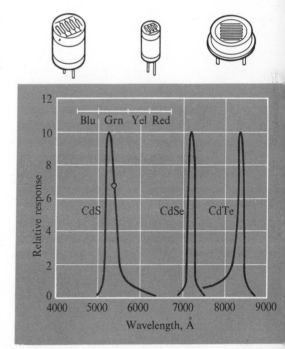

Fig. 2–12 Practical photoconductive cells and the photoconductivity of CdS, CdSe, and CdTe vs. wavelength.

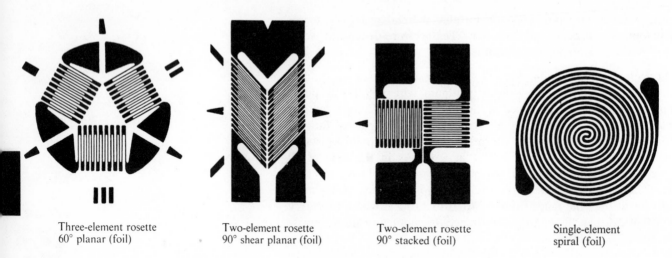

Three-element rosette
60° planar (foil)

Two-element rosette
90° shear planar (foil)

Two-element rosette
90° stacked (foil)

Single-element
spiral (foil)

Fig. 2–13 Practical strain gauges.

light by the material increases rapidly. This results in a marked decrease in the fraction of the light which penetrates to the active region of the device.

Strain-Dependent Resistors

A resistor which is made of a fine wire has a resistance $R = \rho l/a$, where ρ is the resistivity of the metal and l and a are the wire length and cross-section respectively. If the wire is distorted, i.e., stretched, l increases and a decreases. Actually, ρ may also change. The result is a resistor for which the resistance is a function of the strain. Such a device is called a **strain gauge.** Strain gauges have extensive applications in mechanical and biological measurements. Often the fine wire is bonded to a flexible insulating substrate so that the wire dimensions change as the substrate is flexed. Some strain gauges are made by depositing a thin metal film pattern on the substrate material. Several practical strain gauges are shown in Fig. 2–13. Since the resistance change with strain is a very small fraction of the total resistance, very sensitive resistance measuring techniques are required. Another difficulty is separating the changes in resistance due to temperature from those due to strain.

2–1.5 RESISTANCE MEASUREMENT

Resistance measurements in the laboratory tend to fall into two categories: testing or troubleshooting, and measurements using resistive transducers. The ohmmeter function provided in the common multimeters and FET voltmeters fulfill the testing requirement well. They are useful for checking the continuity of a circuit, searching for short circuits, testing components for catastrophic breakdown, and rough resistance measurements. The common ohmmeter circuits are described in this section. However, because of their generally poor accuracy and nonlinear scales, few ohmmeters are useful for laboratory measurements of resistive transducers. Improved accuracy can be obtained from the resistance-to-voltage converter circuit based on the current follower, which can provide an output linear in conductance or resistance. The laboratory standard for resistance measurement is the Wheatstone bridge, which is shown to be a true null resistance comparison technique.

Voltage-Current Ratio

By Ohm's Law the resistance of a device can be determined when there is a current through it by measuring the ratio of the IR drop (voltage) across it to the current through it. This is often the most convenient way to measure the resistance of a device in an operating circuit and under operating conditions. The resistance of an isolated device is often measured by placing it in a circuit with a battery, current meter, and other resistors. The measured current is related to the device resistance. Such a circuit which has the meter face calibrated in ohms is called an ohmmeter. For a linear conversion of resistance or conductance to voltage, an operational amplifier current follower circuit is generally used.

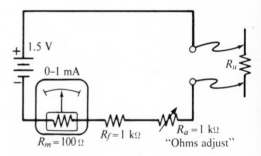

Fig. 2–14 An ohmmeter of simple series type.

The Ohmmeter. A series-type ohmmeter circuit is shown in Fig. 2–14. With the test probes short-circuited ($R_u = 0$), the "ohms adjust" control is turned so that the current I_1 through the total circuit resistance $R = R_m + R_f + R_a$ deflects the meter exactly full scale. Now, by connecting the test probes across the unknown resistance R_u, the current is decreased to a value I_2 which depends on the value of R_u. The 1.5 V battery is across the total resistance in the two cases; so

$$I_1 R = 1.5 \text{ V} \quad \text{and} \quad I_2 R + I_2 R_u = 1.5 \text{ V},$$

from which it follows that

$$R_u = \left(\frac{I_1}{I_2} - 1 \right) R \qquad (2\text{–}12)$$

and

$$I_2 = I_1 \left(\frac{R}{R_u + R} \right). \qquad (2\text{–}13)$$

Since a 1 mA meter movement is shown in Fig. 2–14, and the battery is 1.5 V, the resistance $R = R_m + R_f + R_a$ will have to be set to 1500 Ω

for full-scale deflection; so, from Eq. (2–12),

$$R_u = \left(\frac{I_1}{I_2} - 1\right)1500.$$

With $I_2 = (1/2)I_1$ (midscale deflection), the unknown resistance $R_u =$ 1500 Ω; with $I_2 = (1/3)I_1$, $R_u = 3000$ Ω, etc. It is apparent from Eqs. (2–12) and (2–13) that the transfer function between R_u and I_2 or scale position is not linear, since R_u goes from ∞ to 0 as I_2 goes from 0 to I_1. Values of R_u much higher than 1500 Ω become crowded on the "infinite ohms" end of the scale, and values much lower become indistinguishable from zero. The resistance at midscale could be decreased tenfold to 150 Ω by shunting the meter to make it 10 mA full scale and decreasing the series resistance (with test probes short-circuited) to 150 Ω. The resistance at midscale could be increased to 15 kΩ by increasing the circuit resistance R to 15,000 Ω and then using a 15 V battery or a 100 μA current meter. Clearly the more sensitive the current meter, the higher the value of resistance that can be measured accurately.

A more often used variation on the circuit described above is illustrated in Fig. 2–15. This ohmmeter is referred to as the "voltmeter type" because the voltage is measured across a resistance R_s that is in series with the unknown resistance R_u. The voltmeter branch has a total resistance R_v that is in parallel with R_s, so as to give an effective resistance $R_p = R_s R_v/(R_s + R_v)$, which is in series with R_r and any unknown resistance R_u. The battery voltage V is divided across the voltmeter, R_r, and R_u.

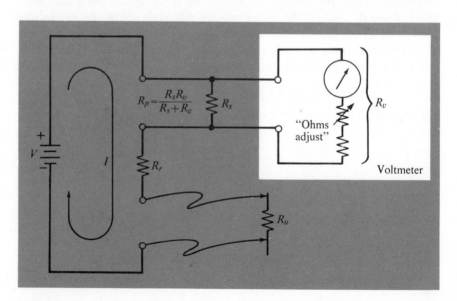

Fig. 2–15   An ohmmeter of voltmeter type.

When the test leads are shorted, the measured voltage is

$$v_1 = V\left(\frac{R_p}{R_p + R_r}\right)$$

and when the test leads are connected to R_u,

$$v_2 = V\left(\frac{R_p}{R_p + R_r + R_u}\right).$$

The relationship among v_1, v_2, and R_u is found by combining the above two equations,

$$v_2 = v_1\left(\frac{R_p + R_r}{R_u + R_p + R_r}\right) \qquad (2\text{–}14)$$

and

$$R_u = (R_p + R_r)\left(\frac{v_1}{v_2} - 1\right). \qquad (2\text{–}15)$$

If the "ohms adjust" resistance is varied until the meter reads full scale with the test leads short-circuited, the meter will have a midscale reading ($v_2 = v_1/2$) when $R_u = R_p + R_r$ and the same nonlinear scale as the meter of Fig. 2–14.

If $R_p = 11.5\ \Omega$ and $R_r = 1.0\ \Omega$, then $R_u = 12.5\ \Omega$ for a midscale meter reading. The meter could be calibrated to read ohms directly. If $R_p + R_r$ is increased to 1250 Ω, the resistance value will equal the meter reading $\times$ 100. In this case $R_u = 1250\ \Omega$ for a midscale reading if the "ohms adjust" was previously varied so that the meter current was full scale with the test leads short-circuited.

As the circuit resistance, $R_p + R_r$, is increased for measuring larger values of R_u, the circuit current decreases. Consequently the fraction of circuit current sent through the meter branch must be increased in order for the meter to read full scale. The range switch, therefore, not only switches the absolute value of $R_p + R_r$ to the desired resistance but also changes relative values of R_s and R_v and, for a very high resistance range, it may switch in a higher battery voltage, in order to get sufficient current through the meter.

When the ohmmeter function is part of a very high input resistance voltmeter such as a vacuum tube or field-effect transistor voltmeter, the simpler circuit of Fig. 2–16 is often used. The voltmeter is used to measure the fraction of V which appears across the unknown resistor R_u. When the test probes are shorted, the measured voltage is zero. The ohms adjust is a gain control on the amplifier which is adjusted to give full-scale de-

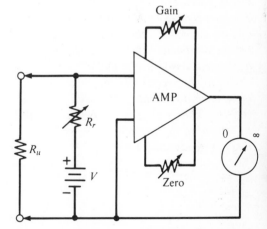

Fig. 2–16 VTVM or FET VM type of ohmmeter.

flection *when the test probes are open* ($R_u = \infty$ Ω). Note that the scale on this type of meter is reversed from the scales of Figs. 2–14 and 2–15. When an unknown resistance R_u is connected between the test probes, the relation between the measured voltage v_2 and R_u is

$$v_2 = V\left(\frac{R_u}{R_r + R_u}\right) \tag{2–16}$$

and

$$R_u = R_r\left(\frac{v_2/V}{1 - v_2/V}\right). \tag{2–17}$$

As Eqs. (2–16) and (2–17) show, the scale is still nonlinear with a range from 0 to ∞ Ω and a half-scale value of R_r. The ohmmeter range is changed simply by switching in various values for R_r up to a maximum value of 0.1% or 1% of the amplifier input resistance.

An ohmmeter is never used while the circuit is in operation,[2] and thus there is no circuit distortion introduced by the measurement. For resistances that depend on circuit conditions, the only solution is to establish normal operating conditions, measure the voltage across the resistance, measure the current through the resistance, and calculate the resistance value from Ohm's Law.

The resistance of devices that might be damaged by moderate currents cannot be measured with an ordinary ohmmeter. Such devices include meter movements and some fuses, lights, relays, tube filaments, diodes, etc. When the danger of damage exists, some other means must be devised to make the measurement.

Linear Resistance-to-Voltage Converters. The nonlinear relationship between the unknown resistance and the current or voltage measured in the above ohmmeter circuit is inconvenient for many types of resistive transducer measurements. From Ohm's Law, $R = V/I$, the voltage across a resistor is proportional to the resistance for a constant value of current. Similarly, the conductance, $G = I/V$, is proportional to the current through a resistor for a constant applied voltage. To take advantage of these linear relationships, a constant current source is required in the first case, and a current measurement with negligible input voltage in the second case. The current follower circuit introduced in Module 1 provides a convenient and accurate means to achieve either of these requirements. In Fig. 2–17a R_u is connected to the voltage source V, and the result is a current $i = V/R_u$ which is proportional to the conductance of R_u. The operational amplifier maintains an output voltage such that point S is at the common potential.

Note 2.   In-Circuit Resistance Measurements.

Even when the power supply for an electronic circuit is off, the circuit and power supply capacitors might remain charged and discharge only slowly as the resistors in the circuit are checked with an ohmmeter. These capacitor voltage sources can cause erroneous resistance measurements or damage to the ohmmeter. Therefore, after the instrument is turned off, all capacitors should be discharged before ohmmeter measurements are attempted. Also, discharged capacitors in a circuit might become charged when the ohmmeter probes are connected across a resistor. This can cause the measured value to drift until the voltages in the circuit reach a steady state.

Therefore, the input current, i, produces a voltage $v_o = -iR_r$ at the current follower output. Also, the connection of the G-to-I converter circuit to point S is equivalent to a connection to common; i.e., essentially no additional voltage or current is added to the input circuit. Since $i = V/R_u$ and $v_o = -iR_r$,

$$v_o = -VR_r/R_u = -VR_rG_u, \qquad (2\text{--}18)$$

resulting in an output voltage directly proportional to the conductance of R_u within the output voltage and current capabilities of the operational amplifier.

In the circuit of Fig. 2–17b, a fixed resistance, R_r, is used in the input current generating circuit. Since this same current passes through R_u and the operational amplifier (OA) maintains point S at the common potential, the output voltage, $v_o = -iR_u$, is proportional to R_u. From $i = V/R_r$ and $v_o = -iR_u$,

$$v_o = -\frac{V}{R_r}R_u. \qquad (2\text{--}19)$$

Of course, both circuits are simply the OA inverting amplifier introduced in Section 1–4.2, for which $v_o = -v_{\text{in}}R_f/R_{\text{in}}$ with $v_{\text{in}} = V$ and R_u taking the place of R_{in} in the first case and R_f in the second.

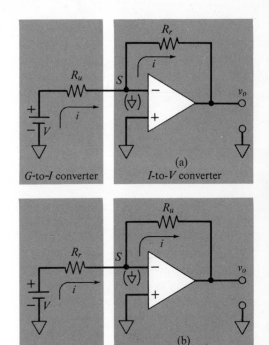

Fig. 2–17 Current follower circuit for resistance measurement: (a) $v_o = -VR_r/R_u$; (b) $v_o = -VR_u/R_r$.

Experiment 2–1 *Light-Sensitive and Temperature-Sensitive Resistive Devices*

A photoconductive cell is used to illustrate how the resistance of a device can change as a function of the illumination, and how the measurement of current through the resistive cell can be related to the illuminance. A thermistor is then used to show the relationship between resistance and temperature for a very useful temperature transducer, again using a current meter to obtain the resistance information.

Experiment 2–2 *Measurement of Thermistor Resistance with a Voltmeter*

In this experiment a voltmeter is used to measure the voltage drop across a thermistor through which a constant current is applied. The temperature is varied and the voltmeter readings are related to resistance values, from which temperature values are calculated.

Experiment 2–3 *Measurement of Thermistor Resistance with an Ohmmeter*

Temperature information was obtained in Experiments 2–1 and 2–2 by measuring current and voltage, respectively, which were then related to resistance values, from which temperature was calculated. It is shown in this experiment that an ohmmeter can be used directly to provide the resistance measurements that are related to temperature. Thus the first three experiments demonstrate the use of three types of conventional meters (current meters, voltmeters, and ohmmeters) for obtaining similar information from a resistive transducer.

Experiment 2–4 *Operational Amplifier Measurement of Thermistor Resistance*

Temperature information is obtained in this experiment from a thermistor whose resistance is calculated from measurements of the output voltage of an operational amplifier circuit containing the thermistor. A constant current is applied to the thermistor by placing it in the feedback loop of an OA inverting amplifier to which a known input current is applied. Output voltage measurements are made and related to the thermistor resistance and thus to temperature. Alternatively the thermistor is used as the input resistor for the OA and a constant input voltage is applied. The thermistor resistance determines the input current, which is then measured by a current comparison method and related back to the temperature of the thermistor.

Null Comparison Measurement

In a null comparison measurement of resistance, the effect of an unknown resistance must be compared with the effect of a variable standard resistance under as near identical conditions as possible. Since a resistance affects the electrical quantities in a circuit, the unknown and standard resistances are placed in identical circuits in such a way that the affected electrical quantity (generally voltage or current) in each circuit can be compared. Several methods for performing this comparison have been devised, of which the Wheatstone bridge is by far the most common. Comparison methods for resistance measurement offer great accuracy, resolution, and relative independence of the electronic sources and detectors used.

The Wheatstone Bridge. The Wheatstone bridge shown in Fig. 2–18 provides the most direct and best-known circuit for comparison of un-

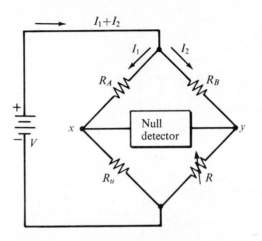

Fig. 2–18 Wheatstone bridge.

known resistances against standard resistances. Resistances R_A, R_B, and R are standard resistance values which are used in the measurement of the unknown resistance R_u. Resistance R is made variable and is adjusted until the null detector indicates that the bridge is balanced.

When the circuit is at balance, there is no current through the null detector and no potential difference between terminals x and y. At balance, four significant conditions exist:

1. The current through R_A and R_u is I_1.
2. The current through R_B and R is I_2.
3. $I_1 R_u = I_2 R$.
4. $I_1 R_A = I_2 R_B$.

Therefore, $R_u/R = R_A/R_B$, and

$$R_u = R\frac{R_A}{R_B}. \qquad (2\text{--}20)$$

It can be seen from Eq. (2–20) that the unknown resistance R_u is determined from the values of the three standard resistances: R, R_A, and R_B. It is common practice to make the ratio R_A/R_B some exact fraction or multiple such as 0.01, 0.1, 1, 10, 100, etc., and to refer to it as the "multiplier." Resistance R is made variable in small increments or continuously, so that the dial reading of R times the multiplier setting equals the unknown resistance R_u.

The main sources of error in a Wheatstone bridge are the inaccuracies of the three standard resistances R, R_A, and R_B. However, these can be made with errors of only about 0.001%. Other factors limiting the accuracy are errors in establishing the null point, thermal emf's, and changes in resistance values due to heating (too high currents). It is important to

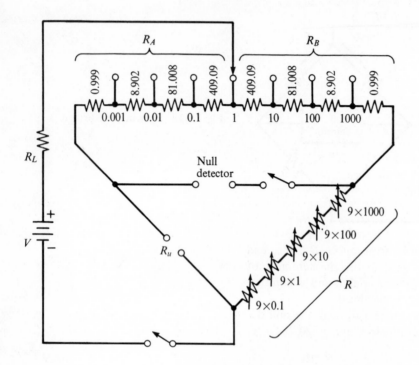

Fig. 2–19   General-purpose Wheatstone bridge.

keep the I^2R loss in the resistors at a safe value, because excessive heating may cause a permanent shift in resistance values. The maximum values specified by the manufacturer should never be exceeded in a bridge measurement.

A schematic of a typical general-purpose Wheatstone bridge is shown in Fig. 2–19. The multiplier R_A/R_B varies from 0.001 to 1000 in seven ranges. The total resistance $R_A + R_B = 1000$ Ω. The resistance R is a decade box varying in steps of 0.1 Ω up to a total resistance of 9999.9 Ω.

The magnitude of the off-balance indication of the null detector of a Wheatstone bridge is sometimes used as a measure of the change in resistance of the unknown resistor R_u. For a derivation of the sensitivity and linearity of the bridge as a resistance-to-voltage converter, the reader is referred to problem 16 at the end of this section. In this application, R_u is frequently a light-, temperature-, or strain-dependent resistor. When matched devices are used for R and R_u, or R_A and R_u, an off-balance output is obtained which is related to the *difference* in the resistance of the two devices and thus the difference in the measured quantity at the two devices. This "differential" resistance measurement with matched devices can tend to cancel the effects of resistance changes due to environmental factors other than the measured quantity.

Experiment 2–5 *Accurate Measurement of Thermistor Resistance with a Wheatstone Bridge*

Unknown resistances are most accurately measured by a comparison measurement, which is conveniently accomplished with the Wheatstone bridge. The resistance bridge is constructed and then used to measure the resistance of a thermistor. Precision resistors and a decade resistance box are used in making the bridge, and a VOM is used on its most sensitive current scale as the null detector. The sensitivity, precision, and accuracy of the resistance (and temperature) measurements with the Wheatstone bridge are compared with those of the methods used in the first three experiments. Some "feel" for the applicability and limitations of several techniques for obtaining resistance information should have been obtained in the first four experiments. It is important to note that the conversion of the resistance information into related electrical signals is necessary so that these signals can be manipulated in the electrical domain so as to provide direct display of temperature, light, or other physical information, or for feedback control information that can be used to hold these parameters constant.

2–1.6 RESISTANCE NETWORK ANALYSIS

Practical circuits are often complex combinations or **networks** of components connected in series and parallel. The methods of analyzing resistance networks to determine the total network resistance or the currents and voltages in various branches of the network will be introduced in this section. The principles involved in these methods are very useful tools, not only for complete network analysis, but also for greatly simplifying the problem of the action of a circuit on a single component or the effect of varying that component on the circuit. This last problem is a matter of concern whenever two devices, circuits, or instruments are connected together.

Successive Reduction

To analyze a complex network of resistances, simple series and parallel branches of the network can often be reduced to single equivalent resistances until only one resistance remains; this single resistance is equivalent to the entire network. If current and voltage values are desired for various network components, they can be calculated as the real network resistances are reconstructed from their equivalent values.

The process of successive reductions is illustrated in Fig. 2–20. The first step is to reduce the simple series and parallel combinations to equivalent values. Thus the parallel combination of R_1 and R_2 is equivalent to a

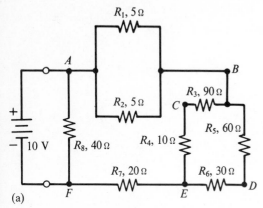

(a)

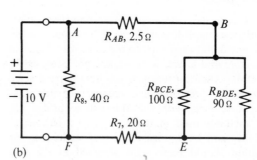

(b)

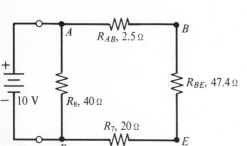

(c)

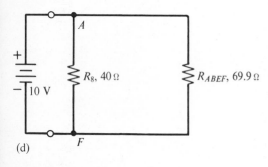

(d)

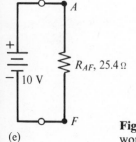

(e)

single 2.5 Ω resistor which we shall call R_{AB}. Similarly, the series combination of R_3 and R_4 is equivalent to a single 100 Ω resistor R_{BCE}, and the combination of R_5 and R_6 is equivalent to a single 90 Ω resistor R_{BDE} as shown in Fig. 2–20b. These steps and subsequent similar steps in the complete reduction are as follows:

From Fig. 2–20a to Fig. 2–20b,

$$R_{AB} = R_1 R_2 / (R_1 + R_2) = 2.5 \ \Omega,$$

$$R_{BCE} = R_3 + R_4 = 100 \ \Omega,$$

$$R_{BDE} = R_5 + R_6 = 90 \ \Omega.$$

From Fig. 2–20b to Fig. 2–20c,

$$R_{BE} = R_{BCE} R_{BDE} / (R_{BCE} + R_{BDE}) = 47.4 \ \Omega.$$

From Fig. 2–20c to Fig. 2–20d,

$$R_{ABEF} = R_{AB} + R_{BE} + R_7 = 69.9 \ \Omega.$$

From Fig. 2–20d to Fig. 2–20e,

$$R_{AF} = R_8 R_{ABEF} / (R_8 + R_{ABEF}) = 25.4 \ \Omega.$$

If the current and voltage values for each resistor are desired, they can now be calculated by "reconstructing" the network. From Fig. 2–20e, for R_{AF}, $V = 10$ V and $I = 0.39$ A. Now from Fig. 2–20d, the current-splitting equation can be used to calculate the current through R_8 and R_{ABEF}.

Fig. 2–20 Reduction of series-parallel network.

These steps and the remaining steps in the complete analysis are tabulated below. Solutions for actual components are indicated by an asterisk (*).

From Fig. 2–20e:

For R_{AF}, $V = 10$ V, $I - 10$ V/25.4 $\Omega = 0.394$ A.

From Fig. 2–20d:

*For R_8, $V = 10$ V, $I = (0.394$ A$)(R_{ABEF})/$
$$(R_8 + R_{ABEF}) = 0.251 \text{ A}.$$

For R_{ABEF}, $V = 10$ V, $I = 0.394$ A $- 0.252$ A $= 0.143$ A.

From Fig. 2–20c:

For R_{AB}, $I = 0.143$ A, $V = (0.143$ A$)R_{AB} = 0.36$ V.

For R_{BE}, $I = 0.143$ A, $V = (0.143$ A$)R_{BE} = 6.77$ V.

*For R_7, $I = 0.143$ A, $V = (0.143$ A$)R_7 = 2.86$ V.

From Fig. 2–20b:

For R_{BDE}, $V = 6.77$ V, $I = (0.143$ A$)R_{BCE}/$
$$(R_{BDE} + R_{BCE}) = 0.075 \text{ A}.$$

For R_{BCE}, $V = 6.77$ V, $I = 0.143$ A $- 0.075$ A $= 0.068$ A.

From Fig. 2–20a:

*For R_1 and R_2, $V = 0.36$ V, $I = 0.143/2 = 0.072$ A.

*For R_3, $I = 0.068$ A, $V = (0.068$ A$)R_3 = 6.10$ V.

*For R_4, $I = 0.068$ A, $V = (0.068$ A$)R_4 = 0.68$ V.

*For R_5, $I = 0.075$ A, $V = (0.075$ A$)R_5 = 4.50$ V.

*For R_6, $I = 0.075$ A, $V = (0.075$ A$)R_6 = 2.25$ V.

Kirchhoff's Laws

Occasionally circuits are encountered where it is not possible to reduce the resistances to simpler circuits by the method of successive reductions illustrated above. To analyze such networks, two simple but very important laws are used. They are called Kirchhoff's Laws, and are often stated as follows:

1) The algebraic sum of all the currents flowing toward a junction is zero. (A junction is a common connection to three or more components.)

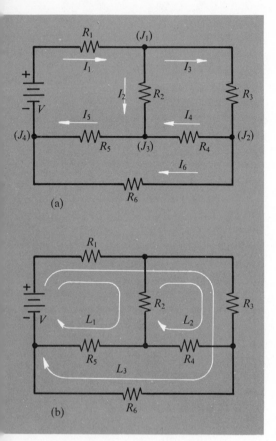

Fig. 2–21 Illustration of Kirchhoff's laws: (a) junction law; (b) loop law.

2) The algebraic sum of all the source voltages is equal to the algebraic sum of all the IR drops in any loop. (A loop is any closed conducting path in the network.)

Kirchhoff's junction law is illustrated in Fig. 2–21a. This network cannot be solved by successive reductions. There are four junctions identified by J's in the figure. A separate equation can be written for each junction as follows:

$$(J_1) \quad I_1 = I_2 + I_3$$

$$(J_2) \quad I_3 = I_4 + I_6$$

$$(J_3) \quad I_4 + I_2 = I_5$$

$$(J_4) \quad I_6 + I_5 = I_1$$

For j junctions there are $j - 1$ independent equations; hence one of the above equations is redundant. Since there are six unknown currents, more equations are needed. These are obtained by applying Kirchhoff's loop law.

The application of Kirchhoff's loop law is illustrated in Fig. 2–21b. Loops are drawn to include every component as shown. Many other combinations could be drawn, but it is seen that the minimum number is three in this case. Summing the voltages around each loop (using the currents and directions from Fig. 2–21a) yields three equations as follows:

$$(L_1) \quad V = I_1R_1 + I_2R_2 + I_5R_5$$

$$(L_2) \quad 0 = I_3R_3 + I_4R_4 - I_2R_2$$

$$(L_3) \quad V = I_1R_1 + I_3R_3 + I_6R_6$$

Thus from Kirchhoff's Laws, six equations have been obtained for this circuit which can be used to solve for six unknown quantities. In general, there will always be enough Kirchhoff's Law equations to solve for all the currents when the voltages and resistances are known.

The simultaneous equations are then solved by successive substitution or by determinants. The Kirchhoff Law method of analysis is thorough, but for complex networks the tedious and mechanical nature of the method invites errors. For this reason, other analysis techniques which are simpler for partial solutions are often used instead. Kirchhoff's Laws, however, are very useful in quickly determining the relationship among currents at a single junction or potentials in a single loop.

Wye-Delta Transformation

A "Y" (wye) connection is the common point of three components such as shown in Fig. 2–22a. The other terminal of each component connected

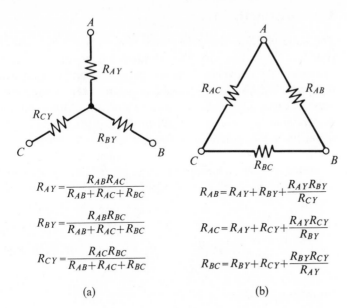

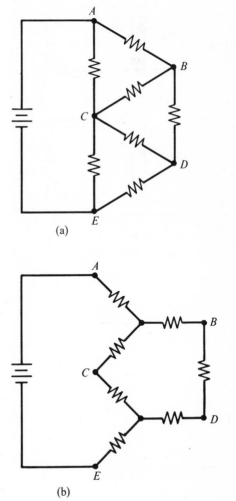

$$R_{AY} = \frac{R_{AB}R_{AC}}{R_{AB}+R_{AC}+R_{BC}}$$

$$R_{BY} = \frac{R_{AB}R_{BC}}{R_{AB}+R_{AC}+R_{BC}}$$

$$R_{CY} = \frac{R_{AC}R_{BC}}{R_{AB}+R_{AC}+R_{BC}}$$

(a)

$$R_{AB} = R_{AY}+R_{BY}+\frac{R_{AY}R_{BY}}{R_{CY}}$$

$$R_{AC} = R_{AY}+R_{CY}+\frac{R_{AY}R_{CY}}{R_{BY}}$$

$$R_{BC} = R_{BY}+R_{CY}+\frac{R_{BY}R_{CY}}{R_{AY}}$$

(b)

Fig. 2–22 Wye-delta transformation: (a) delta to Y; (b) Y to delta.

(a)

(b)

Fig. 2–23 Application of the Δ-Y transformation.

to the Y is labeled A, B, and C respectively. The Y-connected resistors can be shown to be equivalent to a circuit with a resistor between each of the A, B, and C terminals as shown in Fig. 2–22b. This is called the "Δ" (delta) form. Since equivalence relationships exist, a portion of a circuit with Δ-connected components can be transformed into an equivalent Y arrangement of components and vice-versa. The equivalence relationships are given in Fig. 2–22 and are derived in Appendix A.

The complex circuit of Fig. 2–23 can be solved by the use of these transformations and relationships. In the equivalence shown ΔABC and ΔCDE have been transformed into Y's. The circuit can now be analyzed by successive reduction and the voltages at A, B, C, D, and E calculated. Since these voltages must also hold for the equivalent points in the original circuit, the voltage across and the current through any component can be easily calculated. The Y-Δ transformation could also have been used to solve the circuit of Fig. 2–23.

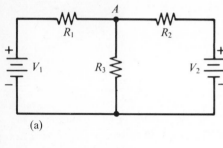

(a)

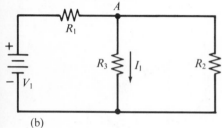

(b)

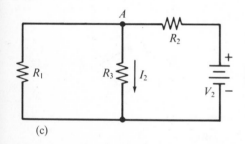

(c)

Fig. 2–24 Superposition theorem application.

Superposition Theorem

The superposition theorem states that each voltage source in a network produces current in any given branch independently of any other voltage sources in the network. Thus the current in any branch will be the sum of the currents from all of the voltage sources taken one at a time. This theorem is especially useful in analyzing circuits with several voltage sources. A common example is the half-bridge shown in Fig. 2–24a. The desired quantity is the potential difference across R_3. When the currents from source V_1 are being considered, V_2 must be replaced by a short circuit as in Fig. 2–24b. The current through R_3 due to V_1 can be obtained by multiplying the total current from V_1 by the current splitting fraction to obtain

$$I_1 = \frac{V_1}{R_1 + \dfrac{R_2 R_3}{R_2 + R_3}} \left(\frac{R_2}{R_2 + R_3} \right) = \frac{V_1 R_2}{R_1 R_2 + R_1 R_3 + R_2 R_3}.$$

From Fig. 2–24c,

$$I_2 = \frac{V_2 R_1}{R_1 R_2 + R_1 R_3 + R_2 R_3},$$

so that

$$I = I_1 + I_2 = \frac{V_1 R_2 + V_2 R_1}{R_1 R_2 + R_1 R_3 + R_2 R_3}$$

and the voltage across R_3 is

$$I R_3 = \frac{R_3 (V_1 R_2 + V_2 R_1)}{R_1 R_2 + R_1 R_3 + R_2 R_3}.$$

Thévenin's Theorem

One of the most useful theorems in circuit analysis is Thévenin's theorem. It states that *any voltage source which has only two output terminals and is composed of resistors and batteries can be represented by a series combination of a resistor R_{Th} and a battery V_{Th}* as in Fig. 2–25. From the Thévenin equivalent circuit it can be seen that V_{Th} is the open-circuit potential difference at the two terminals of the network and R is the resistance between the output terminals when the battery V_{Th} is short-circuited. Thus R_{Th} is the effective source resistance of the network acting as a voltage source of V_{Th} volts.

The network given as an example in Fig. 2–25a is a voltage divider.

The open-circuit voltage V_{Th} is

$$V_{\text{Th}} = V \frac{R_2}{R_1 + R_2} = bV, \qquad (2\text{–}21)$$

where b is the fraction of the total divider resistance across which V_o is taken; this equation shows the proportional relationship between the divider output voltage and the divider fraction. When the battery is replaced by a short circuit, the resistance between the output terminals is the parallel combination of R_1 and R_2:

$$R_{\text{Th}} = \frac{R_1 R_2}{R_1 + R_2}. \qquad (2\text{–}22)$$

If the total divider resistance $(R_1 + R_2)$ is R, $R_2 = bR$ and $R_1 = (1 - b)R$. Substituting these relations in Eq. (2–22), one obtains

$$R_{\text{Th}} = \frac{bR(1 - b)R}{bR + (1 - b)R} = b(1 - b)R. \qquad (2\text{–}23)$$

This equation shows that the output resistance of the divider is a maximum when $b = \frac{1}{2}$, and zero when $b = 1$ or 0.

If $V = 10$ V, $R_1 = 120$ Ω, and $R_2 = 80$ Ω, $V_{\text{Th}} = 4.0$ V and $R_{\text{Th}} = 48$ Ω. This particular voltage divider is thus electrically equivalent to a battery of 4.0 V in series with a resistor of 48 Ω. The actual output of the divider for any load resistance connected to the output terminals can be readily calculated. In fact, through the use of this theorem, the calculation of the output voltage, under load, of *any* voltage source is equally simple. Thévenin's theorem is frequently used to determine the effect of interconnecting two circuits or devices.

Thévenin's theorem is particularly useful when one wishes to calculate the effects of changing or adding one branch in a network. Consider the half-bridge of Fig. 2–26a, which was previously analyzed by superposition. The component for which the voltage and current were sought (R_3) is temporarily removed, and the Thévenin equivalent values are calculated from the remaining network. These calculations are now very simple:

$$V_{\text{Th}} = V_1 + R_1(V_2 - V_1)/(R_1 + R_2) \quad \text{and} \quad R_{\text{Th}} = R_1 R_2/(R_1 + R_2).$$

When R_3 is connected to the equivalent terminals A and B in Fig. 2–26b, the current through R_3 is simply $V_{\text{Th}}/(R_{\text{Th}} + R_3)$.

Thévenin's theorem can also be used to remove a component that makes simple reduction of a network impossible and then add it back to the network after the equivalent circuit has been developed. This technique is illustrated for the Wheatstone bridge of Fig. 2–27a. The 10 Ω resistor

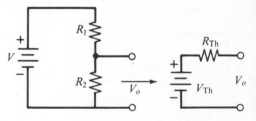

Fig. 2–25 Thévenin's theorem applied to the voltage divider.

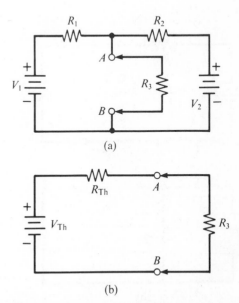

Fig. 2–26 Example of Thévenin's theorem: (a) half-bridge; (b) Thévenin equivalent circuit.

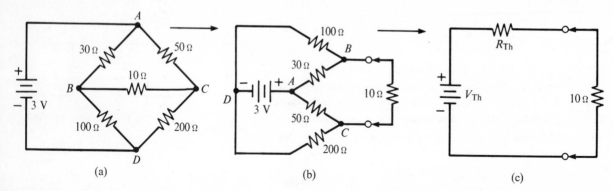

(a)   (b)   (c)

Fig. 2–27 Thévenin's theorem applied to the Wheatstone bridge: (a) original bridge; (b) bridge redrawn; (c) Thévenin equivalent circuit.

from B to C prevents simple analysis of the circuit. (Also, it is the current through this element that is sought in bridge analysis.) The 10 Ω resistor is removed temporarily and the bridge circuit redrawn for clarity in Fig. 2–27b. The Thévenin equivalent voltage is the difference between the voltage at B from the divider ABD and the voltage at C from the divider ACD. Thus

$$V_{Th} = 3 \text{ V} \frac{30}{100 + 30} - 3 \text{ V} \frac{50}{200 + 50} = 0.093 \text{ V}.$$

The Thévenin equivalent resistance, R_{Th} (taken with V shorted), is the parallel combination of the 30 and 100 Ω resistors in series with the parallel combination of the 50 and 200 Ω resistors:

$$R_{Th} = \frac{50 \times 200}{200 + 50} + \frac{30 \times 100}{100 + 30} = 63.1 \text{ Ω}.$$

When reconnected, the current through the 10 Ω resistor is thus 0.093 V/$(63.1 + 10) = 1.26$ mA. The Δ-Y transformation would also have been useful in analyzing the Wheatstone bridge.

Norton's Theorem

A corollary of Thévenin's voltage source theorem is Norton's current source theorem. Norton's theorem states that any two-terminal network of resistors and voltage or current sources is equivalent to a current source with a parallel resistance. The Thévenin and Norton equivalent circuits are shown in Fig. 2–28. In the Norton equivalent circuit, I_N is the current through the terminals with the output shorted. With the half-bridge example of Fig. 2–26a, $I_N = V_1/R_1 + V_2/R_2$. Since an ideal current source has infinite resistance, the equivalent source resistances (R_N and R_{Th}) are the same in both cases. Thus $R_N = R_1 R_2/(R_1 + R_2)$ for the half-bridge.

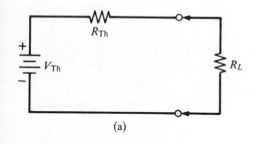

(a)

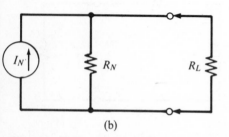

(b)

Fig. 2–28 (a) Thévenin and (b) Norton equivalent circuits.

When the load resistor (R_3 in this case) is connected, the current I_N is split between R_N and the load. Thus for the half-bridge,

$$I = I_N R_N / (R_3 + R_N).$$

All three solutions are thus seen to give identical results.

Since Norton and Thévenin circuits are both equivalent to any two-terminal network, they must be equivalent to each other. The relationships between the equivalent circuit values for the two cases are easily seen from Fig. 2–28. The open-circuit voltage of Thévenin's circuit is V_{Th} and that of Norton's circuit is $I_N R_N$. Therefore,

$$V_{Th} = I_N R_N. \tag{2--24}$$

The short-circuit current of Thévenin's circuit is V_{Th}/R_{Th} and of Norton's circuit is I_N. Now,

$$I_N = V_{Th}/R_{Th}. \tag{2--25}$$

From Eqs. (2–24) and (2–25) it follows that

$$R_N = R_{Th}. \tag{2--26}$$

The choice of circuit analysis technique is largely a matter of personal preference and convenience. However, from the above discussions it is clear that the theorems underlying each method provide important insights into component and circuit interactions. These circuit analysis techniques will be used throughout this and other modules to aid in understanding how circuits perform.

Experiment 2–6 *Linearization of a Transducer Transfer Function*

The relationship between certain desired information in the physical/chemical domain and the output from a specific transducer is often a nonlinear function. In order to make it convenient to display this information on linear current or voltage measurement devices (DVM's, strip chart recorders), it is important to introduce linearization circuits. In this experiment a network is constructed that provides a linear voltage output directly related to temperature even though the resistance values obtained from the thermistor transducer are a nonlinear function of temperature.

PROBLEMS

1. The resistance of a thermistor is to be measured by applying 5 V to a circuit consisting of a thermistor and an ammeter in series. A current of 0.99 mA is measured. (a) If the ammeter resistance is neglected, what is the thermistor resistance? (b) If it is known that the meter has a resistance of 5 kΩ on the 50 μA scale, what resistance does it have on the 1 mA scale used to measure the current through the thermistor? (c) What is the true resistance of the thermistor and what is the relative perturbation error (RPE) caused by inserting the 1 mA full scale meter into the circuit?

2. It is desired to measure the resistance of a thermistor at a temperature near 25°C where the thermistor has a resistance of 10 kΩ. Measurement of the current through the thermistor and of the voltage drop across it are both considered. Only a 1 V source is available. Two meters are available: a current meter with a 50 μA basic sensitivity 5 kΩ resistance and a 100 μA scale; and a voltmeter whose input resistance is 1 MΩ on the 1 V scale. Which of the two meters will cause less loading error (perturbation error)?

3. An ohmmeter of the voltmeter type, similar to that of Fig. 2–15, is used to measure resistance. The voltage source V is a 1.5 V battery, $R_p = 12.5$ Ω, and $R_r = 1.5$ Ω. If full scale is set to 0 Ω with the "ohms adjust" and a reading of two-thirds of full scale is obtained with the unknown resistance R_u, what is the value of R_u?

4. The Wheatstone bridge of Fig. 2–18 is balanced with $R_A = 909.09$ Ω and $R_B = 90.909$ Ω, and the variable standard resistance $R = 628$ Ω. What is the unknown resistance R_u?

5. The circuit shown in Fig. P–5 is constructed. What is the current I_7 through resistor R_7?

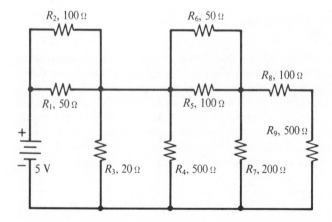

R_2, 100 Ω R_6, 50 Ω

R_8, 100 Ω

R_1, 50 Ω R_5, 100 Ω

R_9, 500 Ω

5 V R_3, 20 Ω R_4, 500 Ω R_7, 200 Ω

Fig. P–5

6. In the circuit of Fig. P–5 what value of source voltage V is needed so that the current through R_3 is 1 mA?

7. The circuit of Fig. 2–21 is constructed in the laboratory; $V = 10$ V, $R_1 = 1$ kΩ, $R_2 = 500$ Ω, $R_3 = 750$ Ω, $R_4 = 1$ kΩ, $R_5 = 2$ kΩ, and $R_6 = 500$ Ω. Find the current through each resistor (currents I_1 through I_6).

8. What is the minimum input resistance required of a voltmeter to measure the IR drop between points A and B in Fig. P–8 with an accuracy of at least 0.01%. Use Thévenin's theorem.

9. In the half-bridge of Fig. 2–26, $V_1 = 10$ V, $V_2 = 15$ V, $R_1 = 1$ kΩ, $R_2 = 5$ kΩ, and $R_3 = 2.5$ kΩ. What is the current through each resistor?

10. In the half-bridge of Fig. 2–26, suppose that the voltage V_2 is variable while V_1 is a fixed voltage. Given that V_2 is changed until the current through $R_3 = 0$ A, develop an equation relating V_2 to V_1 and the resistors R_1 and R_2. Draw an operational amplifier circuit which will give the same transfer function and operates on the current null comparison principle of the half-bridge.

11. In the half-bridge of Fig. 2–26, $V_1 = 14$ V, $V_2 = 9$ V, $R_1 = 20$ Ω, and $R_2 = 5$ Ω. Determine the value of R_3 which will absorb the greatest amount of power and determine the maximum power dissipated.

12. The Wheatstone bridge of Fig. P–12 is to be used to measure resistance. If $R_u = 20$ Ω and the ammeter resistance is 9 Ω, what is the current through the ammeter when the bridge is unbalanced with the standard resistor $R = 10$ Ω? Use Thévenin's theorem.

13. Measurements of resistance change of strain gauges require high sensitivity and are often made using a variation of the Wheatstone bridge shown in Fig. P–13. Resistors R_1 and R_2 provide a multiplier as usual, and resistor R_3 may be only 1–10 Ω for high sensitivity. Resistor R_u is the strain gauge. (a) Consider that ΔR_3 is set to zero and the bridge is balanced with no strain on the gauge, so that $R_u = R_S$. Now consider the gauge to be strained so that $R_u = R_S + \Delta R_S$. Show that when the bridge is rebalanced

$$\Delta R_S = \Delta R_3 (R_2 / R_1).$$

(b) The ratio between the relative resistance change and the relative strain change in the strain gauge is called the gauge factor F and is defined as

$$F = \frac{\Delta R_S / R_S}{\Delta l / l},$$

where l is the length of the wire and Δl is the change in length upon strain. A 120 Ω strain gauge with $F = 2$ is used in an equal-arm bridge ($R_1 = R_2$) that is initially balanced with no strain and rebalanced after straining. Compute the relative strain $\Delta l / l$ that will be present if a change of 0.1 Ω in resistor ΔR_3 is required to rebalance the bridge.

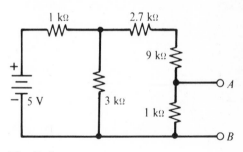

Fig. P–8

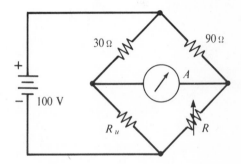

Fig. P–12

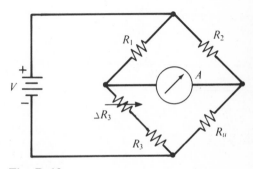

Fig. P–13

14. Consider a precision voltage divider designed to produce 1 V from a 5 V source. (a) If the load to be connected to the 1 V output of the divider has a resistance of 100 kΩ, what are the highest divider resistances that can be used so that the loading error is no more than 0.01%? Use Thévenin's theorem. (b) For a specific divider which satisfies the restrictions of part (a), draw a Norton equivalent circuit and compute I_N, the Norton equivalent current, and R_N, the equivalent shunt resistance.

15. For the Wheatstone bridge of problem 12, recalculate the current through the ammeter using a Norton equivalent circuit instead of a Thévenin equivalent circuit.

16. For many resistance measurement applications it is desirable to record resistance changes directly on a strip chart recorder or other device. Consider the Wheatstone bridge of Fig. 2–18. Consider that the null detector is replaced by a strip chart recorder. (a) Develop a general expression relating the output voltage v_o of the bridge (voltage between points x and y) to the resistance values. (b) Derive a relationship between the change in output voltage and the change in unknown resistance. [*Hint:* use calculus and find dv_o/dR_u.] (c) Under what conditions of bridge resistance will v_o vary linearly with R_u? (d) Under what conditions of bridge resistance will the measurement sensitivity be greatest? [*Hint:* Use calculus and set $d^2v_o/dR_u^2 = 0$.] (e) Is the output voltage linear with R_u under maximum sensitivity conditions?

Capacitive and Inductive Control Devices

Capacitors and inductors play a major role in many instruments for the control of electric charge, current, and voltage. Circuits which are simple combinations of capacitors or inductors and other circuit elements provide a variety of important functions. The characteristics of practical capacitors and inductors and their applications in basic RC and RL circuits are described in this section. Capacitors are used with the versatile operational amplifiers in the widely applied integrator and differentiator circuits. These circuits and several applications in scientific measurements are described.

2–2.1 CAPACITORS

Capacitance is a measure of the quantity of charge required to create a given potential difference between two conductors. Since a finite capacitance exists between every pair of conductors, capacitance, like conductance, is a common property of all electronic devices. The property of capacitance in electric circuits affects the rate at which the circuit voltages can be changed or the duration over which voltage changes can be maintained. Devices which are made specially for their property of capacitance are called **capacitors.**

A capacitor consists basically of two pieces of conducting material separated by an insulator. This construction is represented pictorially in Fig. 2–29 as two metal parallel "plates" of area a separated by a distance d. The capacitor is "charged" by taking charged particles (such as electrons) from one plate and putting them on the other. The result of this separation of charge is a potential difference and an electric field between the two plates. The voltage developed by the charge separation is proportional to the amount of charge separated. The number of coulombs of charge per

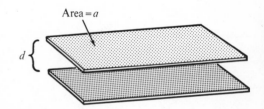

Fig. 2–29 Pictorial representation of a capacitor.

volt of potential difference is the capacitance, C:

$$C = q/V. \tag{2–27}$$

The units of C are coulombs per volt, or **farads** (F). The farad is a measure of the capacity of a capacitor for storing charge at a given voltage.

The separation of charge necessary for the charging of a capacitor requires energy. A charged capacitor (one with a finite potential difference between the plates) stores electric potential energy. This stored energy can be converted to some other form of energy by "discharging" the capacitor, i.e., connecting a conductor or **load** which can produce heat or light or do mechanical work. However, the current through the load occurs at the expense of the charge on the capacitor. As the charge decreases, the capacitor voltage decreases as in Eq. (2–27) so that eventually the charge separation, the voltage, and the current are all zero. When a load is first connected to a capacitor charged to a voltage V, the first electron charge through the load loses an energy of $Q_e V$ joules, where Q_e is the electron charge. The voltage across the capacitor when the last electron charge is passed is essentially zero, so that the energy lost for this charge is also essentially zero. Since the capacitor voltage is proportional to the charge at any time, the average energy lost per electron is the same as the energy lost by the electron passed at half charge. Thus the energy or work obtained in discharging a capacitor completely is

$$W_d = \frac{qV}{2} = \frac{q^2}{2C} = \frac{CV^2}{2}, \tag{2–28}$$

which is also the amount of potential energy (in joules) stored by the capacitor.

By considering the energy stored in the capacitor, one can deduce the geometric factors affecting the capacitance. If the capacitor of Fig. 2–29 is charged to v volts and then the plates are moved apart to a distance $2d$, the energy in the system will be doubled, since the charge has been separated by twice the distance. From Eq. (2–28) it is clear that if W_d is doubled and q is constant, the voltage must be doubled. If the voltage in Eq. (2–27) is doubled for the same charge, the capacitance must be halved. Thus we see that the capacitance is inversely proportional to the conductor separation ($C \propto 1/d$). Again consider the capacitor with separation d charged to v volts. Now we will double the plate area with uncharged pieces of metal. The same charge is now distributed over twice the area and thus produces only half the field. By the same arguments as before, if v is halved when q is constant, C is doubled. Therefore, the capacitance is proportional to the area ($C \propto a$). If the space between the plates of a charged capacitor is at first completely empty (a vacuum), and then insulating material with polarizable molecules (a **dielectric**) is put in that space, the polarizable molecules in the dielectric tend to become aligned under the influence of the electric field. This process uses some of the energy stored in the field. Since q is not changed, v is reduced, which indicates that the capacitance has increased. The factor by which the capacitance is increased by the dielectric over that for a vacuum is called the **dielectric constant** of the material. Thus $C \propto K_d$. If these equations are combined into one,

$$C = \epsilon_0 \frac{K_d a}{d}, \qquad (2\text{--}29)$$

where ϵ_0 is the combined proportionality constant and is called the permittivity of free space. The value of ϵ_0 is 8.854×10^{-12} farads per meter (Fm^{-1}).

If the dimensions and dielectric constant are known for a capacitor, its capacitance can be calculated from Eq. (2–29). For instance, a capacitor made of 2 squares of aluminum foil 10 cm on a side, separated by a cellophane ($K_d \approx 3.5$) sheet 0.02 cm thick, will have a capacitance of

$$C = \frac{(8.9 \times 10^{-12} \text{ Fm}^{-1})(3.5 \times 10^{-2} \text{ m}^2)}{2 \times 10^{-4} \text{ m}} = 16 \times 10^{-10} \text{ F.}$$

It can be seen from this example that the farad is an extremely large unit. For this reason real capacitors are measured in microfarads (1 μF $= 10^{-6}$ F) and picofarads (1 pF $= 10^{-12}$ F). The capacitance in the above example would be given as 0.0016 μF or 1600 pF. Values for some common dielectric constants are given in Table 2–4.

Table 2–4 Common dielectric constants

Material	K_d
Air	1.0006
Glass	3.9–5.6
Plexiglass	2.8
Polyethylene	2.3
Polystyrene	2.6
Teflon	2.1
Mica	5.5
Paper	≈ 3.5
Water	78
Quartz	3.9
Oil	2.2
Paraffin	2.1

Practical Capacitors

The simplest capacitor consists of two sheets of metal foil separated by a dielectric. The dielectric can be paper, oil, mica, ceramic, or plastic film. The whole "sandwich" can be rolled, folded, or otherwise compacted. Then the unit is sealed from the effects of atmosphere. Some common capacitor types are shown in Fig. 2–30. The larger the area of the foil and the thinner the dielectric sheet, the higher the capacitance will be. However, if too high a voltage is applied across the capacitor, the dielectric will break down and a discharge will occur between the two foil sheets. This often results in permanent damage to the capacitor. The breakdown voltage rating may be increased by using a thick dielectric, but the capacitance will be correspondingly decreased. The capacitance of these units, which are the most frequently used capacitors, varies from 0.001 to 1 μF.

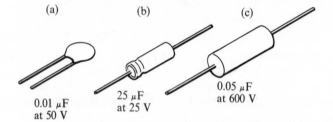

(a)   (b)   (c)

0.01 μF
at 50 V

25 μF
at 25 V

0.05 μF
at 600 V

Fig. 2–30 Common capacitor types: (a) disk ceramic; (b) electrolytic; (c) paper or plastic film.

Capacitors can be obtained in various tolerance ratings, usually 20, 10, or 3%. Capacitors of higher accuracy can be obtained for special applications. Capacitors generally have a high temperature coefficient because of dimensional changes with temperature. In some very critical circuits it is necessary to consider the leakage current through the capacitor dielectric. A capacitor constructed for especially high leakage resistance may have a leakage resistance–capacitance product of 10^6 MΩ-μF.

Electrolytic capacitors are made of aluminum or tantalum foil with a surface that has an anodic formation of metal oxide film. The anodized metal foil is in an electrolytic solution. The oxide film is the dielectric between the metal and the solution. Because the dielectric is so thin, a high capacitance can be obtained in a small space. Most electrolytic capacitors must be used in a circuit where the polarity is always in one direction. If the polarity is reversed, the oxide will be reduced, thus destroying the dielectric. Gas will be evolved at the cathode and the capacitor may explode. Some tantalum electrolytic capacitors are made of two anodized metal foils connected by the electrolyte. Such capacitors are **bipolar** and can be used in ac signals or connected either way in the circuit. The oxide film is a relatively low resistance dielectric which results in significant leakage current.

The oxide in aluminum electrolytics tends to deteriorate with time. Electrolytic capacitors are generally used for filtering and transistor circuit applications where poor dielectric can be tolerated and where large capacitance in a small space is essential. They are generally rated from 1 μF to thousands of microfarads. Naturally, the physical size of the capacitor increases with increased capacitance and voltage rating. Tantalum capacitors are generally much smaller for a given capacitance than aluminum capacitors.

Variable Capacitors.   Variable capacitors are usually small-valued capacitors not exceeding about 1000 pF. When larger variations are desired, a switch is often used to put various fixed-value capacitors into the circuit. There are two common types of variable capacitors, the mica trimmer capacitor and the variable air capacitor.

The mica trimmer capacitor is composed of two sheets of metal sandwiching a sheet of mica. The spacing between the metal sheets, and thus the capacitance, is adjusted by turning a machine screw. This kind of variable capacitor is used when adjustment is only occasionally necessary and when knowledge of the actual value of the capacitance is unimportant.

The variable air capacitors are adjusted from zero to full capacitance by turning a shaft 180°. If the movable plates are semicircular, the capacitance is nearly a linear function of the angular position of the shaft. End effects cause a small deviation from linearity at both limits of rotation.

There are many considerations involved in choosing a capacitor for a given circuit, such as size for a specific capacitance value, applied voltage effects, and stability in all environmental and circuit conditions. For example, the temperature can affect the capacitance by causing variations of the dielectric constant or changing the conductor area or spacing; or it can change the leakage resistance or affect the breakdown voltage. Moisture can make the dielectric more susceptible to breakdown and decrease the insulation resistance and capacitor life unless the capacitor is hermetically sealed.

There are some energy losses in a capacitor. The **capacitor power factor** is the ratio of power lost in the capacitor to the volt-ampere input, and it is usually expressed as a percentage. It is a complex function of frequency, temperature, applied voltage, and dielectric material. The power factor takes into account *all* losses, including leakage losses, dielectric absorption losses, and ohmic losses in the leads and conducting plates.

Many capacitors develop potential differences across their terminals shortly after complete discharge and thus produce unwanted dc voltages that can appear as offset or noise. This phenomenon is due to dielectric polarization effects and is known as **dielectric absorption,** sometimes called the "soaking effect." Capacitors with polystyrene or Teflon™ dielectrics are especially free from this effect and thus are well suited for integrators and other critical instrumentation applications.

Capacitive Transducers

Transducers in which the capacitance varies in response to the measured phenomenon are quite common. Because the capacitor is so sensitive to dimensional changes, it is often used where the parameter to be measured can cause a physical motion. The motion can cause a variation in the gap between the plates, a change in the area of the plates, or a change in the dielectric constant. Thus capacitive transducers exist for displacement, pressure, fluid level, and so on. Capacitive transducers have also been used to measure the dielectric constants of materials.

Combinations of Capacitors

When two or more capacitors are connected together in a circuit, the resulting equivalent circuit capacitance is related to the capacitance of the individual capacitors by simple relationships. Simple series and parallel combinations will be considered first. Then the analysis of a series-parallel network of capacitors will be illustrated.

In the circuit of Fig. 2–31, several capacitors are connected in parallel to a voltage source V. Since the capacitors are in parallel, the voltage V appears across each one. It follows from the definition of capacitance, $C \equiv Q/V$, that the charge on C_1 is C_1V, the charge on C_2 is C_2V, and so on. Thus the total charge Q_T is

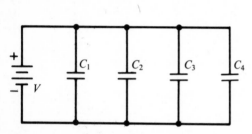

Fig. 2–31 Capacitors in parallel.

$$Q_T = V(C_1 + C_2 + C_3 + C_4). \qquad (2\text{--}30)$$

The equivalent parallel capacitance C_p is Q_T/V, so that

$$C_p = C_1 + C_2 + C_3 + C_4. \qquad (2\text{--}31)$$

Thus we see that the equivalent capacitance of capacitors in parallel is simply the sum of the capacitances. The combined capacitance of capacitors in parallel is always greater than the largest individual capacitance.

A series combination of capacitors connected to a voltage source V is shown in Fig. 2–32. It is assumed that each capacitor was completely discharged before being connected in the circuit. While charging, the current to each capacitor must be identical, since the capacitors are connected in series. When charged, the charge on each capacitor must be identical as indicated in the figure. The voltage v_1 across C_1 must be q/C_1; the voltage v_2 across C_2 must be q/C_2; and so on. The sum of the voltages across the capacitors must be equal to V (Kirchhoff's loop law) so that

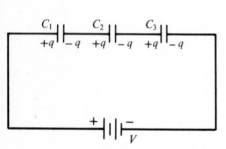

Fig. 2–32 Capacitors in series.

$$V = \frac{q}{C_1} + \frac{q}{C_2} + \frac{q}{C_3}. \qquad (2\text{--}32)$$

The equivalent series capacitance $C_S = q/V$. If this expression is substituted in Eq. (2–32) for V and q is canceled, Eq. (2–33) results:

$$\frac{1}{C_S} = \frac{1}{C_1} + \frac{1}{C_2} + \frac{1}{C_3}. \qquad (2\text{–}33)$$

The equivalent capacitance is the reciprocal of the sum of the reciprocals. For two capacitors, a simpler form of Eq. (2–34) exists:

$$C_S = \frac{C_1 C_2}{C_1 + C_2}. \qquad (2\text{–}34)$$

As Eq. (2–33) or Eq. (2–34) indicates, the capacitance of the series combination will always be *less* than the smallest individual capacitance.

The circuit of Fig. 2–33 is a series-parallel combination of capacitors connected to a voltage source of 16 V. This network can be solved by successive reduction as in the resistive network of Section 2–1.6. The equivalent capacitance of the parallel combination of C_2 and C_3 ($C_{2,3}$) is $C_2 + C_3$; $C_{2,3} = 3\ \mu\text{F} + 5\ \mu\text{F} = 8\ \mu\text{F}$. This equivalent capacitance $C_{2,3}$ in series with C_1 gives an equivalent capacitance $C_{1,2,3} = 8 \times 8/(8 + 8) = 4\ \mu\text{F}$. The series combination of C_4 and C_5 has an equivalent capacitance $C_{4,5} = 2 \times 6/(2 + 6) = 1.5\ \mu\text{F}$. The parallel combination of $C_{4,5}$ and $C_{1,2,3}$ results in $C_T = 1.5 + 4 = 5.5\ \mu\text{F}$. The total charge Q_T on the circuit is $C_T V = 5.5 \times 10^{-6}\ \text{F} \times 16\ \text{V} = 88\ \mu\text{C}$. This charge is split between $C_{1,2,3}$ and $C_{4,5}$: $1.5 \times 10^{-6}\ \text{F} \times 16\ \text{V} = 24\ \mu\text{C}$ on $C_{4,5}$ and 64 μC on $C_{1,2,3}$. Since C_4 and C_5 are in series, they each have a charge of 24 μC. This means the voltage on C_4 is $24\ \mu\text{C}/2\ \mu\text{F} = 12\ \text{V}$ and the voltage on C_5 is 4 V. Now C_1 and $C_{2,3}$ are in series, with 64 μC of charge on each. Therefore, the voltage on C_1 is $64\ \mu\text{C}/8\ \mu\text{F} = 8\ \text{V}$. The voltage on C_2 and C_3 is 8 V. The charge on C_2 is $3\ \mu\text{F} \times 8\ \text{V} = 24\ \mu\text{C}$; and on C_3, 40 μC. In this way the total equivalent capacitance is calculated and then the voltage on each capacitor and the charge across each capacitor are calculated.

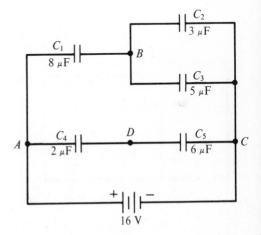

Fig. 2–33 Capacitor network.

Experiment 2–7 *The Capacitor as a Charge Storage Device*

Most measurement and control circuits are dependent in some way on the ability of a capacitor to store charge. Therefore, it is important to obtain laboratory experience on how capacitors perform when charged and when combined in various ways. In this experiment, the leakage characteristics of various types of capacitors are observed, and the distribution of charge in capacitor networks is measured. The voltage values obtained from the theoretical relationships are compared with the measured values in capacitor networks.

2–2.2 *RC* CIRCUITS

When a series combination of a resistor and a capacitor is connected to a voltage source, the values of the resistance and the capacitance affect the rate at which the capacitor can charge. Similarly, when a charged capacitor is discharged through a resistor, the discharge rate is affected by the capacitance and resistance values. Since resistance and capacitance are inherent parts of all circuits and devices, an understanding of the charging and discharging characteristics of *RC* circuits is important in determining the rate with which steady-state conditions will be reached after a change in the applied potential.

Series *RC* Circuit

There are several basic factors that govern the response of the simple *RC* circuit to a sudden change in voltage. In Fig. 2–34 a source of voltage *V* (of negligible internal resistance) can be suddenly impressed across the series *RC* circuit by turning switch *S* to *A*. From inspection of Fig. 2–34, assuming that the switch *S* is at *A*, the following statements can be deduced:

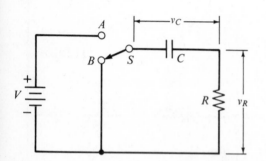

Fig. 2–34   A series *RC* circuit and step signal source.

A. At every instant, the voltage across the capacitor is directly proportional to the stored charge q; that is, $v_C = q/C$.

B. The voltage across the resistance at any instant is $v_R = iR$.

C. The sum of the voltage drops around the circuit must equal the applied voltage at any instant (Kirchhoff's voltage law); that is,

$$V = v_R + v_C = iR + \frac{q}{C}. \qquad (2\text{--}35)$$

D. The current at any instant is the same in all parts of a series circuit and is equal to v_R/R. If Eq. (2–35) is rearranged and solved for the current, the result is

$$i = (V - v_C)/R. \qquad (2\text{--}36)$$

Suppose in the circuit of Fig. 2–34 that $V = 10$ V, $R = 100$ Ω, $C = 1$ µF, and the switch has been at *B* for a sufficient time so that $q = 0$. Therefore, $i = 0$, $v_R = iR = 0$, and $v_C = q/C = 0$. Now the switch is turned to *A*. At the instant of closing, the capacitor has not had a chance to charge, and $v_C = 0$. Therefore, $i = (V - v_C)/R = V/R = 10/100 = 0.1$ A, and $v_R = 10$ V. Immediately the flow of electrons starts to charge the capacitor; so q and v_C increase and i and v_R decrease, each exponentially, as shown in Fig. 2–35. Note that the time scale is calibrated in units of *RC*. The product *RC* has the units of seconds,

$$RC = \left(\frac{V}{i}\right)\left(\frac{q}{V}\right) = \left(\frac{\text{volts}}{\text{coulombs/sec}}\right)\left(\frac{\text{coulombs}}{\text{volts}}\right) = \text{sec},$$

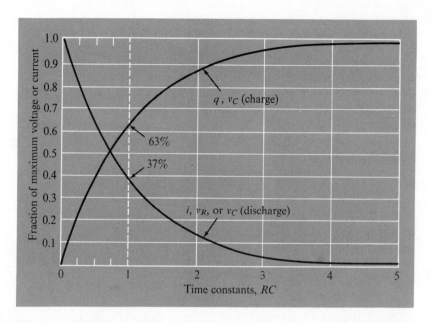

Fig. 2–35 Charge and discharge curves in the *RC* circuit.

and is called the **time constant.** The current *i* at any time *t* after turning the switch to *A* is given by

$$i = \frac{V}{R} e^{-t/RC}. \qquad (2\text{–}37)^3$$

After a time $t_1 = RC$, the current $i_t = (v/R)e^{-RC/RC} = (v/R)e^{-1} = (v/R) \times 0.368$. In other words, at time $t_1 = RC = 100 \times 1 \times 10^{-6} = 10^{-4}$ sec, the current is 36.8% of the current at the instant of impressing voltage V. This means that at time $t_1 = RC$ the voltage across the resistor is only 36.8% of its initial value and the capacitor is charged to 63.2% of the impressed voltage. The time $t_1 = RC$, which is referred to as the time constant of the *RC* circuit, is given the symbol τ. Table 2–5 tabulates values of v_C and v_R for different multiples of τ, both for the charging of the capacitor by impressing voltage V and for discharging a capacitor that is charged to a voltage V. Note that when $t = 4.6RC$ the capacitor is charged to 99.0% of the impressed voltage and v_R is only 1% of its initial value. For practical purposes the capacitor is often considered to be fully charged when $t \approx 5\tau$.

It is important to keep in mind that the voltage across a capacitor cannot change instantly; i.e., the capacitor voltage changes exponentially with time. From one point of view, the capacitor may be considered a short circuit at the first instant of impressing a sudden voltage change.

By watching the output of the series *RC* circuit, first across the resistor and then across the capacitor, various waveshapes similar to those in Fig.

Note 3.   Derivation of Charging Equation.

Equation (2–37) can be derived by substituting dq/dt for i in Eq. (2–35) and solving the resulting differential equation. The result is

$$v_C = V(1 - e^{-t/RC}). \quad (2\text{–}38)$$

To obtain the i vs. t relation, substitute Eq. (2–38) into Eq. (2–35),

$$v_R = V - v_C = Ve^{-t/RC}, \quad (2\text{–}39)$$

and, since $v_R = iR$, it follows that

$$i = \frac{V}{R} e^{-t/RC}. \qquad (2\text{–}40)$$

Table 2–5 Output voltages across capacitor and resistor in series *RC* circuit

	Capacitor charging		Capacitor discharging	
Time	v_c % V applied	v_R % V applied	v_c % V initial	v_R % V initial
RC	63.2	36.8	36.8	36.8
2*RC*	86.5	13.5	13.5	13.5
2.3*RC*	90.0	10.0	10.0	10.0
3*RC*	95.0	5.0	5.0	5.0
4*RC*	98.2	1.8	1.8	1.8
4.6*RC*	99.0	1.0	1.0	1.0

2–36 can be observed if the switch of Fig. 2–34 is turned on and off, or if a rectangular pulse source is used as the input. It is immediately apparent that the output waveform is greatly dependent on the relationship of the *RC* time constant τ to pulse width T_p. It is interesting to observe that the leading edge of the output across the resistor is always steep, so long as the input voltage has a steep leading edge. In contrast, the leading edge of the capacitor output is always changing exponentially. Note that the sum of voltages across the capacitor and resistor equals the input voltage at each instant for a given *RC* time constant. This can be observed by comparing the pairs of curves in Fig. 2–36c, d, and e.

It is worthwhile to observe that sharp positive and negative pulses can be obtained across the resistor when the *RC* time constant is much shorter than the pulse width. This type of response finds application in many circuits. In Fig. 2–36c the voltage across the capacitor is a rather linear sawtooth voltage. This suggests that the circuit can act as an integrator when the time constant is long in comparison with the pulse width. In fact, when the output is taken across the capacitor, the *RC* circuit is often referred to as an "integrator," because

$$v_o = v_c \simeq (1/RC) \int v_i \, dt. \qquad (2\text{–}41)$$

Because *the output is the integral of the input only when the RC time constant is much longer than the pulse width,* such a "passive" *RC* circuit does not provide an exact integration. *RC* circuits in conjunction with operational amplifiers can provide very accurate integration of all waveforms, and are presented in Section 2–2.3.

Parallel *RC* Circuit

To determine the shape and time constant of voltage changes in a circuit, it is only necessary to determine whether the output signal is across the capac-

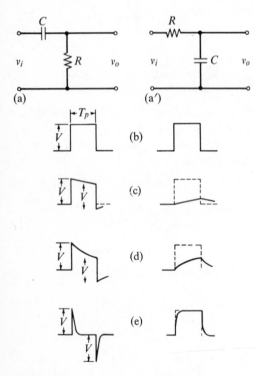

Fig. 2–36 The *RC* series circuit: (a) output taken across resistor; (a′) output taken across capacitor; (b) input voltage; (c, d, e) output voltages across *R* and *C*, respectively, when (c) τ ≫ T_p; (d) τ ≈ T_p; (e) τ ≪ T_p.

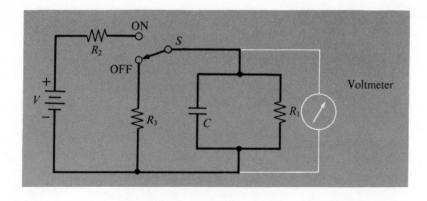

Fig. 2–37 Parallel *RC* switching circuit.

itance or the resistance, and what components determine the *RC* time constant in the case of each change. To take another example, consider the parallel *RC* circuit of Fig. 2–37. When switch *S* is thrown to the **ON** position, *C* will begin to charge toward the final voltage established by the voltage divider (R_1 and R_2), that is, $VR_1/(R_1 + R_2)$. The rate of charge is determined by *C* and by the output resistance of the charging circuit, which consists of the source *V* and the voltage divider. As was pointed out in Section 2–1.6, the equivalent resistance of a voltage divider is the parallel combination of the divider resistors, or $R_1R_2/(R_1 + R_2)$. Thus the charging time constant is $CR_1R_2/(R_1 + R_2)$. When switch *S* is turned **OFF**, capacitor *C* discharges toward zero through the parallel discharge paths R_1 and R_3. Thus the discharge time constant is $CR_1R_3/(R_1 + R_3)$. Since the voltage across the capacitor is being observed, the waveforms will resemble those in the right-hand column of Fig. 2–36.

Such parallel *RC* circuits are of great importance in switching and will be discussed in more detail in Section 2–3.

Experiment 2–8 *Control of Current and Voltage with the RC Network*

The charging and discharging of a capacitor through a resistor provides a current control function that is generally applicable in measurement and control systems. The *RC* network is used for controlling time delay circuits, for controlling the repetition rate of waveform generators, and for integration and differentiation of voltage signals. In this experiment the *RC* circuit is first tested with a step function input signal and then with a repetitive square wave. The types of output waveforms that can be produced by the *RC* circuit with a square wave input are observed and compared with theoretical curves.

2–2.3 ACTIVE *RC* INTEGRATORS, DIFFERENTIATORS, AND LOGARITHMIC CONVERTERS

An active *RC* circuit is one in which the voltage across or current through a section of the circuit is actively controlled. Active *RC* circuits can easily be made by using combinations of resistors and capacitors in the input and feedback circuits of operational amplifiers. Such circuits have a wide variety of applications as integrators, differentiators, oscillators, compensators, and filters. These applications will be explained and illustrated in this and other modules. The active *RC* circuit concept is introduced here with the often used active integrator and differentiator circuits.

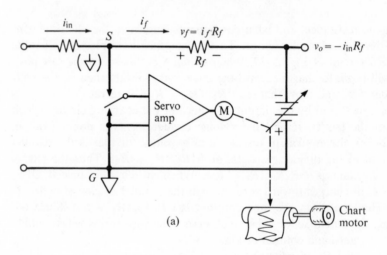

(a)

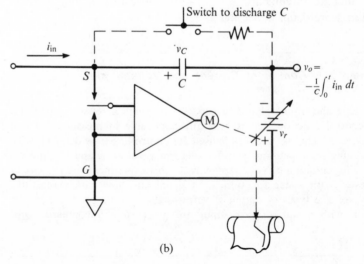

Fig. 2–38 (a) Servomechanical current follower. (b) Servomechanical current integrator.

(b)

Integrator

In many measurement and control applications it is desirable to obtain an accurate electronic integration of an input current or voltage. For example, the quantitative information from many transducers, such as those used in gas or liquid chromatography, is contained in the area of a peak-shaped signal. The use of an electronic integrator allows the quantitative information to be displayed directly without time-consuming manual area measurements. In other applications integration may be used to average the effects of random noise or to produce a linear time-base signal. The basic current follower systems described in Module 1 can perform the integration operation if a capacitor is substituted for the resistor as the feedback element.

The servomechanical current follower with resistive feedback is repeated in Fig. 2–38a. The slidewire output voltage v_o is automatically adjusted to maintain the reference current i_f (normally called the feedback current) equal to the input current i_{in} and thus to maintain the potential at the summing point S equal to that at point G. At balance, $v_o = -i_{in}R_f$, and the pen position, which is proportional to v_o, is thus related to the input current.

In Fig. 2–38b the feedback resistor is replaced with a capacitor C. As before, the servo system adjusts the slidewire output voltage to maintain a current null by keeping the potential at point S very nearly equal to that at point G. The reference or feedback current to point S comes from changing the charge on (and thus the voltage across) the capacitor C. For the servo system to balance, the slidewire output voltage v_o must equal the voltage across the feedback capacitor v_C and must be of opposite polarity. Since $v_C = q/C$, where q is the charge stored on the capacitor,

$$v_o = -v_C = -q/C. \tag{2–42}$$

The charge stored on the capacitor is equal to the integral of the current during the charging period t, and since $i_{in} = i_f$,

$$v_o = -\frac{1}{C} \int_0^t i_{in}\, dt. \tag{2–43}$$

Input voltages as well as currents can be integrated with the servo system by adding an input resistance between the source and the summing points. Just as was the case with the current follower, the servomechanical system can be replaced by an all-electronic operational amplifier for performing current and voltage integration. The OA voltage integrator illustrated in Fig. 2–39 operates on the same principle as the servo current integrator. The input current $i_{in} = i_f = v_{in}/R$. Substituting for i_{in} in Eq.

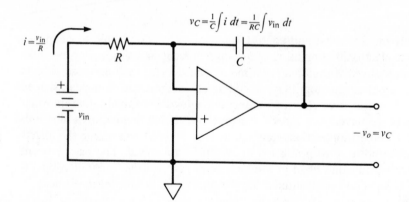

Fig. 2–39 Operational amplifier voltage integrator.

(2–43) gives

$$-v_o = v_C = \frac{1}{C} \int_0^t \frac{v_{in}}{R} \, dt = \frac{1}{RC} \int_0^t v_{in} \, dt. \qquad (2\text{–}44)$$

Thus either currents or voltages can be accurately integrated by the servo-mechanical system or by the all-electronic operational amplifier. For accurate and precise integration, an OA with very low input bias current is highly desirable. Input bias currents can cause large errors because they are also integrated and contribute to v_o.

The integrator circuit can also be used as a capacitance-to-voltage converter. If the integrator circuit is used to place a known amount of charge on the capacitor, the OA output voltage will be inversely proportional to the capacitance as given by Eq. (2–42). The known amount of charge is usually generated by applying a constant voltage V through a known resistance R for a specific time. Thus $q = Vt/R$ and

$$C = -\frac{Vt}{v_o R}. \qquad (2\text{–}45)$$

Experiment 2–9 *An Operational Amplifier Integrator*

When a capacitor is connected as the feedback device in an operational amplifier, the resulting integrating circuit can be used to measure the area under a current or voltage signal, to produce a linear sweep voltage (ramp), and to provide feedback control signals for controlling some other device. In this experiment the integrator is constructed and used as a linear sweep generator and for the integration of various waveforms.

Experiment 2–10 *Determination of Capacitance Using a Voltage Measurement Method*

If a known constant current i is used to charge a capacitor of capacitance C for a known time t, then the charge $q = it$, and the voltage v_C across the capacitor is $v_C = q/C$. Therefore, the capacitance can be determined by measuring the voltage after establishing a known charge on the capacitor. An accurate charge is placed on the capacitor by connecting it as the feedback element in an OA circuit. Since the output voltage v_o of the OA is equal to the voltage v_C across the capacitor, it is simple to measure v_C without disturbing the charge on the capacitor. The precision of the method is determined for various capacitors and the limitations of the method are noted.

Charge-Coupled Amplifier

It is clear from Eq. (2–42) that the integrator circuit is very useful for the measurement of charge, or more precisely, for the conversion of data from the charge domain to the voltage domain. Thus it is particularly useful with transducers having an output in, or easily converted to, the charge domain. Several charge-output transducers were described in Section 1–2.5 of Module 1. Since the charge on a capacitor is a function of the voltage across the capacitor and the capacitance value, both voltage and capacitance variations are easily converted into charge variations.

A common charge generating and measuring circuit is shown in Fig. 2–40. Since one terminal of C_{in} is at voltage v_{in} and the other terminal is held at the common potential by the operational amplifier, the charge on the capacitor is $q = C_{in}v_{in}$. If v_{in} varies by an amount Δv_{in}, the charge on C_{in} will vary by

$$\Delta q = C_{in}\, \Delta v_{in}. \qquad (2\text{--}46)$$

This same change in charge will appear across C_f, so that

$$\Delta v_o = -\Delta q/C_f = -\Delta v_{in} C_{in}/C_f. \qquad (2\text{--}47)$$

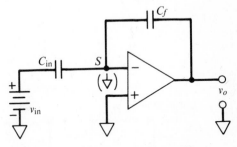

Fig. 2–40 Charge-coupled amplifier.

Thus the input voltage variations appear at the output multiplied by the factor C_{in}/C_f. This circuit is often used with biological transducers which produce voltage variations with a limited charge supply capability. It is necessary to short out C_f occasionally or bypass it with a large resistor to keep the output voltage from drifting off as a result of the integration of leakage currents.

The output voltage of the circuit of Fig. 2–40 will also vary with changes in the value of either C_{in} or C_f. Equation (2–46) can be rewritten to show the change in q due to a change in C_{in}:

$$\Delta q = v_{in}\, \Delta C_{in}. \qquad (2\text{–}48)$$

In this case the output voltage will be

$$\Delta v_o = -\Delta q / C_f = -V_{in}\, \Delta C_{in} / C_f, \qquad (2\text{–}49)$$

proving the circuit to be a linear capacitance-to-voltage converter.

Differentiator

The generation of an output voltage v_o that is proportional to the time rate of change (derivative) of the input voltage, dv_{in}/dt, is often an important operation in electronic measurement and control circuits. The derivative signal is readily obtained with an OA differentiator, as shown in Fig. 2–41. The input voltage is applied to the capacitor C, which is connected to the summing point of the OA. The charging or discharging current for C in the OA input circuit is $i_{in} = dq/dt$. Since $q = Cv$, the input current $i_{in} = C\,(dv_{in}/dt)$. The feedback current i_f through R is equal to i_{in} (that is, $i_f = i_{in}$), and the voltage drop across the feedback element, $i_f R$, is equal to the output voltage v_o, but of opposite polarity. Therefore,

$$v_o = -i_f R = -RC\frac{dv_{in}}{dt}. \qquad (2\text{–}50)$$

The second and higher derivative signals can be obtained by connecting one or more differentiator circuits in sequence with the circuit shown in Fig. 2–41.

Some applications of the differentiator circuit can be seen by referring to Figs. 2–42a, b, and c. In Fig. 2–42a the derivative output signal is pro-

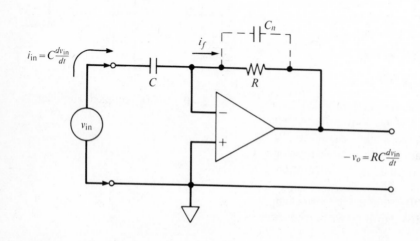

Fig. 2–41   Operational amplifier differentiator.

portional to the positive or negative slope of the input ramp voltage. The sudden transition from a positive to a negative voltage output when the slope reverses from positive to negative can be a very useful control signal. In Fig. 2–42b an information peak is superimposed on a variable background signal. If the second derivative of this signal is taken, there is negligible change in output voltage except when the information signal is present. The second derivative signal is also very useful in measurement and control circuits, and is used in several spectrometric applications as well as in analog computers.

The signals in Fig. 2–42c show how the first and second derivative signals of a sigmoid-type curve are independent of the dc level of the signal. There are several applications where it is necessary to locate the inflection point (maximum rate of change) of the sigmoid curve, and in this case either the first or the second derivative curve can be a valuable control signal. The second derivative output voltage is especially useful as a control signal because the output is zero until just before the inflection point, then rises to a peak in one direction and suddenly reverses and rapidly crosses the zero axis at the inflection point. Therefore, the second derivative signal can easily be used to control a switching circuit that will turn some device ON or OFF exactly at the time of the inflection point of the input signal.

Unfortunately, noise pulses that have a larger dv/dt than the input signal waveform will provide output noise that greatly degrades the useful output information. To minimize the problems of noise, the upper frequency response of the differentiator is generally limited by putting a small capacitor C_n across the feedback resistor, as shown in Fig. 2–41. However, there are many applications where the input signal is too noisy to differentiate in this way without greatly degrading the differentiator response, and it becomes necessary to use a much quieter integration technique to obtain the rate information. A circuit which allows rate information to be acquired in the presence of large amounts of noise is described next under integrating-type differentiators.

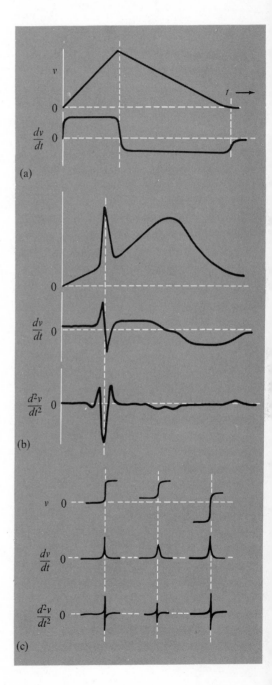

Fig. 2–42 Application of the differentiator. (a) First derivative output signal of a sawtooth input signal. (b) First and second derivative signals for a signal peak superimposed on a variable background. (c) Derivative signals for sigmoid input signals with various dc levels.

(a)

(b)

Experiment 2–11 *An Operational Amplifier Differentiator*

Information about the rate of change of a quantity with respect to time is frequently required for measurement or control purposes. In this experiment the rate of change of a voltage is measured by using an OA differentiator which produces an output voltage v_o that is directly proportional to the input rate of change, dv_{in}/dt. The differentiator circuit is connected, and also the test circuit is easily assembled by connecting an OA integrator which provides a known ramp voltage output as the test signal. The problems of using the conventional differentiator with noisy signals are noted, and corrective measures are tested to improve noise immunity.

Integrating-Type Differentiators

One technique for obtaining rate data utilizes a *comparison method*. An operational amplifier integrator is used to generate a variable reference curve whose output is maintained equal to the unknown signal. When the unknown and reference curves are equal and remain equal as determined by a null detector, the input voltage to the integrator must be equal to the rate of change of the unknown signal. That is, since the output v_o of the integrator is proportional to the integral of the input voltage v_{in}, then dv_o/dt is proportional to v_{in}.

A very good technique for generating reliable rate-of-change signals is based on the integration of two successive segments of the input signal and the subtraction of the resultant areas. At first, this might seem like a cumbersome method but it is relatively easy to implement and does exceptionally well in discriminating against noise present with the input signal. The principle of the method is illustrated in Fig. 2–43a, where the integration is made over two equal time increments Δt which are sequential, and the difference Δa between the resultant areas a_1 and a_2 is the parallelogram $ABCD$. The area of the parallelogram is given by

$$\Delta a = (\Delta t)l. \qquad (2\text{–}51)$$

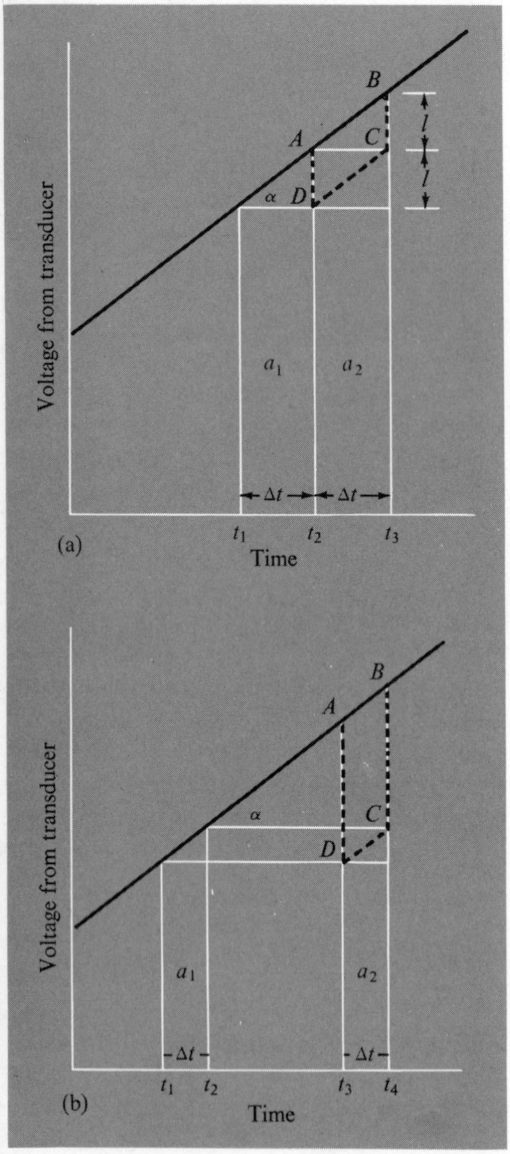

Fig. 2–43 Rate-of-change measurements using an integration method. (a) Typical rate curve showing measurement and subtraction of two sequential areas A_1 and A_2. (b) Measurement and subtraction of two areas a_1 and a_2 separated by time interval $t_3 - t_2$.

The slope S is defined as

$$S = \tan \alpha = \frac{l}{\Delta t}. \tag{2-52}$$

Substitution of Eq. (2–51) into Eq. (2–52) provides an expression relating the slope to the difference in area:

$$S = \frac{\Delta a}{(\Delta t)^2}. \tag{2-53}$$

Since Δt is a constant, the slope is directly proportional to the measured difference Δa. In Fig. 2–43b the two time increments are not consecutive but are separated by a time interval $t_3 - t_2$. In this case Eq. (2–53) can be rewritten as

$$S = \frac{\Delta a}{(t_3 - t_2)\Delta t + (\Delta t)^2}. \tag{2-54}$$

Since $(t_3 - t_2)$ is also kept constant, the slope remains directly proportional to the measured area. The total time for the measurement is $2\Delta t + (t_3 - t_2)$, and can vary from microseconds to seconds or more, depending on the aplication, specific circuits used, and type of noise problems.

Implementation with an Analog Integration System. The integration and subtraction operations illustrated in Fig. 2–43b can be implemented with the relatively simple analog circuit shown in Fig. 2–44a. This circuit uses two operational amplifiers, which are connected to one another and to a signal modifier by a switch network. During the first integration period, the input signal is directed through switches 1 and 2 to the integrator (OA2). If the signal modifier output is as illustrated in Fig. 2–44b, then line BC represents the signal applied to the integrator, which charges capacitor C, causing the output of OA2 to rise to a voltage v_1 as shown in Fig. 2–44c. During the second integration period the signal from the modifier is inverted by the gain-of-one inverter (OA1) and connected to the integrator by switch 3. The voltage represented by line DE in Fig. 2–44b is ap-

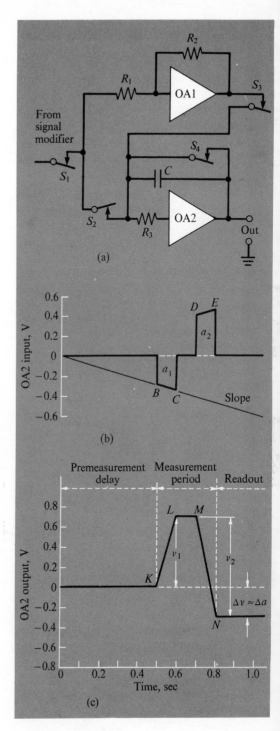

(a)

(b)

(c)

Fig. 2–44 Implementation of integrating rate meter. (a) Integration and subtraction circuit. (b) Rate signal and OA2 input. (c) Output signal of OA2.

plied to the integrator input, which discharges capacitor C, and the integrator output decreases in magnitude by the amount v_2 shown in Fig. 2–44c. The voltage difference Δv, which is read out at the end of the measurement period, is proportional to the difference in area, Δa, and is therefore proportional to the initial slope according to Eq. (2–54).

The switches S_1–S_4 can be activated with pulses generated by a logic circuit. Thus, upon receipt of a suitable trigger pulse, the logic circuit controls the start of the sequence of delay time, the length of the integration periods, the subtraction, and the readout of the result. The integration period is selected to be as long as feasible so as to average the noise most effectively. The shortest practical integration period is governed by the switching and OA response times and at present is about 0.01 to 0.1 msec, as will be discussed under high speed analog switches in Section 2–3.

Logarithmic Converters Using a Capacitor Discharge Technique

There are many measurement and control applications where it is important that the output signal be proportional to the logarithm of an input signal or proportional to the log ratio of two input signals. In spectrometric applications the logarithm of the intensity of a reference light beam P_r relative to an unknown sample light beam P_s [that is, $\log (P_r/P_s)$] is directly proportional to the desired information (concentration, photographic exposure, etc.).

Fig. 2–45 Logarithmic converter using capacitor discharge method. (a) Block diagram of conversion system. (b) Discharge curve showing times at which inputs from log generator are equal to reference and sample voltages.

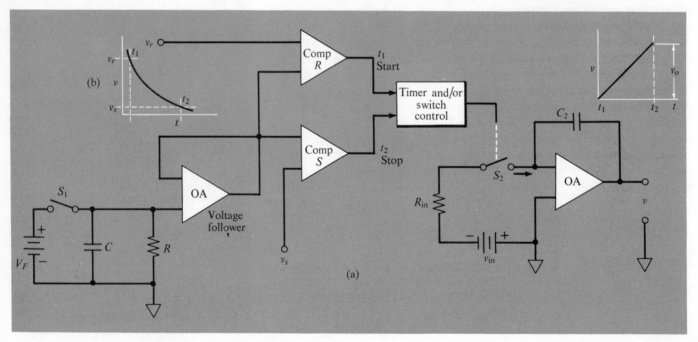

One method that has been used to generate an output signal that is proportional to the log ratio of two input signals is illustrated in Fig. 2–45. It encompasses several of the principles previously described in this section. The technique is as follows. A capacitor C in the log generator circuit is charged, by closing switch S_1 in Fig. 2–45, to a fixed stable voltage V_F that is always greater than the input signal voltages that contain the measurement information. The output of the log generator is fed to one input of both the reference comparator R and the sample comparator S. The input reference voltage, v_r, is fed to the other input of comparator R, and the input sample voltage, v_s, is fed to the other input of comparator S.

When switch S_1 is opened, the capacitor C discharges through R to provide a logarithmic discharge curve. The voltage follower in the log generator unloads the RC network and provides a low output resistance. When the voltage of the log generator decreases to a value equal to v_r at time t_1 (as shown in Fig. 2–45b), the comparator R rapidly changes its output level, starting a readout timer and/or closing switch S_2 at the input of a readout integrator. The capacitor in the log generator continues to discharge until the output is equal to v_s at time t_2. The rapid change in the output level of comparator S stops the timer and/or opens switch S_2. The readout can be from the timer, the integrator output, or both. If the integrator readout is used, switch S_2 closes at time t_1, and the capacitor C_2 is charged by the integrator input current $i_{\text{in}} = v_{\text{in}}/R_{\text{in}}$. At time t_2, switch S_2 opens and the output voltage of the integrator remains constant, so that

$$v_o = \frac{q}{C_2} = \frac{i_{\text{in}}(t_2 - t_1)}{C_2} = k_1(t_2 - t_1). \tag{2-55}$$

We can readily see how this output voltage v_o or the time interval $t_2 - t_1$ is proportional to $\log (v_r/v_s)$ by referring to Fig. 2–45b and considering the following basic equations. The voltage v_C across the capacitor C at any time t is

$$v_C = V_F \, e^{-t/RC}. \tag{2-56}$$

At time t_1 the voltage $v_C = v_r$, so that

$$v_r = V_F \, e^{-t_1/RC}. \tag{2-57}$$

At time t_2, the voltage $v_C = v_s$, so that

$$v_s = V_F \, e^{-t_2/RC}. \tag{2-58}$$

Therefore,

$$t_1 = RC \ln (V_F/v_r) \tag{2-59}$$

and

$$t_2 = RC \ln (V_F/v_s). \tag{2-60}$$

By subtracting Eq. (2–59) from Eq. (2–60) we obtain

$$t_2 - t_1 = RC \ln (v_r/v_s) = k_2 \log (v_r/v_s), \tag{2-61}$$

and substituting Eq. (2–61) into Eq. (2–55) gives the voltage output

$$v_o = k \log (v_r/v_s). \tag{2-62}$$

More recent logarithmic converters utilize the exponential current-voltage transfer functions of certain devices in conjunction with operational amplifiers to provide direct log readout. These converters are described in Module 3, *Digital and Analog Data Conversions*.

Experiment 2–12 *A Logarithmic Converter Using a Capacitor Discharge Method*

It is frequently necessary for a readout device to display linearly the logarithm or log ratio of voltage or current signals. In this experiment the log ratio of two voltage signals is determined by using a capacitor discharge method. The time difference involved in discharging a capacitor from a fixed reference voltage V_F to the two sample signals is measured and shown to be proportional to the log ratio. The precision and limitations of the method are determined.

2–2.4 ELECTRICAL INDUCTANCE

Inductance is that property of a device which resists a change in current through the device. All devices have electrical inductance associated with them to some extent and tend to impede or react against changes in current. Because of this impedance to changes in current, inductance is often an undesirable property in circuits, and it is important to understand the basic principles of inductance in order to minimize its effects in circuits where inductance is unwanted. On the other hand, some devices called **inductors** are purposely introduced in circuits to resist changes in current and thus to serve important control functions. The physical principles underlying the property of inductance are considered in this section, and several basic relationships are introduced. Practical inductors and their characteristics are described, and the effects of inductance in circuits in which a step change in voltage is applied are discussed. The important effects of inductance in alternating current circuits are considered in detail in Module 3.

Principles of Inductance

The basic principle of electromagnetic induction was introduced in conjunction with the generation of voltage in Section 1–3 of Module 1, *Electronic Analog Measurements and Transducers*. Faraday's law of electromagnetic induction is one of the basic laws of electromagnetism and will be reviewed briefly here in order to introduce the concept of **self-inductance** in devices.

In Fig. 2–46a, a magnet is shown moving toward a loop of wire which is connected to a current meter. Ordinarily no current would be expected because the electrical circuit (loop and current meter) contains no voltage source. However, when the magnet is moving, the meter deflects and a current exists in the circuit. When the magnet is at rest with respect to the coil, no current exists in the coil, and the meter shows no deflection. If the poles of the magnet are reversed so that the south pole is moved toward the coil, the meter deflects in the opposite direction. The current that appears in the circuit of Fig. 2–46a is called an **induced current** and is set up by an **induced electromotive force** (emf). In practice it makes no difference whether the magnet is moved toward the coil or vice-versa. Only the relative motion of the magnet and the coil is important.

The presence of an induced emf in a coil of wire can be demonstrated without the presence of a permanent magnet, as in Fig. 2–46b. Here two coils are placed near each other but at rest with respect to each other. If switch S is closed in the right-hand circuit containing the coil, the meter in the left-hand circuit momentarily deflects. When the current in the right-hand coil is steady, no current is measured by the meter. Again when the switch at the right is opened, an opposite, momentary deflection of the meter is observed. These experiments show that the emf in the circuit at the left is induced whenever the current in the other circuit changes. The magnitude of the induced emf depends on the rate of change of current and not on the magnitude of the steady current.

The important parameter that is common to the coils shown in Figs. 2–46a and b is the magnetic flux ϕ of the coil in which the induced emf appears. It makes no difference whether the flux is established by a permanent magnet as in Fig. 2–46a, or by a current loop as in Fig. 2–46b. The magnitude of the induced emf is given by Faraday's law of induction, as shown in Eq. (2–63):

$$v = -\frac{d\phi}{dt}. \qquad (2\text{–}63)$$

Here the rate of change of flux is the parameter which determines the induced emf. The negative sign indicates the sign of the induced emf. If the rate of flux change is expressed in webers/sec, the emf v is in volts. For a

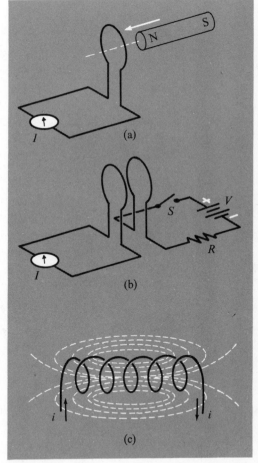

Fig. 2–46 Electromagnetic induction: (a) by a moving magnet; (b) by changing current in a nearby wire; (c) by nearby turns of the same coil.

coil of N turns, the induced emf is proportional to the number of turns, as shown in Eq. (2–64):

$$v = -\frac{N\,d\phi}{dt} = -\frac{d(N\phi)}{dt}. \qquad (2\text{–}64)$$

Here the quantity $N\phi$ is the number of **flux linkages** of the coil.

The presence of two separate coils is not necessary for the inductive effect to appear. If the current through a single coil is changed, an induced emf appears in that same coil. This effect is called **self-inductance,** and the self-induced emf follows Faraday's law (Eqs. 2–63 and 2–64). A practical inductor may be simply a multiple-turn coil as shown in Fig. 2–46c. When current passes through the coil, the magnetic flux established in each turn is approximately the same, and the induced emf is given by Eq. (2–64), which shows the importance of the number of flux linkages, $N\phi$, in determining the induced emf.

In the absence of magnetic materials, such as iron, the number of flux linkages in a given coil is directly proportional to the current, as shown in Eq. (2–65):

$$N\phi = Li, \qquad (2\text{–}65)$$

where the proportionality constant L is called the **inductance** of the coil. The relationship between induced emf and the current in the coil can readily be found from substitution for $N\phi$ in Faraday's law:

$$v = -\frac{N\,d\phi}{dt} = -L\frac{di}{dt}. \qquad (2\text{–}66)$$

The self-induced emf is often called a **counter emf** because it is of opposite polarity to those changes in voltage applied to the coil which cause the changing current di/dt, as the minus sign in Eq. (2–66) indicates. As Eq. (2–66) also indicates, only a changing current causes such a self-induced emf. The counter emf always acts to oppose the current change.

Inductance L has the unit volt-second/ampere, which is given a special name, the **henry** (H). An inductance of 1 henry will induce a counter emf of 1 V when the current is changing at the rate of 1 A/sec. The units millihenry (mH) and microhenry (μH) are also used for many practical inductors.

The inductor reacts against a change in current through itself by storing electrical energy in the magnetic field. Work is required to increase the magnetic field; therefore, *the current in an inductive circuit cannot change instantaneously.* Equation (2–66) shows that an attempt to change the current abruptly will be opposed by an infinite emf induced in the coil. If the applied potential in an inductive circuit decreases, the energy stored in the magnetic field returns to the circuit in the form of an emf which tends to maintain the previous current level. In a perfect inductor, there is no opposition to a steady current (dc) because $di/dt = 0$.

Practical inductors and their characteristics are described below, and the effects of inductive circuits on various signals are discussed in Section 2–2.5 and in Module 3.

Practical Inductors

Inductors are never pure inductances because there is always some resistance in the coil windings and some capacitance between the coil windings. When choosing an inductor (occasionally called a choke) for a specific application, it is necessary to consider the value of the inductance, the dc resistance of the coil, the current-carrying capacity of the coil windings, the breakdown voltage between the coil and the frame, and the frequency range in which the coil is designed to operate.

Inductors are available with values ranging from several hundred henrys down to a few microhenrys. To obtain a very high inductance it is necessary to have a coil of many turns. The inductance can be further increased by winding the coil on a closed-loop iron core or ferrite[4] core. To obtain as pure an inductance as possible, it is desirable to reduce the dc resistance of the windings to a minimum. This can be done by increasing the wire size, which, of course, will increase the size of the choke. The size of the wire also determines the current-handling capacity of the choke. As was pointed out earlier, the work done in forcing a current through a resistance is converted to heat in the resistance. Magnetic losses in an iron core also account for some heating, and this heating restricts any choke to a certain safe operating current. The windings of the coil must be insulated from the frame as well as from each other. Heavier insulation, which necessarily makes the choke more bulky, is used in applications where a high voltage will be present between the frame and the winding. The losses sustained in the iron core increase as the frequency increases. They become so large at about 15 kHz that the iron core must be abandoned. This results in coils of reduced coupling efficiency, but fortunately very large inductance values are not used at high frequencies. The iron-core chokes are restricted to low-frequency applications, but ferrite-core chokes are good at high frequencies. Several practical inductors are illustrated in Fig. 2–47.

The use of precision inductors is rather uncommon. However, they are available and restrictions with respect to frequency, temperature, etc., must be taken into account.

Variable Inductors. In a variable inductor the magnetic coupling between the windings of the coil or the effective number of turns is varied in order to change the total inductance value. The magnetic coupling can be changed

Note 4. Ferrite.

Ferrite is a powdered, compressed, and sintered magnetic material consisting mainly of ferric oxide combined with one or more other metal oxides. It has high resistivity, which makes the eddy-current losses very low at high frequencies. It is used in magnetic core memories because it can be magnetized and demagnetized very rapidly.

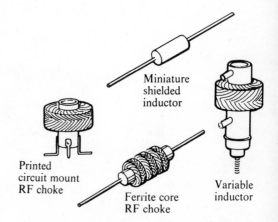

Fig. 2–47 Practical inductors.

by varying the orientation of part of the coil winding, or the effective number of turns can be varied by positioning a silver-plated brass core.

Another type of variable choke is called the swinging choke. The inductance of this choke increases with decreasing current. This type of choke is used to advantage in certain power supply circuits.

Large inductors, rated in henries (H), are used principally in power applications. The frequency in these circuits is relatively low, generally 60 Hz or low multiples thereof. In high frequency circuits, such as those found in FM radios and television sets, very small inductors are frequently used. These frequencies are generally in the 10 to 1000 MHz region and the chokes are of the order of microhenries (μH). In radio-frequency applications (broadcast band), chokes are frequently rated in millihenries (mH).

Inductive Transducers

Various sources of chemical, thermal, mechanical-displacement, or velocity input information can cause changes in inductive devices which can provide readily measured electrical signals. Inductive transducers are based on the relationship from Eq. (2–66),

$$L = N(d\phi/di), \tag{2–67}$$

where L is the inductance in henries, N is the number of turns cut by the flux linkages, and ($d\phi/di$) is the rate of change of flux with current.

The rate of change of flux with current can be changed by varying the length of flux path, the cross-sectional area, or the permeability. A physical movement of the core, a change in the distance between two or more coils, a change in the chemical composition of a portion of the flux path, or a stress on ferromagnetic material in the path, all can cause changes of inductance that can be measured.

2–2.5 *LR* Circuits

It was shown in the introduction to this section that an inductor reacts against—tends to impede—a change in current, and that it is impossible to change the current value instantaneously in a circuit that has inductance. In this respect inductive circuits are analogous to capacitive circuits, in which the voltage cannot be changed instantly. In each case, there will be a time lag between the attempt to change the voltage or current level in the circuit and the attainment of that change. The time of response of *LR* circuits can be analyzed in a way very similar to the method used for *RC* circuit response in Section 2–2.2. This is illustrated for the step signal response of a series *RL* combination below.

Series *LR* Response to Step Signals

A series *LR* circuit with a test step potential source is shown in Fig. 2–48a. When the switch is closed, $v_R = iR$ and $v_L = L\,di/dt$. From Kirchhoff's Law,

$$V = iR + L\frac{di}{dt}. \tag{2–68}$$

Solving this differential equation yields the following:

$$v_L = \overline{V}e^{-tR/L}, \tag{2–69}$$

$$v_R = V(1 - e^{-tR/L}). \tag{2–70}$$

The quantity L/R in these equations is the time constant for *LR* circuits. The curves for v_L and v_R as a function of L/R units are shown in Fig. 2–48b. The voltage v_L approaches zero and v_R approaches the steady-state value V. The curves of Fig. 2–48 verify that the *LR* circuit is an approximate differentiator when the output is taken across L and an integrator when the output signal is v_R.

The storage of energy in the magnetic field in an inductor can readily be seen from Eq. (2–68) for the *LR* circuit. If both sides of the equation are multiplied by i, the rate of doing work, Vi, is obtained:

$$Vi = i^2 R + Li\frac{di}{dt}. \tag{2–71}$$

The rate at which energy is delivered to the circuit, Vi, is seen to be equal to the sum of two terms. The i^2R term is the power dissipated in the resistor or the rate at which energy appears as heat in the resistor. Energy that does not appear as heat is stored in the magnetic field, and the rate of energy storage is given by the last term in Eq. (2–71). If the energy in the field is called U_B, the rate at which energy is stored may be written

$$\frac{dU_B}{dt} = Li\frac{di}{dt}. \tag{2–72}$$

If Eq. (2–72) is integrated, the total stored magnetic energy in an inductor of inductance L, which carries a current i, is given by

$$U_B = \tfrac{1}{2}Li^2. \tag{2–73}$$

This equation can be directly compared to the energy U_E stored in a capacitor which has a charge q, as given earlier in Eq. (2–28) as the work W_d available on discharge:

$$U_E = W_d = \frac{1}{2}\frac{q^2}{C}.$$

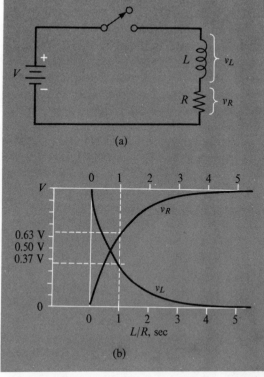

(a)

(b)

Fig. 2–48 *LR* circuits: (a) with step potential source; (b) voltage-time curves after switch closure.

With a capacitor the energy is stored in the electric field, whereas with an inductor the energy is stored in the magnetic field.

Experiment 2–13 *Response Characteristics of an LR Network*

Several different input waveform signals are applied to a series *LR* circuit and the resulting output waveforms are observed. These output signals are then compared with those from an *RC* network.

PROBLEMS

1. A 0.001 µF capacitor was charged to a voltage of 8.32 V. Calculate the work done in completely discharging the capacitor.

2. In the capacitive network of Fig. 2–31, $V = 5$ V, $C_1 = 0.1$ µF, $C_2 = 0.05$ µF, and $C_3 = 0.02$ µF. Assume that capacitor C_4 has been disconnected. Calculate the charge on each capacitor, the total charge q_T, and the equivalent parallel capacitance C_p.

3. The same capacitors and voltage source as in problem 2 are rewired in a series circuit like that of Fig. 2–32. Calculate the charge on each capacitor and the equivalent series capacitance, C_s.

4. The complex capacitive network shown in Fig. P-4 is wired. Calculate the equivalent capacitance of the network, the voltage across each capacitor, and the charge on each capacitor.

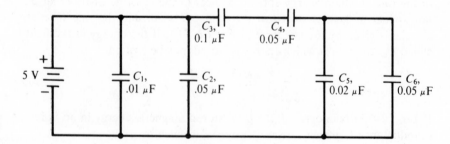

Fig. P–4

5. In the circuit of Fig. 2–34, $V = 1.0$ V, $R = 100$ kΩ, and $C = 0.1$ µF. Calculate the voltage across the capacitor, v_c, and the voltage across the resistor, v_R, at 0.5 msec, 1 msec, 5 msec, 10 msec, 50 msec, and 100 msec after switch S has been turned to position A.

6. In the circuit of Fig. 2–37, $V = 10.0$ V, $R_1 = 10$ kΩ, $R_2 = 40$ kΩ, and $C = 0.22$ µF. (a) How long after switch S is turned to the ON position does it take for capacitor C to become fully charged for practical pur-

poses? (b) If $R_3 = 10 \, \Omega$, how long, for practical purposes, does it take for C to discharge?

7. (a) A 10 V rectangular pulse with a pulse width of 1.0 msec is applied to a series RC circuit at time t_0. If $R = 100 \, \text{k}\Omega$ and $C = 1000$ pF, calculate the voltage across R and C, v_R and v_C respectively, at t_0, $t_0 + 0.1$ msec, $t_0 + 0.3$ msec, $t_0 + 0.5$ msec, $t_0 + 1$ msec, $t_0 + 1.3$ msec, and $t_0 + 1.5$ msec. (b) Repeat part (a) for $R = 1 \, \text{M}\Omega$ and for times t_0, $t_0 + 0.5$ msec, $t_0 + 1.0$ msec, and $t_0 + 1.5$ msec.

8. A constant voltage of 5.0 V is applied to the OA integrator of Fig. 2–39. If $R = 5 \, \text{M}\Omega$ and $C = 0.5 \, \mu\text{F}$, how long will it take for the output voltage to decrease by 1.0 V?

9. Design an OA current integrator with components selected so that a 5.0 V output is equivalent to 1.000 microequivalents of charge (1 μeq = 96,487 μC).

10. The OA of Fig. 2–39 has an output voltage limit of ± 12 V. The input resistor $R = 100 \, \text{k}\Omega$, $C = 0.1 \, \mu\text{F}$, and the inverting $(-)$ input has an offset voltage of $+50 \, \mu$V with respect to the circuit common. The capacitor is initially shorted with a switch of 1 Ω resistance. (a) What is the initial output voltage when the switch is closed? (b) How long will it take the OA to reach its voltage limit after the switch is opened if $v_{\text{in}} = 0.00$ V? (c) What average input voltage is required to give an output integral accurate to 1%? (d) How long can the integration of the input voltage found in part (c) proceed before the OA reaches its output voltage limit? (e) What is the maximum tolerable offset at the inverting input for a 1000 sec integration accurate to 1% if the output voltage at 1000 sec is to be 10 V?

11. (a) Design an OA differentiator which will give an output of -5.0 V for an input rate of charge of $+8$ V/sec. (b) Design a differentiator to give an output of $+1.0$ V for an input rate of charge of $+15$ V/sec.

12. In the logarithmic converter of Fig. 2–45, $V_F = 1.0$ V, $R = 1 \, \text{M}\Omega$, and $C = 0.5 \, \mu\text{F}$. The readout integrator has $v_{\text{in}} = 500$ mV, $R_{\text{in}} = 100 \, \text{k}\Omega$, and $C_2 = 1.0 \, \mu\text{F}$. What will the readout voltages of the integrator be for the following values of input signals v_r and v_s? (a) $v_r = 0.5$ V, $v_s = 0.05$ V. (b) $v_r = 0.5$ V, $v_s = 0.1$ V. (c) $v_r = 0.5$ V, $v_s = 0.25$ V.

13. An unknown inductor produces an induced voltage of 2 V when the current through it is changing at 15 A/sec. What is the inductance?

14. What is the induced voltage produced by a 300 mH choke if the current through it is changed at a rate of 5 A/sec?

15. In the series LR circuit of Fig. 2–48a, $L = 200$ mH, $R = 100 \, \Omega$, and $V = 1.5$ V. Calculate the voltage across the inductor, v_L, and that across the resistor, v_R, at 0.5 msec, 1 msec, 2 msec, 10 msec, and 100 msec after the switch has been closed.

Electronic Switching   Section 2–3

One of the most familiar devices used for the control of electrical signals or power is the switch. A switch is a device with two states, conducting and nonconducting, which is used to ensure that a particular circuit is either complete or open. Since the control of electrical quantities is prerequisite to their application, significant developments in control devices have opened up possibilities for new tools and opportunities for the scientist and instrument designer alike. The recent (and continuing) development of inexpensive semiconductor switches capable of changing states of conductance in the submicrosecond time scale is a major factor in the current computer and instrument revolution.

A switch is composed of two conductors which are joined when the circuit is complete and separated by an insulator when the circuit is open. Because the junction of conductors is basic to all switches, considerable insight into the operation and limitations of practical switches can be gained by first considering the nature of the junctions between conductors, their characteristics, and their mechanism of conduction. After this introduction, the large variety of modern switching devices can then be categorized and discussed according to the type, arrangement, and number of junctions employed. The later portions of this section are devoted to an introduction to the applications of switches in the control of information transmission, in waveform generation, and in the control of ac power.

2–3.1 MECHANISMS OF CONDUCTION AT JUNCTIONS

A conductor has been shown to be a material that has a significant concentration of charge carriers. When two dissimilar conducting materials are joined, the junction between them is not necessarily a conductor. A nonconducting junction occurs when the charge carriers of the two materials interact in such a way as to deplete the concentration of charge carriers in the junction region, or when there is no mechanism for the transport of charge carriers from either material across the junction. On the other hand, when the junction region is populated with charge carriers that can readily move from one material to the other, the junction is conducting. The characteristics of junctions that always conduct are important, since such junctions provide all the electrical connections between components, devices, and instruments. Junctions that conduct under some conditions but not under others provide a useful function as switches which change conducting state in response to a change in the condition(s) controlling the conduction process. The following discussion begins with the metal-metal junction and then proceeds through the pn semiconductor junction, the metal-semiconductor junction, and the metal-gas/vacuum junction.

Metal-Metal Junctions

The energy of conduction electrons in a metal can be considered to be the combination of two types of energy, chemical and electrical. The chemical portion of the electron energy depends on several properties of the material, including the temperature, and is called the **chemical potential energy,** μ_e. The electrical portion of the electron energy results from the electrical potential of the material or electric fields within it and is equal to $-Q_e V$, where Q_e is the electron charge and V is the electrical potential of the material with respect to some arbitrary reference. The combination of chemical potential energy and electrical energy is called the **electrochemical potential energy,** $\bar{\mu}_e$, so that

$$\bar{\mu}_e = \mu_e - Q_e V. \qquad (2\text{--}74)$$

The minus sign appears in Eq. (2–74) because electrons are negatively charged. For positively charged carriers the sign in Eq. (2–74) is $+$.

Contact Potential. When two metals A and B are joined as in Fig. 2–49, conduction electrons pass from one to the other until the electrochemical potential energy of the electrons in both is equal, that is, $\bar{\mu}_e^A = \bar{\mu}_e^B$. Expressing the electrochemical potential energy of the electrons in each material as the sum of the chemical and electrical contributions, we obtain

$$\mu_e^A - Q_e V_A = \mu_e^B - Q_e V_B,$$

from which the difference in potential between A and B can be obtained:

$$V_A - V_B = V_{AB} = \frac{1}{Q_e} (\mu_e^A - \mu_e^B). \qquad (2\text{--}75)$$

From Eq. (2–75) we can see that if metals A and B have the same com-

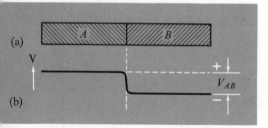

(a)

(b)

Fig. 2–49   (a) Junction of dissimilar metals. (b) Potential profile for $\mu_e^A > \mu_e^B$.

position and temperature ($\mu_e^A = \mu_e^B$), the potential difference between them is zero. On the other hand, if A and B are dissimilar, electrons will cross the junction until a potential difference is developed which depends on (and offsets) the chemical potential difference between the metals. This potential difference, *which occurs at all junctions between dissimilar conductors,* is called the **contact potential.** The contact potential is characteristic of the two materials at any given junction temperature.

An attempt to measure the contact potential is illustrated in Fig. 2–50. The voltmeter leads of material C are connected to A and B respectively. Assume that the entire system is at the same temperature, so that the voltage measured will be simply the sum of the contact potentials for the three dissimilar metal junctions in the circuit:

$$V = V_{CA} + V_{AB} + V_{BC}. \tag{2–76}$$

If we substitute into Eq. (2–78) the equivalent of Eq. (2–75) for each junction,

$$V = \frac{1}{Q_e}(\mu_e^C - \mu_e^A) + \frac{1}{Q_e}(\mu_e^A - \mu_e^B) + \frac{1}{Q_e}(\mu_e^B - \mu_e^C) = 0. \tag{2–77}$$

Thus the contact potentials are shown to cancel exactly. This is reasonable if one considers that the electron energy is constant throughout the system and hence an electron will experience no change in energy as it moves from C on one side of the voltmeter through A and B to C on the other side. Thus the contact potentials in an isothermal (constant temperature) system will always cancel regardless of the number of dissimilar metal junctions in the circuit. The potentials generated in nonisothermal circuits are discussed later in this section.

The contact potential difference between any two metals in electronic equilibrium can be estimated by considering that no energy is lost or gained by moving an electron from A to B across the junction, and so similarly there must be no energy gained or lost by taking an electron from A to a place far from either metal and then placing it in B. The energy required to remove an electron from A is called the **work function,** $-Q_e\Phi_A$, where Φ_A is called the **Richardson potential** for the material. The energy required to remove the electron from the electric field of A is $-Q_eV_A$. Adding the energies required for each step of the process and setting the sum equal to zero, we obtain

$$(-Q_e\Phi_A + Q_eV_A) - (-Q_e\Phi_B + Q_eV_B) = 0,$$

from which we get

$$V_A - V_B = (\Phi_A - \Phi_B). \tag{2–78}$$

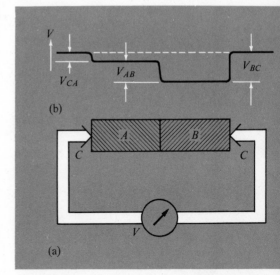

(b)

(a)

Fig. 2–50   (a) Attempt to measure contact potential between A and B. (b) Potential profile showing cancellation of contact potentials.

Equation (2–78) indicates that the contact potential between two metals is simply equal to the difference in their Richardson potentials. Since work function measurements are possible, but greatly affected by the material surface, the tabulated values of work functions should be taken as approximate values in most cases. However, since values of the Richardson potential are from 1 to 5 V for most metals, the contact potential between dissimilar metals is often several volts! Yet, because of the precise cancellation of contact potentials in a complete circuit, such junctions are used routinely to transmit accurately signals only microvolts in amplitude.

Thermal Potentials with Zero Current. When two portions of a conductor are at different temperatures, there will be a flow of heat from the hotter portion to the colder. One of the mechanisms for the transmission of heat energy is the motion of the mobile charge carriers. Thus the transport of heat and the motion of charge are closely related for most conducting materials. The thermoelectric phenomena which are based on the relations between heat and charge flow are significant in several areas of scientific measurements. It was shown above that there is a net flow of electrons under conditions of nonuniform electrochemical energy of conduction electrons. Now we wish to consider also the flow of electrons due to a nonuniform temperature. A general relationship for the total flux of electrons, J_e (particles/sec-cm²), is

$$J_e = -K_\mu \frac{d\bar{\mu}_e}{dx} - K_T \frac{dT}{dx}. \qquad (2-79)$$

The first term to the right of the equality sign indicates that the flux of electrons due to a nonuniform $\bar{\mu}_e$ is proportional to the change in $\bar{\mu}_e$ per increment of distance along the wire. The flux of electrons due to the nonuniform temperature is similarly proportional to the temperature gradient, dT/dx, as given by the right-hand term. The proportionality constants are specific for a given material at a given temperature.

Consider a conducting wire A which is at temperature T_1 at one end and T_2 at the other, as shown in Fig. 2–51. Since there is no electrical connection to the wire, there is no net current; that is, J_e in Eq. (2–79) is zero. Since there is a continuous heat flow through A, heat must be added to the warmer end and removed from the cooler end to maintain the temperatures T_2 and T_1. Setting Eq. (2–79) equal to zero, we obtain

$$\frac{K_T}{K_\mu} = \frac{-d\bar{\mu}}{dx} \cdot \frac{dx}{dT} = \frac{-d\bar{\mu}}{dT}. \qquad (2-80)$$

The ratio of proportionality constants K_T/K_μ has physical significance as the relative response of the electron to gradients of temperature and electro-

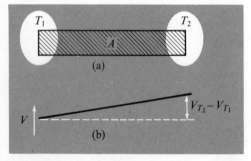

Fig. 2–51 (a) Conductor with a temperature gradient. (b) Potential profile for the conductor.

chemical energy, and has the units J C^{-1} deg^{-1}. This ratio is given the symbol S^*. Thus

$$S^* \equiv \frac{K_T}{K_\mu} \qquad (2\text{–}81)$$

and for $J_e = 0$ (the case of Eq. 2–80),

$$d\bar{\mu}_e = -S^* \, dT, \qquad (2\text{–}82)$$

which tells us that a temperature change will result in a change in $\bar{\mu}_e$ in proportion to the value of S^* for the material at that temperature. Equation (2–82) can be integrated to find the change in $\bar{\mu}_e$ for A between T_1 and T_2:

$$\bar{\mu}_{eT_2} - \bar{\mu}_{eT_1} = -\int_{T_1}^{T_2} S^* \, dT.$$

The $\bar{\mu}_e$ terms can be separated into their chemical and electrical parts from Eq. (2–74):

$$\mu_{eT_2} - Q_e V_{T_2} - (\mu_{eT_1} - Q_e V_{T_1}) = -\int_{T_1}^{T_2} S^* \, dT.$$

This equation can be solved for the potential difference $V_{T_2} - V_{T_1}$:

$$V_{T_2} - V_{T_1} = \frac{1}{Q_e} (\mu_{eT_2} - \mu_{eT_1}) + \frac{1}{Q_e} \int_{T_1}^{T_2} S^* \, dT. \qquad (2\text{–}83)$$

Thus we see that a potential is developed across a wire as a result of the temperature difference. This potential is due partly to the difference in the chemical potential of the electron at the two temperatures (the first term to the right of the equal sign) and due partly to the average value of S^* in the temperature range from T_1 to T_2 (right term of Eq. 2–83).

An arrangement such as that shown in Fig. 2–52 might be used to attempt to measure the thermally generated potential in A; B and b are the voltmeter leads of identical composition but having different temperature gradients. The voltmeter is at a temperature T_R, which may be equal to T_1 or T_2 in some situations. The total measured voltage will be the sum of the potentials generated in the conductors by the thermal gradients and the contact potentials between A and the voltmeter leads. The latter might not cancel in this case, since the junctions are not at the same temperature:

$$V = (V_{bT_R} - V_{bT_1}) + V_{bA} + (V_{AT_1} - V_{AT_2}) + V_{AB} + (V_{BT_2} - V_{BT_R}). \qquad (2\text{–}84)$$

If we substitute the equivalent of Eqs. (2–83) and (2–75) for each tem-

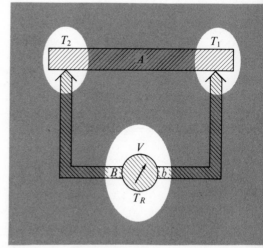

Fig. 2–52 Measurement of thermally generated potential.

perature gradient and junction into Eq. (2–84), we obtain

$$V = + \frac{1}{Q_e} (\mu^b_{eT_R} - \mu^b_{T_1}) + \frac{1}{Q_e} \int_{T_R}^{T_1} S^*_b \, dT + \frac{1}{Q_e} (\mu^b_{eT_1} - \mu^A_{eT_1})$$

$$+ \frac{1}{Q_e} (\mu^A_{eT_1} - \mu^A_{eT_2}) + \frac{1}{Q_e} \int_{T_1}^{T_2} S^*_A \, dT + \frac{1}{Q_e} (\mu^A_{eT_2} - \mu^B_{eT_2})$$

$$+ \frac{1}{Q_e} (\mu^B_{eT_2} - \mu^B_{eT_R}) + \frac{1}{Q_e} \int_{T_2}^{T_R} S^*_B \, dT,$$

in which we find that all the terms in chemical potential cancel ($\mu^b_{eT_R} = \mu^B_{eT_R}$ because it is the same material at the same temperature):

$$V = \frac{1}{Q_e} \int_{T_R}^{T_1} S^*_b \, dT + \frac{1}{Q_e} \int_{T_1}^{T_2} S^*_A \, dT + \frac{1}{Q_e} \int_{T_2}^{T_R} S^*_B \, dT$$

and $S^*_b = S^*_B$ at any given temperature, so

$$V = \frac{1}{Q_e} \int_{T_2}^{T_1} S^*_B \, dT + \frac{1}{Q_e} \int_{T_1}^{T_2} S^*_A \, dT = \frac{1}{Q_e} \int_{T_1}^{T_2} (S^*_A - S^*_B) \, dT.$$

$$(2\text{--}85)$$

We see that the portion of the thermally generated potential which is due to the change in chemical potential with temperature within each material exactly cancels the change in contact potential between materials with temperature. Thus the potential developed when dissimilar conductor junctions are at different potentials (the **Seebeck effect**) is entirely due to the difference in the value of S^* between the two materials.

Of particular interest is the amount of change in the thermally generated voltage for a given change in temperature. Equation (2–85) can easily be solved for dV/dT:

$$\frac{dV}{dT} = \frac{1}{Q_e} (S^*_A - S^*_B) = S_A - S_B = S_{AB}. \qquad (2\text{--}86)$$

The quantity dV/dT is called the **thermopower** or **thermal sensitivity** of A relative to B and is given the symbol S_{AB}. Table 2–6 gives the thermopower of many materials relative to platinum.

The reader will recognize the thermally generated potential difference as being the principle of operation of the thermocouple used to measure temperature difference. It can also be the cause of spurious potentials which can seriously affect the accuracy of low-level voltage measurements. For measurement of temperature, one would choose the metals A and B to have as large a value for S_{AB} as possible. Also, the less S_{AB} varies with

Table 2–6 Thermopower $S = dV/dT$ of thermoelement made of materials listed against platinum (reference junction kept at a temperature of 0°C), µV/°C

Bismuth	−72	Silver	6.5
Constantan	−35	Copper	6.5
Nickel	−15	Gold	6.5
Potassium	− 9	Tungsten	7.5
Sodium	− 2	Cadmium	7.5
Platinum	0	Iron	18.5
Mercury	0.6	Nichrome	25
Carbon	3	Antimony	47
Aluminum	3.5	Germanium	300
Lead	4	Silicon	440
Tantalum	4.5	Tellurium	500
Rhodium	6	Selenium	900

From Kurt S. Lion, *Instrumentation in Scientific Research—Electrical Input Transducers,* McGraw-Hill, New York, N.Y., 1959.

temperature, the more linear the temperature-voltage transfer function will be. It is of interest that the potential developed by the temperature gradients in Fig. 2–52 is independent of the temperature T_R. This is because the potential caused by the temperature gradients $T_R - T_2$ and $T_1 - T_R$ cancel except for $T_1 - T_2$.

In the thermocouple circuit of Fig. 2–52 the voltmeter and its connecting wires are made of material B. It is frequently desirable to make the thermal junctions out of materials other than the material used for the voltmeter connections. This introduces additional dissimilar metal junctions. A typical arrangement is shown in Fig. 2–53. An analysis similar to that used with Fig. 2–52 shows that the potentials generated in the voltmeter leads C and c cancel (providing the wires are homogeneous) and the potential measured depends only on the relative thermopower of A and B and the temperatures T_1 and T_2. It is essential that the CA and Bc junctions be held at the same temperature, which becomes the reference temperature in the measurement of the AB junction temperature T_1.

It should be clear from the above discussion that to avoid thermally generated potentials, all wires should be homogeneous or free of thermal gradients, junctions between dissimilar metals should be avoided where possible, and when dissimilar metal junctions are unavoidable, the junctions should be held at a common temperature. The uniformity of composition of a wire is often tested by measuring the potential between the two ends of the wire with the wire ends held at the same temperature (as in

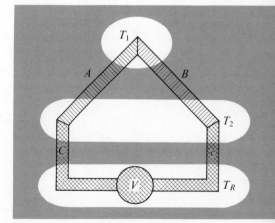

Fig. 2–53 Practical thermocouple measurement arrangement.

Fig. 2–52 with the wire at A and $T_1 = T_2$). Now, by means of a flask of very cold liquid, a cold spot is moved along the length of the wire. If the wire is homogeneous, the effects of the opposing thermal gradients on either side of the cold spot will exactly cancel. Experimenters using thermocouples for exact temperature measurements will test the wires used in the thermocouple circuits this way, rejecting all wire sections giving a measurable potential difference as the cold spot was passed through them.

Thermoelectric Effects with Current. It has been demonstrated above that the motion of charge carriers can be due to a temperature gradient as well as an electrochemical energy gradient. It is also then necessarily true that heat flow can be due to a gradient of the electrochemical energy of charge carriers as well as a gradient of temperature. If there is a current in a conductor, there must be an electrochemical potential gradient. This gradient, in turn, will result in a flow of heat in the wire (and a temperature gradient along it). This is referred to as the **Thomson effect.** The rate at which heat must be added to a section of a current-carrying wire to keep it at a constant temperature is then equal to the heat removed by the Thomson effect minus the heat generated by the energy lost in passing the current through the resistance of the wire (Joule heating). Thus

$$\Delta W = \tau i \, \Delta T - i^2 R, \tag{2–87}$$

where ΔW is the rate of heat addition required, τ is the Thomson coefficient, ΔT is the temperature change in the section of wire, and R is the resistance of the section of wire. The first term to the right of the equals sign represents the Thomson effect, the next the Joule heating. The Thomson coefficient is related to S^* for the material by the equation

$$\tau = \frac{T}{Q_e} \left(\frac{\partial S^*}{\partial T} \right) = T \left(\frac{\partial S}{\partial T} \right), \tag{2–88}$$

where $S = S^*/Q_e$. Thus we see that the Thomson coefficient is proportional to the change in S with temperature for the material. The units of τ are volts/deg. The Thomson coefficient can have either sign. Thus the Thomson effect may tend to heat or cool the wire, depending on the sign of τ *and on the direction of the current* (sign of i). Note that the Joule heating term warms the wire regardless of the sign of i. Thus the Thomson effect is reversible while the Joule heating is not.

Another heating and cooling effect that results from the motion of charge occurs at the junctions of dissimilar materials. It is now convenient to define a term Q^*_{AB} as the heat transported per particle as it traverses the junction from A to B. Q^*_{AB} is related to S^*_{AB} by the equation

$$Q^*_{AB} = TS^*_{AB} + (\bar{\mu}_e^A - \bar{\mu}_e^B), \tag{2–89}$$

where the difference in $\bar{\mu}_e$ between A and B is due to the iR voltage developed across the junction, or

$$Q^*_{AB} = TS^*_{AB} - Q_e iR_j.$$

Thus heat must be added to or taken from a junction between materials of different S^* for every electron passing through the junction if the junction is to remain at a constant temperature. This is called the **Peltier effect.** The rate of heat addition ΔW is obtained by multiplying Q^* by the rate of electrons crossing the junction, or i/Q_e. When this is done,

$$\Delta W = \frac{Q^*_{AB} i}{Q_e} = \frac{-iT}{Q_e} S^*_{AB} - i^2 R_j,$$

in which $S^*_{AB}/Q_e = S_{AB}$ and $-i^2 R_j$ is the irreversible joule heating in the junction. Now,

$$\Delta W = -iTS_{AB} - i^2 R_j, \qquad (2\text{--}90)$$

where $-TS_{AB}$ is called the **Peltier coefficient,** π_{AB}. In its final form, then,

$$\Delta W = i\pi_{AB} - i^2 R_j. \qquad (2\text{--}91)$$

The Peltier effect is reversible and π may have either sign; thus heating or cooling may occur at a junction with current. A simple circuit with two dissimilar materials would have two junctions, one AB and one BA as shown in Fig. 2–54. If π_{AB} is positive, then according to Eq. (2–90) heat must be added to the left-hand junction to keep it at a constant temperature (assuming $i\pi_{AB} > i^2 R_j$). Therefore, the left-hand junction will become cool. The current traverses the right-hand junction from B to A. Since $\pi_{BA} = -\pi_{AB}$, the right-hand junction will become warm. Thus whenever current is forced through a pair of dissimilar junctions, one becomes warmer and the other becomes cooler. This is the basis for **thermoelectric heating/cooling** devices. Since the Peltier effect is reversible, it also follows that if the cold junction were heated by an external source and the hot junction were cooled, a current in the direction shown would be generated. Stacks of such "thermocouple junctions with current" are used as relatively inefficient heat-to-electric-power converters called **thermoelectric generators.**

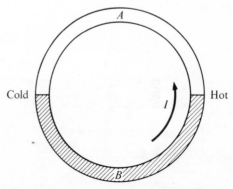

Fig. 2–54 Illustration of Peltier effect.

pn Semiconductor Junctions

A pn semiconductor junction is generally made by changing the dominant dopant from acceptor to donor type in a single crystal of semiconductor. The region where the majority charge carrier changes from electrons to holes is called the pn junction. The holes from the p region and the electrons from the n region are free to cross the junction. Since the chemical energy for electrons is higher in the n region than in the p region, some electrons will cross from the n to the p region. Similarly, some holes will cross to the n region as a result of the higher chemical energy for holes in the p region. The transport of + charges into the n region and − charges into the p region develops a potential difference between the n and p regions. This potential presents an energy barrier sufficient to offset the energy of the majority carriers on either side of the junction, so that at equilibrium there is no net flow of charge across the boundary. A contact potential is thus established with the n region positive with respect to the p region in a manner somewhat similar to the previously discussed metal-metal junction.

The drift of holes across the junction reduces the concentration of electrons in the junction region, since the product of the electron and hole concentrations must remain constant as given in Eq. (2–11). Similarly, the concentration of holes is lower at the junction than in the rest of the p region. The concentration profiles for holes and electrons are shown in Fig. 2–55b. In the example chosen, the dopant concentration is slightly less in the n region than in the p region. As shown by the concentration profiles, the total concentration of charge carriers is greatly depleted in the junction region, decreasing to the intrinsic value at its minimum. This region is called the **depletion region.** Thus the junction region is essentially nonconducting compared to the remainder of the n and p doped regions. The contact potential appears across the insulating junction region as shown in Fig. 2–55c. The magnitude of the contact potential is about 0.3 V for germanium pn junctions and about 0.6 V for silicon.

As shown above, the concentration of charge carriers in the depletion region is essentially that of the intrinsic semiconductor. The charge carriers in intrinsic semiconductor material are electron-hole pairs generated by thermal excitation. However, there is a strong electric field in the depletion region causing the generated holes to move to the p region and the electrons to move to the n region. The current due to the carriers in the intrinsic region is thus from right to left, as shown in Fig. 2–55d. The magnitude of the intrinsic carrier current, I_i, depends on the rate of carrier generation in the depletion region, which is a function of temperature only. The net flow of + charge from right to left due to the intrinsic carrier current would tend to reduce the contact potential across the junction. However, as the potential barrier is decreased, the number of majority carriers on each side with enough energy to cross the junction increases. The current due to majority carriers, I_m, is from left to right and increases

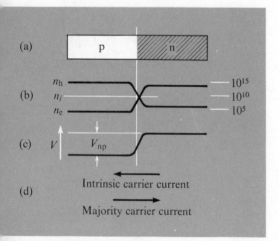

Fig. 2–55 (a) pn junction with (b) carrier concentration, (c) potential profiles, and (d) junction current directions.

exponentially with increasing temperature and decreasing junction potential as given in Eq. (2–92):

$$I_m = K e^{-Q_e V_{np}/kT}, \qquad (2\text{–}92)$$

where k is the Boltzmann constant J/°K and T is the temperature in °K. Since the net current across the junction must be zero, the intrinsic and majority currents must cancel exactly. For any given temperature, there is only one value of the junction potential for which this will be true. One can qualitatively see how the higher intrinsic carrier generation rate of germanium over silicon results in the lower junction potential for germanium.

The Biased pn Junction. A **bias** is an external voltage applied to the material with the pn junction. In order to apply an external voltage source, it is necessary to make metal contacts to the semiconductor. The resulting device is shown pictorially in Fig. 2–56a. At both metal-semiconductor junctions, a contact potential is established. As discussed in the previous sections for a series of metal-metal junctions, the sum of the contact potentials must be zero in any circuit with the same material at both extremes. The potential profile through the contacts and junction is shown in Fig. 2–56b. We will assume in this discussion that the metal-semiconductor contacts are good conductors for current in both directions. Metal-semiconductor junctions are discussed in greater detail in the next section.

Since the connecting wires, the contacts, and the doped semiconductors are all very much better conductors than the depletion region, essentially all of the applied bias voltage will appear as a change in the potential difference across the pn junction. This is shown by the family of potential profiles in Fig. 2–57b. The change in V_{np} will have a great effect on the majority carrier current I_m as given by Eq. (2–92), but will have little effect on the intrinsic current I_i. At zero applied bias, $I_m = I_i$, so

$$I_i = I_m^0 = K e^{-Q_e V_{np}^0/kT}, \qquad (2\text{–}93)$$

where I_m^0 and V_{np}^0 are the values for the majority current and junction potential respectively at zero bias. At nonzero bias, the junction potential is $V_{np}^0 - V$ and a net forward current, I, can cross the junction. The net current is equal to the difference between I_m and I_i. Thus

$$I = I_m - I_i = K e^{-Q_e(V_{np}^0 - V)/kT} - I_i. \qquad (2\text{–}94)$$

If I_i from Eq. (2–93) is substituted in Eq. (2–94) and the equation is rearranged, we obtain the Shockley equation,

$$I = I_i(e^{Q_e V/kT} - 1). \qquad (2\text{–}95)$$

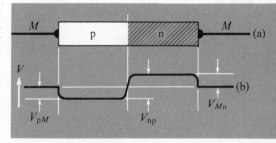

Fig. 2–56 pn junction with metal contacts and potential profile.

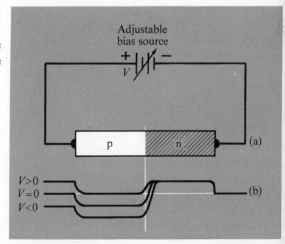

Fig. 2–57 (a) Biased pn junction. (b) Potential profiles showing the effect of bias voltage on the pn junction potential.

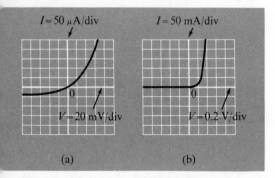

$I = 50\ \mu A/div$ $I = 50\ mA/div$

0 0

$V = 20\ mV/div$ $V = 0.2\ V/div$

(a)   (b)

Fig. 2–58 Current-voltage characteristic of a germanium pn junction: (a) near zero bias; (b) complete forward characteristic.

A plot of the current I through the junction vs. the applied bias voltage V is shown in Fig. 2–58 for a germanium pn junction. The exponential rise in current for increasing positive values of bias voltage is clearly seen in Fig. 2–58a. For many germanium pn junctions the observed I-V curve fits the Shockley equation very well. However, to fit the observed behavior for silicon diodes, it is necessary to add a factor η in the denominator of the exponential. This has been ascribed to charge carrier recombination in the depletion region. In Fig. 2–58b, which shows a larger range of the I-V behavior, it is seen that the forward bias required for substantial forward current is 0.3 to 0.4 V. This corresponds to the bias required to offset the junction contact potential essentially completely. For a silicon pn junction, 0.6 V of forward bias is required for substantial forward currents. As the bias voltage is increased in the reverse direction, the magnitude of I_m decreases rapidly as a result of the increased junction potential. After -0.1 V bias, the current is simply I_i, approximately $-40\ \mu A$ for the germanium pn junction at room temperature shown in Fig. 2–58. A comparable silicon pn junction might have an I_i of -40 nA. Of course, I_i increases rapidly with increasing temperature since it is directly related to the conductivity of the intrinsic semiconductor. Thus the pn junction is a good conductor with a small forward bias applied, but a very poor conductor for even large values of reverse bias voltage. The pn junction is described later as a switch whose state (closed or open) is determined by the polarity of the bias voltage. Such a switch is called a **rectifier** and a junction with these characteristics is called a **rectifying junction.** A device consisting of a rectifying junction with two connectors is called a **diode.**

Reverse-Bias Breakdown. The reverse- or zero-biased pn junction can be thought of as two conductors separated by a thin layer of insulator (the depletion region). Thus the junction has the property of capacitance. As the reverse bias voltage is increased, the electric field strength in the depletion region increases as shown in Fig. 2–57b. This has the effect of increasing the thickness of the depletion layer somewhat. The resulting variation in capacitance is the basis of operation of voltage-controlled capacitors called **varactors.** The electric field strength in the depletion region can be increased to a value where the insulating capability of the junction region breaks down. Breakdown can occur by any of three mechanisms: **thermal instability, avalanche multiplication,** or **tunneling.** The heat generated in the junction region under reverse bias is IV watts. If V increases to a value which produces heat more rapidly than it is conducted away from the junction, the temperature will rise. This causes an increase in the reverse current, which causes a further increase in the heat generated. There is a maximum voltage and temperature beyond which this regenerative process proceeds to the thermal destruction of the device containing the junction. This is breakdown by thermal instability.

The thermally generated (intrinsic) charge carriers in the depletion region are accelerated toward the region where they are majority carriers by the electric field in the junction region. The maximum velocity attained by these charge carriers depends on the electric field strength and the distance traveled between collisions of the carriers with other particles. Thus the momentum of the charge carrier at the instant of collision increases with increasing electric field strength. When the momentum of the charge carriers is equal to the energy of ionization of the semiconductor atoms, electron-hole pairs will be produced upon collision. These charge carriers will also be accelerated to attain the momentum required for further ionizing collisions, and so on. The bias voltage at which the avalanche multiplication breakdown takes place depends primarily on the thickness of the depletion layer. Clearly, the thinner the depletion region, the lower the junction potential needs to be to establish a given electric potential gradient. In the construction of pn junctions, the depletion layer thickness is controlled by the dopant concentrations. The greater the concentration of majority charge carriers, the thinner the depletion region, because a smaller fraction of the charge carriers is required to create the depletion region. The avalanche breakdown voltage varies from several hundred volts to 10 V as the concentration varies from 10^{15} to 10^{17} atoms/cm³. Over this same range of concentration, the depletion layer thickness at breakdown varies from about 100 to 0.1 μm. The avalanche breakdown voltage increases somewhat with increasing temperature. The higher the breakdown voltage, the wider the range of reverse voltages for which the device is useful. Some devices are made to be operated at their breakdown voltage (at a safe current). Breakdown diodes and their applications are discussed in a later section.

At an extremely high electric field strength of about 10^6 V/cm, the valence electrons in the p region have an energy comparable to the conduction electrons in the n region. When the p region valence electron energy exceeds the n region conduction electron energy, it may cross the pn junction directly by a process called **tunneling.** Similarly, an n region conduction electron having a higher energy than a valence band hole can move through the junction to fill the hole. Thus current can be in either direction across the junction by the tunneling mechanism. Sufficiently high fields for tunneling are obtained by heavy doping so that the depletion layer is extremely thin. A depletion layer of 10^{-6} cm (0.01 μm) would have a field strength of 10^6 V/cm with only 1 V of junction potential. This is achieved with a dopant level of the order of 10^{19} atoms/cm³. Thus, as the dopant concentration increases, the breakdown voltage decreases and the breakdown mechanism changes from avalanche (about ~8 V) to tunneling (below ~4 V). This is shown by the I-V curves (a and b) in Fig. 2–59. Between 4 V and 8 V, a combination of both mechanisms

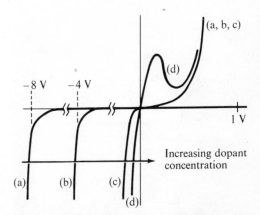

Fig. 2–59 *I-V* curves for several high dopant concentrations: (a) avalanche breakdown; (b) tunnel breakdown; (c) backward diode; (d) tunnel diode.

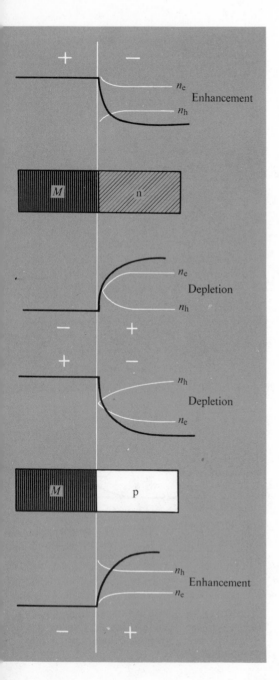

operates simultaneously. The tunnel breakdown voltage decreases with increasing temperature—the opposite of the avalanche temperature effect. Diodes with a breakdown voltage of about 6 V have nearly zero change in breakdown voltage vs. temperature, since the temperature effects of the two mechanisms cancel.

A further increase in dopant concentration could produce tunneling breakdown with only 0.6 V of junction potential. Since this is the zero bias junction potential of a silicon pn junction, the junction is in breakdown with no applied bias! Of course, the tunneling current across the junction is equal in both directions, so that no net current flows. However, only a small value of reverse bias is needed to produce a large reverse current. On the other hand, a forward bias decreases the junction potential and thereby eliminates the tunneling effect. Very little forward current passes until the forward bias offsets the contact potential. Such a device, which conducts in the reverse direction better than in the forward, is called a **backward diode.**

A still further increase in doping produces a junction that is still in tunneling breakdown for small values of forward bias. This produces good conduction in both directions near zero bias. For higher values of forward bias, the tunneling current decreases and the total forward current falls off until the normal majority carrier mechanism causes it to increase once more. A device with this type of junction is called a **tunnel diode.**

Metal-Semiconductor Junctions

The structure and characteristics of metal-semiconductor junctions follow directly from the discussion of metal-metal and pn semiconductor junctions. When a metal and a semiconductor are joined, a contact potential is established as electrons cross the junction to the material in which they have the lower energy. The equilibrium contact potential just offsets the difference in chemical energy of conduction electrons in each material.

The contact potential between a metal and a semiconductor is characteristic of the specific combination at a given temperature and may have either sign. Depending on the sign of the contact potential and the majority carrier of the semiconductor, the junction is either of the **depletion** or **enhancement** type. The four possible combinations of carrier type and contact potential polarity are shown in Fig. 2–60. The top case in the illustration is a junction to an n-type semiconductor with the metal positive. Because the charge carrier concentration in the semiconductor is generally many orders of magnitude less than that in the metal, virtually

Fig. 2–60 Potential and carrier concentration profiles of contacts to n and p semiconductor for each contact voltage polarity.

all of the contact potential gradient occurs in the semiconductor. The junction potential requires an excess of electrons in the semiconductor at the junction. Thus the concentration of majority carriers at the junction is *enhanced* by the contact potential. Since there is a high concentration of charge carriers on both sides of the junction, conduction across the junction occurs readily in both directions. A junction which has a straight-line current-voltage curve passing through the origin is called **ohmic** because it obeys Ohm's Law for both positive and negative voltages.

The opposite effect occurs if the contact potential is of the opposite sign. As shown for the case of the n-type semiconductor which is more positive than the metal, the potential gradient requires a reduction in the electron concentration at the junction. This is a depletion junction similar to the pn semiconductor junction. At normal dopant concentrations, the depletion metal-semiconductor junction is rectifying, i.e., it conducts much better in one direction than the other. Rectifying diodes have been manufactured using depletion metal-semiconductor junctions; these include the copper-oxide, selenium, and point-contact germanium diodes. In most applications pn junction diodes have replaced these earlier types of diodes. A depletion junction can be made ohmic by using a very high level of dopant. This decreases the depletion region thickness to the point where tunneling can occur or, in the extreme case, where it is so thin that it is simply transparent to the electrons. As Fig. 2–59 shows, a tunneling junction is ohmic for small applied voltages. This is generally not a limitation, since the tunneling junction is such a good conductor that the junction bias is always small. An ohmic contact on n-type semiconductors is achieved in practice by evaporating a gold-antimony alloy onto the semiconductor. When this is heated, the antimony diffuses into the semiconductor, creating a region of very high dopant concentration. A contact wire is then soldered or welded to the gold spot.

The case of junctions between metals and p-type semiconductors is exactly analogous to the discussion of the n-type above. When the metal is positive, there is depletion; and when the metal is negative, there is enhancement. Ohmic contacts to p-type semiconductors are made by depositing a gold-arsenic alloy on the material to produce a highly doped junction.

Rectifying metal-semiconductor junctions known as Schottky diodes are used for extremely rapid switching. These diodes are discussed in Section 2–3.3.

Other Junctions

Several other junctions across which charge flows are of interest. One is the junction between a metal and a vacuum. Early in this discussion of junctions, the work required to move an electron from a metal *M* through a vacuum to a distant place was defined as the work function, $Q_e\Phi_M$. Any electron in M having this much energy can theoretically escape from the metal. As the temperature of the material increases, the number of electrons at the surface having sufficient energy to escape increases. Some materials which have a low work function and a high melting point can be heated to a temperature where electrons literally boil out of the metal. When this is done in a vacuum, the metal becomes positively charged and an increasing number of "free" electrons are attracted back to the metal. When a steady state is reached, the rate of evaporation of electrons from the metal is equal to their rate of return and a constant number of electrons exists in a cloud around the metal. This cloud of electrons is called a **space charge.** If another conductor (**electrode**) is placed in the vacuum and made more positive than the hot metal electrode, some of the electrons in the cloud will move in the direction of the positive electrode. Thus a current of electrons is established in the vacuum. If one electrode is hot and the other is cold, the electrons can only go from the hot electrode (**cathode**) to the cold electrode (**anode**) and not the other way. Such a device has rectifying characteristics and is called a **vacuum tube diode.** The **thermionic emission** of electrons and their acceleration by electric fields in a vacuum are the basis of operation of many important scientific devices, including the cathode ray tube.

The junction between the metal and the gas in a **neon lamp** or **gas discharge tube** is also of interest. A neon lamp consists of two cold metal electrodes in a glass envelope with a low pressure of neon gas. At low voltages between the two electrodes, there is virtually zero current because there are essentially no charged particles to act as charge carriers. As the applied voltage is increased, the electric field strength between the electrodes increases. Because of the shape of the electrodes, the electric field is not uniform and eventually, at some voltage V_s, the field strength is large enough to ionize a neon gas atom. The positive ion and electron are then accelerated toward the negative and positive electrodes respectively. By the time these accelerated particles collide with a neutral neon atom, they have acquired more than enough energy to ionize the atom. Thus an avalanche-type breakdown of the neon gas into a conducting plasma of neon ions, electrons, and neon atoms occurs.

Once conduction is initiated, some means of limiting the current to a safe value is required to avoid the destruction of the device by Joule heating. If the current is limited to a safe value, the voltage across the device will adjust itself to the minimum voltage necessary to maintain the applied current. In order to maintain current, it is necessary to maintain a sufficient concentration of ions and electrons in the gas and to transport charge across both metal-plasma junctions. Since the electrons are so much more mobile than the ions, they carry most of the current. The field strength in the gas region will be that required to produce ionizing collisions. At the positive electrode (anode), the current is composed of electrons crossing from the plasma to the metal. The potential across this junction is approximately the Richardson potential for the metal. At the negative electrode (cathode) electrons must cross the junction from the metal to the plasma. Since the metal is not hot, surface electrons must gain the work function energy from some other source. In this case it is from the energy of the positive ion bombardment of the cathode. Thus the electric field in the region of the cathode must be strong enough to accelerate the ions to the work function energy. These effects produce the potential energy profile shown in Fig. 2–61. The electrons emitted from the cathode are given a high energy as they are accelerated by the steep gradient in the cathode region. The collision of these electrons with neon atoms produces highly excited atoms which lose some of their energy by emitting light. Thus a sheath of light is produced about the cathode.

The presence of neon ions in the high field region about the cathode is due to collisions with the emitted electrons. Because of inefficiencies in the energy transfer processes required for ion production and ion bombardment emission, there is a minimum current density which will support this process. Thus the ion bombardment process occurs only on a portion of the electrode area which is small enough to maintain the minimum current density. If the total current is increased, the "active" portion of the electrode area increases as evidenced by an increase in the size of the illumination sheath. The voltage across the cathode-plasma junction is thus essentially independent of the applied current.

The voltage required to maintain conduction, V_m in Fig. 2–61a, is clearly less than that required to start it, V_s. The resulting current-voltage curve is shown in Fig. 2–61b. When the applied voltage reaches V_s, the conduction process begins and the voltage drops to V_m. As the current increases, the maintenance voltage remains almost constant since the junction potentials vary only slightly. The gas discharge tube is seen to switch from nonconducting to conducting with a constant voltage when the applied bias voltage exceeds V_s. A practical circuit includes a source of voltage V greater than V_s connected to a series combination of the discharge tube and a current limiting resistor R. The voltage across the tube will be V_m and the current will be $(V - V_m)/R$. When the bias voltage is reversed, the same processes occur but at opposite electrodes. Differences in electrode material and geometry of the two electrodes will result in different values of V_s and V_m for forward and reverse conduction.

Gas discharge tubes are used as switches, as constant voltage devices, and as indicator lamps. For neon indicator lamps, typical values of V_s and V_m are 87 V and 60 V respectively. Geometry, gas pressure, and gas purity affect these values considerably. Other gases (e.g., argon) are used

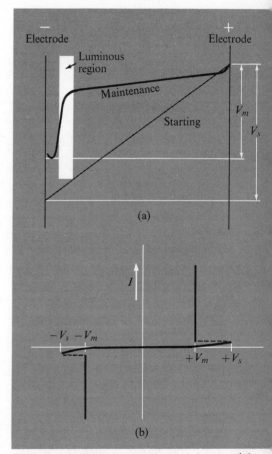

Fig. 2–61 Gas discharge tube: (a) potential profile; (b) current-voltage curve.

for other voltage ranges. At low gas pressures, a higher field in the plasma results in higher energy collisions and the entire plasma region is illuminated as in the all-too-common neon sign.

2–3.2 SWITCHING PRINCIPLES AND MECHANICAL SWITCHES

The switch has been described as a device which determines whether a particular circuit path is conducting or nonconducting. An ideal switch could be placed in any circuit and act as a solid connection when in its closed state and as a perfect insulator in its open state. Since no switch is ideal, there are many kinds of switches available with various combinations of features and compromises. Mechanical switches have a pair of metal contacts which touch when the switch is closed and are apart when the switch is open.

Mechanical switch contacts can be actuated in several ways: manually (toggle and rotary switches), mechanically (lever switches), electro-mechanically (relays and solenoids), and magnetically (magnetic reed relays). Purely electronic devices, including many types of transistors and semiconductor diodes, are also used as switching elements. These solid state switching devices are discussed in the next section (2–3.3). Regardless of the switch type, the basic requirements for a switching device are the same: the device must have two terminals between which an electrical current can be conducted; the device or the circuit must be designed so that there are only two stable states of conductivity in the circuit; and the device must be capable of being "switched" from one state to the other. The appropriate choice of switch depends on the desired method of actuating the switch, the voltage and current levels in the switched circuit, and the connection accuracy and switching speed required. The considerations of accuracy and speed are first discussed in relation to a simple switch operating in a generalized circuit. The descriptions of practical mechanical switches, including manual switches and relays, in this section are intended to aid in the practical application of these basic switch characteristics and limitations.

Switch Resistance

An ordinary mechanical switch (e.g., light switch) has two electrical states: **open** and **closed** (or OFF and ON). The ideal switch has zero resistance between its two connections when closed and zero conductance (infinite resistance) when open. Neither ideal is completely achieved by practical switches.

The schematic symbols that are often used for open and closed switches are shown in Fig. 2–62a. However, in considering the action of a real switch in an electronic circuit it is frequently important to represent

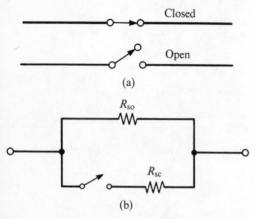

Fig. 2–62 (a) Switch symbol. (b) Schematic representation of nonideal switch.

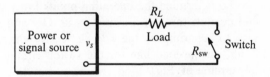

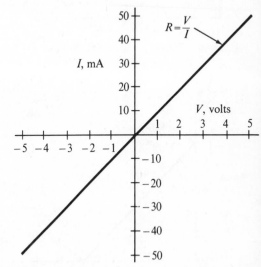

Fig. 2–63 Generalized switching circuit.

the nonideal open and closed switch resistances R_{so} and R_{sc}, as illustrated by the schematic for a nonideal switch in Fig. 2–62b.

In the generalized switching circuit shown in Fig. 2–63, the current in the circuit (through the load) is controlled by the switch. Another way to view the circuit, in consideration of the nonideal switch, is that the source voltage v_s is divided between the resistance of the load and the resistance of the switch. To transfer the maximum signal to the load the closed switch resistance, R_{sc}, should be very small compared to R_L. Similarly, if the open switch resistance is not very much greater than R_L the current will not be effectively turned off by the switch. For purely resistive switches, when R_{so}, R_{sc}, and R_L are known, the effectiveness of a given switch in a desired application can be evaluated using Kirchhoff's laws and the circuit of Fig. 2–63.

Current-Voltage Curves and the Load Line. As the earlier discussion on conducting junctions demonstrated, many junction types used in switches have nonohmic characteristics; i.e., the current through the switch is not a linear function of the voltage across the switch terminals. Therefore, the dc analysis of how the switch performs in a circuit is made most readily by a graphical solution which utilizes the **current-voltage curve** of the device and the **load line** of the circuit. The graphical solution is illustrated in this section for a simple ohmic switch, and in the next section for the pn junction diode and other solid state switching devices. The electrical characteristics of a resistive device for which the resistance is not constant over its operating range are often best shown by a current-voltage curve. A current-voltage (I-V) curve is a plot of the voltage V across a device vs. the current I through it. The I-V curve of a resistor is shown in Fig. 2–64. This line is straight as expected from Ohm's Law. A large resistance has a more horizontal I-V curve, whereas a lower resistance has a more vertical curve. The current-voltage curve of a perfect conductor would coincide with the vertical coordinate, and the line for an infinite resistance would coincide with the horizontal coordinate.

A switch has two I-V curves, one for the resistance of each state. The I-V curves for a perfect switch, which has zero ON resistance and infinite OFF resistance, will fall on the vertical and horizontal coordinates respectively. Exaggerated deviations from the ideal switch characteristics are shown in Fig. 2–65.

Fig. 2–64 Current-voltage (I-V) curve for a resistor.

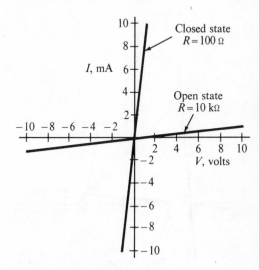

Fig. 2–65 Current-voltage curves for a less-than-ideal switch.

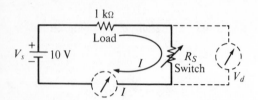

Fig. 2–66   Simple circuit with switch.

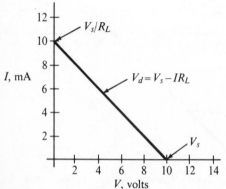

Fig. 2–67   Load line for the circuit of Fig. 2–66.

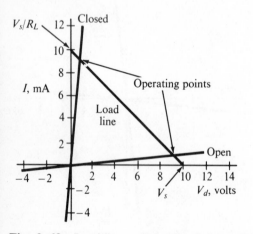

Fig. 2–68   Load line and characteristic curve combined to give operating points.

To determine the **operating points** for a switch, i.e., the voltage across and current through a switch in its ON and OFF states, it is only necessary to superimpose the I-V curves for the switch, such as those in Fig. 2–65, on the load line for the switch circuit. This is illustrated below for the circuit of Fig. 2–66, where a simple switch is used to control the current delivered to a resistive load.

Note that the switch is shown as a device with a variable resistance R_S. From Ohm's Law

$$I = V_s/(R_L + R_S). \qquad (2\text{–}96)$$

Rearranging yields $IR_S = V_s - IR_L$, where IR_S is equal to V_d, the voltage across the device. Therefore,

$$V_d = V_s - IR_L. \qquad (2\text{–}97)$$

This is a linear equation relating I and V_d, and a current-voltage line can be drawn as shown in Fig. 2–67. It is called "load line" because it is determined only by the source voltage V_s and the load resistance R_L. An easy way to obtain the load line is to locate the intercepts and join them by a straight line. From Eq. (2–96) it is seen that if $V_d = 0$, then $I = V_s/R_L$ and if $I = 0$, then $V_d = V_s$. For any value of R_S, the operating values of I and V_d must be located on the load line.

Figure 2–68 shows a superposition of the current-voltage curves of Fig. 2–65 on the load line of Fig. 2–67. Since the switch I-V curve and the circuit load line both define the possible current-voltage relationships for the system, the actual circuit current and switch voltage must be given by the intersection of the I-V curve and the load line. Each intersection is called an **operating point** of the switch *in that circuit*. Since the switch has an I-V curve for each state, there are two operating points as shown in Fig. 2–68. The action of each new switching device can be studied using this graphical method.

Experiment 2–14 *Characteristic Curves and Load Line for an Ideal Switch*

Experience is gained with a curve tracer for obtaining characteristic curves of an ideal switch. The load in a switch circuit is varied to show the variation of the load line and to gain experience with graphical techniques which are also applicable in studying diode and transistor characteristics in subsequent experiments.

Switch Capacitance

In the previous discussion of the nonideal switch we considered only the reduction in amplitude of the voltage across the switch caused by its nonideal ON and OFF resistances. There are also factors inherent in the switch device or circuit which affect the maximum speed of switching. One such factor is the capacitance, which is unavoidable in the construction of the device or circuit. This capacitance results in a certain "inertia" in the switch and affects the time required to go from the ON to the OFF state or vice versa.

To account for the switch capacitance the schematic of the nonideal switch given in Fig. 2–62 can be modified to have an equivalent capacitance C as shown in Fig. 2–69.

The determination of the maximum switching speed of the nonideal switch of Fig. 2–69 in the switch circuit of Fig. 2–66 requires an analysis of the charge and discharge of the capacitance C through the effective circuit resistances.

Switch Inertia. The nonideal switch represented by the schematic diagram of Fig. 2–69 is shown connected in a series circuit in Fig. 2–70, and the effect of the equivalent switch capacitance will now be noted. The resistances R_{so} and R_{sc} are the open and closed values for switch S, and R_L is the load resistance of the circuit.

Assume that the switch S is initially closed, so that the voltage V_{sc} across the switch is

$$V_{sc} = V_s \frac{R_{sc}}{R_L + R_{sc}}$$

as determined by the voltage divider R_L and R_{sc}. Now if the switch is opened, the voltage across the open switch will rise exponentially as the equivalent capacitance C is charged to the maximum open switch voltage determined by the voltage divider R_L and R_{so},

$$V_{so} = V_s \frac{R_{so}}{R_L + R_{so}},$$

as illustrated in Fig. 2–71.

The charging time constant τ_0 will be determined by C and the output resistance of the voltage divider R_L and R_{so}. Thus

$$\tau_0 = CR_LR_{so}/(R_L + R_{so}). \tag{2–98}$$

If $R_{so} \gg R_L$ as is usually the case, then $\tau_0 \approx CR_L$.

The discharge time constant τ_C is determined by C and the parallel discharge paths R_{so}, R_{sc}, and R_L. Since it is assumed that for an effective

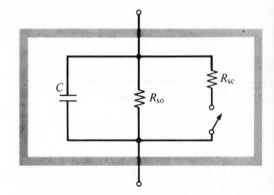

Fig. 2–69 Schematic representation of a nonideal practical switch.

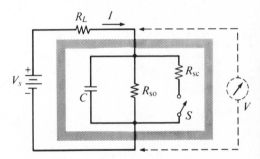

Fig. 2–70 Switch circuit with nonideal switch.

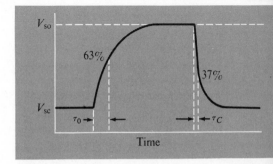

Fig. 2–71 Effect of capacitance on transition between voltage leads.

switch, $R_{so} \gg R_{sc}$,

$$\tau_C = CR_L R_{sc}/(R_L + R_{sc}). \qquad (2\text{--}99)$$

In most cases $R_{sc} \ll R_L$ so that $\tau_C \approx CR_{sc}$.

In the above circuit, if $C = 100$ pF, $R_L = 10$ kΩ, $R_{so} = 1$ MΩ, and $R_{sc} = 100$ Ω, then $\tau_0 = CR_L = 100 \times 10^{-12} \times 10^4 = 10^{-6}$ sec $= 1$ μsec, and $\tau_C = CR_{sc} = 100 \times 10^{-12} \times 10^2 = 0.01$ μsec. Note that $\tau_C \ll \tau_0$, which is true for the typical switch. Therefore, a circuit with significant switch capacitance reaches a steady state more rapidly when the switch is closed than when it is opened. There are other factors inherent in certain devices such as diodes and transistors which determine switching speed, and these will be discussed in the next section.

Actuating Switch Contacts Electrically

When an electrical signal is used to actuate a switch, there are generally two electrical circuits involved: the actuating circuit and the switched circuit, as shown in Fig. 2–72. The degree of interaction between the actuating and switched circuits depends on the type of switch used and the circuit. Some switching devices have inherently one or two connections common between the switch and the actuating element. The three possible degrees of interconnection for simple switches are shown in Fig. 2–73.

Fig. 2–72 Electrically actuated switching circuit.

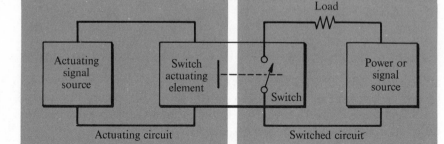

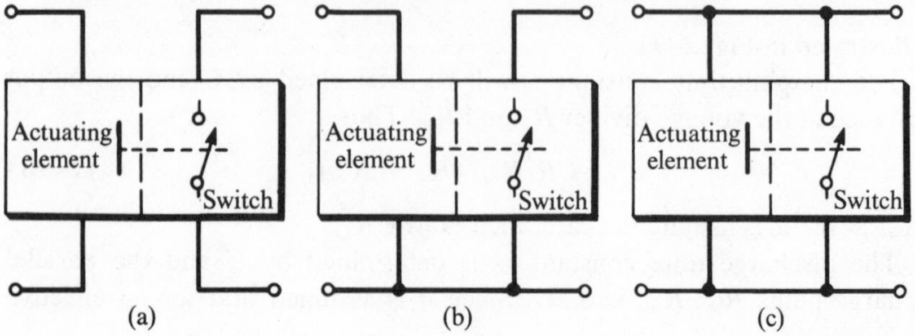

Fig. 2–73 Switching devices: (a) 4-terminal; (b) 3-terminal; (c) 2-terminal.

Examples of four-, three-, and two-terminal devices are the relay, transistor, and diode, respectively. As the descriptions of specific switching devices in this section illustrate, the degree of interaction between the actuating and switched circuits inherent in the switch is a major factor in defining its range of applicability.

Mechanical Switches

Mechanical switches are available in a wide variety of types with characteristics for many different applications. All mechanical switches require some form of pressure to close (make) and open (break) the switch contacts. Mechanical switch contacts can be actuated by the pressure of a human hand, in which case they are known as **manual switches.** Contacts can also be actuated by mechanical pressure, such as the pressure exerted by a cam rotated by a motor. These switches are known as **mechanically actuated switches.** Finally, switch contacts can be actuated electrically by electromagnets or permanent magnets. Such remotely controlled mechanical switches are known as **relays.** In this section a few of the most important types, characteristics, and applications of mechanical switches are presented and some of their advantages and limitations are discussed.

Manual Switch Types. Manual switches are used in applications where complete operator control over opening and closing switch contacts is desired and where switching speed is relatively unimportant. For example, the ON/OFF power switches on most instruments are manually actuated. Likewise manual selector switches are used in instruments for function selection, range changing, introduction of specific circuit elements, and many other control operations.

Manual switches can be classified in a variety of ways. However, it is quite common to classify switches by their actuating mechanism and by their number, type, and arrangement of switch contacts.

Several different types of manual actuating arrangements for switches are illustrated in Fig. 2–74. The choice of which type of actuating mechanism to use for a particular application depends on many factors, including ease of operation, whether repeated actuation is required, whether the switch must interconnect multiple contacts, etc.

Fig. 2–74 Manually actuated switches showing different actuating arrangements.

Pushbutton switch   Toggle switch   Rocker switch   Slide switch   Multiple-wafer rotary switch

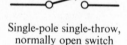

Single-pole single-throw,
normally open switch
(SPST, NO)

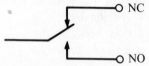

Single-pole double-throw (SPDT)
break-before-make switch
showing normally open (NO) and
normally closed (NC) contacts

Pushbutton switch
(SPST, NO)

Pushbutton switch
(SPST, NC)

Fig. 2–75 Typical manual switch contact arrangements.

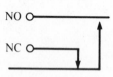

Single-pole double-throw (SPDT)
make-before-break switch

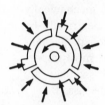

Rotary switch, 3-pole, 3-circuit
with 3 break-before-make contacts

A very important consideration in switches is the switch contacts themselves. Metal-metal contacts are often silver or gold plated for applications where the contact quality is important. Some switches employ metal-liquid contacts, such as mercury-to-dry-metal contacts or mercury-to-wet-metal contacts. Others employ liquid-liquid contacts as in mercury-mercury switches. Mercury switches are actuated by a tilting motion which moves the mercury to make or break the contact.

Some contact arrangements and circuit symbols for typical manual switches are illustrated in Fig. 2–75. Specific circuit applications usually dictate the most appropriate arrangement of contacts. Break-before-make contacts are used when it is permissible to have a momentary open circuit and undesirable to have two circuit components connected together even momentarily. Make-before-break contacts are used when such a momentary open circuit would be undesirable, as in the switch selection of OA feedback resistors for a current follower, where a momentary open circuit would drive the OA to limit.

Manual Switch Characteristics. There are many important characteristics of switches which should be considered for specific applications. Manufacturers of switches usually list **load-life characteristics, contact re-**

sistance, **insulation resistance,** and **dielectric strength** of the insulation in their ratings of switches as well as contact current and voltage ratings. Load and life ratings are given for stringent test conditions established for the industry. These, of course, apply rigidly only to the test conditions. A switch may have a shorter or longer useful life in different applications, depending on the demands made on the switch. All switches deteriorate with age, and the rate at which deterioration takes place varies greatly with switch design.

The contact resistance of a mechanical switch is the resistance between a pair of closed contacts, or the ON resistance. Typical values are in the milliohm range for new switches. However, the contact resistance increases with age and is greatly influenced by the voltage, current, frequency, and environment of the switch.

The insulation resistance is the resistance between open contacts, or the OFF resistance. Typical values for new switches are in the thousand megohm range. Insulation resistance decreases with age, because of surface contamination.

The dielectric strength is a measure of the insulator's ability to withstand high voltages without breakdown. For new switches a typical value might be 1000 V ac.

Current and voltage contact ratings vary widely for different switch designs. Signal switching applications which involve low power switching can be carried out with miniature or subminiature switches, whose nominal ratings might be 5 A at 125 V ac and 24 V dc. Heavy duty power switches for ac power line switching can have typical ratings as high as 20 A at 125 V ac.

Fig. 2–76 Typical actuating mechanisms for mechanically actuated switches.

Pin plunger   Lever   Roller lever

Mechanically Actuated Switches. Switches whose contacts are meant to be operated mechanically are also available in a wide variety of forms. Many of the contact considerations and characteristics are similar to those of the manual switches just described. Devices which transfer the actuating pressure to the switching mechanism are called **actuators,** and are available in many forms, as illustrated in Fig. 2–76.

In addition to contact ratings and load-life ratings, there are other important characteristics for mechanically actuated switches. The **operate**

force is the force required (in ounces) to actuate the switch. The **release force** is the force at the actuator at the instant of the return of the contacts to their normal position. **Pretravel** is the distance from the normal position to the operate position. **Overtravel** is defined as the limit of plunger travel beyond the operate position. Finally, **differential travel** is the distance the actuator moves from the operate position to the release position during its return stroke.

Relay Switch Types. Relays are widely used in electronic circuits as remotely controlled mechanical switches to turn a sequence of events ON or OFF. They are not so fast as transistor or optoelectronic switches, but they have higher open circuit resistance, lower contact resistance, and can generally switch higher loads. The electromagnetic relay is a four-terminal device in which the actuating terminals are electrically isolated from the switched signal terminals. Reed relays can be activated by a permanent magnet or a coil.

Electromagnetic relays utilize a current through a coil to provide a magnetic field that moves the switch contacts, as illustrated in Fig. 2–77 for an armature relay. An armature relay operates by energizing an electromagnet (with a suitable current) which attracts a pivoted lever of magnetic material to a fixed pole. The pivoted lever is called the armature. A switch contact moves with the armature to provide a movable contact. If the current in the coil (and, therefore, the magnetic force) is sufficient, the armature moves the movable contact until it touches the stationary contact. A spring holds the movable contact (and armature) in the open position when the coil is not energized.

There are many types of electromagnetic relays. With a plunger relay, movable contacts are attached to a plunger that moves within a tubular magnetic coil (solenoid). Rotary relays often utilize the rotation of a motor shaft to move switch contacts.

Relays are constructed according to many designs to perform a wide range of functions. Some of the types of relays are interlock, stepping, sequencing, time delay, latch-in, polarized, differential, and general purpose.

The reed relay is a relatively new type of switch. Two or more metal reeds are enclosed in a hermetically sealed glass capsule. A normally open SPST reed relay is shown in Fig. 2–78. The overlapping reeds can be closed or opened by positioning a permanent magnet near or away from the reed contacts. The reed contacts can also be switched by actuating an electromagnet. Some typical arrangements for the use of the reed relay are illustrated in Fig. 2–79.

Relay Contacts and Characteristics. The number of relay contacts and contact arrangements is determined by the application. The nomenclature

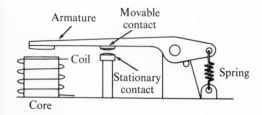

Fig. 2–77 Schematic diagram of single-pole, single-throw, normally open (SPST, NO) relay.

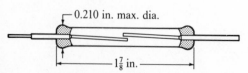

Fig. 2–78 Reed relay, with silver alloy contacts. (Courtesy of Hamlin, Inc.)

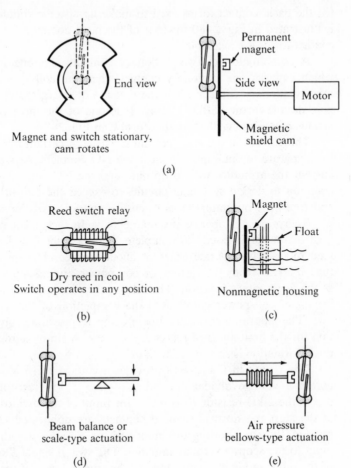

End view

Magnet and switch stationary,
cam rotates

Permanent magnet

Side view

Motor

Magnetic shield cam

(a)

Reed switch relay

Dry reed in coil
Switch operates in any position

(b)

Magnet

Float

Nonmagnetic housing

(c)

Fig. 2–79 Typical reed switch arrangements. (a) Switch operates whenever cut-out portion of shielding metal is between magnet and switch. (b) Switch located in cylindrical coil. (c) Solution level control. (d) Beam position indicator. (e) Magnet positioned by air-driven bellows. (Courtesy of Hamlin, Inc.)

Beam balance or
scale-type actuation

(d)

Air pressure
bellows-type actuation

(e)

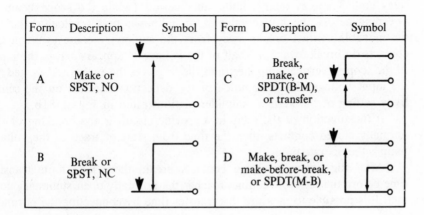

Form	Description	Symbol	Form	Description	Symbol
A	Make or SPST, NO		C	Break, make, or SPDT(B-M), or transfer	
B	Break or SPST, NC		D	Make, break, or make-before-break, or SPDT(M-B)	

Fig. 2–80 Four common forms of relay contacts with designations.

for the basic contact forms used to make up specific contact arrangements is illustrated in Fig. 2–80 for four of the most common forms. Many other contact forms are available.

A combination of two stationary contacts and one movable contact which engages one stationary contact when the coil is energized and the other stationary contact when the coil is not energized is called a single-pole double-throw (SPDT) relay. It is one of the most common contact arrangements and is illustrated as form C in Fig. 2–80.

The current in the relay coil must exceed a certain minimum value for the armature to "pull in" and close the NO contacts. At a somewhat lower current the armature will "drop out" and the NO contacts will open. It is common to design switching circuits to exceed the pull-in current by several times the minimum so as to ensure operation of the relay.

Relays do not operate instantaneously. In fact, most relays require at least a few milliseconds to complete their contact function, although the reed relays have operate times of about 1 msec. The operate time is the time interval from the instant of coil-power application until completion of the last contact function. The release time is the time interval from the instant of coil-power cutoff until the completion of the last contact function. The measurement of the operate and release times gives an appreciation for the limitations of relays as switches. A simple circuit for obtaining this information is shown in Fig. 2–81.

The circuit in Fig. 2–81a provides a voltage V (5 V in the example) across a voltage divider consisting of two 1 kΩ resistors. The voltage across the 1 kΩ resistor R_1 is fed to the input of the oscilloscope. However, at the start the normally closed contacts are connected across the 1 kΩ resistor, thereby shorting the input resistor R_1 and providing zero voltage input to the scope's vertical amplifier. The switch signal S operates the coil of the relay and is also connected to the trigger input of the scope. That is, the $+5$ V signal triggers the sweep and applies the drive current to the relay coil. There is then a finite time period (while the scope beam is sweeping across the horizontal axis at a known preset velocity) before the movable contact breaks away from the stationary contact. When the contacts do break away, one-half of the voltage V appears across the input to the scope's vertical amplifier, and the trace on the scope is deflected by the input voltage. The magnitude of the deflection depends on the sensitivity setting of the vertical amplifier, as illustrated in Fig. 2–81b.

If the function of the relay in a specific circuit is the breaking of the normally closed contacts, then the time from start of trace to the voltage jump is the operate time.

After the normally closed contacts break, there is a finite transfer time before the movable contact reaches the normally open stationary contact. It is possible to measure this transfer time by connecting the normally

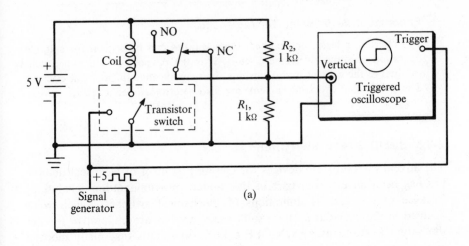

(a)

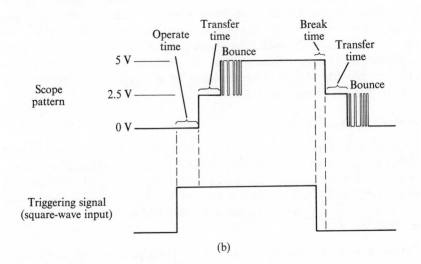

(b)

Fig. 2–81 Circuit for measuring operate, transfer, bounce, and release times.

open contact to the +5 V terminal. Note that the transfer time should not include the bounce time, which is very obvious on the scope trace and of considerable duration.

When the contacts break or make there is always some contact bounce. Although this bounce and its duration can seriously distort the switched signal, it is not included in the operate and release times. The bounce time is also a characteristic of manual switches.

Experiment 2–15   *Relay Characteristics*

The operate, release, and contact bounce times can be critical for suitable operation of some electronic circuits. It is the purpose of this experiment to observe these times and determine their significance in practical circuits. The pull-in and drop-out currents are also determined for a relay.

2–3.3 SOLID STATE SWITCHES

The advent of solid state devices for ON-OFF control of electrical quantities has been an essential part of the modern instrumentation revolution. Because of the various limitations of mechanical switches, which were detailed in the previous section, solid state switches are imperative for the short-time (μsec to nsec) ON-OFF control operations that many modern experiments require. The recent availability of inexpensive diodes, transistors, and thyristors has brought about new short-time measurement and control applications which would have been impossible only a few years ago. In this section the switching characteristics of solid state devices are described, and the graphical analysis of switching circuits is presented in order to provide a basic understanding of the types and properties of modern high speed switches.

The pn Junction Diode

Because of its special current-voltage characteristic the junction diode is often used as a high speed ON-OFF control device. It has the disadvantage of being a two-terminal device, which means that the signal that turns the switch ON and OFF is also the signal that is transmitted. In other words, the actuating signal and the signal to be switched must be the same signal in a junction diode.

Current-Voltage Curves and Load Lines. The circuit of Fig. 2–82 shows the junction diode acting as a switching device in a circuit with a voltage source which alternates between +10 V and −10 V. The switching cir-

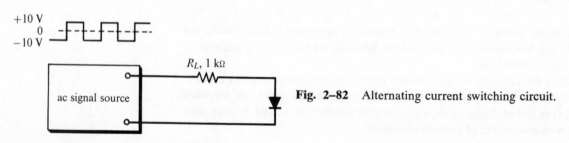

Fig. 2–82 Alternating current switching circuit.

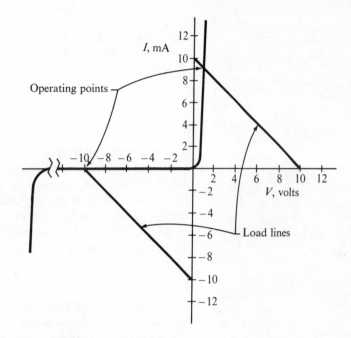

Fig. 2–83 Operating points for circuit of Fig. 2–82.

cuit can be graphically analyzed by the technique introduced in Section 2–3.2. The load line for the circuit is superimposed on the diode current-voltage curve and the operating points are established graphically. The equation for the load line is, as before, $v_d = v_s - iR_L$, where v_d is the voltage across the diode, v_s is the source voltage, and iR_L is the voltage drop across the load resistor. Since v_s has two values in the circuit of Fig. 2–82 ($+10$ V and -10 V), there are two load lines: one for each value of v_s, as shown in Fig. 2–83. The two operating points are found from the intersections of the load lines and the current-voltage curve. From the two operating points shown in Fig. 2–83, it can be seen that the voltage drop across the load resistor will be ≈ 9.4 mA $\times$ 1 k$\Omega = 9.4$ V when $v_s = +10$ V and ~ 0 V when $v_e = -10$ V. Hence from the operating points it is clear that a diode can be a switch which is **OFF** or **ON** depending on the polarity of the applied signal. The voltages across both the diode and the load for each value of v_s can be obtained by the above technique.

If the signal source is a sinusoidally varying ac voltage as shown in Fig. 2–84a, a graphical analysis can still be made although it is somewhat cumbersome, since load lines must be drawn for each value of v_s. As v_s varies with time, beginning with $t = 0$, the load line shifts outward horizontally during the first quarter-cycle until v_s reaches its peak value, V_p, as illustrated in Fig. 2–84b. During the second quarter-cycle, the load line moves

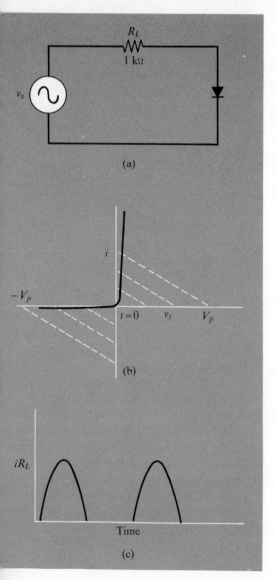

(a)

(b)

(c)

Fig. 2–84 Diode switch with sine wave source: (a) switching circuit; (b) graphical analysis; (c) load voltage vs. time.

inward horizontally as v_s decreases to zero. When v_s goes to negative values, the diode is reversed biased and the load line is drawn in the third quadrant. Again the load line moves outward horizontally until the peak negative value $-V_p$ is reached and moves inward horizontally until the signal source reaches zero, at which point the cycle repeats itself. The operating points for each value of v_s can be determined from the intersection of the i-v curve and the varying load line. The value of current at each operating point can be read from the graph, and the voltage drop across the load can be calculated at each instant in time and plotted as shown in Fig. 2–84c.

Circuit Models. In many situations, particularly those in which the signal source varies continuously with time, the graphical analysis technique described above is rather cumbersome, and it becomes easier to work with a linear circuit model for the diode. For switching applications in which the signal source varies between two fixed voltage values, as in Fig. 2–82, the circuit model used is usually the **static model** of the diode. In the static model, a forward biased diode is modeled as a simple resistor of resistance R_{fs}, the static forward resistance. The value of R_{fs} is found from the ratio of diode voltage to diode current at the ON operating point. Likewise the reversed biased diode is modeled as a simple resistor of resistance R_{rs}, where the reverse static resistance is found from the OFF operating point. Hence the complete linear model for the circuit of Fig. 2–82 consists of two circuits: one for the $+10$ V condition of the signal source and the other for the -10 V condition, as shown in Fig. 2–85a and b. Since these circuits give the same electrical quantities at the two values of v_s as does the diode, they are often known as **equivalent circuits.**

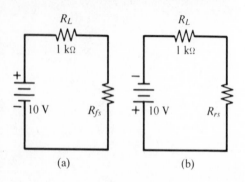

(a)   (b)

Fig. 2–85 Static circuit models for the diode switch of Fig. 2–84: (a) forward biased equivalent circuit; (b) reversed biased equivalent circuit.

Fig. 2–86 Piecewise-linear model of diode: (a) current-voltage characteristic; (b) reverse bias model; (c) forward bias model; (d) simplified model neglecting reverse current.

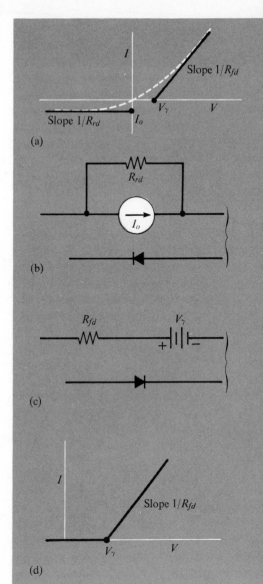

The static model of the diode given above is not particularly useful for continuously varying signals such as sine waves, or for two-state signals when the values of v_s are unknown. Instead a linear model consisting of two straight-line segments is used. Such models are known as **piecewise-linear models.** The most complex piecewise-linear model that is used for the pn junction diode is illustrated in Fig. 2–86. In Fig. 2–86a the diode in reverse bias is approximated by a straight-line current-voltage characteristic whose slope is the reciprocal of the reverse dynamic resistance, R_{rd}. The reverse dynamic resistance is the reciprocal of the slope of the diode I-V curve in the region of the reverse operating point ($R_{rd} = dV/dI$). The resulting circuit model of the diode in reverse bias is shown in Fig. 2–86b and consists of a constant current source and a shunt resistance of value R_{rd}. In forward bias the diode current-voltage characteristic is approximated by a second straight-line segment whose slope is now the reciprocal of the forward dynamic resistance R_{fd}. The forward dynamic resistance is the reciprocal of the slope of the diode I-V curve in the region of the forward operating point ($R_{fd} = dV/dI_m$). The straight-line characteristic intersects the voltage axis at voltage V_γ, which is sometimes called the **threshold voltage** or **cut-in voltage.** For forward currents of up to 10 mA, V_γ is approximately 0.2 V for germanium diodes and 0.6 V for silicon diodes. For larger currents, up to 50 mA, better values of V_γ are 0.3 V for germanium and 0.65 V for silicon. The circuit model which results in the forward bias region is shown in Fig. 2–86c and consists of replacing the diode with a constant voltage source of voltage V_γ and a series resistor R_{fd}.

Because I_i, the reverse current of the diode, is usually very small (of the order of nanoamperes for a silicon diode), it is often completely neglected, as shown by the simplified piecewise-linear model of Fig. 2–86d. Here the diode is considered an open circuit of infinite back resistance for values of $v_s < V_\gamma$; for $v_s > V_\gamma$ the circuit model of Fig. 2–86c applies. This simplified model is most often used in applications where the reverse current can be neglected.

A useful relationship for estimating the dynamic resistance of a forward biased diode can be obtained from the basic diode current-voltage

(*I-V*) equation

$$I = I_i(e^{V/\eta V_T} - 1), \tag{2-100}$$

where V_T, *the voltage equivalent of temperature,* is equal to $(kT/Q_e) = T/11{,}600$, and η is a characteristic of the diode. For germanium $\eta \simeq 1$ and for silicon, $\eta \simeq 2$.

Forward bias corresponds to positive values of V in Eq. (2–100). For values of V greater than 0.25 V for silicon or 0.12 V for germanium, the exponential term is more than 100, so the quantity (-1) is negligible and Eq. (2–100) is approximately

$$I = I_i e^{V/\eta V_T}, \tag{2-101}$$

showing an exponential increase in current as a function of the voltage. Taking the logarithm of both sides of Eq. (2–101) and evaluating at 300°K,

$$\log I - \log I_i = V/(2.3\eta V_T) = V/(0.059\eta).$$

Thus the voltage increases by 59η mV for every tenfold increase in current at room temperature. For real diodes, a plot of $\log I$ vs. V often gives a straight line over several orders of magnitude.

If Eq. (2–101) is differentiated with respect to V, Eq. (2–102) results:

$$\frac{dI}{dV} = \frac{1}{\eta V_T} I_i e^{V/\eta V_T} = \frac{1}{\eta V_T} I, \tag{2-102}$$

or inverting,

$$\frac{\eta V_T}{I} = \frac{dV}{dI} = R_{fd} = \frac{0.026\eta}{I}. \tag{2-103}$$

For example, according to Eq. (2–103), if $I = 0.1$ mA, then the dynamic resistance is 260 Ω for germanium and 520 Ω for silicon, but at 1 mA it is about 26 Ω and 52 Ω, etc. At lower currents, the ohmic resistance of the semiconductor becomes significant.

Classes of Diodes. The two general classes of diodes are signal diodes and rectifier diodes. The signal diodes are used for many functions in digital instrumentation and detector circuits where the forward current and reverse voltage requirements are not high (e.g., 50 mA and 50 V) but where fast response times, low capacitance, and low leakage currents are desirable characteristics.

The rectifier diodes are capable of high current, starting at about 100 mA and ranging to several hundred amperes. The reverse voltage ratings typically vary from about 50 to 600 V, but some newer designs exceed 1000 V. All pn junctions can be damaged by excessive reverse voltages unless the current is limited to a safe value, and consequently they have peak inverse voltage (PIV) ratings.

Diode Switching Times. Most electronic switching devices operate inherently many orders of magnitude more rapidly than mechanical switches. The development of signal switching techniques operable on the microsecond or lower time scale has been a major breakthrough for scientific instrumentation. As we continue in our efforts to measure time more precisely and to resolve events separated by shorter and shorter times, the problem of switching time continues to limit the accuracy and speed of our measurements.

A reverse biased junction diode has very few charge carriers in the junction region. If the diode is suddenly forward biased, the resistance will remain high until majority carriers, accelerated by the new potential gradient, cross the junction boundary. This requires only a few nanoseconds. Having crossed the junction, they are now minority carriers and are neutralized eventually by majority carriers. This reduces the majority carrier concentration near the junction, causing a net diffusion of majority carriers toward the junction from each side. Once established, this diffusion maintains the supply of charge carriers at the junction necessary for the flow of charge across the junction. But until the concentration gradient is established, an electric potential gradient is required to maintain the necessary rate of transport of majority carriers to the junction. Thus, even though forward current is flowing, the forward potential drop across the diode will be higher than its steady-state value. The attainment of steady state is usually accomplished within a fraction of a microsecond. In most applications the turn-on time would be just the time required for the forward current to reach a particular fraction of the steady-state level after the forward bias was applied.

The forward biased junction diode has relatively large concentrations of minority carriers near the junction. These carriers have just crossed the junction but have not yet been neutralized by the majority carriers. When a reverse bias is suddenly applied to a forward biased diode, the minority carriers in the junction region will be attracted back across the junction by the new gradient. That is, electrons near the junction in the p region will be accelerated back toward the nonpositive n region, and similarly for the holes near the junction in the n region. In other words, the junction is still conducting. This gives rise to a substantial reverse current until the minority carriers in the junction region are depleted to their usual low value for the reverse biased diode. This is called the **stored-charge effect** because of the excess of minority charge carriers which are stored in the junction region during conduction and which need to be swept out of the junction region before the diode is OFF. This effect will be discussed again in more detail when the junction transistor is described.

The diode current behavior following the sudden application of a reverse bias is shown in Fig. 2–87. The storage time, t_S, is the time required

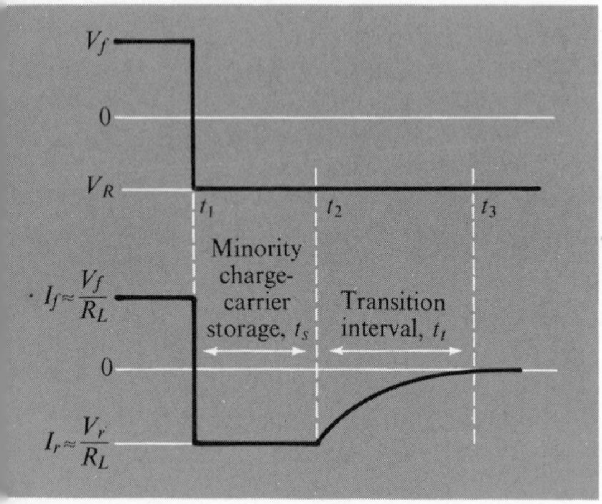

Fig. 2–87 Diode current behavior after application of step reverse bias.

to sweep the minority carriers from the junction. During the transition interval, the diode resistance is increasing. This occurs gradually because excess minority carriers farther from the junction are still crossing the junction and because the capacitance of the reverse biased pn junction must charge through R_L to the reverse bias potential $-V_r$.

Schottky Barrier Diodes. Special high speed switching diodes, known as Schottky diodes, are made which greatly reduce the amount of the stored charge and the width of the junction region. Such devices in appropriate circuits have achieved nanosecond switching times. The Schottky diode is a metal-semiconductor junction rather than a normal pn junction. Since there are no mobile positive charge carriers in the metal, the forward current is essentially all carried by electrons flowing from the semiconductor into the metal. These electrons rapidly reach equilibrium with other electrons in the metal, which eliminates the stored charge effect. Schottky diodes have a lower forward voltage drop across them for a given current than a pn junction diode.

Experiment 2–16 *Characteristic Curves and Resistance of a Diode*

The characteristic curve for a silicon diode is obtained with an oscilloscopic curve tracer, and the significant diode parameters are measured from the curve. The values obtained from the forward biased diode characteristic curve are used to plot log I vs. V so as to determine the logarithmic relationship over a few decades. The dynamic resistance of the diode is also determined and compared with the static resistance at a current of 1 mA. The problems of using an ohmmeter to measure diode resistance are determined.

Reverse Breakdown and Zener Diodes

In Fig. 2–83 it was noted that when the reverse voltage exceeds a certain value the reverse current suddenly increases, and this is called the breakdown voltage. The large reverse bias produces an electric field (greater than 10^6 V/cm) that is sufficient to break the covalent bonds and produce free mobile electron-hole pairs, by the process of tunneling. It occurs in abrupt junctions between highly doped regions. Another type of breakdown can occur called avalanche breakdown, which is caused by carriers that gain sufficient kinetic energy to ionize other atoms with which they collide, thus producing additional mobile electron-hole pairs. The mechanism of breakdown depends on the voltage region of breakdown, and both mechanisms can be present in the same diode. Silicon diodes that break down above 8 V probably depend on the avalanche mechanism, and those in the region below 4 V on the tunneling mechanism. For those diodes that break down between 4 and 8 V both mechanisms are involved, and the dominant mechanism depends on the impurity distribution at the junction. Even though the junction breakdown mechanism originally proposed by C. Zener was the tunneling mechanism, all diodes designed specifically for operation in the reverse breakdown region are called **Zener diodes.** Two common symbols for Zener diodes are shown in Fig. 2–88. The symbol is meant to suggest the shape of the *I-V* curve.

In a similar manner, the reverse bias region before breakdown is sometimes used as a source of very small constant currents. A very high resistance is a desirable characteristic in constant current sources. This application of the diode is not very common, however, because of the large effect of temperature on the reverse current I_i and the limited range of possible currents.

The use of a Zener diode for regulating the output voltage of a power supply is described in Section 2–4.

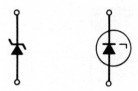

Fig. 2–88 Zener diode symbols.

Experiment 2–17 *Reverse Breakdown and Zener Diodes*

In this experiment the reverse characteristic curve for a Zener diode is observed and the significant parameters of the Zener diode are measured from the curve.

Microcircuit Applications of pn Junctions

The development of microcircuits consisting of hundreds of components (diodes, transistors, capacitors, and resistors) fabricated in a silicon chip has been at the heart of the instrument/automation revolution. Many components can be interconnected right on the chip to form complex circuits called **integrated circuits.** The reverse biased pn junction finds many applications in such microcircuits.

Figure 2–89 illustrates a pn junction and the depletion layer associated with it. As was pointed out in Section 2–3.1, the depletion layer contains very few mobile charge carriers. When the junction is reverse biased, a negligibly small current flows across the junction. Hence a reverse biased pn junction is an excellent insulator and is extensively applied to provide isolation between the p- and n-type semiconductor materials across the junction. In most present integrated circuits the various active components are isolated from each other inside the tiny silicon chip by means of such reverse biased pn junctions. Components are then connected by means of a metal, such as aluminum, which is deposited on the chip surface to form

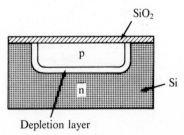

Fig. 2–89 A pn junction used for isolation.

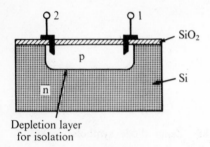

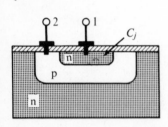

Fig. 2–90 A diffused microresistor with pn junction isolation.

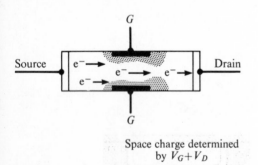

Fig. 2–91 A microcapacitor from a pn junction.

Fig. 2–92 Illustration of channel space charge.

the interconnection pattern. Because there is capacitance associated with the pn junction, the degree of isolation of components by reverse biased junctions falls off at high frequency.

Another common use of the reverse biased pn junction for isolation in microcircuits is illustrated in Fig. 2–90, where a diffused microresistor is shown. The p-type semiconductor is used as the resistor and the reverse biased junction isolates the resistor from any other components on the chip. Terminals 1 and 2 provide the electrical contacts for the resistor. Resistance values ranging from 100 Ω to about 30 kΩ can be obtained in this manner.

Reverse biased pn junctions are also occasionally used as microcapacitors for integrated circuits as illustrated in Fig. 2–91. Here two junctions are shown. One of the junctions provides the desired capacitance C_j, while the other isolates the capacitor from any other components on the chip. Electrical connections to the capacitor are provided by the terminals 1 and 2. Unfortunately the small size of integrated circuits prevents the achievement of large values of capacitance in this manner, and typical maximum capacitances are in the range of tens of picofarads.

Field-Effect Transistor Switches

The field-effect transistor (FET) was briefly introduced in Module 1 with respect to its application as the input stage of a voltmeter. In this application the high input resistance of the FET prevents loading of voltage sources that are measured with the FET voltmeter. Another very important application of the FET is as a high speed switch with very high OFF/ON resistance ratio.

Junction Field-Effect Transistors. Recall that the operation of the unipolar FET as a variable resistance depends on the formation of a "channel" of variable thickness through which the electrons must pass in going from the source to the drain. This is illustrated in Fig. 2–92 for an n-type semiconductor. The channel is formed by introducing p-type impurities into opposite sides of the semiconductor bar, and this type of FET is called a junction FET, abbreviated J-FET. A space charge is developed at the p-type "gate" regions which provides a certain effective channel thickness through which electrons flow. If a reverse bias is applied between the source and gate, the space charge is increased and the effective channel thickness for electron flow is decreased.

When the voltage V_D is applied as shown in Fig. 2–93a, the drain is made positive with respect to the source, and electrons flow from the source through the channel to the drain. The gate voltage V_G controls the current by controlling the effective channel thickness. As would be ex-

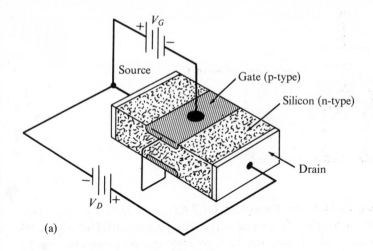

(a)

Fig. 2–93 The junction FET; (a) pictorial; (b) symbols.

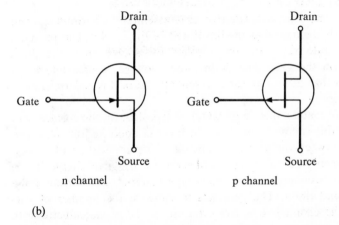

(b)

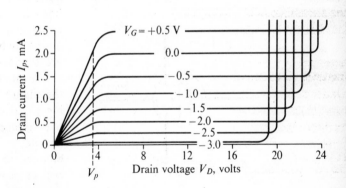

Fig. 2–94 Characteristic curves for junction FET.

pected, there is a critical drain voltage above which there is no increase in drain current. This is called the "pinch-off voltage" V_p and is dependent on the mobility of the electrons in the silicon bar. As shown in Fig. 2–94 a plot of drain current vs. drain voltage provides a family of characteristic curves for different values of gate voltage V_G. The circuit symbols for n- and p-channel FET's are shown in Fig. 2–93b.

When the FET is used as a switch, the gate bias V_G is made sufficiently large to turn the carrier flow **OFF**. The **OFF** resistance is very high, typically greater than $10^9 \, \Omega$. The **ON** resistance for zero gate bias can be estimated at worst to be $R_{ON} = V_p/I_{DSS}$, where I_{DSS} is the current determined from the point where the pinch-off voltage V_p crosses the characteristic curve for $V_G = 0$ in Fig. 2–94.

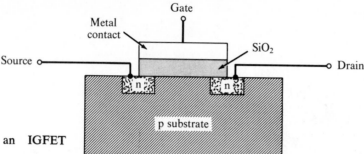

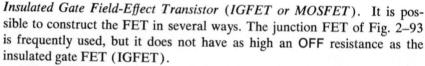

Fig. 2–95 Representation of an IGFET (MOSFET).

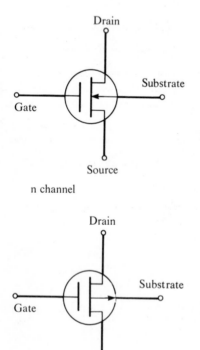

Drain

Substrate

Gate

Source

n channel

Drain

Substrate

Gate

Source

p channel

Insulated Gate Field-Effect Transistor (IGFET or MOSFET). It is possible to construct the FET in several ways. The junction FET of Fig. 2–93 is frequently used, but it does not have as high an OFF resistance as the insulated gate FET (IGFET).

The IGFET can be constructed so that essentially no drain current flows under zero bias conditions, as illustrated in Fig. 2–95. The pn junctions in the substrate very effectively isolate the source from the drain under zero bias on the gate. The drain-source voltage reverse biases the drain junction so that drain current is essentially zero. Typical resistances are 10^{14} Ω or more.

A thin layer of metal oxide (e.g., SiO_2) is deposited on the surface over the channel, and this is covered with a thin layer of deposited metal, forming a capacitor between the gate and substrate. The application of a positive voltage between gate and source will induce a negative charge in the channel so that it becomes in effect an n-type material and current flows between source and drain. The gate acts to enhance the number of current carriers in the channel. It is from the use of the semiconductor with metal and oxide structure that the name MOSFET, frequently used to identify the device, originates. The abbreviation MOST is also used to identify this type of transistor. In those cases where the **insulated gate** is formed by a chemical compound other than the oxide, the appropriate name is IGFET.

The MOSFET can also be fabricated so that with $V_G = 0$ there is a conductive path between source and drain. A negative voltage applied to the gate will then induce a positive charge in the channel which causes a depletion of carriers and can turn the MOSFET OFF.

The symbols for the MOSFET are shown in Fig. 2–96.

Fig. 2–96 MOSFET symbols.

Experiment 2–18 *Field-Effect Transistor (FET) Characteristics*

The characteristic curves for a FET are obtained from a curve tracer and the load line and ON-OFF characteristics are measured.

Junction Transistor Characteristics

Seldom has a component in any apparatus received so much general publicity as has the transistor, a word known to every schoolchild. This remarkable device has brought about major advances in electronics and has pointed the way to the development of other devices and the new integrated circuits.

One of the important applications of the transistor is as a switch to control electrical current and voltage. The transistor has the advantage that it is a three-terminal device and hence offers some isolation between the information signal to be switched and the electrical actuation signal.

The operation of the junction transistor is based on many of the principles described in Section 2–3.1 for the pn junction. In fact, the transistor is essentially made up of two pn junction diodes coupled by a very thin common base, either of p-type or n-type semiconductor material. Therefore, there are two types of junction transistors depending on the material of the base, the npn and the pnp transistors, as shown by the pictorial representation in Fig. 2–97. Because of the two pn junctions, this type of transistor is often called a **bipolar junction transistor** and referred to as a BJT.

In Fig. 2–97 a lead is shown attached to the base. Leads are also shown connected to the two regions of semiconductor material that are adjacent to and of opposite type to the base material. These regions are called the **emitter** and the **collector.** Although it would appear from the figure that the emitter and collector are identical, the fabrication techniques provide each with certain practical characteristics described later in this section. However, in principle, the emitter-base and collector-base junctions are of similar type, and the operation of the transistor depends on the applied voltages between the three terminals.

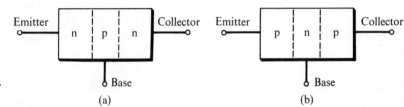

Fig. 2–97 Basic transistor types; (a) npn; (b) pnp.

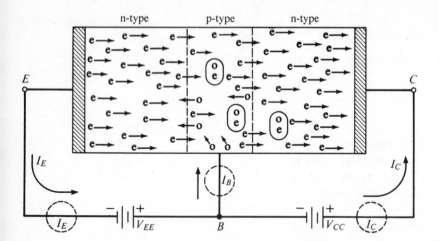

Fig. 2–98 Representation of npn transistor in operation with forward biased emitter-base junction and reverse biased collector-base junction (e = electrons, o = holes, and oe = recombination of holes and electrons).

Principles of Transistor Operation. The internal operation of the npn transistor is represented schematically in Fig. 2–98. It should be noted that the base is very thin, only about 10^{-3} to 10^{-4} cm for a typical transistor. If the emitter-base junction is forward biased the energy barrier is reduced, so that the majority carriers in the emitter (electrons for the npn transistor) cross the barrier and are emitted or ejected into the base, where they become minority carriers. The emitted electrons that cross the junction rapidly diffuse through the very thin base region. When they reach the region of the base-collector junction they are rapidly accelerated toward the collector. Note that because the collector-base junction is reverse biased the n-type collector is positive with respect to the p-type base. The electrons that reach the collector junction are, therefore, attracted to the collector and into the external circuit. In effect, then, the movement of minority carriers through the base (i.e., electrons in the p-type base) primarily determines the collector current I_C.

Some of the electrons combine with holes as they diffuse through the base, and thus a small base current is required to maintain the hole concentration. Also contributing to the base current is the need to compensate for the difference between hole currents across the two junctions. The holes in the base cross the forward biased emitter-base junction, but the number of these holes is much smaller than the number of electrons crossing from emitter to base, because the emitter has a much higher impurity content (charge carrier density) than the base. The number of holes that cross the collector-base junction is usually very small because the collector is heavily doped and the junction is reverse biased. Therefore, the base current I_B is primarily a result of the emission of holes from base to emitter and the electron-hole recombination that occurs in the base.

Because $I_E = I_B + I_C$ from applying Kirchhoff's Law to junction B of Fig. 2–98, and because I_B is usually small compared to I_E, the collector current I_C will be just a little less than I_E. Typical relative values might be $I_C = 99$, $I_E = 100$, and $I_B = 1$. For the configuration shown in Fig. 2–98 it is useful to define a forward current gain α_N as

$$\alpha_N = I_C/I_E, \qquad (2\text{–}104)$$

the fraction of the emitter current that appears in the collector circuit.

The collector current I_C is assumed here to be a fraction θ of the injected electron current I_n,

$$I_C = \theta I_n, \qquad (2\text{–}105)$$

and because the emitter current I_E is the sum of the electron current I_n and the hole current I_p,

$$I_E = I_n + I_p, \qquad (2\text{–}106)$$

it follows from Eqs. (2–104), (2–105), and (2–106) that

$$\alpha_N = \frac{I_C}{I_E} = \frac{\theta}{1 + I_p/I_n}. \qquad (2\text{–}107)$$

Therefore, even if there is no recombination in the base, so that $\theta = 1$, it can be seen from Eq. (2–107) that α_N would be less than unity. As noted previously, the fabrication procedures for the emitter and base regions of the npn transistor result in very few holes crossing the emitter-base junction as compared to the number of electrons that cross, so that the ratio I_p/I_n is small, and from Eq. (2–107) it follows that α_N is approximately equal to θ.

In the discussion above it was assumed that the collector current I_C was a fraction θ of the emitter electron current I_n. However, it is possible that the kinetic energy of the electrons as they are accelerated through the base-collector junction will be sufficient to cause additional electrons to be released from the atoms in the collector. The collector current I_C in the external circuit will thus exceed the rate of electron flow into the collector through the base-collector junction. Therefore, θ can be expressed as the product of two terms: there is a term M to account for any collector-current multiplication, and a term β_t (called the "base transport factor") that is the ratio of carriers which flow from base into collector to the number of carriers entering the base from the emitter, so that $\theta = \beta_t M$. The current gain relationship of Eq. (2–107) can thus be written

$$\alpha_N = \frac{1}{1 + I_p/I_n} \beta_t M = \epsilon \beta_t M. \qquad (2\text{–}108)$$

In practice, transistors are generally designed to keep α_N close to unity.

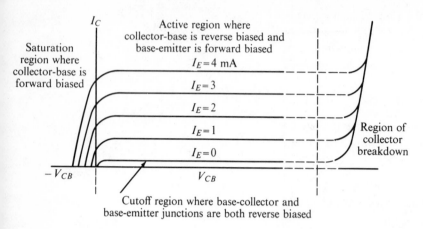

Collector characteristics for common-base transistor configuration.

Common-Base Collector Characteristics. In Fig. 2–98 the connection of one lead of V_{EE}, one lead of V_{CC}, and the base lead to a common point provides the so-called **"common-base"** circuit configuration. If the collector current I_C is plotted against the collector-base voltage V_{CB} for fixed values of emitter current I_E, the characteristic curves of Fig. 2–99 are obtained for the npn transistor described above.

It can be seen from Fig. 2–99 that when $I_E = 0$ there is a collector current, which is a result of collector-base reverse leakage current I_{C0}. The collector current can be written

$$I_C = \alpha_N I_E + I_{C0}. \tag{2–109}$$

If the base-emitter and collector-base junctions are both reverse biased, the collector current can be decreased below I_{C0}. In this case the sign of I_E in Eq. (2–109) is negative. If the collector-base and emitter-base junctions are both forward biased, then the operating characteristics are in the region left of the vertical axis, and this is the saturation region for the common-base connection.

In Fig. 2–99, in the region that is bounded by the two vertical lines and the line for $I_E = 0$, the base-emitter junction is forward biased and the collector-base junction reverse biased. In this region the collector-base voltage V_{CB} has very little effect on the collector current I_C. This is often called the "active" region, which is utilized for small-signal linear circuits.

As V_{CB} is increased, the depletion layer of the collector-base junction increases so that the effective width of the base decreases. This decrease of effective base width means that for the npn transistor the electrons from the emitter need to move an even shorter distance before they enter the collector. Therefore, there is less chance for recombination and less loss of electrons in the base. When the value of V_{CB} gets too high, an effect called

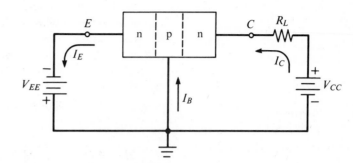

Fig. 2–100 Effect of voltage drop across load resistor on the transistor operation.

punchthrough occurs. The large collector-base voltage causes the effective base width to approach zero so that the emitter electrons (for the npn transistor) move directly to the collector and breakdown occurs. In this case the value of M in Eq. (2–108) increases rapidly and/or Zener breakdown occurs. The breakdown region is shown in Fig. 2–99 to be to the right of the dotted vertical line.

Saturation. The saturation effect is very important for switching circuits. If a load resistor R_L is connected in the collector circuit as shown in Fig. 2–100, then the collector-base voltage $V_{CB} = V_{CC} - I_C R_L$. If the voltage drop across the load, $I_C R_L$, is less than V_{CC}, then the collector-base junction remains reverse biased, and the emitter electrons which cross the collector junction flow into the external circuit. However, when the emitter current I_E is increased sufficiently, the collector current I_C increases proportionately, and the voltage drop $I_C R_L$ across the load can exceed V_{CC}. In this case the collector-base junction becomes forward biased. Now the collector can inject electrons into the base, although at the same time it is receiving emitter-injected electrons to provide the collector current I_C. The net result is that there is a buildup of electron concentration at the collector-base junction, and the transistor is said to be in **saturation.**

If the emitter current is suddenly decreased, the collector current continues to flow for a short time because of the electrons stored at the collector junction. The collector current remains relatively constant for a brief period after cutoff of I_E until the electron concentration at the collector junction returns to zero. This time delay between emitter and collector current changes can be very important in switching circuits, and is often called the **storage-time delay.** It is discussed later in this section. The storage time can be reduced if the collector can be made to act as a poor emitter when forward biased. If the collector is fabricated from a high resistivity

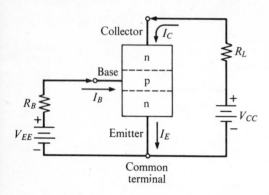

Fig. 2–101 Common-emitter connections for npn transistor.

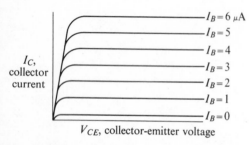

Fig. 2–102 Collector characteristic curves for the common-emitter connection of npn transistor.

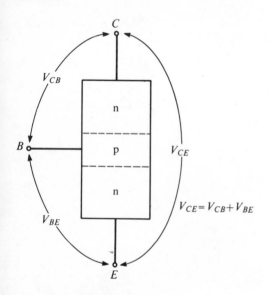

material, its electron injection efficiency is low, and this decreases the storage-time delay.

Common-Emitter Collector Characteristics. When the transistor emitter lead is in common with the base and collector voltage sources, a so-called **"common-emitter"** circuit exists as illustrated in Fig. 2–101. The base-emitter junction is shown forward biased so that electrons that diffuse to the collector junction are accelerated into the collector and move through the external circuit. Again, from Kirchhoff's Law and the direction of the electrode currents shown in Fig. 2–101,

$$I_E = I_C + I_B. \tag{2–110}$$

Typical collector characteristic curves are shown in Fig. 2–102 for the common-emitter connection of the transistor. These curves illustrate the relationship of I_C, V_{CE}, and I_B, and are useful in evaluating several dc parameters and for circuit analysis.

The voltage from collector to emitter, V_{CE}, is equal to the sum of V_{CB} and V_{BE}, as illustrated in Fig. 2–103. If the transistor goes into saturation, the polarity of V_{CB} reverses but V_{CB} always remains smaller than V_{BE}. Since the voltage across a forward biased silicon junction is about 0.6 V, it would be expected that a saturated silicon transistor would have a collector-emitter voltage $V_{CE(sat)}$ less than 0.6 V; it is typically 0.1–0.2 V. The exact value of $V_{CE(sat)}$ depends on both the base current and collector current and can vary from a few millivolts to over one volt.

It should be noted that the relatively high ohmic resistance R_C in the collector material can cause a voltage drop, $I_C R_C$, so that a more accurate expression for the collector-emitter voltage in saturation is

$$V_{CE(sat)} = V_{CB} + V_{BE} + I_C R_C. \tag{2–111}$$

Methods for fabricating transistors are discussed in Appendix B.

Transistor Circuit Symbols and Types. The junction transistor symbols used in schematic diagrams are shown in Fig. 2–104. The arrowhead indicates the direction of conventional collector current flow through the transistor and is always located on the emitter. The symbols may or may not be enclosed in circles, but the circle has no electronic significance.

Fig. 2–103 Voltage relationship for common-emitter connection of npn transistor.

The common-base and common-emitter connections for the transistor were illustrated in Figs. 2–98 and 2–101 respectively. Another method of connecting the transistor is to make the collector lead common to the input and output. The common-collector configuration is usually called an **emitter-follower** circuit. The three basic configurations are illustrated in Fig. 2–105.

If the common connection in each of the circuits is grounded (as indicated by the dotted ground symbol), the circuits in Fig. 2–105 are referred to as grounded-base, grounded-emitter, and grounded-collector circuits respectively.

Switching Circuit Analysis. In the first part of this section, it was shown that a diode could be operated as a switch that is either ON or OFF, depending on the polarity of the applied signal. A junction transistor can act as a switch that is ON or OFF depending on the input current. Since there are three terminals, only one terminal needs to be common to the actuating and switched circuits, as illustrated in Fig. 2–106.

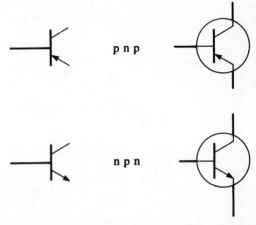

Fig. 2–104 Symbols for pnp and npn transistors.

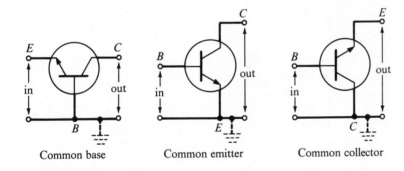

Fig. 2–105 Types of circuit connections for the transistors.

Common base   Common emitter   Common collector

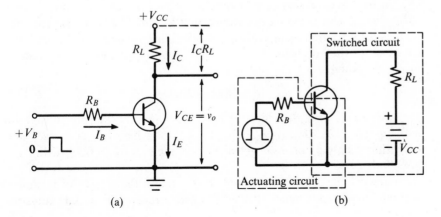

Fig. 2–106 A common-emitter npn transistor switch circuit: (a) typical schematic; (b) outline of actuating and switched circuits.

(a)   (b)

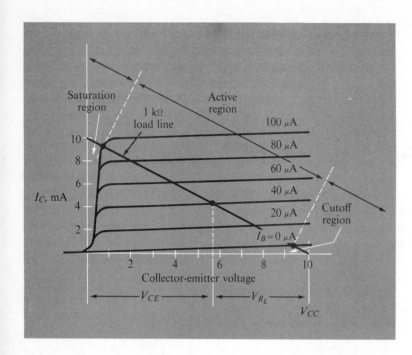

Fig. 2–107 Transistor characteristic curves.

The dc analysis of the common-emitter switch circuit of Fig. 2–106 is done as outlined in Section 2–3.2 by superimposing the load line and *I-V* curves for the collector-emitter circuit (I_C vs. V_{CE}). This is shown in Fig. 2–107 for a typical transistor with a 1 kΩ collector load resistor R_L and $V_{CC} = 10$ V. In the transistor case, there is a separate *I-V* curve for each value of the base current I_B. It is customary to show a few representative curves and to interpolate if any intermediate curves are required. By varying I_B, the operating point can be placed almost anywhere along the load line. In the "active" region, a small change in I_B results in a much larger change in I_C. This is the mode of operation for amplifier circuits. The ratio I_C/I_B is the dc current gain, sometimes called the "dc forward current transfer ratio" or the "dc beta" for short. The symbol for this characteristic is h_{FE} or β_N, where

$$\beta_N = h_{FE} = I_C/I_B. \qquad (2\text{–}112)$$

Because $\alpha_N = I_C/I_E$ and $I_B = I_E - I_C$, it follows that, by substitution into Eq. (2–112), $\beta = I_C/I_B = \alpha_N/(1 - \alpha_N)$, which can be a useful relationship if only β_N or α_N is given. For example, if $\alpha_N = 0.99$, then $\beta_N = 0.99/0.01 = 99$.

In switching circuits one is more interested in the regions of maximum and minimum conductance than in the active region. If the **ON** transistor were a perfect conductor, that is, if $V_{CE} = 0$, the collector current I_C would have the maximum possible value,

$$I_{C(\max)} = V_{CC}/R_L. \qquad (2\text{--}113)$$

If the OFF transistor were a perfect insulator the collector current $I_C = 0$, and

$$V_{CE(\max)} = V_{CC}. \qquad (2\text{--}114)$$

Two points ($V_{CE} = 0$, $I_C = V_{CC}/R_L$) and ($I_C = 0$, $V_{CE} = V_{CC}$) are used to draw the load line. Since $I_C R_L + V_{CE} = V_{CC}$, the collector current can be expressed as

$$I_C = -\frac{1}{R_L} V_{CE} + \frac{V_{CC}}{R_L}. \qquad (2\text{--}115)$$

That is, the collector current I_C can be determined from the slope $(-1/R_L)$ and intercept (V_{CC}/R_L) of the load line. As shown in Figs. 2–106 and 2–107, the sum of V_{CE} and V_{R_L} equals V_{CC} at any point along the load line.

If a voltage of V_B is applied to the input of the circuit in Fig. 2–106a, the base current I_B will flow, and the switch is turned ON if I_B is sufficiently large. Whether the switch can be considered ON can be determined by the point where a specific value of I_B crosses the load line in Fig. 2–107. Usually it is desired that the switch go into saturation, and thus V_{CE} is very small (about 0.1 to 0.2 V).

To ensure that the transistor goes into saturation and the switch is truly ON, the base network should provide a turn-on base current I_{B1} such that

$$h_{FE}I_{B1} > I_{C(\max)}, \qquad (2\text{--}116)$$

where

$$I_{C(\max)} = \frac{V_{CC} - V_{CE(\text{sat})}}{R_L} \qquad (2\text{--}117)$$

and

$$I_{B1} = \frac{V_B - V_{BE(\text{ON})}}{R_B}. \qquad (2\text{--}118)$$

The voltage $V_{BE(\text{ON})}$ is the voltage drop across the forward biased base-emitter junction.

To ensure that the transistor is cut off ($I_E = 0$) and the switch is truly OFF, the base network must prevent the base-emitter junction from being forward biased. Referring to Fig. 2–106, in a cut-off transistor, the base current I_B is equal to the reverse bias collector-base leakage current I_{C0}.

This leakage current is in the opposite direction to the I_B arrow shown in the figure. The actual base-emitter voltage will be equal to the applied actuating voltage (OFF value) plus $I_{C0}(R_B + R_s)$, where R_s is the voltage source resistance. If the OFF value of the actuating voltage is 0 volts as shown and $(R_B + R_s)$ is 10 kΩ, and $I_{C0} = 10$ μA, the base-emitter voltage will be $+0.1$ V. In some applications, this amount of forward bias may produce an unacceptably large emitter current for the OFF state. To avoid this problem, it is necessary to apply a negative value of actuating voltage such that the resulting base voltage is actually slightly negative. Obviously high values of R_B and R_s are to be avoided. If the base circuit is left open $(R_s = \infty)$, the charge leaking into the base from the collector raises the base voltage until the base-emitter junction is forward biased enough to conduct current from the base at the rate I_{C0}. This is an effective base-emitter current source of I_{C0} which results in a collector current of $\beta_N i_B = \beta_N I_{C0}$ according to Eq. (2–112).

In Fig. 2–106 it can be seen that $v_o = V_{CC} - I_C R_L$. When the transistor is cut off the output voltage v_o is at its maximum of nearly $+V_{CC}$ volts. When the base input signal goes to its maximum of $+V_B$ volts so that the transistor saturates, then v_o drops to a minimum value of $V_{CE(\text{sat})}$, about 0.1 to 0.2 V. In other words, the switch of Fig. 2–106 inverts the input signal. It is, therefore, often called an inverter. The grounded-emitter switch always provides an output voltage change of opposite direction with respect to the input voltage change.

Experiment 2–19 *Transistor Characteristics*

The collector characteristic curves for an npn transistor that is connected in a common-emitter circuit are obtained with an oscilloscopic curve tracer. If a scope camera is available, the curves should be photographed. The recorded curves are to be used to evaluate several dc parameters. The values of $V_{CE(\text{sat})}$ and $V_{CE(\text{OFF})}$ are measured and compared with the expected values. The dc beta β_N, or h_{FE}, is determined from the characteristic curves and compared with typical values for the specific type of transistor.

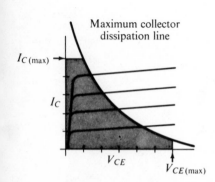

Maximum collector dissipation line

Fig. 2–108 Limiting regions for a transistor.

Voltage and Current Limitations in Transistors. Transistor switching circuits are routinely used in switched circuits where the supply voltage is hundreds of volts or where the current switched can be several hundreds of amperes. Despite this great versatility, there are a number of ways of destroying a transistor through careless circuit design. Three limitations encountered under normal bias conditions are shown in Fig. 2–108. The maximum collector dissipation in watts is $I_C V_{CE}$. The safe (with adequate cooling) region is the shaded area. The load line should normally stay within the safe boundaries. It is possible, in switching applications, for the load line to go through the area outside the maximum dissipation line if both operating points are well within the maximum limitations and the transition time is very short.

In addition to the above limitations, there are two breakdown voltages that must not be exceeded unless there is some protection against the damaging currents which could result from the breakdown. The breakdown voltages are BV_{CBO}, the maximum inverse collector-base voltage, and BV_{CEO}, the maximum collector-emitter voltage.

Transistor Switching Time. The transistor switch cannot respond instantaneously to a turn-on or a turn-off actuating signal. In many applications it is important to be aware of the magnitude of possible errors caused by a time delay in the switching circuit and how to minimize the time lags. In Fig. 2–109 the response of a transistor switching circuit is shown when a turn-on and turn-off actuating signal V_B is applied to the input circuit of Fig. 2–106.

Each segment of the turn-on and turn-off times illustrated in Fig. 2–109 will be considered briefly. First, there is the delay time, t_d. The base-emitter input capacitance must charge through R_B before the base-emitter junction is actually forward biased; there is a finite transit time for the first minority carriers to cross the base region; and some time is required for the collector current to reach 10% of its final value.

Next, there is the rise time, t_r. This is the time needed to establish the concentration of minority carriers in the base region (i.e., the electron density in the p-type base of the npn transistor) which is required to carry 90% of the final value of I_C. The combination of t_d and t_r is the turn-on time, t_{ON}. The rise time can be substantially reduced if the base current is larger than the minimum needed for saturation. The collector current rises toward $h_{FE}I_B$ with a characteristic time constant, but cannot, of course, exceed $I_{C(max)}$. However, increasing I_B greatly beyond the current required to saturate the transistor substantially increases the steady-state excess minority carrier concentration in the base region during saturation.

Assuming now that the actuating signal was of sufficient duration and magnitude to saturate the transistor, it is important to consider the turn-off behavior. First, there is the storage time, t_s. This is the time necessary to clear the collector-base junction of excess minority carriers, as discussed under saturation earlier in this section, and to decrease I_C to 90% of its maximum value. Since the saturation of the transistor occurs as soon as the collector-base junction becomes forward biased, any further increase of base current causes excess charge to accumulate in the base region. Then, when the forward bias on the base-emitter junction is suddenly re-

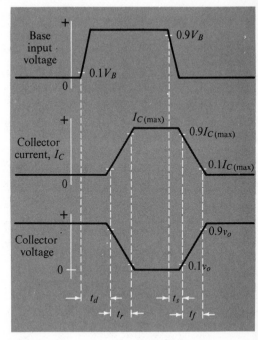

Fig. 2–109   Transistor switching time.

moved, the excess base charge flows into the collector. At the instant the potential gradient at the collector-base junction reaches zero, the collector no longer emits carriers into the base region, and the transistor comes out of saturation.

The fall time, t_f, is the time required for the output current to fall from 90% to 10% of its maximum value, and depends on the time required to discharge the collector-emitter capacitance and the time required for the minority carriers in the base to be collected so that there is zero density gradient at the emitter junction as well as at the collector junction. This is illustrated in Fig. 2–110, where the charge density gradients in the base between the emitter and collector junctions are drawn.

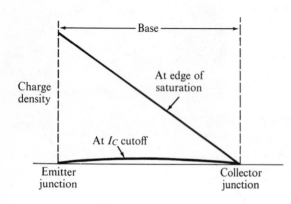

Fig. 2–110 Charge density gradients in the base.

To summarize then, the important switching times are as follows:

t_d: Delay time or "turn-on delay," the finite time that elapses between application of the base input voltage and the start of collector current flow in the transistor. The time t_d is measured at values of input voltage and output current that are 10% of the maximum values.

t_r: Rise time, the time required for the collector current to increase from 10% to 90% of its total change.

t_s: Storage time, the time required for the collector current to decrease to 90% of its maximum value after the input has decreased to 90% of its maximum value.

t_f: Fall time, the time required for the output waveform to fall from 90% to 10% of its maximum value.

t_{ON}: Turn-on time, the sum $t_d + t_r$.

t_{OFF}: Turn-off time, the sum $t_s + t_f$.

It should be noted that the various transistor switching times are usually measured by using the V_{CE} output waveform because it is easy to display on the oscilloscope. This is valid only for cases where the load is resistive.

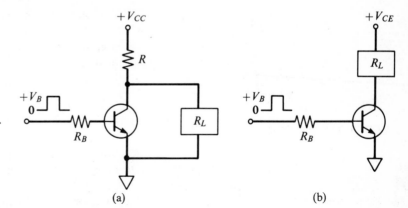

Fig. 2–111 Connection of a load to a BJT switch.

(a)   (b)

Connection of Load. A basic application of the transistor switch is to control current in an external load. There are two ways to connect the load, as illustrated in Fig. 2–111. When the load is placed in *shunt* with the transistor, a resistance R is connected in series between the parallel network and the supply voltage. When the switch is OFF, current flows through the load and the series resistance R. When the switch is ON, the load is effectively short-circuited and the current is steered from the load into the transistor. When the external load is in series with the transistor, the current through the load is the same as the collector current I_C.

If the external load requires a larger current than can be supplied by one transistor that is controlled by a current-limited base control voltage source, it is possible to cascade transistors as shown in Fig. 2–112.

The base-emitter junction of Q_2 is the shunt load of Q_1. The base current of Q_2 is dependent on the supply voltage V_{CC}, the Q_1 load resistor R_{L_1}, and $V_{BE(ON)}$. The current $I_{B(Q_2)} = (V_{CC} - V_{BE(ON)})/R_{L_1}$.

Whenever Q_1 is ON the base current for Q_2 is cut OFF, and when Q_1 is OFF there is a base current $I_{B(Q_2)}$ for Q_2. The current through the load is equal to the collector current for Q_2 because it is in series with Q_2. If resistor R_{L_1} is chosen such that the base current of Q_2 causes saturation (see Eqs. 2–116 through 2–118), the current in the load $I_{C(Q_2)} = (V_{CC} - V_{CE(sat)})/R_{load}$.

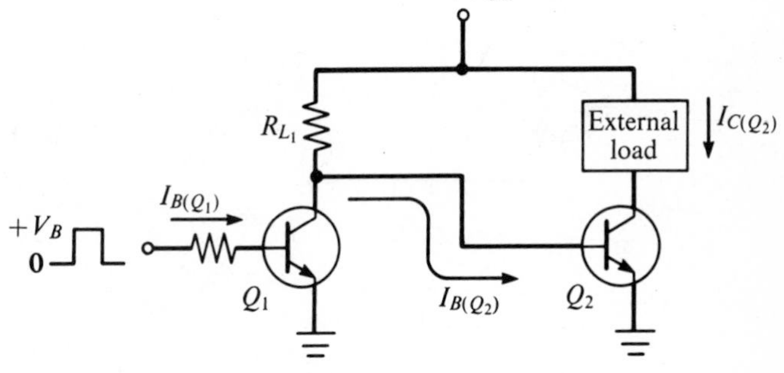

Fig. 2–112 Cascaded transistors for large loads.

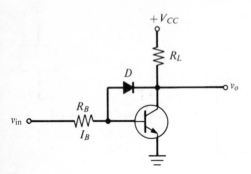

Fig. 2–113 Nonsaturating transistor switch with diode clamp.

Note 5. Clamping Diode.

A germanium or a Schottky diode is best to use because it conducts sooner and ensures that the transistor does not get close to saturation. A silicon diode could be used because the clamping diode carries only a low current compared with the transistor base-emitter junction, and consequently the forward voltage drop across the clamping diode is less than V_{BE}.

Note that with the circuit of Fig. 2–112 the current in the load is OFF when the base voltage of transistor Q_1 is positive, and vice versa. If it is desirable to have the current in the load go ON when the input goes positive, it is necessary to add a transistor which acts as an inverter between the two transistors in Fig. 2–112 or connect the emitter of Q_1 to the base of Q_2. In the latter case, Q_1 acts to switch ON the base current of Q_2.

Nonsaturating Switching Circuits. The storage time caused by allowing a transistor to saturate significantly limits the switching speed. The problem of storage time does not arise if the transistor is not allowed to saturate, and thus an improvement in switching speed is possible.

One method of providing a nonsaturating transistor switch is to clamp the collector in such a way that the collector-base junction cannot become forward biased. One method of clamping is shown in Fig. 2–113. The diode[5] D becomes forward biased as the transistor begins to leave the active region, at the edge of saturation. Therefore, the collector never becomes sufficiently negative with respect to the base for the collector-base junction to become forward biased, and the transistor cannot saturate. The excess base current is conducted away through the diode. The Schottky diode is especially effective in this application. Its forward voltage drop is less than that of the collector-base junction, thus keeping the transistor farther from saturation. In addition, the Schottky diode has essentially no stored charge itself. For these reasons, the storage time is almost completely eliminated in a Schottky clamped transistor.

Another method of providing nonsaturated switches is to control the collector current. A circuit utilizing transistors operated in the common-base configuration is shown in Fig. 2–114. When the input signal is negative, transistor Q_1 is cut off, but the base-emitter junction of Q_2 is forward biased, and thus Q_2 is ON and the emitter current $I_E \approx V_{EE}/R_E$. Now when the input signal goes positive the transistor Q_1 goes ON and the voltage drop across R_E causes the base-emitter junction of Q_2 to become reverse biased and Q_2 goes OFF. The current is in effect "steered" to either Q_1 or Q_2 by the input signal. The steering depends on both the polarity

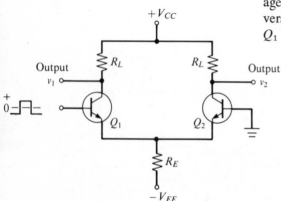

Fig. 2–114 Nonsaturating transistor switches.

and the amplitude of the input signal. The collector-base junction will always be at some value of reverse or zero bias if $I_C R_L \leqslant V_{CC}$, and the transistor will thus stay out of saturation.

The outputs v_1 and v_2 are complementary, with v_2 in phase with the input signal. A change of input signal of only a few tenths of a volt is required to switch states.

Experiment 2–20 *The Basic Bipolar Junction Transistor (BJT) Switch*

The basic BJT switch circuit is connected and tested so that its important characteristics can be observed. The inverting action of the transistor is first observed. If a high speed oscilloscope is available, the switching times are measured. Then the transistor switch is connected to a load so as to control the current in an external load. The nonsaturating transistor switch circuit is also connected and tested and its characteristics are compared with those of the saturated switch.

The Unijunction Transistor

The unijunction transistor (UJT) has some unique features as compared to the two-junction npn and pnp transistors. It is a three-terminal device with only one pn junction. The inherent stability of the characteristic parameters provides advantages for certain circuit applications.

Characteristics and Biasing. The UJT is conveniently considered as a bar or strip of silicon to which a pn junction is made, as illustrated in Fig. 2–115a, although present structures of the UJT deviate from the bar structure for simplicity and economy of production. The leads from the two ends are called base 1 and base 2. The rectifying contact is called the emitter, and it is represented as being between base 1 and base 2 on the bar of silicon. The schematic circuit symbol is shown in Fig. 2–115b. The

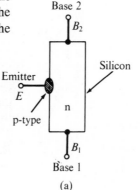
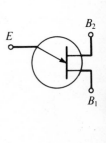

Fig. 2–115 Unijunction transistor: (a) structure representation; (b) circuit symbol.

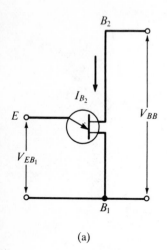

(a)

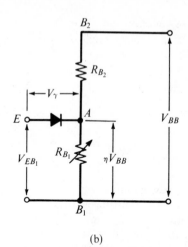

(b)

Fig. 2–116 Unijunction transistor: (a) nomenclature; (b) simplified equivalent circuit.

operation of the UJT is readily illustrated by consideration of the equivalent circuit in Fig. 2–116.

If a voltage V_{BB} is applied so that base 2 is positive, a current will flow between base 2 and base 1 as determined by the interbase resistance R_{BB} (typically 7 kΩ) of the silicon bar, so that

$$I_{B2} = V_{BB}/R_{BB}. \qquad (2\text{–}119)$$

The resistance R_{BB} can be considered as consisting of the resistances R_{B1} and R_{B2} in series, acting as a voltage divider. A fraction η (typically 0.6) of the applied voltage (that is, ηV_{BB}) is divided between the emitter contact and base 1. Therefore, the pn junction at the emitter is reverse biased and only a small leakage current normally flows in the emitter lead.

When the voltage V_E that is applied between the emitter and base 1 is increased, a voltage is reached where V_{EB_1} is equal to the sum of the voltage across R_{B1} and the forward voltage drop of the pn junction V_γ. This voltage is known as the peak-point voltage V_p,

$$V_p = V_\gamma + \eta V_{BB}. \qquad (2\text{–}120)$$

For emitter-base voltages greater than V_p the pn junction is forward biased, so that holes are injected from the emitter into the silicon bar. Because base 1 is negative with respect to the emitter, the electric field is such that most holes move toward the base 1 terminal. An equal number of electrons are injected from base 1 to maintain electrical neutrality in the bar. The increase in current carried in the silicon bar decreases the value of R_{B1}. This causes the fraction of voltage between point A and base 1 to decrease, and this in turn causes a further increase of emitter current I_E and a lower resistance for R_{B1}.

A plot of V_E vs. I_E is shown in Fig. 2–117, which shows the region of negative resistance where the voltage across R_{B1} decreases as I_E increases. The region of negative resistance occurs between the peak-point voltage V_p and the so-called valley-point voltage V_v. At the valley-point voltage and higher voltages the density of charge carriers is so high that the lifetime of the carriers is decreased. This decrease counteracts the effect of new carriers being generated, and thus the emitter voltage V_E increases gradually at currents above I_v. The region to the right of the valley point is known as the saturation region, where the dynamic resistance is positive.

The emitter characteristic curve given in Fig. 2–117 is not drawn to scale, in order better to show the various operating regions. A typical emitter characteristic curve drawn to scale is shown in Fig. 2–118, where only a small portion of the cutoff region is shown. The peak-point voltage is about 16 V and is reached at a forward emitter current of about 10 pA. The voltage then remains constant until about 0.1 μA, where it starts to decrease. The valley-point voltage of 1.6 V is reached at a valley current of about 8 mA. The saturation resistance can be determined by the slope of the I-V curve in the saturation region, and is seen to be about 5 Ω.

Peak-Point Voltage. The peak-point voltage V_p is the most important characteristic of the unijunction transistor because it determines the trigger or switch point in its circuit applications. The value of V_p is expressed in Eq. (2–120) as a function of the intrinsic standoff ratio η.

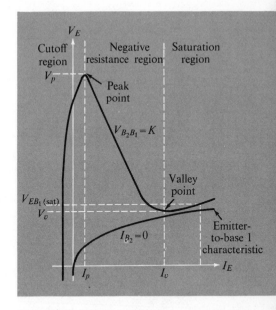

Fig. 2–117 Static emitter characteristic curves for unijunction transistor. (Courtesy of General Electric Co.)

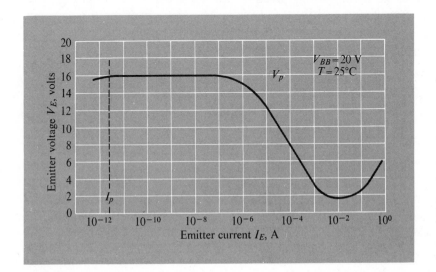

Fig. 2–118 Static emitter characteristics for unijunction transistor. (Courtesy of Motorola Inc.).

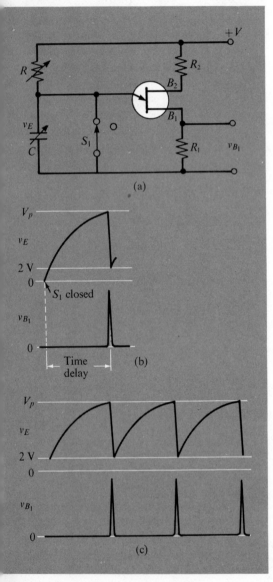

(a)

(b)

S_1 closed

V_p

v_E

2 V

0

v_{B_1}

0

Time delay

V_p

v_E

2 V

0

v_{B_1}

0

(c)

Intrinsic Standoff Ratio. A rearrangement of Eq. (2–120) provides the definition of the intrinsic standoff ratio,

$$\eta = \frac{V_p - V_\gamma}{V_{BB}} = \frac{R_{B1}}{R_{BB}}. \qquad (2\text{–}121)$$

This ratio is not completely independent of temperature but the temperature coefficient is quite small, about $0.06\%/°C$. The intrinsic standoff ratio is also slightly dependent on the supply voltage V_{BB}, but the variation is generally negligible for practical purposes.

UJT Trigger Circuit. One of the most important uses of the UJT is to provide a precise time-delayed trigger pulse for activating other devices. The basic UJT trigger circuit is shown in Fig. 2–119a. The emitter–base 1 voltage is provided by the voltage across the timing capacitor C. At the start, capacitor C is shorted by switch S_1. When S_1 is opened, the capacitor charges through resistor R from the power supply. The emitter voltage v_E thus increases exponentially toward the supply voltage $+V$ with a time constant RC. When the emitter voltage reaches the peak-point voltage V_p, the UJT turns ON and the capacitor discharges through R_1. When v_E reaches about 2 V, the UJT turns OFF. Since R_1 is usually a small resistance value ($<100\ \Omega$), the discharge of timing capacitor C is usually quite rapid, and a sharp trigger pulse is obtained across R_1 as shown in Fig. 2–119b. If only a single timing pulse is desired, capacitor C must be shorted by closing switch S_1 when the UJT turns OFF. The time delay is approximately given by the RC time constant and can be varied within limits by varying R and/or C. The upper limit on the time delay depends on the desired timing precision, the V_p of the UJT, the leakage current of capacitor C, and the temperature. The upper limit for R is determined by the requirement that the current to the UJT emitter be large enough for it to fire (larger than I_p).

If switch S_1 is omitted or left open, capacitor C will begin to recharge after the emitter voltage falls to about 2 V and the UJT turns OFF. Again C will charge until v_E reaches V_p, at which time discharge through R_1 occurs. This process keeps repeating itself as shown by the waveforms in Fig. 2–119c. Because of the repetitive nature of the waveforms at the emitter and across R_1, this circuit is known as a relaxation oscillator, and has many important applications.

Fig. 2–119 Basic UJT trigger: (a) circuit; (b) waveforms for single pulse; (c) waveforms for repetitive operation (relaxation oscillator).

Experiment 2–21 *Unijunction Transistor (UJT) Trigger Circuit*

A unijunction transistor is combined with an *RC* network to provide a simple time delay and trigger circuit. The range of useful time delay is investigated and the circuit signals are observed.

The SCR, Triac, and Other Thyristors

There are a variety of important semiconductor switches which are pnpn four-layer devices, called **thyristors.** These devices find numerous applications as switches for controlling ac and dc power, as will be discussed in Section 2–3.7. The term thyristor applies to any pnpn semiconductor switch. The most common of these devices is the silicon controlled rectifier (SCR), which is a unidirectional three-terminal switch (current can flow in one direction only). Such unidirectional three-terminal switches are often called **reverse blocking triode thyristors.** Other important reverse blocking triode thyristors are the silicon unilateral switch (SUS), the gate turn-off switch (GTO), the programmable unijunction transistor (PUT), and several light-actuated switches which will be described in Section 2–3.5. There are other unidirectional or reverse blocking thyristors such as the Shockley or four-layer diode, which is classified as a reverse blocking diode thyristor. The silicon controlled switch, which has two control inputs, is a reverse blocking tetrode thyristor. Other thyristors can conduct current in both directions and are known as **bidirectional thyristors.** The most important of these are bidirectional triode thyristors, such as the triode ac switch (triac) and the silicon bilateral switch (SBS). The characteristics and general mechanisms for switching thyristors are first considered. Then specific pnpn devices are briefly described.

Two-Transistor Representation of pnpn Devices. In Fig. 2–120a the pnpn device is shown as a four-layered pellet of alternating p- and n-type semiconductor material (usually silicon). The four-layer diode has leads

Fig. 2–120 Illustration of external lead connection for various pnpn devices.

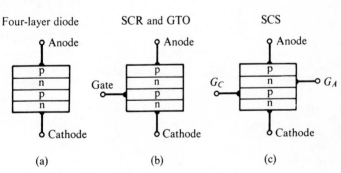

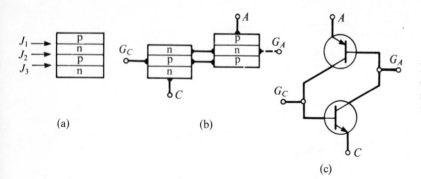

(a)

(b)

(c)

Fig. 2–121 Two-transistor representation of pnpn device. (a) Junctions indicated. (b) Hypothetical separation into two transistors. (c) Schematic of two-transistor equivalent for pnpn device.

connected only to the end layers of semiconductor. The SCR has three layers accessible by external leads, and the SCS has all four layers accessible, as shown in Fig. 2–120b and c respectively. Each layer interacts with its adjacent layers as in the transistor. It can be seen in Fig. 2–120 that, starting from one end of the stack, three adjacent layers form a pnp transistor, and from the other end an npn transistor. This two-transistor concept is illustrated in Fig. 2–121, where it is apparent that the pnpn structure can be considered as a complementary (npn-pnp) transistor feedback pair. The collector of each transistor is direct-coupled to the base of the other, providing a positive feedback loop. That is, a small change in base current of the npn transistor causes a larger change in its collector current, which is the base current of the pnp transistor. The collector current of the pnp transistor also contributes to the base current of the npn transistor. Therefore, what begins as a small change in the base current of the npn transistor becomes a large change because of positive feedback.

If the cathode of the pnpn device is connected to the negative terminal and the anode to the positive terminal of the power supply, the center pn junction (J_2 of Fig. 2–121a) is reverse biased. Therefore, the device does not conduct unless the current gain around the feedback loop is approximately unity. This is seen more clearly by considering the equations expressing the total anode-to-cathode current I_A.

If the current gains of the npn and pnp transistors are h_{FE1} and h_{FE2} respectively, then the current gain G_l of the internal feedback loop is

$$G_l = h_{FE1} h_{FE2}. \tag{2-122}$$

With the base leakage current of the npn designated as I_{C01} and that of the pnp as I_{C02}, it follows that

$$I_{C1} \text{ (for the pnp)} = h_{FE1}(I_{C2} + I_{C01}) + I_{C01} \tag{2-123}$$

and

$$I_{C2} \text{ (for the npn)} = h_{FE2}(I_{C1} + I_{C02}) + I_{C02}. \tag{2-124}$$

The total anode-to-cathode current I_A is

$$I_A = I_{C1} + I_{C2}. \qquad (2\text{--}125)$$

Combining Eqs. (2–123), (2–124), and (2–125), we obtain

$$I_A = \frac{(1 + h_{FE1})(1 + h_{FE2})(I_{C01} + I_{C02})}{1 - h_{FE1}h_{FE2}}. \qquad (2\text{--}126)$$

When the anode is positive with respect to the cathode, the center section is reverse biased and h_{FE1} and h_{FE2} are very small compared to 1, and thus $I_A \approx I_{C01} + I_{C02}$, that is, the sum of only the junction leakage currents. Therefore, the device is in its high resistance OFF state. However, it can be seen from Eq. (2–126) that as $G_l \rightarrow 1$, $I_A \rightarrow \infty$. In other words, as the loop gain approaches unity the circuit feedback becomes sufficient for each transistor to drive the other into saturation, and the device goes to its low resistance or ON state. When the device is ON, the anode current is limited only by the external circuit.

Mechanisms for Switching pnpn Devices. Any mechanism that increases h_{FE} so that $G_l \rightarrow 1$ will turn the device ON. These mechanisms are all based on the emitter-current dependence of h_{FE} as illustrated in Fig. 2–122. When the emitter current is very low the gain h_{FE} is low, but if anything temporarily increases I_E sufficiently, then the pnpn device is rapidly switched ON. Some important mechanisms for producing the necessary emitter current to switch the device ON include anode-to-cathode voltage, rate of voltage change, temperature, transistor action, and radiant energy.

When the applied voltage across the device gets high enough there is avalanche breakdown, and when the avalanche current causes $G_l \rightarrow 1$ the device is switched ON. This is the normal method used to switch the pnpn diodes into conduction. These diodes can be constructed to switch ON at certain selected anode-to-cathode voltages.

As previously discussed, any pn junction has capacitance. Therefore, when a step function is impressed across the pnpn device a charging current $i = C(dV/dt)$ will flow from anode to cathode to charge the device capacitance. If the current i is sufficient for $G_l \rightarrow 1$ the device switches ON; this mechanism is known as "the dV/dt effect."

Because the leakage current in a silicon pn junction doubles for every 8°C increase in temperature, it is possible for $G_l \rightarrow 1$ and the device to switch ON from high junction temperature.

A relatively small base current can turn a transistor ON. Likewise, for the pnpn device the injection of current carriers into either "gate" (base region) can switch the device ON. This is the normal method of switching the SCR and the SCS.

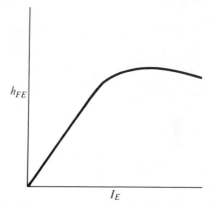

Fig. 2–122 Dependence of h_{FE} on emitter current in a silicon transistor.

The pnpn devices can be turned OFF by the removal of the anode-to-cathode dc supply voltage, or if the supply voltage is ac the device will turn OFF when the voltage is decreased sufficiently. To maintain the device in conduction after a gate signal is removed, it is necessary to exceed a certain load current called the **holding current.**

The junctions in the SCR are all forward biased when the device is conducting, and the base regions are heavily saturated with current carriers. Therefore, to turn the SCR OFF in a minimum time, it is necessary to apply a reverse voltage. This causes the electrons and holes in the region of the two end junctions, J_1 and J_3 in Fig. 2–121, to diffuse to these junctions so as to cause a reverse current in the external circuit. Therefore, the voltage across the SCR will remain at about 0.7 V. When the holes and electrons are removed from the region of the junctions the reverse current ceases and the junctions are again in the blocking state. The reverse voltage across the SCR is now determined by the applied supply voltage. However, the recovery of the SCR is not complete until the center junction is cleared of excess charge carriers, and this depends primarily on the recombination process. This process is quite independent of external applied voltage. When the excess charge carriers at the center junction have been reduced to a low value, it is again safe to apply a forward voltage to the SCR without danger of turn-on. It usually requires about 10 μsec after cessation of the forward current flow before the forward voltage may be safely reapplied. This delay time is known as the **turn-off time.**

Reverse Blocking Thyristors. In Fig. 2–123 the circuit symbols are shown for the common types of unidirectional thyristor switches. All of these devices have current-voltage characteristics which are quite similar to that shown in Fig. 2–124 for the SCR. As can be seen from the characteristic curve, the reverse blocking type thyristors conduct in only one of the four quadrants unless the reverse voltage exceeds the reverse breakdown voltage, which is highly undesirable. The major differences between the devices are in the number of control gates (number of terminals) and in the gate construction, which can govern the turn-off characteristics.

As can be seen from Fig. 2–123d the four-layer diode (Shockley diode) has only two terminals and must be actuated by a signal applied at the anode. As such it is not nearly as useful as other pnpn devices, which can be actuated by signals applied to control gates.

The SCR has one cathode control gate, as does the gate turn-off switch (GTO). The GTO has a special gate construction that allows it to be turned OFF by a small reverse gate current. The GTO can also be turned OFF in the same way as a conventional SCR. For dc application the GTO enables the turn-off circuitry to be simpler than that for the SCR. However, when GTO's are conducting high currents, there is difficulty in

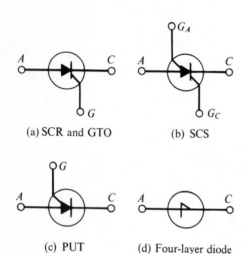

(a) SCR and GTO   (b) SCS

(c) PUT   (d) Four-layer diode

Fig. 2–123 Circuit symbols for common reverse blocking thyristors: A = anode, C = cathode, G = gate.

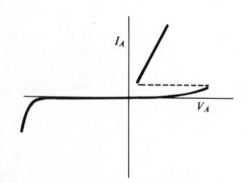

Fig. 2–124 Anode current-voltage characteristic for reverse blocking thyristors.

maintaining turn-off control by the gate. Consequently GTO devices are operated at much lower currents than SCR's. In recent years GTO devices have been replaced in many applications by high voltage–high gain silicon power transistors.

The silicon controlled switch (SCS) is a four-terminal device as shown in Fig. 2–120c and Fig. 2–123b. The control gate connections are made to both the lower p-base (G_C) and the upper n-base (G_A). Hence the SCS can be turned ON by either injecting current at G_C or withdrawing current at G_A. Triggering a thyristor by withdrawal of current at the anode gate requires more current than triggering a conventional SCR because the upper n-base area is constructed of high resistivity silicon. Also n-base triggering necessarily removes current from the loop and reduces the loop gain G_l.

The programmable unijunction transistor (PUT) is a thyristor with an anode gate rather than the usual cathode gate of the SCR. The PUT conducts when the anode voltage exceeds the anode gate voltage by the forward voltage drop across one pn junction. At this voltage the peak-point voltage of the PUT is reached and conduction occurs. The major difference between the PUT and the normal UJT is that the peak-point voltage can be controlled in the PUT by controlling the anode gate voltage, whereas the UJT turns ON at a constant fraction η of the supply voltage. PUT's have important applications in time-delay circuits, trigger circuits for high current SCR's, and relaxation oscillators. One important characteristic that makes the PUT more suitable for triggering SCR's than normal UJT's is the higher peak pulse current output of the PUT. Switching times are normally in the 10 nsec region, which makes the PUT quite valuable.

The final reverse blocking thyristor to be discussed in this section is the silicon unilateral switch (SUS). The circuit symbol for the SUS is shown in Fig. 2–125, along with an equivalent circuit and the current-voltage characteristic. The SUS is a miniature SCR with an internal low voltage avalanche diode between the gate and cathode and an anode gate.

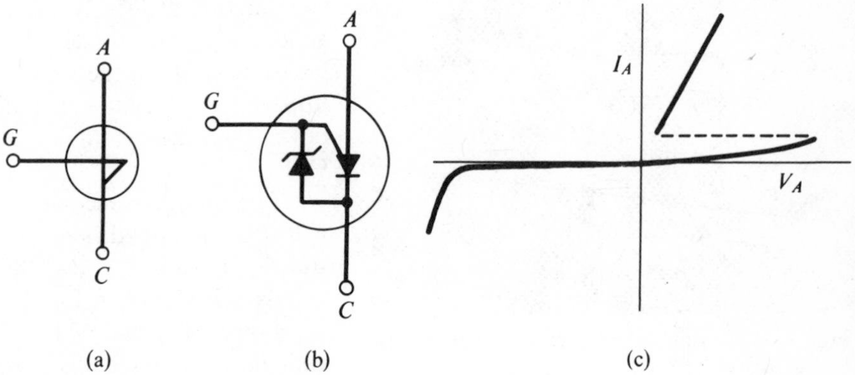

Fig. 2–125 Silicon unilateral switch (SUS): (a) circuit symbol; (b) equivalent circuit; (c) current voltage characteristic.

(a)   (b)   (c)

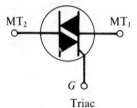

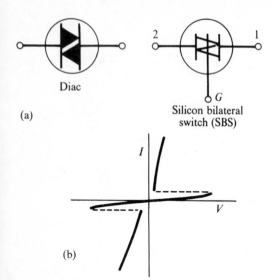

Diac

Silicon bilateral
switch (SBS)

Triac

(a)

(b)

Fig. 2–126 Common bidirectional thyristors:
(a) circuit symbols; (b) current-voltage characteristic.

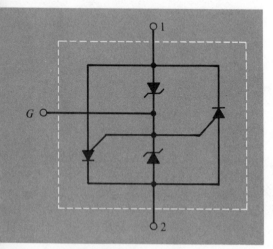

Fig. 2–127 Equivalent circuit of silicon bilateral switch (SBS).

The SUS is usually used in relaxation oscillator circuits. The SUS functions similarly to the UJT except that it switches at a fixed voltage determined by the breakdown voltage of the built-in avalanche diode, rather than at a fraction (η) of the supply voltage as does the UJT.

Bidirectional Thyristor Switches. Circuit symbols for the most common bidirectional thyristors are shown in Fig. 2–126a. These devices all have current-voltage characteristics similar to that shown in Fig. 2–126b for the triac. As can be seen, these devices can conduct current in either direction and hence are quite valuable in ac switching circuits. The diac is a bi-directional diode thyristor since it has only two terminals, while the triac and SBS are bidirectional triode thyristors. Since the diac has limited application, this discussion will focus on the operating principles of the SBS and the triac. In Section 2–3.7, applications of triacs in power control circuits are discussed in detail.

The SBS is very similar to the SUS except for its bidirectional current conduction characteristics. The equivalent circuit of the SBS is shown in Fig. 2–127, and illustrates that the SBS is equivalent to two SUS's con-nected in reverse-parallel fashion. The control gate is at the anodes of the two internal PUT's, while avalanche diodes are connected between the anodes and the cathodes. The SBS finds its major application in triggering triacs, since it can be switched with voltages of either polarity. The basic SBS trigger circuit is a relaxation oscillator with an ac supply voltage rather than the dc supply voltages used with unidirectional thyristors and UJT's.

The triac has become such an important ac power control device that it will be described here in some detail. The basic pellet structure of the triac is shown in Fig. 2–128a, while a more complete current-voltage char-acteristic is shown in Fig. 2–128b. The terminals in the triac are normally numbered MT_1 and MT_2 (for main terminals 1 and 2) since anode and cathode terminology is not applicable. Terminal MT_1 is taken as the reference terminal for voltage measurements; current measurements are taken at terminal MT_2 and gate G. Also identified in Fig. 2–128b are two quadrants I and III. As can be seen, conduction with the triac occurs in quadrants I and III. In quadrant I the voltage at MT_2 is positive with respect to MT_1, while the voltage at MT_2 is negative with respect to MT_1 in quadrant III. Also identified in Fig. 2–128b are two important voltages $+V_{BO}$ and $-V_{BO}$, known as breakover voltages. In normal triac use it is desirable to induce conduction by signals applied to the control gate. Thus the ac voltage to be switched should always be less than V_{BO}, since a voltage exceeding the breakover voltage will cause conduction. This relinquishes control over conduction by the gate. Even transient voltages in excess of V_{BO} will cause conduction. The triac remains in its conducting state until the current falls below the holding current, I_H, marked in Fig. 2–128b. One desirable aspect of this feature is that the triac is self-protected against transient voltages exceeding V_{BO}. Hence triacs are not damaged by high voltage transients, although switching the triac into conduction may cause damage to circuits being controlled by the triac.

The triac is an extremely versatile switch and can be triggered by either positive or negative gate currents so long as V_{BO} is not exceeded. Triggering can be accomplished with dc signals, ac signals, or pulses. UJT's, neon lamps, and SBS relaxation oscillators are commonly used. The triggering and conduction characteristics are usually identified by specifying the quadrant in which conduction takes place and the sign of the gate current. Thus the symbol I (+) signifies that the triac is switched in quadrant I by a positive gate current, while I (−) signifies triggering in quadrant I with a negative gate current. The other two possibilities are III (+) and III (−) triggering.

The most sensitive modes of triggering are the I (+) and III (−) modes, although the I (−) mode is only slightly less sensitive. Triggering in the III (+) mode is rather insensitive and often avoided, although specially selected triacs are available for this purpose.

The versatility and simplicity of triac switching have brought about many important ac power control applications, which are discussed in Section 2–3.7. Triacs are commercially available at present with maximum current ratings of up to 40 A and maximum voltages of up to 600 V.

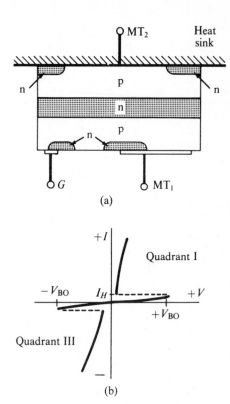

Fig. 2–128 The triac: (a) pellet structure; (b) current-voltage characteristic.

Experiment 2–22 *pnpn Device Characteristics*

A curve tracer is constructed and the current-voltage curves of two common pnpn devices, the SCR and the triac, are obtained. This experiment also demonstrates the relationship between the turn-on time of the thyristor and the gate current.

2–3.4 INFORMATION TRANSMISSION GATES

In an electronics-aided measurement, the information about the quantity being measured is first converted into an electrical quantity. This quantity is an **electrical signal.** In a complete measurement, the signals representing the measurement data may proceed through an electronic signal modifier (such as an amplifier) and then through one or more data domain converters (such as a current-to-voltage converter and a recorder). In instruments of even modest complexity there are a number of signals which represent different characteristics of the system being measured. Examples of related multiple signals include the current through a device and the voltage across it, the voltages from 10 thermocouples measuring a temperature profile, and voltages representing the pH of a solution and its temperature. As the instrument performs the measurements on such signals, certain signal (information) paths must be opened and others closed. Also, many instruments provide several different measurement functions which are achieved by altering the signal paths through the various data modification and transformation elements in the instrument. Circuits used to direct and interrupt the paths of electrical signals are called **information transmission gates.**

The basic elements in information gates are the switching devices described earlier. The application of several switching devices in gating circuits is discussed in this section. While all information transmission gates operate on the same basic switching concepts, there is a significant variation in design and implementation depending on the accuracy of signal transmission required. Two extremes of accuracy requirement are considered in this discussion: analog transmission gates and binary information gates. In the former, any distortion of the electrical signal quantity represents a proportional distortion of the data represented by the signal. Binary information.gates are used with signals representing simple two-state information, such as "the temperature is or is not greater than 42°." With such signals, one range of voltage (or current) values represents one signal state (such as "greater than") and another quite different range of voltage values represents the other state (such as "*not* greater than"). With such binary information signals, considerable distortion of signal level can occur with no degradation of the information content of the signal.

Analog Transmission Gates

A basic transmission gate is shown in Fig. 2–129. When the gate is **open** or ON, the signal at the gate input appears at the output. When the gate is **closed** or OFF, the output signal is zero. The signal level at the gate control input determines whether the gate is open or closed. An analog gate is designed to transmit from its input to its output an exact reproduction of the

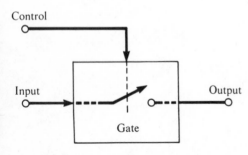

Fig. 2–129   Basic gate.

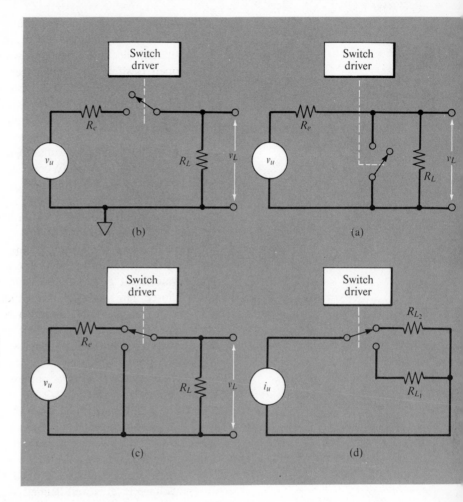

Fig. 2–130 Basic types of analog switches: (a) shunt; (b) series; (c) series-shunt voltage switch; (d) series-shunt current switch.

input waveform during the selected interval when it is open and to have a zero output when it is OFF. Other terms used for such a device are analog switch, linear switch, sampling gate, transmission gate, and linear gate. The simplest analog gate could be a manual switch, but the term gate generally refers to an automatically actuated switch. The first analog gates were made from relays. The modern reed relay form of the mechanical contact gate is still one of the most accurate and simplest analog gates to implement.

The actual switching devices can be arranged in the gate circuit in several different ways. The most common four arrangements are shown in Fig. 2–130 along with the signal sources and loads. The choice among these possibilities depends on the switch characteristics and the gate application requirements. The simple series switch gate is shown in Fig. 2–130b where R_e is the equivalent source resistance. When made from an ideal switch, the output voltage would be $v_u R_L / (R_e + R_L)$ when the gate is open (switch closed) and zero volts when the gate is closed. In order to approach this ideal, it is necessary that the switch's ON resistance, R_{ON}, be very much less than R_L and that its OFF resistance, R_{OFF}, be very much greater than R_L. With the optimum value of R_L it is possible that a transmission accuracy of 0.1% would require a switch with an R_{OFF}/R_{ON} ratio of 10^6 or more. Note that in the closed state, the output terminal is "floating." That is, its only connection to common is through the load. This is an advantage when other signal sources are to be gated to this same load when this gate is closed, but it is a disadvantage if a well-defined zero voltage output is required for the closed-gate state.

The gate circuit of Fig. 2–130a employs a shunt switch. The gate is open when the switch is open, producing the same ideal output voltage as the series switch above. When the switch is closed the switch shorts the gate input and output terminals. To provide a good zero output in the closed-gate state requires that $R_{ON} \ll R_e$. Practical signal sources and switches generally require a resistor in series with the source to increase the effective R_e. At the same time, R_{OFF} must be very much larger than R_L to avoid distorting the signal voltage.

Better switching efficiency with fewer demands on the source and load resistance values are obtained with a combination of series and shunt switches as in Fig. 2–130c. When the gate is open, the circuit acts like the series switch; and when the gate is closed, the source is disconnected (as in the series switch) but the load is shorted (as in the shunt switch). Thus accurate transmission requires that $R_{ON} \ll R_L$ and, for a good zero output, that $R_{OFF} \gg R_{ON}$. Thus the R_{OFF}/R_{ON} ratio requirement for 0.1% accuracy is 1000 times less than for the series switch and the closed output voltage level is well defined at zero. One common application of the series-shunt switch is as a **chopper.** A chopper is a device that is used to alternately transmit and interrupt a signal at regular intervals. The chopping operation is generally performed on a signal so that the input signal level can be accurately compared with a known reference level (often common). For this reason, it is important that the gate circuit used for chopping have a well-defined closed-gate output voltage.

A series-shunt current switch is shown in Fig. 2–130d. This switch alternately directs the signal current i_u through two different loads, R_{L_1} and R_{L_2}. A current switch always has the signal current i_u pass through the switch regardless of the value of the load R_L. The voltage that is switched is a secondary consideration and is equal to $i_u R_L$.

There are other important considerations in classifying analog switches. For instance, it is important whether or not the signal and control paths are isolated and how the switch is controlled. A **galvanically isolated switch** has no currents that flow between signal and control circuits, whereas a **direct-coupled switch** has no galvanic isolation. A **voltage-controlled switch** requires a control voltage to turn it ON, whereas a **current-controlled switch** depends on a control current to turn it ON.

The applications of several solid state switches in analog gating circuits are described below.

Diode Analog Gate Circuits. Even though the diode is a two-terminal device, it is possible to use a diode circuit to gate signals ON and OFF in response to an actuating source. The circuit of Fig. 2–131a shows a

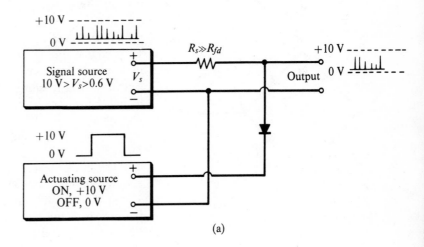

Fig. 2–131 Diode analog gates: (a) shunt switch diode; (b) series-shunt switch.

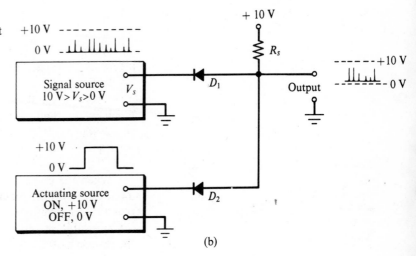

shunt gate circuit using a diode. When the actuating source output is 0 V, the diode acts as a closed shunt switch shorting the output terminals for any signal source voltage greater than 0.6 V.

When the actuating source output is +10 V, the diode becomes reverse biased and will not affect the transmission of signals of 10 V or less to the output terminals.

The switching circuit of Fig. 2–131a puts a heavy demand on both the signal and actuating sources. In order to maintain no variation in output in the OFF position, the actuating source must be capable of accepting the full signal current at its output without changing the output voltage; i.e., output resistance of the actuating source must be nearly zero in the 0 V state. Another circuit which provides better isolation between the signal and actuating sources uses two diodes as shown in Fig. 2–131b. When the actuating source output is 0 V, diode D_2 conducts the current from the 10 V source through R_s to ground. The output potential is held at 0.6 V, the forward drop across D_2. Positive signals from the signal source give D_1 a reverse bias resulting in no change in current through D_2 or output potential. However, when the actuating source potential is more positive than the signal source, D_1 is conducting and D_2 is reverse biased. The output will now follow V_s and will be equal to V_s plus the forward drop across D_1.

The diode analog switches shown in Fig. 2–131 can be improved by adding two more diodes as shown in Fig. 2–132. If the control voltages V_c and $-V_c$ of this circuit are balanced, there will be no voltage offset at the output of the type that resulted from the forward voltage drop across the switching diodes in the series-shunt switch. The required magnitude of the control voltages depends on the amplitude of the input voltage v_u, and is determined by the requirement that the current be in the forward direction in diodes D_1, D_2, D_3, and D_4 when V_c and $-V_c$ are applied as shown in Fig. 2–132.

When the control voltages are suitable all four diodes conduct, and a very low resistance path exists between the input signal v_u and the output load resistor R_L. Therefore the output voltage v_o will faithfully follow the input voltage v_u. It is, of course, important that the load resistor R_L be much larger than the diode resistances and the output resistance of the signal source. It can also be seen in Fig. 2–132 that the voltage drops across the diode pairs D_1-D_3 and D_2-D_4 are in opposition, so that when $v_u = 0$ the output voltage $v_o = 0$.

When the control voltages are reversed to $(-V_c)$ and (V_c), the four diodes are reverse biased and the resistance in series with the output is much larger than the load resistor R_L. Therefore the output voltage is zero for any value of input signal up to the magnitude of the control signal in the reverse biased condition.

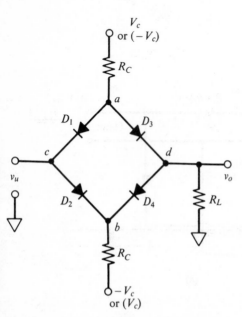

Fig. 2–132   A four-diode analog switch.

Transistor Analog Gate Circuits. The transistor switch shown in Figs. 2–106 and 2–111 can be used in analog gating circuits as shown in Fig. 2–133. In the series switch circuit (Fig. 2–133a), the gate is closed when the transistor base is positive with respect to v_u. The reverse biased base-collector junction makes an open output when the gate is OFF. A negative pulse is used to turn the gate (and the transistor) ON. The output voltage will be 0.1 to 0.2 V less than the input voltage because of $V_{CE(sat)}$ and will follow any input voltage which is more than 0.6 V positive with respect to the base voltage. When negative signal levels are to be switched, an npn transistor is used with a positive-going turn-on signal.

The shunt switch shown in Fig. 2–133b uses an npn transistor to short the output terminals for positive signals when the switch base is driven positive. This switch gives better transmission accuracy for the open gate and a gate-closed output voltage of $V_{CE(sat)}$. The series-shunt combination is shown in Fig. 2–133c.

For precision analog switching BJT's are frequently used in an inverted or common-collector mode as illustrated in the shunt switch gate of Fig. 2–134. It is known that the voltage offset is smaller when a BJT is operated in the inverse mode. Also, the ON resistance can be as low as 2 Ω, and many BJT's can be turned ON and OFF in 10 nsec or less when they are operated in the inverse mode.

The control signal that is applied to the base should be sufficiently more positive than ground potential so that the base-collector junction becomes forward biased when the transistor is to be turned ON. When the transistor is ON, the output voltage v_o is equal to the voltage drop between the collector and emitter, $V_{CE(sat)}$, which can be only 0.2–0.4 mV for base currents greater than 0.5 mA. Therefore, when the transistor is ON the output voltage is close to zero and the gate is closed. When the transistor is OFF the output voltage v_o is equal to the input signal v_u. To turn the transistor OFF and transmit the signal v_u to the output, both the base-collector and base-emitter junctions must be reverse biased.

Field-Effect Transistor Analog Gate Circuits. As might be expected, the field-effect transistor family (JFET and IGFET) make excellent switches for analog gates. They are used in the now-familiar series, shunt, and series-shunt configurations as shown in Fig. 2–130. The FET switch offers high analog transmission accuracy because it does not introduce a junction potential in series with the signal. It also offers the versatility of being

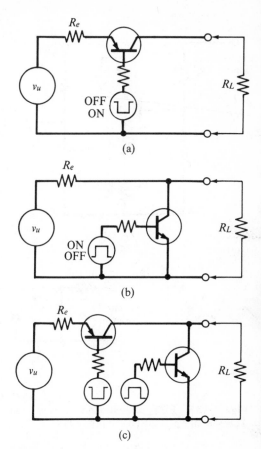

Fig. 2–133 Transistor gate circuits: (a) series; (b) shunt; (c) series-shunt and driver.

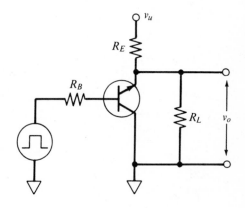

Fig. 2–134 Inverted-mode BJT switch.

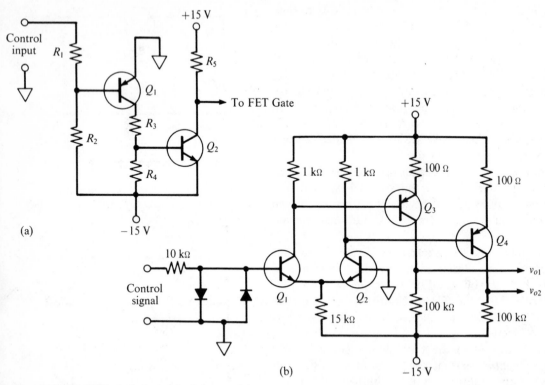

(a)

(b)

Fig. 2–135 FET switch and driver: (a) Tobey et al., p. 329; (b) Smith, p. 191.

able to conduct current in either direction. Its disadvantages with respect to the BJT switch are higher ON resistance and generally slower switching speed.

The gate voltage of the field-effect transistor switch is provided by a **FET switch driver circuit** which converts an ON-OFF control signal with convenient voltage levels into a signal suitable for turning the FET completely ON and OFF. Two such driver circuits are shown in Fig. 2–135a and b. In Fig. 2–135a, R_1 and R_2 are chosen so that for the more positive value of the control signal (say +4 V), transistor Q_1 will be OFF, and for the more negative value (say +0.5 V), it will be ON. Thus for the more negative control signal value, Q_1 is ON; this causes Q_2 to turn ON and produce ~−15 V at the FET gate driver output. For the more positive control input level, Q_1 and Q_2 are OFF and the FET gate output voltage is ~+15 V. This ±15 V gate voltage swing is enough to switch signal levels of ±10 V with most FET devices.

The driver circuit of Fig. 2–135b is convenient for a series-shunt gate circuit or for two series switch gates in an SPST configuration because of the complementary gate outputs provided. Transistors Q_1 and Q_2 have

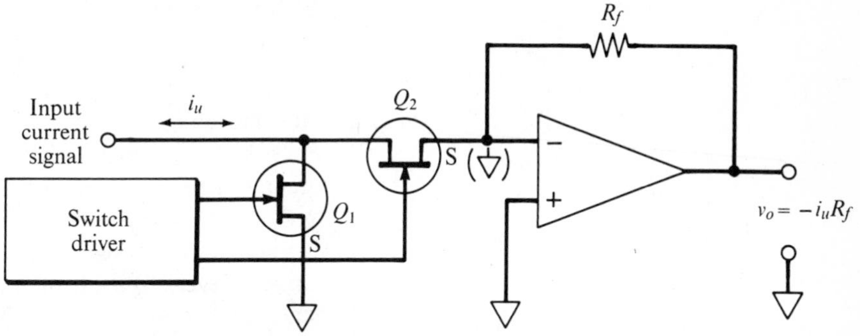

Fig. 2–136 A JFET series-shunt current switch.

the same emitter potential while the base of Q_2 is connected to common. If the control signal is greater than $+0.6$ V, Q_1 is ON; this turns Q_3 ON and produces $\sim +15$ V at the v_{o1} output. At the same time, Q_2 is OFF; therefore Q_4 is OFF and the v_{o2} output is ~ -15 V. When the control input is less than -0.6 V, the opposite transistors are conducting and the output voltage levels are reversed. When these "complementary" output voltages are used to drive the gates of two FET switches, the switches will always be in opposite states.

When JFET's are used as current switches, they do not require large amplitude driver voltage changes which reduce the switching speed. For a FET current switch, the drain and source voltages are always zero and the voltage across the switch is zero. Since the driver voltage needs to vary only between zero and the pinch-off voltage (<5 V), the current switch is faster than a voltage switch. A series-shunt current switch is shown in Fig. 2–136 with the input current signal feeding into the summing point of an OA.

When JFET Q_1 is ON, Q_2 is OFF, so that the signal current i_u passes through Q_1 to common. When the driver reverses so that Q_2 is OFF and Q_1 is ON, then the current i_u is connected to the summing point of the OA.

Experiment 2–23 *Analog Switching Circuits*

An analog diode switching circuit is first constructed in this experiment and then tested to determine whether the switch faithfully transmits analog current and voltage signals from input to output when it is ON and effectively disconnects them from the output when it is OFF. Then a JFET switch and pnp transistor driver circuit are constructed and tested for analog transmission. Finally, a commercial analog switch is employed and tested. The experiment is thus designed to show the principles of analog switching and some of the circuits for accomplishing high-accuracy analog transmission.

Binary Information Gates

A binary information signal represents a piece of information for which there are only two possible states. Examples of such information are: the laboratory animal is pressing (not pressing) the button; it is after (before) 8:26 a.m.; the conductivity of the distilled water is less than (equal to or greater than) 1 μmho; or, in general, any question that can be answered yes or no, or any statement that is either true or false. Binary information is encoded electrically by assigning a separate range of signal voltage or current levels to each of the two states, for example, 0 to +1 V for one state and +3 to +5 V for the other. The two signal level ranges are called **logic levels.** The more positive logic level is called HI and the more negative, LO. The logic levels are also sometimes called 1 and 0 for the two numerals in the binary number system, or TRUE and FALSE, though in these latter names there is no implication regarding the relative signal values.

In transmitting a binary information signal, variation of the signal level can occur within the given logic level range without any change in the information content of the signal. Thus binary information gates require a much lower signal amplitude transmission accuracy than do analog gates. However, a signal level which is not within either of the two logic level ranges conveys no information at all (if the information is truly binary) and is to be avoided. Therefore, the circuits of some binary information gates are designed to have only two stable signal levels, which are within the logic level ranges of the system. The discussion below introduces several types of binary information gates, or **logic gates.** For a more thorough discussion of logic gates and applications, the reader is referred to Module 3, *Digital and Analog Data Conversions*.

Diode Logic Gates. It was shown in the previous section that diodes can be used as switches in analog gate circuits. The same basic circuit as that of Fig. 2–131b can also be used to switch logic level signals, as illustrated in Fig. 2–137. This circuit is called an AND logic gate. The reason it is called a "logic gate" can be understood by following the circuit action in Fig. 2–137. The logic levels of the signal source are +4.0 V for a HI logic

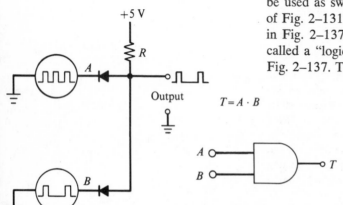

Fig. 2–137 Diode AND gate and symbol.

level and 0 V for a LO logic level. When one input is held at the HI logic level, the output will follow the level changes at the other input; i.e., the gate is open. On the other hand, if one of the inputs is LO, the output will be LO regardless of the input signal; i.e., the gate is closed. For an AND gate all inputs must be HI in order to have a HI level at the output.

Note in Fig. 2–137 that during the first pulse both inputs A and B are at $+4.0$ V. The diodes are still forward biased between the $+5$ V supply and the $+4$ V logic level input signals. The output voltage is thus 4.0 V $+ V_\gamma$ (the voltage drop across the forward biased diode), which is clearly a HI logic level. There is a LO level logic pulse at B during the second pulse at A so that the output is connected through the diode to common; this provides a LO level output. As stated above, a LO level at either input will keep the gate closed.

When diodes are connected as shown in Fig. 2–138, they perform the so-called OR function. That is, a HI level at input A OR B OR C will cause a HI level at the output. During the period when any one input is HI, the logic levels at all other inputs can change without having the changes appear at the output. A common application of the OR gate is for preventing digital signal sources which need to be connected to one point in a circuit from interfering with each other. If input A is held at a HI logic level, then the A-input diode conducts and there is a voltage drop across the output resistor which reverse biases the other diodes so as to isolate the other inputs.

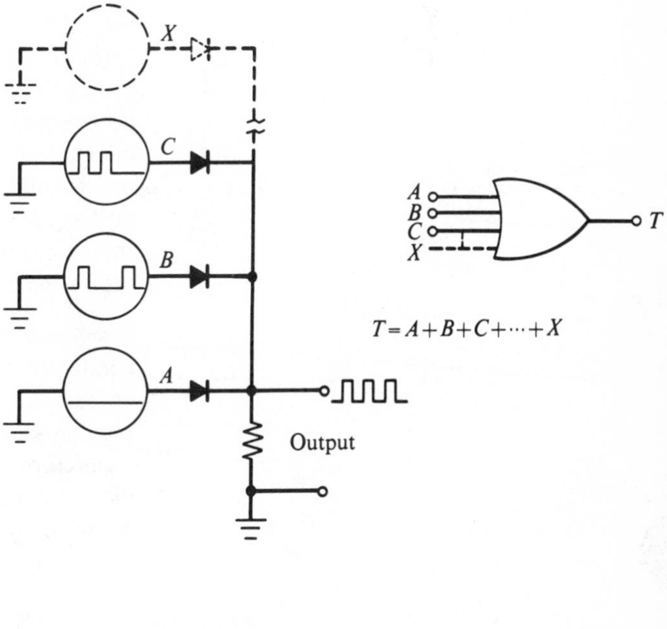

Fig. 2–138 Diode OR gate and symbol.

$$T = A + B + C + \cdots + X$$

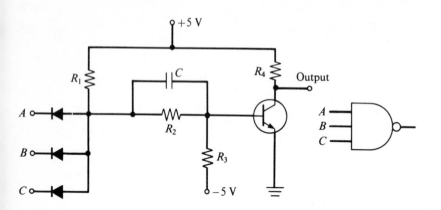

Fig. 2–139   Positive AND gate with inverting amplifier.

Diode-Transistor Logic Gates. A natural development from the diode gates was the addition of a transistor shunt switch to the gate output to reduce the output resistance as shown in Fig. 2–139. This is a diode-transistor logic or DTL gate. The input diodes and R_1 are the AND circuit shown in Fig. 2–137. When the AND output is near 0 V, the transistor is cut off and the output is $+5$ V, a HI logic level. A HI output from the AND gate is sufficient to saturate the transistor, making the output very near a LO 0 V. The gate function is now the AND function with an inverted (opposite logic level) output. This is called a NOT-AND or NAND gate. The symbol for a NAND gate is shown in Fig. 2–139. When a LO level signal source is connected to one of the inputs, a current of $5/R_1$ A must pass to ground through the signal source without raising the input voltage beyond the upper limit of the LO level. Therefore a signal source for the gate of Fig. 2–139 must have a very low output resistance when its output level is LO. When the output of the NAND gate is LO, the transistor is saturated and conducting heavily. Its output resistance is so low that it can be connected to many gate inputs. The number of gate inputs which can be connected to a gate output without jeopardizing the logic level is called the **fan-out,** an important gate characteristic. Very little current is required of a signal in the HI state, and the amplifier is easily designed to supply enough current at the HI logic level to equal the fan-out limitation set by the LO state conditions.

Integrated Circuits. Many of the currently practiced techniques of semiconductor fabrication involve diffusion of dopants and chemical etching on selected areas of a silicon crystal "chip." These techniques allow for hundreds of transistors and/or diodes to be made on a single chip. Many transistors and diodes can then be interconnected right on the chip to form more complex circuit units called "integrated circuits." Resistors can also be made by careful control of doping area and depth. The doping and etch-

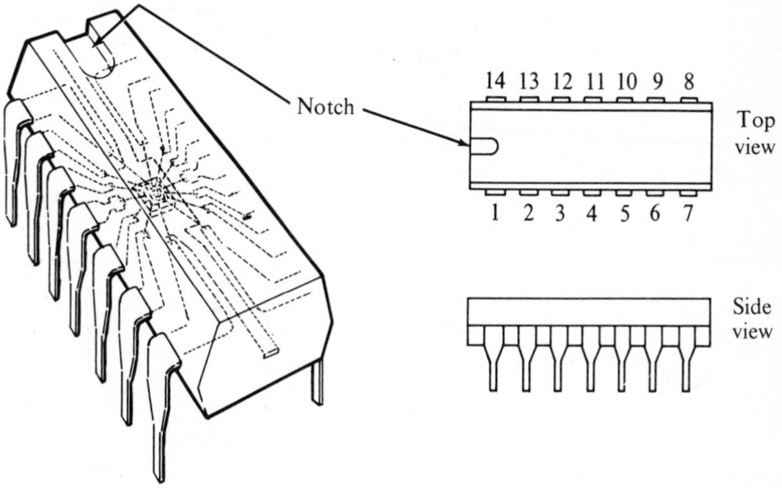

Fig. 2-140 Integrated circuit plastic dual in-line package.

ing operations are carried out first to make the desired array of transistors, diodes, and resistors. Then an oxide insulating layer is put over the chip. This is followed by selective etching through the insulation where the connections to the components are to be made. Next aluminum is deposited on the chip in the form of the required interconnections. The chip is then mounted in a metal, ceramic, or plastic body and connections are welded from the connector pins of the body to aluminum pads on the interconnection pattern where the external connections are to be made. Finally, the body is sealed and the device is given a final check. An example of an integrated circuit is shown in Fig. 2-140.

As semiconductor technology has improved, the number of active devices and components that can be economically put on a single silicon chip has increased greatly. Several complete gates of considerable complexity are now available in a single tiny package, as are many prewired combinations of gates that perform special functions. Indeed, it is the availability of such versatile circuits in convenient and inexpensive form that has sparked a new revolution in scientific instrumentation and many other areas of electronics.

TTL Gates. Available in integrated circuit form are the DTL gates described in the previous section and several other types in many variations. Along with the development of these circuits came the realization that some circuit improvements were possible and economical with IC's which were not practical with discrete components. An example is the TTL gate shown in Fig. 2-141, where Q_1 is a multiple-emitter transistor. Grounding any one or more inputs forward biases transistor Q_1; this puts the collector of Q_1

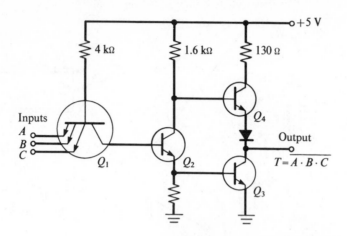

Fig. 2–141 TTL positive NAND gate. (Courtesy of Texas Instruments.)

at a low potential and turns Q_2 OFF. This, in turn, turns Q_4 ON and Q_3 OFF and results in a HI output. If, however, all inputs are HI (or unused) the base-collector junction of Q_1 will conduct, forward biasing the base-emitter junction of Q_2. When Q_2 is ON, Q_3 is ON and Q_4 is OFF, resulting in a LO level at the output. This corresponds to the inverted AND or NAND function.

The circuit is actually very similar to that of the DTL gate, with Q_1 replacing the gate and coupling diodes. Transistors Q_4 and Q_3 are a series-shunt switch providing a larger current capability in the 1 state, and Q_2 provides gain for increased speed. Since the connecting leads inside the integrated circuit (IC) gate are extremely short, lead capacitance and inductance are minimized and very high speed is possible. The first commercial TTL gates had only a 13 nsec propagation delay, but even faster TTL gates are now available. Schottky TTL gates have only about a 3 nsec propagation delay. The fan-out is about 10 and the noise immunity is typically 1.0 V, making the TTL a very convenient general purpose binary gate.

Experiment 2–24 *Binary Information Gates*

It is shown in this experiment that the diode switching circuit of the previous experiment can also be used to transmit binary information, but the considerations for successful operation are different. The diode binary information gate is constructed and tested, and limitations are noted. Then a diode-transistor binary gate is constructed and tested in order to show how a transistor inverter following the diode gate can improve performance. Then an integrated circuit binary information gate is used. The input and output levels are determined for a TTL NAND gate and advantages and limitations over the diode and diode-transistor gates are noted.

2–3.5 PRINCIPLES AND APPLICATIONS OF LIGHT-ACTIVATED SWITCHES

In recent years there have been numerous new devices introduced which are light activated. These devices convert radiant energy into electrical current or voltage or a variable resistance. In Module 1 and Section 2–1 of this module light-sensitive devices were discussed as transducers for measurements of light intensity. They also have important uses as switches for power control and for electrical isolation. In this section the basic principles of light-activated semiconductors are first discussed with emphasis on photodiodes, phototransistors, and light-activated thyristors, such as the light-activated SCR (LASCR). Then the most commonly used radiation sources, such as tungsten lamps, neon lamps, and light-emitting semiconductors, are briefly discussed. This section concludes with a presentation of source-detector combinations for achieving electrical isolation and for accomplishing several important switching functions.

Light-Activated Switches

A wide variety of light-sensitive devices is available. The cadmium sulfide, cadmium selenide, and cadmium telluride photocells convert radiant energy into a variable resistance. These devices were discussed in Section 2–1 of this module. Other devices such as photovoltaic cells, which were described in Module 1, produce voltage and current from incident radiant energy. This discussion will focus on the newer semiconductor junction light-sensitive devices, which include light-activated diodes, transistors, and thyristors.

The Photodiode. As explained in Section 2–3.1, the current through a reverse biased pn junction is due to the energetic creation of electron-hole pairs in the depletion region at the junction. In addition to thermal generation of electron-hole pairs, electromagnetic radiation of the proper energy (wavelength), when incident on the depletion region, can be absorbed by the material and create electron-hole pairs as illustrated in Fig. 2–142a. Provided the rate of charge carrier production by absorbed electromagnetic radiation is very much larger than the rate of thermal generation, the reverse bias current is directly proportional to the photon flux (light intensity). The characteristic curves shown in Fig. 2–142b demonstrate this linear relation for a wide range of reverse bias voltages.

If the photodiode is operated with enough reverse bias to prevent the recombination of charge carriers in the depletion layer, the photocurrent is given by

$$-I_p = b_\lambda Q_e P_k,$$

where b_λ is the quantum efficiency, Q_e is the electron charge, and P_k is the incident photon flux (photons/sec) on the junction.

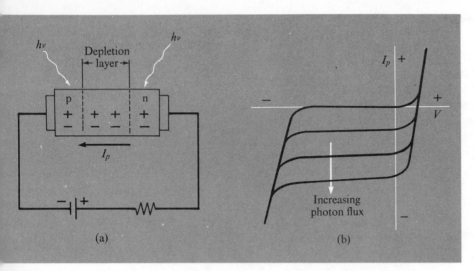

Fig. 2–142 The photodiode: (a) pictorial showing electron-hole pair production; (b) current-voltage characteristics for different photon fluxes.

The photodiode can provide currents in the μA range and is extremely fast-responding. Transit times of charge carriers across the junction are in the subnanosecond range; frequency response is therefore in the gigahertz region.

Phototransistors. The ordinary phototransistor is similar to the photodiode except that internal gain is provided. A simplified pictorial of the phototransistor is shown in Fig. 2–143a, along with an equivalent circuit and the common circuit symbol. In the equivalent circuit (Fig. 2–143b) the collector-base junction, which serves as the photodiode, is separated from the transistor for ease in understanding its operation.

Light striking the photodiode gives rise to a photocurrent I_p which supplies the base current of the transistor. The base current is amplified by the dc current gain, h_{FE}, of the transistor so that the emitter current I_E is given by

$$I_E = I_p(h_{FE} + 1). \qquad (2\text{--}127)$$

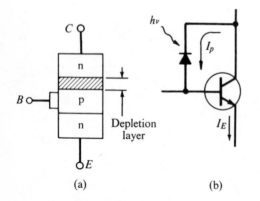
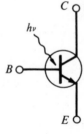

Fig. 2–143 The phototransistor: (a) pictorial; (b) equivalent circuit; (c) circuit symbol.

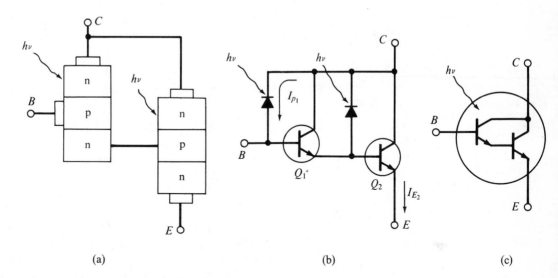

(a)   (b)   (c)

Typical phototransistors can amplify the photodiode current by 10–100 times. Switching speeds of phototransistors must be carefully considered for some applications. Typical delay times are 1–2 μsec and rise times are of the order of 4–8 μsec.

A newer type of phototransistor called a photodarlington amplifier is illustrated in Fig. 2–144. The equivalent circuit (Fig. 2–144b) illustrates that the photodarlington consists of two coupled phototransistors. The emitter current from Q_2, I_{E_2}, is approximately given by

$$I_{E_2} \approx I_{p_1} h_{FE1} h_{FE2}. \tag{2-128}$$

The fact that the current is the product of two dc current gains accounts for the high sensitivity of the photodarlington amplifier. Photodarlingtons can provide output currents of the order of hundreds of milliamperes. However, their switching speeds are somewhat slower than those of normal phototransistors.

The Light-Activated SCR (LASCR). The LASCR is a relatively recent device, whose basic operating principles can be understood with the aid of Fig. 2–145. If forward voltage is applied from the anode to the cathode of the LASCR, junctions J_1 and J_3 are forward biased. Conduction does not occur, however, because J_2 is reverse biased in the absence of light. When light strikes the silicon, free electron-hole pairs are created as in the photodiode. These pairs created near the depletion layer can move across J_2. If sufficient light is present, the current in the reverse biased diode becomes large enough to increase the net current gain of the two transistors above unity and the SCR goes into conduction. Thus the basic firing cri-

Fig. 2–144 Photodarlington amplifier: (a) pictorial representation; (b) equivalent circuit; (c) circuit symbol.

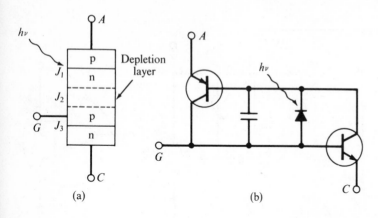

(a)

(b)

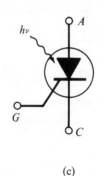

(c)

Fig. 2–145 The light-activated SCR (LASCR): (a) pictorial representation; (b) two-transistor equivalent; (c) circuit symbol.

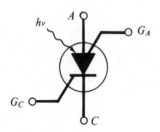

Fig. 2–146 Circuit symbol of light-activated silicon-controlled switch (LASCS).

terion is the same as for a normal SCR with the exception that there is an additional light-generated current which can cause conduction. Like normal SCR's, the LASCR can also be switched by the injection of current into the gate. Hence the LASCR is a very versatile switching device.

LASCR's are available which can switch amperes of current and block several hundred volts in the reverse direction.

The Light-Activated Silicon Controlled Switch (LASCS). Perhaps the most versatile of all pnpn devices is the LASCS, which has both anode and cathode control gates in addition to being sensitive to light. The circuit symbol for the LASCS is shown in Fig. 2–146. The same techniques used in switching the SCR and LASCR can be used with the LASCS. However, the presence of an anode gate allows turn-on to be accomplished with negative pulses at the anode, and turn-off with positive pulses.

Light Sources

A variety of light-emitting devices are used in conjunction with light-activated switches to provide numerous important control and isolation applications. The emitted radiation from the light source as "seen" by the photodetector depends on many variables, including the relative sensitivity of each as a function of wavelength, the geometry, use of light filters and light pipes, and others. It is important that the source have a long stable life, and where many sources are to be used the cost is important. Where distance between the source and light-activated device must be great, the intensity and focusing are especially important. Incandescent lamps, neon bulbs, electroluminescent devices, light-emitting injection diodes, ruby and gas lasers, and others have been used.

Incandescent Lamp. The tungsten filament lamp is inexpensive and has a wide spectral output which is similar to blackbody radiation as shown in

Fig. 2–147. However, the rise time of the output light is poor (typically 100 msec), and tungsten bulbs are susceptible to vibration breakage and burnout. They are used primarily with systems that have a low pulse rate.

Neon Lamp. The neon lamp is very inexpensive, has long life, and the required input power is much less than for the incandescent bulb. Response time is in the microsecond range. Most of the light output is limited to two orange spectral lines, the maximum available output is quite limited, and a high voltage supply is needed. This limits the flexibility of the optical link, and a detector must be chosen that is sensitive to the orange lines.

Semiconductor Sources. Light-emitting diodes (LED's) and junction lasers are dependent on semiconductor pn junctions which emit photons as a result of hole-electron recombinations in the structure. In silicon and germanium diodes the energy from recombinations is delivered to the crystal primarily as heat, but for junction diodes made from gallium arsenide (GaAs), gallium phosphide (GaP), and gallium arsenide phosphide (GaAsP) the recombination energy is converted mostly into emitted photons. This light is either incoherent and concentrated in a narrow bandwidth, as illustrated in Fig. 2–147 for a GaAsP LED, or it can be coherent and concentrated primarily in a narrow line. When emitting coherent light it is called a **junction laser.**

These diodes have very fast response times in the nanosecond range and have good dynamic range and linearity. Present diode devices have their light output typically in the infrared region (about 900 nm), the red region (650 nm), or the green region (570 nm) so that a detector must be used that is sensitive in this region.

The circuit symbol for a LED is illustrated in Fig. 2–148, along with a plot of relative intensity vs. current input for a GaAsP LED. For a LED it is customary to define the quantum efficiency or electrical-to-optical conversion efficiency as the ratio of the number of photons out to the num-

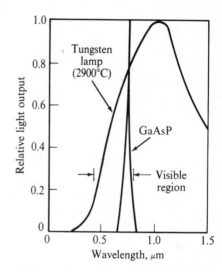

Fig. 2–147 Relative output of sources vs. wavelength.

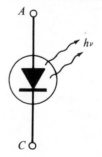

Fig. 2–148 Light-emitting diodes: (a) circuit symbol; (b) power output vs. forward current for a GaAsP LED.

(a)

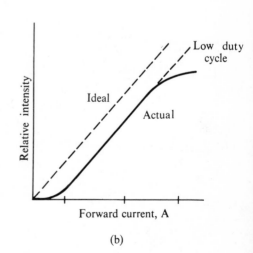

(b)

ber of electrons in. LED's are often used in a pulsed mode to increase the output intensity, since the LED can withstand higher peak currents under low duty cycle pulsing conditions.

Optical Coupling for Electrical Isolation

The combination of a light source with a photosensitive device inside an opaque enclosure can perform electronic functions that are difficult by other means. The information from the light source and associated devices can be transferred to the photodetector without any electrical coupling. Light is the coupling link. Because the light source terminals and photodetector terminals are isolated, the pair forms an almost ideal four-terminal network in which there is essentially no interaction between input and output circuits (although the 1–2 pF capacitance typical between terminals of some pairs can sometimes be significant).

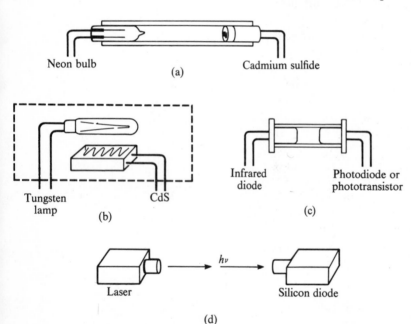

Neon bulb   Cadmium sulfide

(a)

Tungsten   CdS
lamp

(b)

Infrared   Photodiode or
diode   phototransistor

(c)

Laser   Silicon diode

(d)

Fig. 2–149 Source-detector pairs for opto-electronic devices.

Some typical optical links matching source detector pairs are illustrated in Fig. 2–149. In recent years completely encapsulated optical couplers have become available commercially. Figure 2–150a shows an optical coupler made from a LED and a phototransistor. When input current is applied to the LED, radiation is produced and transduced by the phototransistor back into an electrical current. Complete isolation of the input and output circuits is accomplished by optical coupling and no feedback can occur. Figure 2–150b shows a typical plot of output current vs. input current for the optical coupler of Fig. 2–150a.

Optical coupling has several other important advantages. It eliminates ground loops and isolates noise sources. Switching can be accomplished on a very short time scale. Also, optical coupling can be used to shift voltage levels as in binary information transmission.

Applications of Optoelectronic Devices

The applications of semiconductor optoelectronic devices could be classified into two groups: one group in which the device acts as a continuously variable resistor (rheostat), and another group that depends only on the low ON resistance and high OFF resistance for switching applications. A few examples are presented to illustrate some of the many uses of source-detector pairs.

The use of the rheostat characteristics of a CdS detector together with a tungsten source is illustrated in Fig. 2–151 for an illumination control. A portion of the light radiation is made to fall on the CdS cell, which is in series with resistor R_{ai}. The voltage drop across R_{ai} is fed to an amplifier which provides an output signal to a series power element. The effective resistance R_s of the series element depends on the input to the amplifier, which depends on the light incident on R_c. The amplifier and power series element are connected in such a way that an increase in lamp intensity for any reason will cause a decrease in R_c. This causes an increase in the amplifier input voltage $v_{ai} = V_r R_{ai}/(R_{ai} + R_c)$, which causes the resistance R_s of the series element to increase, thus decreasing the lamp current. The resulting optoelectronic feedback, therefore, brings the light illumination down toward its original value, and tends to hold illumination constant, as "seen" by the CdS detector.

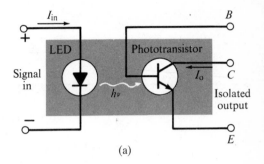

(a)

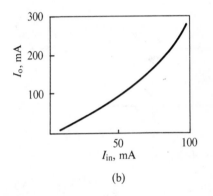

(b)

Fig. 2–150 Optical coupler: (a) source-detector pair; (b) output current vs. input current.

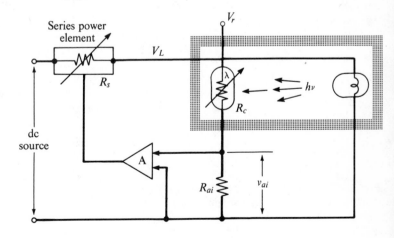

Fig. 2–151 Illumination control using a tungsten source CdS detector.

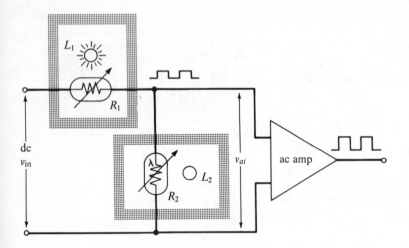

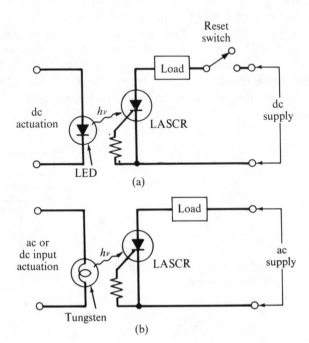

Fig. 2–152 Neon source–cadmium selenide chopper.

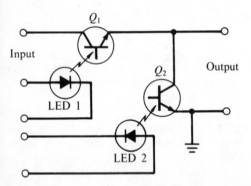

Fig. 2–153 A series-shunt chopper circuit using light-emitting diodes (LED's) and phototransistors.

Fig. 2–154 Light-actuated relays: (a) dc latching relay; (b) ac nonlatching relay.

A widely used switch application of the neon source CdSe detector is a chopper (i.e., a modulator to convert dc to ac voltage). This is illustrated in Fig. 2–152. Lights L_1 and L_2 are alternately turned ON and OFF so that when L_1 is ON, L_2 is OFF. Therefore, when R_1 is LO (e.g., 1 kΩ), R_2 is HI (e.g., 1 MΩ) and amplifier input $v_{ai} = v_{in}R_2/(R_1 + R_2) \approx v_{in}$; and when R_1 is HI, R_2 is LO, so that $v_{ai} \approx 0$. The amplifier input voltage is therefore a chopped signal proportional in amplitude to the input dc signal v_{in}. This chopped signal can now be advantageously amplified or manipulated as an ac signal. A similar series-shunt chopper using LED's and phototransistors is shown in Fig. 2–153.

A semiconductor analog of an electromechanical relay can be simply accomplished with a LED and a LASCR as illustrated in Fig. 2–154a and b. In addition to offering complete isolation, the solid state relay has the advantages of microsecond response, no contact bounce, and very small size. The dc latching relay shown in Fig. 2–154a requires a dc supply. When the lamp is off, the LASCR is nonconducting and no current flows through the load. When the lamp is turned on by a control signal, the LASCR conducts and current passes through the load. The LASCR remains in conduction even after the lamp is turned off until it is reset by removing the supply voltage. A nonlatching relay is illustrated in Fig. 2–154b. It uses an ac supply for the LASCR. The LASCR resets during the negative half-cycle of the ac supply voltage. Current will flow through the load only when the lamp is energized during the positive half-cycle of the LASCR supply voltage.

Only a few of the important applications of optoelectronic switches have been discussed in this section. Other applications include the reading, converting, or translation of known information from punched cards or paper tape. Major applications are being found in communication systems and other areas beyond the scope of this module.

Experiment 2–25 *Optoelectronic Switching Circuits*

Several different light sources and detectors are characterized in this experiment for applications in optoelectronic switching. Experience is gained in wiring a BJT driver circuit for operating a tungsten lamp, and neon lamps and light-emitting diodes (LED's) are also characterized. The photoconductive cell and phototransistor are used as detectors. A photoconductive cell optoelectronic chopper is connected as a modulator for converting a dc signal to an ac signal. A series chopper, a shunt chopper, and a series-shunt chopper are studied.

2–3.6 PRECISE SWITCHING FOR SIGNAL LEVEL CONTROL AND WAVEFORM GENERATION

Switches which are actuated when a preset signal level is exceeded are very widely used in electronic instrumentation. Switch devices which are signal level actuated include the breakdown diode, the unijunction transistor, the gas-discharge lamp, and any switching device controlled by the output of a comparator. Level-actuated switches can be used to keep signals within safe or desired limits when the switch actuating levels are set at the boundaries of the desired range. Signal limiting circuits of this type, called clipping circuits, using a variety of switch types, are described in this section.

Another very interesting and useful application of level-actuated switches is in the increasingly popular function generator. The basis of most generators which produce a variety of waveforms, including triangular and square waves, is integrating circuits controlled by level-actuated switches. A discussion of representative techniques and circuits used in the generation of sawtooth, triangular, square, sine, and pulse waveforms is presented in this section.

Level-actuated switches are also used to convert an analog-encoded signal (such as the voltage level from a strain gauge bridge) to a binary information signal (such as the information that the strain on gauge A is greater, or less, than the strain on gauge B). Many applications of this type appear in Module 3, *Digital and Analog Data Conversions.*

Clipping Circuits

Clipping or limiting circuits are used to prevent voltages from exceeding desired positive or negative maximum values, to cut off either positive or negative pulses, and to isolate or eliminate various sections of waveforms. This is accomplished by a variation of the analog transmission gate, which is open when the signal is within the set limits and which provides the limit voltage at the output when the input exceeds the limit value. Biased pn junction diodes and breakdown diodes are the most commonly used switching devices in clipping circuits.

The pn junction diode can be thought of as a voltage-actuated switch with the actuation voltage threshold at 0.6 V. The clipping circuits which result when the diode is used in the series and shunt switching positions are shown in Fig. 2–155a and b respectively. In the series circuit, the diode is OFF (gate closed) except when v_{in} is more positive than +0.6 V. Signal voltages on the positive side of the threshold are followed at the output, while voltages less than 0.6 V are disconnected from the load by the reverse biased diode. Thus the negative input voltages are "clipped off." The shunt diode in Fig. 2–155b becomes conducting for v_{in} less than −0.6 V, thus cutting the output voltage off at −0.6 V. In each case, if the diode

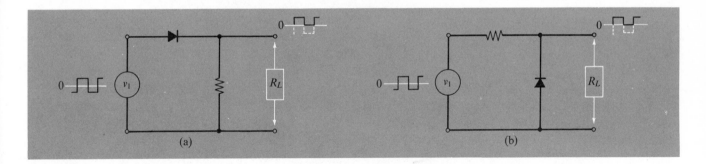

were reversed, the negative signal voltages would be transmitted and the positive values cut off.

It is often desirable to set the clipping threshold at some voltage other than 0 V. This is accomplished with pn junction diodes by placing in series with the diode a bias voltage source which is equal to the desired threshold voltage. Examples for a series diode limiter are shown in Fig. 2–156. In the circuit of Fig. 2–156a, the diode will conduct only when v_{in} is greater than the bias voltage, V. When the diode is not conducting, the output voltage is V V. Thus the circuit clips off all input voltage excursions on the negative side of V V. This will be true for any value or polarity of bias voltage V. The effect of reversing the diode is shown in Fig. 2–156b. In this case, $v_o = V$ when v_{in} is greater than V and the diode is **OFF**. However, when v_{in} is less than V, the diode conducts and $v_o = v_{in}$. Thus voltages greater than V are clipped off. Shunt clipping circuits using biased diodes are just

Fig. 2–155 Limiters: (a) series; (b) shunt.

Fig. 2–156 Clipping above and below a fixed voltage level: (a) clipping below; (b) clipping above (component positions for equivalent shunt circuits shown in parentheses).

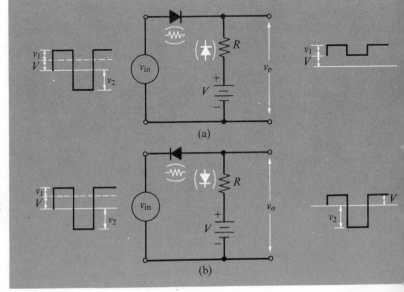

like the circuits of Fig. 2–156 except that the positions of the diode and resistor in the circuit are interchanged. When the diode is oriented as shown, the clipping operation is analogous to that of the series circuit.

A limiter that clips signals above and below fixed voltage limits is shown in Fig. 2–157a. It is sometimes called a **slicer,** because it slices out a section of the input voltage. The same type of clipping can be obtained by using two Zener diodes in a back-to-back configuration, to replace both the regular diodes and bias batteries, as shown in Fig. 2–157b. During the positive half-cycle, Zener diode Z_1 is forward biased and Z_2 conducts only when the breakdown potential V_z is exceeded. During the negative half-cycle Z_2 is forward biased and clipping occurs at the Z_1 breakdown voltage. Perfectly symmetrical clipping using the circuit of Fig. 2–157b would require the breakdown potentials of Z_1 and Z_2 to be perfectly matched. This requirement is avoided by using a single Zener diode in the diode bridge circuit shown in Fig. 2–157c. On the positive half-cycle, D_1 and D_3 are

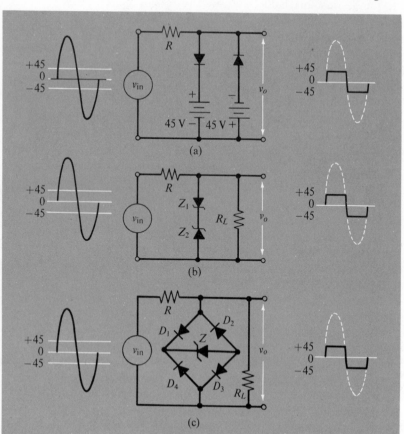

Fig. 2–157 Slicer circuits: (a) use of diodes and bias batteries; (b) use of Zener diodes; (c) use of single Zener diode in diode bridge.

forward biased, placing the input voltage v_{in} as a reverse bias across the Zener diode. When $v_{in} > V_z + 2V_f$, the diode bridge circuit acts as a shunt switch clipping the output voltage at that value. On the negative half-cycle, D_2 and D_4 are forward biased and v_{in} again appears as a reverse bias on the Zener diode.

The four-diode analog switch of Fig. 2–132 (repeated in Fig. 2–158) is also used as a limiting circuit. The output voltage, v_o, will follow v_{in} as long as all the diodes are forward biased. Assume that there is a maximum positive voltage for point A in the circuit, $V_{A(max)}$. If v_{in} exceeds $V_{A(max)}$, D_1 becomes reverse biased. Since D_4 and D_2 are still ON, $V_B = v_{in}$ and $v_o = V_{A(max)}$, making D_3 reverse biased. Now we can see that v_o is the $+V_C$ supply voltage divided by R_1 and R_L. Therefore,

$$v_{o(+max)} = +V_C R_L / (R_1 + R_L).$$

By similar arguments

$$v_{o(-max)} = -V_C R_L / (R_2 + R_L),$$

which results in a clipping of the input voltage at these values. The clipping points are easily adjusted by varying the supply voltages or the series resistors. It is easy to see how the "window" of voltages between clipping levels decreases as the supply voltages decrease and how it disappears completely when the supply voltage polarities are reversed.

Clipping Circuits with Operational Amplifiers. Clipping circuits are frequently used in conjunction with operational amplifiers to keep the input signals within safe bounds, to keep the amplifier output from limiting, or to provide a precision limiting function. When the ideal output voltage of an operational amplifier is greater than its maximum practical value (for instance, when an OA with a $v_{o(max)}$ of $+12$ V is connected as a gain-of-100 amplifier to an input of 0.2 V), the amplifier limits. Some types of OA's recover from a limit state very slowly in comparison to their normal response. Thus, when an OA is used for an application in which it is likely to limit, it is desirable to provide a circuit that will automatically switch to a lower gain as the amplifier output voltage approaches its maximum value. This is accomplished by putting a shunt limiting circuit in the negative feedback path of the OA circuit as shown in Fig. 2–159. Since the output voltage must equal the voltage across the feedback element when the amplifier is in control, the output voltage cannot exceed the turn-on

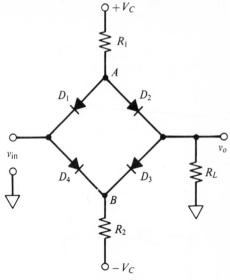

Fig. 2–158   Diode bridge limiter.

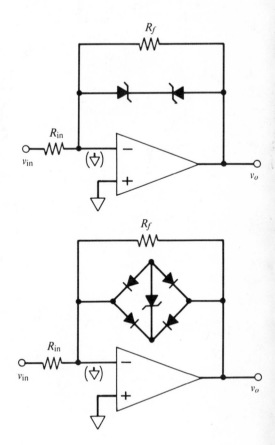

Fig. 2–159   Fast recovery limiting of OA output voltage.

threshold voltage for the shunt switches connected from the output to the inverting input. These OA output limiting circuits may be used with any OA circuit with a virtual ground at the inverting input. When used with the inverting amplifier circuit shown in Fig. 2–159, the resulting circuit is called a **bounded** amplifier. Such an amplifier has its normal gain (R_f/R_{in}) between the limits and breaks cleanly to zero gain at the set output voltage limits.

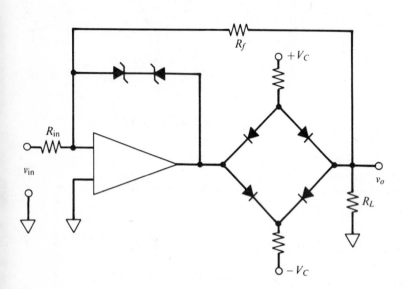

Fig. 2–160   Precision bounded amplifier.

A precision limiting circuit is shown in Fig. 2–160. This circuit uses the diode bridge limiter of Fig. 2–158 in the feedback circuit. Since the fed-back voltage is taken after the diode bridge, any offset or loss in the bridge is compensated for by the amplifier output. The voltage limits are the same as those for Fig. 2–158 except that R_f must be considered to be in parallel with R_L. When the output voltage is at a limit, the normal feedback path is open and the OA itself limits and goes out of control. This is prevented by the Zener diodes, for which the breakdown voltage should be between the desired output voltage boundary and the maximum OA output voltage.

Clamping Circuit

The clamping circuit is usually designed to "clamp" the top or bottom of a waveform to a fixed dc level, which may be zero, while preserving its shape and amplitude. It is sometimes called a "dc restorer," because it can restore the dc level that may have been shifted in amplifier or coupling circuits. A simple dc restorer is illustrated in Fig. 2–161 together with the positive and negative applied voltages and the initial waveforms during dc restoration.

The input voltage is a 500 Hz square wave with amplitude varying from +10 to −10 V, or 20 V peak to peak. Assume that the capacitor is uncharged and that the square wave is suddenly applied as shown by the input waveform in Fig. 2–161. The +10 V is applied across the diode so that it conducts. The effective resistance is now the diode ON resistance in parallel with R_L. Therefore the time constant for charging the capacitor is short. Assuming that the period between alternations of the square wave is equal to one charging time constant, the capacitor will charge to 63% of the applied voltage, or about 6.3 V, before the input voltage drops to −10 V. The instant the voltage drops, the diode stops conducting and the effective resistance is R_L. Now the time constant for the discharge of the capacitor is much longer and it does not discharge significantly during the negative half-cycle.

On the next positive half-cycle the diode does not conduct until the voltage has exceeded 6.3 V, because this is the voltage retained on the capacitor from the previous cycle, and it opposes the applied input. The net voltage applied across the diode at the start of the second cycle is 10 − 6.3 = 3.7 V. During the second positive half-cycle the capacitor will charge some more and will add 2.3 V (63% of 3.7 V) to the previous 6.3 V, to give a total of 8.6 V. Again during the negative half-cycle the discharge of the capacitor is insignificant, so that the 8.6 V on the capacitor opposes the positive swing of the third half-cycle. The capacitor continues to charge to 63% of the remaining voltage until the charge is essentially 10 V.

After the capacitor is charged to 10 V, it is apparent that, during each positive half-cycle of the input, the output will be 0 V. During each negative half-cycle of the input, the output will be −10 − 10 = −20 V. In effect the entire input waveform has shifted downward at the output so that the top is on the zero axis. The bottom of the waveform could be clamped to the zero axis instead by simply reversing the diode. In the case of narrow pulses it is necessary to clamp to the base, not the peak, to provide a sufficient portion of the cycle for charging.

A dc restorer can be biased to clamp the bottom or top of the waveform at some preselected voltage by placing a voltage source in series with the diode as in the shunt limiting circuit.

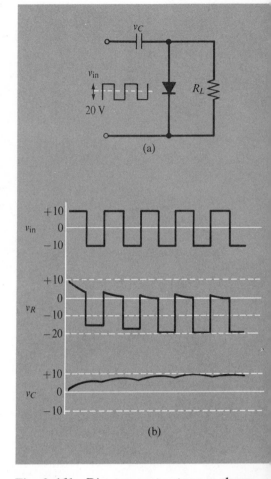

Fig. 2–161 Direct current restorer or clamping circuit: (a) clamping circuit; (b) waveforms.

Experiment 2–26 *Diode Clipping and Clamping Circuits*

Several diode clipper circuits are connected and tested so as to illustrate how they can slice off sections of a waveform. A diode clamp is also connected, and its ability to restore the dc component in an input square wave after it has been coupled through an *RC* circuit is investigated.

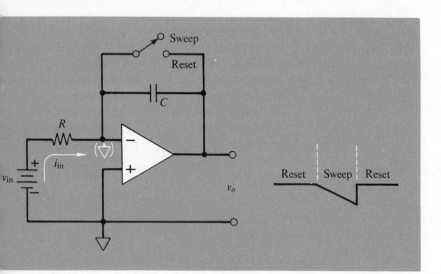

Fig. 2–162   OA integrator connected as a simple sweep generator.

Linear Sweep Generators

A signal that changes at a constant rate is generally obtained by integrating a constant input signal. The circuit most often used for this purpose is the operational amplifier integrator introduced in Section 2–2.3 and shown in Fig. 2–162. The constant voltage v_{in} supplies a current $i_{in} = v_{in}/R$ to the summing point of the OA. The amplifier output voltage v_o is always such that the summing point S is maintained at nearly 0 V (virtual ground) while the current i_{in} passes through the feedback circuit (switch or C). When the switch is closed (reset position), there is no resistance between point S and the output, so they must be at the same potential and $v_o = 0$ V. When the switch is opened (sweep position), the current i_{in} causes a continuous accumulation of charge on C. The charge q at any instant is $i_{in}t$, where t is the time in seconds since the sweep began. The voltage across the capacitor $v_C = q/C = i_{in}t/C$. The voltage $v_o = v_C$ (since S is at 0 V), so

$$v_o = -v_{in}t/RC \qquad (2\text{--}129)$$

and the sweep rate dv_o/dt in volts per second is the value of v_o after $t = 1$ sec, or

$$\frac{dv_o}{dt} = \frac{-v_{in}}{RC}. \qquad (2\text{--}130)$$

For example, if $v_{in} = +5$ V, $R_{in} = 100$ kΩ, and $C = 0.1$ µF, the sweep rate would be $-5/(10^5 \times 10^{-7}) = -5 \times 10^2$ or -500 V/sec. If the OA has a maximum output voltage of ± 10 V, the sweep will be linear for

$10/500 = 1/50$ sec before limiting at -10 V. When the switch is closed, the capacitor discharges through the switch and v_o returns quickly to zero. This simple sweep circuit is useful for many laboratory applications for sweep times in the range of microseconds to minutes.

To produce a repetitive, periodic sweep signal, it is necessary to provide an automatic periodic reset for the circuit of Fig. 2–162. This can be done in two ways. One is to use the output of an oscillator to close the reset switch momentarily at regular intervals. With this method, the values of R and C affect the sweep rate and the maximum amplitude, but not the frequency, because that is controlled by the oscillator. The other method is to use a level-actuated switch to momentarily close the reset switch when the desired maximum sweep voltage is reached. This technique is illustrated in Fig. 2–163. A generalized comparator-operated switch is shown. As the sweep output, v_o, increases toward the comparison voltage, v_c, the comparator output is positive. A fraction of the comparator output provides v_c. When this level is reached, the comparator output becomes negative, turning the switch ON and charging v_c to 0 V. Thus the switch will remain closed until the capacitor is discharged, and then it will open again. In the level-control method of automatic resetting, the R- and C-values affect both the sweep rate and the frequency, but not the amplitude, because that is controlled by the comparator.

A very simple sort of sweep generator/oscillator which operates on this same principle uses a neon bulb as a level-controlled switch to discharge the capacitor. The circuit is shown in Fig. 2–164. When the capacitor is charged through R to the starting voltage of the neon lamp, the lamp conducts, discharging C to the maintenance voltage v_m. The resistor R is too large to allow the minimum current required for discharge in the lamp; so it stops conducting and the charge on the capacitor starts to increase once again. The sweep is not perfectly linear because the charging rate is not constant. A flash of light is produced during each discharge. Another very similar circuit is the unijunction relaxation oscillator of Fig. 2–119, described in Section 2–3.3.

Triangular and Square Wave Generators

Another way to discharge the capacitor is to reverse the integrator input voltage polarity so that the current is reversed. This will cause the capacitor to discharge at the same rate it was charged and produce a triangular wave

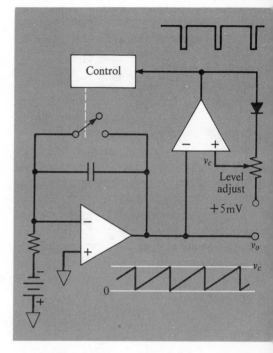

Fig. 2–163 Sweep generator with automatic level switch reset.

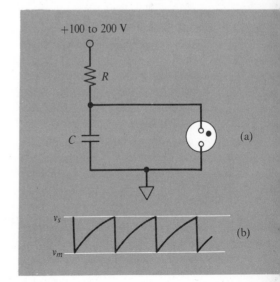

Fig. 2–164 Neon bulb sweep generator/oscillator: (a) circuit; (b) waveform.

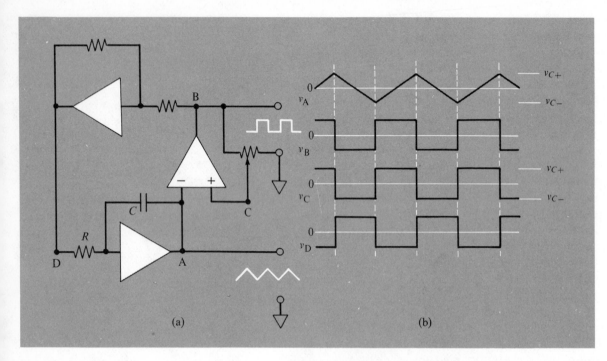

Fig. 2–165   Triangular waveform generator: (a) circuit; (b) waveforms.

at the output. If the triangular waveform is to be repeated, the input voltage polarity should be reversed again when the capacitor has been discharged to its starting voltage. Thus an automatic reversal of the input voltage is required at each limit of the sweep voltage. This can be accomplished with two comparators, one with the threshold set at the positive limit and one at the negative limit. Another way is to change the threshold setting to the other limit each time the threshold is crossed. This technique is shown in Fig. 2–165. At the start of the waveforms shown, the input voltage D is negative, causing the integrator output voltage A to sweep in the positive direction. Because the threshold voltage C is more positive than A, the comparator output is positive. A fraction of the comparator output voltage B provides the threshold voltage C. When the integrator output A just exceeds the threshold voltage *C*, the comparator output B, and consequently the threshold voltage C, become negative. This change in state at B is often used to operate a polarity reversing switch for the voltage at D. In the circuit shown, the B voltage is simply inverted by the third operational amplifier to provide a voltage at D of the proper polarity. As the waveforms show, the signal at A is a triangular wave and the signal at B is a square wave. It is this basic triangular waveform generator which is used in the popular laboratory **function generator,** which has triangular, square, and sometimes sine wave outputs. The value of R is varied to change the frequency (and sweep rate) over one or two orders of magnitude. Different values of C are switched in to cover the entire frequency range.

The basic generator concept described above can be adapted to provide several additional waveforms and capabilities. If the integrator input voltage is supplied by an external source, the frequency of oscillation will be proportional to that voltage. Thus the function generator becomes a **voltage controlled oscillator** (VCO). Of course, the comparator output must operate a reversing switch on the input voltage. If the positive and negative integrator input voltages are not equal, the capacitor will charge and discharge at different rates and produce asymmetrical waveforms. The result would be a sawtooth instead of a triangle and pulses instead of a square wave. This is often done intentionally to produce those waveforms.

Sine Wave Synthesis. In the function generator, the basic waveform is the triangular wave and the sine wave must somehow be derived from it. This is usually accomplished by a diode wave-shaping circuit. The triangular waveform and the desired sine wave are shown in Fig. 2–166. In order to obtain the sine wave from the triangular wave, an attenuator is needed which will attenuate the larger input voltages much more than the smaller ones.

A nonlinear voltage divider to serve this purpose can be made with diodes and resistors as shown in Fig. 2–167. The single diode in Fig. 2–167a is connected as a shunt clipping switch with a resistor in series. Therefore, for $v_{in} < v_b$, the output signal is the unattenuated input signal.

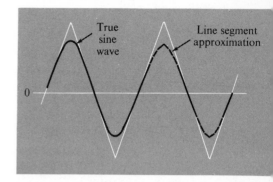

Fig. 2–166 Synthesizing a sine wave from a triangular wave by line segment approximation.

Fig. 2–167 Diode wave shaping: (a) single break; (b) multiple break.

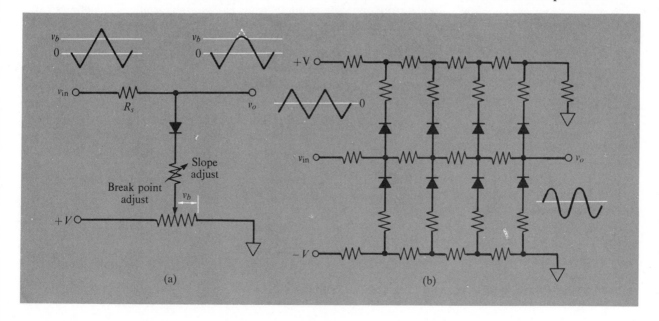

When v_{in} is greater than v_b, the diode will conduct a current proportional to $v_{in} - v_b$. This current causes an *IR* drop in R_s proportional to $v_{in} - v_b$. Thus the slope of the output signal changes at v_b. The point of the slope change is set by the **breakpoint,** v_b, adjustment. The amount of slope change depends on the value of the slope-adjust resistor. Additional diodes and resistors can be added to provide additional breakpoints and slope changes until the desired waveform is approximated by a series of line segments. The line segment approximation is shown in Fig. 2–166, and a multiple-break circuit (with four positive and four negative breakpoints) is shown in Fig. 2–167.

The sine wave output produced by diode wave shaping in a good function generator is generally of high quality. The line segments are not observable with an oscilloscope. The function generator approach is the best way to produce low-frequency (below 20 Hz) sine waves, since they are difficult to produce with quality by standard harmonic oscillator circuits.

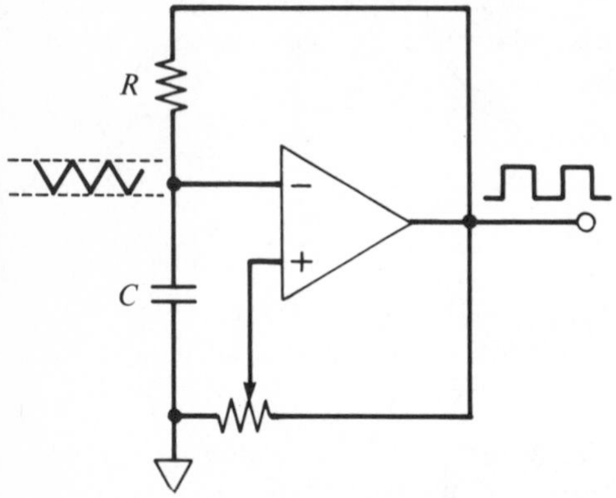

Square Wave Generator.   The function generator of Fig. 2–165 produced a square wave simultaneously from the level switching comparator output. If the linearity of the integrating circuit is not important, that is, if the square wave is the desired output waveform, the circuit can be simplified considerably. A simple passive *RC* integrating circuit can take the place of the active OA integrator and, since the *RC* circuit does not invert as the OA integrator does, the inverting amplifier is also unnecessary. The resulting circuit is shown in Fig. 2–168. A circuit of this type, which essentially snaps from one state to another but has no stable state, is sometimes called an **astable multivibrator.**

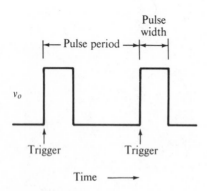

Fig. 2–168   OA square wave generator.

Pulse Generators

A pulse is a waveform in which the voltage or current changes quickly from its normal steady value and then returns to the normal value after a period of time called the **pulse width.** Such a circuit has two states, only one of which is stable, and is thus called a **monostable multivibrator.** Most laboratory pulse generators produce "square" pulses of the type shown in Fig. 2–169. The pulse is initiated by some sort of "trigger" signal. The pulse width is determined by the time constant of an *RC* circuit in the pulse generator. The astable circuit of Fig. 2–168 can be made monostable by putting a diode shunt clipping switch across the capacitor. The resulting circuit is shown in Fig. 2–170. Because of the diode, the capacitor cannot charge in the positive direction. Thus the stable state occurs when the output is positive and the threshold voltage at the + input of the OA is positive. A trigger circuit is added which can convert a sharp negative-going transition of a trigger signal into a negative pulse of sufficient magnitude

Fig. 2–169   Output signal from a pulse generator.

to make the threshold voltage negative. When the trigger pulse occurs, the output goes negative, holding the threshold negative and charging the capacitor in the negative direction. When the capacitor voltage reaches the threshold, the output returns to its stable positive state. The pulse width depends on, and is approximately equal to, the RC time constant.

Monostable multivibrators are available in convenient integrated circuit form, which makes the assembly of such circuits from discrete components increasingly rare. If two monostable circuits can be connected to trigger each other, the resulting circuit is astable with a total period of oscillation equal to the sum of the two monostable pulse widths. This type of astable circuit has the advantage of providing independent control of the width of the positive and negative portions of the resulting waveform.

Experiment 2–27 *Waveform Generators*

This experiment illustrates that a general purpose function generator can be constructed from a basic operational amplifier linear sweep generator. The function generator produces variable-frequency sine, square, and triangular waveforms. A linear sweep generator with automatic reset is first constructed and tested. Then the circuit is modified to produce the various other waveforms. Some characteristics and limitations of the function generator are then studied.

Experiment 2–28 *Multivibrators*

In this experiment various multivibrator circuits are constructed and tested. The basic astable, monostable, and bistable multivibrators are constructed using operational amplifier circuits. The limitations and advantages of these circuits are noted.

2–3.7 POWER CONTROL WITH SOLID STATE SWITCHES

The solid state devices discussed in previous sections find many important applications for power switching in addition to the low-level signal switching applications discussed in Section 2–3.4. Power transistors are often used for **ON-OFF** switching of dc power to various loads. Since the switching considerations are identical to those discussed in Section 2–3.3, BJT power switches will not be considered in this section. Instead this section will be devoted to the use of SCR's and triacs in power control applications. Power switching can be classified into two categories: in the first, **static switching,** the controller is used to turn the power applied to the load completely **ON** and **OFF** In the second category of power switching, often called

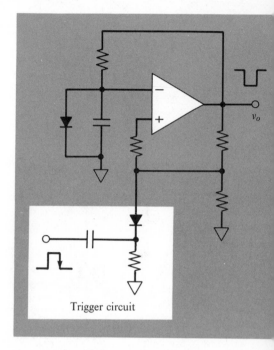

Trigger circuit

Fig. 2–170 Pulse generator.

phase control, the ac supply is switched to the load for a precisely controlled fraction of each cycle so as to control the magnitude of the average power applied to the load.

This section discusses some of the basic considerations in static and phase control and presents several important applications. The reader is referred to the GE *SCR Manual** for a more exhaustive discussion.

Static Switching with Thyristors

The SCR and the triac are important controllers for a variety of static switching applications. Switching of both ac and dc can be accomplished, although with the latter the SCR must be turned off by one of the methods discussed in Section 2–3.3. The uses of SCR's and triacs in static ac switching, static dc switching, protective circuits and time delay relay circuits are considered below.

AC Switching Circuits. High-speed switching of ac power to a load can be readily accomplished with a triac circuit similar to that shown in Fig. 2–171. A control device such as a reed switch or light-activated switch is used to trigger the triac into conduction and thus to apply the ac voltage to the load. Resistor R is used to limit the gate control current to the triac to a safe value. The triac has many advantages over electromechanical relays for simple static switching applications. Contact bounce is completely eliminated with triacs and contact wear is no longer a factor. Hence the lifetime of such a solid state switch is considerably greater than that of an ordinary relay. The amount of power that can be switched in this manner is determined by the triac ratings, which should be carefully considered in power switching. If loads containing inductance are to be switched, the *RC* network shown in dotted lines in Fig. 2–171 should be included to protect the triac from voltage transients (recall that a rapidly changing current through an inductor produces a large counter emf). A similar but more complex ac switching circuit can be constructed with two SCR's connected in inverse-parallel fashion. This latter circuit should be used if the ac supply has a frequency greater than about 400 Hz, since triacs are normally limited to lower frequencies.

DC Switching Circuit. A static SCR switching circuit for control of dc through a load is shown in Fig. 2–172. The circuit is basically a bistable multivibrator or flip-flop. A turn-on control signal applied to the gate of SCR1 triggers the SCR into conduction and applies the dc voltage to the load. The application of the supply voltage begins to charge capacitor C through resistor R. When the turn-off signal is applied to the gate of SCR2,

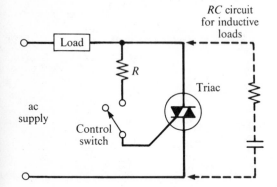

Fig. 2–171   Basic ac static switch.

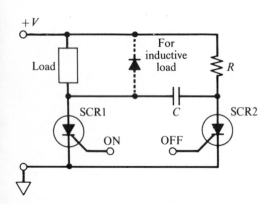

Fig. 2–172   DC static switch.

---

* *SCR Manual,* 1972, fifth edition, General Electric Company, Syracuse, N.Y.

SCR2 conducts (acts like a short circuit) and capacitor C is essentially connected across SCR1. Since the right-hand plate of C was charged positively with respect to the left-hand plate, this reverse biases SCR1 momentarily and turns it OFF. The current through the load then decreases exponentially to zero.

SCR Protection Circuits. Because SCR's are capable of rapid power switching, they are often used for protection against transient voltages, overvoltages, and current shorts.

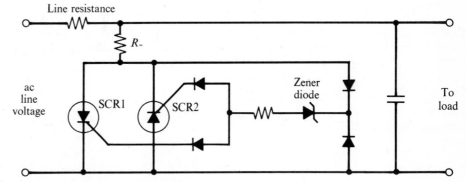

Fig. 2–173 Transient line voltage protection circuit with SCR's.

Figure 2–173 shows an SCR circuit that is useful in protecting loads from transient line voltage surges. Any transient surge which exceeds the Zener voltage of the Zener diode causes either SCR1 or SCR2 to trigger. Which SCR triggers depends on the line voltage polarity at the instant of the transient. Resistor R is a current limiting resistor which protects the SCR. The loading effect of the SCR conduction causes the transient voltage to be dropped across the resistance of the ac line rather than across the load. When the line voltage reverses polarity, the SCR which was conducting turns OFF.

An interesting circuit used in power supply protection is the so-called "crowbar" circuit shown in Fig. 2–174. This circuit provides protection against both line voltage surges and possible short circuit loads. Two UJT's are used to trigger the SCR's. UJT1 provides a trigger pulse if overvoltages are present, while UJT2 triggers the SCR if load currents exceed the maximum value. Potentiometer P_1 sets the maximum desired value of the dc supply. If the supply should exceed the value set, the peak-point voltage of UJT1 is exceeded; this supplies a trigger pulse to the SCR. The entire dc supply voltage is then applied to the circuit breaker, which opens the main supply line. Short circuit protection is provided by UJT2. The voltage across resistor R_1 is proportional to the load current. When the

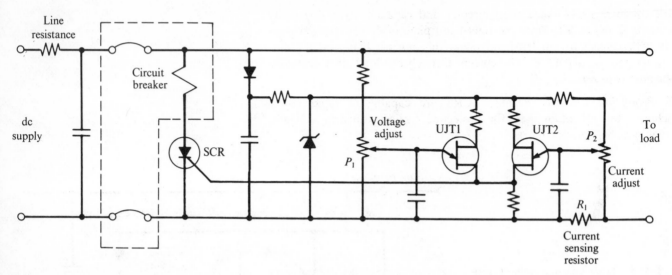

Fig. 2–174 Crowbar overvoltage and over-current protection circuit.

voltage across R_1 exceeds the maximum desired value set by potentiometer P_2, UJT2 triggers the SCR into conduction. As before, this shuts down the dc supply by operating the circuit breaker. Many supplies are protected against only overvoltage or overcurrent instead of both. Hence, for only-overvoltage protection UJT2 and the circuitry associated with it are often left out. Likewise for only-short-circuit protection UJT1 and its associated circuitry are deleted.

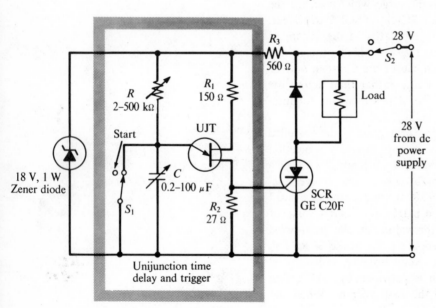

Fig. 2–175 Precision solid state time delay circuit.

Time-Delay Relay Circuits. The SCR is often combined with UJT, neon lamp, or PUT relaxation oscillators to provide precision time delays. A typical application of the unijunction transistor is shown in Fig. 2–175. The UJT is used to provide a precise time-delayed trigger pulse to turn the SCR power control circuit ON. The emitter-base 1 voltage for the UJT is supplied by the voltage across the capacitor C. At the start, C is shorted by switch S_1. When S_1 opens, the capacitor C charges through R from the stable 18 V Zener-regulated supply. The voltage across C is determined by the RC time constant and Zener voltage source. When the capacitor charges to the UJT peak-point voltage, V_p, the UJT fires. This provides a pulse across R_2 which triggers the SCR. The full supply voltage (less the small voltage drop across the SCR) is then applied across the load terminals.

The voltage across the emitter-to-base 1 of the UJT drops to less than 2 V when the SCR triggers, so that a low voltage is maintained on capacitor C, but this varies as a function of time.

The upper limit on the time delay depends on the specified accuracy, the V_p of the UJT, the leakage current of capacitor C, and the temperature. The upper limit for R is determined by the requirement that the current to the UJT emitter be large enough for it to fire (larger than I_p).

To reset the circuit for another timing cycle, the dc supply must be removed from the SCR by opening switch S_2 and the capacitor C should be shorted by closing S_1.

Phase Control Switching Circuits

In phase control switching, as opposed to the static switching discussed in the previous section, the ac supply is switched to the load for a precisely controlled fraction of each cycle. This is the basic way in which power is controlled in lamps, motors, heaters, etc. The thyristor is a basic component in this mode of switching. Control of the average power applied to a load is obtained by triggering the thyristor at a specific point in time (phase) on the ac waveform. The thyristor then conducts for the remainder of the half-cycle.

Phase control can take many forms. Only one half-cycle of the ac wave might be used (half-wave control), or both half-cycles might be used (full-wave control), depending on the application. Although SCR's are used in many phase control applications, the triac is a more versatile controller since it conducts in both directions and is self-protecting against transients.

SCR and Triac Power Controllers. An example of a circuit that can be used for control of power in a load is shown in Fig. 2–176. The section of the circuit enclosed in the block acts as a switch to provide current through the load only during a discrete selected fraction of each half-cycle. The RC circuit determines the portion of each half-cycle required to charge

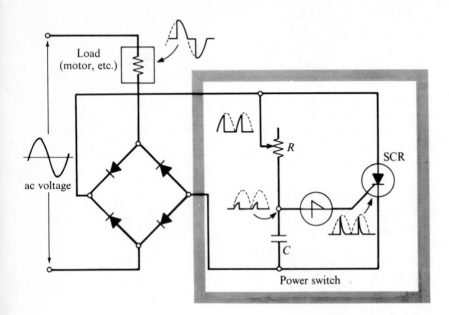

Fig. 2–176 Use of SCR to control power in a load.

the capacitor C to a sufficient voltage for the four-layer diode to turn ON, which provides the necessary gate signal for the SCR to turn ON. Because the power delivered to the load in each half-cycle depends on the portion of each half-cycle required to charge the capacitor to the firing potential of the diode and the SCR, it is easy to change the power to the load by varying resistor R or capacitor C. The voltage waveforms shown in the various portions of the circuit of Fig. 2–176 as related to the input supply voltage illustrate the circuit action. Note that because the effective resistance of the ON SCR is very small, the circuit current is essentially determined by the load resistance. This circuit uses full-wave power control.

A similar circuit designed to operate a 60 Hz motor with full-wave phase control using a triac is shown in Fig. 2–177. The diac is used to trigger the triac and resistor R_1 adjusts the trigger point (phase).

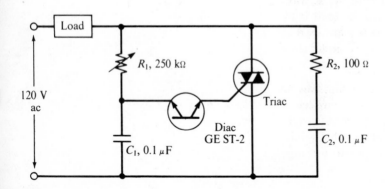

Fig. 2–177 Full-wave phase controller for motors.

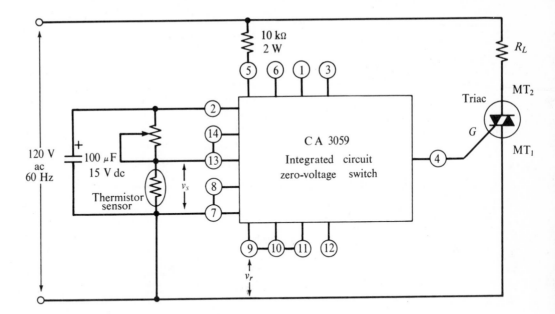

Fig. 2–178 Zero voltage switched triac circuit.

Zero Voltage Switching

Whenever power is switched **ON** and **OFF** very rapidly as with solid state switches, high-frequency electromagnetic interference (EMI) is generated. Although thyristors eliminate interference problems caused by rapid current interruptions in inductive loads, special circuits are required to eliminate EMI during turn-on. Experimentally it has been shown that the minimum amount of EMI is generated in ac power switching if the thyristor is turned **ON** when the ac waveform crosses zero voltage. Likewise the minimum EMI is created when the switch turns **OFF** the instant the current through it drops to zero.

A variety of circuits have been proposed for sensing the ac waveform zero crossing and providing a trigger pulse for the SCR or triac. However, recently zero voltage switches have become available in integrated circuit packages and their use in power switching is increasing at a rapid rate. A triac on-off temperature controller which uses an IC zero voltage switch is shown in Fig. 2–178. The integrated circuit is a combination zero voltage threshold sensor and a trigger circuit. Hence the triac is turned **ON** very near the zero crossing by a trigger pulse from the IC to the triac gate whenever the voltage v_s across the thermistor sensor exceeds the reference voltage v_r. Such a zero voltage switched triac circuit almost completely eliminates generated EMI. In the past few years encapsulated modules have become commercially available which include a zero voltage switch and triac. These solid state ac relays are rugged and simple to use.

Experiment 2–29 *Control of Power in a Load with Solid State Switches*

Several circuits employing solid state power switches are constructed. This experiment illustrates how SCR's and triacs can be used to accomplish power supply protection, and several modes of light source control. A circuit is studied as an overvoltage protector. Then a lamp output is controlled by a triac circuit. A circuit is constructed to turn the lamp off after a precise amount of radiation has been received by the detector, and to control the power applied to the lamp in order to keep the intensity constant.

PROBLEMS

1. In the nonideal switch circuit of Fig. 2–70, the switch has an open resistance $R_{so} = 900$ kΩ, a closed resistance $R_{sc} = 80$ Ω, and an equivalent capacitance $C = 400$ pF. It is desired to switch a 5 V source V_s to a 50 kΩ load R_L. (a) What are the voltage drops across the switch and load for both states of the switch? (b) What are the two time constants, τ_o on opening the switch and τ_c on closing the switch?

2. A silicon pn junction is described by Eq. (2–95) with $I_i = 10^{-10}$ A at $300°$K. Calculate and plot the *I-V* curve of the diode for voltages of 0.30 0.40, 0.50, and 0.60 V.

3. The diode switching circuit of Fig. 2–82 was graphically analyzed and the diode found to have a forward static resistance $R_{fs} = 65$ Ω, and a reverse static resistance $R_{rs} = 5$ MΩ at the operating points. Calculate the voltage across the load for the two values of the voltage source.

4. Find the theoretical forward dynamic resistance R_{fd} of a diode with a forward current of 0.8 mA.

5. In the transistor switching circuit of Fig. 2–106, $h_{FE} = 100$, $V_{CE(sat)} = 0.2$ V, and $V_{BE(ON)} = 0.6$ V. The input signal varies between 0 and $+1$ V, $R_B = 4$ kΩ, $R_L = 1$ kΩ, and $V_{CC} = 5$ V. (a) When the input signal is $+1$ V, is the transistor in saturation? (b) Calculate the collector current, I_C, when the transistor is ON, and the ON resistance of the transistor.

6. The transistor switch circuit in Fig. P–6 is used to control current through a relay coil. The relay coil has a resistance of 200 Ω and requires 20 mA to operate. (a) Does the transistor have to be in saturation to energize the relay coil? Explain. (b) If a $+4$ V signal at the control input is to turn the relay ON, what current will be drawn from the source ($V_{BE(ON)} = 0.6$ V)? (c) What is the minimum value of h_{FE} the transistor can have and still energize the relay coil?

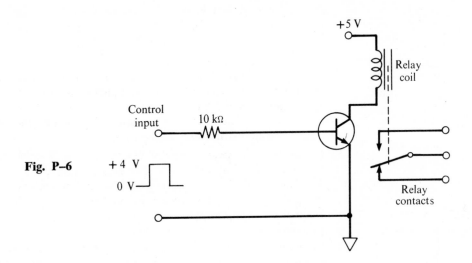

Fig. P–6

7. A switching transistor is characterized by a collector-emitter cutoff current $(\beta_N I_{CO})$ of 20 μA at $V_{CE} = 15$ V and $V_{CE(\text{sat})} = 0.4$ V at $I_C = 50$ mA and $I_B = 500$ μA. For a switching application with $V_{CC} = 15$ V and $R_L = 1$ kΩ, estimate the **ON** and **OFF** resistances of the transistor.

8. A unijunction transistor (UJT) has a peak-point voltage of 15 V at a value of $V_{BB} = 19$ V. If $V_\gamma = 0.6$ V, what is the intrinsic standoff ratio of the UJT?

9. (a) For stabilization of the peak-point voltage against temperature variations, an external resistor R_2 can be added in series with B_2 of the UJT circuit of Fig. 2–116. If $R_2 \ll R_{BB}$, show that the peak-point voltage can be approximated by

$$V_p \approx V_\gamma + \eta V_B - \frac{\eta V_B R_2}{R_{BB}},$$

where V_B is the total supply voltage between R_2 and B_1. (b) Show what value R_2 is necessary to cancel the temperature dependence of V_γ, which is the major contributor to temperature variations in V_p.

10. The series-type analog switch of Fig. 2–130b is constructed using a FET switch. The **ON** and **OFF** resistances of the switch are 50 Ω and 100 kΩ respectively. If the source v_u has an output of 5 V and an equivalent resistance R_e of 100 Ω, and $R_L = 1$ kΩ, calculate v_L for the two states of the switch.

11. Consider the four-diode analog gate of Fig. 2–132. The voltages V_c and $-V_c$ are determined by the condition that the current in each of the four diodes must be in the forward direction. If the diode forward resistances

are negligible, show that the minimum value of V_c required is

$$(V_c)_{\min} = v_u \left(2 + \frac{R_c}{R_L} \right).$$

12. (a) A reverse biased photodiode has a quantum efficiency of 0.3 at the wavelength of an incident light beam of photon flux 10^{12} photons/sec. What is the photocurrent I_p? (b) A phototransistor has a dc current gain of 50. If it is operated under the same conditions as the photodiode in part (a), what is the emitter current I_E?

13. (a) In the diode series limiter of Fig. 2–155a, v_{in} is a square wave which has values of -5 V and $+5$ V. The diode has an ON static resistance of 50 Ω and an OFF static resistance of 500 kΩ at the operating points. If the load resistance is 5 kΩ, what are the output voltages for the two states of the source? (b) Repeat the calculation for the shunt limiter of Fig. 2–155b and the same conditions as in part (a).

14. Design diode clipping circuits which will limit the output voltage v_o to the following limits regardless of the range of input voltages: (a) $0 < v_o < +5$ V, (b) -10 V $< v_o < +10$ V, (c) $+50$ V $< v_o < +75$ V.

15. The shunt limiter of Fig. P–15 was analyzed graphically, and R_{fs} was found to be 50 Ω, while $R_{rs} = 1$ MΩ. Calculate v_o for the two values of v_{in}.

16. The OA linear sweep generator of Fig. 2–162 has $v_{\text{in}} = +1.5$ V, $R = 250$ kΩ, and $C = 0.5$ μF. (a) What is the sweep rate of the generator? (b) If the maximum output voltage of the OA is ±12 V, for how many seconds will the sweep remain linear, before limiting?

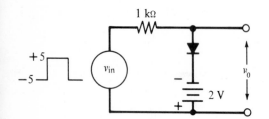

Fig. P–15

Instrumentation Power Supplies

It was shown in previous sections that electronic circuits, devices, and systems all require specific voltages V and the transfer of charge Q between the various circuit elements and networks so as to perform the functions for which they were designed. In other words, a supply of electrical energy W is required for suitable operation, and W (joules) $= Q$ (coulombs) $\times V$ (volts). However, it is not sufficient for an energy source to supply the required charge over an arbitrary period of time. It is necessary that the power P (the time rate of supplying the energy) be sufficient for continuous operation of the many circuits and devices in a system; the electrical power P (watts) $= W$ (joules)$/t$ (seconds). Since the voltage V (volts) is equal to W (joules)$/Q$ (coulombs), and the current I (amperes) is equal to Q (coulombs)$/t$ (seconds), the power $P = VI$. One watt of electrical power is thus one volt-ampere or one joule per second.

The sources of electrical power that are used to meet the specific voltage and current requirements for electronic measurement and control systems are called **instrument power supplies.** The major type of instrument power supply uses the ac electrical power that is fed into buildings from power companies and converts it into the required voltages, ensuring that these voltage levels are maintained in accordance with the current demands. By combining diodes, resistors, capacitors, and other devices discussed in the previous sections, it is possible to convert the ac line power into dc power at the required voltages. Therefore, the line-operated dc power supply circuit is a form of control circuit for manipulation of the ac line power so as to maintain the desired dc output voltages and currents for which it was designed.

The power company that supplies the ac electrical line power might have developed its electrical power in any of several ways. In the classical methods a water turbine is used to drive an electrical turbine (hydroelectric power plant), or a fossil fuel (coal, oil, gas) is burned to convert water into steam and then steam turbines move electrical conductors through a magnetic field so as to generate the ac voltage and current. More recently atomic energy (at nuclear power plants) is used to convert water into steam to drive the steam turbines which operate the electrical generators. However, when we plug instruments into a wall socket it does not make any difference what was the original type of energy used to provide the electrical power; the only important thing is that the correct ac electrical power is available at that socket. Therefore, the first part of the discussion in this section will assume the availability of ac electrical line power and will concentrate on its conversion to the dc voltages and currents required by modern electronic measurement and control systems. For example, transistor and integrated circuits usually require dc operating voltages of about 3 to 30 V and currents of a few milliamperes to several amperes. Transducers such as photomultiplier tubes and ionization detectors require dc power supplies that provide hundreds to thousands of volts but very low currents, usually less than 1 mA. The stability of the instrument power supply is frequently very important because it can determine the stability of the entire instrument. Therefore, the general types of power supply regulation devices are discussed in this section.

In recent years portable electronic instruments that do not require ac electrical line power have become increasingly important (remote pollution monitors, space flights) and popular (portable television). Most of these instruments utilize some form of electrochemical cell or batteries to provide the necessary voltages and currents. For these instruments, it is important to know something about the basic energy source in order to recognize the range of usefulness and limitations of many types of portable instruments with self-contained electrical energy sources. Both the disposable dry cells and rechargeable batteries are discussed. Also, the principles of solar cells and fuel cells are presented.

2–4.1 CONVERSION OF AC LINE VOLTAGE TO DC VOLTAGE

The block diagram in Fig. 2–179 illustrates the basic components and functions usually required to convert an ac to a useful dc voltage. First, a transformer is used to convert the ac line voltage to about the desired voltage. Next, a rectifier circuit converts the ac to a pulsating dc voltage. To smooth out the pulsating dc voltage, a filter network is employed. In modern instruments some type of regulator circuit is usually added to improve the stability and regulation and to reduce further the ripple of the output voltage. The critical characteristics of a dc power supply are its rated output voltage, its internal resistance, its output voltage variation (noise), and its current limits. Any voltage source, being made of imperfect conductors, has some internal resistance. Since the current in the voltage source is the same as

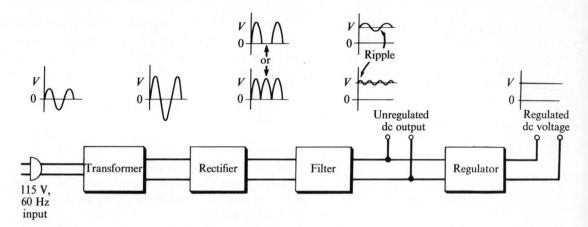

the current in the external circuit (for power supplies, the external circuit is called the load), an *IR* potential drop will be developed in the voltage source. Since this potential drop is in opposition to the no-load voltage source output potential difference, the output voltage will decrease as the load current increases.

Fig. 2–179 Block diagram showing conversion of ac input to dc output voltage.

An ideal dc voltage power supply would have (1) a constant output voltage regardless of variations in the current required by the load (good regulation or low output resistance); (2) a constant output voltage regardless of variations in temperature, ac line voltage, age of power supply, etc. (good stability); and (3) no noise voltage of line or other frequency superimposed on the dc output (low ripple). In addition to these characteristics, the dc output voltage and the current capability of the power supply must meet the operational requirements of the electronic devices.

Transformers

A transformer is used to provide an efficient multiplication or division of the ac power line voltage. The line voltages in the United States are usually 110 V or 220 V, 60 Hz, and in Europe 220 V, 50 Hz. A schematic diagram and pictorial of a transformer are given in Fig. 2–180. The operation of the

Fig. 2–180 Pictorial and schematic representation of transformer.

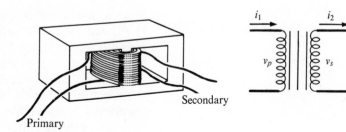

transformer is based on the principles of inductance introduced in Section 2–2.4. A changing current in the primary winding produces a changing magnetic flux in the secondary coil, which produces an induced emf across that coil. The voltage produced in the secondary winding is equal to the number of turns, N_s, times the rate of change of magnetic flux ϕ. The relationship, earlier given as Eq. (2–64), is repeated here:

$$v_s = - N_s \frac{d\phi}{dt}. \tag{2–131}$$

Since the primary and secondary are intimately wound on a core of high magnetic permeability, the magnetic flux in the primary and the magnetic flux in the secondary windings are equal. According to Eq. (2–66) the rate of change of flux is

$$\frac{d\phi}{dt} = \frac{-v_p}{N_p}, \tag{2–132}$$

where v_p and N_p are the voltage on the primary and the number of turns in the primary, respectively.

Combining Eqs. (2–131) and (2–132) to eliminate $d\phi/dt$, we get

$$\frac{v_s}{v_p} = \frac{N_s}{N_p}. \tag{2–133}$$

Therefore, the voltage induced in the secondary winding is proportional to the voltage applied to the primary. The voltages referred to are, of course, ac voltages because the transformer depends on a constantly changing flux. By choosing a transformer with an appropriate turns ratio, one can convert the ac line voltage to a higher or lower value. A single transformer can have several windings or several taps in one winding to satisfy several different voltage requirements. The transformer also provides a degree of isolation of the secondary load from the primary supply, since there is no direct connection between the two. Isolation transformers are used where there is a problem of conflicting common voltage levels between the primary and secondary circuits.

Rectifier Circuits

Since a diode conducts current effectively in only one direction, the current in a circuit which is in series with a diode must be dc. The simplest rectifier circuit is shown in Fig. 2–181. On the positive half-cycle of the ac voltage, the diode can conduct, allowing current to pass through R_L. The resistor R_L is the "load" or the circuit which is to be supplied with direct current. On the negative half-cycle, the diode is reverse biased and, therefore, nonconducting. This rectifier circuit is called a "half-wave" rectifier because only half of the ac current wave is present in the load circuit.

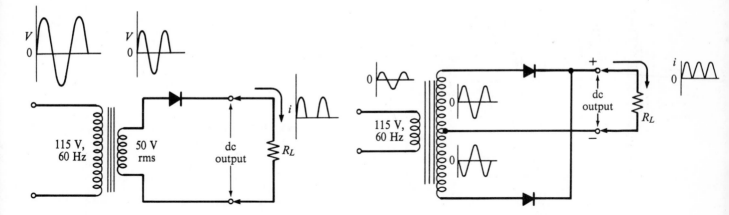

Fig. 2–181 Half-wave rectifier circuit.   **Fig. 2–182** Full-wave rectifier circuit.

In considering rectifier circuits, it is important to consider the ratings of the rectifiers, which include: (1) the maximum average forward current rating, which is approximately $\frac{1}{2} \times V_{av}/R_L$, because the diode conducts only half the time; and (2) the **peak inverse voltage** (PIV), which is the maximum voltage of the reverse or nonconduction polarity that should be applied to the rectifier. The diode of Fig. 2–181 has to withstand a peak inverse voltage of $50 \times 1.4 = 70$ V. If a capacitor is connected across the output of the rectifier circuit in Fig. 2–181 the diode PIV must be 2×70 V $= 140$ V. The effective resistance of a conducting diode is not constant but depends on the current, but an approximate knowledge of the forward resistance allows one to calculate the power loss in the rectifier (I^2R_{fs}).[6]

Full-Wave Rectifiers. For many applications it is desirable to have a rectifier circuit which supplies current during both half-cycles of the ac power, and thus provides a more continuous current to the load. A "full-wave" rectifier circuit is shown in Fig. 2–182. This circuit is essentially two half-wave rectifiers in parallel with inputs that have a phase difference of 180°. The voltage output of the full-wave rectifier is equal to the voltage developed by each half of the transformer secondary. For a 30 V peak output from a full-wave rectifier, one would use a $(60/1.4) + 1 = 44$ V center-tapped transformer. The extra 1 V is to compensate for the drop in the rectifiers. Note that the rectifiers must withstand an inverse voltage of twice the peak value of the source. For the case above, the peak inverse voltage would be $1.4 \times 44 = 62$ V.

A way to obtain full-wave rectification which does not require a center-tapped transformer is shown in Fig. 2–183. This circuit is called the "bridge rectifier." Trace the current path through the rectifiers. On the positive half-

Note 6. Peak, Average, and RMS Current and Voltage.

In calculating power the rms values are used. The relationships between peak current, I_p, peak-to-peak current, $I_{p\text{-}p}$, average current, I_{av}, and root mean square current, I_{rms}, are

$$I_{p\text{-}p} = 2I_p,$$

$$I_{rms} = 0.707I_p,$$

$$I_{av} = 0.637I_p.$$

See Appendix C for the derivation of these basic relationships.

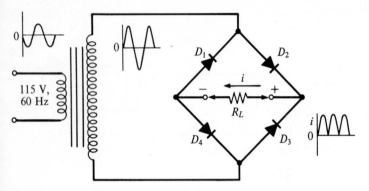

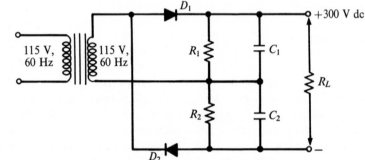

Fig. 2–183   Bridge rectifier circuit.

Fig. 2–184   Voltage doubler.

cycle, D_2 and D_4 conduct. On the negative half-cycle, D_1 and D_3 conduct. In each case, the direction of electron flow through the load R_L is the same.

Having two rectifiers in series with the load drops the peak inverse voltage that each rectifier must withstand to the peak value of the supply voltage.

Voltage-Doubler Rectifier.   Two rectifiers can be connected to a single ac source and wired so that their outputs are in series as in Fig. 2–184. The output voltage available from such a circuit is twice that available from the ac source with a half-wave or bridge rectifier. For this reason, this kind of circuit is called a "voltage-doubler" rectifier. On the positive half-cycle, capacitor C_1 is charged to the peak value of the supply voltage (in this case $115 \times 1.4 = 160$ V). On the negative half-cycle, C_2 is charged to the same potential. Since C_1 and C_2 are in series across the load, the output voltage is twice the peak voltage of the ac source. The capacitors are essential to the operation of the circuit, because they maintain the voltage developed during one half-cycle so that the voltage developed during the next half-cycle can be added to it. The capacitors C_1 and C_2 have a filtering action which is described in the next section. Since current is drawn from the transformer during both half-cycles, this voltage-doubler circuit is considered to be full-wave. The peak inverse voltage applied to each rectifier is twice the peak value of the supply voltage, in this case 320 V. Where the peak inverse voltage rating of the diode is insufficient, two diodes may be put in series so that their peak inverse voltage ratings are additive.

Filtering the Rectified Voltage

The output voltages of the rectified circuits discussed in the previous section vary with time as shown in Fig. 2–185. The output can be considered as a voltage that varies about the average dc voltage. The average dc voltage is called the dc component of the output. For the half-wave rectifier output, the lowest and predominant frequency of the ac component is the frequency of the ac line as shown by the dotted line in Fig. 2–185a. From Fig. 2–185b it can be seen that the lowest and predominant frequency of the output from a full-wave rectifier is twice the ac supply frequency. For most applications, it is necessary to reduce the magnitude of the ac component of the rectifier output to a value which is very small compared with the average dc potential. The electrical device that accomplishes this task is called a **filter.** The effectiveness of the filter is called the **ripple factor,** r, which is defined as the rms value of the ac component or ripple divided by the average dc voltage; that is, $r = I_{ac}/I_{dc} = V_{ac}/V_{dc}$.

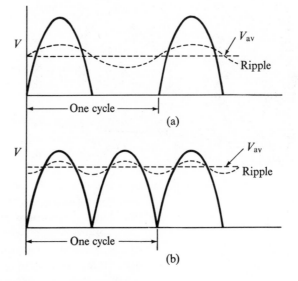

Fig. 2–185 Rectified voltage and fundamental ripple frequency: (a) half-wave and (b) full-wave rectifier outputs.

A rather effective filter is simply a large capacitance in parallel with the load R_L. The capacitor may be thought of as storing charge when the ac component is positive and as discharging through the load when the ac component is negative. Another way to think of the capacitor filter is to consider

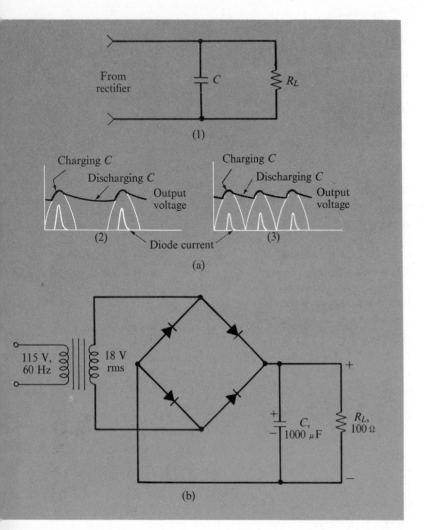

Fig. 2–186 (a) Capacitor filter: (1) circuit; (2) half-wave input; (3) full-wave input. (b) Low-voltage, capacitor-filtered power supply.

that its resistance for the ripple frequency is very low compared with R_L, and thus it diverts most of the alternating current away from the load circuit.

A more detailed picture of the action of a capacitor filter is presented in Fig. 2–186a. The capacitor charges to the peak value of the input voltage. If R_L were infinite, the voltage across the capacitor would quickly reach a constant value equal to the peak value of the alternating current supplying the rectifier. In the practical case where R_L is not infinite, the capacitor begins to discharge through R_L as soon as the input voltage decreases to that voltage to which the capacitor has charged. This results in the output waveshapes shown in Fig. 2–186a.

It can be seen from Fig. 2–186a that the magnitude of the ripple voltage will be decreased if R_L, C, or the frequency is increased. The expression for the ripple factor, $r = 1/(2\sqrt{3}fCR_L)$, bears this out. Here f is the frequency of the main ac component, equal to the line frequency for half-wave rectifiers and twice the line frequency for full-wave rectifiers. It can also be seen that, as the ripple increases in magnitude, the average dc output will decrease and be approximately

$$V_{dc} = 1.4V_{rms} - \frac{I_{dc}}{2fC}, \tag{2–134}$$

where V_{rms} is the rms value of the rectifier supply voltage and I_{dc} is the average dc current through R_L. The regulation of the capacitor filter improves also with larger values of C and f.

Since $I_{dc} \approx V_{dc}/R_L$, by substituting for I_{dc} in Eq. (2–134), we obtain

$$V_{dc} = 1.4\, V_{rms}\, [1 - 1/(2fCR_L)]. \tag{2–135}$$

Equations (2–134) and (2–135) are based on assumptions with respect to waveshape and the load. They are not very suitable if the load is sufficient to cause the output voltage to decrease by more than about 20% from the peak voltage ($1.4\, V_{rms}$).

The output voltage and ripple voltage are determined here for the low voltage power source in Fig. 2–186b. Note that, since this is a full-wave rectifier, $f = 2 \times$ line frequency $= 120$ Hz. Inserting the specific values for f, C, and R_L, we obtain

$$V_{dc} = 1.4 \times 18\, [1 - 1/(2 \times 120 \times 1000 \times 10^{-6} \times 100)] = 24.2 \text{ V}. \tag{2–136}$$

The ripple factor r is

$$r = \frac{1}{2\sqrt{3}fCR_L}$$

$$= \frac{1}{3.5 \times 120 \times 1000 \times 10^{-6} \times 100} = 0.024. \tag{2–137}$$

The rms ripple voltage is

$$V_{ripple} = rV_{dc} \approx 0.6\, V_{rms} \quad (1.7 \text{ V peak to peak}). \tag{2–138}$$

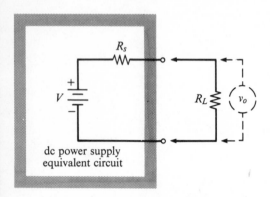

Fig. 2–187 Power supply equivalent circuit.

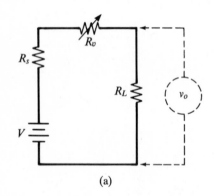

(a)

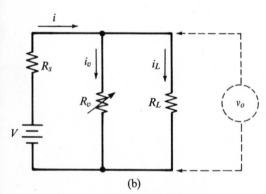

(b)

Fig. 2–188 Types of voltage regulators: (a) series type; (b) shunt type.

Another filter device is simply an inductance in series with the load. The inductance, by opposing the changes in current through it, tends to maintain a more constant current through the load. This results in a more constant output voltage. The current tends toward the average value under the influence of the inductance. The voltage across the load then becomes the average voltage output of the rectifier. For a full-wave rectifier and a choke filter, $V_{dc} = 2/\pi V_{peak} = 2\sqrt{2}/\pi V_{rms} \approx 0.9 V_{rms}$. This expression assumes that the dc resistances of the transformer, rectifier, and choke are negligible compared with R_L. Note that the output voltage is quite independent of the load resistance, making the choke-filtered supply well regulated. The output voltage, however, is considerably lower than with the capacitor filter.

To be effective with a half-wave rectifier, the inductance value must be rather large. The sharp increase in current which occurs at the beginning of the conduction cycle in a half-wave rectifier causes such a large voltage to be induced in the coil that a severe strain is put on the insulation of the windings. For this reason, it is not practical to use chokes to filter the output of a half-wave rectifier.

Combinations of chokes and capacitors make the best passive filters, but the current trend is toward active filtering and regulating circuits for power supplies in precision instruments. Passive and active filters are considered in detail in Modules 3 and 4.

Voltage Regulation

Any power supply can be represented by the Thévenin equivalent circuit (see Section 2–1) shown in Fig. 2–187 with the load R_L connected at its output. The voltage v_o across the load is

$$v_o = V - iR_s. \qquad (2\text{--}139)$$

Therefore, anything that changes the current i changes the voltage v_o. Obviously, if the load changes, the current changes, and the effect on voltage v_o depends on the relative values of R_L and R_s. Also, anything that changes V will change v_o. Therefore, if the line voltage changes, the voltage across the load changes. In other words, to regulate fully (control) the voltage v_o, it is necessary to compensate for changes of V and/or changes of load current i.

One method of providing voltage regulation is to introduce a variable resistor R_v in series with the load, as shown in Fig. 2–188a. The output voltage is now

$$v_o = V - i_L(R_s + R_v), \qquad (2\text{--}140)$$

and in order to provide regulation of the output voltage v_o, the resistance R_v is varied automatically to compensate for changes of V or i_L. There are various control circuits that can automatically change the effective resistance of a series element by monitoring the output voltage and comparing it to a fixed reference voltage. This type of **series regulator** can provide excellent voltage regulation. With the recent developments in integrated circuits the cost of providing series regulation is low. Because of the low cost and excellent voltage regulation, the series type regulator is now very widely used. Specific circuits for series regulators are discussed in Section 2–6.

Another type of voltage regulator is the **shunt regulator,** illustrated in Fig. 2–188b. One of its most important applications is for low-power voltage reference sources where Zener diodes are used as the automatically variable shunts. In the shunt regulator, part of the total supply current i goes through the load, $R_L(i_L)$, and part through a variable resistance, $R_v(i_v)$. Since $v_o = V - iR_s$ and since $i = i_L(R_v + R_L)/R_v$, it follows that

$$v_o = V - \left(i_L \frac{R_v + R_L}{R_v} \right) R_s. \qquad (2\text{–}141)$$

It can be seen that a device that would effectively vary R_v so as to maintain v_o constant at all times would be a valuable regulator. This is the role of a Zener diode in certain regulated supplies. A Zener-regulated supply is shown in Fig. 2–189.

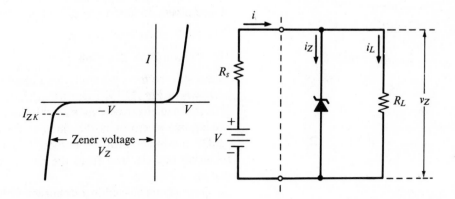

Fig. 2–189 Zener-regulated supply and characteristic curve.

Since the Zener voltage remains essentially constant for a wide range of currents, its effective resistance $R_v = V_Z/i$ changes to compensate for changes of supply voltage V or load current. The diode continues to regulate until the circuit conditions require the Zener current to fall to I_{ZK} in the region of the knee on the diode I-V curve. The power dissipation rating determines the upper limit for the Zener diode current $(P_{\max} = I_{Z(\max)}V_Z)$.

2–4.2 BATTERIES

Despite the availability of high-quality units for the conversion of line-voltage alternating current to direct current in all voltage and current ranges, batteries find a widespread application in modern scientific instruments. The use of batteries as power sources for instruments is essential in remote places, such as outer space, underseas, and in mines, where a central line voltage is not available. Batteries are called for when mobility, portability, and extremely high reliability are required. For some applications, a battery is less expensive and simpler than a power supply that would have the equivalent stability. Also the battery is free from line frequency noise.

A battery consists of several electrochemical "cells" connected to provide the necessary voltage and current capacity. However, it is not unusual to hear the term "battery" used in reference to a single cell. For many years there were only two kinds of batteries readily available that were suitable for light- or moderate-duty applications. These were the familiar carbon-zinc dry cell and the lead-acid storage battery. Today the designer or user of batteries has a choice of half a dozen different kinds of batteries, all commercially available in a wide variety of sizes and voltages. Each of these battery types has characteristics which make it particularly valuable for certain uses. Some new kinds of batteries come in the same case sizes and voltage ratings as the familiar carbon-zinc dry cell, thus allowing the user to choose the best type for his particular application. In this section, the construction, characteristics, and general areas of usefulness in instruments will be discussed for each of the common battery types.

The Carbon-Zinc Dry Cell

This is still the most generally used of all the so-called dry cells. Its structure is shown in Fig. 2–190. During discharge, the zinc (Zn) metal of the can goes into the electroylte as zinc chloride, hydroxide, amine, or oxychloride, leaving two electrons in the zinc can for every atom of zinc dissolved. The MnO_2 in contact with the carbon (C) electrode changes to $Mn_2O_3 \cdot H_2O$, requiring one electron from the carbon for every molecule of MnO_2 converted.

This chemical action establishes a voltage between the two electrodes of about 1.5 to 1.6 V. The carbon electrode is positive, and electrons flow in an external circuit from the zinc can to the carbon rod. Thirty individual cylindrical cells could be connected in series to provide a standard 45 V battery, but this would waste space. Therefore, dry batteries are generally made by stacking a carbon plate, a layer of electrolyte paste, and a zinc plate, alternately, as many times as necessary to give the desired voltage. The most common dry batteries have voltages of 1.5, 3, 6, 7.5, 22.5, 45, 67.5, and 90 V.

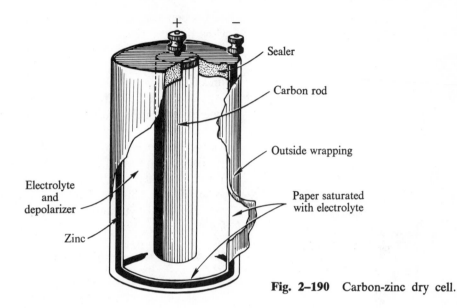

+ −

Sealer

Carbon rod

Outside wrapping

Electrolyte
and
depolarizer

Paper saturated
with electrolyte

Zinc

Fig. 2–190 Carbon-zinc dry cell.

The carbon-zinc dry cell is one of several kinds of primary batteries, that is, nonrechargeable, "one-shot" batteries.

The service life of a battery is the number of operating hours during which a fresh battery will satisfactorily operate the actual circuit under normal operating conditions. The service life of a battery can vary widely depending on the following factors: the quality of the battery, the length of time it has been stored before use, its temperature during the storage period, the rate at which it is discharged, the number and duration of the off periods, its temperature during discharge, and the lowest voltage for satisfactory operation of the circuit. After actuation all batteries discharge at some rate internally. This limits the time a dry cell can be stored before use. At room temperature a dry cell could be stored for about a year, but increasing the temperature shortens the shelf life considerably.

It is very important to use a battery which is physically large enough to handle the current drain comfortably. The discharge rate is commonly measured by the number of operating hours necessary to wear out the battery, that is, the service life. For a carbon-zinc cell, a service life of less than 10 hours is considered a heavy drain. In addition to the nuisance of frequent battery changing, the number of milliampere-hours (the capacity) of the battery falls off severely as the rate of discharge increases. Furthermore, the voltage of a battery under heavy drain is low because of the *IR* drop in the cell and the polarization at the electrodes. If a carbon-zinc dry cell must be used under conditions of heavy drain, periodic rest periods will definitely extend the capacity and thus the service life of the battery.

Carbon-zinc dry cells do not work very effectively at temperatures below freezing. Even at temperatures where the cell can still supply a current, the service life may be very low. The potential of the carbon-zinc cell decreases continuously during use. This decrease is due partly to the formation of compounds which interfere with the operation of the battery and partly to the fact that substances in the electrolyte may be involved, to some extent, in the electrode reaction. When the latter is true, the change in electrolyte composition during discharge has a direct effect on the electrode potentials.

The dissolution of the zinc tends to weaken the structure of the cell. Furthermore, during discharge or storage a pressure of evolved hydrogen gas builds up. This can lead to a rupture of the zinc and leakage of the corrosive electrolyte into the instrument. *Instruments using these dry cells should therefore not be stored with the batteries installed.*

The Alkaline-Manganese Battery

Commonly called the alkaline battery, the alkaline-manganese battery is quite similar in mechanism to the carbon-zinc dry cell. The electrode reactions are basically the same. The arrangement of the electrodes is different, and the electrolyte is strongly basic. The structure is shown in Fig. 2–191. The changes have resulted in quite a number of improvements over the ordinary carbon-zinc cell. The capacity has roughly doubled. The shelf life has increased. The capacity does not decrease under heavy drain. The internal resistance is lower, and the available current is higher. The low-temperature operating limit has been reduced to 40°F. At roughly four times the price, the alkaline battery will not replace the carbon-zinc cell completely. It does have obvious advantages for heavy drain applications, where the capacity of the conventional cell is greatly decreased. Its longer shelf life and lower operating temperature make it a much better choice for an emergency power source. Its low internal impedance gives it an advantage for circuits where large fluctuations in the current demand occur.

The Mercury Battery

This is a third common kind of primary cell. It has a zinc amalgam for one electrode and mercuric oxide–carbon for the other. During discharge the zinc oxidizes to zinc oxide, yielding two electrons per atom, while the mercuric oxide is reduced to mercury by accepting two electrons per molecule from the carbon electrode. The structure is shown in Fig. 2–192a. The potential developed by this cell is 1.35 V and is remarkably reproducible from one cell to the next. The capacity of this cell is somewhat greater than even that of the alkaline cell, but this capacity is not maintained under heavy drain or at low temperatures (only 58% capacity at 40°F). The

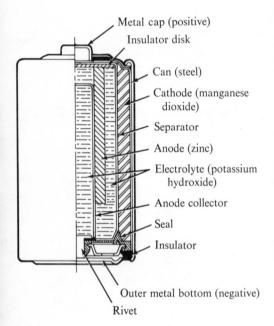

Metal cap (positive)

Insulator disk

Can (steel)

Cathode (manganese dioxide)

Separator

Anode (zinc)

Electrolyte (potassium hydroxide)

Anode collector

Seal

Insulator

Outer metal bottom (negative)

Rivet

Fig. 2–191   Alkaline-manganese cell.

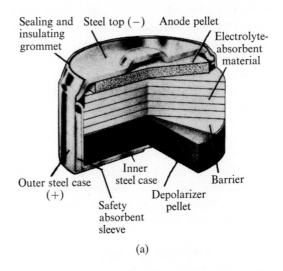

Sealing and   Steel top (−)   Anode pellet
insulating
grommet

Electrolyte-
absorbent
material

Outer steel case
(+)

Inner
steel case

Barrier

Safety
absorbent
sleeve

Depolarizer
pellet

(a)

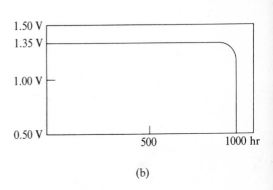

(b)

internal resistance is very low, as with the alkaline battery. In addition to the reproducibility of cell voltages, the great virtue of the mercury battery is its virtually constant voltage during discharge. This is achieved because of the electrode structure and composition and because the electrolyte does not change composition during discharge. The voltage discharge curve (Fig. 2–192b) is so flat that the mercury battery has been used in some older instruments as a secondary voltage standard. No gas is evolved in the mercury battery, and its structure does not readily deteriorate. Because of its high price (about five times that of the alkaline battery for a comparable size), the mercury battery is likely to be used only where a constant potential during discharge is important. In many applications the circuit will not operate satisfactorily at reduced voltages. This requirement shortens the service life of the less expensive batteries so much that the mercury battery is practical economically. Mercury batteries are available in many voltages from 1.35 to 42 V.

Fig. 2–192 Mercury battery: (a) pictorial; (b) discharge curve.

The Lead-Acid Storage Battery

The familiar car battery, this is a series combination of several $Pb/H_2SO_4/PbO_2$ cells. Under discharge the lead (Pb) in the lead electrode is converted to insoluble lead sulfate, releasing two electrons to the lead electrode per molecule of lead sulfate created. The lead dioxide in the PbO_2-Pb electrode is reduced to lead sulfate by accepting two electrons per molecule of lead sulfate formed. The chemical reaction which provides the electrical energy is

$$Pb + PbO_2 + 2H_2SO_4 \rightarrow 2PbSO_4 + 2H_2O.$$

When fully charged, the potential of each cell is 2.06 to 2.14 V. The lead-acid battery is a **secondary** cell; i.e., it can be recharged. When a reverse or charging current is forced through the cell, the $PbSO_4$ is converted to Pb and PbO_2 respectively, so that the cell is returned to very nearly its original state.

The reaction shows that sulfuric acid is converted to water on discharge. This dilution of the sulfuric acid causes the cell voltage to decrease during discharge. The decrease in voltage is slow at first but becomes quite rapid during the last one-third of the service life. The internal resistance of the lead-acid battery is so low that it can be ignored at normal discharge rates. The high current free from ac ripple makes it the most economical power source for certain applications. However, the lead-acid battery is bulky and heavy and requires a considerable amount of care if it is to give proper service.

The Nickel-Cadmium Battery

The nickel-cadmium battery has come into widespread use in recent years. One reason for its growth in popularity has been the development of the sealed nickel-cadmium cell. This is a completely sealed unit which requires no attention other than charging. Two forms of this cell are shown in Fig. 2–193a. Under discharge the cadmium (Cd) is oxidized, supplying electrons, while at the positive electrode, nickel oxide is reduced to a lower oxidation state by accepting electrons. The open-circuit potential of this cell is 1.3 V. The electrolyte is not involved in the electrode reaction, and thus the potential is fairly constant over the service life. The capacity of the nickel-cadmium battery is reduced very little even at very high discharge rates. The batteries can be stored charged or discharged without harm. The sealed units will not leak. They have the lowest self-discharge rate of any secondary cell. Sealed nickel-cadmium batteries are made in many sizes and voltages which make them interchangeable with the primary batteries. To all the areas of the application of conventional batteries, nickel-cadmium batteries offer the advantages of high current capability, long service life, reasonably constant potential, and the possibility of recharge. These characteristics are particularly suited to instruments where heavy drain and frequent use would require frequent battery replacement. The price is about 25 times that of a similar-sized carbon-zinc dry cell, and there is some concern about cadmium pollution in the environment.

During the charging process, the nickel oxide is reoxidized to its higher oxidation state and the cadmium oxide is reduced. The cell is constructed with some oxidized cadmium so that the nickel electrode is the first electrode to reach full charge. If the cell is overcharged, the oxygen gas produced at the nickel electrode diffuses to the cadmium electrode, where it is reduced.

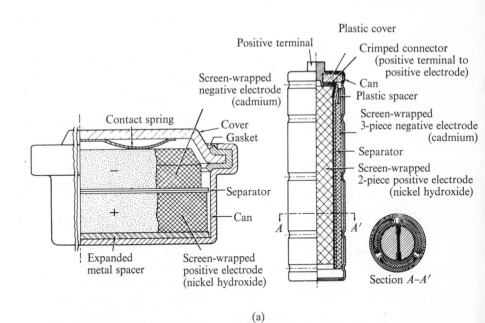

(a)

Fig. 2–193 Rechargeable nickel-cadmium cell: (a) sealed button cell at left and cylindrical cell at right; (b) battery charger for rechargeable cells.

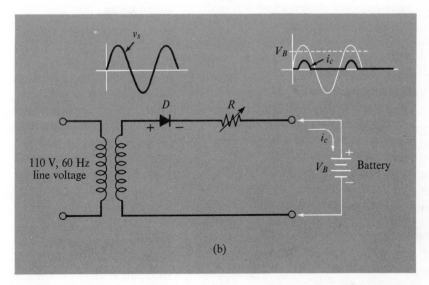

(b)

This provides a "chemical short circuit" for excess charge and prevents the buildup of a large gas pressure.

Battery Charger

Rechargeable batteries such as the nickel-cadmium battery can be readily charged from the ac power line by use of a half-wave rectifier circuit as shown in Fig. 2–193b. The battery is charged by reversing the electrode reactions which occur when the battery is a voltage source. A dc charging current, i_c, is obtained by introducing a diode in series with the battery and applying a sufficiently high voltage, v_s, at the secondary of the transformer so that the diode conducts during part of the ac cycle. When the magnitude v_s of the sine wave exceeds the battery voltage, V_B, the diode conducts and the charging current, i_c, flows in the correct direction so as to force the electrochemical reactions to occur which recharge the battery. When the magnitude of v_s is less than V_B the diode is reverse biased, and this effectively opens the circuit so that no discharge current occurs. By varying the resistance R, the rate of charging the battery can be adjusted.

2–4.3 SOLAR AND FUEL CELLS

Although not generally applicable now, both solar cells and fuel cells hold considerable promise for the direct conversion of sunlight and chemical energy, respectively, into electrical power. These cells would be especially useful for instrumentation in remote locations.

Solar Cells

The radiation from the sun to the earth each day represents an amount of energy that is many times our yearly consumption of electrical energy. It is estimated that on a summer day the solar radiant power is 1 kW/m² at the earth's surface. Unfortunately there is no efficient way of directly converting this radiant power into electrical power. Of course, the solar radiation is used indirectly to obtain electrical power. The sun's radiation causes evaporation of water from the earth's oceans and other major bodies of water, and the rain returns the water to the mountains, where it can be utilized by hydroelectric power plants. Also, the sun's radiation was involved in the chemical and biological processes that resulted over many years in the formation of coal and oil, which are used in fossil fuel power plants.

In recent years, **semiconductor solar cells** have been used with fair efficiency for direct conversion of solar radiation into electrical power. These cells provide a high power capacity per unit weight and have been used successfully on space vehicles and communication satellites. A representation of a silicon photovoltaic cell is shown in Fig. 2–194a. A 1×2 cm silicon wafer has been used to build up single-panel arrays of many thousands of wafers so as to provide a capacity of several hundred watts.

A very thin layer of p-type semiconductor is formed on an n-type silicon strip (or vice versa) to form a pn junction and to provide the light-sensitive face. A narrow conducting strip serves as the collector term-

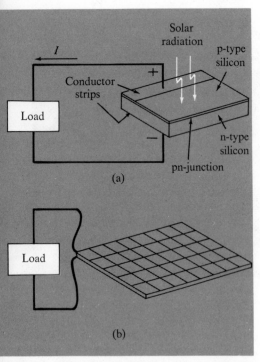

Fig. 2–194 Solar cells: (a) silicon cell; (b) an array of forty 1×2 cm silicon cells.

inal. The bottom of the cell is nickel plated and tinned. The mechanism of conversion from solar radiation to electrical power can be understood by recalling the discussion of the pn diode in Section 2–3. The diffusion of majority carriers from both the n- and p-type materials causes a combination of holes and electrons that sets up a barrier potential across the junction. The p- and n-type materials contain many valence electrons that are not affected by the impurity doping. However, a photon of light can interact directly with a valence electron and provide sufficient energy to promote it to a conductance electron. The free electron and the hole are now attracted by the barrier potential at the pn junction, and will travel in the opposite direction to the majority semiconductor carrier. That is, the photoelectrons will travel toward the positively charged n-type silicon and the holes toward the negatively charged p-type silicon so as to provide a current I through a load as shown in Fig. 2–194a.

A series-parallel array of forty 1×2 cm cells as shown in Fig. 2–194b could provide about 1 W electrical power for 10 W of incident sunlight. The satellite arrays generally provide about 10 W output per pound. In view of the huge supply of silicon and sunlight, there should be increased application of solar cells as the technology improves.

Fuel Cells

The direct conversion of chemical energy in a fuel into electrical power is now practical on a small scale by operation of hydrogen-oxygen fuel cells. The Gemini spacecraft utilized H_2-O_2 fuel cells of the type shown in Fig. 2–195. There are two separate chambers and two porous electrodes separated by an electrolyte. The hydrogen gas in the top chamber diffuses through electrode A and reacts in the electrolyte (in the presence of a catalyst) to produce hydrogen ions and free electrons ($2H_2 \rightarrow 4H^+ + 4e^-$). The hydrogen ions move through the electrolyte to electrode B and there they combine with oxygen and electrons to form water ($4H^+ + 4e^- + O_2 \rightarrow 2H_2O$). The reactions at electrodes A and B cause a flow of current I through the load, as shown in Fig. 2–195.

The theoretical efficiency of greater than 90% and already attained efficiencies of over 80% with laboratory cells, low cost, and quiet operation without exhaust are major reasons for the great interest in this type of direct chemical-to-electrical converter.

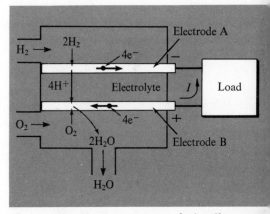

Fig. 2–195 Hydrogen-oxygen fuel cell.

2–4.4 ELECTRICAL POWER DISSIPATION

When charge moves through a resistive element, the power loss in the element appears as heat. The product of the voltage drop across the resistive element and the current through it is equal to the power dissipated in the device, as expressed in Eq. (2–142):

$$P \text{ (watts)} = V \text{ (volts)} \times I \text{ (amperes)}. \qquad (2\text{--}142)$$

Since $V = IR$, Eq. (2–142) can be rewritten as

$$P = I^2 R \qquad (2\text{--}143)$$

or

$$P = \frac{V^2}{R}. \qquad (2\text{--}144)$$

In building power supplies or in connecting loads at the output of power supplies, it is important not to exceed the wattage rating of resistors or other devices. For example, suppose that a 47 kΩ resistor is rated at 1 W. The maximum current through this resistor is thus $\sqrt{1/47{,}000} = 5 \times 10^{-3}$ A, or 5 mA. The maximum voltage that can be applied to the resistor is $\sqrt{1 \times 47{,}000} = 230$ V. A resistor that is dissipating its **maximum rated power** can be hot enough to cause burns and must be well ventilated. For cooler and therefore more stable resistance values, a maximum rated power two to ten times greater than the power dissipation is recommended.

Some devices, such as light bulbs, are rated according to the **power normally consumed** rather than the maximum rated power. For these, the power rating reveals other electrical characteristics.

Experiment 2–30 *Instrumentation Power Supplies*

This experiment illustrates how a Zener-regulated power supply can be built from a rectifier circuit, a filter circuit, and a Zener regulator circuit. Half-wave and full-wave rectifier circuits are first studied and then a capacitor filter is added to a full-wave bridge rectifier. The effects of the *RC* time constant of the filter on the output waveform are noted. A Zener regulator is then added and its characteristics and limitations are studied.

PROBLEMS

1. By what percentage will the secondary voltage of a transformer be reduced if the line voltage decreases from 115 V ac to 95 V ac?

2. A transformer on a power pole transforms the voltage from an efficient transmission value of 4400 V to a safe working value of 220 V. What turns ratio must the transformer have?

3. What is the minimum peak inverse voltage (PIV) rating for a diode to be used in a half-wave rectifier circuit if a transformer is used which has 10 V rms on the secondary, and a large filter capacitor is placed across the load?

4. (a) What are the minimum PIV ratings for diodes to be used in a full-wave rectifier circuit if a transformer is used which has 5 V rms on either side of the center tap and if a large filter capacitor is placed across the load? (b) What are the minimum PIV ratings for the diodes and the voltage rating for the transformer to be used in a bridge rectifier circuit which supplies the same voltage to the same filtered load as in part (a) above?

5. Various rectifier circuits are being considered for use with a transformer which has 25 V rms on the secondary. If it is assumed that the transformer and the diodes are ideal and the load is negligible, what are the maximum output voltages available for the following rectifier circuits with a capacitor filter: (a) half-wave; (b) full-wave bridge; (c) voltage doubler?

6. Consider the power supply of Fig. P–6 and assume that the diodes and transformer are ideal. Calculate (a) the output voltage; (b) the current through the 5 kΩ load; (c) the rms ripple voltage.

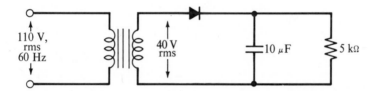

Fig. P–6

7. The power supply of Fig. 2–186b is considered for various values of the filter capacitor C. Calculate the output voltage and ripple voltage for $C = 10$ μF, 100 μF, 500 μF, and 1000 μF.

8. A Zener diode is used in the low-voltage power supply shown in Fig. 2–189. The Zener voltage is 5.1 V and the Zener diode can dissipate a maximum power of $P_{max} = 400$ mW. Calculate the required value of R_s if $V = 10$ V, $R_L = 100$ Ω, and the power dissipation is to be 100 mW.

9. A 50 Ω resistor has a maximum power rating of $\frac{1}{2}$ W. What is the maximum current through this resistor and the maximum voltage which can be applied without exceeding the power rating?

Gain Control of Voltage and Current

The sequential or simultaneous control of events by electronic switching, often at fantastically high speeds, and the control of electrical signals with resistive, capacitive, and inductive networks, were described in the previous sections of this module. Combinations of these basic control operations were shown to provide important functions such as electronic waveform generation and shaping, integration, differentiation, and log conversion. For many of these previously described applications, amplifiers were assumed to be available with the required characteristics, and were used as **gain blocks.** That is, in those cases where voltage and/or current gain of an electrical signal were necessary, an amplifier functional block was inserted that could meet the gain requirements and it was represented in diagrams by a triangular symbol. Fortunately, there are now on the market many inexpensive general purpose amplifiers that come in small prepackaged units and fulfill specific amplification (current and/or voltage gain) needs. Unfortunately, however, all amplifiers have limitations and these limitations can cause problems unless they are recognized and allowed for.

From our own experience we have found that the misuse of amplifier functional blocks can be most easily avoided by the application of some basic and practical understanding of how the control of voltage and/or current gains is achieved with amplifiers. Therefore, many of the basic considerations of ideal and practical amplifiers are presented. In Modules 3 and 4, certain of these considerations, such as frequency response of amplifiers, are discussed in more detail. Some of the most frequently encountered basic amplifiers are presented here so as to acquaint the reader with the terminology and methods of circuit modeling. Several of these basic amplifiers are then combined to form a dc instrumentation amplifier or gain block of the general type that has been used (and represented with a triangular symbol) to perform many of the measurement and control functions previously reported.

2–5.1 IDEAL AMPLIFIERS

Both voltage and current amplification are important in measurement and control applications in instrumentation systems. Often the voltage signal from a transducer or other source is so small that it needs to be increased in magnitude by a **voltage amplifier** before subsequent information processing. In other cases the magnitude of the voltage is sufficient but the current signal needs to be increased by a **current amplifier,** which also provides the necessary **power gain,** so that the signal can control a motor, relay, meter, or other device. Although gain blocks with ideal characteristics are not available, it is often helpful to consider what constitutes an ideal device for a specific purpose. Therefore, in this section both ideal voltage and current amplifiers will be considered and illustrated with circuit models.

An Ideal Voltage Amplifier

First let us examine the characteristics of an ideal voltage amplifier (voltage-controlled source), in which the output voltage v_o is directly proportional to the input voltage v_{in} at every instant, and μ is the proportionality constant (called the voltage amplification factor). The input and output of the amplifier are related as shown in Eq. (2–145), and represented in the circuit model of Fig. 2–196.

$$v_o = \mu v_{in}. \tag{2–145}$$

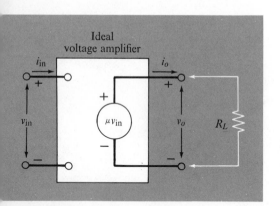

Fig. 2–196 Circuit model of an ideal voltage amplifier.

There are several significant conclusions that can be reached from considering the ideal voltage amplifier. The input terminals are isolated from the output terminals. That is, the electrical connections at the output are completely independent of the electrical connections at the input, as illustrated in Fig. 2–196. The input resistance is infinite (open circuit), so that the input current $i_{in} = 0$. Therefore the input power is zero and there is no loading of the input voltage source. The output power, P_o, is

$$P_o = \frac{v_o^2}{R_L} = \frac{(\mu v_{in})^2}{R_L}. \tag{2–146}$$

Therefore in the ideal voltage amplifier the ratio of output power to input power (i.e., power gain) is infinite.

Ideally, as is also implied in Eq. (2–145), there are no restrictions on the output current. In other words, as the load changes from no load (open circuit) to a short circuit the output voltage remains constant, and this implies that the output resistance is zero.

When the output load is fixed at R_L, it can be seen from Eq. (2–146) that the output power depends on the input voltage. Therefore, if μ is fixed and v_{in} is small, the output power and the voltage might be insufficient for operation of a specific load. In this case several voltage amplifiers can be used in cascade, the output of the first amplifier being connected to the input of the second amplifier, and the load connected to the output of the second amplifier.

An Ideal Current Amplifier

A current-controlled current source or **current amplifier** would be ideal if the output current i_o remained proportional to the input current i_{in} at every instant. The proportionality constant β is called the current amplification factor, and the relationship is expressed in Eq. (2–147):

$$i_o = \beta i_{\text{in}}. \qquad (2\text{–}147)$$

The circuit model of an ideal current amplifier is shown in Fig. 2–197. The input current is applied through a zero input resistance so that the input voltage is zero and the input power ($P_{\text{in}} = i_{\text{in}}^2 R_{\text{in}}$) is zero. Also, the input terminals are perfectly isolated from the output terminals. It can be seen from Eq. (2–147) that the output current is independent of the output voltage and load, and that the output power ($P_o = i_o^2 R_L$) varies with the load. Therefore the output power gain of an ideal current amplifier is infinite.

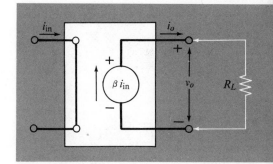

Fig. 2–197 Circuit model of an ideal current amplifier.

Having considered the basic characteristics of ideal voltage and current amplifiers, we will now turn our attention to various amplifier types and become familiar with basic amplifier terminology. Although practical amplifiers fall short of being ideal, field-effect transistor amplifiers approximate ideal voltage amplifiers, while bipolar junction transistor amplifiers can often be treated as ideal current amplifiers.

2–5.2 BASIC AMPLIFIERS

There are many types of basic amplifiers, only a few of which are presented in this section. However, by describing a basic junction field-effect transistor (JFET) amplifier and a basic bipolar junction transistor (BJT) amplifier it is possible to introduce much of the electronic amplifier terminology and circuit modeling concepts that are important in reading the electronics literature. These basic presentations are followed by discussions of the emitter-follower amplifier and the difference amplifier, both of which are frequently found in all types of instrumentation systems. With this brief

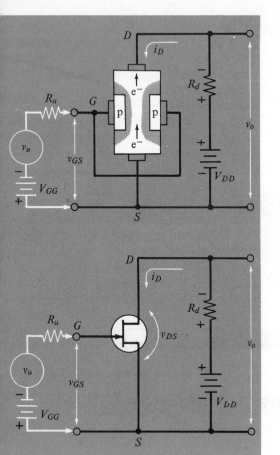

Fig. 2–198 Basic n-channel JFET amplifier: (a) pictorial diagram; (b) schematic diagram.

background in amplifiers it is possible to understand how the basic devices are combined to form a general purpose operational amplifier or gain block that has almost unlimited application for measurement and control functions. This combination of devices to form a general purpose gain block is the subject of the next section.

A Basic JFET Amplifier

The basic structure of a JFET was described in Module 1, and its application as a switch was presented in Section 2–3 of this module. In Fig. 2–198 a silicon n-channel JFET is shown connected in a basic common-source amplifier configuration. The term **common-source** implies that the source terminal of the JFET is common to both the input and output terminals of the amplifier.

The majority carriers (electrons for the n-channel silicon FET shown in Fig. 2–198) flow from the source S to the drain D because of the applied voltage V_{DD}. The gate (G) junctions are usually reverse biased (in this example by a voltage source V_{GG}) and only very small currents (10^{-8} to 10^{-13} A) exist at the gate terminals. Therefore, the voltage drop across the resistance R_u of the input source is usually negligible, and the instantaneous voltage between gate and source, v_{GS}, is essentially equal to the sum of the signal input voltage and the bias voltage, $v_{GS} = v_u + V_{GG}$. The gate voltage v_{GS} controls the current, i_D, between the drain and source of the JFET. Thus v_{GS} also controls the voltage drop, $i_D R_d$, across the drain load resistor R_d, which is connected between the drain terminal of the JFET and the positive lead of the drain voltage supply V_{DD}.

The output voltage, v_o, is equal to the difference between the supply voltage V_{DD} and the voltage drop across R_d caused by the drain current, i_D, so that

$$v_{DS} = v_o = V_{DD} - i_D R_d. \qquad (2\text{--}148)$$

The linear equation (2–148) relates the output voltage and device current, and this can be drawn as a current-voltage line on a set of JFET characteristic curves. This method of drawing a load line on a set of characteristic curves was first introduced in Section 2–3 for switching circuits. In this case it provides a means of graphically observing how an input signal is amplified by a specific JFET, as illustrated in Fig. 2–199. It should be noted that the input signal varies about an average value of V_{GS}, so that when input signal $v_u = 0$, there is an average dc drain current I_D which causes an average dc drain-source voltage V_{DS}. These average values are known as the **quiescent voltages and current**, and they describe an average or **quiescent operating point** about which the signal varies. The quiescent values are designated by capital letters.

A JFET Small-Signal Circuit Model

In analyzing the operation of practical amplifier circuits it is convenient to use circuit models. In developing these models for amplifiers one generally assumes that a linear equivalent circuit is a valid approximation over small portions of the operating range. That is, the circuit model will describe the relationship of the small changes in the amplifier current and voltage caused by a small input signal.

For the JFET amplifier the application of an input signal causes the instantaneous circuit values of v_{GS}, i_D, and v_{DS} (Fig. 2–199) to differ from the quiescent values V_{GS}, I_D, and V_{DS}. The differences between the actual values and the quiescent values are called the signal values and are given the symbols v_{gs}, i_d, and v_{ds}. In other words $v_{gs} = v_{DS} - V_{DS}$, where v_{gs} is the change in gate-to-source voltage which causes a change in drain current $i_d = i_D - I_D$, which causes an output signal $v_{ds} = v_{DS} - V_{DS}$.

By noting that the actual circuit quantities v_{GS}, i_D, and v_{DS} are interdependent variables, and choosing i_D as the dependent variable, the relationship among i_D and v_{GS} and v_{DS} can be written as a general function as shown in Eq. (2–149):

$$i_D = f(v_{GS}, v_{DS}). \qquad (2\text{–}149)$$

If the differential[7] of Eq. (2–149) is taken, Eq. (2–150) results:

$$di_D = \left(\frac{\partial i_D}{\partial v_{GS}}\right)_{v_{DS}} dv_{GS} + \left(\frac{\partial i_D}{\partial v_{DS}}\right)_{v_{GS}} dv_{DS}. \qquad (2\text{–}150)$$

The quantity $(\partial i_D/\partial v_{DS})_{v_{GS}}$ is the reciprocal of the drain-to-source dynamic resistance, or $1/r_{ds}$. The quantity $(\partial i_D/\partial v_{GS})_{v_{DS}}$ is called the **mutual conductance** or **transconductance** and given the symbol g_m. It has the units 1/ohms, or mhos. Since the quantities di_D, dv_{GS}, and dv_{DS} are the small-parameter changes from quiescent values of drain current, gate-to-source voltage, and drain-to-source voltage, respectively, they can be represented by i_d, v_{gs}, and v_{ds}. Therefore Eq. (2–150) can be written as

$$i_d = g_m v_{gs} + \frac{v_{ds}}{r_{ds}}. \qquad (2\text{–}151)$$

Equation (2–151) suggests that the signal drain current i_D is the sum of the currents in two current paths between the source and the drain. The main current is $g_m v_{gs}$, and the other is a small current caused by a slight decrease or increase in channel length with increasing or decreasing drain voltage that results in a small change in drain current equal to v_{ds}/r_{ds}. The slope m of a drain characteristic curve in the region of pinch-off is equal to i_d/v_{ds}, which is $1/r_{ds}$ and is given the symbol g_d. The circuit model can thus be drawn as shown in Fig. 2–200a or b. It is called a **current-**

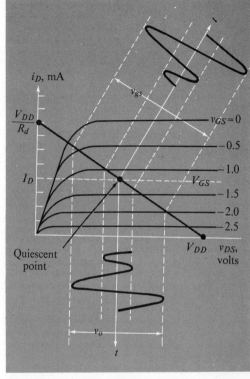

Fig. 2–199 Graphical illustration of amplification by JFET amplifier.

Note 7. Interpretation of Eq. (2–150).

A knowledge of differential equations is NOT necessary to understand this formula or the derivation. Equation (2–150) reads simply: The change in i_D is equal to the change in i_D resulting from a change in v_{GS} at constant v_{DS} times the change in v_{GS}, plus the change in i_D resulting from a change in v_{DS} at constant v_{GS} times the change in v_{DS}.

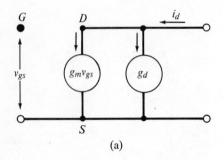

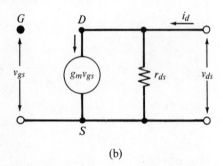

Fig. 2–200 Circuit models for JFET amplifier.

source circuit model because the input signal v_{gs} results in an output current $g_m v_{gs}$.

For the amplifier circuit of Fig. 2–201a, a circuit model can be drawn as shown in Fig. 2–201b. This circuit model includes the input source v_u and its internal resistance R_u and the output load resistance R_d. Since $i_d \approx g_m v_{gs}$, the output voltage $v_o \approx -g_m v_{gs} R_d$. The negative sign is used, as previously, to indicate a 180° phase reversal. That is, when the input voltage increases, the output voltage decreases proportionately, and vice versa, as readily seen in Fig. 2–199.

The voltage gain, A_v, is thus

$$A_v = \frac{v_o}{v_{gs}} = -g_m R_d. \qquad (2\text{–}152)$$

The concept of the small-signal model is very useful in analyzing electronic circuits and it is used extensively. The model represents the device and/or circuit in terms that can be readily manipulated and visualized.

Fig. 2–201 Basic JFET amplifier with input voltage source: (a) circuit schematic; (b) circuit model.

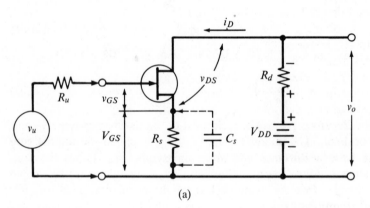

(a)

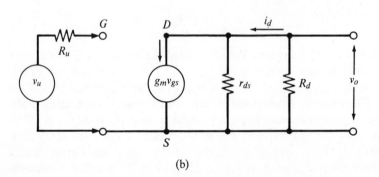

(b)

However, it must be remembered that these models are restricted to a limited range of operating conditions. It should be noted that the terms **circuit model** and **equivalent circuit** are often used interchangeably, but that in recent literature the term circuit model is gaining preference because it shifts emphasis from equivalence to approximation of the circuit.

Self-Bias of JFET Amplifier

In the amplifier circuit shown in Fig. 2–198 the gate-to-drain operating point was established with a bias voltage source V_{GG}. Although it is possible to use a bias source, it is not convenient or economically practical. Instead a resistance R_s is generally put in the source circuit as shown in Fig. 2–201a. The current through R_s with no input signal applied is then I_D, because the gate-to-drain current i_G is negligible. Therefore there is a voltage drop $I_D R_s$ across the JFET source resistor which provides the quiescent bias voltage V_{GS} between gate and source. When an input signal voltage v_u is applied to a self-biased amplifier, the drain current, $i_D = I_D + i_d$, changes with the input signal. This would cause a change in the bias voltage and decrease the effective input signal. To maintain a relatively constant quiescent bias voltage a capacitor C_s is connected across the resistor R_s. The capacitor C_s is normally of very high capacitance so that it acts as a very low impedance[8] for even relatively low frequency signals. Thus the signal current, i_d, is effectively bypassed or short-circuited around R_s, so that the current in R_s is I_D and the current in C_s is i_d. The bias voltage across the parallel combination is then essentially constant, $V_{GS} = I_D R_s$.

When the circuit model is drawn for ac signals, the resistor R_s is omitted and the circuit analysis is the same as for Fig. 2–201a except for the added restriction that the signal must be ac. The circuit model for the JFET basic amplifier is shown in Fig. 2–201b.

A Basic BJT Amplifier

A basic BJT transistor common-emitter amplifier is shown in Fig. 2–202. One great difference between the BJT and the FET amplifiers is the

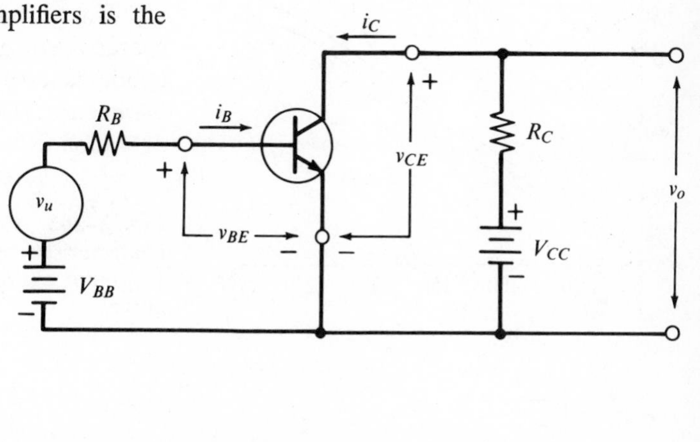

Note 8. Capacitive Reactance.

Capacitive reactance, X_C, *is the opposition or* **impedance** *of a capacitor to alternating current and is measured in ohms. It is dependent on frequency f and the capacitance, so that* $X_C = 1/2\pi f C$. *A thorough discussion of impedance concepts and measurements is included in Module 3.*

Fig. 2–202 Basic BJT amplifier.

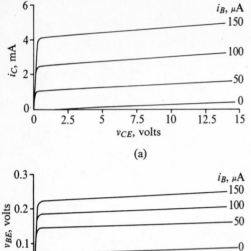

(a)

(b)

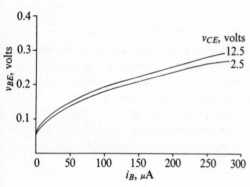

(c)

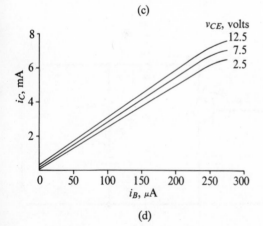

(d)

appreciable base current for the BJT circuit, which is required for control of the emitter-collector current; in contrast, the gate current for the FET is extremely small. The base current, I_B, and base-emitter voltage, V_{BE}, influence the charge distribution and thus the operating characteristics of the BJT. Hence the characteristics of the output circuit (emitter to collector) are dependent on certain parameters of the input circuit. Similarly, conditions in the output circuit will affect the characteristics of the input circuit. The interdependence of the input and output circuits complicates the analysis of the BJT in comparison to the FET described in the previous section.

The characteristic curves of the BJT are readily obtained, as discussed previously. Theoretically, eight different three-parameter characteristic curves can be determined for the four quantities i_B, v_{BE}, i_C, and v_{CE} shown in Fig. 2–203. At least two types of curves are necessary for a complete characterization. Four types of curves are shown in Fig. 2–203.

Circuit Model for BJT Amplifier

To derive a circuit model for the BJT, we must consider all four variables i_B, v_{BE}, i_C, and v_{CE}. Two variables are chosen as independent. It is more logical to choose the two most easy to control, which are i_B and v_{CE}. The other two variables are dependent. Thus we have the equations

$$v_{BE} = f_1(i_B, v_{CE}) \tag{2-153}$$

and

$$i_C = f_2(i_B, v_{CE}). \tag{2-154}$$

The differentiation of the first will yield the input equivalent circuit, and the differentiation of the second the output equivalent circuit. From the first equation,

$$dv_{BE} = \left(\frac{\partial v_{BE}}{\partial i_B}\right)_{v_{CE}} di_B + \left(\frac{\partial v_{BE}}{\partial v_{CE}}\right)_{i_B} dv_{CE}. \tag{2-155}$$

The partial differential $(\partial v_{BE}/\partial i_B)_{v_{CE}}$ is the slope of the v_{BE} vs. i_B characteristic curve (Fig. 2–203c) and is defined as h_{ie}. The symbol h is for hybrid, since the parameters developed by these equations are called hybrid parameters. The i indicates input, and e is for the common-emitter configuration. Note that h_{ie} has the units of resistance and is sometimes referred

Fig. 2–203 Common-emitter characteristic curves: (a) collector characteristic curves; (b) reverse transfer characteristic curves; (c) input characteristic curves; (d) forward transfer characteristic curves.

to as the input resistance. The partial differential $(\partial v_{BE}/\partial v_{CE})_{i_B}$ is the slope of the v_{BE} vs. v_{CE} characteristic curve (Fig. 2–203b) and is defined as h_{re}. This parameter indicates the effect of the output voltage on the input voltage and hence is a sort of amplification factor in reverse. Here h stands for hybrid parameter, r is for reverse, and e represents the common-emitter configuration.

With the substituted parameters, the input circuit equation is

$$v_{be} = h_{ie}i_b + h_{re}v_{ce}, \qquad (2\text{–}156)$$

where v_{be}, i_b, v_{ce}, and other symbols with lower-case subscripts are defined as the signal values. This equation suggests a resistance h_{ie} in the base lead which develops an iR drop of $i_b h_{ie}$ in series with a voltage source $h_{re}v_{ce}$. The schematic representation of the transistor input circuit model is shown as part of Fig. 2–204. The term $h_{re}v_{ce}$ in Eq. (2–156) shows the dependence of the input circuit on the conditions existing in the output circuit (v_{ce}). The reverse amplification factor h_{re} is a measure of the degree of interaction between the input and output voltages.

Differentiating the output equation, Eq. (2–154), we obtain

$$di_c = \left(\frac{\partial i_C}{\partial i_B}\right)_{v_{CE}} di_B + \left(\frac{\partial i_C}{\partial v_{CE}}\right)_{i_B} dv_{CE}. \qquad (2\text{–}157)$$

The partial differential $(\partial i_C/\partial i_B)_{v_{CE}}$ is the slope of the i_C vs. i_B characteristic curve (Fig. 2–203d). It is the current amplification factor symbolized by h_{fe}, the f standing for forward and e for the common-emitter configuration. The partial differential $(\partial i_C/\partial v_{CE})_{i_B}$ is the slope of the i_C vs. v_{CE} characteristic curve (Fig. 2–203a). It is a conductance, has the units mhos, and is symbolized by h_{oe}. Again h is for hybrid, and the o indicates output and e the common-emitter configuration. Therefore

$$i_c = h_{fe}i_b + h_{oe}v_{ce}. \qquad (2\text{–}158)$$

The collector current is the sum of two current paths, a conductance h_{oe} between the emitter and collector, and the amplified input current $h_{fe}i_b$. In the equivalent circuit the conductance h_{oe} is shown as a resistance of $1/h_{oe}$ ohms. The resulting equivalent circuit is shown in Fig. 2–204. Other equivalent circuits and other parameters result from a different choice of the independent variables.

A BJT Common-Emitter Amplifier

A BJT amplifier which has the emitter lead in common with the input and output circuits is shown in Fig. 2–205. The dc analysis of this circuit is very similar to that of the JFET. First the load line is determined by

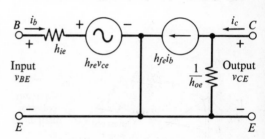

Fig. 2–204 BJT amplifier circuit model with hybrid parameters.

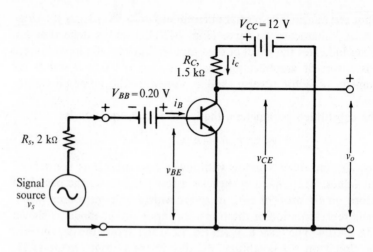

Fig. 2–205 Transistor amplifier circuit.

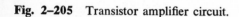

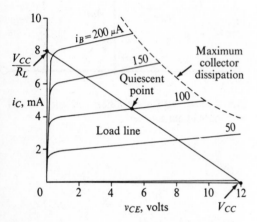

Fig. 2–206 DC analysis for the circuit of Fig. 2–205.

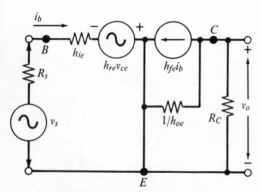

Fig. 2–207 Common-emitter amplifier circuit model with hybrid parameters.

connecting V_{CC} on the voltage axis with V_{CC}/R_L on the current axis of the collector-characteristic curve for the transistor used (Fig. 2–206). Note that the load line should fall well within the maximum collector dissipation line. The maximum collector dissipation $P_{\max}$ is a characteristic of the transistor and is usually given in the transistor manuals or catalog descriptions. The maximum collector dissipation hyberbola is drawn by connecting all the points satisfying the equation $i_C v_{CE} = P_{\max}$. The quiescent base current can be estimated by considering the input circuit of Fig. 2–205. Assume that the base-to-emitter junction is simply the transistor input resistance h_{ie}. If $v_s = 0$ (quiescent state), $i_B = V_{BB}/(R_s + h_{ie})$. Often R_s will be very much greater than h_{ie}, so that $i_B \approx V_{BB}/R_s$. In this case, then, $i_B \approx 0.20 \text{ V}/2 \text{ k}\Omega = 100 \text{ μA}$. Thus the choice of bias voltage will depend on the nature of the signal source. Other methods of biasing are used, but none of them is independent of the input signal circuit to the extent that the JFET is.

Transistor Amplifier Characteristics. The common-emitter amplifier circuit model is shown in Fig. 2–207. Equations for the various amplifier characteristics can be derived from the equivalent circuit and the mathematical relations used to obtain that circuit. The derivations are more involved than those for the JFET amplifier because of the greater complexity of the transistor circuit model. Some of the characteristics of the common-emitter amplifier are given below. The actual derivations can be found in some of the books listed in the bibliography.

The voltage gain is

$$A_v = \frac{-h_{fe}R_C}{(h_{ie}h_{oe} - h_{fe}h_{re})R_C + h_{ie}}. \tag{2–159}$$

For the transistor amplifier circuit shown in Fig. 2–205, the gain can be calculated if the parameters for the transistor are known. Assume that

$$h_{fe} = 55,$$

$$h_{ie} = 2720 \ \Omega,$$

$$h_{oe} = 14 \times 10^{-6} \ \Omega^{-1},$$

$$h_{re} = 3.23 \times 10^{-4}.$$

Then

$$A_v = \frac{-55 \times 1500}{(2720 \times 14 \times 10^{-6} - 55 \times 3.33 \times 10^{-4})1500 + 2720}$$

$$= \frac{-8.3 \times 10^4}{(3.8 \times 10^{-2} - 1.8 \times 10^{-2})1500 + 2720}$$

$$= \frac{-8.3 \times 10^4}{30 + 2720} = -30.$$

The term $h_{ie}h_{oe} - h_{fe}h_{re}$ is found so often in transistor equations that it is given a special symbol, Δh_e. For this transistor, $\Delta h_e = 0.02$. It can be seen from the above equation that the term $\Delta h_e R_C$ in the denominator does not influence the result significantly. Thus the voltage gain could be approximated by the equation

$$A_v \approx \frac{-h_{fe}R_C}{h_{ie}} \tag{2-160}$$

unless Δh_e or R_C is unusually large. The minus sign indicates that the output voltage is 180° out of phase with the input voltage. Recall that phase inversion is also a characteristic of the analogous JFET circuit (the common-source amplifier). Assume that R_C of Fig. 2–205 is a much better conductor than $1/h_{oe}$, that is, $R_C \ll 1/h_{oe}$. Assume also that the generator voltage $h_{re}v_{ce}$ is small compared with the voltage i_bh_{ie}. Now the model could be drawn as in Fig. 2–208. From this circuit it can be seen that $v_{in} = i_bh_{ie}$ and that $v_o = -h_{fe}i_bR_C$, so that the gain $A_v = v_o/v_{in} = -h_{fe}i_bR_C/i_bh_{ie} = -h_{fe}R_C/h_{ie}$, which is Eq. (2–160). Thus the assumptions leading to Eq. (2–160), which were not obvious previously, are actually $R_C \ll 1/h_{oe}$ and $h_{re}v_{ce} \ll i_bh_{ie}$.

The current gain A_i is given by the formula

$$A_i = \frac{-h_{fe}}{h_{oe}R_C + 1}. \tag{2-161}$$

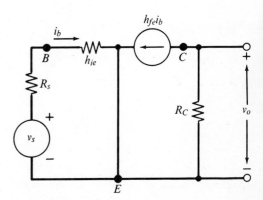

Fig. 2–208 Transistor amplifier simplified equivalent circuit.

The current gain for the amplifier of Fig. 2–205 is then

$$A_i = \frac{-55}{14 \times 10^{-6} \times 1500 + 1} = \frac{-55}{2.1 \times 10^{-2} + 1} = -55.$$

The term $h_{oe}R_C$ is usually so small compared with 1 that it can be neglected. Thus

$$A_i \approx -h_{fe}. \tag{2–162}$$

Again the minus sign is an indication of the phase reversal between the input and output signals. The assumption made in Eq. (2–162), that is, $h_{oe}R_C \ll 1$, is the same as the previous assumption, $R_C \ll 1/h_{oe}$. In calculating this current gain, the output current is defined as the current through R_C, not the current which can be drawn from the output terminals. If an additional load R_L is connected to the output terminals of the amplifier, the output current will be split between R_C and R_L; thus the effective value of the parallel combination of R_C and R_L should be used in the above equations.

The power gain A_p is simply the product of the current and voltage gains. When the assumptions made in Eqs. (2–160) and (2–162) are valid,

$$A_p \approx \frac{(h_{fe})^2 R_C}{h_{ie}}. \tag{2–163}$$

For the amplifier under analysis,

$$A_p \approx (-30)(-55) \quad \text{or} \quad \frac{(-55)^2(1500)}{2720} = 1660.$$

The power gain does not include power losses in the signal source or in transferring the output signal to a load other than R_C. It is simply the ratio of the signal power dissipated in R_C to the signal power dissipated in h_{ie}. Equation (2–163) verifies this definition.

The input resistance R_{in} is found from the formula

$$R_{in} = \frac{h_{ie} + (h_{oe}h_{ie} - h_{fe}h_{re})R_C}{1 + h_{oe}R_C} = \frac{h_{ie} + \Delta h_e R_C}{1 + h_{oe}R_C}. \tag{2–164}$$

Substitution of the parameters shows R_{in}, in this case, to be

$$R_{in} = \frac{2720 + 0.02 \times 1500}{1 + (14 \times 10^{-6}) \times 1500} = \frac{2720 + 30}{1 + 2.1 \times 10^{-2}} = 2700 \ \Omega.$$

Since $\Delta h_e R_C \ll h_{ie}$ and $h_{oe}R_C \ll 1$,

$$R_{in} \approx h_{ie}. \tag{2–165}$$

It is obvious that the assumption $h_{re}v_{ce} \ll i_b h_{ie}$ made in Fig. 2–208 leads to this same conclusion. Note also that $R_{in} = R_C(A_i/A_v)$.

The output resistance R_o can be calculated from

$$R_o = \frac{h_{ie} + R_s}{h_{oe}h_{ie} - h_{re}h_{fe} + h_{oe}R_s} = \frac{h_{ie} + R_s}{\Delta h_e + h_{oe}R_s}. \qquad (2\text{–}166)$$

For the circuit of Fig. 2–205,

$$R_o = \frac{2720 + 2000}{0.02 + 14 \times 10^{-6} \times 2000} = \frac{4720}{0.048} = 10^5 \ \Omega.$$

In this case the terms in the numerator and the denominator are both of the same order of magnitude. It can be seen that, if R_s is small compared with h_{ie},

$$R_o \approx \frac{h_{ie}}{\Delta h_e} \qquad (2\text{–}167)$$

or that, if R_s is much larger than h_{ie},

$$R_o \approx \frac{1}{h_{oe}}. \qquad (2\text{–}168)$$

As Eq. (2–168) indicates, the output resistance, R_o, is in this case the resistance that the transistor as a power source presents to the load R_C. The output resistance of the entire amplifier circuit as seen by a device attached to the output terminals would actually be R_o in parallel with R_C.

The common-emitter BJT amplifier is thus shown to possess the properties of voltage gain, current gain, phase reversal of the input signal, low input resistance, and high output resistance. Similar analyses can be made for the common-collector and common-base amplifiers.

Emitter-Follower Amplifier

A very practical and important circuit that has the basic common-collector configuration is shown in Fig. 2–209. The input and output terminals are connected to the transistor collector through the collector supply voltage V_{CC}. However, the three terminals are in common for ac signals because, from the viewpoint of the varying signal levels, the collector voltage is a fixed constant value just as surely as if it were grounded. It is considered to be in common with the common input and output terminals because the potential difference is a constant.

This circuit has many desirable characteristics such as high input impedance, low output impedance, and excellent linearity over a wide dynamic range, and is therefore frequently used as a buffer stage or an

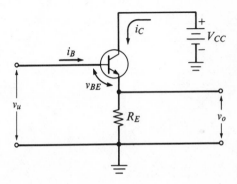

Fig. 2–209 Emitter-follower circuit.

output circuit. The wide applicability of this simple linear amplifier circuit makes it worthwhile to gain an understanding of its characteristics.

When a positive voltage v_u is applied to the base, the transistor conducts, causing a voltage drop v_o across R_E which increases the transistor emitter voltage. If the input voltage v_u then decreases, i_C decreases, and so does the emitter potential. Thus the emitter voltage, v_o, "follows" the input voltage. Therefore this type of common-collector configuration is usually called an "emitter follower." The characteristics of this circuit are analyzed in greater detail as follows.

When the input signal voltage, v_u, is greater than the turn-on voltage for the base-emitter junction, v_{BE}, the output voltage, v_o, is

$$v_o = v_u - v_{BE} \tag{2-169}$$

and the ac voltage gain A_v of the circuit is

$$A_v = \frac{dv_o}{dv_u} = \frac{d(v_u - v_{BE})}{dv_u} = 1 - \frac{dv_{BE}}{dv_u}. \tag{2-170}$$

Since dv_{BE}/dv_u is always positive, it can be seen from Eq. (2-170) that the voltage gain is always less than 1, but it approaches unity in the linear amplification region, where dv_{BE}/dv_u is small. However, the input v_u cannot exceed the supply voltage V_{CC} by much before the collector-base junction becomes forward biased and the voltage between collector and base $v_{CB(ON)}$ is only a few tenths of a volt. The input would thus be clamped at $V_{CC} + v_{CB(ON)}$.

If the transistor of Fig. 2-209 is not saturated,

$$i_C = h_{fe}i_B. \tag{2-171}$$

Since $i_E = i_C + i_B$, it follows that

$$i_E = h_{fe}i_B + i_B = i_B(h_{fe} + 1). \tag{2-172}$$

Therefore, from Eq. (2-172) the current gain A_i is

$$A_i = \frac{di_E}{di_B} = h_{fe} + 1. \tag{2-173}$$

Input Resistance. One of the important characteristics of the emitter follower is its relatively high input resistance, which means that it will not "load" the input signal source significantly. In this section the effective resistance to an ac signal is considered. It is defined as

$$R_{\text{in}} = \frac{dv_u}{di_B}. \tag{2-174}$$

If the input voltage signal, v_u, is applied directly to the base terminal, it can be seen from Eq. (2–169) that $v_u = v_o + v_{BE}$, and since $v_o = i_E R_E$ it follows that

$$v_u = i_E R_E + v_{BE}. \qquad (2\text{–}175)$$

By substituting the equivalent of i_B from Eq. (2–172), $i_B = i_E/(h_{fe} + 1)$, and of v_u from Eq. (2–175) into Eq. (2–174), it follows that

$$R_{\text{in}} = \frac{d(i_E R_E + v_{BE})}{d(i_E/(h_{fe} + 1))} = \left(R_E + \frac{dv_{BE}}{di_E} \right)(h_{fe} + 1). \qquad (2\text{–}176)$$

If the emitter current i_E is larger than a few milliamperes, dv_{BE}/di_E will typically be negligible compared to R_E, and R_{in} will be

$$R_{\text{in}} \approx R_E(h_{fe} + 1) \qquad (2\text{–}177)$$

Equation (2–177) shows that the dc input resistance, R_{in}, is greater than the emitter load, R_E, by the factor $h_{fe} + 1$.

Output Resistance. The emitter-follower stage is often connected between an inverter circuit and the external load, because of its ability to act as an impedance transformer. This is illustrated by the circuit of Fig. 2–210a. A low output resistance for a voltage source means that relatively heavy loads can be connected to the output of the source without changing the output voltage. The low output resistance of the emitter follower is one of its most important characteristics. The output resistance of the emitter follower is seen to be only a fraction $[1/(h_{fe} + 1)]$ of the input source resistance $(R_u = dv_u/di_B)$ by considering the following equations obtained by substitution from Eqs. (2–169) and (2–172):

$$R_o = \frac{dv_o}{di_E} = \frac{d(v_u - v_{BE})}{di_B(h_{fe} + 1)} = \frac{1}{h_{fe} + 1} \times \frac{d(v_u - v_{BE})}{di_B}$$

and, if v_{BE} is negligible compared to v_u,

$$R_o = \frac{R_u}{h_{fe} + 1}. \qquad (2\text{–}178)$$

Since the maximum output voltage $v_{o(\text{max})}$ can be given as $v_{o(\text{max})} = V_{CC} - i_B R_C - v_{BE}$, it can also be readily shown, by substitution for $i_B = i_E/(h_{fe} + 1)$ and for $I_E = v_{o(\text{max})}/R_L$, that

$$v_{o(\text{max})} = \frac{V_{CC} - v_{BE}}{1 + R_C/[R_L(h_{fe} + 1)]}. \qquad (2\text{–}179)$$

If v_{BE} is assumed to be 0.6 V and the supply voltage $V_{CC} = 5$ V, $R_C = 1$ kΩ, and $h_{fe} = 50$ for the selected transistor Q_2 in Fig. 2–210a,

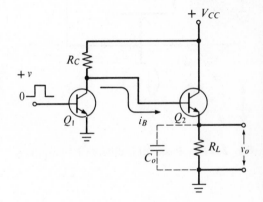

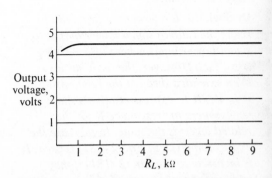

Fig. 2–210 Emitter follower as resistance transformer between inverter and output load: (a) circuit; (b) plot showing small change of output voltage vs. load resistance.

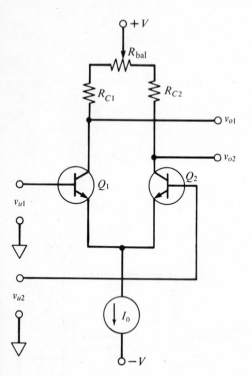

Fig. 2–211   Basic BJT differential amplifier.

then it can be seen that the second term in the denominator in Eq. (2–179) is negligible for a wide range of load resistance R_L. A plot of output voltage vs. load resistance for this example is shown in Fig. 2–210b.

A Basic Differential Amplifier

A difference or differential amplifier is one whose output is a function of the potential difference between its two inputs, and is relatively nonresponsive to voltages of equal magnitude (common-mode voltages v_{cm}) present at both inputs. It is a very useful amplifier because it is capable of rejecting spurious common-mode signals that are coupled to both inputs either from pickup of radiated noise or conduction from the input source. Also, a well-designed differential amplifier is relatively insensitive to environmental changes of temperature or other conditions.

It was shown in Section 1–3.4 of Module 1, *Electronic Analog Measurements and Transducers,* that the very important voltage comparator is a differential amplifier designed specifically for null comparison measurements. It has well-balanced differential inputs and controlled output limits. Although the ideal differential amplifier would provide an output proportional only to the difference in voltage at the two inputs regardless of the magnitude of the input voltages, it is necessary to settle for something less than ideal. The commonly used measure of the perfection of input balance was also shown in Section 1–3.4 to be the **common mode rejection ratio** (**CMRR**):

$$\text{CMRR} = \frac{\Delta v_o / \Delta v_d}{\Delta v_o / \Delta v_{cm}}. \tag{2–180}$$

The numerator of Eq. (2–180) is the change in output voltage Δv_o for a difference signal Δv_d at the inputs, and the denominator is the change in output voltage for a change Δv_{cm} in the common mode signal voltage that is applied to both inputs.

Often the common mode voltage rejection (CMRR) is expressed as a certain number of decibels (dB),[9] where

$$\text{CMRR} = 20 \log (\text{CMRR}) \text{ dB}. \tag{2–181}$$

For example, if the CMRR is 100,000, then CMRR = 20 log 10^5 = 100 dB.

A modern comparator amplifier is generally a combination of several types of amplifiers, and the combination is usually represented by the triangular symbol with two inputs. The following description is of the basic differential amplifier shown in Fig. 2–211, which is made with two matched BJT amplifier stages. The emitters of the two transistors are connected together and to a constant current source I_0.

To understand the operation of the difference amplifier, consider first how the output at v_{o2}, which is influenced by both input signals, originates. The input signal at v_{u1} gives rise to an output at v_{o2} which is amplified, but not inverted, since transistor Q_1 acts as an emitter-follower stage for this signal and Q_2 acts as a common-base stage. Transistor Q_2 provides voltage gain for the signal at v_{u1}, while Q_1 provides high input resistance and low output resistance to drive the common-base stage. The input signal at v_{u2} gives rise to an output at v_{o2} which is both amplified and inverted, because transistor Q_2 acts as a normal common-emitter stage for v_{u2}. Thus the output at v_{o2} is a function of the difference between the two input voltages and can be expressed as

$$v_{o2} = K_1 v_{u1} - K_2 v_{u2},$$

where K_1 and K_2 are the voltage gains experienced by v_{u1} and v_{u2} respectively. Similar reasoning applied to the output at v_{o1} leads to the conclusion that the signal at v_{u2} gives rise to an output which is amplified but not inverted, while the signal at v_{u1} gives rise to an output at v_{o1} which is amplified and inverted. Thus the output at v_{o1} can be expressed as

$$v_{o1} = K_2' v_{u2} - K_1' v_{u1},$$

where again K_2' and K_1' are the gains experienced by v_{u2} and v_{u1} respectively at v_{o1}.

There are several modes in which the difference amplifier can be used. In the **balanced** or **differential** mode, the output voltage v_o is taken from collector to collector so that

$$v_o = v_{o2} - v_{o1} = A_1 v_{u1} - A_2 v_{u2}, \qquad (2\text{--}182)$$

where $A_1 = K_1 + K_1'$ and $A_2 = K_2 + K_2'$. If the circuit is perfectly balanced, the voltage gains experienced by the two signals are equal ($A_1 = A_2 = A$) so that

$$v_o = A(v_{u1} - v_{u2}). \qquad (2\text{--}183)$$

Perfect balance requires that the amplification factors of the two transistors and the collector load resistors, R_{C1} and R_{C2}, be perfectly matched and track with any changes in temperature or other operating conditions. Also, any differences between the base-emitter voltages of the two conducting transistors can lead to an offset voltage at the output, which changes with temperature. For these reasons the two transistors are often fabricated on the same chip, encapsulated in a single case, or placed together on a single heat sink. A balance potentiometer (R_{bal} in Fig. 2–211) is often used in the collector circuit to compensate for any voltage offsets and drifts. A well-balanced difference amplifier is highly independent of supply voltage fluctuations and ripple, since these affect both outputs similarly and do not influence the difference output very strongly.

To a large extent the CMRR depends on the quality of the constant current source I_0. A perfect current source would maintain the gains and dc bias points of Q_1 and Q_2 constant and independent of the input signals. If the current source is nonideal, however, the gains and bias points will shift with a common mode input, which leads to a poor CMRR. The simplest current source is a common mode resistor connected between the negative supply voltage $-V$ and the two emitters. Large values of the common mode resistor make the current relatively independent of the input signals. Improved values of CMRR can be obtained by replacing the common mode resistor with a transistor constant current source or by using feedback to control the current. Almost all high quality difference amplifiers use these latter methods for high CMRR.

The difference amplifier can be used in a **single-ended output** mode. In this mode the input is differential as in Fig. 2–211, but only one output, referenced to the $+V$ supply, is employed. With a single-ended output only one of the collector load resistors is necessary.

A third mode in which the differential amplifier can be operated is the **phase inverter** mode. In this mode, only one of the inputs is used and each output is taken separately. If input 2 in Fig. 2–211 is shorted to the common, $v_{o2} = K_1 v_{u1}$ and $v_{o1} = -K_1' v_{u1}$. If $K_1 \approx K_1'$, the outputs at the two collectors are of equal magnitude, but always of opposite phase.

A differential amplifier made with matched JFET's is illustrated in the next section as the input stage for a general purpose operational amplifier.

Summary

The basic amplifiers presented in this section all illustrate that amplification is essentially a control operation. That is, various electronic devices such as BJT's and JFET's are used as control elements in various configurations to control a power supply, as illustrated in the block diagram, Fig. 2–212.

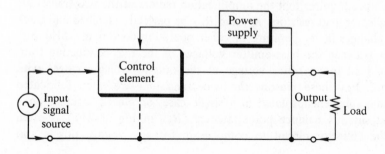

Fig. 2–212   Method of amplification.

The control element (BJT, FET, etc.) is actuated by the input signal, and it controls the magnitudes of the voltage and current that are taken from the power supply and made available at the amplifier output terminals. The controlled output voltage and current from the power supply will follow some known function related to the input signal. The amplification is a result of the small voltage and/or current required by the control element in order to control relatively large voltages and currents from the power supply.

The many amplifier configurations used in practice are aimed at meeting certain design requirements such as linearity (direct proportionality between input and output), high-frequency response (ability to provide the necessary control of power supply voltage and current even when the input signal varies at extremely high rates such as 100 MHz or more), or rejection of noise (as with the differential amplifier). The amplifier concepts, circuits, models, and terminology presented in this section should enable the reader to understand the basic operation of amplifiers and appreciate at least some of the limitations of amplifier circuits. Several basic amplifiers are usually combined in the production of amplifiers with special characteristics. An example of this is illustrated in the next section, where a differential JFET amplifier, a cascaded high-gain BJT amplifier, and an emitter-follower amplifier are combined to make a general purpose operational amplifier.

Experiment 2–31 *Difference Amplifiers*

A difference amplifier is constructed and studied as a differential balanced amplifier, as a phase inverter, and as a single-ended differential amplifier. The amplifier is constructed with both npn transistors and matched JFET's. Characteristics of the amplifiers are determined, and the common mode rejection ratios are determined.

2–5.3 INSTRUMENTATION DC AMPLIFIERS

Economical general purpose dc amplifiers, referred to as operational amplifiers (OA's), are now available which have outstanding characteristics for instrumentation applications. Operational amplifiers are utilized in hundreds of different ways to provide elegant solutions to measurement and control problems. Many applications have been presented in this module and others in the series.

The many functions that can be performed by using OA's with suitable feedback elements and in various combinations make it feasible for anyone with just a little experience to construct some very sophisticated instruments. It is, of course, not necessary to know what is inside the OA gain block (represented simply by a triangular symbol) in order to use an OA profitably. However, most users of OA's feel more comfortable in using them when they have some idea of how they are designed, and certainly the user can appreciate specifications and limitations better after investigation of the internal operation of a typical OA. Therefore the first part of this section is devoted to a look inside the triangular symbol. Then the combination of several OA's to form a very high quality dc instrumentation amplifier is presented, and the section concludes with a summary of specification considerations for OA's.

Integrated Circuit Operational Amplifier

The OA that we have chosen to investigate is constructed on a micro semiconductor chip. This integration of several basic amplifier circuits into a monolithic OA has resulted in units no larger than and costing no more than a single transistor. As previously indicated, the general purpose OA is a gain block with dc integrity, high input impedance, low output impedance, quite wideband frequency response with negative feedback, and other characteristics that will be described later. The ways by which these characteristics are obtained can be seen by referring to Fig. 2–213, which is a somewhat simplified circuit diagram of an integrated circuit OA.

The input circuit is a differential amplifier using matched p-channel JFET's as source followers. An external 10 kΩ potentiometer is used to balance the differential amplifier to compensate for any offset. The differential input provides for the inverting and noninverting inputs and excellent CMRR. The use of JFET's also provides input impedances of 10^9 Ω or higher and very low offset currents. This very high input impedance makes the OA generally applicable as a high quality integrator, for analog memory, or for other applications where high input impedance is necessary.

Two BJT transistors are coupled together to provide a second stage with high current gain. When two transistors are coupled together in this fashion, the pair is known as a **Darlington circuit.** The current gain of the pair is approximately equal to the product of the amplification factors $(\beta_1 \times \beta_2)$ for the two transistors.

The output stage is a cascaded complementary emitter-follower circuit. It has the necessary very low output impedance (< 1 Ω with negative feedback) for the output stage, as described in the previous section.

Fig. 2–213 Integrated circuit operational amplifier.

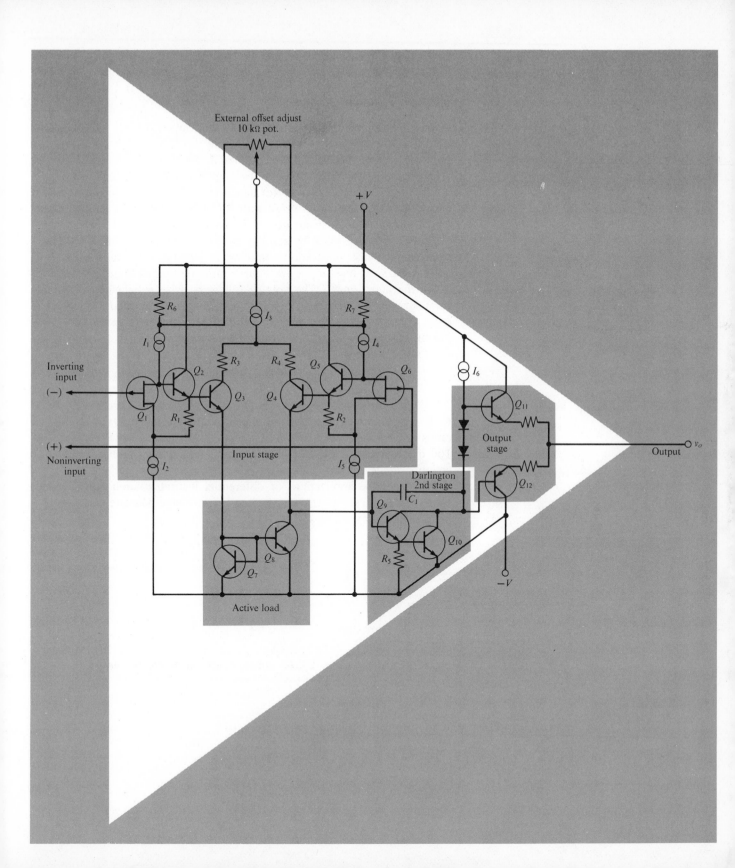

DC Instrumentation Amplifiers

When a dc amplifier has especially high input impedance ($> 10^9 \, \Omega$), very low offset currents and voltages, negligible drift, very high CMRR, high gain over a relatively wide frequency band, and low output impedance, it is frequently called an **instrumentation amplifier.** Obviously, such an amplifier is capable of excellent performance in critical measurement systems. Although many differential OA's with FET inputs fulfill the general needs for many accurate instruments, it is possible to improve certain characteristics such as CMRR by combining three or more OA's so as to form a differential dc instrumentation amplifier, as illustrated in Fig. 2–214. Two OA's with very high input impedance are used as voltage followers to buffer the input signal. The voltage followers drive a balanced difference amplifier which can provide gain and reject common mode voltages. If $R_1 = R_2 = R_3 = R_4$, the gain is unity and the output voltage $v_o = v_2 - v_1$. For the case where $R_3/R_1 = R_4/R_2$ the output voltage is

$$v_o = (v_1 - v_2) \, \frac{R_3}{R_1}. \qquad (2\text{--}184)$$

With this type of amplifier it is necessary to keep the ratio R_3/R_1 equal to R_4/R_2 to obtain the best common mode rejection. By trimming the ratios to be better than 0.1% it is possible to obtain a CMRR greater than 100 dB. With the typical ± 15 V power supplies the amplitude limit for common mode voltage is about ± 10 V.

Some of the applications of differential instrumentation amplifiers include bridge, strain gauge, or any low level voltage measurement where a high CMRR is essential.

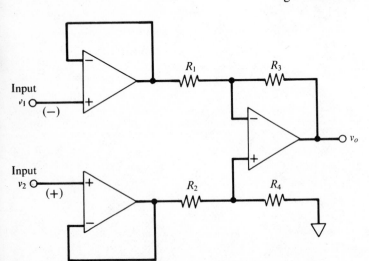

Fig. 2–214 Instrumentation differential amplifier.

Summary of OA Specifications

Every commercial OA has a set of specifications, and although many of these have been described in previous sections, they are summarized here as a useful reference. In addition, some of the other specifications and considerations that would not have meant much earlier can now be briefly described in light of the foregoing discussion of amplifiers.

Open-Loop Gain. The open-loop gain is the OA gain without any feedback. It is typically given in dB. For example, since gain (dB) = 20 log (v_o/v_u), an open-loop gain of 100 dB means that

$$\log \frac{v_o}{v_u} = \frac{100}{20} = 5,$$

so that

$$A = \frac{v_o}{v_u} = 100{,}000.$$

Frequency Response. It is important to realize that the open-loop gain, A, is a function of frequency and is relatively constant from dc to 10 Hz for most OA's and then falls off. As was previously shown in descriptions of various applications, it is important that A be very large for the basic equations to be reliable. Therefore, it is important to check the manufacturer's frequency response characteristics when using OA's for applications at frequencies above 10 Hz. Amplifier frequency response is discussed in detail in Module 3.

Rated Output. The power supply voltages used with many modern OA's are ±15 V with a range of tolerance from about ±12 to ±18 V. Therefore, the maximum available output voltage is often greater than ±10 V. Some IC OA's operate at voltages about ±5 V so that the voltage output swing is less than 5 V. The current output rating varies considerably with the specific OA, but it is typically 2–10 mA for general purpose OA's. Consequently, if the output voltage goes as high as 10 V and the output current rating is 10 mA, then the lowest value for the load resistor for rated output is 1 kΩ. If the current rating is 2 mA, the lowest possible value for the output load is 5 kΩ if the output voltage is to swing to its rated maximum.

Input Offset Voltage. As explained in the previous section, it is improbable that the input differential amplifier will be perfectly balanced, although perfect balance is a design goal of the manufacturer of dc amplifiers. The small, relatively constant, but temperature-dependent voltage that exists between the input terminals is called **input offset voltage.** It can be considered as a small voltage in series with one of the input leads. It will be amplified by the same factor as the input signal, and unless it is balanced out, the output v_o will not be zero with zero input. Usually an external pot is required to balance out the offset voltage. However, the offset voltage changes with temperature, supply voltage, and time. For example, manufacturer's specifications might be ± 50 µV/°C, ± 500 µV/V, and ± 100 µV/day respectively.

Input Bias Current.   Even when the input voltage is zero, there is an input current in each input terminal. For JFET input stages the bias currents are the gate currents of the JFET's and any leakage currents within the amplifier. The manufacturer attempts to make this bias current as small as possible by selecting appropriate input control elements and by careful design and component layout. Typical bias current for a good instrumentation dc amplifier might be in the range 0.1–10 pA at 25°C. The bias current can be balanced by supplying a balance current to the summing point of an OA. However, it should be remembered that the bias current is quite sensitive to temperature changes. For example, for FET OA's the bias current doubles for about each 10°C change in temperature. The difference between the two input bias currents is called the **input offset current.**

Input Resistance.   Modern FET amplifiers have extremely high input impedances. The value depends greatly on the design and will be generally in the range from 10^9 to 10^{14} Ω. An amplifier with an input resistance of 10^{14} Ω can keep the input current in the subpicoampere range for full common mode voltage swings of ± 10 V. The input capacitance in parallel with the input resistance is usually about 5 pF.

Input Noise.   The input current noise of the amplifier is the limiting factor on signal resolution. It varies as a function of source impedance and frequency. Therefore, the specification of input noise is generally given as a function of frequency, and graphs of typical changes of input noise with source resistance are available from the manufacturer.

CMRR.   The common mode rejection ratio of amplifiers classed as dc instrumentation amplifiers should be very good, typically 100 dB at ± 5 V and about 80 dB at ± 10 V.

Leakage Currents.   When using FET amplifiers it is important to take special precautions in the layout of external wiring and feedback elements. That is, unless the circuit boards are free of fingerprints and other contamination, the leakage paths between the 15 V power supply conductors to the amplifier inputs might decrease the insulation resistance well below 10^{12} Ω. With a leakage path of 10^{12} Ω between either input and the 15 V supply there would be an input current of 15 pA. This can be much larger than the amplifier input bias current. Leakage paths are likely to undergo great variations with environmental changes and could cause erratic behavior.

Because of the difficulty in guarding input terminals from leakage paths when the OA's are soldered on printed-circuit boards, it is sometimes preferable to use **hardwiring.**

Shielding. The high resistance levels at which FET instrumentation amplifiers may be operated makes it important to shield the input terminals so as to reduce noise pickup. Shielded input cables are recommended and the feedback elements should be enclosed within a shield. When a shielded cable is flexed, it can generate input noise because of the cable conductor rubbing against the insulation. However, special low noise cables are available that have a graphite coating or conductive tape between the cable conductor and its insulation.

Any changes in capacitance at the input because of the input wiring shifting position can cause extraneous voltages at the input. The obvious remedy for this is to provide rigid wiring with leads as short as possible and spaced as far from other conductors as possible.

Slew Rate. Within the OA there are capacitances inherent in the semiconductor devices and leads used in its construction. The rate of change of voltage at each point in a circuit is limited by the available current to charge the capacitance at that point. The consideration is the same as for the response of the RC circuit to a step function, as discussed in Section 2–2. The maximum rate of change of voltage $(\Delta V/\Delta t)_{max} = I_{max}/C$. Therefore, for example, if a capacitance of 10 pF is to change voltage by 10 V in 1 nsec, then the available current at that point must be $I = (10/10^{-9})10^{-11} = 100$ mA.

Settling Time. In today's world of high-speed data acquisition and processing, many measurements are made in a microsecond or less. Therefore the time required for an electronic feedback control circuit to perform its function with a specified accuracy is important and is referred to as **settling time.** For OA feedback circuits it is the time that elapses between the application of an ideal instantaneous step input and the instant at which the closed-loop amplifier output has entered and remained within a specified error band, usually symmetrical about the final value. The settling time includes any propagation delay in the circuit plus the time required to slew to the final value, recover from any overshooting of the final value associated with slewing, and finally settling within a certain specified accuracy for the final value.

Note that the settling time is associated with a closed-loop OA system, i.e., an OA circuit with feedback, and it cannot be predicted from open-loop specifications for the OA, such as slew rate or small-signal bandwidth.

Experiment 2–32 *Operational Amplifier Characteristics*

Two important characteristics of operational amplifiers which are easily measured are the common mode rejection ratio (CMRR) and the closed-loop output resistance (R_o). In this experiment these values are measured for two different types of operational amplifiers in order to compare and contrast these characteristics. Other OA characteristics can be measured if desired.

PROBLEMS

1. Circuit models of two semi-ideal amplifiers are shown in Fig. P–1. Fig. P–1a corresponds to a FET amplifier, while Fig. P–1b corresponds to a BJT amplifier. Note that the FET amplifier has a finite input resistance, as opposed to the ideal voltage amplifier, while the BJT amplifier has a non-zero input resistance. (a) Calculate the voltage gain of the FET amplifier of Fig. P–1a and the current gain of the BJT amplifier of Fig. P–1b. (b) Calculate the power gains of the two amplifiers. (c) Calculate the voltage gain of the BJT amplifier of Fig. P–1b.

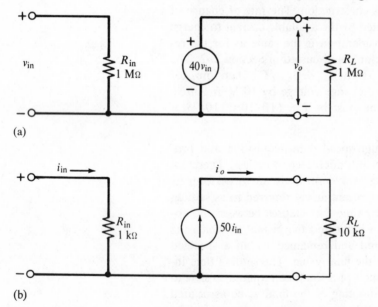

(a)

(b)

Fig. P–1

2. In the JFET amplifier of Fig. 2–201a, $R_s = 1$ kΩ, $R_d = 10$ kΩ, $C_s = 20$ μF, and $V_{DD} = 30$ V. For the particular JFET used, the transconductance $g_m = 2 \times 10^{-3}$ Ω^{-1} and the drain-source dynamic resistance $r_{ds} = 100$ kΩ. (a) Assuming that the current v_{ds}/r_{ds} is negligible, calculate the amplifier

voltage gain. (b) For $v_{gs} = 1$ V, determine if v_{ds}/r_{ds} is significant compared to the current $g_m v_{gs}$.

3. Draw the small-signal model of the JFET amplifier of problem 2 including actual values of the small-signal components. Compare this model to the ideal voltage amplifier and describe how the JFET amplifier falls short of being ideal.

4. Determine the hybrid parameters h_{fe}, h_{ie}, h_{re}, and h_{oe} of the transistor whose characteristic curves are given in Fig. 2–203a through d.

5. The transistor whose hybrid parameters were found in problem 4 above is used in the amplifier circuit of Fig. 2–205. (a) Calculate the current gain, voltage gain, and power gain of the amplifier. (b) Calculate the amplifier input and output resistances.

6. Draw the small-signal model of the common-emitter amplifier of problem 5. Include specific values for all the small-signal circuit components. Compare the resulting model with that of the ideal current amplifier and describe how the BJT amplifier falls short of ideality.

7. The transistor used in the emitter-follower amplifier of Fig. 2–209 has an h_{fe} value of 60. A source is used with an equivalent resistance $R_u = 100$ Ω. The supply voltage $V_{CC} = 5$ V and $R_E = 5$ kΩ. (a) If $dv_{BE}/dv_u = 0.05$, calculate the emitter-follower voltage, current, and power gain. (b) Calculate the input and output resistance of the emitter-follower amplifier.

8. The output voltage of a difference amplifier changes by 5 V for an input difference signal of 250 μV, while the output voltage changes by 1 V for a common mode input signal of 500 mV. (a) What is the common mode rejection ratio (CMRR) of the difference amplifier? (b) Express the CMRR in decibels.

9. Using the differential input of an oscilloscope, it is desired to measure a 50 mV signal to within 1% accuracy. If the CMRR of the oscilloscope amplifier is 10^4, what is the maximum common mode signal which can be tolerated?

10. One circuit model of a difference amplifier is shown in Fig. P–10. The parameters are $\mu = 1000$, $R_{in} = 50$ kΩ, $R_o = 200$ Ω, and $R_L = 2$ kΩ. If the signal $v_{u2} = 0$, calculate the voltage gain, current gain, and power gain from the circuit model.

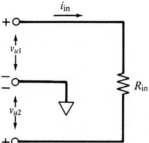
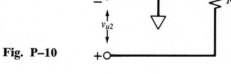
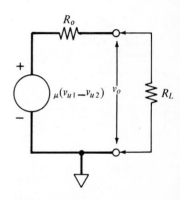

Fig. P–10

11. The same amplifier as in problem 10 is operated with a sine wave input at v_{u1} of 5 mV rms amplitude and $v_{u2} = 0$. Consider the load resistor R_L to be variable. (a) What is the maximum output power that can be drawn from the amplifier by adjustment of R_L? (b) What is the power gain at the value of R_L which gives maximum power?

Automatic Feedback Control

The basic concepts of feedback control (servo) systems were described in Section 1–4 of Module 1, *Electronic Analog Measurements and Transducers.* The potentiometric recorder, potentiometric amplifier, and operational amplifier voltage and current followers have all been presented as part of a family of automated null comparison systems that are usually referred to as **feedback control systems.** The same basic automated null comparison concepts are inherent in all servo systems whether they are used in controlling a ship, spacecraft, manufacturing process, or laboratory experiment.

The applications of the feedback control concept are rather overwhelming, yet the underlying idea is simple and elegant. The servo system recognizes a specific command (such as a command to move an object to a certain position), responds to the command (by comparing the actual position to the desired position of an object), and then takes the necessary steps to correct any difference.

The amplifiers, power supplies, motors, gears, and other hardware can be important parts of the servo system, but it is the total system that must operate so as to recognize and fulfill a specific command by feedback control. Obviously, system design is very important, and as one looks at real servos certain practical system problems become apparent.

No system can respond instantaneously to a command, and unfortunately the more quickly the system reacts to obey the command, the greater are the chances of instability. Consider, for example, a position servo. No mass can be moved instantaneously to fulfill a command for change of position because this would require infinitely rapid acceleration and deceleration and therefore infinite force. Nor can any electronic circuit respond instantaneously, because of propagation delays in moving charge carriers and because of the capacitance and inductance that are inherent in the devices and leads used in circuit construction (as discussed under slew rate in the previous section).

All systems that detect a difference between the actual and the desired states of a controllable quantity (e.g., voltage, temperature, chemical composition, position) and then feed the difference information back to a controlling device, which then reacts and causes the difference to become essentially zero, are classified as **servo systems.** A jet airliner or spacecraft is controlled by many types of mechanical, electromechanical, and electronic servos. A human is controlled by many types of servos, and one type of human servo will be used in the next paragraph as an illustration for focusing on the functional blocks of all servo systems. Then some of the servo response characteristics will be considered and the section will conclude with a discussion of programmable power supplies operated in current-regulated and voltage-regulated modes.

2–6.1 BASIC SERVO COMPONENTS

An understanding of the basic components in a servo system can be gained by considering how a human operates in response to an order to position an object. As shown by the block diagram (Fig. 2–215a), the command is picked up by the eye, ear, or touch, and a signal is transmitted to the human brain, which controls the body muscles that exert a force on the load to move it toward the desired location. The eye or other sensor acts as an information feedback device that describes the actual position. The mind compares the command and feedback information and controls the body so as to change the load position in accordance with the error signal, i.e., the difference between the desired and actual positions of the object. The man-made servo operates in a similar way, as shown in Fig. 2–215b. The command v is picked up by a device that puts it in the form of an acceptable command signal r, and the information on the actual position of the load is put in the form of a feedback signal b. These are usually voltage signals that can be compared to give a difference voltage e that is fed to an electronic signal controller. The controller uses the error signal e to regulate the electrical power to the motor (or other positioner), and the motor operates to move the load to the desired position. When the load is at the desired position, the error (actuating) signal e is essentially zero. Note that the terms "error signal" and "actuating signal" have the same meaning here and are used interchangeably.

The AIEE (American Institute of Electrical Engineers) has recommended standard symbols for feedback control systems; these are shown with a general block diagram (Fig. 2–216) that contains the basic elements and variables. The definitions are as follows:

The **command** v is the input, which is established by some means external to and independent of the feedback control system.

The **reference input elements** a produce a **reference input** r which is proportional to the command.

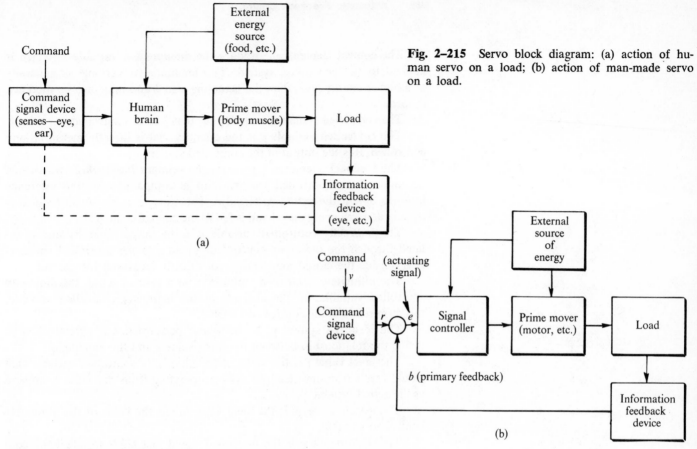

Fig. 2–215 Servo block diagram: (a) action of human servo on a load; (b) action of man-made servo on a load.

(a)

(b)

Fig. 2–216 General block diagram for the feedback control system.

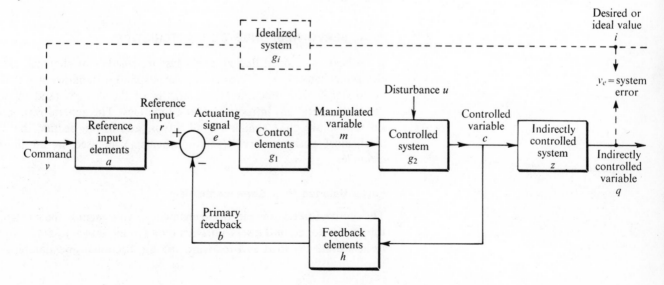

The **control elements** g_1 produce the **manipulated variable** m which is applied to the controlled system. The manipulated variable is generally at a higher energy level than the actuating signal and may also be modified in form.

The **controlled system** g_2 is the device that is to be controlled.

The **controlled variable** c is the quantity that is directly measured and controlled. It is the output of the controlled system.

The **feedback elements** h produce the **primary feedback** b, which is a function of the controlled variable and is compared with the reference input to obtain the **actuating signal** e. Therefore, $e = r - b$; this signal is usually at a low energy level and is fed to the control elements.

The **indirectly controlled variable** q is the output quantity and is related through the **indirectly controlled system** z to the controlled variable. It is outside the closed loop and is not directly measured for control.

The **ultimately controlled variable** is, in a general sense, the indirectly controlled variable. In the absence of the indirectly controlled variable, this term refers to the controlled variable.

The **idealized system** g_i is one whose performance is agreed upon to define the relationship between the **ideal value** i and the command.

The **ideal value** i is the value of the ultimately controlled variable that would result from an idealized system operating from the same command as the actual system.

The **system error** y_e is the ideal value minus the value of the ultimately controlled variable.

The **disturbance** u is the unwanted signal that tends to affect the controlled variable. The disturbance may be introduced into the system at many places.

2–6.2 SERVO RESPONSE CHARACTERISTICS

It has been mentioned that no servo system, whether mechanical, chemical, electromechanical, or all-electronic, can respond instantaneously to a command signal. The inherent characteristics of the devices used to perform the servo operations introduce certain time lags. The general response considerations are quite similar for all servos, and in this section the electromechanical position servo (servomechanism) is used as an illustrative example.

Force Balance in a Servomechanism

The motive force in the electromechanical servo system is the electromotive torque developed in the servo drive motor (prime mover). This torque acts to accelerate the load and to overcome the frictional and damping forces in the system. For most instrument servomotors the torque is roughly proportional to the voltage in the control winding. If the error-signal amplifier is linear, the output voltage of the amplifier is proportional to the error-signal input. Finally, if the system is completely linear, the error signal is proportional to the difference between the actual and the command positions of the load. Therefore the relationship between the load position c and the motor torque T_m is

$$T_m = K'(c_0 - c), \tag{2-185}$$

where c_0 is the command position of the load, or that load position which would give no error signal to the amplifier, and K' is a constant with units of torque per load displacement.

For Eq. (2–185) to be true at all times, it is necessary that the amplifier gain, which affects K', not be a function of frequency. For chopper amplifiers, then, the chopper frequency must be high compared with the highest frequency command that the servo is expected to follow. The signal delay in the amplifier (phase distortion) should also be negligible for signals of the fastest expected response speed.

In considering the linear translation of a load, such as the pen in a recording potentiometer, it is more convenient to use the unit "force" than "torque." The force is simply the torque divided by the lever arm. When the load is positioned by a drive pulley concentric to the motor shaft, the lever arm l is a constant and the force developed in the motor, F_m, as applied to the system, is

$$F_m = T_m/l = (K'/l)(c_0 - c) = K(c_0 - c). \tag{2-186}$$

The force that actually accelerates the load, F_a, is the motor force F_m minus the force lost to damping and frictional forces, F_d:

$$F_a = F_m - F_d. \tag{2-187}$$

The force F_a is related to the acceleration a of the load by

$$F_a = Ma, \tag{2-188}$$

where M is the mass of the load. The damping and frictional forces are generally proportional to the velocity u of the load. The constant of proportionality will be given the symbol D:

$$F_d = Du. \tag{2-189}$$

We assume that all frictional surfaces are lubricated so that the starting friction and unlubricated sliding friction are negligible. Substituting Eqs. (2–186), (2–188), and (2–189) into Eq. (2–187), we obtain

$$Ma = K(c_0 - c) - Du \tag{2-190}$$

or

$$Kc_0 = Ma + Du + Kc. \qquad (2\text{–}191)$$

Expressing the acceleration and the velocity as the second and first time derivatives of c, we can write

$$Kc_0 = M\,\frac{d^2c}{dt^2} + D\,\frac{dc}{dt} + Kc. \qquad (2\text{–}192)$$

Response to Step Command

To characterize the response of the system, one of the traditional tests is to apply a step signal at the input, commanding that the load, at standstill, be located at a new position instantly. From Eq. (2–190), it can be seen that, the greater the command displacement, the greater the accelerating force, and the greater the mass, the less the acceleration.

To consider the effect of the damping term, consider the step response of a system with no damping, i.e., one in which the damping and frictional forces are negligible ($D = 0$). The load accelerates at a rate of $K(c_0 - c)/M$. As c approaches c_0, the acceleration decreases, but until $c = c_0$, the load has continued to gain speed. At this point the load ceases to accelerate and would just maintain its acquired velocity except that, in going beyond c_0, an accelerating force in the opposite direction is developed by the system. Finally, at some position on the other side of c_0 from the initial position, the load is stopped by the accelerative force, which always acts toward c_0. Now the load, accelerated toward c_0, increases its velocity in that direction until it passes c_0 again. This sequence is shown as load position plotted against time in Fig. 2–217. If there are no damping or frictional forces, the load will continue to oscillate about c_0.

When damping or viscous-friction forces are present, the accelerative force is decreased by an amount proportional to the velocity. Again consider Eq. (2–190) as applied to a load approaching c_0. As the velocity u increases and the error $(c_0 - c)$ decreases, Du will, at some c, exceed $K(c_0 - c)$ and the net force will become decelerative (negative Ma). The deceleration of the load before c_0 is attained reduces the overshoot of the load. The overshoot will decrease each half cycle until the load comes to rest at c_0. This action is called a damped oscillation and is shown in Fig. 2–218. If the damping is increased still further, the overshoot can be eliminated altogether but the approach to c_0 is much slower. This is shown in Fig. 2–219.

Both overshoot and slow response are distortions of the desired response to the command signal, and therefore they are important parameters of the servo system. The overshoot is often described as a fraction of the

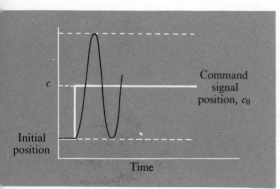

Fig. 2–217 Servo response to step command, no damping.

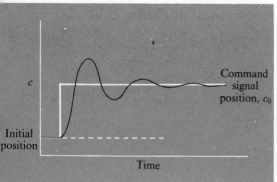

Fig. 2–218 Response to step command, some damping.

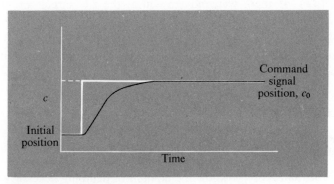

Fig. 2–219 Response to step command, damped to eliminate overshoot.

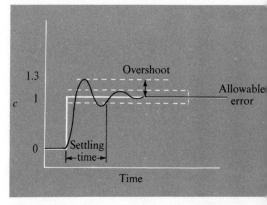

Fig. 2–220 Servo response characteristics.

command displacement. That is, if the command was to move the load 10 cm and the initial overshoot was 1 cm, the overshoot would be 0.1, or 10 percent. A measure of the response speed is the "settling time," the time required for the load to come to the command position within a prescribed error tolerance. The measure of overshoot and settling time is shown in Fig. 2–220. This is the same type of consideration as for the settling time of OA feedback circuits discussed in the previous section.

An *RLC* Analog of Servo Response

It can be shown that the undamped servo system is an oscillator. This is easiest to do by showing that the positioning of the load according to Eq. (2–192) is exactly analogous to the movement of charge in a series-resonant *LC* circuit. When the switch of Fig. 2–221 is closed, the command is to transfer charge to capacitor C through inductance L and resistance R. The sum of the voltages across the circuit elements is equal to the battery voltage. The voltage across the resistor is iR; across the conductance, $L(di/dt)$; across the capacitor, q/C, where q is the charge on the capacitor. The summation of the voltages in the loop of Fig. 2–221 may be written

$$V = L\frac{di}{dt} + iR + \frac{q}{C}$$

or, since $i = dq/dt$,

$$V = L\frac{d^2q}{dt^2} + R\frac{dq}{dt} + \frac{q}{C}. \qquad (2\text{–}193)$$

Thus we see that L is analogous to M, R to D, and $1/C$ to K of Eq.

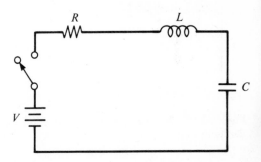

Fig. 2–221 Electrical analog of servo system.

Note 10.   Resonant Frequency.

The resonant frequency, f, of an RLC circuit occurs when the capacitive reactance, X_C, is equal to the inductive reactance, X_L, and is $f = 1/(2\pi\sqrt{LC})$.

(2–192). If there were no resistance, the LC circuit excited as in Fig. 2–217 would oscillate at its resonant frequency[10] theoretically forever. The introduction of resistance results in a damped oscillation, or **ringing.** The resonant frequency is $f_0 = 1/2\pi\sqrt{LC}$ (as will be explained in Module 3). It follows that the natural frequency of the undamped servo system will be

$$f_0 = \frac{1}{2\pi}\frac{\sqrt{K}}{M}. \qquad (2\text{–}194)$$

Damping Factor

It is advantageous to know the amount of damping necessary to reduce the overshoot of the system to a tolerable level. The amount of damping required will be larger for a larger mass, because a larger mass acquires less velocity in a given time for the same accelerating force. Also, in order to bring about a deceleration for a particular value of $c_0 - c$, D must be larger for larger values of the motive force K. The following formula for the damping factor ξ shows the relationship between D, K, and M for a constant damping effect:

$$\xi = \frac{D}{2\sqrt{KM}}. \qquad (2\text{–}195)$$

Figure 2–222 shows the step response as a function of the damping factor. Note that, when the damping factor is less than unity, there are overshoot and damped oscillations. An unnecessarily long time is required to reach command position when the damping factor is greater than 1. A system with a damping factor of unity is said to be critically damped. Most servo systems are operated with a damping factor between 0.6 and 1.0. The lower damping factors give a quicker response and, in spite of the overshoot, a smaller settling time than critical damping. If the damping factor is much below 0.6, the damped oscillations take too long to subside and the settling time is again increased.

Note also from Fig. 2–222 that the damping factor has an effect on the period of the damped oscillation. The actual frequency is given by the following equation when $\xi < 1$:

$$f = f_0\sqrt{1 - \xi^2} = \frac{1}{4\pi M}\sqrt{4KM - D^2}. \qquad (2\text{–}196)$$

Saturation. The assumption of the preceding arguments that the motor torque is proportional to the error in the load's position cannot be true for very large errors. There is a maximum torque for any given motor, and any amplifier has a maximum output capability. When one of these maxima is

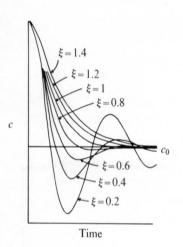

Fig. 2–222 Servo response as a function of damping factor.

reached, the drive system is saturated and a further increase in error signal cannot increase the restoring torque. Today most servo systems are designed to saturate with an error which is only a small percentage of the span. For large displacement, therefore, the term $K(c_0 - c)$ in Eq. (2–190) has a maximum value regardless of c:

$$Ma = K(c_0 - c) - Du.$$

This maximum force will be called K_{max}. The initial acceleration of the load will be $a = K_{max}/M$. As the velocity increases, the acceleration decreases until finally the damping force is equal to the driving force and the velocity is constant and maximum. When this is true, $K_{max} = Du_{max}$. Solving for the maximum velocity, we obtain

$$u_{max} = \frac{K_{max}}{D}. \qquad (2-197)$$

This demonstrates that the maximum velocity in saturation is inversely proportional to the damping; so, here again, excess damping must be avoided if response speed is an objective.

   One of the advantages of operating the system in saturation is that the system always responds with its maximum response speed. When the load comes close enough to the command position to bring the system out of saturation, the driving torque decreases and the load decelerates as it approaches c_0. The response of the system operating for a time in the saturated region is shown in Fig. 2–223. In this case, the settling time is a linear function of the displacement, and *the overshoot is a constant fraction of the control-system span*, rather than a constant fraction of the input step.

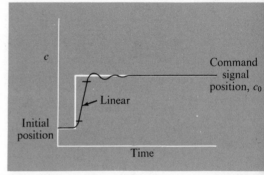

Fig. 2–223 Step response of saturated servo system.

Position Errors

Errors in servo positioning arise from several sources. One of these errors is that the load position does not immediately coincide with the command position. This difficulty has been discussed above. Since it cannot be overcome completely, it is certainly of concern when the load must accurately follow the command signal.

   Another source of position error is the "dead zone" of the system. This is the position error for which the error signal is too small to result in a rotation of the servo motor. An obvious way to correct this difficulty is to increase the amplifier gain. This also has the effect of driving the system into saturation for smaller displacements. The limit of practical amplification is reached when the amplifier is so sensitive that it cannot distinguish between noise and the error signal.

   Backlash, due to a looseness in the coupling between the load and the motor, leads to a kind of oscillation. This is due to the momentum built up by the motor rotating through the backlash region. Backlash between the load and the position indicator will cause a hysteresis in the positioning.

An oscillation can also occur as the result of a time delay in the feedback control loop. That frequency for which the delay is exactly a half cycle will be fed back in phase rather than out of phase.

The Ramp Input Test

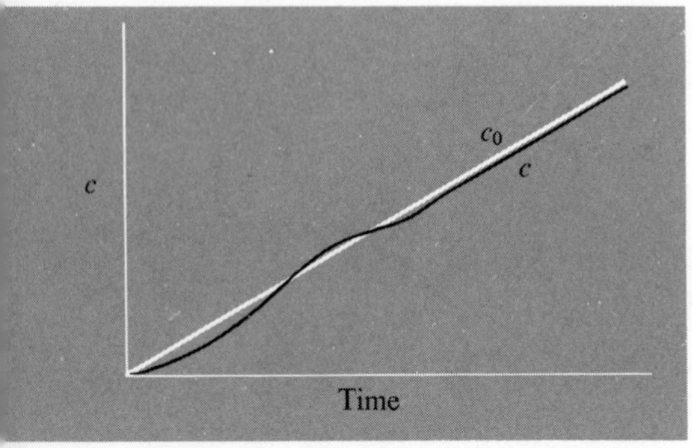

Another standard test command signal is the ramp, a linearly increasing potential. The usual response of the linear servo system to the ramp is shown in Fig. 2–224. The servo should move the load at a constant velocity from the initiation of the ramp. The load cannot be accelerated to that velocity instantly, and so it lags. As the load accelerates to catch up, its velocity exceeds the command velocity and overshoot occurs. The system settles down to follow the command velocity, but the command signal is always a little ahead of the load at any given time. The reason is that, to have the servo motor rotating at a constant velocity, there must be an error signal at the input. It can be seen from Eq. (2–190) that the steady-state displacement error will be proportional to D and u and inversely proportional to K. As the command signal approaches u_{max}, the servo loses the ability even to reproduce the slope of the command signal.

Fig. 2–224 Servo response to ramp command.

2–6.3 OPERATIONAL POWER SUPPLIES

Several power supplies are usually found in each modern measurement or control instrument. For example, a relatively simple instrument for color measurement requires a current-regulated supply at 2 A for a nominal 12 V light source, and voltage-regulated supplies of $+5$ V at 1 A for the TTL circuits, $+15$ V and -15 V at 100 mA for the OA's, and 1000 V at 1 mA for the photomultiplier transducer. Other instruments not only require regulated supplies at fixed current or voltage levels, but also require that specific supplies be programmable and respond rapidly to control signals provided either by an internal program or by remote signals from a computer or other signal controller.

These modern regulated and programmable power supplies are electronic servos. They can recognize a command to provide a specific current or voltage output, make a comparison between the actual and desired values, and take action either to make the actual value equal to a preset command value (regulated supply) or to respond rapidly and track accurately a remote command value (programmable supply).

Many types of discrete component circuits have been used for power supplies, but most of the newest supplies utilize integrated operational amplifier circuits to provide the feedback control. The regulated supply can be described in terms of OA analysis, and a supply with an OA controller is frequently called an **operational power supply.**

The great usefulness of accurate and precisely controlled voltage and current power supplies, and the high accuracy that can be obtained by electronic feedback control, make these supplies ideal examples to conclude this section on automatic feedback control.

Voltage-Regulated and Controlled Power Supply

When the ac line power is adjusted with a transformer, rectified, and filtered to provide a dc output voltage, the magnitude of the output from the raw dc supply is dependent on the line voltage and the attached load. The principles of a series regulator that will adjust and hold the output voltage at a command value, even with rather large changes in both line voltage and load, were introduced in Section 2–4.1. The discussion of these concepts is now enlarged in terms of feedback control, and a practical circuit is developed.

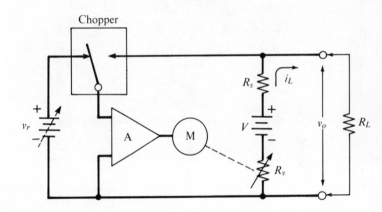

Fig. 2–225 Voltage-regulated supply with electromechanical feedback.

Electromechanical Feedback. The raw dc supply as obtained by a transformer, rectifier, and filter is represented simply by V with an internal resistance R_s in Fig. 2–225. In series with the raw supply is a variable power resistor, R_v, which is varied by mechanical coupling to a motor (M) which operates from the output of a servo amplifier. At the input of the servo amplifier the output voltage v_o is compared with the reference voltage v_r by using a chopper to look alternately at v_r and v_o. If there is any difference between v_r and v_o, the output of the high-gain amplifier provides the necessary power output for the motor shaft to turn, changing R_v and varying v_o until $v_o = v_r$. Since

$$v_o = V - i_L(R_s + R_v), \qquad (2\text{--}198)$$

it is readily seen that any change in the raw supply voltage V because of line voltage changes, or in the term $i_L(R_s + R_v)$ because of load current changes, will cause a change in v_o. However, this error (difference between v_o and v_r) is quickly sensed at the input to the servo amplifier, and the motor shaft rotates and changes R_v so as to bring v_o back to the stable preset reference voltage v_r. If v_o is greater than v_r, the motor shaft turns in the direction to increase R_v, and vice versa. Obviously, the servo amplifier must recognize whether $v_o > v_r$ or $v_o < v_r$ so that the motor shaft will turn in the correct direction to increase or decrease R_v.

This type of electromechanical servo voltage regulator is seldom used anymore because of its high cost and slow response, but it does illustrate the basic principles of all series voltage regulators.

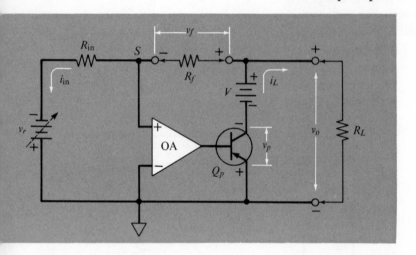

Fig. 2–226 Voltage-regulated supply with electronic feedback.

Electronic Feedback. An electronic servo voltage regulator is illustrated in Fig. 2–226. A pnp power transistor that is capable of carrying the load current i_L is connected in series with the raw supply V. The power transistor is called the **pass transistor** and acts essentially as a variable resistance similar to R_v of Fig. 2–225. It provides the variable voltage drop v_p which compensates for any fluctuations in V or i_L.

As in all OA feedback circuits, the sum of the voltages around the feedback loop from the summing point S to the noninverting input is zero. Therefore,

$$v_o = -v_f \qquad (2\text{–}199)$$

and, since $v_f = i_f R_f$, $i_f \cong i_{in}$, and $i_{in} = v_r/R_{in}$, it follows that

$$v_o = -i_{in}R_f = -v_r\left(\frac{R_f}{R_{in}}\right). \qquad (2\text{–}200)$$

The circuit in Fig. 2–226 to the right of the summing point, including the raw supply, control amplifier, and pass transistor (but not the feedback resistor R_f), can be considered as a high-powered OA. The combination can be represented by a triangular symbol and used as an OA for typical operational functions such as adding and integrating, but with the advantage of higher power output. Of course, the output voltage for the circuit shown in Fig. 2–226 will swing only in the positive direction. This is sufficient in many cases. It is necessary to add an npn power transistor and oppositely connected raw supply to the circuit of Fig. 2–226 if the OA output is to be bipolar. If this is done, the output v_o will swing either positive or negative depending on the direction of the input current.

Voltage Control and Programming. If the reference voltage v_r and R_{in} are fixed values, then R_f can be used as the **voltage control.** It can be seen from Eq. (2–200) that v_o is directly proportional to R_f, and the volts/ohm ratio for R_f is equal to v_r/R_{in}. For example, if a 5 V Zener supply is used for v_r and $R_{in} = 5000 \ \Omega$, then v_o changes by 1 mV per ohm change of R_f. From Eq. (2–200) it is also apparent that a variable reference source could be used for linear control of the output v_o, and R_{in} could be used for an inverse relationship.

The output voltage can be electromechanically or electronically programmed. A timing motor with a cam selector switch can switch in different values of v_r, R_f, or R_{in}. Much higher-speed programming can be accomplished by using electronic switches driven by logic circuits.

Voltage Output Errors. The voltage-regulated power supply is evaluated and specified by its changes in output v_o versus changes in line fluctuations, load variations, temperature variations, and changes over long time periods. In Section 2–5 the input bias current (i_b) and offset voltage (v_{off}) of OA's were described and their variation with temperature was indicated. If the OA gain is very high and i_b and v_{off} are significant, the expression for output voltage is

$$v_o = -v_r \left(\frac{R_f}{R_{in}} \right) \pm i_b R_f \pm v_{off} \left(1 + \frac{R_f}{R_{in}} \right) \qquad (2\text{--}201)$$

and the variation in the output voltage is

$$\Delta v_o = -\Delta v_r \left(\frac{R_f}{R_{in}} \right) \pm \Delta i_b R_f \pm \Delta v_{off} \left(1 + \frac{R_f}{R_{in}} \right). \qquad (2\text{--}202)$$

If the accuracy of the voltage output is to be reliably specified, it is important to include data on Δv_r, Δi_b, and Δv_{off} as a function of temperature, time, line voltage, and load. Also, any changes in R_{in} or R_f should be considered, but it is assumed that resistors with low temperature coefficients are used and that the input and feedback currents are within bounds so that self-heating of the resistors does not cause significant change in the resistance ratio R_f/R_{in}. The value of v_o can be regarded as the output signal and the largest $|\Delta v_o|$ is the worst-case error. The maximum relative error for specified conditions is

$$\text{percent relative error} = 100 \, \frac{\Delta v_o}{v_o}. \qquad (2\text{--}203)$$

A very well-regulated voltage supply would have a relative error of $<0.01\%$ for line variation of 105–125 V, a change from no load to full load, a change of $\pm 5°$C from ambient temperature, and over an eight-hour period.

Integrated Circuit Regulators.   Inexpensive integrated circuit voltage regulators are now available for many applications that provide 0.01 to 1% regulation. These regulators contain both a control amplifier and a Zener supply on the semiconductor chip. If the pass transistor is included on the same semiconductor chip, it must dissipate considerable heat. Even with a good heat sink, the changes in V and i_L cause significant changes in the temperature-dependent parts of the circuit and cause regulator errors. Therefore, at present the high accuracy reference supplies do not include the pass transistor on the regulator chip.

Current-Regulated and Programmable Power Supply

There are many transducers, light sources, and circuits that require a constant current or a programmable sequence of accurate current signals for operation. Again OA feedback circuits can provide accurate and programmable current control. Two types of current-regulated supplies are shown in Fig. 2–227.

The current-regulated supply of Fig. 2–227a is basically the voltage-follower-with-gain configuration. The current i_L in the load R_L is supplied from the raw power supply and the power transistor Q_p. A current-sensing resistor R_{cs} is used to monitor the load current. If the amplifier bias current is negligible, the current in R_{cs} is equal to i_L. The amplifier maintains the voltage v_u at the inverting input equal to the reference voltage v_r. Since $v_u = i_L R_{cs}$, it follows that $i_L = v_r/R_{cs}$ within the limitations imposed by the current capabilities of the power supply. The load current can be adjusted very precisely by appropriate choices of v_r and R_{cs}. Also remote programming is possible by controlling v_r. This type of current-regulated supply is limited to rather low-voltage applications since the output voltage of the OA is approximately equal to the voltage drop across the load.

A second type of current-regulated supply that has been frequently used is illustrated in Fig. 2–227b. The current i_L through the load is equal to the

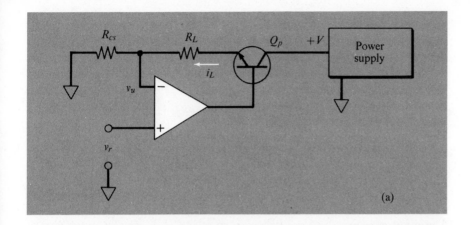

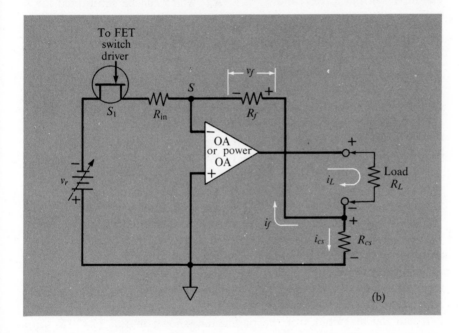

Fig. 2–227 Programmable current-regulated power supplies: (a) noninverting OA; (b) inverting OA.

sum of the current i_{cs} through the current-sensing resistor R_{cs} and the current i_f through the feedback resistor R_f. That is,

$$i_L = i_{cs} + i_f. \qquad (2\text{--}204)$$

The voltage drop v_{cs} across the current-sensing resistor must equal the voltage drop v_f across R_f to maintain the sum of the voltages between common and the summing point S equal to zero; and since

$$v_{cs} = i_{cs}R_{cs} \qquad \text{and} \qquad v_f = i_f R_f,$$

it follows that

$$i_{cs}R_{cs} = i_f R_f \qquad \text{and} \qquad i_{cs} = i_f \left(\frac{R_f}{R_{cs}}\right). \qquad (2\text{--}205)$$

If Eq. (2–205) is substituted in Eq. (2–204),

$$i_L = i_f \left(\frac{R_f}{R_{cs}}\right) + i_f = i_f \left(\frac{R_f}{R_{cs}} + 1\right). \qquad (2\text{--}206)$$

The input current $i_{\text{in}} = i_f$ and $i_{\text{in}} = v_r/R_{\text{in}}$, which can be substituted for i_f in Eq. (2–206), so that

$$i_L = \frac{v_r}{R_{\text{in}}} \left(\frac{R_f}{R_{cs}} + 1\right). \qquad (2\text{--}207)$$

It can be seen from Eq. (2–207) that the load current can be set at a desired value by choosing precise values for v_r, R_f, R_{in}, and R_{cs}. The current can be controlled over a range by varying any of these values. The current can also be programmed by remote control of v_r and/or the resistors R_f, R_{in}, and R_{cs} with logic level signals driving FET switches. The FET analog switch shown in Fig. 2–227 can be used to turn the load current ON and OFF at very high speeds. When the switch S_1 is ON, the load current will be as specified by Eq. (2–207), but when S_1 is OFF, the load current will be zero. The current through the load will be controlled accurately according to the parameters in Eq. (2–207) and can be programmed in a precise ON-OFF sequence as determined by logic level pulses to the FET driver.

This type of programmed current source can be used for low voltage or high voltage applications and the programmed current range will be determined by the power capability of the raw supply and the operational amplifier, and the magnitudes of the parameters shown in Eq. (2–207). For loads such as hollow-cathode lamps that require high voltage and high current when operated in a switched mode, the triangular symbol in Fig. 2–227 would have to be a power OA similar to the combination of OA, Q_p, and V in Fig. 2–226, but with R_f in the circuit as shown in Fig. 2–227.

Switching Regulators

Voltage regulators which utilize high-frequency ON-OFF switching can offer several advantages over conventional continuously controlled, series-

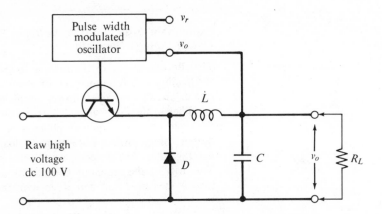

Fig. 2–228   Switching regulator.

type regulators. There are several different types of switching regulators, one of which is illustrated in Fig. 2–228. In this type of regulator the series transistor is turned ON and OFF by an oscillator whose pulse width is variable in proportion to the voltage difference between a reference voltage and the regulator output. The oscillator has a constant repetition rate (typically 20 kHz), but the ON time of the oscillator pulse is variable. The LC filter at the output smooths the pulsed output from the switching transistor. When the transistor conducts, current is delivered to the load from the raw supply. When the transistor is OFF, diode D conducts and current is supplied to the load from the energy stored in the inductor and capacitor.

The major advantage of the switching regulator is its higher efficiency compared to conventional series regulators. Efficiencies of a switching regulated supply can be as high as 80%, while typical series-pass regulated supplies have efficiencies of 50% or less.

Another type of switching-regulated supply makes use of direct rectification of the 60 Hz line voltage. The relatively high dc voltage is then converted to a high-frequency ac (20 kHz) where a stepdown in voltage can take place with a high-frequency transformer. The resulting high-frequency signal can then be rectified and applied to a switching regulator of the type shown in Fig. 2–228. This type of supply can offer considerable size reduction over conventional supplies because the high-frequency transformer can be much smaller than a 60 Hz transformer and less heat sink area is required because of higher efficiency. In addition, the size of the filter capacitor is reduced and its efficiency improved.

At present, because of the number of components needed for a switching-regulated supply, the costs are usually higher than for a conventional supply. However, this type of supply is expected to become less expensive as integrated circuit techniques provide cheaper components.

Experiment 2–33 *An OA Voltage Follower Source and Integrated Circuit Voltage Regulator*

The effectiveness of an OA voltage follower with gain for greatly decreasing the loading of a reference cell or electronic voltage reference source is first tested. The OA voltage follower circuit is also modified to include a booster amplifier in the OA feedback loop to show that the booster reduces drift of the input offset voltage with changes of output load and provides regulated output voltage over a wide range of loads. Then an integrated circuit voltage regulator that includes a reference source is tested with various loads. The drift of the regulator because of temperature change with change of load is measured and compared with the manufacturer's specifications.

Experiment 2–34 *A Programmable Current-Regulated Power Supply*

In many instrumentation systems it is important either to maintain the current in a device at a fixed value or to program the current accurately according to some function such as a precise rectangular pulse or a ramp signal. In this experiment an electronic OA feedback system is constructed which provides accurate programmable control of the current. A booster amplifier is included in the first part to provide bidirectional control of current. In an optional part a circuit can be constructed using a power transistor which provides only unidirectional current control.

Experiment 2–35 *An Electronic Servo for Control of Illuminance*

In this experiment the illuminance from a light source is regulated by an electronic servo system. The response characteristics of the feedback control element are first varied to illustrate their effect on the response of a servo system to a command signal when the controlled device (a light-emitting diode) responds very rapidly (10^{-8} sec). Then a tungsten source with a relatively slow response time is controlled, and it is shown how the servo system can cause continuous oscillation in the controlled device. It is then shown how suitable changes in the feedback control circuit can damp the oscillation and enable the controlled device to follow the command signal.

PROBLEMS

1. In operational amplifier feedback regulators, as in other OA applications, the input bias current can cause an output voltage error. Consider the OA inverting voltage amplifier shown in Fig. P–1. Assume that the voltage offset

$v_{\text{off}} = 0$. In the inverting amplifier only the bias current into or out of the inverting input $i_{b(-)}$ is important. (a) In which resistor will the bias current exist? (b) If $i_{b(-)} = 15$ nA, what will be the output voltage v_o due to $i_{b(-)}$?

2. Input offset voltage in OA's can also cause output voltage errors. The offset voltage can be considered a voltage source in series with one of the inputs, as shown in Fig. P–1. Assume that the bias currents shown in Fig. P–1 are negligible and that $v_{\text{off}} = 1$ mV. What is the magnitude of the output voltage v_o due to v_{off}?

3. In the OA inverting amplifier of Fig. P–1 consider a 50 mV signal to be applied between R_{in} and common. (a) What is the maximum value of $i_{b(-)}$ which can be tolerated for a 1% or less output voltage error (assume ideal resistors and $v_{\text{off}} = 0$)? (b) What is the maximum value of v_{off} that can be tolerated for the same error (assume $i_b = 0$)?

4. The voltage-regulated supply of Fig. 2–226 is operated with $R_{\text{in}} = 5$ kΩ, $R_f = 10$ kΩ, and $v_r = -2$ V. If $i_{b(-)} = 1$ μA into the noninverting input and $v_{\text{off}} = +2$ mV into the inverting input, what is the output voltage? What is the percentage error due to $i_{b(-)}$ and v_{off}?

5. In the voltage-regulated supply of problem 4, v_{off} drifts with time at a rate of ± 100 μV/day. The input bias current drift rate is ± 1 μA/day and v_r drifts at a rate of ± 1 mV/day. What is the maximum drift rate of the regulated output v_o?

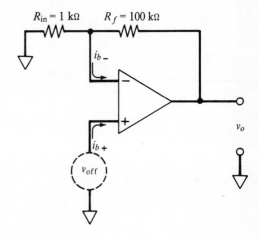

Fig. P–1

Derivation of the Delta-Y, Y-Delta Transformation

Delta to Wye

From Fig. 2–22a the resistance from A to B is $R_{AY} + R_{BY}$, and from Fig. 2–22b the resistance is $R_{AB}(R_{AC} + R_{BC})/(R_{AB} + R_{BC} + R_{AC})$. Therefore,

$$R_{AY} + R_{BY} = \frac{R_{AB}R_{AC} + R_{AB}R_{BC}}{R_{AB} + R_{BC} + R_{AC}}. \qquad (A1)$$

In the same way,

$$R_{BY} + R_{CY} = \frac{R_{BC}R_{AC} + R_{BC}R_{AB}}{R_{AB} + R_{BC} + R_{AC}} \qquad (A2)$$

and

$$R_{AY} + R_{CY} = \frac{R_{AC}R_{AB} + R_{AC}R_{BC}}{R_{AB} + R_{BC} + R_{AC}}. \qquad (A3)$$

Subtract (A2) from (A1) to obtain

$$R_{AY} - R_{CY} = \frac{R_{AB}R_{AC} - R_{BC}R_{AC}}{R_{AB} + R_{BC} + R_{AC}}. \qquad (A4)$$

Now subtract (A4) from (A3) to get

$$R_{AY} = \frac{R_{AB}R_{AC}}{R_{AB} + R_{BC} + R_{AC}}. \qquad (A5)$$

The expressions for R_{BY} and R_{CY} are obtained similarly.

Wye to Delta

From the delta-Y transformation equations,

$$\frac{R_{AY}}{R_{BY}} = \frac{R_{AC}}{R_{BC}} \qquad \text{and} \qquad R_{AC} = \frac{R_{BC}R_{AY}}{R_{BY}}. \tag{A6}$$

Similarly,

$$\frac{R_{AY}}{R_{CY}} = \frac{R_{AB}}{R_{BC}} \qquad \text{and} \qquad R_{AB} = \frac{R_{BC}R_{AY}}{R_{CY}}. \tag{A7}$$

Now substitute (A6) and (A7) for R_{AC} and R_{AB} in (A5) to get

$$R_{BC} = R_{BY} + R_{CY} + \frac{R_{BY}R_{CY}}{R_{AY}} \tag{A8}$$

and obtain the expressions for R_{AB} and R_{AC} similarly.

Fabrication of Transistors

<h1>Appendix B</h1>

It was indicated in the previous discussions that the method of fabrication of the transistor influences its properties. To obtain the desired properties, the location and concentration of the n- and p-type impurities must be carefully controlled. The most popular ways of forming the desired transistor pn junction are by the alloy-junction and diffused-base (mesa and planar) techniques.

Grown-Junction Transistors

The first commercial transistors were produced by the grown-junction technique. With this method a rather large semiconductor crystal is grown as it is continuously pulled from the molten semiconductor material, e.g., silicon. As the crystal grows the composition of the silicon melt is changed from n-type to p-type and back to n-type by doping the melt while continuously turning and pulling the crystal from the melt. By cutting the doped crystal into short cylinders and then into small bars, many segments of npn material are obtained as illustrated in Fig. B–1. Leads are connected to the two n- and the p-sections of the semiconductor bar, which is then mounted on a header as shown in Fig. B–2.

When using the grown-junction technique it is impossible to control the base-layer thickness very well, and the base thickness is relatively large, about 10^{-3} cm. Because of the base thickness, the transistors made by this process do not have good high-frequency response, and the process is quite complicated. Therefore, the grown-junction process has generally been superseded by other fabrication techniques.

Alloy-Junction Transistors

A thin slice (about 20 mils) of semiconductor material, e.g., germanium, can have emitter and collector dopants (indium) alloyed to either side of

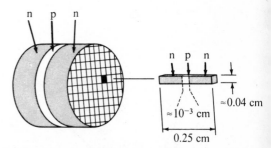

Fig. B–1 Sectioning of grown-junction doped semiconductor to obtain small bars of npn material for transistors.

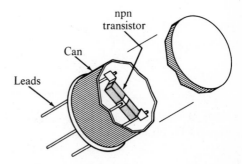

Fig. B–2 Mounting for a grown-junction transistor.

the wafer, as shown in Fig. B–3. When the indium melts, it dissolves some of the germanium and during the subsequent cooling the dissolved germanium recrystallizes on the base with many indium atoms incorporated in the crystal structure. Therefore, the recrystallized germanium is now doped as p-type, and a pnp transistor results.

The larger size of the collector pellet as compared to the emitter enables the carriers injected at the emitter to be collected more efficiently. The dimensions of the junction regions are also hard to control by the alloy process, so that the high-frequency response is generally not good.

Diffused-Base Transistors

It is possible to control the junction fabrication process precisely by diffusion. By a suitable sequence of masking and exposing silicon semiconductor material to a hot gas of antimony or boron dopants and etching, it is possible to produce the diffused silicon mesa or planar transistors. Cross sections of these two types are shown in Fig. B–4.

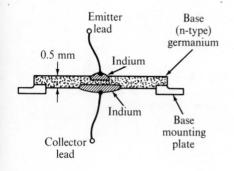

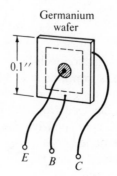

Fig. B–3   A pnp alloy transistor.

Fig. B–4   Cross-sectional view of diffused silicon transistors: (a) mesa; (b) planar.

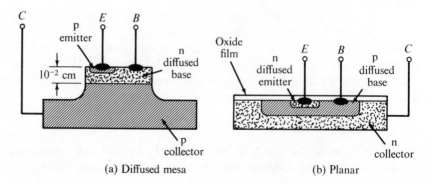

(a) Diffused mesa

(b) Planar

Measures of Sine Wave Amplitude Appendix C

Sine wave voltage and current amplitudes are commonly described in four ways: the peak value, the peak-to-peak value, the average value, and the rms (or *root-mean-square*) value. The particular description used depends on the application. Measuring instruments such as vacuum-tube voltmeters and oscilloscopes *respond* to the peak-to-peak value, which for a sine wave is merely twice the peak value. Moving-coil meters *respond* to the average value (effective rectified dc value), and the rms value is mostly used for the measurement of *power* in an ac circuit. Figure C–1 graphically summarizes the four commonly used measures.

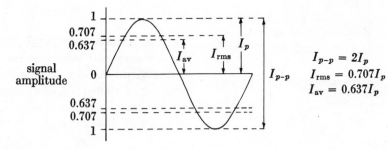

Fig. C–1 Common measures of sine wave amplitude.

$$I_{p-p} = 2I_p$$
$$I_{rms} = 0.707I_p$$
$$I_{av} = 0.637I_p$$

Average Value

Moving-coil meters respond to the average value of the current through them. The mechanism by which the meters operate is discussed in Module 1. The average value is that current above and below which the current-time areas of a half-cycle are equal. This is illustrated graphically in Fig. C–2.

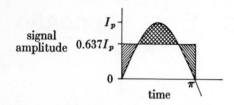

signal amplitude $0.637I_p$

Fig. C–2 Determination of the average value of a sinusoidal signal.

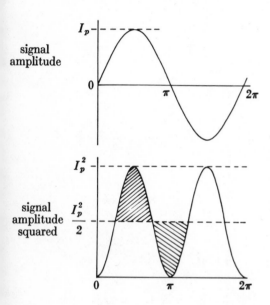

signal amplitude

signal amplitude squared $\dfrac{I_p^2}{2}$

Fig. C–3 Determination of the rms value of a sinusoidal signal.

The average value may be obtained by integrating the area over a half-cycle:

$$I_{av} = \frac{1}{\pi}\int_0^\pi I_p \sin \omega t\, dt$$

$$= \frac{2I_p}{\pi} = 0.637I_p. \tag{C1}$$

In practice, the sine wave must be rectified (full-wave) to obtain meter readings corresponding to Eq. (C1).

RMS Value

The rms value is the *effective* value needed in power calculations. The rms value of a sine wave current will produce the same heating (power) in a resistor as an identical dc current. In other words, 1 rms ac ampere produces the same amount of heat in a resistor in a given time as 1 ampere of dc current. The rms value is obtained by taking the *square root* of the *mean* (average) of the *squared* values of the current. This operation is shown graphically in Fig. C–3. The square of the sine wave ($I_p^2 \sin^2 \omega t$) is plotted directly below the sine wave, because the heating effect of a current is directly proportional to the square of the current. The result is another "sine wave" which is completely above zero and twice the frequency. The mean or average value of this "sine wave" is clearly $I_p^2/2$. On taking the square root,

$$I_{rms} = \sqrt{\frac{I_p^2}{2}} = \frac{I_p}{2}\sqrt{2} = 0.707I_p. \tag{C2}$$

The relation between the effective emf and the peak emf is the same as for effective and peak current. Thus, $V_{rms} = 0.707V_p$. Calculations of voltages and currents in ac circuits provide self-consistent values; e.g., if rms voltages are used, the currents will have rms values, etc. Note that peak values are obtained by multiplying the rms values by the reciprocal of 0.707 (1.414), so that

$$I_p = 1.414I_{rms},$$
$$V_p = 1.414V_{rms}.$$

Bibliography

Angelo, E. J., Jr., *Electronics: BJT's, FET's and Microcircuits,* McGraw-Hill, New York, N.Y., 1969.

A book intended for a first course in electronics for electrical engineering students. It contains an excellent basic description of amplifiers.

De Waard, H., and D. Lazarus, *Modern Electronics,* Addison-Wesley, Reading, Mass., 1966.

The first chapter of this book is an excellent introduction to the mechanisms of conduction in solids.

Haase, R., *Thermodynamics of Irreversible Processes,* Addison-Wesley, Reading, Mass., 1969.

A helpful reference for mathematical treatment of thermoelectric phenomena.

Halliday, D., and R. Resnick, *Physics, Parts I and II,* Wiley, New York, N.Y., 1966.

A standard college physics text that includes good discussions of basic electrical quantities.

Lion, K. S., *Instrumentation in Scientific Research—Electrical Input Transducers,* McGraw-Hill, New York, N.Y., 1959.

Many basic discussions of transducers are included in this classic text.

Malmstadt, H. V., C. J. Delaney, and E. A. Cordos, "Instruments for Rate Determinations," *Analytical Chemistry 44,* 1972, p. 79A.

A description of the principles and circuits for quantitative measurement of rate of change of electrical signals, with special emphasis on integrating types of differentiators.

Malmstadt, H. V., and C. G. Enke, *Digital Electronics for Scientists,* W. A. Benjamin, Menlo Park, Calif., 1969.

The integrated presentation of electronic switching devices and applications first presented in Chapters 2 and 3 of this book has been greatly modified and updated and introduced in Section 2–3 of this new text.

Malmstadt, H. V., C. G. Enke, and E. C. Toren, Jr., *Electronics for Scientists,* W. A. Benjamin, Menlo Park, Calif., 1962.

Some of the basic amplification and servo control concepts presented in this earlier text have been modified and presented in the present text in a new perspective, with an emphasis on the overall control of electrical quantities in instrumentation.

Millman, J., and H. Taub, *Pulse, Digital and Switching Waveforms,* McGraw-Hill, New York, N.Y., 1965.

A standard source book for switching and wave shaping. Contains detailed discussions of diode and transistor switches, including gating.

Motorola, *The Semiconductor Data Book,* 1969.

Contains specifications for many transistors, FET's, IC's, OA's, diodes, etc., and also descriptive material on FET's, phototransistors, unijunction transistors, and their applications.

Schmid, H., *Electronic Analog/Digital Conversions,* Van Nostrand Reinhold Co., 1970.

Includes an excellent section on electronic switches.

SCR Manual, 5th ed., General Electric Company, Syracuse, N.Y., 1972.

Principles and applications of pnpn devices, including light-activated devices. Contains many practical circuits for use of thyristors.

Seely, S., *Electronic Circuits,* Holt, Rinehart and Winston, New York, N.Y., 1968.

A well-known text in which electronic devices and basic semiconductor and tube-driven circuits are presented.

Smith, R. J., *Circuits, Devices and Systems,* 2nd ed., Wiley and Sons, New York, N.Y., 1971.

A basic text that provides an understanding of the fundamental principles of electrical engineering, a familiarity with the viewpoint and terminology of electrical engineers, and an introduction to some modern techniques and many good illustrations and problems.

Sze, S. M., *Physics of Semiconductor Devices,* Wiley-Interscience, New York, N.Y., 1969.

The electrical characteristics of modern semiconductor devices are developed from basic semiconductor theories in a thorough mathematical treatment.

Texas Instruments Inc., *Transistor Circuit Design,* McGraw-Hill, New York, N.Y., 1963.

A good basic description of transistors and their applications.

Tobey, G. E., J. G. Graeme, and L. P. Huelsman (Eds.), *Operational Amplifiers,* McGraw-Hill, New York, N.Y., 1971.

A modern book on operational amplifiers that contains many applications of OA's as well as basic design considerations of the amplifiers.

Solutions to Problems

SOLUTIONS TO PROBLEMS IN SECTION 2–1

1. a) If $R_m = 0$, the measured thermistor resistance R'_T is given by

$$R'_T = 5 \text{ V}/0.99 \times 10^{-3} \text{ A} = 5.05 \text{ k}\Omega.$$

b) $R_m = 250 \ \Omega.$

c) With the 1 mA meter the true resistance is

$$R_T = \frac{V}{I} - R_m = 4800 \ \Omega,$$

and

$$\text{RPE} = \frac{R_T - R'_T}{R_T} = 0.052, \qquad \text{or } 5.2\%.$$

2. a) For the current meter a series circuit consisting of the 1 V battery, the thermistor, and the current meter will be considered. The theoretical current is $10^{-4} \text{ A} = 100 \ \mu\text{A}$. The measured current is

$$I = \frac{1 \text{ V}}{R_T + R_m} = \frac{1}{10^4 + 2.5 \times 10^3} = 0.8 \times 10^{-4} \text{ A},$$

and

$$\text{RPE} = 0.20, \qquad \text{or } 20\%.$$

b) For voltage measurements, a series circuit consisting of the 1 V battery, a 10 kΩ resistor, and the thermistor is considered. The voltmeter is in parallel with the thermistor. The theoretical voltage V_T may be calculated from

$$V_T = \frac{VR_T}{R_T + 10 \text{ k}\Omega} = 0.5 \text{ V}.$$

The voltage measured may be calculated in the same manner, substituting R_{eq} for R_T, where

$$R_{eq} = \frac{R_m R_T}{R_m + R_T} = 9.9 \times 10^3 \ \Omega.$$

Then

$$V_m = 0.497 \ V$$

and

$$RPE = 0.006, \qquad \text{or } 0.6\%.$$

3. The total current in the circuit may be written as

$$I_t = \frac{V}{R_p + R_r + R_u}.$$

If $R_u = 0$, $I_t = 0.107$ A. When R_u is present, $I = \frac{2}{3}I_t = 0.0715$ A. Hence

$$R_u = \frac{V}{I} - (R_p + R_r) = 7 \ \Omega.$$

4. From Eq. (2–20),

$$R_u = R(R_A/R_B) = 6.28 \ k\Omega.$$

5. By successive reductions, the current through $R_7 = 7.05$ mA.

6. $V = 0.71$ V.

7. Use the Kirchhoff junction and loop equations and solve by determinants. $I_1 = 6.22$ mA, $I_2 = 1.87$ mA, $I_3 = 4.35$ mA, $I_4 = -0.15$ mA, $I_5 = 1.72$ mA, $I_6 = 4.50$ mA.

8. The Thévenin equivalent voltage is 0.279 V, $R_{Th} = 925 \ \Omega$. If $V_m = V_{Th}$, IR_{Th} must be

$$IR_{Th} \leqslant 0.0001(V_{Th}) = 2.79 \times 10^{-5} \ V.$$

Then

$$I = 3.02 \times 10^{-8} \ A,$$

and

$$R_{meter} = \frac{V_{Th}}{I} - R_{Th} \cong 9.25 \ M\Omega.$$

9. Analyzing by superposition, we obtain:

	1 kΩ	2.5 kΩ	5 kΩ
I_1, mA	3.75	2.5	−1.25
I_2, mA	−1.87	0.75	2.72
I_t, mA	1.88	3.25	1.37

10. If $IR_3 = 0$, then at the junction of the three resistors,

$$I_1 = -I_2 \quad \text{or} \quad \frac{V_1}{R_1} = \frac{-V_2}{R_2},$$

$$V_2 = -V_1 \frac{R_2}{R_1}.$$

The OA circuit is the inverting voltage amplifier.

11. Since R_3 is to be varied to determine the maximum power P_{max}, it is best to remove R_3 and form a Thévenin equivalent circuit of the rest of the network. When R_3 is removed, the Thévenin equivalent voltage between the two terminals created, V_{Th}, is

$$V_{Th} = 14 \text{ V} - 20I = 9 \text{ V} + 5I.$$

Since

$$I = \frac{5 \text{ V}}{25 \text{ }\Omega} = 0.2 \text{ A},$$

$$V_{Th} = 14 \text{ V} - 4 \text{ V} = 10 \text{ V}.$$

R_{Th} is the parallel combination of the 20 Ω and 5 Ω resistors, or

$$R_{Th} = \frac{20 \times 5}{20 + 5} = 4 \text{ }\Omega.$$

The simple series circuit after reinserting R_3 is thus a 10 V source, a 4 Ω resistance, and R_3. The power in the circuit is given by

$$P = I^2 R_3 = \left(\frac{10 \text{ V}}{4 \text{ }\Omega + R_3}\right)^2 R_3.$$

The maximum power occurs when $dP/dR = 0$,

$$\frac{dP}{dR_3} = \frac{(4 + R_3)^2 100 - 100R_3 \times 2(4 + R_3)}{(4 + R_3)^4} = 0,$$

from which $R_3 = 4 \text{ }\Omega$.

$$P_{max} = \left(\frac{10 \text{ V}}{8 \text{ }\Omega}\right)^2 \times 4 \text{ }\Omega = 6.25 \text{ W}.$$

12. The Thévenin equivalent voltage is

$$V_{Th} = 30 \text{ V},$$

and

$$R_{Th} = 21 \text{ }\Omega.$$

Thus the current through the ammeter, I_m, is

$$I_m = \frac{V_{\text{Th}}}{R_{\text{Th}} + 9\ \Omega} = \frac{30\ \text{V}}{30\ \Omega} = 9\ \text{A} \quad 1\ \text{A}$$

13. a) R_s must equal the setting on ΔR_3 times the multiplier:

$$R_u = R_s = \Delta_1 R_3\ (R_2/R_1),$$

$$R_s + \Delta R_s = (\Delta_1 R_3 + \Delta_2 R_3)(R_2/R_1),$$

or

$$\Delta R_s = \Delta_2 R_3 (R_2/R_1),$$

where $\Delta_2 R_3$ is the change in resistor ΔR_3 required to rebalance the bridge.

b) If $\Delta R_3 = 0.1\ \Omega$, $\Delta R_s = 0.1\ \Omega$, then

$$\Delta l/l = \frac{\Delta R_s / R_s}{F} = \frac{0.1/120}{2} = 0.000416.$$

14. a) R_{Th} should be no greater than 0.01% of 100 kΩ, or 10 Ω. $R_{\text{Th}} = R_1 R_2/(R_1 + R_2) = 10\ \Omega$ and since this is a 4/1 divider, $R_1 = 4R_2$. Therefore, $R_1 = 50\ \Omega$, $R_2 = 12.5\ \Omega$.

b) $I_N = 5\ \text{V}/50\ \Omega = 0.1\ \text{A}$, $R_N = 50\ \Omega \times 12.5/62.5 = 10\ \Omega$.

15. $R_N = 21\ \Omega$, $I_N = I_{30} - I_{20} = -(I_{90} - I_{10}) = 1.45\ \text{A}$, $I_{\text{meter}} = I_N R_N/(R_N + R_m) = 1\ \text{A}$.

16. a) $v_o = v_x - v_y = V\left(\dfrac{R_u}{R_A + R_u} - \dfrac{R}{R_B + R}\right).$

b) $\dfrac{dv_o}{dR_u} = V\left[\dfrac{(R_A + R_u) - R_u}{(R_A + R_u)^2}\right] = V\left[\dfrac{R_A}{(R_A + R_u)^2}\right].$

c) The bridge output will be linear if the derivative in part (b) is a constant. This will occur if $R_A \gg R_u$. Thus

$$\frac{dv_o}{dR_u} = V\frac{R_A}{R_A^2} = \frac{V}{R_A}.$$

d) $\dfrac{d^2v_o}{dR_u^2} = V\left[\dfrac{-R_A \times 2(R_A + R_u)}{(R_A + R_u)^4}\right] = 0.$

The above condition will hold only if $R_A \ll R_u$.

e) No, since when $R_A \ll R_u$ the slope determined in part (b) is not constant.

SOLUTIONS TO PROBLEMS IN SECTION 2–2

1. Substituting into Eq. (2–28), $W_d = CV^2/2 = 10^{-9}\,F \times (8.32)^2/2 = 3.46 \times 10^{-8}\,J$.

2. a) Since the voltage on each capacitor is 5 V and $q = CV$, $q_1 = 0.5 \times 10^{-6}\,C$, $q_2 = 0.25 \times 10^{-6}\,C$, $q_3 = 0.1 \times 10^{-6}\,C$.

   b) Since the charges on parallel capacitors add, $q_T = 0.85 \times 10^{-6}\,C$.

   c) $C_p = 0.17\,\mu F$.

3. Substituting into Eq. (2–33), $C_s = 0.0125\,\mu F$,

$$\frac{1}{C_s} = \frac{1}{C_1} + \frac{1}{C_2} + \frac{1}{C_3}.$$

The charge on each capacitor $= q_T = CV = 6.25 \times 10^{-8}\,C$.

4. a) $C_{eq} = 0.083\,\mu F$.

   b) V_1 and $V_2 = 5$ V. The charges on C_3, C_4, and the parallel combination of C_5 and C_6 are equal. The equivalent capacitance of that branch of the network is $0.023\,\mu F$. The charge on C_3 and $C_4 = 1.13 \times 10^{-7}\,C$. $V_{C3} = q_3/C_3 = 1.13$ V, $V_{C4} = 2.26$ V, $V_{C5} = V_{C6} = 1.61$ V.

   c) The charge on each capacitor is:

	C_1	C_2	C_3	C_4	C_5	C_6
q	5	25	11.3	11.3	3.2	$8.1 \times 10^{-8}\,C$.

5. Using Eq. (2–37), $i = (V/R)e^{-t/RC}$ or $V_R = Ve^{-t/RC}$, $V_C = V - V_R$.

t, msec	V_R, V	V_C, V
0.5	0.95	0.05
1	0.90	0.1
5	0.61	0.39
10	0.37	0.63
50	0.0067	0.9933
100	4×10^{-5}	0.99996

6. a) The charging time constant is

$$\tau_c = \frac{CR_1R_2}{R_1 + R_2} = 1.76\ \text{msec}.$$

The discharge time constant is

$$\tau_D = \frac{CR_1R_3}{R_1 + R_3} = 2.18 \text{ µsec.}$$

For practical purposes, 5τ is sufficient time for charging or discharging, so $5\tau_c = 8.8$ msec, $5\tau_D = 10.9$ µsec.

7. a)

t, msec	V_R, V	V_C, V
0	10	0
0.1	3.68	6.32
0.3	0.50	9.50
0.5	0.067	9.933
1.0	0.0045	9.9955
1.3	−0.50	0.50
1.5	−0.067	0.067

b)

t, msec	V_R, V	V_C, V
0	10	0
0.5	6.06	3.94
1.0	3.68	6.32
1.5	−3.83	3.83

8. $v_o = -\dfrac{1}{RC} \displaystyle\int v_{\text{in}}\, dt,$

$$\frac{dv_o}{dt} = -\frac{v_{\text{in}}}{RC} = -\frac{5 \text{ V}}{2.5 \text{ sec}} = -2 \text{ V/sec.}$$

The output changes by -1 V in 0.5 sec.

9. Using Eqs. (2–27) and (2–45),

$$C = \frac{Vt}{V_o R} = \frac{q}{V} = \frac{0.096 \text{ C}}{5 \text{ V}}, \qquad C = 0.019 \text{ F.}$$

The resistor value is unimportant.

10. a) This is a follower with gain.

$$v_o = v_{\text{in}} \left(\frac{R_f}{R_{\text{in}}} + 1 \right) = 5 \times 10^{-5} \text{ V} \left(\frac{1}{10^5} + 1 \right), \qquad v_o \approx 50 \text{ µV.}$$

b) $v_o = -\dfrac{1}{RC} \displaystyle\int v_{\text{in}}\, dt,$

$$\Delta t = -\frac{v_o RC}{v_{\text{in}}} = - \left(\frac{+12}{-5 \times 10^{-5}} \right) (10^5)(10^{-7}),$$

$$\Delta t = 2.4 \times 10^3 \text{ sec.}$$

c) It must be at least 100×50 μV or 5 mV.

d) 24 sec $\pm 1\%$.

e) For $v_{in} = 0$, Δt must be greater than 10^5 sec.

$$-v_{in} = \frac{v_o}{\Delta t} RC = -10^{-6} \text{ V} = 1 \text{ μV offset.}$$

11. a) $v_o = -RC \dfrac{dv_{in}}{dt}$, $RC = -\dfrac{(-5.0) \text{ V}}{8 \text{ V/sec}} = 0.625$ sec.

   Any R and C with a product of 0.625 sec may be used. Example: $R = 10^6 \text{ Ω}$, $C = 0.625$ μF.

   b) $RC = 0.067$ sec. Example: $R = 1$ MΩ, $C = 0.067$ μF. The differentiator can be followed with an inverting unity-gain OA.

12. a) Using Eq. (2–62),

   $$v_o = k \log \frac{v_r}{v_s} K = K_1 K_2,$$

   $$K_1 = \frac{i_{in}}{C_2} = \frac{v_{in}}{R_{in} C_2}, \qquad K_2 = RC(2.3),$$

   $$v_o = \frac{0.5}{(10^5)(10^{-6})} [(2.3)(10^6)(0.5 \times 10^6)] \log 10,$$

   $$v_o = 5.75 \text{ V.}$$

   b) $v_o = 5.75 \log 5 = 1.82$ V.

   c) $v_o = 0.91$ V.

13. Substituting into Eq. (2–66),

   $$L = \frac{V}{di/dt} = \frac{2}{15} = 0.133 \text{ H.}$$

14. $V = 1.5$ V.

15. From Eqs. (2–69) and (2–70),

   $$V_L = V e^{-tR/L}, \qquad V_R = V(1 - e^{-tR/L}).$$

t, msec	V_L, V	V_R, V
0.5	1.17	0.33
1	0.91	0.59
2	0.55	0.95
10	0.010	1.49
100	~0	1.50

SOLUTIONS TO PROBLEMS IN SECTION 2–3

1. a) From the equations preceding Eq. (2–98) the voltage drops across the closed switch and open switch, V_{sc} and V_{so}, can be calculated:

$$V_{sc} = 5 \left(\frac{80}{50 \times 10^3 + 80} \right) \text{ V} = 0.008 \text{ V},$$

$$V_{so} = 5 \left(\frac{900 \times 10^3}{50 \times 10^3 + 900 \times 10^3} \right) \text{ V} = 4.737 \text{ V}.$$

The voltage drop across the load is the source voltage minus the voltage drop across the switch. For the closed switch, $V_L = 5 - 0.008 = 4.992$ V, and for the open switch $V_L = 5 - 4.737 = 0.263$ V.

b) For τ_o use Eq. (2–98),

$$\tau_o = \frac{400 \times 10^{-12} \times 50 \times 10^3 \times 900 \times 10^3}{50 \times 10^3 + 900 \times 10^3} = 18.9 \ \mu\text{sec},$$

and for τ_c use Eq. (2–99),

$$\tau_c = \frac{400 \times 10^{-12} \times 50 \times 10^3 \times 80}{50 \times 10^3 + 80} = 0.032 \ \mu\text{sec}.$$

2. At 300°K, $Q_e/kT = 39$, so that Eq. (2–95) can be written

$$I = I_i(e^{39V} - 1) = 10^{-10}(e^{39V} - 1).$$

For $V = 0.30$, $I = 10^{-10}(e^{11.7} - 1) = 15 \ \mu$A.

For $V = 0.40$, $I = 10^{-10}(e^{15.6} - 1) = 0.6$ mA.

For $V = 0.50$, $I - 10^{-10}(e^{19.5} - 1) = 30$ mA.

For $V = 0.60$, $I = 10^{-10}(e^{23.4} - 1) = 1.45$ A.

Note that the final value of 1.45 A would not be reached in practice because of the ohmic resistance of the junction.

3. For the source at $+10$ V, the voltage across the load

$$V_L = +10 \times \frac{1 \ \text{k}\Omega}{1 \ \text{k}\Omega + 65\Omega} = 9.39 \text{ V}.$$

For the source at -10 V,

$$V_L = -10 \times \frac{1 \ \text{k}\Omega}{1 \ \text{k}\Omega + 5 \ \text{M}\Omega} = -0.002 \text{ V}.$$

4. Substituting in Eq. (2–103), $I_m = 0.8$ mA, we find

$$R_{fd} = \frac{26}{0.8} = 32.5 \; \Omega.$$

5. a) From Eq. (2–118),

$$I_{B1} = \frac{1 \text{ V} - 0.6 \text{ V}}{4 \text{ k}\Omega} = 100 \; \mu\text{A}.$$

The transistor is saturated if the condition of Eq. (2–116) holds. Since $h_{FE}I_{B1} = 100 \times 100 \; \mu\text{A} = 10$ mA, and

$$I_{C(\text{max})} = \frac{5 \text{ V} - 0.2 \text{ V}}{1 \text{ k}\Omega} = 4.8 \text{ mA},$$

the transistor is in saturation.

b) As calculated above, $I_{C(\text{sat})} = 4.8$ mA.

c) $R_{\text{ON}} = \dfrac{V_{CE(\text{sat})}}{I_{C(\text{sat})}} = \dfrac{0.2 \text{ V}}{4.8 \text{ mA}} = 42 \; \Omega.$

6. a) The minimum voltage across the coil for turn-on, $V_{\text{coil(min)}} = 200 \; \Omega \times 20$ mA $= 4$ V. Hence V_{CE} must be less than 1 V for turn-on, and the transistor need not be in saturation.

b) The source current $I_s = (4 \text{ V} - V_{BE(\text{ON})})/10 \text{ k}\Omega = 3.4 \text{ V}/10 \text{ k}\Omega = 0.34$ mA.

c) $h_{FE(\text{min})} \geqslant I_{C(\text{min})}/I_B$ or $h_{FE(\text{min})} \geqslant 20 \text{ mA}/0.34 \text{ mA} = 59.$

7. The ON resistance R_{ON} is $R_{\text{ON}} = V_{CE(\text{sat})}/I_{C(\text{sat})} = 0.4 \text{ V}/50 \text{ mA} = 8 \; \Omega$. When the transistor is in cutoff, the circuit can be considered to be R_L in series with the transistor OFF resistance, R_{OFF}. Hence,

$$I_{C0} = \frac{V_{CC}}{R_L + R_{\text{OFF}}}$$

and

$$R_{\text{OFF}} = \frac{15 \text{ V} - 20 \; \mu\text{A} \times 1 \text{ k}\Omega}{20 \; \mu\text{A}} = 749 \text{ k}\Omega.$$

8. Substituting in Eq. (2–121), $V_p = 15$ V, $V_\gamma = 0.6$ V, and $V_{BB} = 19$ V, we have

$$\eta = \frac{15 - 0.6}{19} = 0.758.$$

9. a) At the peak point, $V_B = I_p R_2 + V_{BB}$, where I_p is the peak-point current. From Eq. (2–120),

$$V_p = V_\gamma + \eta V_{BB} = V_\gamma + \eta V_B - \eta I_p R_2.$$

The peak-point current I_p is

$$I_p = \frac{V_B}{R_{B1} + R_{B2} + R_2} \approx \frac{V_B}{R_{BB}}.$$

Hence

$$V_p = V_\gamma + \eta V_B - \frac{\eta V_B R_2}{R_{BB}}.$$

b) If R_2 is chosen such that $R_2 \approx V_\gamma R_{BB}/\eta V_B$, then $V_p = \eta V_B$ and will be highly independent of temperature.

10. Use $v_L = v_u R_L/(R_L + R_e + R_s)$, where R_s is the switch resistance. For the closed switch,

$$v_L = \frac{5\ \text{V} \times 1\ \text{k}\Omega}{1\ \text{k}\Omega + 100\ \Omega + 50\ \Omega} = 4.35\ \text{V}.$$

For the open switch,

$$v_L = \frac{5\ \text{V} \times 1\ \text{k}\Omega}{1\ \text{k}\Omega + 100\ \Omega + 100\ \text{k}\Omega} = 0.049\ \text{V}.$$

11. The current in each diode is composed of two components, one due to the supply voltage $\pm V_c$ and the other due to the input signal v_u. If each diode is considered to be ideal (zero resistance), the current due to the supply voltage in Fig. 2–132 is $V_c/2R_c$. This current is in the forward direction in each diode. The current due to v_u is in the reverse direction in diodes D_1 and D_4. The reverse current in D_1 is $v_u/R_c + v_u/2R_L$, while that in D_4 is $v_u/2R_L$. Hence the larger reverse current is in D_1. The forward current $V_c/2R_c$ must thus be greater than $v_u/R_c + v_u/2R_L$. Hence

$$(V_c)_{\min} = \frac{v_u}{R_c} \times 2R_c + \frac{v_u \times 2R_c}{2R_L} = v_u \left(2 + \frac{R_c}{R_L} \right).$$

12. a) Substituting into the equation preceding Eq. (2–127), $b_\lambda = 0.3$, $Q_e = 1.603 \times 10^{-19}$ C, and $P_k = 10^{12}$ photons/sec, we obtain

$$-I_p = 0.3 \times 1.603 \times 10^{-19} \times 10^{12} = 4.81 \times 10^{-8}\ \text{A}.$$

b) Substituting $I_p = -4.81 \times 10^{-8}$ A, and $h_{FE} = 50$ into Eq. (2–127), we obtain

$$I_E = -4.81 \times 10^{-8}(51) = -2.45 \times 10^{-6}\ \text{A}.$$

13. a) For $v_{in} = +5$ V, $v_s = +5$ V $\times \dfrac{5 \text{ k}\Omega}{5 \text{ k}\Omega + 50 \text{ }\Omega} = 4.95$ V.

For $v_{in} = -5$ V,

$$v_o = -5 \text{ V} \times \frac{5 \text{ k}\Omega}{5 \text{ k}\Omega + 500 \text{ k}\Omega} = -0.049 \text{ V.}$$

b) For $v_{in} = +5$ V,

$$v_o = +5 \text{ V} \times \frac{500 \text{ k}\Omega}{500 \text{ k}\Omega + 5 \text{ k}\Omega} = 4.95 \text{ V.}$$

For $v_{in} = -5$ V,

$$v_o = -5 \text{ V} \times \frac{50 \text{ }\Omega}{5 \text{ k}\Omega + 50 \text{ }\Omega} = -0.049 \text{ V.}$$

14. See Fig. S–14 for three clipping circuits which will solve the problem.

15. When $v_{in} = +5$ V, the diode is forward biased. The current I in the series circuit is

$$I = \frac{5 \text{ V} + 2 \text{ V}}{1 \text{ k}\Omega + 50 \text{ }\Omega} = 6.66 \text{ mA.}$$

The output voltage v_o is

$$v_o = 6.66 \text{ mA} \times 50 \text{ }\Omega - 2 \text{ V} = -1.67 \text{ V.}$$

When $v_{in} = -5$ V, the diode is reverse biased. Now

$$I = \frac{5 \text{ V} - 2 \text{ V}}{5 \text{ M}\Omega + 1 \text{ k}\Omega} = 6 \times 10^{-7} \text{ A}$$

and $v_o = -2 \text{ V} - 5 \text{ M}\Omega \times 6 \times 10^{-7} \text{ A} = -5$ V.

16. a) Substituting in Eq. (2–130) $v_{in} = 1.5$ V, $R = 250$ kΩ, and $C = 0.5$ μF, we have

$$\frac{dv_o}{dt} = -\frac{1.5 \text{ V}}{250 \text{ k}\Omega \times 0.5 \text{ μF}} = -12 \text{ V/sec.}$$

b) 1 sec.

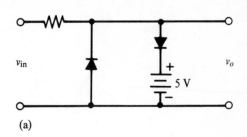

(a)

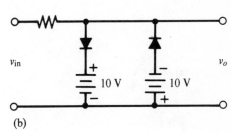

(b)

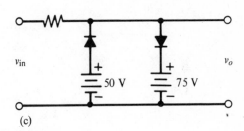

(c)

Fig. S–14

SOLUTIONS TO PROBLEMS IN SECTION 2–4

1. Since by Eq. (2–133) the secondary voltage is directly proportional to the primary voltage, it will decrease by the same percentage, or $(115 - 95)/115 = 17\%$.

2. Substituting in Eq. (2–133) $v_s = 220$ V, $v_p = 4400$ V, we have $N_s/N_p = 220/4400 = 1/20$.

3. When the capacitor is present, it charges to $+1.4 \times 10 = 14$ V during the transformer's positive half-cycle. During the negative half-cycle, the diode must withstand the sum of the negative transformer voltage and the positive voltage on the capacitor. Thus $\text{PIV} = 2 \times 1.4 \times 10 = 28$ V.

4. a) By the same reasoning as in problem 3, $\text{PIV} = 2 \times 1.4 \times 5 = 14$ V.

   b) During the positive half-cycle the capacitor charges to $5 \times 1.4 = 7$ V. Two diodes conduct, while the other two must each withstand 7 V of reverse voltage. Likewise during the negative half-cycle the other two diodes must withstand 7 V. Hence the PIV rating is 7 V. The transformer required for the bridge circuit must have a 5 V rms secondary, whereas the transformer in part (a) must have a 10 V rms secondary with a center tap.

5. a) $V_{dc} = 1.4 \times 25$ V $= 35$ V.

   b) 35 V.

   c) 2×35 V $= 70$ V.

6. a) From the discussion preceding Eq. (2–135), it can be shown that

   $$V_{dc} = 1.4\,V_{rms}[1 - 1/(2fCR_L)].$$

   Hence

   $$V_{dc} = 1.4 \times 40\,[1 - 1/(2 \times 60 \times 10 \times 10^{-6} \times 5 \times 10^3)] = 46.7 \text{ V}.$$

   b) Since $I_{dc} = V_{dc}/R_L$ and $V_{dc} = 48$ and $R_L = 5$ kΩ, we have $I_{dc} \approx 47/5000 = 9.3$ mA.

   c) Substituting in Eq. (2–137) $f = 60$ Hz, $C = 10$ µF, $R_L = 5$ kΩ, we find for the ripple factor

   $$r = \frac{1}{2 \times 1.75 \times 60 \times 10 \times 10^{-6} \times 5 \times 10^3} = 0.095.$$

From Eq. (2–138)

$$V_{\text{ripple}} = rV_{\text{dc}} = 0.095 \times 47 = 4.4 \text{ V}_{\text{rms}}.$$

7. For V_{dc} use Eq. (2–135). With $C = 100 \text{ μF}$,

$$V_{\text{dc}} = 1.4 \times 18 \, [1 - 1/(2 \times 120 \times 100 \times 10^{-6} \times 100)] = 14.7 \text{ V}.$$

Note that Eq. (2–135) is not reliable here because of the heavy load. With $C = 500 \text{ μF}$,

$$V_{\text{dc}} = 1.4 \times 18 \, [1 + 1/(2 \times 120 \times 500 \times 10^{-6} \times 100)] = 23.1 \text{ V}.$$

With $C = 1000 \text{ μF}$, $V_{\text{dc}} = 24 \text{ V}$. For the ripple voltages use Eqs. (2–137) and (2–138). For $C = 500 \text{ μF}$,

$$r = \frac{1}{2 \times 1.75 \times 120 \times 500 \times 10^{-6} \times 100} = 0.048,$$

$$V_{\text{ripple}} = 0.048 \times 23.1 = 1.11 \text{ V}_{\text{rms}}.$$

For $C = 1000 \text{ μF}$, $r = 0.024$, $V_{\text{ripple}} = 0.6 \text{ V}_{\text{rms}}$.

8. The current through the Zener diode $i_Z = P/v_Z = 100 \text{ mW}/5.1 \text{ V} \approx 20 \text{ mA}$. The load current $i_L = v_Z/R_L = 5.1 \text{ V}/100 = 51 \text{ mA}$. Hence the maximum supply current $i = i_Z + i_L = 71 \text{ mA}$. To stay within this limit, $R_s = V/i = 10 \text{ V}/71 \text{ mA} = 141 \text{ Ω}$.

9. From Eq. (2–143) the maximum current is

$$I = \sqrt{P/R} = \sqrt{0.5 \text{ W}/50} = 0.1 \text{ A}.$$

From Eq. (2–144) the maximum voltage is

$$V = \sqrt{PR} = \sqrt{0.5 \times 50} = 5 \text{ V}.$$

SOLUTIONS TO PROBLEMS IN SECTION 2–5

1. a) For the amplifier of Fig. P–1a, $v_o = 40v_{\text{in}}$, $A_V = 40$, while for the amplifier of Fig. P–1b, $i_o = 50i_{\text{in}}$, $A_i = 50$.

   b) The power gain A_P is the ratio of the output power to the input power, $A_P = P_o/P_{\text{in}}$. From Eq. (2–146) for the voltage amplifier, $P_o = (\mu v_{\text{in}})^2/R_L$ and $P_{\text{in}} = v_{\text{in}}^2/R_{\text{in}}$. Thus

$$A_P = \frac{(40v_{\text{in}})^2(10^6)}{v_{\text{in}}^2 (10^6)} = 1600.$$

For the current amplifier of Fig. P–1b, $P_o = (\beta i_{in})^2 R_L$, and $P_{in} = (i_{in})^2 R_{in}$. Thus

$$A_P = \frac{(50i_{in})^2 \times 10 \text{ k}\Omega}{i_{in}^2 \times 1 \text{ k}\Omega} = 25{,}000.$$

c) $v_o = i_o R_L = 50i_{in} \times 10 \text{ k}\Omega$, $v_{in} = i_{in} R_{in} = i_{in} \times 1 \text{ k}\Omega$,

$$A_v = \frac{v_o}{v_{in}} = \frac{50i_{in} \times 10 \text{ k}\Omega}{i_{in} \times 1 \text{ k}\Omega} = 500.$$

2. a) Substituting $g_m = 2 \times 10^{-3} \ \Omega^{-1}$ and $R_d = 10$ kΩ into Eq. (2–152), we have $A_v = (2 \times 10^{-3})(10 \times 10^3) = 20$.

   b) Since $v_{ds} \approx 20v_{gs}$, $v_{ds} \approx 20$ V. The current $v_{ds}/r_{ds} \approx 20 \text{ V}/100 \text{ k}\Omega$ $= 2 \times 10^{-3}$ A. Hence v_{ds}/r_{ds} is 10% of the main current.

3. The small-signal model would be as in Fig. 2–201b, with $v_{gs} = 1$ V, $g_m = 2 \times 10^{-3} \ \Omega^{-1}$, $r_{ds} = 100$ kΩ, and $R_d = 10$ kΩ. This model differs from the ideal voltage amplifier in that the input resistance is not infinite and the output resistance is not zero. However, the JFET amplifier is a fairly good approximation to the ideal voltage amplifier at low signal frequencies.

4. From the slopes of the characteristic curves in Fig. 2–203, it is found that $h_{fe} \approx 25$, $h_{ie} \approx 700$ Ω, $h_{re} \approx 6 \times 10^{-4}$, $h_{oe} \approx 50 \times 10^{-6} \ \Omega^{-1}$.

5. a) For A_v, use Eq. (2–159):

$$A_v = \frac{-25 \times 1.5 \times 10^3 \ \Omega}{(700 \ \Omega \times 50 \times 10^{-6} \ \Omega^{-1} - 25 \times 6 \times 10^{-4}) \times 1.5 \times 10^3 \ \Omega + 700 \ \Omega}$$

$$A_v = \frac{-3.75 \times 10^4}{(0.035 - 0.015) \times 1.5 \times 10^3 + 700} = -51.$$

For A_i, use Eq. (2–161):

$$A_i = \frac{-25}{50 \times 10^{-6} \ \Omega^{-1} \times 1.5 \times 10^3 \ \Omega + 1} = -23.2.$$

For A_p, use $A_p = A_v \times A_i = 51 \times 23.2 = 1183.$

b) To find R_{in}, use Eq. (2–164):

$$R_{in} = \frac{700 \ \Omega + (0.02) \times 1.5 \times 10^3 \ \Omega}{1 + 50 \times 10^{-6} \ \Omega^{-1} \times 1.5 \times 10^3 \ \Omega} = 679 \ \Omega.$$

To find R_o use Eq. (2–166):

$$R_o = \frac{700\ \Omega + 2 \times 10^3\ \Omega}{0.02 + 50 \times 10^{-6}\ \Omega^{-1} \times 2 \times 10^3\ \Omega} = 22.5\ \text{k}\Omega.$$

6. The small-signal model would be as in Fig. 2–207 with the appropriate values. This differs in several ways from the ideal current amplifier. First, the input resistance is not zero (see problem 5b above). Second, the output resistance is finite and dependent on h_{oe} or $h_{ie}/\Delta h_e$ (see Eqs. 2–167 and 2–168). Third, the current gain is not strictly h_{fe}. Also, the output of the BJT influences the input through the voltage generator, $h_{re}v_{ce}$. However, the BJT can be considered an ideal current amplifier for many purposes.

7. a) For A_v, use Eq. (2–170):

$$A_v = 1 - \frac{dv_{BE}}{dv_u} = 1 - 0.05 = 0.95.$$

For A_i, use Eq. (2–173):

$$A_i = h_{fe} + 1 = 60 + 1 = 61.$$

For A_p, use $A_p = A_i \times A_v = 61 \times 0.95 = 58.$

   b) R_{in} can be found from Eq. (2–177) with $R_E = 5\ \text{k}\Omega$ and $h_{fe} = 60$:

$$R_{\text{in}} = 5 \times 10^3\ (60 + 1) = 305\ \text{k}\Omega.$$

R_o can be found from Eq. (2–178) with $R_u = 100\ \Omega$ and $h_{fe} = 60$:

$$R_o = \frac{100\ \Omega}{61} = 1.6\ \Omega.$$

8. a) Substitute into Eq. (2–180)

$$\frac{\Delta v_o}{\Delta v_d} = \frac{5\ \text{V}}{250\ \mu\text{V}} = 20{,}000 \quad \text{and} \quad \frac{\Delta v_o}{\Delta v_{\text{cm}}} = \frac{1\ \text{V}}{500\ \text{mV}} = 2.$$

   Then

$$\text{CMRR} = \frac{20{,}000}{2} = 10{,}000.$$

   b) Use Eq. (2–181) to find CMRR in dB $= 20 \log 10^4 = 80.$

9. To measure a 50 mV signal to 1% accuracy, we need to measure to 0.5 mV. With a CMRR of 10^4 the maximum common mode voltage which can be tolerated is 0.5 mV $\times 10^4 = 5$ V.

10. Since R_L and R_o form a voltage divider for μv_{u1},

$$v_o = \mu v_{u1} \frac{R_L}{R_L + R_o} = 1000 v_{u1} \times \frac{2 \text{ k}\Omega}{2.2 \text{ k}\Omega} = 909 v_{u1}.$$

Hence $A_v = v_o/v_{u1} = 909$. To find A_i, we need i_{in} and i_o. The input current $i_{\text{in}} = v_{u1}/R_{\text{in}} = v_{u1}/50 \text{ k}\Omega$. The output current i_o is

$$i_o = \frac{\mu v_{u1}}{R_o + R_L} = \frac{1000 v_{u1}}{2.2 \text{ k}\Omega}.$$

Hence

$$A_i = \frac{i_o}{i_{\text{in}}} = \frac{1000 v_{u1}/2.2 \text{ k}\Omega}{v_{u1}/50 \text{ k}\Omega} = 22{,}727.$$

Since $A_p = A_v \times A_i$, $A_p = 909 \times 22{,}727 = 20.6 \times 10^6$.

11. a) The output power $P_o = v_o^2/R_L$, or since $v_o = \mu v_{u1}[R_L/(R_L + R_o)]$,

$$P_o = \frac{\mu^2 v_{u1}^2 R_L^2}{(R_L + R_o)^2 R_L} = \frac{\mu^2 v_{u1}^2 R_L}{(R_L + R_o)^2}.$$

The maximum power occurs when $dP_o/dR_L = 0$. This is found where $R_L = R_o$. The maximum power is then

$$P_{o(\text{max})} = \frac{1000^2 \times (5 \times 10^{-3})^2}{4 \times 200} = 31 \text{ mW}.$$

b) The input power $P_{\text{in}} = v_{u1}^2/R_{\text{in}} = (5 \times 10^{-3})^2/5 \times 10^4 = 5 \times 10^{-10}$ W.

$$A_p = \frac{P_o}{P_{\text{in}}} = \frac{31 \times 10^{-3} \text{ W}}{5 \times 10^{-10} \text{ W}} = 6.2 \times 10^7.$$

SOLUTIONS TO PROBLEMS IN SECTION 2–6

1, a) The bias current will exist in the feedback resistor R_f, since the input resistor is connected between the circuit common and virtual common.

b) Since there is no current in R_{in}, $v_o = i_{b(-)}R_f = 15 \times 10^{-9} \text{ A} \times 100 \times 10^3 \ \Omega = 1.5 \text{ mV}$. The sign of v_o will depend on the direction of $i_{b(-)}$.

2. The circuit as shown is a follower with gain to the offset voltage. Hence $v_o = v_{\text{off}}(1 + R_f/R_{\text{in}}) = 10^{-3} \text{ V} (1 + 100) = 101 \text{ mV}$.

3. a) The output voltage should be

$$v_o = \frac{-R_f}{R_{in}} v_{in} = -100 \times 50\,\text{mV} = -5\,\text{V}.$$

For 1% or less output error, 50 mV at the output due to $i_{b(-)}$ is all that can be tolerated. Hence,

$$i_{b(-)} = \frac{v_o}{R_f} = \frac{50\,\text{mV}}{100\,\text{k}\Omega} = 5 \times 10^{-7}\,\text{A}.$$

b) Again 50 mV error is all that can be tolerated. So

$$v_{off} = \frac{v_o}{\left(1 + \dfrac{R_f}{R_{in}}\right)} = \frac{50\,\text{mV}}{101\,\text{k}\Omega} = 495\,\mu\text{V}.$$

4. Substituting into Eq. (2–201), we have

$$v_o = -\left(-2\,\text{V} \times \frac{10\,\text{k}\Omega}{5\,\text{k}\Omega}\right) \pm 10^{-6}\,\text{A} \times 10\,\text{k}\Omega$$

$$\pm 2 \times 10^{-3}\,\text{V}\left(1 + \frac{10\,\text{k}\Omega}{5\,\text{k}\Omega}\right)$$

$$= +4\,\text{V} \pm 10^{-2}\,\text{V} \pm 6 \times 10^{-3}\,\text{V} = +4\,\text{V} \pm 16\,\text{mV}.$$

So the percentage error is 0.4%.

5. Use Eq. (2–202) as follows:

$$v_o = \pm 1\,\text{mV/day} \times \frac{10\,\text{k}\Omega}{5\,\text{k}\Omega} \pm 1\,\mu\text{A} \times 10\,\text{k}\Omega$$

$$\pm 1\,\text{mV}\left(1 + \frac{10\,\text{k}\Omega}{5\,\text{k}\Omega}\right)$$

$$= \pm 2\,\text{mV/day} \pm 10\,\text{mV/day} \pm 0.3\,\text{mV/day}.$$

So the maximum drift rate is 12.3 mV/day.

DIGITAL AND ANALOG DATA CONVERSIONS

Module 3

Introduction

When electronic devices are used to aid a measurement process, the quantity to be measured is converted by an input transducer into an electrical signal which is related to the measured quantity in some known way. Many popular input transducers convert the measured parameter into a related charge, voltage, current, or resistance. The resulting electrical signal or resistance is then measured by the techniques described in Modules 1 and 2, *Electronic Analog Measurements and Transducers* and *Control of Electrical Quantities in Instrumentation.*

In many input transducer applications an electrical quantity is produced that is directly related to the measured parameter. However, there is an increasing number of input transducer applications for which the measured parameter is represented by the manner of the *variation* of the signal, such as the amplitude, rate, time, or number of the variations in the signal quantity. An example is a Geiger tube, which produces a pulse of current for each measured particle of radiation. There are, in fact, a great many ways by which measurement information (data) can be represented (encoded) as electrical signal variations. Each method of electrically encoding the data has advantages in particular applications. Many electronic data encoding techniques have been impractical in common instruments or have been associated primarily with special areas of electronics such as communications. However, integrated circuits, which provide an increasing variety of complex electronic functions inexpensively, have opened up many new encoding methods to scientific instrumentation. Additional impetus to adopt new data coding methods has come from the desire to use convenient "digital" readouts and to combine scientific measurement devices with modern data processing equipment.

The many ways in which data can be encoded electrically in an instrument can be grouped in three major types of electrical data domains: analog, time, and digital. The first section of this module introduces the characteristics of signals in the time and digital domains and illustrates the application of time and digital domain encoding and conversions in several laboratory instruments. Sections two, three, and four develop the applications of analog, time, and digital data encoding, respectively, in scientific instrumentation. Included in the discussion of each of the three groups of electrical data domains are some representative input transducers and generators which produce signals in that domain, the techniques of measuring the quantity encoded in the signal, and the related interdomain conversions. The module ends with several examples of the applications of digital data encoding in measurement and control instrumentation.

Electrical Data Domains   Section 3–1

Each characteristic or property used to represent measurement data is a **data domain.** Devices which convert the data representations from one characteristic or property to another are called **data domain converters.** Electronic measurement systems employ a sequence of data domain conversions beginning with the physical or chemical quantity being measured and ending with a readout from which a number can be obtained. An input transducer is used to convert the measured quantity or property into an electrical signal, some characteristic of which is related to the measured quantity. In general, it is necessary to modify this signal or combine it with other data before it is suitable for operating the readout device from which a number related to the desired quantity can be obtained. In the course of modifying the electrical signal, the way, or electrical domain, in which the data are encoded in the signal may be changed several times.

Understanding the functions of the electronic parts of the instrument is greatly simplified if the attention is kept on the changes in the domains in which the data are encoded as they proceed through the instrument rather than on the details of the electronic circuits which perform the data domain conversions. This is simply a study of the measurement process, which is sufficient for many purposes. Once the specific data domains used in an instrument are known, a data flow chart or block diagram for the instrument can be drawn in which the data domain conversion function of each section of the instrument is identified. The study of the electronic circuits themselves can then be undertaken in terms of the data domain conversions they perform, a manner of study now very relevant to the needs of the experimenter. Thus a knowledge of the electrical data domains is prerequisite to a study of the data domain converters, the building blocks of scientific instruments.

3–1.1 ELECTRICAL DOMAIN CLASSES

The study of data domains is simplified by categorizing into three major groups the ways in which measurement data are represented by an electrical signal. These groups are: **analog** (A), in which the magnitude of an electrical quantity (charge, current, voltage, or power) is related to the data; **time** (T), in which the time relationship between signal level changes is related to the measurement data; and **digital** (D), in which an integer number is represented by binary level signals. Each of these groups includes many data domains which are related by the general nature of the data encoding. The general characteristics of each group and examples of data domains within each group are given in this section.

Analog Domains

The measurement data in the analog domain are represented by the magnitude of one of the four electrical quantities charge, current, voltage, or power. Most input transducers used today convert the measurement data from a physical or chemical domain (P) to one of the analog (A) domains. Among the many examples of P-to-A converters described in Modules 1 and 2 are photodetectors which convert light intensity into an electrical charge or current, a thermistor bridge which converts temperature difference to an electrical voltage, and a flame ionization detector which converts the concentration of ionizable molecules in a gas into an electrical current.

At each instant in time, the measured quantity is represented by an electrical signal amplitude. Because the smallest unit of charge, the charge on an electron, is so small compared to the total charge stored or transferred by most signal sources, the amplitude of an analog signal is essentially continuously variable. That is to say, the amplitude of an analog signal is variable in essentially infinitesimal increments. By means of the measurement techniques described in Module 1, the amplitude of an analog signal can be measured continuously with time. When continuous measurements are made, the variations in the signal amplitude may be plotted against time, wavelength, magnetic field strength, temperature, or other experimental parameters as shown in Fig. 3–1. From such plots, additional information can often be obtained from a correlation of amplitudes measured at different times. Such information would include simple observations like peak height, peak position, number of peaks, or more complex correlations such as peak area, peak separation, signal averaging, and Fourier transformation. The techniques of correlating data taken at different times must be distinguished from the techniques of converting the data taken at any given time into a usable form. This module includes a discussion of the techniques of measuring the amplitudes of analog signals

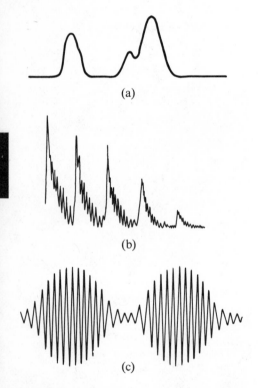

Fig. 3–1 Variation of analog signal sources. (a) Flame ionization detector current vs. time in gas chromatography. (b) Photomultiplier tube current vs. wavelength in spectroscopy. (c) Antenna voltage vs. time for a single AM radio signal.

at particular instants in time. Once the required instantaneous data points have been converted to a useful form and stored, the correlation of the resulting data is handled by various processing techniques.

Signals in the analog domains are susceptible to electrical noise sources contained within or induced upon the circuits and connections of the instrument and input transducer. The resulting signal amplitude at any instant is the sum of the data signals and the noise signals. Data encoded as signals in either a time or digital electrical domain are less affected by electrical noise. The problems of noise in electrical signals are discussed in Module 4, *Optimization of Electronic Measurements.*

Time Domains

In the time (T) domains the measurement data are contained in the time relationship of the signal variations, not the amplitude of the variations. Typical T domain signals are shown in Fig. 3–2. The most common T domain signals represent the data as the **frequency** of a periodic waveform (a), the time duration of a pulse or **pulse width** (b), or as the **time** or average **rate** of pulses (c). These are logic level signals; i.e., their signal amplitude is either in the HI (or **1**) logic level region or the LO (or **0**) logic level region. The data are contained in the time relationship between the logic level transitions, such as the time between successive **0→1** transitions (**period**), the time between **0→1** and **1→0** transitions (**pulse width**), and the number of **0→1** transitions per unit time (**frequency** or **rate**). Examples of converters producing signals in the T domains from physical domain quantities are as follows: a crystal oscillator that produces a temperature-dependent frequency because of the temperature characteristics of the quartz crystal, an oscillator which has an output frequency dependent on the value of the capacitance used in the oscillator circuit, and the Geiger tube which converts the level of radioactivity to a pulse repetition rate. An example of a domain converter between an A domain and a T domain is a voltage-controlled oscillator or voltage-to-frequency converter which provides an output frequency related to an input voltage.

The greater the slope (dV/dt) of the signal through the logic level threshold region, the more precisely the transition time can be defined. Because the data encoding in T domain signals is less amplitude dependent than in A domain signals, it is less affected by electrical noise sources. A common example of this is the FM radio signal (frequency domain) vs. the more noise-susceptible AM radio signal (an amplitude domain). The greater the difference between the average **1** or **0** signal level amplitude and the logic level threshold, the less susceptible the signal will be to noise-induced error. In these respects, the signal shown in Fig. 3–2b is better than those of Figs. 3–2a and 3–2c. The logic level transitions of signals

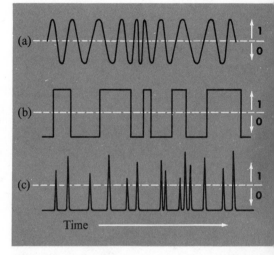

Fig. 3–2 Time domain signals. (a) Frequency. (b) Pulse duration. (c) Pulse rate.

like Figs. 3–2a and 3–2c are generally sharpened to those like Fig. 3–2b before the significant time relationship is measured. This is accomplished by a **comparator** or **Schmitt trigger circuit.**

Signals in the T domains, like signals in the A domains, are continuously variable, since the frequency or pulse width can be varied infinitesimally. However, the encoded variable of signals in the T domains cannot be measured continuously with time or at any instant in time. The minimum time required for conversion of T domain signals to any other domain is necessarily at least one period or one pulse width.

Digital Domain

In the digital domain, the measurement data are contained in a two-level signal (HI/LO, **1/0**, etc.) which is coded to unambiguously represent a specific integer (or character). The signal in the digital domain may be a coded series of pulses in one channel (serial form) or a coded set of signals on simultaneous multiple channels (parallel form). Since a digitally coded signal represents a specific number exactly, the data are actually in numerical form. The unit for the signal characteristic encoding the data in the digital domain is thus *number*. Since this is true for all the many varieties of digital encoding, all digital signals are in a single domain called **digital.** *Data in the digital domain need no further data domain transformations to complete the measurement; it is only necessary to decode the signal and display the number.* Representative digital signal waveforms are shown in Fig. 3–3.

The **count** waveform (Fig. 3–3a) is a series of pulses with a clearly defined beginning and end. The number represented is the number of pulses in the series. The count waveform of Fig. 3–3a might represent, for instance, the number of photons of a particular energy detected during a single spark excitation. The count form is simple but not very efficient. To provide a resolution of one part per thousand, the time required for at least 1000 pulses to occur must be allowed for each series of pulses.

The most efficient **serial digital signal** is the **binary-coded serial signal** shown in Fig. 3–3b. In this signal, each pulse time in the series represents a different binary digit, or **bit** position in a binary number. The appearance of a pulse at a time position indicates a 1; the absence of a pulse, a 0. The data are not represented by the exact time of the pulse as in the T domains, but by the signal logic level present within a given time range. The binary number represented by the waveform shown is 10110101, which is 181 in the decimal system. A series of n pulse times has a resolution of one part in 2^n. Thus a 10-bit series has a resolution of one part in $2^{10} = 1024$, and a 20-bit series has a resolution of better than one part per million.

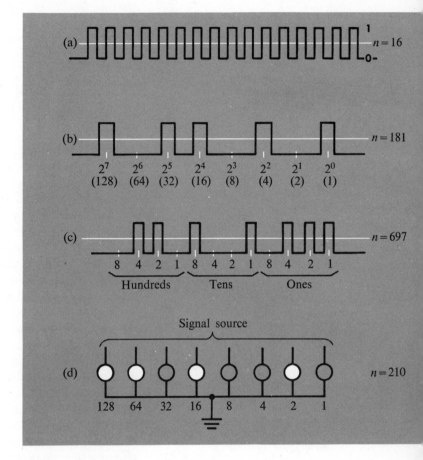

Fig. 3–3 Digital domain signal forms. (a) Count.
(b) Binary-coded serial. (c) Binary-coded decimal.
(d) Eight-bit parallel.

The binary-coded decimal serial form is somewhat less efficient but very convenient where a decimal numerical output is desired. Each group of four bits represents one decimal digit in a number. Twelve bits can thus represent three decimal digits and provide a resolution of one part in one thousand.

A **parallel digital signal** uses a separate wire for each bit position instead of a separate time on a single wire. The principal advantage of parallel digital data connections is speed. An entire **word** (group of bits) can be conveyed from one circuit to another in the time required for the transmission of one bit in a serial connection. An eight-bit parallel data source is shown in Fig. 3–3d connected to indicator lights to show the simultaneous appearance of the data logic levels on all eight data lines. Binary coding (shown), binary-coded decimal coding, and others are used for parallel digital data. Parallel data connections are used in all modern,

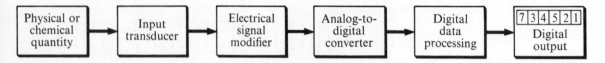

Fig. 3–4 A common form of digital measurement.

fast computers. Serial data connections are often used for telemetry and slow computer peripherals such as teletypes.

Digital Measurement. A common form of digital measurement system is shown in block form in Fig. 3–4. At some point after the measurement data have been converted into electrical amplitudes, an analog-to-digital data domain converter is used. If the digitization has been performed to take advantage of the great accuracy, power, and versatility of digital data processing, that will be done next. Finally, the numerical binary level signal is decoded into a number which is displayed, or recorded.

Because so many advantages are claimed for digital techniques, many techniques have claimed to be "digital." In fact, any type of device which has dial settings or outputs which are numerals in a row is likely to be called digital. By that standard a decade resistance box would be a digital instrument. Since the end result of any measurement is a number, all instruments could be called digital, but the meaning of the word in that sense becomes trivial. Some confine the use of the words "digital instrument" to those instruments which contain binary level electronic logic circuits of the type developed for digital computers. As was shown in the discussion of time domains, it is common for measurement data to be represented by a binary level electronic signal and still not be "digitized" or numerical. Therefore, for our purposes, a **digital measurement** or instrument will be defined as one that uses a digital domain electrical signal to represent the measurement data somewhere within the instrument.

3–1.2 DATA DOMAIN CONVERSIONS

The data domains map that was first presented in Module 1, *Electronic Analog Measurements and Transducers,* can now be expanded to include the time domains and digital domain which were described in the previous section. The resulting map is shown in Fig. 3–5. The specific domain of any signal can be determined by stating the units for that signal characteristic which represents the data. For instance, the characteristic of the signal from a photodiode which is related to the light intensity is the amplitude of the current. Therefore the signal is in the analog current domain. The frequency of the signal from a temperature-dependent crystal oscillator contains the temperature information. The units are cycles per *time*. Thus the data are encoded in the time frequency domain. If each

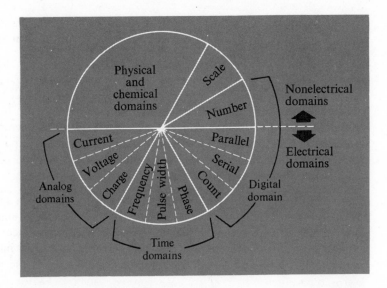

Fig. 3–5 Data domains map.

time a laboratory animal presses a lever, a simple circuit produces a voltage pulse, the information lies in the number of pulses occurring during a given phase of the experiment. The unit is just *number* and the domain is clearly digital.

If the **units** of the data-encoding characteristic of a signal can be used to define the data domain for that signal, then *signals in each different domain represent the data by different units.* A data domain converter is required whenever there is a change in the units used to represent the signal. You will recall from Module 1 that a data domain conversion has all the characteristics of a measurement. Therefore a data domain converter must have a **difference detector** and a **reference standard** with the same characteristic (units) as the input signal, and either the difference detector and/or the reference standard provides the converted output. It is very useful in assessing measurement errors to analyze in this way each domain converter involved in the measurement.

One often encounters circuits which alter the measurement signal in some way without changing the units of data representation. Such circuits include a voltage amplifier (input and output units are both volts) and a digital counter (count input and coded parallel output; both units are number). Such devices perform **intradomain conversions** and do not necessarily have all the measurement characteristics or introduce error. Since all digital signals have the units of number, they are all in the same domain and devices converting one form of digital signal to another are potentially error-free intradomain converters. This is one of the primary attractions of the digital encoding of measurement data.

In the domains map of Fig. 3–5 the analog power domain is not included because it would complicate the map and is not used in any of the discussions in this module.

Mapping Domain Conversion

To illustrate the application of the variety of electrical domains in scientific measurements, the domain conversions of several instruments will be described and mapped below. Note in these examples the possibly unexpected encountering of signals in the time domains and the distinctions between **interdomain** and **intradomain** conversions.

Digital pH Meter. A digital pH meter (Fig. 3–6) differs from an ordinary pH meter in that the meter is replaced by an analog-to-digital converter (ADC) and a digital display. An electrode transducer converts solution pH to a related voltage and thus performs a P-to-A interdomain conversion. A frequently used ADC for this application is the dual-slope

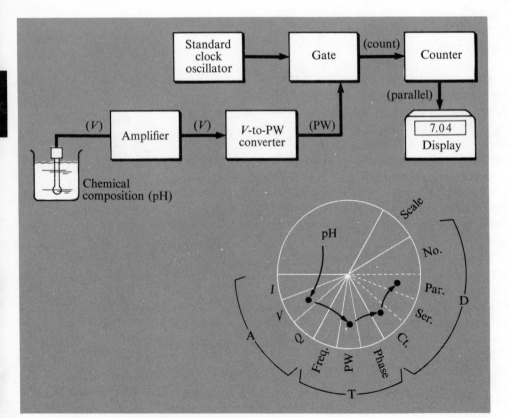

Fig. 3–6 Digital pH meter.

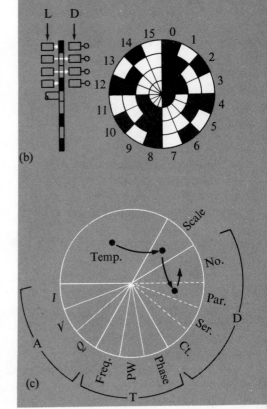

Fig. 3–7 Digital temperature measurement using a bimetallic coil and an encoded wheel. (a) Pictorial. (b) Encoded wheel detail. (c) Data domain path.

converter. As is often the case, this ADC does not convert directly from the A to D domains. The dual-slope circuit produces a pulse which has a duration proportional to the input signal voltage—i.e., a T pulse width signal. The pulse width is converted to a digital signal using the pulse to turn an oscillator ON and OFF, generating a count digital signal. The count signal is in turn "counted" or converted to a parallel digital signal for the display by a counter. From the domains map it is seen that three interdomain (P-to-V, V-to-PW, and PW-to-D) and two intradomain (count–to–parallel digital, and parallel digital–to–displayed number) conversions are involved in the measurement.

Digital Temperature Measurement.   Three approaches to a digital thermometer are compared here. In the first example, the temperature is converted to a shaft rotation by a bimetallic coil as shown in Fig. 3–7a. The shaft rotation is converted to a parallel digital signal by means of the encoded wheel illustrated in Fig. 3–7b. The wheel is divided into 16 radial segments and four concentric rings. Each radial segment has a combination of light and dark sections which is different from all the others. This combination is thus a code representing that specific position. In order to convert this code to an electrical digital signal, a row of light-photodetector pairs is placed radially from the shaft with their light path interrupted by the wheel. The photodetectors provide logic level outputs that depend on whether the wheel is opaque or transparent in that position. A four-bit parallel digital signal results. In this case, the position of the wheel is known within 1/16 of 360° or 22.5°. The temperature can then be known to within whatever temperature change causes a 22.5° shaft rotation. For greater accuracy, additional rings and photodetectors are required. The data domains path for this measurement is shown in Fig. 3–7c. Note the directness of the path and that the only electrical domain used is the digital domain. This is quite unusual, but as the use of the digital domain becomes more common, the invention of more input transducers with digital domain outputs is to be expected.

Instruments do not always follow the most direct domain path in providing the desired measurement. For instance, the direct digital converter described above might not be sufficiently accurate, fast, sensitive, or economical for the desired application. If there are no good interdomain converters available for direct conversion to the digital domain, an input transducer to another domain can be used, followed by additional interdomain conversions. For instance, the resistance of a thermistor or fine platinum wire is related to the temperature. When the thermistor is used in a Wheatstone bridge circuit, as shown in Fig. 3–8, a voltage is produced

Fig. 3–8 Thermistor digital thermometer.

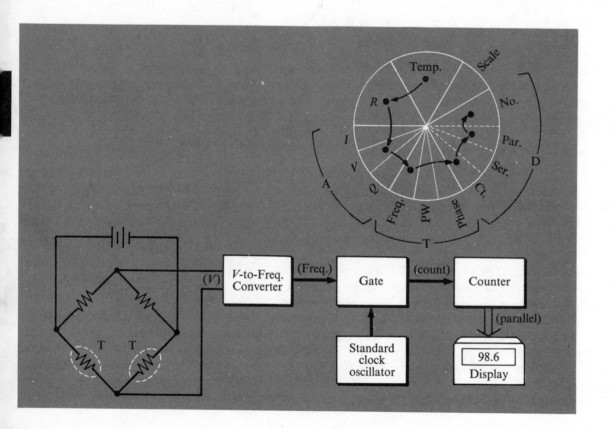

at the bridge output that is related to the temperature. A digital voltmeter (DVM) can then be used to convert the bridge voltage to a digital form. The particular digital voltmeter shown uses a voltage-to-frequency converter followed by a digital frequency meter. Thus, from the measured temperature to the DVM readout, as shown in Fig. 3–8, there are several interconversions: temperature–resistance–voltage–frequency–digital. Not the most direct route, indeed! It is possible to make an A-to-D conversion without traversing a T domain. Several ADC techniques are described in Section 3–4.

A transducer that converts temperature to frequency is the basis of an accurate instrument for temperature measurement. The input transducer is a quartz crystal with a resonant frequency that is linearly dependent on temperature. The quartz crystal is used in a crystal oscillator, the frequency of which is measured with a digital frequency meter. The measurement system, a temperature–frequency–digital instrument, is shown in Fig. 3–9. Because the frequency of the quartz crystal changes about 1 kHz/°C out of 28 MHz, the frequency f_t is mixed with a standard frequency f_s to provide a difference frequency $f_t - f_s$ that can be measured and calibrated more

Fig. 3–9 Crystal oscillator digital thermometer.

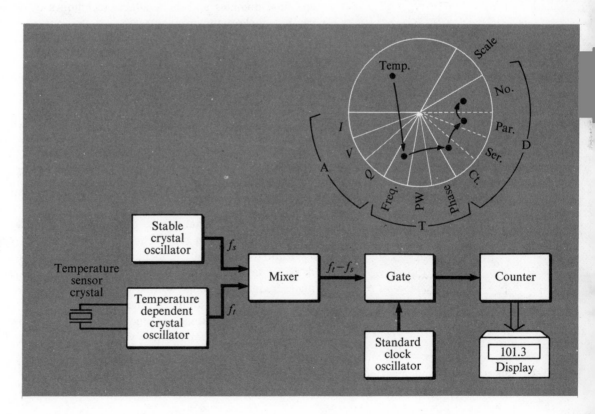

easily. This is an example of signal modification occurring within the frequency domain. Note that this digital thermometer requires one less interdomain conversion than that of Fig. 3–8. Whether this simplification would result in greater accuracy, however, depends on the accuracy of the converters involved in each case.

Even though the signals f_t or $(f_t - f_s)$ may be logic level signals, they are *not* in the digital domain. The information is contained in the frequency and the frequency is a continuous function of the temperature, not a specific number represented by the succession of logic levels. However, the standard clock opens the gate to the counter for a precise time interval, and it is this operation that provides an absolute number of pulses at the counter input. The signal at this point, being a specific number of logic level changes, is in the count digital form. The counter converts the count digital signal to a parallel digital signal at the counter outputs [binary-coded decimal (BCD) or binary]. Thus the clock and gate perform the T-to-D conversion. A clear understanding of the difference between the T and D domains is important to an understanding of the various forms, applications, and limitations of digital instrumentation. In addition, a careful analysis of the required initial and final domains and the available techniques of domain interconversion can reveal shortcuts in many cases.

Conversions of Varying Quantities

The examples used in the previous section are measurements of steady-state quantities which are not expected to vary perceptibly over the interval of measurement. When the time variation of quantities in the various data domains is considered, a third dimension (real time) needs to be added to the data domain map of Fig. 3–5, as shown in Fig. 3–10. Here each interdomain conversion is shown as a slice across the real-time continuum. If a quantity that varies continuously with time is to be converted to the digital domain, the resulting number can only be true for a specific instant in time. It is not possible, therefore, to make a truly continuous digital record of a varying quantity. What can be done is to measure the varying quantity at successive instants in time. The numerical result of each measurement is then stored sequentially in memory registers, or recorded on punched cards or paper tape, or by magnetic recording devices. If the measurements are made frequently enough so that the varying quantity changes only slightly between each measurement time, the digital record can quite accurately represent the amplitude-vs.-time behavior of the measured quantity. Of course, the maximum frequency of measurement is limited by the time required to convert the measured quantity into the digital domain and store or record it. Successive data domain conversions will be illustrated by the following three examples of digitized measurement systems.

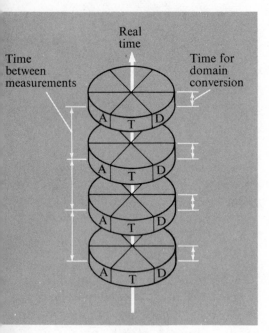

Fig. 3–10 Successive data domain conversions in real time.

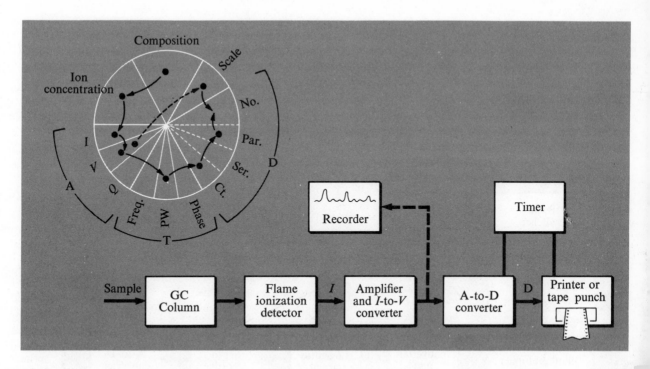

Gas Chromatography. A digital recording gas chromatograph is shown in Fig. 3–11. The components in the sample mixture are separated by the column; the result is a flow of gas of varying composition through the flame ionization detector. The flame in the detector converts the hydrocarbon concentration into an ion concentration, which is then converted by the electric field in the detector to an electrical current. The current signal is amplified and converted to a voltage amplitude suitable for A-to-D conversion and/or recording. The desired information is a record of detector current vs. time. The time relationship of the printed or punched values of the current amplitudes is generally obtained by having the successive A-to-D conversions performed at precise and regular time intervals. This is accomplished by the timer shown in the block diagram. The domain path for a single conversion (one time slice) is also shown.

Absorption Spectrophotometry. The block diagram of a digitized recording double-beam spectrophotometer is shown in Fig. 3–12. A narrow wavelength range of light from the light source is selected by the monochromator and passed on to the beam switcher and cell compartment. The beam switcher alternately directs the monochromatic beam through the reference and sample cells to the photomultiplier tube detector. This produces an electrical current (analog *I* domain) which has an amplitude alter-

Fig. 3–11 Digital recording gas chromatograph.

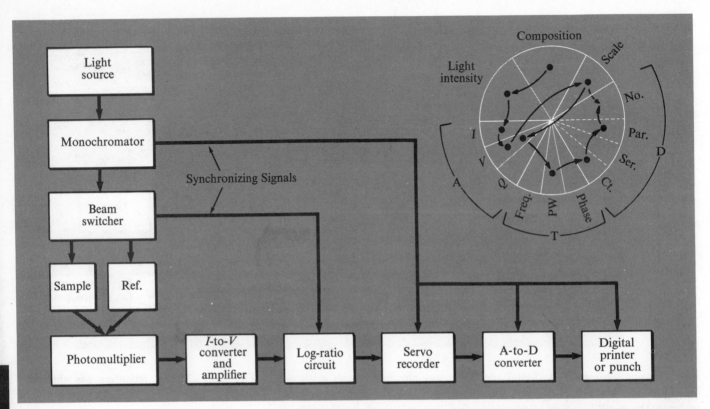

Fig. 3–12 Digitized double-beam spectrophotometer.

nating between sample and reference beam intensities P and P_0. The desired output signal for the recorder is absorbance $A = \log_{10}(P_0/P)$, which is accomplished in the log-ratio circuit. This circuit performs a correlation between signal levels measured at two different times. It must, therefore, have a memory and a synchronizing connection to the beam switcher. The recorder plots the absorbance vs. the wavelength of light from the monochromator. The recorder chart drive thus requires a synchronizing connection to the monochromator wavelength drive mechanism.

The output of this standard spectrophotometer was later digitized by putting a retransmitting pot assembly on the servo recorder. This converts the recorder pen position to a voltage amplitude which is connected to an ADC or A-to-D converter and printer or punch. Since the absorbance value recorded for precise *wavelength* (rather than time) intervals is desired, the ADC or A-to-D converter and printer are synchronized to the wavelength drive mechanism rather than to a timer.

The data domains map for the resulting instrument is shown in Fig. 3–12. It contains nine conversions: seven interdomain and two intradomain. The excursion into the scale position domain is unnecessary to the

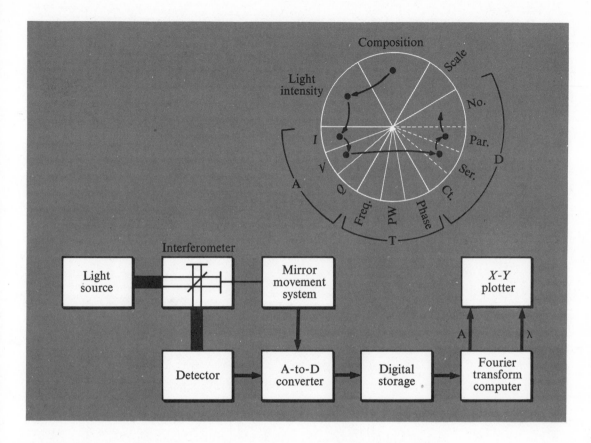

Fig. 3–13 Fourier transform spectrometer.

digitizing process and suggests that the ADC or A-to-D converter input would have been better connected to the log-ratio circuit output, if possible. It is interesting to note that if photon counting were used to measure the relative intensities, the number of interdomain conversions would be reduced to three and a digital log-ratio circuit would be required.

Fourier Transform Spectroscopy. A Fourier transform spectrometer is an example of a conceptually very simple data acquisition system connected to a complex data correlating and processing system. The block diagram is shown in Fig. 3–13. To obtain the interference pattern, the ADC converts the detector output signal at constant increments of movement of the reference beam mirror. Each piece of data is stored for use in the Fourier transform calculation. The data domains map as shown is complete for each piece of data as it is acquired and stored. A direct type of ADC has been assumed in this map. There are four interdomain and two intra-domain conversions in all. After the transform calculation (which involves an intercorrelation of all the measured data points) is complete, a plot of absorbance vs. wavelength (λ) can be made.

This example demonstrates that Fourier transformation from amplitude vs. time or distance to amplitude vs. frequency (or the reverse) is really an intercorrelation of analog signals which have already been "measured." It is essential to distinguish between the data domains involved in the methods of acquiring each data point and the methods of correlating and displaying groups of data points after they have been acquired.

Measurements of Varying Analog Signals

Section 3–2

Electrical quantities in the analog domains are voltage, charge, current, and power. The definitions of these quantities and the methods of measuring their magnitudes under steady-state conditions were discussed in Module 1. In this section, we will consider the measurement of these quantities as they vary with time. First the conditioning of the signal by conversion to another analog domain, amplification, or altering the form of the variation is described. Then the characterization of varying signals in terms of their frequency composition is presented. The frequency-dependent transfer functions of reactive elements in measurement circuits are seen to affect the form of the signal variation being measured. The applications of the oscilloscope, sampling devices, and various ac meters are discussed. Which of the measurement devices is best suited to a particular application is shown to depend on the form and rate of the variation with time, whether the waveform is repetitive (reoccurs in essentially the same form), and how the desired information is encoded in the amplitude-time variation. The discussion in this section is organized by the kind of waveform being measured. It is assumed, in every case, that the desired information is encoded in the *amplitude* of the signal at some time or times in its variation and *not* in the time of particular amplitude variations. Measurements of signals with time-encoded information are discussed in Section 3–3.

3–2.1 ANALOG SIGNAL CONDITIONING

Conditioning or modification of an analog signal is frequently required to bring it to the particular analog domain, signal level, or functional relationship with the measured quantity which is best suited for the readout (analog-to-number) system chosen. For instance, the output voltage of a combination pH electrode changes from approximately −420 mV to +420 mV as the solution pH changes from 0 to 14 respectively. If the solution pH is to be displayed directly by a digital voltmeter with a 0–2 V full scale range, the signal from the electrode must be offset (shifted positive) and amplified so that solutions of 0 to 14 pH produce voltages of 0 to 1.4 V at the digital voltmeter input. This section begins with a brief review of the analog interdomain conversions of current and charge to voltage which were described in Module 1. Several operational amplifier voltage amplification circuits are then reviewed with emphasis on the addition or subtraction of signals for offset or combination. Other mathematical operations which are very useful in signal conditioning, such as logarithmic response and multiplication, are also described.

Current-to-Voltage Conversion

The most common current-to-voltage converter device is a resistor. The resistor is introduced into the circuit in such a way that it is in the path of the current i_{in} to be measured. From Ohm's Law, the voltage across the resistor is then $i_{in}R$. This simple technique is not suitable when the introduction of the *IR* drop into the circuit affects the value of the current to be measured. The operational amplifier (OA) current follower circuit introduced in Module 1, Section 1–4.2, and shown in Fig. 3–14, is frequently used to convert the signal from a current-producing input transducer to a proportional voltage. The OA is a very high-gain dc amplifier which produces an output voltage v_o equal to $-Av_s$, where A is the amplifier gain (typically 10^3 to 10^7) and v_s is the potential difference between its − and + inputs. From the high gain, it follows that a value of v_s of only a few millivolts or microvolts is necessary to produce a v_o of several volts. The output voltage is applied to R_f to generate a current i_f which follows the input current i_{in}.

From the currents at point S, it can be seen that

$$i_{in} = i_f + i_b, \tag{3–1}$$

where i_b is the input bias current of the OA. By adding the voltage drop $i_f R_f$ across the feedback resistor to v_s, we obtain the following equation for v_o:

$$v_o = v_s - i_f R_f. \tag{3–2}$$

Substituting for i_f from Eq. (3–1) and v_s from the inverting amplifier relationship $v_o = -Av_s$ into Eq. (3–2), we obtain

$$v_o = -i_{in}R_f + i_bR_f - \frac{v_o}{A} \quad \text{or} \quad v_o = -R_f(i_{in} - i_b)\left(\frac{A}{1+A}\right) \tag{3–3}$$

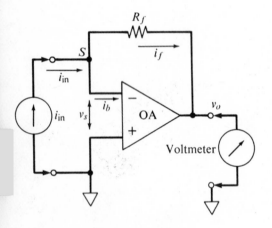

Fig. 3–14   OA current follower.

From either form of Eq. (3–3) it can be seen that when A is very large and i_b is very much less than i_{in}, v_o is proportional to i_{in}:

$$v_o \approx -i_{\text{in}}R_f \quad \text{(for } A \gg 1 \text{ and } i_b \ll i_{\text{in}}\text{).} \qquad (3\text{–}4)$$

Thus i_f follows i_{in} and produces an output voltage proportional to i_{in}.

The current follower output voltage is the same as would be produced across R_f if it were just placed in the current measurement circuit. However, the current follower has a very low input resistance (ideal for current measurement) and a very low output resistance (ideal for a voltage source) even for very large values of R_f. The effective input resistance R_{cf} of the current follower circuit as seen by the unknown current source is the input voltage, v_s, divided by the input current, i_{in}:

$$R_{cf} = v_s/i_{\text{in}}. \qquad (3\text{–}5)$$

Substituting for v_s from $v_o = -Av_s$ and for i_{in} from Eq. (3–4), we have

$$R_{cf} = R_f/A, \qquad (3\text{–}6)$$

which is a factor-of-A improvement over placing R_f in the measurement circuit directly.

If more than one current source is connected to point S of the current follower circuit as in Fig. 3–15, summing the currents to point S yields

$$i_1 + i_2 + i_3 = i_f + i_b, \qquad (3\text{–}7)$$

which has the same form as Eq. (3–1), where the sum of the applied current sources is equivalent to i_{in}. According to the same arguments which led to Eq. (3–4), the output voltage is

$$v_o \approx -(i_1 + i_2 + i_3)R_f \quad \text{(for } A \gg 1 \text{ and } i_b \ll i_1 + i_2 + i_3\text{).}$$
$$(3\text{–}8)$$

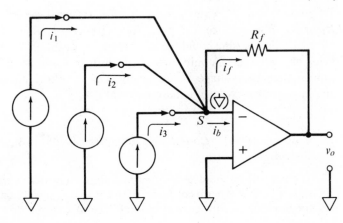

Fig. 3–15 Summing current follower.

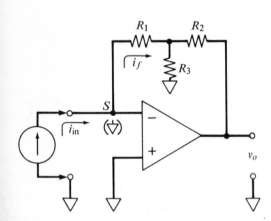

Fig. 3–16   Circuit for high equivalent R_f.

From Eq. (3–8), it is seen that the current i_f and the output voltage v_o are proportional to the sum of the input currents. Since the OA maintains point S at the common potential, it is an ideal point to sum (combine) the currents from several sources without interference with each other. For this reason, point S is called the **summing point.** This circuit is very useful for measuring the sum of two or more current sources or for off-setting a constant base current from the measured current source.

The range of currents that can be converted to voltages by the OA current follower is limited on the high end by the output current limit of the OA, which must supply i_f and the current to the output load. On the low end, it is limited by the OA input bias current, i_b, as shown in Eq. (3–3), by v_s as shown by Eq. (3–2), or by the maximum practical value for R_f. The best way to reduce i_b is by the proper choice of OA, although a steady value of i_b can be compensated for by applying a constant source of current opposite to i_b to point S. The value of v_s, compared to $i_f R_f$, will be sufficient to cause an appreciable error according to Eq. (3–2) only when a low-gain amplifier is used or when the amplifier voltage offset or offset voltage drift is not very much less than the expected measurement v_o.

The maximum practical value of R_f is limited by the state of the art in producing and using stable, accurate, high-value resistors. A circuit which has the effect of a high R_f but is made of lower-value resistors is shown in Fig. 3–16. The output voltage is divided by R_2 and R_3 and the resulting fraction of v_o is applied to R_1 to provide i_f. The resistance R_1 is a load on the R_2-R_3 divider. If point S is assumed to be at the common voltage, R_1 and R_3 are effectively in parallel and the equivalent resistance between the output and point S is derived to be

$$R_{f(\text{eq})} = \frac{R_1 R_2 + R_2 R_3 + R_1 R_3}{R_3}. \tag{3–9}$$

To produce a larger equivalent resistance, R_3 must be smaller than either R_1 or R_2. For instance, for $R_1 = R_2 = 10^6 \ \Omega$ and $R_3 = 10^3 \ \Omega$, $R_{f(\text{eq})} = 10^9 \ \Omega$. This circuit should be used with the caution that the maximum value of v_s due to OA gain, offset, and drift limitations must remain much smaller than the voltage at the junction of the three resistors. In the example given, this voltage is only $v_o/1000$.

Experiment 3–1   *Operational Amplifier Current-to-Voltage Converter*

The operational amplifier current-to-voltage converter (current follower) is shown in this experiment to be a highly useful signal conditioning circuit for transducers with current outputs. The OA current-to-voltage converter is constructed, and the output voltage–input current relationships (transfer function) are studied for a general purpose OA and a FET input OA. Deviations from ideality are noted. The OA current-to-voltage converter is compared to the load resistor method for current-to-voltage conversions.

Charge-to-Voltage Conversion

Since the voltage across a capacitor is directly proportional to the charge on it, the capacitor is the basic device used in charge-to-voltage converter circuits. The measurement of charge and current are related, since current is the rate of charge transfer and charge is the total charge transferred by a current during a particular event or time. The basic operational amplifier charge-to-voltage converter is shown in Fig. 3–17a. A positive charge applied to the input would tend to make point S positive, which in turn makes the output voltage negative by the amount $-Av_s$. Because of the very large gain, A, the change in v_o is very much greater than that of v_s. Since the change in voltage at point S is negligible, the total change in input charge appears on capacitor C. The voltage across C is

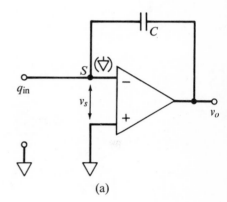

(a)

$$v_s - v_o = q_{in}/C. \qquad (3\text{-}10)$$

When Eq. (3–10) is combined with $v_o = -Av_s$ to eliminate v_s,

$$v_o = \frac{-q_{in}}{C}\left(\frac{1+A}{A}\right) \approx \frac{-q_{in}}{C} \quad \text{for } A \gg 1, \qquad (3\text{-}11)$$

showing that v_o is proportional to q_{in} when the OA gain is much greater than 1. The value of C is chosen for the range of charge values to be measured. Equation (3–11) would be perfectly accurate for any amount of charge over any time period if it were not for the leakage of charge to point S from sources other than the input. The major source of leakage current is often the amplifier input bias current i_b. The total charge measured must be large enough so that the charge leaked by i_b during the measurement time is small by comparison.

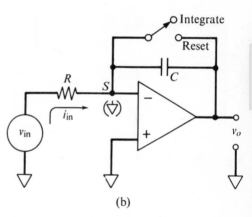

(b)

Fig. 3–17 (a) Charge-to-voltage converter. (b) OA voltage integrator.

Integration. Since the total charge passing a point over a given time is the integral or summation of the instantaneous currents during that time, the charge-to-voltage converter can be used to perform the mathematical function of integration on a signal in the current domain. If q_{in} is applied to the circuit of Fig. 3–17a as an input current, i_{in} over time t,

$$q_{in} = \int_0^t i_{in}\, dt \qquad (3\text{-}12)$$

and, combining Eqs. (3–12) and (3–11),

$$v_o \approx -\frac{1}{C} \int_0^t i_{\text{in}} \, dt. \tag{3–13}$$

A signal encoded in the voltage domain can be integrated if a resistor is used to convert the input voltage to a proportional current. The resulting circuit is shown in Fig. 3–17b. Since point S is maintained at the common voltage, $i_{\text{in}} = v_{\text{in}}/R$. Substituting for i_{in} in Eq. (3–13),

$$v_o \approx -\frac{1}{RC} \int_0^t v_{\text{in}} \, dt. \tag{3–14}$$

A switch (often electronic) is used to discharge C and reset the integrator to begin another measurement as described in Module 2.

Differentiation.   The time derivative of a signal is proportional to its rate of change. The rate of change of charge on a capacitor, dq/dt, is proportional to the rate of change of voltage across the capacitor by

$$\frac{dq}{dt} = C\frac{dv}{dt}, \tag{3–15}$$

where dq/dt is the rate at which charge must be brought to the capacitor, or, in other words, the current, i. Thus the change in voltage on a capacitor produces a current proportional to the rate of change. The measurement of this current is then a measurement of the time derivative of the voltage signal. A derivative-measuring circuit using an OA current follower for the current measurement is shown in Fig. 3–18. Since $i_{\text{in}} = C(dv_{\text{in}}/dt)$ and $v_o \approx -Ri_{\text{in}}$,

$$v_o \approx -RC\frac{dv_{\text{in}}}{dt}. \tag{3–16}$$

Derivative circuits are useful for sharpening transitions of signals for easier identification of regions of change as well as for measurements of rate. Because the output voltage is proportional to the rate of input voltage change, a small but very rapid pulse could produce a larger output signal than a slower but larger change in signal level. Practical derivative circuits include a small resistor in series with the input capacitor to limit its rate of charge and/or a small capacitor in parallel with the feedback resistor to limit the response rate of the amplifier.

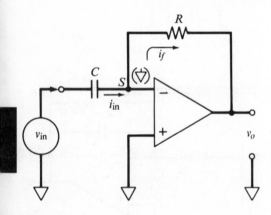

Fig. 3–18   OA differentiator.

Experiment 3–2   *Integration and Differentiation*

When a capacitor is used in the feedback loop of an OA, the resulting integrator is highly useful in signal conditioning applications. The OA integrator is often used to obtain the area under transient current or voltage signals, to average noise present with steady signals, and to provide various control functions. Information about the rate of change of a quantity with respect to time is frequently required for measurement or control purposes and readily obtained with an OA differentiator. In this experiment an integrator is first constructed and tested with various waveforms. Then an OA differentiator is constructed and used to obtain rate information from a linear sweep test signal.

Voltage Amplification

A voltage amplifier is not an interdomain converter, since the unit describing the data-encoding quantity is volts at both the input and the output. However, the transfer function, i.e., the input/output voltage relationship, is very important in the measurement process. A typical transfer function is shown in Fig. 3–19a. Ideally, the change in v_o resulting from a given change in v_{in} (the gain) is absolutely constant throughout the linear region. As v_o approaches its positive and negative limits, the gain changes and the input/output voltage relationship becomes nonlinear. In the limiting region, the output voltage is independent of the input voltage. For amplification of signals in the voltage domain, it is important to know the gain, the gain variation with signal level and time, and the limits of the linear range. The effect of changing gain on the transfer function is shown in Fig. 3–19b.

The transfer functions shown in Fig. 3–19a and b are for amplifiers with zero offset, i.e., those in which v_o is zero when v_{in} is zero. Figure 3–19c shows the transfer function for several positive and negative values of offset. Most amplifiers have some offset, either unintentional, due to the circuit characteristics or component instability, or intentional, to compensate for an offset inherent in the voltage source. A "zero" or "balance" adjustment is often provided to reduce offset to within the offset stability of the amplifier. "Offset" controls on amplifiers cover a wide range, sometimes exceeding the full scale range of input voltages.

It is important for a voltage amplifier to avoid distorting the input signal voltage by loading the voltage source. To avoid this, it must have an input resistance very much greater than the signal source resistance. On the other hand, the amplifier output voltage should not be loaded when connected to common voltage measurement devices. Thus the amplifier should have a low equivalent source resistance at its output. An amplifier with high input and low output resistance may be used between a high resistance voltage source and a low resistance voltmeter simply to eliminate the loading which would occur if the source were connected to the voltmeter directly.

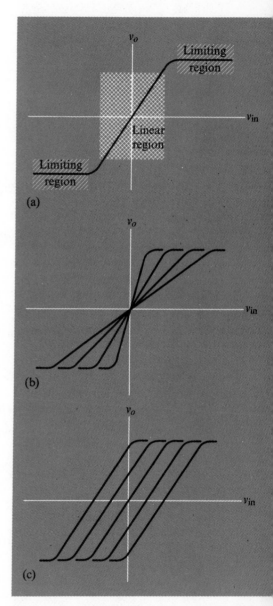

Fig. 3–19 Amplifier transfer functions. (a) Showing linear and limiting regions. (b) For various gains. (c) For various offsets.

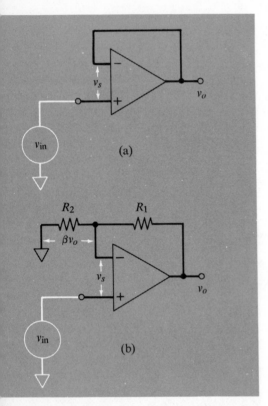

Fig. 3–20 (a) Operational amplifier voltage follower. (b) Follower with gain.

Voltage Follower Amplifier. The most popular type of voltage amplifier for precision applications is the voltage follower. In this type of amplifier, the amplifier output voltage is compared to the input voltage and corrected if it is not the same. In this way, the output voltage is made to "follow" the input voltage exactly. An operational amplifier voltage follower circuit is shown in Fig. 3–20a. The amplifier output voltage, v_o, equals $-v_s A$ and $v_s = v_o - v_{in}$. When these equations are combined to eliminate v_s,

$$v_o = v_{in} \left(\frac{A}{A+1} \right) \approx v_{in} \quad \text{for } A \gg 1. \tag{3–17}$$

In other words, so long as the gain is high, the difference v_s between v_o and v_{in} is very small. The gain of the follower amplifier is thus almost exactly unity. Its great effectiveness lies in its very high input resistance (10^9 to 10^{15} Ω) and its very low output resistance (< 1 Ω), which make it an excellent intermediary between any voltage source/load combination. An assumption made in the voltage follower circuit is that the amplifier offset will remain zero (that is, zero v_o for zero v_s) over the entire range of v_{in} applied to the $+$ input. To achieve this, the amplifier must have an excellent **common mode rejection** characteristic, a figure of merit for difference amplifiers.

The voltage follower circuit can provide voltage gain with little sacrifice to its input and output resistance characteristics if only a fraction of v_o is compared with the input voltage. The resulting circuit is shown in Fig. 3–20b. The resistors R_1 and R_2 divide the output voltage to produce a voltage βv_o where $\beta = R_2/(R_1 + R_2)$. Following the same arguments that led to Eq. (3–17),

$$v_o = v_{in} \left(\frac{A}{\beta A + 1} \right) \approx v_{in} \left(\frac{1}{\beta} \right) \quad \text{for } A \gg 1, \tag{3–18}$$

which gives the amplifier a gain of $1/\beta$.

Summing Amplifier. The linear combination of voltages from two or more sources is accomplished by a summing amplifier. The most frequently used summing amplifier circuit is the summing current follower of Fig. 3–15, to which input resistors have been added to convert the input voltages to proportional currents. The resulting circuit is shown in Fig. 3–21. Substituting the values for i_1, i_2, and i_3 in Eq. (3–8), we obtain

$$v_o \approx - \left(v_1 \frac{R_f}{R_1} + v_2 \frac{R_f}{R_2} + v_3 \frac{R_f}{R_3} \right) \quad \text{for } A \gg R_f/R_{in}, \tag{3–19}$$

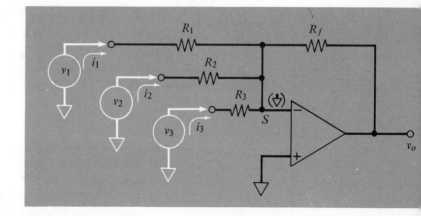

Fig. 3–21 Summing amplifier.

which shows that the output voltage is the negative of the sum of the input voltages, each multiplied by its own R_f/R_{in} ratio. If the simple (unweighted) sum is desired, all R_{in} values are made equal. Then

$$v_o \approx \frac{-R_f}{R_{in}} (v_1 + v_2 + v_3) \quad \text{for } A \gg R_f/R_{in}. \qquad (3\text{–}20)$$

Summing amplifiers are used in instruments and in analog computers to perform the addition function on data in the voltage domain. A summing amplifier can also be used to add a constant to a signal voltage, thus introducing (or eliminating) an offset in the voltage transfer function. The summing amplifier is often used with just one input source when the polarity-inverting characteristic of this amplifier circuit is desired. The input resistance of the summing amplifier is R_{in} at each input. Where this presents an excessive load on the voltage source, a voltage follower circuit (Fig. 3–20a) is used between the voltage source and the summing amplifier input.

Difference Amplifier. As the name implies, a difference amplifier produces an output voltage that is proportional to the difference between two input voltages. The basic difference amplifier circuit is shown in Fig. 3–22. The circuit is a combination of the inverting amplifier and the follower with gain. A fraction of the input voltage v_2 is applied to the noninverting OA input while the same fraction of the difference between v_1 and v_o is applied to the inverting OA input. The difference between v_- and v_+, the $-$ and $+$ OA input voltages, is v_o/A, which is negligibly small for amplifiers with high gain. The relation between the output voltage and the two input voltages can be solved by substituting the expressions for v_- and

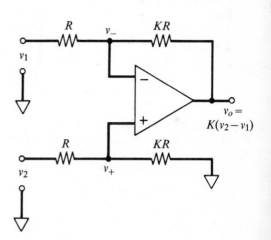

Fig. 3–22 Difference amplifier.

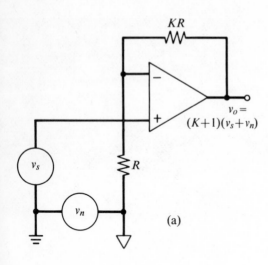

(a)

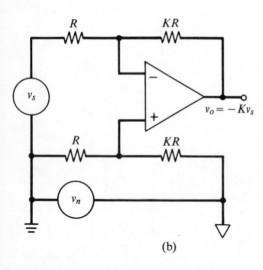

(b)

Fig. 3–23 Demonstrating the use of the difference amplifier to eliminate noise generated between source and amplifier commons.

v_+ into $v_o = A(v_+ - v_-)$ as follows:

$$v_+ = v_2 \frac{KR}{R + KR},$$

$$v_- = v_1 + (v_o - v_1)\frac{R}{R + KR},$$

$$v_0 = \frac{AK(v_2 - v_1)}{1 + K + A} \approx K(v_2 - v_1) \quad \text{for } A \gg K + 1. \quad (3\text{–}21)$$

As Eq. (3–21) shows, the output voltage is equal to the input voltage multiplied by K.

The difference amplifier is very convenient where the desired measurement is the difference between two transducer outputs. It is also the type of amplifier most often used for low level signal sources to eliminate noise signals occurring between the commons of the source and amplifier. This application is illustrated in Fig. 3–23. The "single-ended" follower with gain amplifier shown in (a) amplifies v_s, the desired signal, *plus* v_n, the voltage difference between the commons of the source and amplifier. However, the signal input to the difference amplifier connected as shown in Fig. 3–23b does not include v_n.

The ideal difference amplifier has perfectly balanced inputs; that is, equivalent voltage changes applied to both inputs will have exactly the same effect on the output voltage. Thus the output will respond only to differences in the input voltages. To achieve this ideality in the circuit of Fig. 3–22, it is necessary that the R's and KR's be perfectly matched and that the response of the − and + inputs of the OA be identical. A measure of the degree of balance in a difference amplifier is the ratio of the response of the amplifier to a signal applied between the difference inputs to its response to the same signal applied between both inputs and common. This ratio is called the **common mode rejection ratio** (CMRR), since the portion of the input signal which is applied identically to both inputs is called the **common mode signal.** The common mode response of a difference amplifier can be checked by connecting the output of a signal generator to both inputs in parallel and observing the amplifier output. If the amplifier has a balance control, it should then be adjusted to give zero output. A balance adjustment for the amplifier of Fig. 3–22 could be provided by a trimming adjustment on any one of the resistors.

A complete **instrumentation amplifier** would combine the advantages of the difference input with the high input resistance of the voltage follower. This is frequently achieved by simply putting a voltage follower amplifier before each input of the difference amplifier as shown in Fig. 3–24. The temptation to introduce gain in the follower amplifiers of

Fig. 3–24 is tempered by the requirement that the amplifier gains be well matched in order to achieve a high CMRR. The resulting circuit would then require four precisely matched pairs of stable resistors. A very clever circuit which cross-couples the two follower-with-gain circuits so that they track each other is shown with the difference amplifier in Fig. 3–25. The gain and the cross-coupling are provided by the three resistors between the two follower outputs. The follower amplifiers OA1 and OA2 keep the feedback points equal to v_1 and v_2 respectively. This results in a current through the three resistors of $i = (v_1 - v_2)/R_1$. From this information, the follower output voltages can be calculated by adding the feedback voltage to the iR drop through the feedback resistor. Thus

$$v_{o1} = v_1 + a(v_1 - v_2)$$

and

$$v_{o2} = v_2 - b(v_1 - v_2),$$

showing that each follower amplifies its input signal by 1 and the difference signal by $+a$ or $-b$. The difference gain, G_d, of OA1 and OA2 is thus

$$G_d = \frac{v_{o1} - v_{o2}}{v_1 - v_2} = 1 + a + b \qquad (3\text{–}22)$$

and the common mode gain, G_{cm}, is

$$G_{cm} = \frac{(v_{o1} + v_{o2})/2}{v_{cm}} = 1, \qquad (3\text{–}23)$$

where v_1 and v_2 are both equal to v_{cm}, the common-mode input signal. In normal applications, a and b would be approximately equal, but these

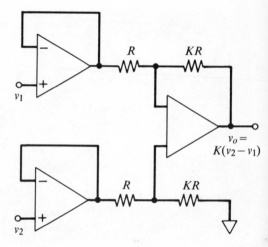

Fig. 3–24 Instrumentation amplifier using the difference amplifier with follower inputs.

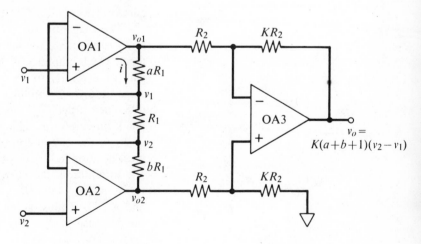

Fig. 3–25 Instrumentation amplifier using cross-coupled differential follower input.

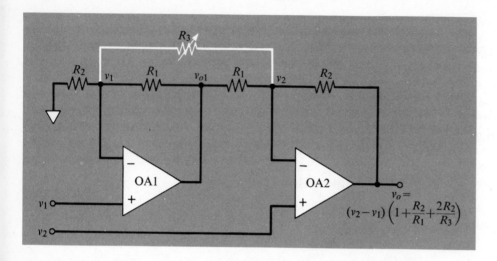

The equation shown in the figure:

$$v_o = (v_2 - v_1)\left(1 + \frac{R_2}{R_1} + \frac{2R_2}{R_3}\right)$$

Fig. 3–26   Instrumentation amplifier.

equations were derived with separate symbols to show that the common-mode gain is exactly unity without matched resistors. This stage then provides a CMRR equal to the difference gain, which is generally between 10 and 1000. The gain of this stage can be adjusted by changing the single resistor R_1. In general, the value of R_1 for a difference gain of 10 is wired into the circuit and other resistors are switched in parallel with R_1 when higher gains are desired. The CMRR of the cross-coupled input amplifiers is then multiplied by the CMRR of the difference amplifier to produce an instrumentation amplifier with excellent CMRR, high input impedance, and stable, easily adjustable gain. For instance, if the gain of the input stage is set at 100 and the gain and CMRR of the difference stage are 10 and 10^4 respectively, the resulting amplifier would have a gain of 1000 and a CMRR of 10^6.

The elegantly simple instrumentation amplifier circuit shown in Fig. 3–26 is adequate for many applications. Ignoring resistor R_3 for the moment, OA1 amplifies v_1 by the factor $1 + R_1/R_2$ while OA2 provides a gain of $-R_2/R_1$ to v_{o1} for a net gain for v_1 of $-(1 + R_2/R_1)$. At the same time OA2 amplifies v_2 by $+(1 + R_2/R_1)$ to provide a gain for the difference input $(v_2 - v_1)$ of $1 + R_2/R_1$. The CMRR of this circuit depends on careful matching of the R_1 and R_2 pairs of resistors. The addition of R_3 to the circuit causes a current of $(v_2 - v_1)/R_3$ in R_3. This current must be supplied by OA2 through R_2, which increases v_o by $R_2(v_2 - v_1)/R_3$. In addition, the current through R_3 decreases the OA output voltage v_{o1} by $R_1(v_2 - v_1)/R_3$. To maintain a feedback voltage of v_2 between R_1 and R_2, OA2 must supply an additional iR drop in R_1 to compensate for the decrease in v_{o1}. This additional current through R_2 increases v_o by the quantity $R_2(v_2 - v_1)/R_3$. The result is an increase in difference gain of $2R_2/R_3$ for a complete output voltage expression of[1]

$$v_o = (v_2 - v_1)\left(1 + \frac{R_2}{R_1} + 2\frac{R_2}{R_3}\right). \qquad (3\text{-}24)$$

Again, the gain can be adjusted by a single resistor without affecting the CMRR. Typical values for R_1 and R_2 are 1 kΩ and 9 kΩ to give a gain of 10 with no R_3. A value of 200 Ω for R_3 would then provide a gain of 100.

Experiment 3–3 *OA Voltage Follower*

This experiment illustrates that a voltage follower is highly useful in isolating voltage sources from loads because it can supply fairly large amounts of current to a load while drawing only small amounts of current from a source. The input-output relationship of a unity-gain follower is determined, and the output current capabilities of the follower are observed. The very low input current of the follower is estimated from the voltage drop across a large input resistor. A follower with gain is then constructed and the experimental gain compared to the gain expected.

Experiment 3–4 *OA Inverting and Summing Amplifiers*

This experiment demonstrates the use of an operational amplifier as an inverting voltage amplifier and as a summing amplifier. The inverting amplifier is tested by measuring the output voltage as a function of input voltage, of feedback resistance, and of input resistance. The summing amplifier is used to provide an output voltage related to the sum or difference of two or more input currents. Voltage sources are connected to the summing point of an OA inverting amplifier with different input resistances. Addition and subtraction of input currents are demonstrated. Results obtained with the inverting and summing amplifiers are compared to those expected from theoretical expressions.

Experiment 3–5 *OA Difference Amplifiers*

The difference amplifier is a highly useful signal conditioning circuit for measurements where the desired result is the difference between two transducer signals. This experiment demonstrates the use of operational amplifiers as differential amplifiers. The basic OA difference amplifier is first investigated. The output voltage is measured over a wide range of input

Note 1. Derivation of Gain for the Circuit of Fig. 3–26.

OA1 controls the junction of R_1, R_2, and R_3 to be v_1. This requires a current through R_1 equal to the currents through R_2 and R_3, which are v_1/R_2 and $-(v_2 - v_1)/R_3$ respectively. Therefore

$$v_{o1} = v_1 + \frac{v_1}{R_2}R_1 - \frac{(v_2 - v_1)}{R_3}R_1.$$

Similarly OA2 maintains its feedback voltage at v_2 by supplying a current through R_2 equal to the currents through R_1 and R_3:

$$v_o = v_2 + \frac{(v_2 - v_{o1})}{R_1}R_2$$
$$+ \frac{(v_2 - v_1)}{R_3}R_2.$$

When the equation for v_{o1} is substituted into the equation for v_o and the terms are collected, the result is

$$v_o = (v_2 - v_1)\left(1 + \frac{R_2}{R_1} + \frac{2R_2}{R_3}\right).$$

voltages, and limitations are noted. Then voltage follower inputs are added to the basic difference amplifier to make a complete instrumentation amplifier. In the final part of the experiment an instrumentation amplifier with cross-coupled follower inputs is investigated.

Nonlinear Operations

The transfer function for a complete measurement system is the mathematical relationship between the magnitude of the numerical output and the number of units of the measured quantity. The transfer function can be determined experimentally by plotting the observed numerical output vs. the magnitude of the measured quantity for several known values of input quantity. Such a plot is the familiar "calibration curve" for a measurement. A linear transfer function (a straight-line input/output relationship) is often sought as the most convenient. As explained in Section 3–1, each measurement system can be considered to be a sequence of inter-domain converters, each of which has its own input-quantity-vs.-output-quantity transfer function. The overall transfer function for the system is the product of all the separate interdomain converter transfer functions. For instance, the transfer function for a thermocouple temperature meter is the product of the voltage/temperature relationship of the thermo-couple, the current/voltage relationship of the amplifier and voltage-to-current converter, the scale position/current relationship of the meter, and the scale numbers/scale position relationship of the meter face markings. When all of these transfer functions are multiplied together, the intermediate units (domains) cancel and the overall transfer function (scale numbers/temperature relationship) is obtained. If all the interdomain transfer functions are linear, the overall transfer function will also be linear.

Many occasions arise where an interdomain converter with a non-linear transfer function is chosen for a system. If all the other system converters are linear, the value of the measured quantity cannot be read directly from the output device. In some cases the measured quantity must then be obtained by consulting a graph or table of output readings vs. measured quantities. To avoid this difficulty, a second nonlinear transfer function is frequently introduced into the system. If the additional transfer function is the reciprocal of the first one, the product of all the transfer functions will again be linear. One simple place for a manufacturer to introduce the compensating nonlinear transfer function is in the meter scale markings, that is, the scale numbers/scale position relationship. However, when a recorded output is produced, the need to use a specially printed nonlinear recording paper is undesirable. No similar nonlinear scaling is possible when the readout device is an inherently linear device such as a digital voltmeter. For such readout units, the compensating nonlinear transfer function must be introduced while the data are still in the electrical analog domain.

Several approaches have been used in the development of nonlinear transfer function circuits, most of them involving the use of devices with nonlinear current-voltage relationships. These can be divided into two categories: those that have piecewise-linear transfer functions, composed of two or more line segments, and those that have smooth, continuous-curve transfer functions. We will consider the piecewise-linear category first.

Limiter Function. The ideal limiter circuit provides a linear transfer function between the input and the output over a limited range. Beyond this range, the output voltage will respond much less or not at all to changes in the input voltage. This circuit has many applications in keeping signals within a safe range as illustrated in the discussion of bounded amplifiers in Section 2–3.6 of Module 2. The basic limiter function is also very useful in nonlinear analog signal conditioning, as we shall see. The heart of the limiter circuit is the pn junction diode, which we will consider temporarily as an ideal diode—zero resistance when forward biased and zero conductance when reverse biased.

A simple series limiter is shown in Fig. 3–27a. So long as v_{in} is less positive than V_R, the diode is OFF and v_o is zero. When v_{in} is more positive than V_R, the diode conducts and the voltage $v_{in} - V_R$ appears at the output. The transfer function is two line segments with a breakpoint at $v_{in} = V_R$ and $v_o = 0$ V. The output voltage is seen to be limited to positive values. The effect of varying the sign or magnitude of V_R is to move the breakpoint back and forth along the zero v_o line. The slope remains the same. The effect of reversing the diode is to produce a zero v_o for all values of v_{in} more positive than V_R and a v_o of $v_{in} - V_R$ for all other values of v_{in}. Thus v_o is limited to negative values, but the slope and breakpoint of the transfer function remain unchanged. A more practical circuit which avoids the need of the series-connected bias source V_R is shown in Fig. 3–27b. In this circuit, the diode will conduct when the voltage at the junction of R_S and R_R is positive. This occurs when v_{in} is more positive than $-V_R R_S / R_R$. The slope during conduction approaches

Fig. 3–27 Series limiting circuits and their transfer functions.

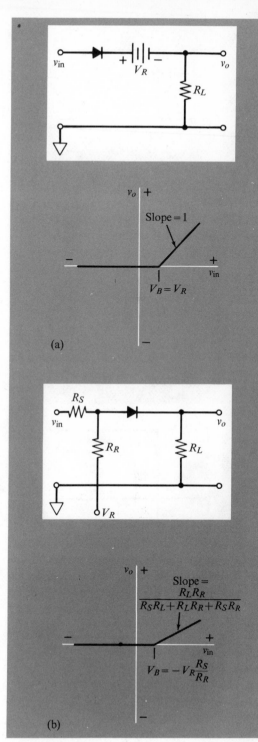

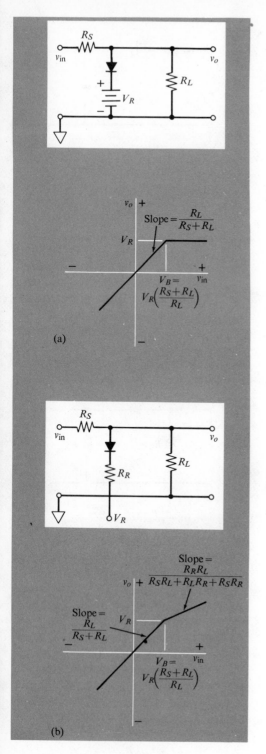

Fig. 3–28 Shunt limiting circuits and their transfer functions.

unity as R_L approaches infinity. The series limiter is thus seen to limit v_o to voltages of a single polarity while allowing the range of effective input voltages to be adjusted by setting V_B.

The shunt limiting circuits of Fig. 3–28 have somewhat different transfer functions. In the simple circuit of Fig. 3–28a, v_{in} is divided between R_L and R_S until the divided fraction becomes more positive than V_R. At this point the diode conducts, holding v_o at the voltage V_R. The linear region passes through the origin and has a slope approaching unity when R_L is very much larger than R_S. The output voltage is limited to values negative of V_R. The breakpoint voltage can be set in any polarity and magnitude by adjusting V_R. Reversing the diode would make V_R the negative limit of the output voltage range. A resistance R_R in series with the diode and the source of V_R affects the slope of the transfer function after the breakpoint, as shown in Fig. 3–28b. This provides two line segments of adjustable slope and breakpoint. As seen from the formula, the slope after the breakpoint is 0 for $R_R = 0$ and is $R_L/(R_S + R_L)$ for $R_R = \infty$; thus the slope after the breakpoint must always be less than that in the region where the diode is not conducting. Note that in the case of the shunt limiter, varying V_R moves the breakpoint back and forth along the $v_o = v_{in}R_L/(R_S + R_L)$ line, affecting *both* the output voltage limit and the effective input voltage range.

The series and shunt limiting circuits shown operate essentially as the models indicate for signal levels of many volts. For signal levels less than about 10 V, the assumption of an ideal diode causes errors. If the forward conduction bias of 0.6 V required by a silicon diode is taken into account, the models can be extended to signals of the order of 1 V. Below 1 V, the logarithmic curvature of the diode current/voltage curve becomes important and destroys the two-line-segment circuit function model.

Precision Limiter.   A precise limiter circuit that performs well even at signal levels in the millivolt range or less is shown in Fig. 3–29. The circuit is very similar to the summing amplifier of Fig. 3–21 except that there are two feedback paths, one for each possible direction of feedback current. If the sum of the currents from V_R and v_{in} is positive, v_A will be sufficiently negative to cause D_1 to conduct the feedback current required to keep point S at virtual ground. The output voltage is exactly zero in this case, since the output is connected to ground through R_L, to virtual

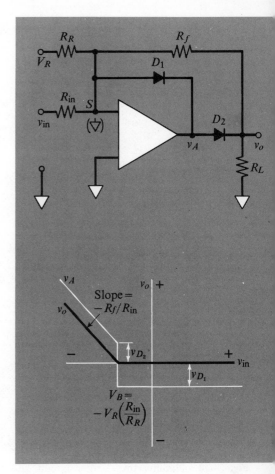

Fig. 3–29 Precision limiter circuit.

ground through R_f, and diode D_2 is reverse biased. Note that the voltage required to make D_1 conduct does not appear at the output. On the other hand, when the sum of the currents through R_R and R_{in} is negative, the only possible feedback path is through diode D_2 and R_f. The amplifier output v_A must be sufficiently positive to provide the required forward bias for D_2 *and* the iR drop across R_f as shown by the dashed-line curve in Fig. 3–29. Since point S is still maintained at virtual ground, v_o is simply $-i_f R_f$, where i_f is equal to the total input current $(v_{in}/R_{in}) + (V_R/R_R)$. Thus

$$v_o = -R_f \left(\frac{v_{in}}{R_{in}} + \frac{V_R}{R_R} \right) \quad \text{for } v_{in} < -V_R \frac{R_{in}}{R_R}. \quad (3\text{–}25)$$

The inclusion of the forward bias of D_2 in the control loop reduces the error due to the forward voltage across D_2 by the loop gain of the OA circuit. It is important to include an actual load resistor R_L such that significant current (≈ 1 mA) will be drawn through D_2 when v_o is positive. This will ensure good control of v_o. It should also be pointed out that if the output is required to sink current, the feedback current through R_f will be excessive and the amplifier output v_A will go negative to supply a compensating current through D_1. In such a case, D_2 is reverse biased and the circuit can no longer control v_o. Therefore, the load circuit should be passive.

The transfer function of this extremely accurate and useful circuit resembles the "zero-bounded" transfer function of the series limiter except that it is inverted by the inverting amplifier configuration. As the figure shows, both the slope and the breakpoint are easily adjusted. When V_R is zero and $R_{in} = R_f$, the precision limiter acts as an ideal diode providing unity transfer for one polarity of v_{in} and zero transfer for the other. Reversing both diodes in the circuit makes zero the upper bound of v_o and changes the condition of validity of Eq. (3–25) to $v_{in} > -V_R R_{in}/R_R$.

A similar limiter circuit based on the noninverting voltage follower circuit is shown in Fig. 3–30. The OA, through the feedback paths, works to keep the voltage difference between its $-$ and $+$ inputs negligibly small. Thus, when v_{in} is positive, the feedback path through D_2 and the 10 kΩ isolating resistor enable the voltage v_{in} to be established at the $-$ input. The voltage difference between v_o and the $-$ input voltage is the iR

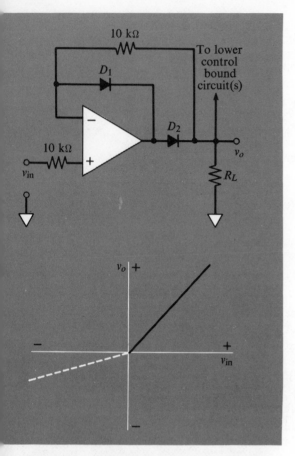

Fig. 3–30 Noninverting limiter circuit and transfer function.

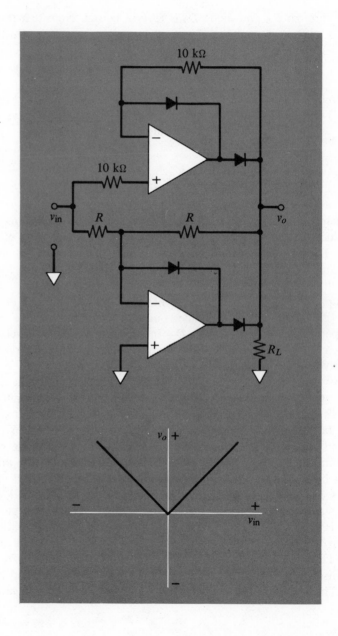

Fig. 3–31 Absolute value circuit and transfer function.

drop across the feedback resistor due to the OA input current. Since this is negligibly small, $v_o = v_{in}$ for all positive values of v_{in}. For negative values of v_{in}, the feedback path is D_1, and D_2 is reverse biased. This would at first appear to essentially disconnect the circuit from R_L, as in the case of the inverting limiter of Fig. 3–29. However, in this case the OA — input is not held at virtual ground; it follows v_{in} negative and the voltage v_{in} will be divided between R_L and the 10 kΩ feedback resistor. Thus this circuit must be used only in conjunction with some other circuit that will control v_o at zero volts should it tend to become negative. This lower bound v_o control function could be performed by connecting the output of a second noninverting limiter to the v_o terminal. If the v_{in} of the second limiter circuit is connected directly to common, it will establish 0 V as the negative limit of v_o. In fact, several noninverting limiters can be connected together to produce an output voltage equal to the most positive of the input voltages. The 10 kΩ feedback resistor in this circuit limits the feedback current that must be supplied by the lower bound control circuit. The input resistor is present to match OA input source resistances. The actual values of the resistors are not critical.

These basic limiter circuits have been used in a variety of ingenious combinations to produce many useful nonlinear functions, several of which are described below.

Absolute Value Function.   One of the most useful of the two-line-segment functions is the absolute value function. As its name suggests, this circuit produces a positive output voltage equal in magnitude to the input voltage regardless of the sign of the input voltage. A simple implementation of this function combines the inverting and noninverting limiter circuits of Figs. 3–29 and 3–30, as shown in Fig. 3–31. The upper, noninverting limiter circuit establishes the voltage v_{in} at the output for all positive values of v_{in}. Similarly, the lower inverting limiter establishes a voltage of $-v_{in}$ at the output for all negative values of v_{in}. The two resistors in the lower circuit should be carefully matched to ensure unity gain for both polarities of v_{in}. The same circuit will produce a negative v_o for either polarity of v_{in} if all the diodes are reversed.

The absolute value circuit is actually a precise full-wave rectifier circuit for use with analog signals. When the measurement information is encoded as the magnitude of voltage variations of both polarities, the absolute value circuit will convert the signal variations to a proportional unipolar signal suitable for measurement with a dc voltage measurement system.

Multiple-Line-Segment Functions.   As suggested above, precision limiter circuits can be combined to produce transfer functions of more than two

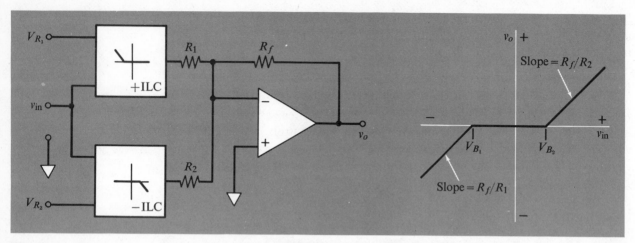

Fig. 3–32 Dead space circuit and transfer function.

line segments. In Fig. 3–32, the outputs of two inverting limiter circuits (ILC), as shown in Fig. 3–29, are summed. One of the limiters has zero as its lower bound and the other has zero as its upper bound. If the breakpoints are adjusted so that there is a range of v_{in} where both limiters are zero bound, the transfer function shown in Fig. 3–32 results. This function is called the dead-space function because there is a space in the range of v_{in} for which the output voltage is zero. Thus a window of inactivity can be produced at any point in the input/output transfer function. This can be useful if one wishes to avoid using a significant fraction of the span of the readout device for a portion of the signal of no interest. Note that the slopes and breakpoints are easily and independently adjustable.

An extremely versatile multiple-line-segment circuit results from summing the outputs of one or more inverting limiting circuits with the input signal itself. Such a circuit and a representative transfer function are shown in Fig. 3–33. Throughout the range of v_{in} where the ILC is zero bound, the circuit acts as an ordinary inverting amplifier, producing a gain of $-R_f/R_1$. For input voltages exceeding the breakpoint, a voltage change of opposite slope to v_{in} is summed with v_{in}. If the input summing resistors R_1 and R_2 are equal, the input slopes cancel, producing an output that limits at V_R. Thus the circuit can provide an output limiting function. If a zero-upper-bound ILC were used in addition, a negative output limit would also be obtained.

When the summing resistance from the ILC is made larger or smaller than the summing resistance from the signal, the slope of the input signal will be under- or overcompensated respectively. Thus the slope and the breakpoint of the second line segment can be easily adjusted. Of particular interest is the case where $V_R = 0$ and $R_2 = R_1/2$. In this case, the breakpoint comes at exactly 0 V and the slope of the inverting amplifier

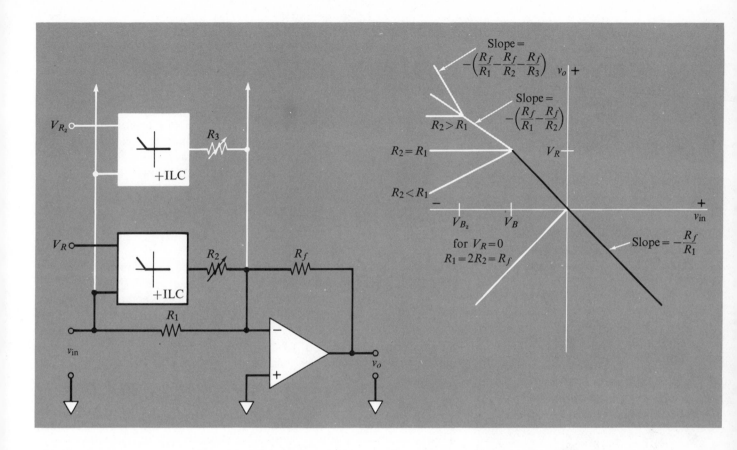

Fig. 3–33 Limiter circuit and transfer function.

v_o/v_{in} curve in the $-v_{in}$ region is doubly compensated. The resulting transfer function, shown in dashed lines in Fig. 3–33, is the absolute value function with negative v_o. This means of obtaining the absolute value function, though given in several sources, is less attractive than that of Fig. 3–31 because it requires three additional precision resistors.

Additional upper or lower bound ILC circuits can be added to provide additional line segments as shown in the white portion of the circuit diagram. As shown, each additional ILC provides another breakpoint and slope modification. By summing the outputs of many such ILC's (6 to 24 units are not unusual) it is possible to approximate virtually any desired transfer function with a number of line segments. The accuracy and stability of this circuit along with the neatly independent breakpoint and slope adjustments make this line-segment function circuit better for precise low level signal conditioning than the diode ladder circuit described in Section 2–3.6 of Module 2.

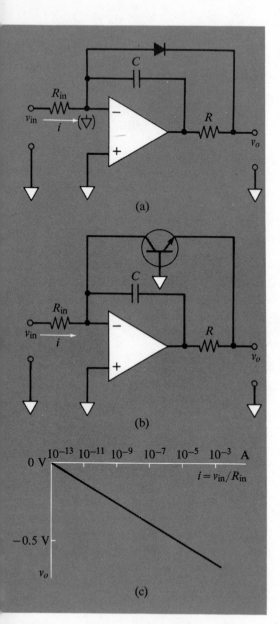

(a)

(b)

(c)

Fig. 3–34 Simple logarithmic function circuit. (a) Diode. (b) Transistor. (c) Transfer function for $\eta = 1$, $I_0 = 10^{-13}$ A.

Logarithmic Function. The logarithmic transfer function is one which finds many uses, including compensating for logarithmic function transducers, compression of especially wide dynamic range signals, and nonlinear arithmetic function modules such as multipliers and dividers. The logarithmic transfer function, $v_o = \log v_{\text{in}}$, is generally achieved in either of two ways, by approximation with a multiple-line-segment function generator or by taking advantage of the approximately logarithmic relationship between current and voltage in a semiconductor pn junction. The latter technique, which can provide good accuracy and a smooth transfer function over a wide dynamic range, is described below.

The current-voltage relationship of a forward biased pn junction was shown in Module 2 to be

$$i = I_0 e^{v/\eta V_T} - I_0 \tag{3–26}$$

where I_0 is a temperature-dependent term related to the reverse bias current, V_T is the temperature equivalent of voltage $= kT/Q_e = T/11,600$, and η is an empirical parameter related to the device composition, having a value of ≈ 1 for germanium and ≈ 2 for silicon devices. For forward bias values greater than 0.25 V for silicon and 0.12 V for germanium, the exponential term in Eq. (3–26) is greater than 100. Therefore, for all but very small forward bias voltages,

$$i \approx I_0 e^{v/\eta V_T} \quad \text{for } i > 100 I_0. \tag{3–27}$$

Taking the $\log_{10}$ of both sides of Eq. (3–27) and solving for v, we obtain

$$v = 2.3 \eta V_T (\log i - \log I_0), \tag{3–28}$$

showing that the voltage across a pn junction varies as the log of the current through it. Since the factor $2.3 V_T = 0.059 \eta$ at 300°K, the voltage will change by 59η mV for each tenfold change in current. If an input voltage is converted to a proportional current and that current is passed through a pn junction, the voltage across the junction will vary as the log of the input voltage. A simple circuit which implements this concept is shown in Fig. 3–34a. The voltage v_{in} is converted to a proportional current by resistor R_1 connected to the virtual common at the summing point of the operational amplifier. The output voltage establishes the exact diode bias voltage required to pass the input current through the diode. The resistor R and capacitor C are present only for circuit stability. Their values, which are typically 1 kΩ and 10^{-2} µF respectively, do not affect the static accuracy of the circuit. Thus the output voltage v_o is $-v$, the forward bias voltage of the diode. Substituting $v_o = -v$ and $i = v_{\text{in}}/R_{\text{in}}$ in

Eq. (3–28), we obtain

$$v_o = -\gamma_T \left(\log \frac{v_{\text{in}}}{R_{\text{in}}} - \log I_0 \right), \qquad (3\text{--}29)$$

where $\gamma_T = 2.3\eta V_T$.

The range of currents over which Eq. (3–29) is valid is limited on the low end by the minimum bias voltage for the approximation of Eq. (3–27) to be valid, and on the high end by the assumption that the iR drop across the semiconductor material, contacts, and leads is negligible compared to v. For most diodes, Eq. (3–29) is valid over at least a two-orders-of-magnitude change in current. Some diode types provide a logarithmic function over four or five orders of magnitude. The trend recently is to use the base-emitter junction of a transistor in place of the diode as shown in Fig. 3–34b. For a transistor, the collector current has the same relationship to the emitter-base junction voltage as the diode. The OA applies the exact emitter-base bias voltage required to maintain a collector current of i. In practice, the transistor has been found to have a better logarithmic accuracy, a wider dynamic range, and a value of η of very nearly unity. Figure 3–34c shows a typical transfer function for the simple logarithmic function circuit.

The output voltage of the circuits of Fig. 3–34 is the sum of two terms, as shown by expanding Eq. (3–29):

$$v_o = -\gamma_T \log \frac{v_{\text{in}}}{R_{\text{in}}} + \gamma_T (\log I_0). \qquad (3\text{--}30)$$

In the first term, the desired log v_{in} relationship is multiplied by the temperature-dependent factor γ_T (which varies about 0.3%/°C at room temperature). The second term is a temperature-dependent offset term in which γ_T and I_0 are temperature dependent. At room temperature, the variation in the offset term with temperature is about −2.5 mV/°C.

One of the greatest difficulties with logarithmic function circuits has been maintaining reasonable accuracy in circuits with such a large temperature dependence. The earliest solution was to enclose the log element, resistors, and even the OA in a carefully temperature-controlled "oven." More recently, effective temperature compensation circuits have been developed. The most common of these is shown in Fig. 3–35. The circuit surrounding OA1 is identical to that of Fig. 3–34b, which yields an output voltage v_{o1} as given by Eq. (3–30). A second transistor interposed between v_{o1} and a follower-with-gain output amplifier is used to counteract the temperature variation in the second term of Eq. (3–30). A source of constant current i_R (obtained from V_R and the very large resistor R_R)

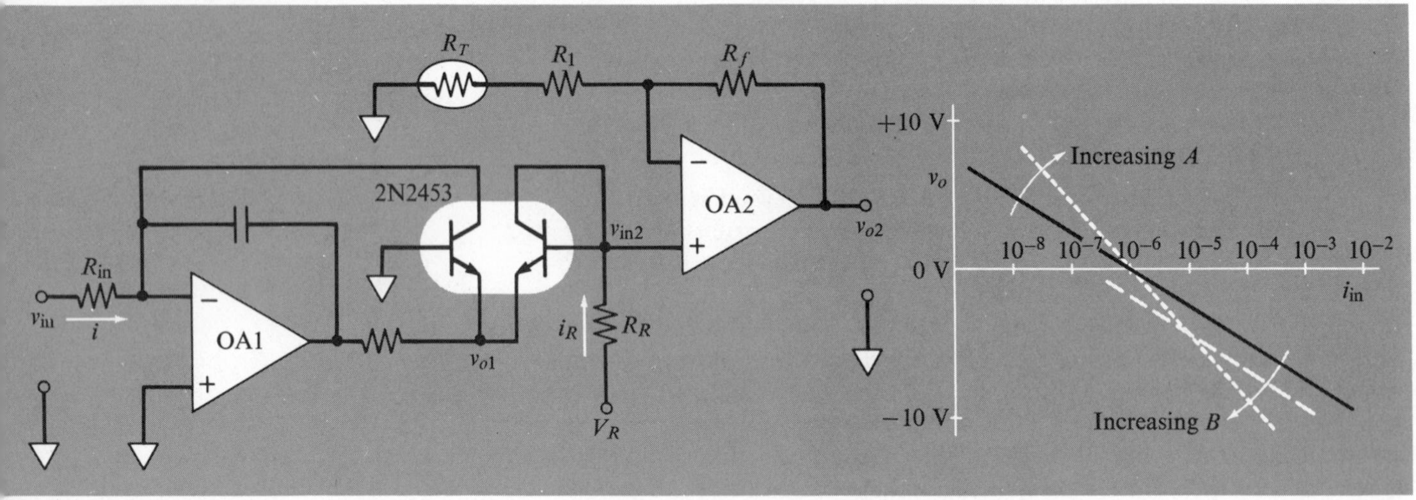

Fig. 3–35 Temperature-compensated logarithmic function circuit and its transfer function.

is applied to the emitter-base junction of the compensating transistor. The emitter-base voltage across this transistor is given by Eq. (3–28). If the two transistors are identical, the η, V_T, and I_0 terms for both transistors will be equal. The matching of transistors and their temperatures is greatly improved by using matched transistor pairs that are fabricated on the same silicon chip. Assuming a perfect match, the resulting voltage v_{in2} can be obtained by summing the voltages in Eqs. (3–30) and (3–28):

$$v_{in2} = -\gamma_T \left(\log \frac{v_{in}}{R_{in}} + \log I_0 + \log i_R - \log I_0 \right)$$

$$= -\gamma_T \log \frac{v_{in}}{R_{in} \, i_R}. \tag{3–31}$$

Thus only the temperature dependence of the scaling factor γ_T remains. This is compensated by using the temperature-dependent resistor R_T in the gain-determining circuit of the follower with gain. Since the gain of this amplifier is $(R_f + R_1 + R_T)/(R_1 + R_T)$, the circuit output voltage v_{o2} is

$$v_{o2} = -\frac{(R_f + R_1 + R_T)\gamma_T}{R_1 + R_T} \log \frac{v_{in}}{R_{in} \, i_R}$$

$$= -A\gamma_T \log B v_{in}, \tag{3–32}$$

where A is the gain of the follower amplifier and B is determined by R_{in} and i_R to provide complete flexibility of range adjustment as shown in the ideal transfer function of Fig. 3–35. A useful dynamic range of 10^7

has been reported for a circuit of this type. For a wide dynamic range of voltage inputs, very low values of offset voltage and current are required of the input operational amplifier.

Of course, the circuits of Figs. 3–34 and 3–35 are restricted to positive input signal voltages. Negative input signal voltages can be used by reversing the diode in Fig. 3–34a or using pnp transistors in Figs. 3–34b and 3–35. In such a case, the resulting equation for the output voltage of Fig. 3–35 is $v_{o2} = A\gamma_T \log -Bv_{\text{in}}$.

Logarithmic function circuits are now available as complete functional modules including temperature compensation. A number of very useful variations on and applications of the logarithmic function circuit are described in the remaining portions of this section.

Log-Ratio Amplifier.   If the outputs of two logarithmic function circuits are combined at the inputs of a difference amplifier, the difference amplifier output will be the difference of the logs of the two input voltages, which is the same as the log of the ratio of the input voltages. A log-ratio amplifier circuit is shown in Fig. 3–36. The output voltages v_{o1} and v_{o2} of OA1 and OA2 are given by Eq. (3–30). When these are combined in

Fig. 3–36 Log-ratio amplifier.

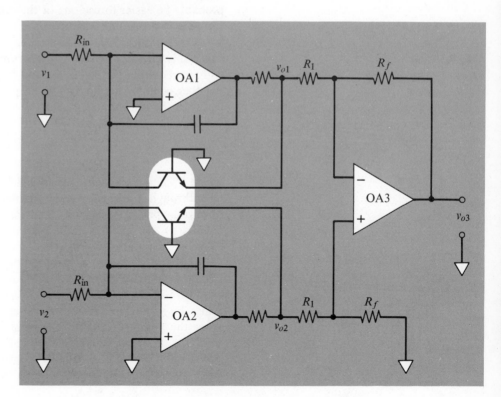

the difference amplifier, OA3, the resulting output voltage is

$$v_{o3} = \frac{R_f}{R_1} \left(-\gamma_T \log \frac{v_2}{R_{in}} + \gamma_T \log I_0 \right)$$

$$- \frac{R_f}{R_1} \left(-\gamma_T \log \frac{v_1}{R_{in}} + \gamma_T \log I_0 \right)$$

$$= \frac{\gamma_T R_f}{R_1} \left(\log \frac{v_1}{R_{in}} - \log \frac{v_2}{R_{in}} \right)$$

$$= \frac{\gamma_T R_f}{R_1} \log \frac{v_1}{v_2}. \tag{3--33}$$

Note that the $\log I_0$ terms cancel in the difference operation assuming matched transistors. The resulting expression for v_{o3} contains the $0.3\%/°C$ temperature-dependent scaling factor γ_T. The difference amplifier shown could be made to have a compensating temperature-dependent gain, but this would require matched temperature-dependent resistors. It would probably be easier to use one of the difference amplifier circuits for which the gain is adjustable with a single resistance.

The log-ratio circuit has many applications. If either v_1 or v_2 is constant, this circuit provides temperature compensation of the $\log I_0$ term without requiring the stable V_R and high-valued resistor R_R of Fig. 3–35. Another common application is the production of a signal voltage proportional to the optical absorbance A_0 of a sample:

$$A_0 = \log \frac{P_R}{P_S},$$

where P_S and P_R are the flux of light through the sample and the flux of light through a reference under similar conditions respectively. A voltage or current proportional to P_R and P_S is obtained from the light detection circuits(s) and applied to the log-ratio circuit.

Antilog Function. A simple antilog or exponential function circuit is obtained by interchanging the resistor and log elements in the simple circuits of Fig. 3–34. Temperature compensation of the $\log I_0$ term is obtained by supplying a temperature-dependent bias voltage as in Fig. 3–37. Since the antilog function is a signal amplitude expander, it is necessary to divide the input voltage with R_1 and R_2 so that the output voltage will stay within the range of the OA output. The voltage follower amplifier, OA1, provides a voltage of v_{o1} equal to the divided input voltage plus the

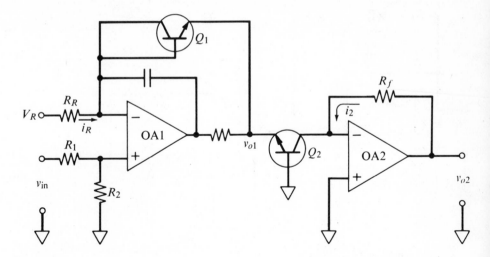

Fig. 3–37 Antilog circuit.

forward bias voltage of Q_1. Thus

$$v_{o1} = v_{in} \frac{R_2}{R_1 + R_2} - \gamma_T(\log i_R - \log I_0). \qquad (3\text{–}34)$$

This voltage is applied to Q_2, resulting in a current i_2 through Q_2 and R_f. The current-voltage relationship for Q_2 is

$$v_{o1} = -\gamma_T(\log i_2 - \log I_0). \qquad (3\text{–}35)$$

Combining Eqs. (3–34) and (3–35) to eliminate v_{o1}, we obtain

$$v_{in} \frac{R_2}{R_1 + R_2} = \gamma_T \log \frac{i_R}{i_2}. \qquad (3\text{–}36)$$

Now, since $v_{o2} = i_2 R_f$,

$$-v_{in} \frac{R_2}{R_1 + R_2} = \gamma_T \log \frac{v_{o2}}{R_f i_R}$$

and

$$v_{o2} = R_f i_R \log^{-1} \left[-v_{in} \frac{R_2}{(R_1 + R_2)\gamma_T} \right]$$

$$= C \log^{-1} (-Dv_{in}/\gamma_T), \qquad (3\text{–}37)$$

where $C = R_f i_R$ and D is the input divider fraction. The remaining temperature-variant term, γ_T, appears in the scaling factor for v_{in}. This could be compensated for by introducing an approximate temperature dependence in the input voltage divider.

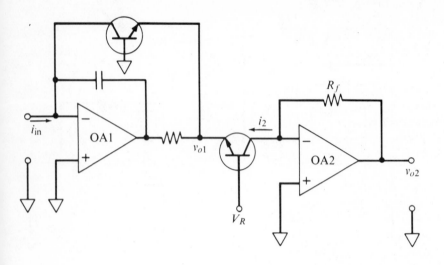

Fig. 3–38 Wide dynamic range current follower.

The choice of npn transistors in the circuit of Fig. 3–37 restricts the input voltages to negative values, as Eq. (3–37) shows. If pnp transistors are used instead, positive input voltages must be used and the output voltage equation is

$$v_{o2} = -C \log^{-1} (Dv_{\mathrm{in}}/\gamma_T).$$

One possible application for the antilog circuit is in conjunction with the log-ratio circuit to perform a multiplier function, as will be described below. A very interesting application uses a log and antilog circuit to obtain wide dynamic range in a current-follower circuit without the problems of range-switching very large valued feedback resistors. The basis of the design is shown in Fig. 3–38. The output voltage v_{o1} of OA1 is given by Eq. (3–30). The current-voltage relationship for the antilog transistor at the OA2 input is

$$V_R - v_{o1} = \gamma_T(\log i_2 - \log I_o). \tag{3–38}$$

Combining Eqs. (3–38) and (3–30) to eliminate v_{o1}, we obtain

$$V_R = \gamma_T \log \frac{i_2}{i_{\mathrm{in}}}. \tag{3–39}$$

Substituting $v_{o2} = i_2 R_f$ for i_2 in Eq. (3–39) and solving for v_{o2}, we find

$$v_{o2} = i_{\mathrm{in}} R_f \times 10 e^{V_R/\gamma_T}, \tag{3–40}$$

from which we see that the scaling factor between the output voltage and the input current is an exponential function of V_R. A change of 59η mV in V_R causes a tenfold change in the current-to-voltage gain factor. The actual value of R_f used can be moderate (≈ 10 kΩ) and the range changing accomplished by stepping V_R. Temperature compensation of γ_T could be achieved by introducing a suitable temperature dependence of V_R.

Analog Multipliers. An analog multiplier circuit produces an output voltage that is proportional to the product of two input voltages. As analog multiplier function modules have improved in accuracy, speed, versatility, and economy, their use has spread rapidly. There is already such a large selection of multiplier modules available that few experimenters will find it necessary (or advantageous) to construct their own. However, a knowledge of the basis of operation of the several types is helpful in understanding the particular advantages and limitations of each. Several of the most important types of multipliers are based on the nonlinear function circuits discussed above. These types (the logarithmic, the quarter-square, the triangle-averaging, and the transconductance) are described below. One other type, the time-division multiplier, is described in Section 3–3.3.

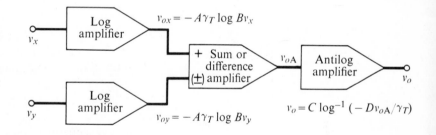

Fig. 3–39 Logarithmic multiplier, divider, or power circuit.

Logarithmic multiplier. The technique of using logarithms to obtain the product of two numbers can be implemented electronically by a combination of log and antilog function circuits, as shown in Fig. 3–39. This circuit can perform several functions depending on the choice of combining amplifier. For the multiplication function, the amplifier is a summing amplifier with a gain of $+1$. (Note that the output of the usual inverting summing amplifier must be inverted unless the antilog circuit is built to accept positive input voltage.) The output voltages of the log amplifiers given in the figure are obtained from Eq. (3–32). From these, the amplifier output voltage, v_{oA}, is

$$v_{oA} = -A\gamma_T \log Bv_x - A\gamma_T \log Bv_y$$

$$= -A\gamma_T \log B^2 v_x v_y. \tag{3–41}$$

Using the expression for v_{oA} in the antilog function of Eq. (3–37), the circuit output voltage, v_o, is obtained:

$$v_o = C \log^{-1} (AD \log B^2 v_x v_y)$$

$$= C(B^2 v_x v_y)^{AD}. \tag{3-42}$$

Thus the output voltage is proportional to the product of v_x and v_y raised to the power AD. Since A is the gain of the amplifier in the log circuit and D is the divider fraction in the antilog circuit, a unity value for AD is easily obtained by using a simple follower in the log circuit and omitting the divider in the antilog circuit. The temperature-dependent γ_T terms cancel, eliminating the need for temperature-dependent gains in the circuit. The factor CB^2 is generally adjusted to be 0.1 so that the voltage level of the product will not exceed the maximum output voltage of the antilog circuit.

When the amplifier is a difference amplifier with a gain of $+1$, the log of v_y is subtracted from the log of v_x and the division function results. By the same steps used in obtaining Eq. (3–42), the result for the difference amplifier is

$$v_o = C(v_x/v_y)^{AD}. \tag{3-43}$$

Again, the log and antilog circuits are simplified to provide unity values of A and D. Notice that the B term cancels out. The value of C is normally 10 to bring the output voltage range nearer to that of the antilog amplifier.

This same circuit is also used to provide a power function through the adjustment of the product AD. Either the multiplier or divider circuit can be used. To obtain the power function for a single input, the other input of a general purpose multiplier-divider must be held constant. To obtain the power function only, one of the input log amplifiers and the sum/difference amplifier could be eliminated.

The logarithmic circuit will accept input signals of a single polarity only. When incorporated into the multiplier or divider circuit shown in Fig. 3–39, the function will be performed for only a single combination of input polarities, e.g., both positive. There are four possible combinations of input polarities: $++$, $+-$, $-+$, and $--$, corresponding to the four quadrants defined by the $X = 0$ and $Y = 0$ coordinates in an ordinary Euclidean graph of X vs. Y. The basic logarithmic multiplier is thus said to be a **single-quadrant multiplier.** It is possible to make a two- or four-quadrant multiplier from a single-quadrant multiplier by introducing enough offset at both inputs to keep the net signal level in the allowed quadrant. If the offset voltages are V_{ox} and V_{oy}, respectively, the resulting product is proportional to $(v_x + V_{ox})(v_y + V_{oy})$, which expanded is $v_x v_y + v_x V_{oy} + v_y V_{ox} + V_{ox} V_{oy}$. Since the last three terms are either constant or single variables multiplied by a constant factor, signals propor-

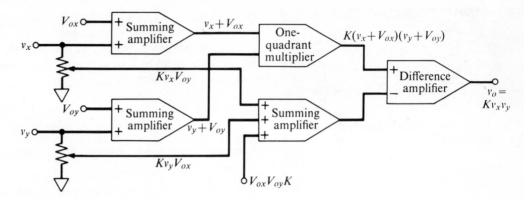

Fig. 3–40 Four-quadrant multiplication using a single-quadrant multiplier.

tional to these quantities can be easily obtained and subtracted from the product. The resulting circuit is shown in Fig. 3–40.

Quarter-square multiplier. A popular multiplication technique which works in all quadrants makes use of the fact that the square of the sum of two numbers less the square of their difference equals four times their product.[2] The analog implementation of this identity is shown in block form in Fig. 3–41. The absolute value function is obtained by the circuit of Fig. 3–31 and the square function is obtained by a multiple-segment function circuit such as that shown in Fig. 3–33 or the less accurate diode ladder described in Section 2–3.6 of Module 2. The major source of error in this approach is the approximation of the square function by line segments.

Triangle-averaging multiplier. A cleverly conceived and quite simple four-quadrant multiplier is based on the fact that the area of a triangle is one-half the product of its base and its altitude. If a symmetrical bipolar triangular waveform V_w of peak amplitude V_p is summed with a voltage v, the output waveform is as shown in Fig. 3–42. A voltage proportional to the triangular area positive of the zero line can be obtained by using a pre-

Note 2. The Quarter-Square Identity.

$$\frac{(x+y)^2 - (x-y)^2}{4}$$

$$= \frac{(x^2 + 2xy + y^2) - (x^2 - 2xy + y^2)}{4}$$

$$= xy$$

Fig. 3–41 Quarter-square multiplier.

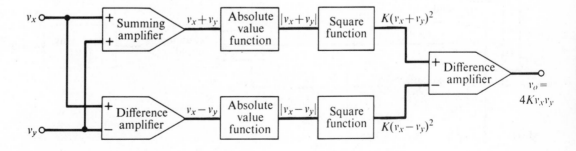

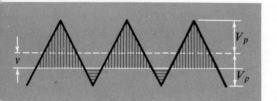

Fig. 3–42 Positive and negative areas of sum of symmetrical triangular wave and voltage v.

cision limiter to clip off the negative portion and then filtering the resulting waveform to obtain the average voltage. The average voltage of the positive portion of the waveform $\overline{V}_+$ is one-half the base b times the altitude h, where the base is given as a fraction of the period of V_w. Thus $h = V_p + v$ and $b = (1/2 + v/2V_p)$ and $\overline{V}_+$ is

$$\overline{V}_+ = \frac{1}{2}\left(\frac{1}{2} + \frac{v}{2V_p}\right)(V_p + v). \tag{3-44}$$

Similarly, the average voltage of the negative portion of the waveform $\overline{V}_-$ is

$$\overline{V}_- = -\frac{1}{2}\left(\frac{1}{2} - \frac{v}{2V_p}\right)(V_p - v). \tag{3-45}$$

It can be seen that these products contain a term in v^2 which allows a variation on the quarter-square identity to be implemented. In practice, the voltage summed to obtain $\overline{V}_+$ is $v_x - v_y$, and that summed to obtain $\overline{V}_-$ is $-(v_x + v_y)$. Thus

$$\overline{V}_+ = \frac{1}{2}\left(\frac{1}{2} + \frac{v_x - v_y}{2V_p}\right)(V_p + v_x - v_y) \tag{3-46}$$

and

$$\overline{V}_- = -\frac{1}{2}\left(\frac{1}{2} + \frac{v_x + v_y}{2V_p}\right)(V_p + v_x + v_y). \tag{3-47}$$

If $\overline{V}_+$ and $\overline{V}_-$ are summed, the result is

$$\overline{V}_+ + \overline{V}_- = -v_y - v_x v_y/V_p. \tag{3-48}$$

The undesired $-v_y$ term in the result is easily removed by subtracting it away. A block diagram of the triangle-averaging multiplier is shown in

Fig. 3–43 Triangle-averaging multiplier.

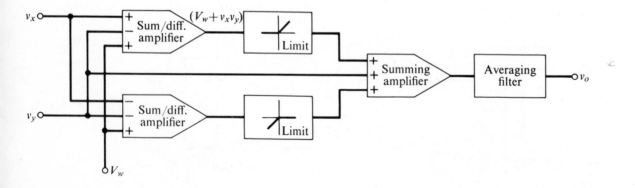

Fig. 3–43. This type of multiplier can be made to be very accurate. Its primary disadvantage is that its rate of response is limited by its output filter to frequencies much less than the frequency of the triangular wave generator.

Transconductance multiplier. The transconductance multiplier type is most responsible for the great increase in multiplier applications. Since it is relatively simple, it has been possible to produce it in integrated circuit form. It is based on the proportional relationship between the current through a transistor and its transconductance (ratio of output current to input voltage). One input signal is used to control the voltage applied to a matched pair or transistors in a differential amplifier circuit, and the other input signal is used to control the collector current. The difference in collector current is then approximately proportional to the product of the two input voltages. The technique depends on carefully matched transistors for temperature stability and is accurate over a very narrow range of input voltages and currents. Thus the input signals are scaled down before application to the transconductance circuit and the differential output signal is amplified afterward. A more sophisticated circuit based on the same principle is called the current-ratioing multiplier. It incorporates a differential current control circuit which provides better stability and accuracy. The transconductance multiplier also has very good response speed, and with added circuitry can be made into a four-quadrant multiplier.

Analog Multiplier Applications. The analog multiplier, now available in convenient modular form, is an exceptionally versatile building block for a variety of analog signal processing applications. Several of the most common applications are described below.

Division is accomplished by using an OA to adjust one of the multiplier inputs so that the product is equal to the numerator. The resulting circuit is shown in Fig. 3–44. The OA controls v_o so that $v_x v_y / 10 = -v_z$ at the summing point. Since $v_o = v_y$,

$$v_x v_o / 10 = -v_z$$

Fig. 3–44 Analog divider using multiplier circuit.

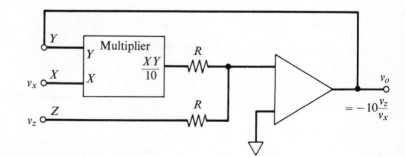

and

$$v_o = -10v_z/v_x. \tag{3-49}$$

As Eq. (3–49) shows, v_o becomes indeterminately large as v_x approaches zero; hence low values of the denominator must be used with caution. With the circuit of Fig. 3–44 negative values of v_x must be avoided because the multiplier will then invert the sense of the Y signal, and the feed-

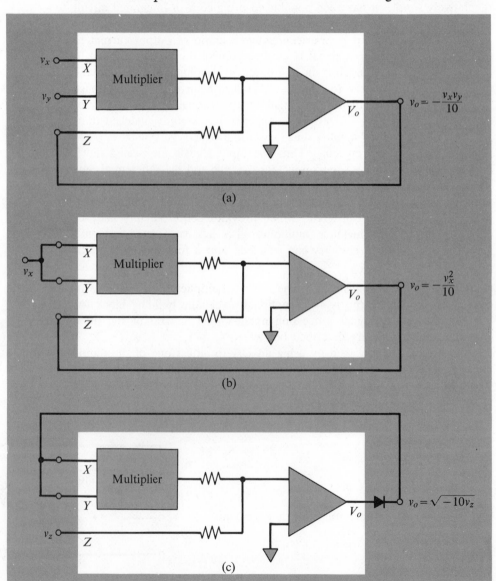

Fig. 3–45 Multiplier applications. (a) Multiplier. (b) Square. (c) Square root.

back control loop will become positive and immediately go to limit. The OA circuit shown in Fig. 3–44 is useful for so many applications that it is frequently included as part of the multiplier circuit module. Several applications in which the multiplier-OA combination is used are shown in Fig. 3–45. In Fig. 3–45a, the OA is used as an inverting output buffer amplifier by connecting the v_o and Z terminals of the multiplier module. The square circuit of Fig. 3–45b is identical to the multiplication circuit except that the same signal is connected to both inputs. The square root circuit of Fig. 3–45c is closely related to the divider of Fig. 3–44. The diode is present to maintain positive voltages at Y and X. From the output relationship (and the fact that $XY/10$ must be positive) it is clear that v_z must be negative.

One interesting computation using a combination of multiplying circuits is the calculation of a vector magnitude from its orthogonal components (such as calculating impedance from the magnitude of reactance and resistance). The function to be calculated is $v_o = \sqrt{v_x^2 + v_y^2}$ and the implementation is shown in Fig. 3–46. The undesirable $\sqrt{10}$ factor in the output expression could be eliminated by the proper choice of summing resistors in the OA circuit of the square root circuit.

Fig. 3–46   Vector magnitude computation.

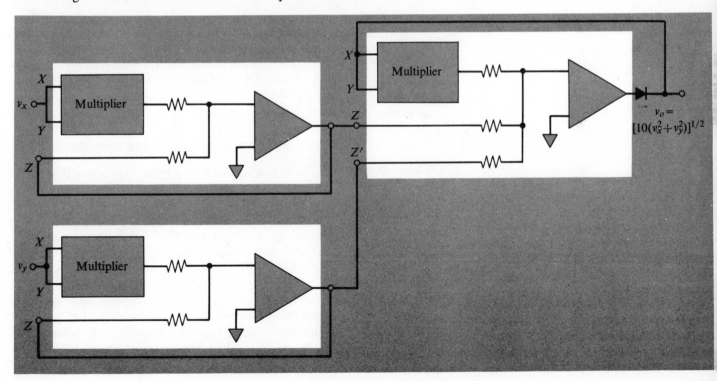

Experiment 3–6 *Absolute Value Circuit*

A circuit which produces a unipolar output voltage equal in magnitude to the input voltage regardless of its polarity is investigated in this experiment. The absolute value circuit is first tested with dc input voltages and its transfer function is verified. Then an ac input voltage is connected, and the circuit is shown to be a precision full-wave rectifier for low level signals.

Experiment 3–7 *Logarithmic Amplifiers*

By using a diode or transistor as the feedback element in an OA circuit, a logarithmic amplifier is obtained. The log amplifier finds many applications for analog signal conditioning. In this experiment, the basic log amplifier is constructed and its response over several decades is tested. Then a temperature-compensated log amplifier is constructed and tested. As an optional part of the experiment an OA circuit which responds to the logarithm of the ratio of two input signals can be constructed and tested.

Experiment 3–8 *The Analog Multiplier*

The analog multiplier is shown in this experiment to provide the important functions of multiplication, squaring, division, and taking the square root of signals. Two voltage signals are first applied to the multiplier and the accuracy of the resulting product is obtained. Then the multiplier is tested in its other modes of operation. Subsequent experiments in this module and Module 4 demonstrate many other uses of analog multipliers.

3–2.2 FREQUENCY COMPOSITION OF ELECTRICAL SIGNALS

Electrical signals in the analog domains often vary with time. With time-varying signals the amplitude at a particular time or the entire amplitude-time relationship may be the desired information. For simple time-varying signals, such as sinusoidal waveforms, the amplitude-time function can be completely described by a simple equation, which includes the frequency of the waveform, the time, and the peak amplitude. In general, signals which result from scientific experiments are not simple sine waves, composed of single frequencies, and their description is necessarily more complex. However, all signals, no matter how complex, may be represented as a sum of simple sine waves. It is the purpose of this section to illustrate how the amplitude-time relationships of periodic waveforms and transient signals can be represented by the combination of simple sinusoidal waveforms.

A finite series of elementary sine waves of the fundamental frequency and multiples of the fundamental frequency is known as a **Fourier series** and can be used to describe periodic waveforms. More complex signals can be represented by a continuous function of composite frequencies known as a **Fourier integral.** Such a representation is often used to *transform* amplitude-time information into amplitude-frequency information. This type of transformation is known as a **Fourier transformation.** The amplitude-time function and the corresponding amplitude-frequency function are known as **Fourier transform pairs.** It will be shown in this section how the Fourier transmission of signals can reveal the frequency components of signals or their **frequency spectrum.** The frequency content of signals is a very important factor in designing measurement systems which will preserve the quality of the desired analog domain information. In Module 4, *Optimization of Electronic Measurement,* the relation between frequency response and signal-to-noise ratios in measurement systems is described.

Sine Wave Signals

A sine wave signal is, of course, a voltage or current which varies sinusoidally with time as illustrated in Fig. 3–47. The sine wave is of great importance in physics and electronics. *It is the simplest periodic waveform.* Objects whose displacement is a sinusoidal function of time are undergoing "simple harmonic motion" and are very common in nature. Sinusoidal current is produced by rotating a wire loop in a uniform magnetic field, as in a power generator. The sine wave signal is the only true single-frequency periodic waveform. The repetition interval is the **period,** as shown in Fig. 3–47. Each repetition is called a **cycle.** The period is thus expressed in units of seconds per cycle. The reciprocal of the period is the number of cycles per second, or the **frequency.** The unit **cycles per second** has been given the name **Hertz** (Hz).

The generation of a sine wave from a rotating vector of magnitude A_p is illustrated in Fig. 3–48. The sine wave results from the projection of

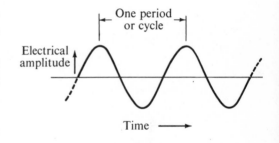

Fig. 3–47 A sinusoidal signal.

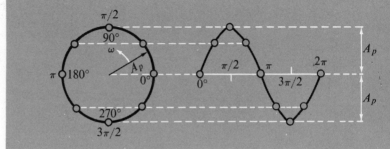

Fig. 3–48 Development of a sine wave from the projection of a rotating vector.

the vector in the vertical axis as the rector rotates counterclockwise with a uniform **angular velocity** ω. If the vector is rotating at the rate of one revolution per second, the sine wave repeats itself periodically once every second. One revolution of the vector produces one cycle every second. The period is the time interval, t_{per}, required to produce one cycle. Since a cycle is generated each time the vector sweeps through 360°, or 2π radians, the time axis is conveniently expressed as the angle of the vector. Since the vector sweeps one cycle every t_{per} seconds, the frequency f is

$$f = 1/t_{per}. \tag{3–50}$$

The vector is rotating at a rate ω or 2π radians per t_{per} seconds; thus

$$\omega = 2\pi/t_{per} = 2\pi f. \tag{3–51}$$

If the amplitude of the sine wave is at any instant of time t designated a_t and the maximum or peak amplitude is designated A_p, the equation for the sine wave is written

$$a_t = A_p \sin \omega t = A_p \sin 2\pi ft. \tag{3–52}$$

If the vector represents a current or voltage, the instantaneous current i or voltage v can be given as

$$i = I_p \sin \omega t, \tag{3–53}$$

$$v = V_p \sin \omega t, \tag{3–54}$$

where I_p and V_p are the peak current and voltage, respectively.

Fourier Series Waveform Analysis

Any periodic waveform such as those shown in Fig. 3–49 can be represented by a **Fourier series expansion.** These waveforms have a fundamental

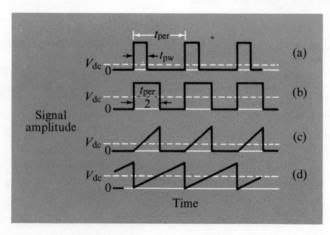

Fig. 3–49 Nonsinusoidal, periodic signals. (a) Rectangular wave. (b) Square wave. (c) Ramp. (d) Sawtooth.

frequency, or repetition rate, and contain information at other frequencies. The Fourier series is a summation of the fundamental frequency and its harmonics in various amplitude and phase relationships. The general features of composite waveforms made up of sine waves of similar and different frequencies are presented in this section. A complex waveform, a square wave, is constructed from a Fourier series expansion of sine waves in order to illustrate the frequency components of various portions of the square wave.

Combinations of Sine Waves of Similar Frequency.   When two sine waves of the same frequency are added to each other, the result is another sine wave of the same frequency. The amplitude of the resulting sine wave depends on the individual amplitudes of the elementary sine waves as well as on the relative time at which the two elementary sine waves reach similar points in their oscillations. Any difference in time between similar points such as the peak amplitudes or zero crossings of two sine waves of the same frequency is known as a **phase difference.** A phase difference, or lack of phase difference, can cause a combination sine wave to be of greater or lesser amplitude than the elementary sine waves. For example, if the two sine waves reach their maxima and minima at exactly the same times (zero phase difference), the resultant combination sine wave would have a peak amplitude which is the simple sum of the peak amplitudes of the individual sine waves. At the opposite extreme, if one sine wave reaches its maximum amplitude at the same time the other reaches its minimum (completely opposite in phase), partial or complete cancellation occurs. Complete interference would occur if each sine wave had exactly the same peak amplitude and the two waves were exactly opposite in phase.

The mathematical expression for the amplitude and phase of the combined sine wave can be readily obtained from the amplitude and phase of each elementary sine wave. Consider two voltage signals which vary sinusoidally as shown in Fig. 3–50. The two signals v_1 and v_2 have the same frequency, but cross zero at different times, and are thus out of phase. The difference in time can be conveniently expressed in terms of an angle (fraction of a cycle) called the **phase angle,** θ. If the two sine waves are shown as rotating vectors as in Fig. 3–50, the phase angle can be seen to arise from the fact that one vector **leads** or **lags** the other. A sine wave which is 90° ($\pi/2$ radians) out of phase with another sine wave could be called a "cosine wave." However, the waveform is still sinusoidal and may be written as $\cos \omega t = -\sin (\omega t - \pi/2)$.

A generalized voltage sine wave signal can be expressed as

$$v = V_p \sin (\omega t + \theta), \tag{3–55}$$

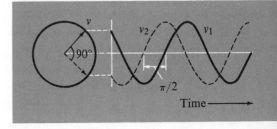

Fig. 3–50 Two sine waves with a 90° phase difference.

where θ is the phase angle between the sine wave under consideration and a reference sine wave.

The equation for the waveform which results from combining the two sine wave voltage signals v_1 and v_2 of Fig. 3–50 can be readily derived. If the voltage signal v_1 is considered to be the reference signal, the equation which describes its variation with time is $v_1 = V_{p1} \sin \omega t$, where V_{p1} is the peak amplitude of signal v_1. The equation which describes signal v_2 is $v_2 = V_{p2} \sin (\omega t + \theta)$, where V_{p2} is the peak amplitude of signal v_2 and θ is the phase angle with respect to signal v_1. The simple sum of the two signals is given by Eq. (3–56):

$$v_1 + v_2 = V_{p1} \sin \omega t + V_{p2} \sin (\omega t + \theta). \qquad (3\text{–}56)$$

From the trigonometric identity $\sin (x + \theta) = \sin x \cos \theta + \cos x \sin \theta$, the sum can be written

$$v_1 + v_2 = (V_{p1} + V_{p2} \cos \theta) \sin \omega t + (V_{p2} \sin \theta) \cos \omega t. \qquad (3\text{–}57)$$

If Eq. (3–57) is rearranged, the sinusoidal nature of the combination waveform is revealed as shown in Eq. (3–58):

$$v_1 + v_2 = V_{p1,2} \sin (\omega t + \theta_{1,2}), \qquad (3\text{–}58)$$

where

$$V_{p1,2} = [(V_{p1} + V_{p2} \cos \theta)^2 + (V_{p2} \sin \theta)^2]^{1/2},$$

and

$$\tan \theta_{1,2} = \frac{V_{p2} \sin \theta}{V_{p1} + V_{p2} \cos \theta}.$$

Thus the peak amplitude of the combination sine wave is $V_{p1,2}$ and the phase angle with respect to v_1 is $\theta_{1,2}$.

If the elementary sine waves are in phase ($\theta = 0°$) the peak amplitude is $V_{p1,2} = V_{p1} + V_{p2}$, the expected simple sum. If $\theta = 180°$ (elementary sine waves completely opposite in phase), $v_1 + v_2 = (V_{p1} - V_{p2}) \sin \omega t$. For the special case where the peak amplitudes of the two elementary signals are equal and $180°$ out of phase, complete destructive interference occurs and $v_1 + v_2 = 0$. Another interesting case occurs when $\theta = 90°$ as in Fig. 3–50. For this case,

$$V_{p1,2} = (V_{p1}^2 + V_{p2}^2)^{1/2} \quad \text{and} \quad \tan \theta_{1,2} = V_{p1}/V_{p2}.$$

For the special case of $\theta = 90°$ and $V_{p1} = V_{p2}$, the resultant peak amplitude of the combination wave is $V_{p1,2} = V_{p1}\sqrt{2}$, and the resultant phase angle of the combination wave is $\theta_{1,2} = 45°$.

Fourier Series Expansion. The method of combining sine waves shown in the previous section can be extended to include sine waves of different frequencies and phase relationships. A **Fourier series** of sine waves is a special combination of sine waves which is a summation of multiples of a single fundamental frequency. The multiple frequencies, which may have various amplitudes and phase angles, are known as **harmonics.** Any single-valued, periodic voltage waveform $v(t)$ can be represented by a Fourier series expansion, as shown in Eq. (3–59):

$$v(t) = V_{dc} + V_1 \sin (\omega t + \theta_1) + V_2 \sin (2\omega t + \theta_2)$$
$$+ \ldots + V_n \sin (n\omega t + \theta_n). \quad (3\text{–}59)$$

The term V_{dc} is called the **dc level** of the signal, and is the average value about which the signal varies. The terms $V_1, V_2, \ldots, V_n$ are the peak amplitudes of the fundamental, second harmonic, and so on through the nth harmonic, while the terms $\theta_1, \theta_2, \ldots, \theta_n$ are the corresponding phase angles.

As an example of how a Fourier series can be used to represent the waveform of a periodic signal, a square wave can be constructed as shown in Fig. 3–51. Curves A and B in Fig. 3–51a show the fundamental and the third harmonic, while curve C shows the graphical addition of the fundamental and the third harmonic. Additional odd harmonics are shown in Fig. 3–51b and c. Curve G shows the graphical addition of the fundamental plus the third, fifth, and seventh harmonics and is a fairly good approximation to the square wave. The complete Fourier series representation of the square wave of Fig. 3–51 would include higher odd harmonics, and the equation can be written

$$v(t) = \frac{4V}{\pi} (\sin \omega t + \tfrac{1}{3} \sin 3\omega t + \tfrac{1}{5} \sin 5\omega t + \tfrac{1}{7} \sin 7\omega t + \ldots).$$
$$(3\text{–}60)$$

As Eq. (3–60) shows, the square wave is made up of sine waves of the fundamental period and its odd harmonics. A plot of the relative amplitude of the signal at each frequency vs. the frequency is shown in Fig. 3–52. This is a plot of the **frequency spectrum,** $F(\omega)$. As can be seen

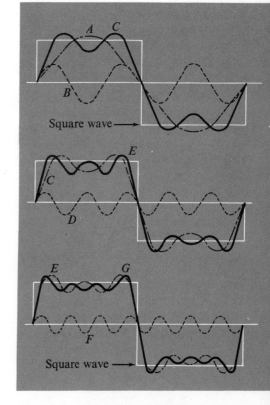

Fig. 3–51 Harmonic composition of a square wave. (A) Fundamental. (B) Third harmonic. (C) Fundamental plus third harmonic. (D) Fifth harmonic. (E) Fundamental plus third and fifth harmonics. (F) Seventh harmonic. (G) Fundamental plus third, fifth, and seventh harmonics.

Fig. 3–52 Frequency spectrum of square wave.

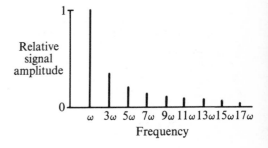

in Fig. 3–51, the higher frequency components are required to produce the abrupt changes in the signal at the leading and trailing edges of the square wave. Thus the high frequency components are contained in the steep rising and falling portions. Lower harmonic components are present in the flat top and bottom portions of the square wave. Because the square wave contains low and high frequency components, it is a very useful waveform in testing the frequency response of measurement systems.

Experiment 3–9   *Combinations of Sine Waves*

This experiment demonstrates the combination of sine wave signals of various frequency and phase relationships. Two sine waves of equal frequency, whose phase relationship can be varied, are first applied to an OA adder circuit and the resulting combination sine wave is observed on the oscilloscope. Then sine wave signals from two different generators are applied to the adder and the resulting combination waveforms are observed for equal frequencies and frequencies which are multiples of each other. The observed combination waveforms are explained in terms of the frequency and phase relationships of the two sine waves being combined.

Fourier Transform Analysis

The Fourier series analysis of the square wave demonstrates that a periodic nonsinusoidal signal is made up of many discrete frequency components. However, most electronic signals are not periodic but have a finite duration. A single square pulse waveform is an example of this more common type of waveform. The frequency composition or spectrum of such a signal is a continuous function of frequency rather than a series of discrete frequencies. The process of Fourier transformation can be used to obtain the frequency spectrum of a signal. The basic properties of Fourier transformations and the frequency spectra of several common signals are presented in this section.

Fourier Integral.   The Fourier integral is a mathematical equation that can be used to calculate the frequency spectrum of an amplitude-time waveform. The form of the Fourier integral for a voltage signal $v(t)$ is shown in Eq. (3–61):

$$F(\omega) = \int_{-\infty}^{\infty} v(t) \, [\cos \omega t - j \sin \omega t] \, dt, \qquad (3\text{–}61)$$

where j is the complex operator $\sqrt{-1}$. The convenience of using complex notation to represent vector quantities is illustrated in Section 3–2.3. For every amplitude-time waveform [$v(t)$] such as an isolated square pulse, there exists a related function $F(\omega)$ known as the frequency spectrum of that signal. The two functions $v(t)$ and $F(\omega)$ are mathematically related and are called Fourier transform pairs. Thus the frequency spectrum of any real waveform or signal can be calculated from Eq. (3–61). This calculation is normally done on a digital computer.

Frequency Spectra of Signals.   If the frequency spectrum of a signal from a transducer is known, an electronics-aided measurement system can be designed to amplify or modify the signal and provide appropriate data treatment and data display without the loss of information. Also, if the frequency spectrum of undesired electrical noise is known along with that of the signal, systems can be designed which have high signal-to-noise ratios. In many techniques for improvement of signal-to-noise ratio, the measurement system is designed to pass a limited band of frequencies such that the signal integrity is preserved while as much of the noise as possible is rejected. Thus it is very important to have an intuitive feel for the frequency spectra of common waveforms and signals. This can easily be obtained by studying some simple pictorial Fourier transform pairs.

Several pictorial Fourier transform pairs are shown in Fig. 3–53. The rigorous application of Eq. (3–61) to the calculation of frequency spectra results in negative frequencies. For essentially all practical purposes these can be ignored. The frequency spectra depicted in Fig. 3–53 show only the positive frequencies.

An infinitely sharp amplitude-time signal as shown in Fig. 3–53a has a frequency spectrum which contains equal amplitudes at all frequencies. This type of frequency spectrum is often called a **white spectrum** and it is also typical of a random noise amplitude-time waveform. This Fourier transform pair indicates that an instrument would have to have an infinitely wide frequency response in order to transmit a sharp impulse signal without distortion. As the pulse gets wider the frequency spectrum becomes narrower and in the specific case of a rectangular pulse the frequency spectrum has the form of the function $(\sin x)/x$. This is shown in Figs. 3–53b and 3–53c. In both cases the major component is at zero frequency (dc) but there are also frequency components extending to very high frequencies. As with the periodic square wave, an electronic circuit must pass these high frequencies in order to avoid distortion of the rising and falling edges of the rectangular pulse.

The form of the frequency spectrum is determined by the shape of the pulse. If the pulse is triangular, the functional form of its frequency spectrum is $(\sin^2 x)/x^2$. This is shown in Fig. 3–53d. Other common pulse and signal forms are Gaussian, Lorentzian, and exponential. These

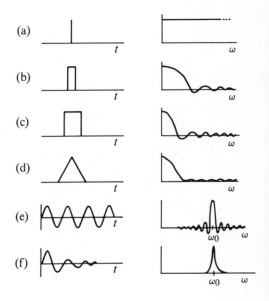

Fig. 3–53   Pictorial Fourier transform pairs.

have frequency spectra that have Gaussian, exponential, and Lorentzian functional forms respectively.

The frequency spectrum of a sine wave with a finite length is shown in Fig. 3–53e. The frequency spectrum is a $(\sin x)/x$ function centered on the frequency axis at the frequency of the sine wave. In contrast to the Fourier series representation of a sine wave, this frequency spectrum is not an infinitely narrow spike. This simply reflects the fact that the Fourier series representation of signals is applicable only to periodic waveforms and assumes that they are infinitely long. The Fourier series can usefully be thought of as a limit of the Fourier transform in the case of infinitely long waveforms. The longer the sine wave in Fig. 3–53e, the narrower would be the $(\sin x)/x$ frequency spectrum. The Fourier transform pair shown in Fig. 3–53f illustrates that the manner by which the sine wave is truncated (exponential) also alters the frequency spectrum in a manner exactly analogous to the various pulse shapes discussed above. In this case the frequency spectrum of the amplitude-time function is a Lorentzian peak.

Overall, two important properties of amplitude-time waveforms and their corresponding frequency spectra are illustrated in Fig. 3–53: namely, the inverse dependence of the broadness of the frequency spectrum on the duration of the amplitude-time waveform, and the direct dependence of the form of the frequency spectrum on the form of the amplitude-time waveform. These ideas and concepts are very useful in understanding the frequency response characteristics required for the measurement of various time-variant analog signals.

Experiment 3–10   *Frequency Composition of Electrical Signals*

It is important to know the frequency composition of varying analog signals in order to use measurement systems which have the appropriate frequency response for low distortion, high precision and accuracy, and other desired characteristics. This experiment illustrates the frequency components of several common analog waveforms. The frequency spectrum of a signal waveform is obtained by applying the waveform as one input to an analog multiplier. A sine wave whose known frequency can be varied linearly is applied as the second multiplier input. The multiplier output is filtered so that a dc voltage is obtained whenever the sine wave frequency reaches the fundamental or a harmonic of the waveform to be analyzed.

3-2.3 IMPEDANCE AND FREQUENCY RESPONSE OF REACTIVE CIRCUITS

When a varying voltage is applied across a conductor, one must consider not only the current which results from the varying field at any instant, but also the reaction of the conductor to the *variation* in voltage. For ac signals the term **impedance,** given the symbol Z, is defined to take both effects into account. Impedance in ac circuits is analogous to dc resistance, and its magnitude is $Z = V_p/I_p$, where V_p and I_p are the peak values of the voltage and current respectively. As the rate of signal variation (its frequency) approaches zero, the impedance of a device approaches its resistance.

With simple resistors, the current is always proportional to the instantaneous applied voltage as given by Ohm's Law. Since action and reaction are simultaneous in a pure resistance, there is no special reaction to signal variations and $Z = R$ for all frequencies. Thus the application of a sine wave voltage of the form $v = V_p \sin (\omega_t + \theta)$ to a pure resistor gives rise to a sine wave current of the form $i = I_p \sin (\omega_t + \theta)$. Note that the phase θ does not change; thus the current and voltage sine waves are in phase for all frequencies with a pure resistance. Pure resistance is an ideal which is closely approached by many practical resistors. However, the physical characteristics of conducting devices make some reaction to the rate of signal change unavoidable.

Reactions to the rate of change of a signal can be classified as **inductive reactance** or **capacitive reactance.** In some circuits reactance may be considered a deviation from ideality. However, it is often used intentionally in ac circuits to produce a variety of important effects as described in this section.

Capacitive Reactance

The potential across a capacitor cannot change without a change in charge. Therefore a current i equal to the rate of charge flow dQ_e/dt is required. Since the charge on a capacitor is equal to the product of the capacitance C and the voltage v_C across the capacitor, the current may be written

$$i = \frac{dQ_e}{dt} = C \frac{dv_C}{dt}. \tag{3-62}$$

If Eq. (3-62) is rewritten as

$$\frac{dv_C}{dt} = \frac{i}{C},$$

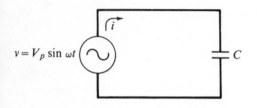

Fig. 3–54 Capacitor with ac voltage source.

it becomes clear that *the potential across a capacitor cannot change instantaneously* because an infinite current would be required.

When a capacitor is connected to a source of alternating voltage $v = V_p \sin \omega t$ as shown in Fig. 3–54, the resulting current is given by

$$i = C \frac{dv_C}{dt} = \omega C V_p \cos \omega t = I_p \cos \omega t. \qquad (3\text{–}63)$$

The resulting current and voltage waveforms are shown in Fig. 3–55. As the sine and cosine relationship suggests, there is a 90° phase difference between the current and voltage waveforms. The current through the capacitor is said to *lead* the voltage across it by 90°.

Since the impedance of a device $Z = V_p/I_p$, the impedance of a pure capacitor, called the capacitive reactance and given the symbol X_C, is

$$X_C = \frac{V_p}{\omega C V_p} = \frac{1}{\omega C} = \frac{1}{2\pi f C}. \qquad (3\text{–}64)$$

The units of capacitive reactance are ohms. Note that the capacitive reactance is frequency dependent. The higher the frequency, the less reactance a capacitor offers to the flow of current. Because of the way in which its reactance depends on frequency, a capacitor is said to be an open circuit to the flow of direct current ($X_C \to \infty$ as $f \to 0$).

The reactance of capacitors in series (X_{Cs}) is the sum of the separate reactances:

$$X_{Cs} = X_{C_1} + X_{C_2} + X_{C_3} + \ldots. \qquad (3\text{–}65)$$

The resulting capacitance of a series combination of capacitors can be found by substituting $1/(2\pi f C)$ for X_C in the above equation:

$$\frac{1}{C_s} = \frac{1}{C_1} + \frac{1}{C_2} + \frac{1}{C_3} + \ldots. \qquad (3\text{–}66)$$

The capacitance of a series combination of capacitors must be smaller than that of the smallest capacitor in the series.

The instantaneous power p supplied to a capacitor is $p = v \times i$. For the circuit of Fig. 3–54 the power is given by

$$p = \omega C V_p^2 \sin \omega t \cos \omega t = \tfrac{1}{2} C V_p^2 \sin 2\omega t. \qquad (3\text{–}67)$$

The relationship of the power waveform to the current and voltage waveforms is shown in Fig. 3–55. When the voltage across C increases, p is positive and energy is being stored in the capacitor. When the applied voltage decreases, p is negative and the energy stored in the capacitor is released to the circuit to be dissipated in the resistive elements. The

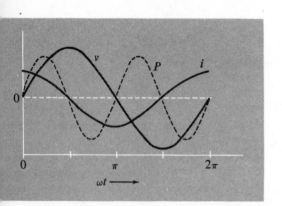

Fig. 3–55 Current, voltage, and power waveforms in circuit of Fig. 3–54.

average power supplied to the capacitor over a full cycle of the applied voltage is zero.

Experiment 3–11 *Current and Voltage Relationships in a Capacitive ac Circuit*

The relationship of current and voltage in a purely capacitive ac circuit is determined in this experiment by oscilloscope observations. A sine wave voltage source which is floating (not connected to chassis or line common) is required. The current passing through a capacitor and the voltage across it are observed simultaneously on an oscilloscope. To measure the current, a small resistor whose resistance is negligible compared to the capacitive reactance is placed in series with the capacitor. The iR drop across the resistor is displayed along with the voltage across the capacitor on a dual trace or dual beam oscilloscope.

Inductive Reactance

Inductors can also offer reactance to alternating current, although the effect of inductance is quite different from that of capacitance. A varying current through an inductor gives rise to a varying magnetic field. Because energy is stored in the magnetic field, an inductor reacts against rapid changes in current. The inductance L is defined by Eq. (3–68):

$$v_L = L \frac{di}{dt}, \tag{3–68}$$

where v_L is the voltage across the inductor. When v_L is in volts and i is in amperes, the unit of L is the henry (H).

When an inductor is connected to a source of alternating voltage of the form $v = V_p \sin \omega t$, as shown in Fig. 3–56, the rate of change of current can be found from Eq. (3–69):

$$\frac{di}{dt} = \frac{v_L}{L} = \frac{v_p}{L} \sin \omega t. \tag{3–69}$$

If Eq. (3–69) is integrated, the instantaneous current is given by

$$i = \frac{-1}{\omega L} V_p \cos \omega t = -I_p \cos \omega t. \tag{3–70}$$

From the sign in Eq. (3–70), it is clear that the current through an inductor *lags* the voltage across it by 90°. Note that this is opposite to the current and voltage waveforms in a capacitive circuit.

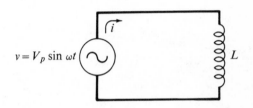

Fig. 3–56 Inductor with ac source.

Again the impedance of a pure inductor, called the inductive reactance and symbolized X_L, can be found from V_p/i_p as shown in Eq. (3–71):

$$X_L = \frac{V_p}{V_p/(\omega L)} = \omega L = 2\pi f L. \qquad (3\text{--}71)$$

Note that inductive reactance, like capacitive reactance, is frequency dependent. For inductors, however, the reactance increases with increasing frequency. An inductor is said to be a short circuit to direct current, since $X_L \rightarrow 0$ as $f \rightarrow 0$.

The reactances of inductors in series are simply additive. That is,

$$X_{Ls} = X_{L_1} + X_{L_2} + X_{L_3} + \cdots. \qquad (3\text{--}72)$$

Thus the total inductance for inductors in series is

$$L_s = L_1 + L_2 + L_3 + \cdots. \qquad (3\text{--}73)$$

Experiment 3–12   *Current and Voltage Relationships in an Inductive ac Circuit*

The relationship between current and voltage in a purely inductive ac circuit is determined in this experiment by oscilloscopic observations. A high frequency sine wave source which is floating with respect to chassis or line common is required. A resistor, whose resistance is negligible with respect to the inductive reactance, is placed in series with the inductor in order to measure the current through the inductor. The iR drop across the resistor is displayed simultaneously with the voltage across the inductor in order to obtain the current and voltage waveforms.

Complex Notation

When resistors, capacitors, and inductors are combined in reactive circuits, trigonometric and graphical means can be used to obtain the resulting circuit impedance and the phase angle between current and voltage waveforms. However, with circuits which have several reactive or resistive components, it is much easier to deal with complex numbers in the solutions for impedance and phase angle. In complex notation the **imaginary operator** $j = \sqrt{-1}$ is used to indicate a phase shift of 90°. This is quite similar to multiplying a signal by -1 to indicate a 180° phase shift, as is commonly done with amplifiers. Since reactive elements shift the phase of signals by 90°, the use of the operator j allows quick and accurate algebraic treatment of reactive circuits. The impedance of an inductor is thus given by

$$\mathbf{Z}_L = jX_L = j\omega L \qquad (3\text{-}74)$$

and the impedance of a capacitor is given by

$$\mathbf{Z}_C = -jX_C = \frac{-j}{\omega C}, \qquad (3\text{-}75)$$

where the boldface symbol for the impedance indicates a complex quantity. The sign difference in the complex notation for $\mathbf{Z}_L$ and $\mathbf{Z}_C$ indicates the phase difference between the voltage across and the current through the reactive element. Thus $+j$ for $\mathbf{Z}_L$ indicates that the signal voltage across the inductor leads the current by 90°, while $-j$ for $\mathbf{Z}_C$ indicates that the signal voltage lags the current by 90°.

The method of complex notation can be readily illustrated with the aid of the vector diagram of Fig. 3–57. Here a complex quantity A is specified by its component along the real axis (a_1) and its component along an imaginary axis at right angles to the real axis (ja_2). The complex quantity A is written as the sum of these two components, as shown by Eq. (3–76):

$$\mathbf{A} = a_1 + ja_2, \qquad (3\text{-}76)$$

where $+j$ is used to show that the imaginary component has been rotated by $+90°$. A rotation of $-90°$ would be indicated by $-j$ in Eq. (3–76) and would correspond to the imaginary component $-ja_2'$ shown in dotted lines in Fig. 3–57 and the complex number A'.

Complex notation is very useful for analyzing circuits with reactive and resistive components because the impedance $\mathbf{Z}$ can be represented as a complex number with a real (resistive) component and an imaginary (reactive) component as shown in Eq. (3–77) for a series combination of resistance and reactance:

$$\mathbf{Z} = R + jX. \qquad (3\text{-}77)$$

For parallel combinations of R and X, the reciprocal of the complex impedance is the sum of the reciprocal resistance and reciprocal imaginary reactance.

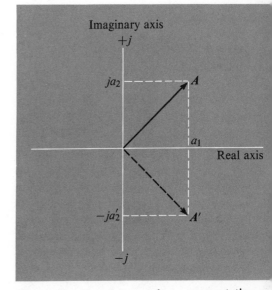

Fig. 3–57 Complex number representation as vectors.

RC Circuits

Simple circuits consisting of resistors and capacitors are of great importance in treating ac signals. The series RC circuit is often intentionally used to attenuate low frequency or high frequency signal components. In other cases unintentional RC combinations which arise from stray capacitance can limit frequency response. A discussion of parallel RC circuits and the frequency-compensated voltage divider follows the analysis of series *RC* circuits.

Series RC Circuits.   A series *RC* circuit with a sinusoidal voltage source is shown in Fig. 3–58. The source voltage $v_s = V_p \sin \omega t$ causes a sinusoidal current i of unknown magnitude and phase angle to flow in the circuit. The magnitude of the resulting current and the phase difference between current and voltage could be solved by applying Kirchhoff's voltage law to the circuit, $v_s = v_R + v_C$. If v_R is replaced by iR and v_C by $Q_e/C = 1/C \int i\, dt$, Eq. (3–78) results:

$$v_s = iR + \frac{1}{C} \int i\, dt. \tag{3–78}$$

This equation can be readily solved by substituting $v_s = V_p \sin \omega t$ and differentiating all terms with respect to time. However, complex notation allows the desired result to be obtained by ordinary algebraic methods.

The impedance of the series combination of R and C is given by

$$\mathbf{Z} = R - jX_C = R - \frac{j}{\omega C}. \tag{3–79}$$

The impedance is again complex, with a real part R and an imaginary component X_C. An impedance diagram can be drawn with real and imaginary components as shown in Fig. 3–59. The magnitude of the complex impedance is written Z, not boldface, and from the geometry of the diagram is

$$Z = V_p/I_p = \sqrt{R^2 + X_C^2}. \tag{3–80}$$

The phase angle θ is

$$\tan \theta = X_C/R. \tag{3–81}$$

The above relationships can be illustrated by a simple example. Suppose that in Fig. 3–58, $R = 10$ kΩ, $C = 0.02$ µF, and the signal source has an output voltage of 25 V peak and a frequency of 500 Hz. The capacitive reactance is thus $X_C = 1/2\pi \times 500 \text{ sec}^{-1} \times 2 \times 10^{-8} \text{ F} = 1.59 \times 10^4 \ \Omega$. The magnitude of the impedance is

$$Z = \sqrt{(10 \text{ k}\Omega)^2 + (15.9 \text{ k}\Omega)^2} = 18.8 \text{ k}\Omega.$$

The peak current in the circuit is $I_p = 25 \text{ V}/18.8 \text{ k}\Omega = 1.33$ mA. The phase angle θ between the current and the applied voltage is $\theta = \arctan 15.9 \text{ k}\Omega/10 \text{ k}\Omega = 57.8°$ with the current leading the voltage. The peak

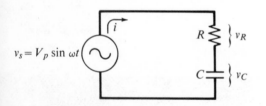

Fig. 3–58   Series *RC* circuit.

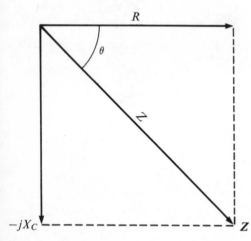

Fig. 3–59   Impedance vectors for series *RC* circuit.

voltage across the resistor is

$$V_{pR} = I_p R = 1.33 \times 10^{-3}\,\text{A} \times 10^4\,\Omega = 13.3\,\text{V}$$

and across the capacitor it is

$$V_{pC} = I_p X_C = 1.33 \times 10^{-3}\,\text{A} \times 15.9 \times 10^3\,\Omega = 21.2\,\text{V}.$$

Note that the sum of the peak voltages across R and C does not equal the applied V_p. Kirchhoff's law is obeyed for the instantaneous values. The difference in phase prevents simple addition for the peak values.

The power dissipated in the resistor, p_R, is simply $p_R = I_{\text{rms}}^2 R = (0.707 \times 1.33 \times 10^{-3}\,\text{A})^2 \times 10^4\,\Omega = 8.85 \times 10^{-3}\,\text{W}.$[3] As was shown earlier, the power dissipated by the capacitor is zero.

The **apparent power** supplied by the voltage source is

$$\text{Apparent power} = V_{\text{rms}} \times I_{\text{rms}}. \tag{3–82}$$

In this case, the apparent power is $(0.707 \times 1.33 \times 10^{-3}\,\text{A}) \times 0.707 \times 25\,\text{V} = 1.66 \times 10^{-2}\,\text{W}$, but the **actual power** dissipated in the circuit as shown above is only $8.85 \times 10^{-3}\,\text{W}$. Another way to find the actual power dissipated is to take the *vector* product of the current and voltage supplied by the source:

$$\text{Actual power} = V_{\text{rms}} \times I_{\text{rms}} \times \cos\theta, \tag{3–83}$$

where θ is the phase angle as given in Eq. (3–81). For the above example the actual power, calculated from Eq. (3–83), is $1.66 \times 10^{-2} \times \cos 57.8° = 8.85 \times 10^{-3}\,\text{W}$. The ratio of the actual power to the apparent power is called the **power factor.** From Eqs. (3–83) and (3–82),

$$\text{Power factor} = \frac{\text{actual power}}{\text{apparent power}}$$

$$= \frac{V_{\text{rms}} \times I_{\text{rms}} \times \cos\theta}{V_{\text{rms}} \times I_{\text{rms}}} = \cos\theta. \tag{3–84}$$

Frequency Response of Series RC Circuits. The series RC circuit of the above section is an ac voltage divider in much the same way that a series resistive circuit is a dc voltage divider. The division of voltage between resistor and capacitor is, however, frequency dependent. Figure 3–60a shows one configuration of the RC voltage divider, which is, of great importance in ac circuits. The output of the circuit is taken across the resistance R. The ratio of the voltage across R to that from the source is the attenuation of the divider and is called the **network transfer function,** $H(j\omega)$. The transfer function contains both frequency and phase information and is thus the ratio of the resistance R to the total complex im-

Note 3. RMS Value for Varying Signals. The power being dissipated in a resistor at any given time is proportional to the square of the current or voltage as given by the relations $i^2 R$ and v^2/R. When a repetitively varying signal is applied to the resistor, the average power dissipated over the repetition period is proportional to the average of the square of the current or voltage during that period. From the standpoint of power, then, the effective voltage or current of the signal is the square root of that average value. This is called the root-mean-square or rms value of the signal. As derived in Section 3–2.4, the rms value of a sinusoidal signal is 0.707 times the peak value.

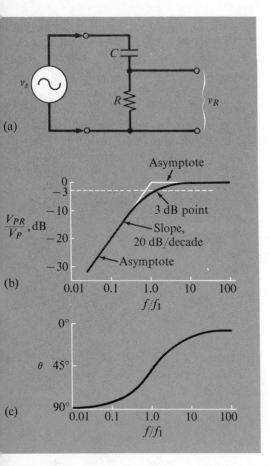

Fig. 3-60 Frequency response of high pass circuit. (a) High pass network. (b) Bode diagram of magnitude ratio vs. f/f_1. (c) Phase shift vs. f/f_1.

pedance **Z** as shown in Eq. (3-85):

$$H(j\omega) = \frac{v_R}{v_s} = \frac{R}{Z} = \frac{R}{R - jX_C}. \qquad (3\text{-}85)$$

Quite often the magnitude of the attenuation is of primary interest. In that case the amplitude of the transfer function, symbolized $|H(j\omega)|$, is written. Note that this function contains no information about the phase relationships between the signals across R and the source voltage. The magnitude function $|H(j\omega)|$ depends on the ratio of R to the magnitude of the network impedance as shown in Eq. (3-86):

$$|H(j\omega)| = \frac{V_p R}{V_p} = \frac{R}{Z} = \frac{R}{\sqrt{R^2 + X_C^2}}. \qquad (3\text{-}86)$$

The frequency dependence of the transfer function can be easily seen from Eq. (3-86). At high frequencies X_C becomes very small compared to R, and $V_p R$ is nearly equal to V_p. Hence the transfer function is approximately unity. As the frequency decreases, X_C increases and the fraction of V_p found across R approaches zero. Thus $|H(j\omega)|$ decreases to zero at low frequencies. Such a circuit is called a **high pass filter,** because only high frequency signals pass unattenuated. Note also that any dc component of the source will not be passed by C and cannot appear in the output across R.

The voltage across R leads the signal voltage by the phase angle $\theta = \arctan(X_C/R)$. Thus the phase difference between v_R and v_s increases with decreasing frequency.

With a signal frequency such that $X_C = R$, the power dissipated in R is exactly half the apparent power, the output voltage $V_p R = 0.707 V_p$, and the phase angle $\theta = 45°$. This frequency is called the **lower cutoff frequency,** f_1, and is given by

$$f_1 = \frac{1}{2\pi RC}. \qquad (3\text{-}87)$$

It is common practice to use a graphical representation to express the frequency dependence of filters and other reactive circuits. The **Bode diagram** is the most common method for expressing the frequency dependence of circuits. Equation (3-86) can be easily rearranged to give $|H(j\omega)|$ as a function of frequency. Usually $|H(j\omega)|$ is plotted in decibels[4] vs. the logarithm of the ratio f/f_1 or ω/ω_1. If the numerator and denominator of Eq. (3-86) are divided by $R = 1/(2\pi f_1 C) = 1/(\omega_1 C)$, Eq. (3-88) results:

$$|H(j\omega)| = \frac{1}{\sqrt{1 + (X_C^2/R^2)}} = \frac{1}{\sqrt{1 + (f_1/f)^2}} = \frac{1}{\sqrt{1 + (\omega_1/\omega)^2}}. \qquad (3\text{-}88)$$

The Bode diagram has two limiting slopes at very low and very high frequencies. The actual plot approaches these limiting slopes asymptotically. At very high frequencies where $f \gg f_1$, $|H(j\omega)| \approx 1$ and a plot of $|H(j\omega)|$ in dB vs. log f/f_1 has a slope of 0. At very low frequencies, where $f \ll f_1$, $(f/f_1)^2 \gg 1$ in the denominator of Eq. (3–88), and the transfer function reduces to $|H(j\omega)| = (f/f_1)$ or, in decibels, is 20 log (f/f_1). For each tenfold (decade) decrease in f, at low frequencies, $|H(j\omega)|$ decreases by 20 dB. For each twofold (octave) decrease in f, $|H(j\omega)|$ decreases by 6 dB. Thus the response of the high pass filter is said to roll off below f_1 at a rate of 20 dB/decade or 6 dB/octave.

The diagram in Fig. 3–60b shows a Bode plot for the high pass filter. Note that the two asymptotic slopes shown white represent quite accurately the true response curve except for the region very near f_1. Hence for very quick construction of Bode diagrams, two slopes can be drawn which intersect at f_1. For the high pass filter the low frequency slope increases at 20 dB/decade until the attenuation reaches 0 dB at f_1. Note that the actual attenuation factor at f_1 is 0.707, which represents an attenuation of -3 dB. Hence f_1 is often referred to as the **lower 3 dB frequency.** Since it can also represent the turning point of the two limiting slopes, f_1 is also called the **corner frequency.**

The phase angle, θ, between v_R and v_s can also be plotted against log (f/f_1) as shown in Fig. 3–60c. Hence both amplitude and phase information can be conveniently plotted by the Bode method to obtain a complete picture of the transfer response of a network. Note that $\theta = 45°$ at the lower 3 dB frequency.

In the circuit of Fig. 3–61a, the same series RC network is rearranged so that the output is taken across the capacitor. The network transfer function is

$$H(j\omega) = \frac{v_C}{v_s} = \frac{-jX_C}{\mathbf{Z}} = \frac{-jX_C}{R - jX_C} \qquad (3\text{–}89)$$

Note 4.   Decibel.

Decibel, dB, is a term used primarily to express the ratio of two power levels such as the signal power output and signal power input where the circuit power gain in dB $= 10$ log $(P_\text{out}/P_\text{in})$. The term has been extended through the relationship $P = V^2/R$ to express voltage level ratios where the circuit voltage gain in dB $= 20$ log $(V_\text{out}/V_\text{in})$. Technically, the voltage ratio relationship should only be used when the impedances across which the voltages appear are equal. However, it is common practice to express voltage ratios in dB's regardless of the impedances involved.

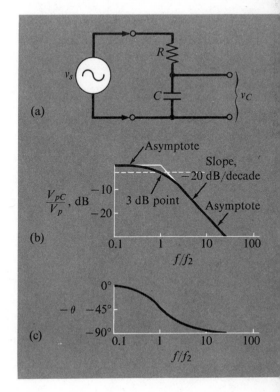

Fig. 3–61 Frequency response of low pass circuit. (a) Low pass network. (b) Bode diagram of magnitude ratio vs. f/f_2. (c) Phase shift vs. f/f_2.

and the amplitude of the transfer function is

$$|H(j\omega)| = \frac{X_C}{Z} = \frac{X_C}{\sqrt{R^2 + X_C^2}}. \quad (3\text{-}90)$$

This RC circuit is a **low pass filter** because only low frequencies appear unattenuated. The voltage across C lags the signal voltage by $90° - \theta$, where $\theta = \arctan (R/X_C)$. The frequency at which $X_C = R$ is called the **upper cutoff frequency**, f_2, and is given by the same relationship as that in Eq. (3-87):

$$f_2 = \frac{1}{2\pi RC}.$$

Equation (3-90) is easily arranged to express $|H(j\omega)|$ in terms of the ratio f/f_2 as shown in Eq. (3-91):

$$|H(j\omega)| = \frac{1}{\sqrt{1 + (f/f_2)^2}}. \quad (3\text{-}91)$$

At low frequencies the transfer function is approximately unity, while at frequencies above f_2, a rolloff slope of -20 dB/decade or -6 dB/octave occurs. Figure 3-61b and c shows Bode plots for the amplitude of the transfer function and the phase shift of the network.

The high pass circuit is used extensively to attenuate components of a signal below f_1, while the low pass circuit is used to remove those above f_2. Both networks also find use in applications where a known phase shift between input and output waveforms is desired.

Parallel RC Circuits. A parallel combination of resistance and capacitance is shown in Fig. 3-62. The reciprocal complex parallel impedance $1/\mathbf{Z}$ is equal to the sum of $1/R$ and $1/{-jX_C}$ as shown in Eq. (3-92):

$$\frac{1}{\mathbf{Z}} = \frac{1}{R} - \frac{1}{jX_C} = \frac{1}{R} + j\omega C. \quad (3\text{-}92)$$

The reciprocal of the complex impedance is given a separate name, the **complex admittance**, and the symbol $\mathbf{Y}$. The reciprocal of capacitive reactance is given the name **capacitive susceptance** and the symbol B_C. Since $1/R$ is the conductance G, Eq. (3-92) is often written

$$\mathbf{Y} = G + jB_C. \quad (3\text{-}93)$$

All three reciprocal quantities have the units ohms^{-1} or mhos.

An admittance diagram for the parallel RC circuit of Fig. 3-62 is shown in Fig. 3-63. The magnitude of the admittance is simply

$$Y = \sqrt{B_C^2 + G^2}, \quad (3\text{-}94)$$

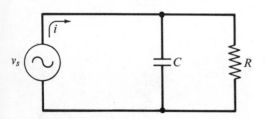

Fig. 3-62   A parallel RC circuit.

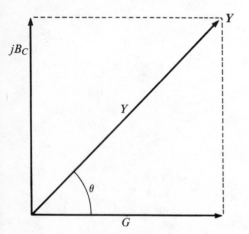

Fig. 3-63   Admittance diagram for a parallel RC circuit.

while the phase angle θ is given by

$$\theta = \arctan (B_C/G) = \arctan (R/X_C). \qquad (3\text{–}95)$$

Frequency-Compensated Voltage Divider. The frequency response of resistive circuits can often be limited by unintentional stray capacitance, as shown by the voltage divider of Fig. 3–64a. The stray capacitance C_2 is shown in dotted lines. Often by adding a second capacitor C_1 in parallel with R_1, the effect of stray capacitance can be neutralized as shown in Fig. 3–64b. Such compensation capacitors are often used in the attenuator probes of oscilloscopes and the attenuators of amplifiers to compensate for the input capacitance.

The circuit consists of two parallel RC networks in series. Components C_2 and R_2 can be replaced by an equivalent impedance

$$\mathbf{Z}_2 = \frac{R_2}{1 + j\omega C_2 R_2},$$

while R_1 and C_1 can be replaced by

$$\mathbf{Z}_1 = \frac{R_1}{1 + j\omega C_1 R_1}.$$

The fraction v_o/v_{in} appearing across the parallel combination of R_2 and C_2 is the ratio of the impedance $\mathbf{Z}_2$ to the total impedance $\mathbf{Z}_1 + \mathbf{Z}_2$ as shown in Eq. (3–96):

$$H(j\omega) = \frac{v_o}{v_{\text{in}}} = \frac{\mathbf{Z}_2}{\mathbf{Z}_1 + \mathbf{Z}_2}$$

$$= \frac{R_2/(1 + j\omega C_2 R_2)}{R_1/(1 + j\omega C_1 R_1) + R_2/(1 + j\omega C_2 R_2)}. \qquad (3\text{–}96)$$

If a compensating capacitor is not used ($C_1 = 0$), Eq. (3–96) reduces to

$$H(j\omega) = \frac{R_2}{R_1(1 + j\omega C_2 R_2) + R_2},$$

showing the normal divider transfer function of $R_2/(R_1 + R_2)$ at dc ($\omega = 0$) and a decrease in the transfer function at high frequencies. The Bode diagram of the circuit is the same as Fig. 3–61 with f_2, the 3 dB point, occurring where $X_{C_2} = R_1 R_2/(R_1 + R_2)$. If compensating capacitor C_1 is chosen such that $R_1 C_1 = R_2 C_2$, then Eq. (3–96) reduces to

$$v_o/v_{\text{in}} = R_2/(R_1 + R_2) \qquad (3\text{–}97)$$

and the voltage divider fraction is independent of frequency.

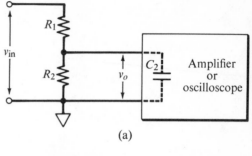

(a)

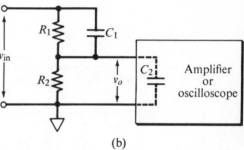

(b)

Fig. 3–64 Voltage dividers. (a) Uncompensated. (b) Frequency-compensated.

Experiment 3–13   *Voltage and Phase Angle Relationships for a Series RC Circuit*

A series *RC* circuit is studied in this experiment in order to obtain information about the voltage and phase angle relationships. A sine wave signal of fixed frequency is applied to the *RC* circuit, and the peak voltages across *R* and *C* are measured with an oscilloscope for various values of the resistance *R*. From each value of *R* and *C*, the total impedance of the *RC* circuit is calculated and used to obtain expected values of the peak voltages. The expected phase angles between the capacitor and resistor voltages are also calculated.

Experiment 3–14   *RC Filters*

This experiment demonstrates the use of series *RC* circuits as high and low pass filters. The high pass filter is first studied and its response is measured as a function of the input frequency. Results are compared to those expected from the ac voltage divider equations. Then the low pass filter is similarly characterized.

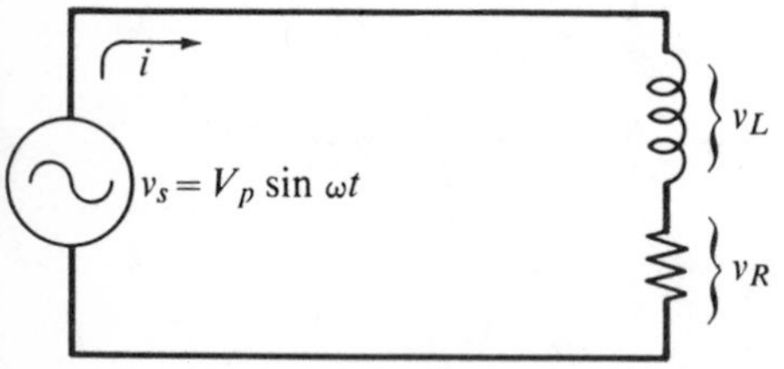

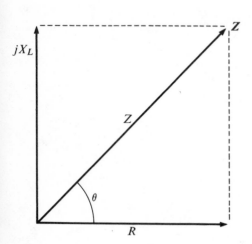

Fig. 3–65   Series *RL* circuit.

Fig. 3–66   Impedance diagram for series *RL* circuit.

RL and *RLC* Combinations

Several important circuits result when inductors and resistors are placed in series or parallel. AC voltage dividers can be made from series *RL* circuits to form high and low pass filters. The combination of *R*, *L*, and *C* can provide the important series and parallel resonant circuits which are the heart of some oscillators and frequency-selective circuits.

Series RL Circuits.   A series *RL* circuit is illustrated in Fig. 3–65. Since the two components are in series, the complex impedance **Z** is the sum of the resistance and the complex capacitive reactance as shown in Eq. (3–98):

$$\mathbf{Z} = R + j\omega L. \qquad (3\text{–}98)$$

The magnitude of **Z** can be found from the impedance diagram of Fig. 3–66:

$$Z = \sqrt{R^2 + X_L^2}. \qquad (3\text{–}99)$$

The phase angle θ is given by

$$\tan \theta = X_L/R. \qquad (3\text{–}100)$$

The discussion of power in the series *RC* circuit applies directly to the *RL* case. The only actual power dissipated is in the resistance of the circuit.

The discrepancy between power dissipated and the product of rms current and rms source voltage is accounted for by the power factor $\cos \theta$.

The series RL circuit is also useful as an ac voltage divider, as shown in Fig. 3–67. If the output voltage is taken across R, a low pass filter results, while a high pass filter is obtained when the output is taken across L.

For the low pass filter of Fig. 3–67a, the fraction of V_p to appear at the output is given by the ratio of R to the magnitude of the total impedance as shown by Eq. (3–101):

$$|H(j\omega)| = \frac{R}{Z} = \frac{R}{\sqrt{R^2 + X_L^2}}. \tag{3–101}$$

Only at low frequencies, where $X_L \ll R$, will the full source output appear across R. The upper cutoff frequency, f_2, is reached when $R = X_L$ and is given by Eq. (3–102):

$$f_2 = R/2\pi L. \tag{3–102}$$

At this frequency $V_{pR}/V_p = 0.707$. The phase difference between the input voltage v_s and the output voltage v_R is θ as given in Eq. (3–100). The output voltage leads the input voltage.

For the high pass network of Fig. 3–67b, the fraction appearing across L is

$$|H(j\omega)| = \frac{X_L}{Z} = \frac{X_L}{\sqrt{R_2 + X_L^2}}. \tag{3–103}$$

Since X_L increases with increasing frequency, the full source voltage will appear across L only at high frequencies, where $X_L \ll R$. The lower cutoff frequency, f_1, where $X_L = R$ and $V_{pL}/V_p = 0.707$, is given by $f_1 = R/2\pi L$.

Series Resonant Circuit. When a resistor, an inductor, and a capacitor are placed in series as shown in Fig. 3–68, a **series resonant circuit** results. The series LRC circuit is a simple combination of the series RC and RL circuits already discussed. The complex impedance of the circuit may be written

$$\mathbf{Z} = R + j\omega L - \frac{j}{\omega C}. \tag{3–104}$$

If the inductive reactance is greater than the capacitive reactance, the circuit has a net inductive effect and the impedance diagram is as shown in Fig. 3–69. The magnitude of the impedance is given by

$$Z = \sqrt{R^2 + (X_L - X_C)^2} \tag{3–105}$$

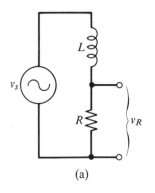

(a)

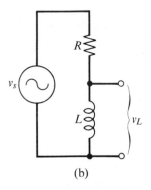

(b)

Fig. 3–67 RL voltage dividers. (a) Low pass filter. (b) High pass filter.

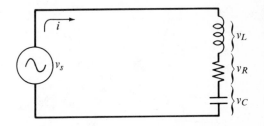

Fig. 3–68 Series resonant circuit.

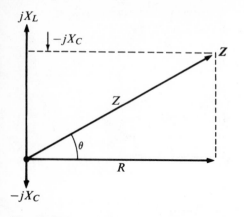

Fig. 3–69 Impedance diagram for series *LRC* circuit with $X_L > X_C$.

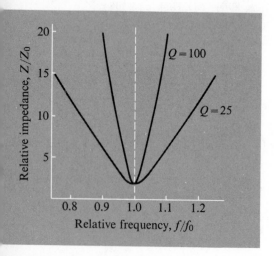

Fig. 3–70 Relative impedance of series resonant circuit vs. relative frequency.

and the phase angle between the voltage applied to the circuit and the current is given by

$$\tan \theta = (X_L - X_C)/R. \qquad (3\text{--}106)$$

There is a special case of the series resonant circuit worth noting; that is when $X_C = X_L$. From Eq. (3–105) it is apparent that the impedance of the circuit at the frequency where $X_C = X_L$ is simply the resistance R and that this is the minimum impedance the circuit can show. The frequency at which $X_C = X_L$ is called the **resonant frequency,** f_0, and is given by

$$f_0 = 1/2\pi\sqrt{LC}. \qquad (3\text{--}107)$$

At the resonant frequency the source voltage is in phase with the current, since the impedance is just R. Thus the current is simply v_s/R. The magnitude of the voltage across the inductor is

$$v_L = iX_L = \frac{v_s}{R} X_L. \qquad (3\text{--}108)$$

Since $X_C = X_L$, the magnitude of the voltage across C is equal to v_L but $180°$ out of phase. If R is much less than X_L, the voltage across the inductor (or capacitor) will be much greater than the signal voltage. Kirchhoff's voltage law is obeyed, because the voltage across L is offset by an equally high voltage across C which is out of phase. In other words, at the resonant frequency f_0, a large single-frequency voltage can be obtained across L or C while the voltage source supplies only the power dissipated in R.

The series resonant circuit is thus a selective frequency filter, since the magnitude of the voltage across L or C will decrease rapidly on either side of the resonant frequency.

To increase the selectivity or the response of the resonant circuit, R should be minimized. Usually R can be reduced to the resistance of the wire in the inductor. The **quality factor,** Q, of the circuit is a measure of the frequency selectivity of the circuit and is given by the ratio of the inductive reactance of the coil to its resistance, and is usually measured at the resonant frequency:

$$Q = \frac{X_L}{R} = \frac{2\pi f_0 L}{R}. \qquad (3\text{--}109)$$

The effect of the Q factor of the inductor on the frequency selectivity is shown in Fig. 3–70, where the ratio of the impedance of the circuit to the impedance at resonance Z_0 is plotted vs. the relative frequency f/f_0. Note that as Q increases the selectivity increases, because the impedance increases more rapidly from the resonance impedance as f deviates from f_0. Common values for the Q factor are from 10 to 100. Since the inductive reactance of a given inductor increases with frequency, coils with high Q values are more easily constructed for high frequency resonant circuits.

Parallel Resonant Circuit. A parallel resonant circuit is shown in Fig. 3–71. The circuit is seen to be a series-parallel combination of components. Again R is usually the resistance of the inductor. The impedance of the series RL combination $\mathbf{Z}_{RL}$ is $R + jX_L$. The total circuit impedance is the parallel combination of $\mathbf{Z}_{RL}$ and X_C as shown in Eq. (3-110):

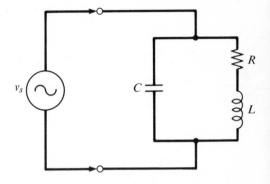

Fig. 3–71 Parallel resonant circuit.

$$\frac{1}{\mathbf{Z}} = \frac{1}{X_C} + \frac{1}{\mathbf{Z}_{RL}}, \qquad (3\text{–}110)$$

$$\mathbf{Z} = \frac{(-jX_C)(R + jX_L)}{R + jX_L - jX_C}. \qquad (3\text{–}111)$$

At the resonant frequency, f_0, the capacitance reactance $X_C = X_L = X$ and Eq. (3–111) can be written

$$\mathbf{Z}_0 = \frac{-jX(R + jX)}{R} = \frac{X^2}{R} - jX, \qquad (3\text{–}112)$$

where $\mathbf{Z}_0$ is the complex circuit impedance at resonance. Since the Q factor of the coil is X_L/R, the impedance at resonance can be written

$$\mathbf{Z}_0 = QX - jX. \qquad (3\text{–}113)$$

For high-Q coils the complex reactive term in Eq. (3–113) can be neglected and the impedance at resonance is approximately $\mathbf{Z}_0 = QX$.

The resonant frequency is the same as that for the series resonant circuit, $f_0 = 1/2\pi\sqrt{LC}$. On either side of f_0 the impedance of the parallel resonant circuit decreases from its maximum value at resonance. Note that this is the opposite behavior to that of the series resonant circuit. Again the value of Q determines the sharpness of the tuning, as illustrated in Fig. 3–72.

The parallel resonant circuit is very common in tuned amplifier circuits and frequency selective filters. These narrowband circuits are discussed in Module 4.

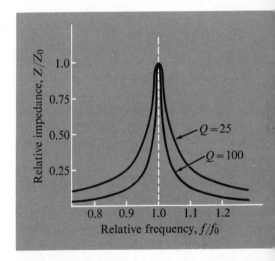

Fig. 3–72 Impedance of parallel resonant circuit vs. frequency.

Impedance Characteristics of Transmission Lines

The lines used to transmit electrical signals from one point to another in laboratory measurements are known as transmission lines. Some of the common lines used in measurement applications are parallel-wire lines, single-wire lines over a grounded plane, and coaxial cables. These three

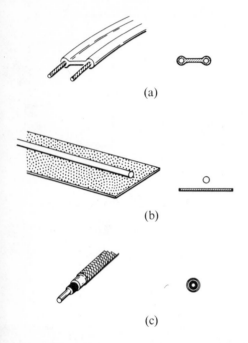

(a)

(b)

(c)

Fig. 3–73 Typical transmission lines. (a) Parallel-wire line, top view and cross section. (b) Single-wire above ground plane, top view and cross section. (c) Coaxial line, side view and cross section.

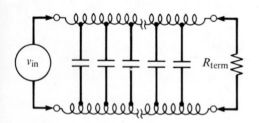

Fig. 3–74 Equivalent circuit of a transmission line.

transmission lines are illustrated in Fig. 3–73. The parallel-wire configuration of Fig. 3–73a usually consists of equal diameter wires held apart by an insulation material. The familiar antenna lead-in for a television set is an example of a parallel-wire transmission line. The single-wire line appears in many electrical circuits where the interconnection of components is made by a single wire. The components of such circuits are normally held on a grounded metal chassis which completes the circuit and is sometimes referred to as an infinite ground plane. The coaxial line is very often used for input and output connections between laboratory instruments and has the advantage that the signal on the cable is shielded from external radiation sources. Likewise, external sources are shielded from the signal transmitted by the line.

Transmission lines are characterized by a series inductance L per unit length and a parallel capacitance C per unit length, which act to impede changes in the voltage applied to the transmission line. The load which the transmission line presents to a varying signal source is determined by the inductance and capacitance of a short section at the input end. As the equivalent circuit of Fig. 3–74 suggests, the effect on the input signal of the sections of series and parallel reactance farther from the input is increasingly less, so that a limiting **characteristic impedance** Z_0 is soon reached which is given by

$$Z_0 = \sqrt{L/C}. \qquad (3\text{–}114)$$

Note that Z_0 is independent of the length of the line and the frequency of the signal. For lines where losses due to the resistance of the wire and leakages between the conductors can be neglected, Z_0 is a pure resistance but it is clearly *not* the dc resistance of the line. Since L and C are functions of the line geometry type and dimensions, lines with various characteristic impedances are available. For parallel-wire lines Z_0 is typically about 300 Ω, while for coaxial and twisted-pair lines values from 50 to 100 Ω are most common. The 50 Ω characteristic impedance is most common in commercially available coaxial cables. The popular RG58/U coaxial cable has a Z_0 of 50 Ω, and L and C of 212 nH/m and 85 pF/m respectively. A change in v_{in} applied to the transmission line travels down the line at a rate given by

$$t_d = \sqrt{LC}, \qquad (3\text{–}115)$$

where t_d is the transmission delay time per unit length of line. When the change in signal level reaches the end of the transmission line, any change in the impedance causes a phase shift between current and voltage which results in a reflection of the change back along the line toward the input. To avoid reflections, transmission lines must be terminated with a resistance equal to the characteristic impedance of the line. Transmission lines are sometimes used as intentional delay elements in signal transmission. Applying the characteristics of the RG58/U cable to Eq. (3–115) results in a t_d of 4.2 nsec/m.

Impedance Measurements

The impedance of a device or circuit can be measured in several ways which are entirely analogous to the techniques of resistance measurement described in Section 2–1 of Module 2. If only the magnitude of the impedance is desired, one can apply a sinusoidal voltage or current to the circuit and measure the ratio V_p/I_p. However, impedance is a complex quantity and its complete determination would require that the phase angle also be determined. The techniques of phase angle measurement are described in Section 3–3. Another possible technique is to separately measure the resistive and reactive components of the impedance. When a sinusoidal current is applied to a circuit, the resulting sinusoidal voltage signal can be considered as the sum of two sine wave components of different phase angle. The magnitude of the component in phase with the applied current is proportional to the resistive component of the impedance, and the magnitude of the component 90° out of phase with the applied current is proportional to the reactive component of the impedance. Knowing the resistive and reactive components allows direct calculation of the magnitude of the impedance and the phase angle. The technique of measuring the components of signals at particular phase angles is described in Section 3–3. A closely related technique is the impedance comparison bridge method described below.

Null Comparison Impedance Measurement. The comparison of impedances is accomplished in much the same way as the comparison of resistances, i.e., with a bridge. A bridge for the measurement of impedance is shown in Fig. 3–75. If R_A is equal to R_B, then at balance the impedance $\mathbf{Z}_s$ of R_s and C_s must equal the unknown impedance $\mathbf{Z}_u$, but there is only one combination which will yield the same impedance *and* phase angle as the unknown. To demonstrate that phase balance is necessary for a null, consider the left arms (a,d,c) as an ac voltage divider for the source voltage. The right arms (a,b,c) also form a divider for the same source. When the impedance $\mathbf{Z}_s$ is equal numerically to the impedance $\mathbf{Z}_u$, the magnitudes of the voltages across $\mathbf{Z}_u$ and $\mathbf{Z}_s$ will be identical. However, these two voltages may be out of phase. The detector sees the potential difference between points d and b. The difference between two sine waves of equal amplitude V_p and different phase is

$$v = V_p \sin \omega t - V_p \sin (\omega t + \theta), \qquad (3\text{–}116)$$

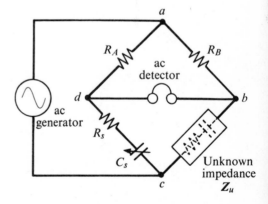

Fig. 3–75 Impedance bridge.

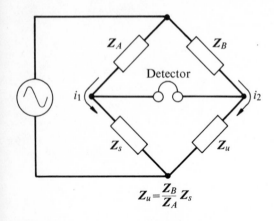

$$Z_u = \frac{Z_B}{Z_A} Z_s$$

Fig. 3–76 Impedance bridge at balance.

which, following the solution developed for Eq. (3–56), is

$$v = 2V_p(1 - \cos \theta) \sin \left(\omega t - \tan^{-1} \frac{\sin \theta}{1 - \cos \theta} \right). \quad (3\text{–}117)$$

This shows that the difference is a sinusoidal signal of magnitude $2V_p \times (1 - \cos \theta)$ and that only when $\theta = 0$ will the detector observe a true null.

It is possible to derive the conditions of balance for the impedance bridge just as was done for the Wheatstone bridge. To make the derivation completely general, the possibility of having a complex impedance in all four arms has been considered, as in Fig. 3–76. At balance, the detector current is zero, and i_1 and i_2 flow in the impedances as shown. The two equations resulting from the fact that there is no potential across the detector are

$$i_1 \mathbf{Z}_A = i_2 \mathbf{Z}_B \quad \text{and} \quad i_1 \mathbf{Z}_s = i_2 \mathbf{Z}_u.$$

From the above,

$$\frac{i_2}{i_1} = \frac{\mathbf{Z}_A}{\mathbf{Z}_B} = \frac{\mathbf{Z}_s}{\mathbf{Z}_u}.$$

Solving for $\mathbf{Z}_u$, we obtain

$$\mathbf{Z}_u = \frac{\mathbf{Z}_B}{\mathbf{Z}_A} \mathbf{Z}_s. \quad (3\text{–}118)$$

If this equation is applied to the bridge of Fig. 3–75, the following solution for $\mathbf{Z}_u$ results:

$$\mathbf{Z}_u = \frac{R_B}{R_A} \left(R_s - \frac{j}{\omega C_s} \right). \quad (3\text{–}119)$$

$\mathbf{Z}_u$ may be separated into its resistive and reactive parts according to Eq. (3–119). The resistive component of $\mathbf{Z}_u$ is $R = (R_B/R_A)R_s$, and the reactive component is $X = -(R_B/R_A)(1/\omega C_s)$. The impedance is thus

$$\mathbf{Z}_u = \frac{R_B}{R_A} \sqrt{R_s^2 + \left(\frac{1}{\omega C_s} \right)^2} \quad (3\text{–}120)$$

and the phase angle

$$\theta = \cos^{-1} \frac{R_s}{\sqrt{R_s^2 + (1/\omega C_s)^2}}. \quad (3\text{–}121)$$

If it is known that the unknown impedance is a series combination of a resistance R_u and a capacitance C_u, the actual values of R_u and C_u can be determined. Substitute $R_u - j/\omega C_u$ for $\mathbf{Z}_u$ in Eq. (3–119):

$$R_u - \frac{j}{\omega C_u} = \frac{R_B}{R_A}\left(R_s - \frac{j}{\omega C_s}\right).$$

Separating the above equation into its "real" and "imaginary" parts yields

$$R_u = \frac{R_B}{R_A}R_s, \qquad\qquad (3\text{–}122)$$

$$C_u = \frac{R_A}{R_B}C_s. \qquad\qquad (3\text{–}123)$$

This bridge is called the series-capacitance bridge, and Eqs. (3–122) and (3–123) above are its conditions of balance. Note that the ratio R_A/R_B appears as a multiplier, as in the Wheatstone bridge. Note also that the conditions of balance do not contain a frequency term. Theoretically, then, this type of bridge could be operated from a generator of any waveform, or even from pulses or switching transients.

Suppose that the unknown impedance $\mathbf{Z}_u$ is composed of a parallel combination of resistance R_u and capacitance C_u. The admittance $\mathbf{Y}_u$ of the parallel combination is

$$\mathbf{Y}_u = \frac{1}{\mathbf{Z}_u} = \frac{1}{R_u} + j\omega C_u. \qquad\qquad (3\text{–}124)$$

Equation (3–117) may be rewritten as

$$\mathbf{Y}_u\mathbf{Z}_s = \frac{\mathbf{Z}_A}{\mathbf{Z}_B} \qquad\qquad (3\text{–}125)$$

because it is easier to work with admittances in a parallel circuit. Substituting the complex equivalents of $\mathbf{Y}_u$ and $\mathbf{Z}_s$ into Eq. (3–125), we obtain

$$\left(\frac{1}{R_u} + j\omega C_u\right)\left(R_s - \frac{j}{\omega C_s}\right) = \frac{R_A}{R_B}.$$

Expansion of this equation yields

$$\frac{R_s}{R_u} + j\omega C_u R_s - \frac{j}{\omega C_s R_u} + \frac{C_u}{C_s} = \frac{R_A}{R_B}.$$

Separating the equation into the real and imaginary terms yields the fol-

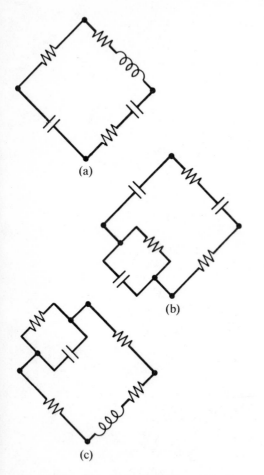

(a)

(b)

(c)

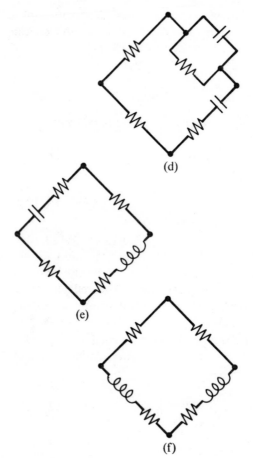

(d)

(e)

(f)

Fig. 3–77 Impedance bridges. (a) Owen. (b) Schering. (c) Maxwell. (d) Wein. (e) Hay. (f) Inductance.

lowing equations of balance:

$$\frac{R_s}{R_u} + \frac{C_u}{C_s} = \frac{R_A}{R_B},$$ (3–126)

$$\omega^2 C_u C_s R_s R_u = 1.$$ (3–127)

When the above equations are solved for C_u and R_u in terms of the known values,

$$C_u = \frac{R_A}{R_B} \frac{C_s}{(R_s^2 C^2 \omega^2 + 1)},$$ (3–128)

$$R_u = \frac{R_B}{R_A} \frac{(\omega^2 C_s^2 R_s^2 + 1)}{\omega^2 C_s^2 R^2}.$$ (3–129)

Equation (3–127) shows the bridge balance to depend on the frequency, and Eqs. (3–128) and (3–129) confirm this. A bridge with a series combination in one arm and a parallel combination in an adjacent arm is called a Wein bridge. This sort of bridge is useful for measuring resistance and capacitance in terms of standard components and frequency, or it can be used to measure frequency. When a frequency term enters the calculations, as above, it is necessary to have a very stable and accurate oscillator.

There are many other types of bridges which are used for various measurements. Some of these are shown in Fig. 3–77. The choice of bridge arrangement will depend on the approximate value of the components to be measured, the availability of accurate standards, and the necessity of an accurate oscillator. The analysis of each of the bridge types begins with Eq. (3–118) and follows the pattern of the two illustrations exactly.

3–2.4 AMPLITUDE MEASUREMENTS OF PERIODIC SIGNALS

The information contained in a regularly recurring (periodic) waveform must be in the shape of the waveform, its amplitude, or its frequency. The encoding of information as a signal frequency is discussed in Section 3–3. The amplitude of a repetitive waveform may be expressed as the peak value, the peak-to-peak value, the average value, or the rms value, depending on the application. The discussion of periodic signal measurement devices in this section is organized according to the amplitude characteristic that the device responds to directly. It is also shown in this section how the relationships among the several amplitude characteristics depend on the waveshape and thus how serious measurement errors can occur if the waveshape, desired amplitude characteristic, and device function are not carefully considered. The measurement devices themselves often have insufficient sensitivity to effectively measure many of the signals we would like to measure. For this reason, periodic signal measuring systems frequently include an amplifier to increase the signal amplitude. Ideally, the signal will be amplified linearly and without distortion of the waveshape. However, the frequency response limitations of practical amplifiers are very often the limiting factor in the useful range of application of the measurement system. This section begins with a discussion of the source and effects of frequency limitations in amplifiers.

Frequency Response of Amplifiers

All amplifiers have limited frequency response, which is sometimes desirable, but for amplifying broadband signals (signals with a broad frequency spectrum) such limitations are undesirable. Bandpass limitations in amplifiers cause the gain to be frequency dependent. For example, an ac amplifier, such as an audio amplifier, might have a constant gain of 1000 for frequencies from 20 Hz to 20 kHz, but at frequencies outside this range the gain may rapidly decrease. A dc amplifier may have a constant gain from dc to 1 kHz, but have decreasing gain for signals of higher frequencies. The basic reactive circuits presented in the last section are a major cause of frequency limitations in amplifiers. In this section the gain-vs.-frequency characteristics of wideband ac and dc amplifiers are analyzed along with the phase and waveform distortion which accompanies limited frequency response.

Ac amplifiers may be classified in a variety of ways. Quite often the frequency range over which the amplifier is to operate is a common criterion for differentiating among types of amplifiers. Audio amplifiers are normally designed to provide voltage and/or power gain in the audio frequency range, that is, in the frequency range of the human ear, which extends from a lower limit of about 15–20 Hz to an upper limit of about 20 kHz. Video amplifiers, used for television, must be capable of even broader frequency response, often from as low as 30 Hz to as high as 100 MHz. Radio frequency amplifiers used in AM radio broadcasting must handle signals from 535 kHz to 1800 kHz, and in FM broadcasting from 88 to 108 MHz.

Low Frequency Response.   Consider the amplifier input circuit shown schematically in Fig. 3–78. The coupling capacitor C_C is added to ensure that the average input voltage is zero regardless of any dc component from the source. In other words, the coupling capacitor blocks any dc component from the amplifier input and ensures that only ac signal components appear at the amplifier input.

The amplifier with a blocking capacitor input represented schematically in Fig. 3–78 has a certain bandwidth over which the amplification is relatively constant. At low frequencies and at high frequencies the amplification decreases. The voltage amplification of an amplifier over the frequency range where constant gain is achieved is called the **midfrequency amplification** and often symbolized A_v.

The major reason for the decrease in the gain of an *RC*-coupled amplifier at low frequencies is that the coupling capacitor C_C and the parallel combination of R_b and the input resistance of the amplifier form an ac voltage divider (high pass network). The capacitive reactance, $X_C = 1/2\pi f C_C$, increases as the frequency decreases, and an increasing fraction of v_{in} appears across the coupling capacitor. Thus the low frequency response of a capacitance-coupled amplifier is merely that of the high pass filter discussed in Section 3–2.3.

The low frequency cutoff, f_1, is designed as that frequency at which the amplification falls to 0.707 times the midfrequency value, A_v. This

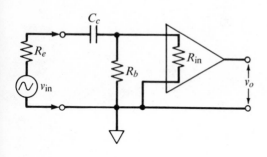

Fig. 3–78   Capacitor-coupled amplifier input.

corresponds to a reduction in gain of 3 dB. In a high pass network a signal is attenuated to this extent when $X_c = R$. If the low frequency attenuation is due entirely to the input coupling network, f_1 is given by

$$f_1 = \frac{1}{2\pi C_C(R_p + R_e)}, \tag{3–130}$$

where $R_p = R_{\text{in}}/(R_b + R_{\text{in}})$ and R_e is the equivalent source resistance. Since the attenuation of a high pass filter is equal to $1/[1 + (f_1/f)^2]^{1/2}$, as shown in Section 3–2.3, the magnitude of the ratio of the gain A_l at low frequency to the midfrequency gain A_v is given by

$$\left| \frac{A_l}{A_v} \right| = \frac{1}{[1 + (f_1/f)^2]^{1/2}}. \tag{3–131}$$

Thus if the amplifier 3 dB point and A_v are known, the gain at any low frequency can be determined from Eq. (3–131).

High Frequency Response.   The mechanisms responsible for an amplifier's loss in gain at high frequencies are not so simple as those responsible for the low frequency loss. In vacuum tube amplifiers, the high frequency limitations are caused by shunt capacitances which are not intentionally part of the circuit and effectively short-circuit the high frequency signal to ground. Stray wiring capacitance and internal capacitance between elements in the tube both contribute to the effective shunt capacitance. In addition, however, the transit time for charge carriers to move from one element to another becomes an important consideration in most amplifiers at high frequencies. In transistor amplifiers there are also capacitive effects. The transit time effects usually result in a variation of the current gain with frequency. Both transit time effects and capacitive effects are usually combined to develop a high frequency equivalent circuit of the transistor amplifier. New designs continue to extend the frequency response of solid state devices. Integrated circuits have greatly reduced the capacitance due to the connections between transistors.

With both transistor and vacuum tube amplifiers, the equivalent circuits at high frequencies contain a low pass network which leads to a reduction in gains. Although the exact gain-loss mechanisms are quite different in the two amplifier types, an upper cutoff frequency, f_2, can be defined in both cases which is the upper cutoff frequency of the low pass network:

$$f_2 = \frac{1}{2\pi R_s C_s}, \tag{3–132}$$

where R_s is the equivalent series resistance of the amplifier and C_s is the equivalent shunt capacitance. The upper cutoff frequency f_2 is again that frequency at which the amplification falls to 0.707 of its value A_v at mid-frequencies or the upper 3 dB frequency. Since the attenuation of a low pass network can be written as $1/[1 + (f/f_2)^2]^{1/2}$, the magnitude of the ratio of the amplification at an arbitrary high frequency A_h to that at midfrequency can be expressed as

$$\left| \frac{A_h}{A_v} \right| = \frac{1}{[1 + (f/f_2)^2]^{1/2}}. \tag{3–133}$$

Amplifier Bandwidth.   The combined high, low, and midrange frequency dependence of an amplifier may be summarized by the log-log plot of Fig. 3–79. Here the magnitude of the relative gain A/A_v, in dB, is plotted vs. log frequency. In the midfrequency range, where $A = A_v$, the relative gain is 0 dB. At $f = f_1$ and $f = f_2$, the relative gain is —3 dB. Beyond the 3 dB points the gain rolls off with frequency with an asymptotic slope of 6 dB/octave or 20 dB/decade. Such rolloff characteristics are merely those of the low and high pass networks. The bandwidth of an ac amplifier is usually taken to be $f_2 - f_1$, and although there is appreciable gain outside this frequency range, it should also be remembered that the attenuation at f_1 and f_2 is already 30%.

It seems reasonable to assume that if the frequency spectrum of a signal falls within the bandpass of the amplifier, it will be amplified without distortion. This assumption can be tested by the square wave for which the frequency spectrum is given by Eq. (3–60) and Fig. 3–52. It has been demonstrated that the discontinuities of the square wave are made up of the highest frequency components. Since the Fourier expansion of the square wave is a series which extends indefinitely to the high harmonics, a square waveform with an instantaneous transition from one voltage level to the other would contain frequency components to infinite frequency. A practical amplifier with a finite high frequency limit will distort the signal as shown in Fig. 3–80. The ability of an amplifier to respond to instantaneous signal changes is measured in terms of the **rise time,** t_r, the time required to go from 10% to 90% of the applied change. It is a useful rule-of-thumb that the rise time of an amplifier is approximately related to f_2 by

$$t_r \approx 1/(3f_2). \tag{3–134}$$

An even more dramatic example of square wave distortion is provided by the low frequency limitation of the amplifier. The lowest frequency term in the Fourier series expansion for a 1000 Hz square wave is the 1000 Hz fundamental. A high pass filter with an f_1 of 1000 Hz has

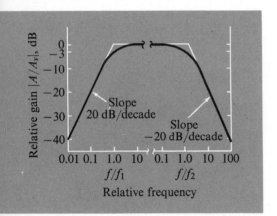

Fig. 3–79   Amplifier frequency response.

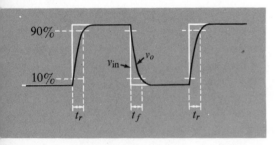

Fig. 3–80   Distortion of square wave due to high frequency response limit.

an RC time constant of $1/2\pi f_1$ (from Eq. 3–87) $= 160$ μsec. However, a half-cycle is 500 μsec long and the amplitude of a square wave subjected to such a filter would drop to a small fraction of the initial step as shown in Fig. 3–81. Even when $f_1 = 30$ Hz and $RC = 5$ msec there is a definite slope or a "droop" to the square wave. The percentage droop, D, can be determined from the expression

$$D = 100\pi \frac{f_1}{f}, \tag{3–135}$$

where f is the frequency of the square wave. For a 10% droop on a 1000 Hz square wave,

$$f_1 = \frac{1000 \times 10}{100\pi} = 32 \text{ Hz}.$$

If only 1% droop can be tolerated, the low frequency response of the circuit must be extended to 3.2 Hz. This somewhat surprising result is due to the fact that the fundamental frequency of the square wave contains a majority of the energy of the signal and that a signal of frequency f_1 is subject to substantial attenuation and phase distortion. To predict the effect of bandpass limitations on complex waveforms, the Fourier integral (Eq. 3–61) of the waveform is multiplied by the complex transfer function $H(j\omega)$ of the amplifier, and the product transformed back to the resulting amplitude-time function.

Phase Distortion.   Phase distortion occurs when the signal undergoes a phase change in passing through a high pass or low pass network. As was shown in Section 3–2.3, the extent of the phase change in RC networks depends on frequency. A complex waveform can be severely distorted by phase distortion. The phase change in a high pass or low pass network is significant only in regions of attenuation. Therefore, as the gain of the amplifier falls off, the phase distortion increases. Figure 3–82 shows the gain and phase shift of an amplifier as a function of f/f_1 or f_2/f. This graph shows that the frequency must be nearly $10f_1$ or $0.1f_2$ before the midfrequency gain is reached. The phase shift θ between the output and input can be expressed in terms of f_1 or f_2 as shown in Eq. (3–136):

$$\theta = \arctan (f_1/f) \quad \text{or} \quad \arctan (f/f_2). \tag{3–136}$$

Thus the phase distortion is small within the region of good frequency response $(f_1/f$ or $f/f_2 \ll 1)$. For virtually undistorted amplification of nonsinusoidal signals, it may be necessary for the signal frequency to be 100 times greater than f_1 to avoid phase distortion. Note that at the 3 dB points the phase shift is 45°. At f_1 the output signal leads the input, while at f_2 it lags the input signal.

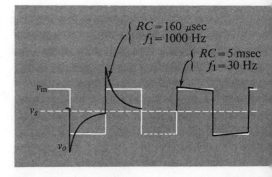

Fig. 3–81   Response of a high pass circuit to a square wave signal.

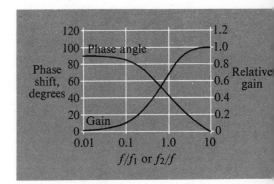

Fig. 3–82   Universal amplification curve.

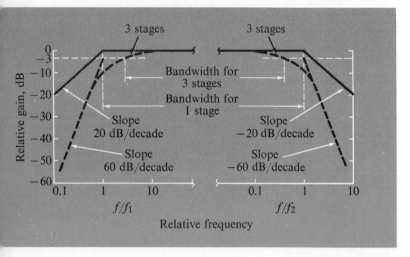

Fig. 3–83 Bode diagram showing attenuation of a one-stage and a three-stage amplifier.

Cascaded Amplifier Bandwidth. It is quite common to amplify a signal successively through several stages of amplification. The gain of the resulting amplifier is the product of the gain of each stage. The bandwidth of the entire amplifier can be no better than the bandwidth of the worst stage.

In fact, if all the stages of an amplifier have identical bandwidths, the overall bandwidth is somewhat worse than that of each stage. This is shown by the frequency response curve of Fig. 3–83. The upper cutoff frequency of an n-stage amplifier, f_{2n}, can be expressed by

$$f_{2n} = \sqrt{2^{1/n} - 1}\, f_2, \qquad (3\text{--}137)$$

where f_2 is the 3 dB point of each stage. The lower cutoff frequency, f_{1n}, is also related to the 3 dB point of each stage f_1 by Eq. (3–138):

$$f_{1n} = \frac{1}{\sqrt{2^{1/n} - 1}}\, f_1. \qquad (3\text{--}138)$$

For a three-stage amplifier, for example, $f_{23} = 0.26f_2$ and $f_{13} = 3.8f_1$, which shows clearly the reduction in bandwidth caused by cascading stages.

Response of dc Amplifiers. The outputs of many transducers are dc voltages which must be amplified electronically to produce a sufficient output signal to operate a meter, recorder, or other readout device. Dc amplifiers also find widespread use in oscilloscopes and in the operational amplifier.

For the amplification of dc signals, basically two different methods are used. In the first method, the dc signal is converted to an alternating signal by means of a mechanical or solid state chopper. The ac signal thus produced is amplified to the appropriate level by an ac amplifier. Such chopper amplifiers have been described for potentiometric measurements in Module 1. The ac amplifier can be a relatively narrowband amplifier, since most of the signal information occurs in a small band of frequencies around the chopping frequency. The resulting amplified ac signal is then converted back to dc by rectification techniques. Such modulation-demodulation methods are described further in Module 4. The alternative method for amplifying dc signals is to use an amplifier which has a frequency response bandwidth which extends to 0 Hz. Since the RC method of coupling ac amplifier stages reduces low frequency response, true dc amplifiers utilize *direct coupling* between stages.

The frequency response curve of the amplification A of a dc amplifier, such as an operational or instrumentation amplifier, is shown in Fig. 3–84. The amplification at dc and low frequencies is called the open-loop gain and often symbolized A_0. The amplifier shown in Fig. 3–84 has an open-loop gain of 10^4 or 100 dB. At higher frequencies A begins to decrease and typically rolls off with a slope of -6 dB/octave just as does a low pass network. The upper cutoff frequency f_2 (3 dB point) is located on the diagram. The bandwidth of the dc amplifier is just f_2, since the low frequency response extends to 0 Hz. If the OA is used with feedback to provide an amplifier with a voltage gain of 100 (40 dB), the value of f_2 for the resulting amplifier will be 10 kHz, as shown in Fig. 3–84. The gain of an amplifier times the bandwidth at that gain is a constant called the **gain-bandwidth product.** For the amplifier of Fig. 3–84, this product is 10^6. Another important frequency located on Fig. 3–84 is the frequency at which the open-loop gain has decreased to unity (0 dB), f_t. The open-loop unity-gain bandwidth, f_t, is often quoted in commercial literature, and is a useful figure when the dc amplifier is used as a unity-gain or feedback control amplifier. The amplifier of Fig. 3–84 has a unity-gain bandwidth of 1 MHz, which is typical of many operational amplifiers.

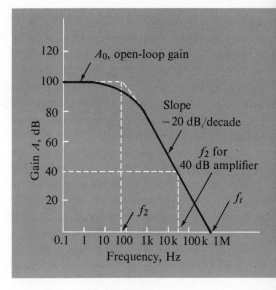

Fig. 3–84 Bode diagram for an OA.

Experiment 3–15 *Measurement of Amplifier Frequency Response*

The frequency response characteristics of an ac and a dc amplifier are measured in this experiment in order to illustrate the way in which amplifier gain and phase shift vary with the input frequency. The oscilloscope is used as an *x-y* plotter in order to obtain gain and phase shift information simultaneously. An ac amplifier is first evaluated and the frequency response curve is plotted to obtain the amplifier upper and lower cutoff frequencies (3 dB points). Then a dc amplifier is characterized and its upper cutoff frequency is obtained.

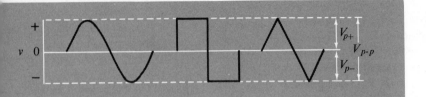

V_{p+}
$V_{p\text{-}p}$
V_{p-}

Fig. 3–85   Peak measures of common waveforms.

Measures of Periodic Signal Amplitude

The amplitudes of periodic signals are generally described in one of four ways: the peak value, the peak-to-peak value, the average value, or the rms value. In this section, these amplitude measures are defined and applied to the common waveforms sine, square, and triangle.

Peak Values.   The peak value is the voltage or current at the maximum point in the waveform. If the signal has both positive and negative polarity during a cycle, there will be a positive peak value, V_{p+}, and a negative peak value, V_{p-}, as shown in Fig. 3–85. If the waveform is centered about zero volts, V_{p+} and V_{p-} are equal in magnitude. This measure is useful where the maximum positive or negative excursions are of principal interest. Where the total magnitude of the change independent of the zero position is of interest, the peak-to-peak value is used. For a waveform symmetrical about the zero point, $V_{p\text{-}p} = 2V_p$. The concept of an ac signal as one in which the charge flow actually reverses direction periodically is too confining. In signal descriptions, it is useful to define an ac signal as that part of the signal current or voltage which varies with time. The average value about which the ac signal varies is called the dc signal level. Figure 3–86 shows a signal with sinusoidal variation of V_p volts about an average signal value of V_{dc}. The total signal may actually be considered as the sum of the steady or dc voltage and the variant or ac voltage. If V_{dc} were zero, the flow of charge would reverse direction and the voltage would change sign, as in the conventional concept of alternating current.

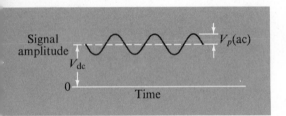

Signal
amplitude
V_{dc}
$V_p(ac)$
0
Time

Fig. 3–86   Ac and dc signal components.

Average Values.   The average value of a waveform is the average of the absolute magnitude of the signal during an entire period. This is illustrated in Fig. 3–87 for several waveforms. For the sine wave symmetrical about zero, the average value is obtained by integrating the waveform over a half-period and dividing by the time of one-half period:

$$V_{av} = \frac{1}{\pi} \int_0^{\pi} V_p \, (\sin \omega t) \, d\omega t = \frac{2V_p}{\pi} = 0.637 V_p. \quad (3\text{–}139)$$

The average value of a unipolar square wave is clearly equal to $0.5V_p$. If the wave is symmetrically bipolar, the signal maintains a constant amplitude of V_p, making $V_{av} = V_p$. On the other hand, the average value of the triangular waveforms is $0.5V_p$ whether they are unipolar or bipolar. The average value is of particular interest when the integral of the absolute value of the signal amplitude is the effective quantity. This is often the case when the frequency of the variation is greater than f_2 of the system used to make the measurement.

RMS Values. The rms value of a waveform is the effective value used in power calculations. The rms value of a waveform will produce the same heating (power dissipation) in a resistor as an identical dc current. In other words, 1 rms ac ampere produces the same amount of heat in a resistor in a given time as 1 ampere of dc current. The power dissipated in a resistor at any time is proportional to the square of the current through the resistor ($P = I^2R$) or of the voltage across the resistor ($P = V^2/R$). Therefore the average power dissipated by a varying signal will be proportional to the average of the square of the current or voltage amplitude over one complete cycle. The voltage or current value equivalent to the varying voltage in power dissipation is then the square root of the average of the squared amplitude value. This is called the **root-mean-square** or **rms** value.

The rms value for a sine wave signal is thus obtained by solving the equation

$$V_{rms} = \left(\frac{1}{\pi} \int_0^\pi V_p^2 (\sin^2 \omega t) \, d\omega t \right)^{1/2} = \frac{V_p}{\sqrt{2}} = 0.707V_p. \tag{3-140}$$

This same result is shown graphically in Fig. 3–88. The square of the sine wave ($V_p^2 \sin^2 t$) is plotted directly below the sine wave. The result is another sine wave which is completely positive and has twice the frequency. The mean or average value of the squared sine wave is clearly $V_p^2/2$. On taking the square root, we confirm that $V_{rms} = V_p/\sqrt{2} = 0.707V_p$.

Calculations of voltages and currents in ac circuits provide self-consistent values; thus, if rms voltage magnitudes are used, the currents will have rms values, etc. Note that peak values for sine waves are obtained by multiplying the rms values by $\sqrt{2} = 1.414$, so that

$$V_p = 1.414V_{rms}. \tag{3-141}$$

The rms values for several other waveforms shown in Fig. 3–87 are calculated below. For the unipolar square wave, the squared voltage is V_p^2 for one half-cycle and 0 for the other, giving a mean square of $V_p^2/2$ and a $V_{rms} = V_p/\sqrt{2}$, as for the sine wave. However, the bipolar square wave has a mean square voltage of V_p^2, which means that for that wave-

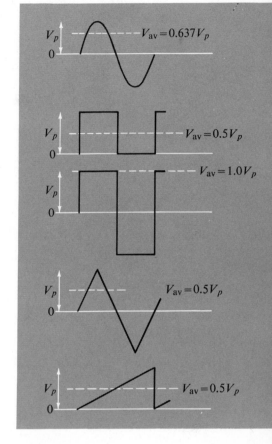

Fig. 3–87 Average measures of common waveforms.

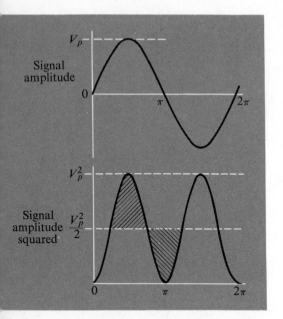

Fig. 3–88 Determination of the rms value of a sinusoidal signal.

form $V_p = V_{av} = V_{rms}$. The rms value of a triangular wave is obtained by taking the root mean square of the linear voltage relation $v = kt$ over the time from 0 to V_p/k:

$$V_{rms} = \left(\frac{1}{t} \int_0^{V_p/k} k^2 t^2 \, dt \right)^{1/2} = \frac{V_p}{\sqrt{3}} = 0.576 V_p. \quad (3\text{–}142)$$

This relation will apply to all continuously varying linear waveforms, triangular or ramp, unipolar or bipolar. Thus we see that the relationships among the several useful measures of periodic signal amplitude are highly dependent on the waveshape. The importance of this fact is made clear in the manner of application of the measurement devices discussed below.

Average-Responding Meters

The force exerted by the coil against the spring in the moving coil meter is proportional to the magnitude of the current through the coil. If the current amplitude is fluctuating rapidly, the inertia (damping) of the meter movement tends to average the amplitude variations. Thus the moving coil movement responds to the *average* value of a rapidly varying signal. The average value of a sine wave alternating current is zero, since the direction of the current changes sign each half-cycle. Therefore, alternating (bipolar) current must be converted to direct current (unipolar) to be measured by a moving coil meter. This is usually accomplished by using diodes to limit the current through the meter movement to the proper direction.

A diode and moving coil meter circuit is shown with an ac source in Fig. 3–89. The waveforms for the source voltage and circuit current are also shown. During the positive half-cycles the current passes through the meter in the correct direction with a peak value of V_p/R, where R is the circuit resistance including diode, meter, and source resistance. During the negative half-cycle the current amplitude (everywhere in the circuit) is negligible. The average current amplitude during the positive half-cycle is $0.637 V_p/R$ and zero during the negative half-cycle. The average over the entire cycle is thus $0.318 V_p/R$. This "half-wave" circuit is not practical as an ac current meter because its insertion into the circuit stops the circuit current half the time. The bridge rectifier circuit, which does not have this effect on the measured circuit, is shown in Fig. 3–90. In this case there is a low resistance path through the diodes and meter for either polarity of V_{ac} but the current through the meter is always in the correct direction. The average value of I_m in this case is $0.637 I_p$.

When a moving coil meter is used for ac current measurements with the bridge rectifier, the meter deflection is proportional to the average

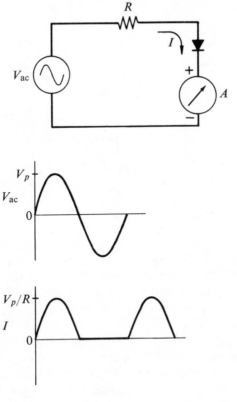

Fig. 3–89   Moving coil meter with diode rectifier.

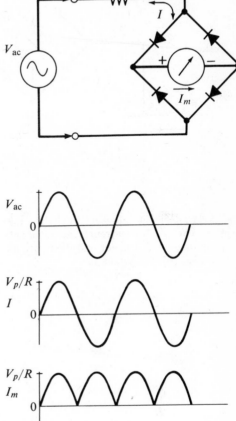

Fig. 3–90   Moving coil meter with full-wave rectifier.

value of the ac current, so that a 1 mA dc full scale meter movement would deflect to full scale for a 1 mA average ac current. However, the average ac current value is generally of less interest than the rms value, since the rms value is related to the work available from the current.

For a sine wave, the rms value = $0.707I_p$ and the average value = $0.637I_p$. Therefore, the rms value = $(0.707/0.637)I_{av}$. By adjusting the current meter shunt resistors and/or the meter scale markings, the meter can be made to *read out* in rms current even though it responds to the average current. Since the calibration factor is the same for all sine wave signals, there is no problem so long as sine wave signals are being measured. Clearly the calibration factor is not the same for other waveforms and the average-responding, rms-calibrated meter will thus not read the actual rms amplitude for most nonsinusoidal signals.

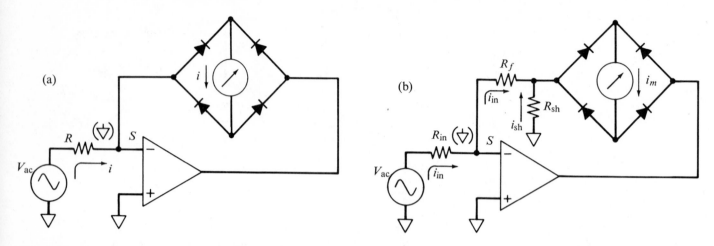

Fig. 3–91 Current follower circuits for low level ac voltage (or current) measurements with the moving coil meter and diode bridge. (a) Simple. (b) With gain.

As in the case of dc voltage measurements with the moving coil meter, ac voltage measurements are made by placing a resistor in series with the meter movement. Recall that the full scale voltage $= I(R_{\text{series}} + R_m)$. For a 100 μA (average) meter movement, $V_{\text{av}} = 10^{-4}R_{\text{series}}$ when $R_{\text{series}} \gg R_m$. Again, users prefer a meter calibrated in V_{rms}. Since $V_{\text{rms}} = (0.707/0.637)V_{\text{av}}$, $V_{\text{rms full scale}} = (0.707/0.637)10^{-4}R_{\text{series}}$. The value of R_{series} is adjusted to give even-unit full scale values for V_{rms} even though the meter responds to the average voltage value.

Another source of error in the circuit of Fig. 3–90 is the several tenths of a volt of forward bias required by practical diodes for conduction. Only that portion of the source voltage in excess of the forward diode bias voltage (times two, since there are two diodes in series) is effective in causing a current in the meter. For this reason, practical ac voltmeters frequently have nonlinear scale markings for full scale voltage ranges of 10 V and less, and are essentially useless below 1 V. For low level ac voltage measurements, the operational amplifier current follower is used with the diode bridge as shown in Fig. 3–91a. Recall that the current follower maintains point S at a virtual common voltage by holding the output voltage to the exact voltage which appears across the feedback circuit (diode bridge and meter in this case) when the input current i is passing through it. Thus the input current i is determined by V_{ac}/R only, and this same current passes through the diode bridge and meter. The OA output supplies the necessary diode bias voltages and the IR drop across the meter movement.

The sensitivity of this circuit is increased by adding the resistors R_f and R_{sh} as shown in Fig. 3–91b. The current i_m from the OA output is divided with a portion i_{in} in R_f and the remainder in R_{sh}. Since R_f and R_{sh} are connected to virtual common and common respectively, they are effectively in parallel. From the current splitting theorem for parallel resistors, $i_{in} = i_m R_{sh}/(R_f + R_{sh})$, from which we get the current gain for the circuit:

$$\frac{i_m}{i_{in}} = \frac{R_f + R_{sh}}{R_{sh}}. \qquad (3\text{--}143)$$

Such circuits have reportedly provided accurate results with gains up to 1000. However, for high-gain applications, an amplifier with low input voltage and current offset drift should be chosen.

An excellent alternative to the moving coil meter and diode circuit for average signal measurement is the absolute value circuit of Fig. 3–31 and a low pass filter followed by any voltage readout device, such as a recorder or digital voltmeter.

Experiment 3–16   *Oscilloscopic Measurements of Sine Wave Amplitudes*

The various descriptions of sine wave amplitudes are studied in this experiment by oscilloscopic measurement. The relationships between peak-to-peak, peak, rms, and average voltages for sine waves are illustrated. A multi-tapped transformer is used as a 60 Hz source with several switch-selected output amplitudes. Peak-to-peak voltage measurements with the oscilloscope are converted to rms and average values. An antenna is used to pick up noise in the air from 60 Hz power lines, and the amplitude of the noise is determined.

Experiment 3–17   *Sine Wave Measurements with the Voltohmmeter*

The moving coil meter is used in this experiment to measure sine wave amplitudes in order to illustrate the principles and limitations of ac meters. A rectifier circuit is first wired to demonstrate the conversion of ac to dc. Then the meter is used to measure several voltages from a multitapped transformer, and the results are compared with those obtained with the oscilloscope.

Experiment 3–18   *Measurements of Composite Sine Waves and dc Signals*

A sine wave voltage with a dc level is generated in order to illustrate combinations of ac and dc signals. The dc and ac input coupling modes of the oscilloscope are used to illustrate how small ac signals can be measured in the presence of large dc components.

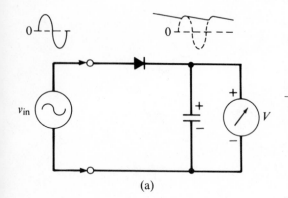

(a)

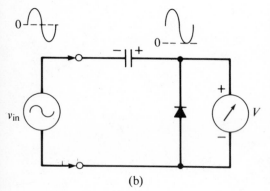

(b)

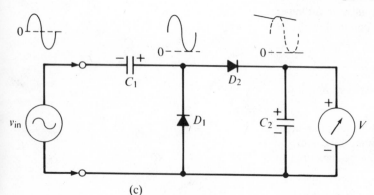

(c)

Experiment 3–19   *Square Wave Characteristics*

The characteristics of square waves are studied in this experiment by displaying these waveforms on an oscilloscope and by measuring their amplitudes on various meters. Square waves with net dc levels are introduced from a square wave generator, and the operation of the input coupling switch of the oscilloscope is observed. Then the rms amplitudes of square waves are obtained from peak-to-peak oscilloscopic measurements and compared to measurements made with a moving coil multimeter.

Experiment 3–20   *Triangular and Sawtooth Waveforms*

Various nonsinusoidal waveforms are observed on an oscilloscope. Triangular and/or sawtooth waves are applied to the scope from a signal generator. Amplitude measurements are made on these signals.

Peak-Responding Voltmeters

Some ac voltmeters are designed to respond to the peak value of the measured voltage. This is accomplished by using a diode circuit to charge a capacitor to the peak signal voltage and then measuring the resulting capacitor voltage with a dc voltmeter. Three diode-capacitor circuits used in peak-responding voltmeters are shown in Fig. 3–92. Circuit (a) is the ordinary half-wave rectifier, which charges the capacitor to the positive peak voltage during the positive half-cycle. The diode is forward biased and conducting only when v_{in} is more positive than the capacitor voltage. Thus the capacitor voltage follows the positive peaks of v_{in}. The voltmeter measures the capacitor voltage. The time constant of the capacitor and volt-

Fig. 3–92 Peak-responding voltmeter circuits.

meter resistance must be large enough that negligible discharge occurs between the positive signal peaks, but small enough for the meter to respond to decreasing peak voltage. This circuit measures only the positive voltage extremes. If the signal is asymmetrical or has a dc component, different readings will be obtained if the input connections are reversed.

In the circuit of Fig. 3–92b, the diode and capacitor positions are reversed. The capacitor will charge to the negative peak voltage during the negative half-cycle. The signal applied to the voltmeter is thus the peak-to-peak signal, with the most negative voltage being 0 volts. The meter responds to the average value of the resulting waveform. For any symmetrical wave, the average voltage applied to the voltmeter is $V_{p\text{-}p}/2$. Any dc component in the input signal is blocked by the capacitor. However, asymmetry in the signal waveform will result in different readings when the input connections are reversed.

The circuit of Fig. 3–92c is seen to be a combination of the other two circuits. Capacitor C_1 charges to the negative peak value as in Fig. 3–92b. The remainder of the circuit (D_2, C_2, and the voltmeter) is the positive peak-responding voltmeter of Fig. 3–92a. The resulting voltage applied to the voltmeter is the peak-to-peak value of the input waveform. The readings of this circuit are affected by neither asymmetry nor a dc component in the input signal.

Most peak-responding voltmeters use a vacuum tube or FET voltmeter. The high input impedance of the voltmeter allows capacitance values of the order of 0.01 μF to be used even for signal frequencies as low as 20 Hz. The higher values of capacitance required by lower resistance meters present a substantial load to the input signal. Because the peak-responding circuit is the best choice when a variety of waveforms are to be measured, and because it is so simple, most FET voltmeters with ac scales use the peak-to-peak responding circuit. Again, the scale is often calibrated in rms volts by applying a scaling factor of 0.707/2 to the voltage measurement for the ac scales. Since 0.707/2 is not the rms/peak factor for most nonsinusoidal waveforms, the rms scale is, in general, only accurate for sine wave voltages. Some meters also have a peak-to-peak or peak scale. Since this scale is calibrated in the characteristic the meter responds to, it will be accurate for all waveforms within frequency response limits and continuously varying (not for low duty cycle pulses).

RMS-Responding Voltmeters

Even though almost all ac meters are calibrated in rms volts or current, very few actually respond to the rms value of the measured signal. When a true rms measurement is needed and the relationship between the average or peak value and the rms value of the measured waveform is not known,

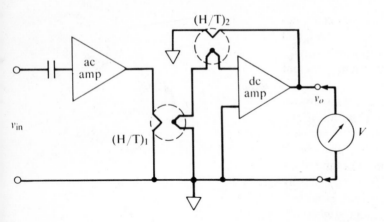

Fig. 3–93 True rms voltmeter using null-balanced heater/thermocouples.

an rms-responding meter must be used. The rms voltage or current has been defined as the value of the steady dc voltage or current which would dissipate an equivalent amount of power in a resistor. One rms measurement technique actually makes a null comparison of the power dissipated by the measured signal and that of a variable dc reference signal. This is accomplished by the circuit shown in Fig. 3–93. The ac component of the input signal (amplified if necessary) is applied to a heating element in a sealed heater/thermocouple unit. Another matched heater/thermocouple unit is connected in opposition to the first. A difference in temperature between the two thermocouples will cause a potential difference at the dc amplifier input. This difference is amplified and applied to the heater in the second heater/thermocouple unit in such a way as to reduce the potential difference from the thermocouples. When the system has reached steady state, the thermocouple temperatures differ only by the amount necessary to keep an input voltage of v_o/A at the dc amplifier input, where A is the dc amplifier gain. If A is very large, the temperatures of $(H/T)_1$ and $(H/T)_2$ are essentially the same and the power dissipated by v_o is equal to that of the amplified input signal. Thus the value of v_o is equal to the rms value of the input signal (divided by the gain of the ac amplifier). The dc amplifier employed in such a system is generally a chopper-input type because of the very high gain and low offset and drift requirements.

Another way to measure the true rms value of a signal is to calculate it. This is done by using an analog squaring circuit to generate a signal which is the square of the input signal. Next the squared signal is averaged over at least one period of the input signal. Finally, an analog square root circuit is used to produce an output voltage equal to the root mean square of the input voltage. The block diagram of a calculating rms voltmeter is shown in Fig. 3–94. Analog multiplier/divider circuits are used for the square and square root functions as shown in Fig. 3–45. The averaging

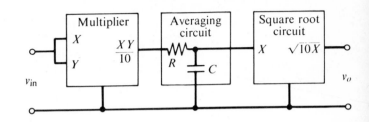

Fig. 3–94   Calculating rms voltmeter.

circuit is often just an *RC* low pass filter with an *RC* time constant greater than the period of the measured waveform. The accuracy of calculating meters is of the order of 0.1% to 1%. A major source of error is in the measurement of pulse-type signal sources. High level pulses or "spikes" can contribute greatly to the rms value of a signal. However, if the amplitude of the pulse exceeds the linear dynamic range of the signal amplifier or the squaring circuit, much of its contribution to the rms value will be lost.

The following measurement guideline is offered as a summary to this section: For accurate measurements of periodic nonsinusoidal signals, consider first whether the information desired is the average, rms, or peak value of the waveform. Then, if possible, use a meter that responds to and is calibrated in that characteristic. If this is not possible and the waveform is constant (except in amplitude), the proportionality constant between the reading and the desired amplitude characteristic can be determined experimentally. However, in such cases, it is generally safer and easier to observe and measure the actual waveform amplitude with an oscilloscope.

3–2.5 MEASUREMENTS OF APERIODIC OR NONREPETITIVE SIGNALS

Time-variant signals which are singular (occur only once, or have no repetitive pattern) or which repeat the amplitude-time variation of interest at irregular and unpredictable intervals pose special measurement problems. One solution is to make a continuous record of the quantity as a function of time so that the portion of interest, whenever it occurs, is sure to be recorded. The measurement of the entire amplitude-time function is the most complete measurement of a signal; that is, it records all the information required to interpret the signal regardless of how the information is encoded.

Strip Chart Recorder

For relatively slowly changing signals, the amplitude-time function is often recorded with a strip-chart recorder. The position of the pen varies with the signal amplitude while the paper moves at right angles to the pen motion at a constant rate. The result is a continuous plot of the signal amplitude vs. time. The data recorded on such plots can later be analyzed to obtain particular instantaneous amplitudes, peak amplitude values and times, maximum rates of change, etc. The disadvantage of recording the entire amplitude-time function is that the regions of interest must be located on the chart and then be interpreted even for the simplest measurements, such as peak amplitude or the amplitude at some particular time. However, there is an advantage in recording complete curves of difficult or unrepeatable experiments even when the measurement is expected to be simple. The advantage is that *all* the information is recorded so that additional interpretations of the curve can later be made without repeating the experiment. The recorder, with its permanent record and its almost unlimited time axis, would be the complete amplitude-time measurement device if it were not restricted to rates of signal change slower than a few tenths of a second. The amplitude-time function of more rapidly changing signals is often measured with an oscilloscope.

The Oscilloscope

The amplitude-time function of a varying voltage signal can be displayed with an oscilloscope. The measured signal is applied to the vertical amplifier and cathode ray tube (CRT) deflection plates while a linearly increasing voltage is applied to the horizontal amplifier and CRT deflection plates. The result is that the position of the electron beam on the CRT phosphor is displaced vertically in proportion to the signal amplitude while it is being displaced horizontally (to the right) at a constant rate by a linear sweep signal. A trace of a sinusoidal voltage variation such as might appear on an oscilloscope is shown in Fig. 3–95. The oscilloscope is capable of displaying signal variations that occur in times as short as fractions of a microsecond to several minutes. However, it has two obvious limitations in comparison with the strip chart recorder for recording amplitude-time functions. One is that the screen is of limited size and therefore the measurement time is very limited. The second is that the phosphorescent trace is a far from permanent record.

The limitation that the CRT screen size puts on the measurement time span is alleviated, in some cases, by displaying the amplitude-time function only during intervals when the signal is of interest. This requires timing the beginning of the horizontal sweep signal to coincide with the features of the measured signal waveform that indicate the beginning of the region of interest. In order to display the desired portion of the signal, the sweep signal generator circuit should be **triggered** to start its sweep just before the desired portion of the signal is to occur. A comparator is used to convert a specified signal variation into a sudden voltage change which can be used

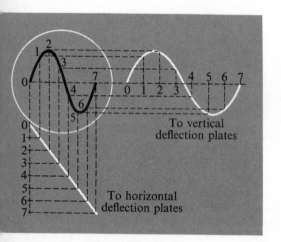

Fig. 3–95 Sawtooth vs. sine wave of the same frequency.

To vertical deflection plates

To horizontal deflection plates

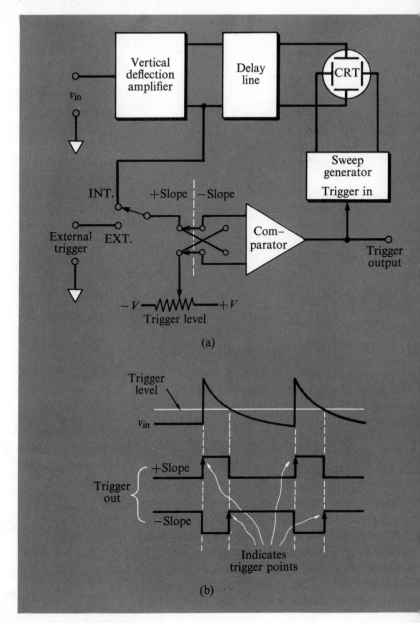

Fig. 3–96 Triggering the oscilloscope sweep generator. (a) Circuit. (b) Waveforms.

to trigger the sweep generator. The oscilloscope trigger circuit is shown in Fig. 3–96a. One of the comparator inputs is an adjustable **trigger level** voltage source. The other input is the signal which is to cause the sweep to trigger. When the trigger input signal waveform crosses the threshold voltage value, the comparator output voltage changes value suddenly, as shown in Fig. 3–96b. The trigger input signal can be the signal to be observed (generally amplified first as shown and called "internal") or an "external" timing signal. The comparator output will trigger the start of the sweep signal on the positive-going transition (marked by arrows in the figure). The triggering transition will occur on the positive-going slope of v_{in} or the negative-going slope, depending on the position of the slope switch at the comparator input. In this example, the proper choice is probably the $+$ slope position. Oscilloscopes designed for observing fast transient signals use a transmission line to delay the vertical deflection signal between the triggering circuit and the vertical deflection plates for a fraction of a microsecond, as shown. This allows the sweep to begin before the portion of the signal which triggered the sweep appears at the vertical deflection plates.

For the recording of singular signals, a **single sweep** provision may be included. This allows the sweep to be triggered only once before being reset. In such a case the very ephemeral nature of the phosphorescent display may cause the display of the signal to be too brief to observe or interpret. This problem can be solved in two ways: (1) photograph the CRT while the waveform is being displayed; (2) use a "storage" type CRT in which the trace phosphorescence is maintained by a "wash" of low-energy electrons.

If the sweep trigger is set to trigger reproducibly on successive repetitions of a repeated waveform, the display from each sweep will be superimposed. The repetitive, superimposed electron beams reinforce the CRT phosphorescence to produce the appearance of a steady display if the repetitions are frequent enough. Because the sweep is triggered by the observed signal feature itself, it does not matter whether the triggering portion of the signal occurs at regular intervals or not. This capability is often used to advantage in observing the result of random events such as emissions from radioactive materials.

Good linearity of the sweep is a major requirement for a reliable oscilloscope. In many oscilloscopes the sweep time is adjustable over a range from 10 μsec to 10 sec or more. At the end of the sweep the electron beam is moved back across the screen to the starting point by returning the sweep voltage to its initial value. The finite time required to change the sweep voltage back to the starting value can cause a visible trace of low intensity, known as the **retrace** or **flyback.** It is customary to turn off or **blank** the electron beam during the retrace by applying a negative voltage to the CRT intensity control element during this short time interval.

Sampling Measurements

Often the desired information in a varying signal is the signal amplitude at specific instants in time. In other words, the signal amplitude is to be **sampled** at those specific instants. Sometimes the desired sampling times come at regular intervals, e.g., a temperature reading every 10 minutes. However, in other cases, the sampling times are the times of events which occur in other parts of the system, such as an electrode potential reading each time a 0.1 milliliter increment of titrant is added to a solution. One way to obtain such data is to record the entire amplitude-time curve and later obtain the amplitude at the desired times from the recorded curve. When sampling from a recorded curve, it is obviously necessary to know the desired sampling times relative to the time axis of the recorded curve. To aid this process, an indication of the instants in time that are of particular significance is sometimes recorded along with the signal. In a strip chart recorder, this is accomplished by an **events marker,** a second pen located at one edge of the paper. The movement of this pen is activated by a solenoid which is pulsed when a time mark is to be made. Time marks can be added to oscilloscope traces by applying the timing signal to the second vertical input of a dual-trace or dual-beam oscilloscope. Another technique is to apply the timing pulses to the **Z-axis,** or intensity control input, of the oscilloscope. The time marks would then be bright or dim spots in the trace of the amplitude-time curve.

Sampling from a recorded curve when the only desired information is the amplitude at specified instants in time is inefficient in two respects: the measurement system spends most of its time measuring data that are not going to be used, and the recording system uses most of its storage capacity storing unnecessary data. An alternative to sampling from recorded curves is to sample the desired points in the analog signal while they are occurring. This **real-time sampling** requires activating the measurement/recording apparatus only at the desired sampling times. The resulting data points can be displayed on a meter, oscilloscope, or recorder for manual digitization, or they can be electronically digitized and recorded by a digital printer or stored in a digital memory. Sampling measurements are employed for three reasons: to reduce the quantity of stored data to the most relevant, to share a single measurement system among several relatively slowly varying signal sources, or to measure specific instants in a rapidly varying signal with a relatively slowly responding measurement system. The sampling techniques applied in each of these three cases are described in order below.

Storage and Display of Sampled Data. If a succession of sampled measurements is to be recorded or stored, it is necessary to record or store each point in such a way that it can be identified with the particular sampling instant to which it corresponds. When the sampling is done at constant intervals of time, volume of reactant, wavelength selected, etc., the correspondence of sampled value with sample number is made simply by advancing the chart paper or the x-axis of the recorder, the horizontal deflection of the oscilloscope, the paper of the printer, or the storage address of the digital memory by one unit before recording each new sampled value.

When sampling occurs at irregular intervals and when the correspondence of the sampled amplitude with some other variable is important, two values must be recorded for each data point. This is done with oscilloscopes and *x-y* recorders by using the *y*-axis deflection for the sampled variable and the *x*-axis deflection for the variable related to the sampling time. The *x-y* data points are displayed as intensified dots on the CRT or as a point plot on the *x-y* recorder. When recording or storing the data points digitally, it is necessary to store or print two numbers (one *y* and one *x*) for each data point.

Multiplexing of Input Sources.   The sharing of a single analog-input measurement and recording system by more than one input signal is accomplished by a signal switching technique called **multiplexing.** The analog multiplexer is simply a controlled selector switch, as shown in Fig. 3–97. In the switch position shown, the input connector pair labeled "CH. 2" is connected to the output. As the switch position changes, the connector pairs of the other input channels are connected, in turn, to the output. The switch position is controlled by a channel control circuit which can be set to any channel or instructed to increment (change to the next channel) by external signals. In response to four successive increment instruction pulses, beginning at channel 2, the active channels would become 3, 0, 1, and 2 in that order. A digital output is generally provided for an external indication of the active channel. The multiplexer switch can be an electromagnetically

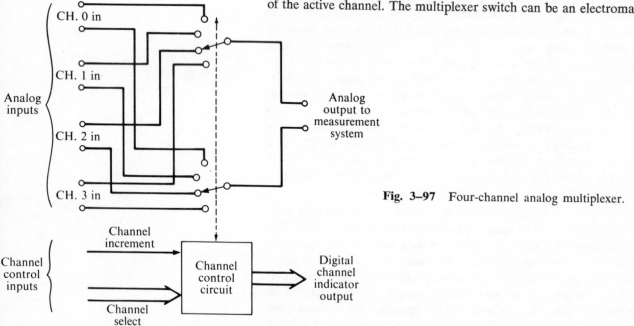

Fig. 3–97 Four-channel analog multiplexer.

operated rotary selector switch or a set of relays or solid state switches controlled in such a way that only one channel is connected at one time. The signal errors introduced by multiplexing depend entirely on the switches used, as described in Module 2. The multiplexer in Fig. 3–97 has two connections per channel, which will allow the use of floating signal sources when a differential input is available on the measurement system. If all the sources are known to have one connection in common with one terminal of the measurement system, the lower pole of the switch in Fig. 3–97 could be replaced by a common connection of all the points at that switch. Multiplexing is most frequently used with groups of similar signal sources such as identical thermocouples monitoring the temperatures in different locations. If disparate sources which would require different range or offset adjustments of the measurement system are multiplexed, they must be conditioned to a common range and offset before they are multiplexed.

Analog multiplexing is frequently used with recorders, oscilloscopes, digital voltmeters with printers, and analog-to-digital converter inputs to computers and data loggers. A multiplexer used with a strip chart recorder often increments the channel selector switch at a fixed time interval which is a little longer than the maximum balancing time of the recorder. At the end of each sampling interval, the recorder pen is dropped briefly to make a mark on the paper. If the chart drive speed is slow enough so that the paper advances only a few millimeters during each complete cycle of the input channels, the recorded trace of each channel will appear essentially continuous. The identification of each recorded trace with the corresponding input channel is often marked by hand on the chart. However, some recorder pens can plot points in various shapes or colors in response to the channel indicator output of the multiplexer. Recorder multiplexers have from four to sixteen channels. Some point-plotting recorders drop the pen as soon as the recorder comes to balance after a new channel selection. A "point-plotted" pulse from the plotter can then be used to increment the multiplexer. In this way, the points are plotted at the maximum rate for the most continuous appearance. Clearly, multiplexing multiplies the storage capacity of the recorder by the number of channels and allows more convenient comparison of corresponding data among the several channels.

Multiplexing the input signals to an oscilloscope is done for the same reasons, but in a somewhat different way. An oscilloscope with a two-channel multiplexer is called a "dual-trace" oscilloscope. A block diagram of the input and multiplex portion of a dual-trace oscilloscope is given in Fig. 3–98. Each input channel has its own preamplifier with separate range (gain) and offset (position) controls in order to allow the dual-trace display of widely different signal levels. The multiplexer is a two-position solid state switch with four modes of control. In the CH. A or CH. B positions, the

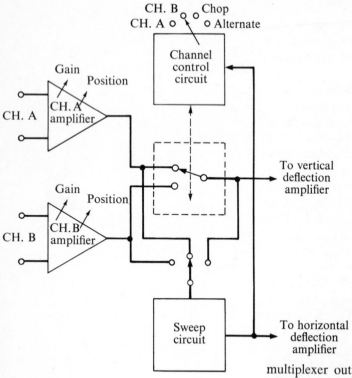

Fig. 3–98 Multiplexer in a dual-trace oscilloscope.

multiplexer output is continuously connected to channel A or B respectively. In the "chop" mode, the multiplexer switch is alternated back and forth between channels A and B at a rate of 100 kHz to 1 MHz. With a horizontal sweep rate of 100 μsec/cm or slower, the traces for both signals will appear to be essentially continuous. As the sweep rate is made less than 100 μsec/cm, the chopping of each signal becomes apparent. At still faster sweep speeds, the chop mode becomes unusable. In the "alternate" mode, the channel selector switch is alternated after each sweep. In other words, channel A is displayed for one entire sweep, and then channel B is displayed during the next sweep, and so on. This mode is practical when the observed signals are repetitive and when alternate sweeps are frequent enough to appear continuous to the eye. Sweep rates faster than 1 msec/cm will generally repeat each trace 50 times or more each second. The retrace portion of the sweep signal is used to alternate the channel selector switch. The sweep trigger selector switch on a dual-trace oscilloscope includes provision for triggering from the channel A signal, the channel B signal, or the multiplexed signal. The latter trigger source is only used in the "alternate" mode. It allows each signal to determine its own trigger point, but the time correlation between the two traces is lost. Some oscilloscope models have a four-channel input which is simply an expansion of the two-channel multiplexer shown in Fig. 3–98. The dual-trace oscilloscope should not be confused with the dual-beam oscilloscope, which has two separate electron guns and separate pairs of vertical deflection plates for each beam. In a dual-beam oscilloscope, the trace of both beams is continuous over the entire range of sweep speeds.

Multiplexers are often used with digital data loggers to acquire data which are to be correlated from several data sources or to share the expense of the data logger among several sources. Typical data logging systems which will be discussed in Section 3–4 are a digital voltmeter or analog-to-digital converter with a printer, paper-tape, or magnetic tape recording attachment. The analog multiplexer of Fig. 3–97 can be used with digital data logging systems. It is generally necessary to record or print the channel number along with the digitized data for each point. When a digital computer is part of the data logging system, the multiplexer channel control circuit is generally put under computer control.

High-Speed Analog Sampling. The third major application for sampling measurements is measuring the instantaneous amplitude of a signal which can vary significantly during the response time of the measurement system. For this application, a **sample-and-hold circuit,** a kind of analog memory, shown in Fig. 3–99, is employed. As the waveforms show, the instant a sample-and-hold circuit receives a "hold" command signal, the voltage level present at its analog input is stored and maintained at its output. The now-steady voltage may then be measured by a relatively slow device. After the value of the held voltage is measured and recorded in some way, the circuit is released from the hold condition so that it can again respond to the input voltage changes. The sample-and-hold function is generally accomplished by charging a capacitor with the signal value during the sample interval, then measuring the voltage across the capacitor with a high input impedance amplifier during the hold period.

A reliable sample-and-hold circuit that uses the voltage follower amplifier is shown in Fig. 3–100. The input signal charges capacitor C through resistance R when the circuit is in the "sample" mode. The time constant

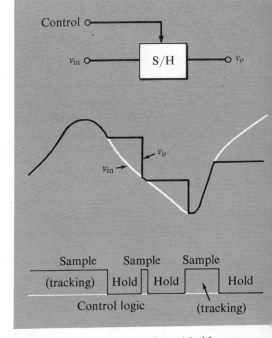

Fig. 3–99 Typical sample-and-hold waveforms.

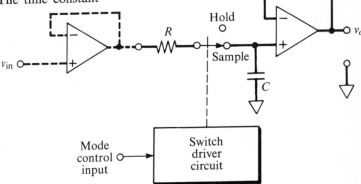

Fig. 3–100 Voltage follower sample-and-hold circuit.

RC and the response time of the amplifier must be short compared to the rate of change of the input signal, so that the follower amplifier input and output will follow the input signal variations. At the desired sampling instant, the switch is changed to the "hold" position, isolating the input signal and leaving the voltage across capacitor *C* at that instant at the amplifier input. Ideally, this voltage will be maintained (held) at the amplifier noninverting ($+$) input (and consequently, at the output) until the circuit is returned to the sample mode.

From the circuit of Fig. 3–100, some of the deviations from ideal behavior of sample-and-hold circuits can be recognized and characterized. These will be discussed briefly below in terms of the two states of the circuit and the transitions from one state to the other. First, the sample mode: Since the response rate of the circuit is limited by the *RC* time constant, *R* is kept very low. Consequently, the input impedance of the circuit is very low. A second voltage follower is almost always used between the signal source and the sample-and-hold input. Other characteristics of concern during the sampling mode are the following: **offset,** the output voltage at zero input voltage; **gain error,** the deviation of the stated output-to-input voltage ratio (unity for the circuit of Fig. 3–100); and **settling time,** the time required for the output voltage to come to within a given percent of its final value after an instantaneous change of some specified fraction in the input voltage. The sampling characteristics are clearly determined by the operational amplifiers used in the voltage follower circuits and the values of *C* and R_{ON} for the switch. Second, the transition from the sample to the hold mode includes errors of the following types: **aperture time,** the total time between the hold command and the actual opening of the hold switch, including the average delay *and* the delay uncertainty where aperture time is frequently as small as 10–100 nsec; **switching offset,** the change in voltage on *C* due to a charge loss or gain during switching; and **settling time,** the time after opening of the sample switch for the output voltage to settle to within a specified difference from its final value. These errors have to do with the switch and drive circuits and the response speed of the output follower amplifier. Third, during the hold mode the problem of output drift is of greatest concern. The rate of change in the output voltage in volts per second is called the **droop.** It is caused by finite currents at the amplifier input or through the sampling switch which cause the voltage across *C* to change with time. This drift in the hold voltage can be minimized by using an amplifier with very low input current and a switch with very high **OFF** resistance. Using a larger capacitance at *C* will also reduce drift, though this might be incompatible with keeping the *RC* time constant small enough for the $+$ input to follow the input signal accurately. Finally, in the hold-to-sample transition, one is concerned with the **acquisition time,** the minimum

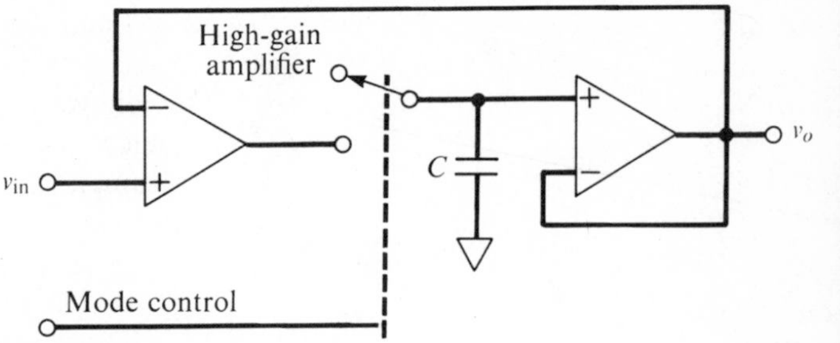

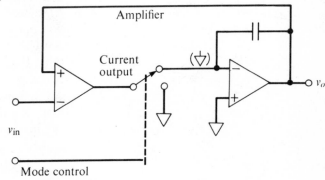

Fig. 3–101 Track-and-hold   circuit   with feedback.

sampling time to acquire the input voltage to within a given accuracy. In most cases this is the same as the sample mode settling time. The critical characteristics in actual applications are the minimum sampling time (given by the acquisition time), the sampling error (given by the offsets, gain error, droop, and overall analog accuracy in percent of full scale signal), and the sampling time uncertainty (the variation in or unknown portion of the aperture time).

The simple circuit shown in Fig. 3–100 is the current choice for the highest speed sample-and-hold circuits. Greater accuracy in sampling can be achieved by variations on this circuit at the expense of response speed. The circuit of Fig. 3–101 is essentially the same as Fig. 3–100 except that the sampling amplifier is included in the feedback loop of the input voltage follower. In this way, the charge on C is forced to be whatever is required to produce the exact input voltage at the output when in the sample mode. In the hold mode, the output amplifier is a voltage follower which will maintain the output voltage at its sampled value. The inclusion of the sampling switch and capacitor in the feedback loop eliminates offset and common mode errors. The circuit of Fig. 3–102 maintains the feedback advantage and uses an integrator as the output holding amplifier. During the sampling mode, the input amplifier supplies current through the switch to the integrator circuit to keep the output voltage exactly equal to the input voltage. In the hold mode, the current supply to the integrator is cut

Fig. 3–102 Feedback track-and-hold with an integrator.

off and the integrator maintains its last output voltage. The advantage over the circuit of Fig. 3–101 is that all three switch connections are at ground voltage in both modes. With essentially zero volts across the switch contacts, switch leakage is greatly reduced. For both examples of feedback sample-and-hold amplifiers, the acquisition time is the time for the whole circuit to settle on the input voltage, not just the capacitor voltage as in Fig. 3–100.

The sample-and-hold circuits of the types shown in Figs. 3–100 to 3–102 are available in convenient circuit modules with characteristics to suit a very wide variety of applications. Minimum sampling times as short as 100 nsec, overall accuracy to 0.005%, and sampling time uncertainty as low as 5 nsec are readily obtained. However, even with improved circuits and high quality switches and OA's, a compromise between acquisition and holding times must be made in most cases. The value of the holding capacitor C is adjusted to give the required holding time and the acquisition time is chosen accordingly. If the measurement requires a ratio of acquisition-to-hold time that is too small to be achieved in an actual circuit, two or more sample-and-hold circuits can be cascaded. The first is set for the required acquisition time and provides an output that holds for the longer acquisition time of the second circuit. A delaying circuit would be used to actuate the switches of the two sample-and-hold circuits in sequence.

Because the desired measurement, in applications using sample-and-hold circuits, is the amplitude of the signal, the emphasis is often on the sampling and holding accuracy of the circuit. However, the accuracy of the sampling time is also often critical. If the signal amplitude is rapidly changing, it stands to reason that the value measured could depend greatly on the exact instant of sampling. In fact, a fairly common reason for using sample-and-hold circuits is to achieve a precise sampling time which is independent of the remainder of the measurement system.

The acquisition time for all the above sample-and-hold circuits is the time for the storage capacitor to charge to the input signal value within the desired accuracy. The acquisition time is thus 4τ seconds, where τ is the RC time constant for charging the storage capacitor. To reduce the acquisition time, it is then necessary to reduce τ. The lower limit for the time constant is the lowest reasonable signal source and sampling switch resistance (e.g., 50 Ω) times the stray input capacitance of the sampling amplifier (e.g., 10 pF). Thus the shortest practical acquisition time for such a circuit would be about 2 nsec. If an acquisition time shorter than 4τ is used, a sampling error will result. In the **error-sampled** sample-and-hold circuit, shown in Fig. 3–103, a second sample-and-hold amplifier with gain is used to correct that error and make sampling measurements with picosecond acquisition times possible.

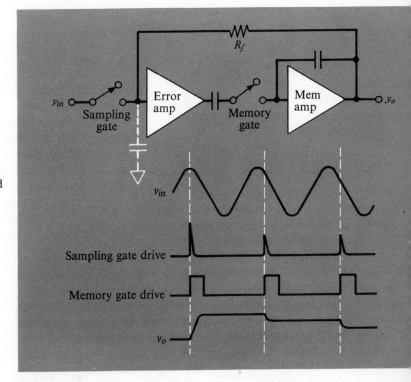

Fig. 3–103 Error-sampled sample-and-hold amplifier.

Assume that the memory output is equal to the input signal voltage during the previous sampling interval. This voltage is applied to the input amplifier through the feedback resistor R_f. The amplifier input capacitance is held at the previous signal voltage. The input potential change between sampling intervals will be called the **error voltage.** When the sampling gate opens, the amplifier input voltage begins to change by the amount of the error voltage. However, when only a fraction of this change has taken place, say 0.1, the sampling gate closes. The voltage change at the amplifier input, representing only one-tenth the desired change, is amplified and capacitor-coupled to the memory gate. The memory gate is open longer than the sampling gate, so that there is sufficient time for the amplified fraction of the error voltage to be applied to the memory amplifier. The overall gain of the amplifier, memory gate, and memory is adjusted to compensate exactly for the fraction of the error voltage change lost as a result of the short sampling aperture. Thus the memory output voltage correctly represents the input voltage. The output voltage now charges the input capacitor up to the hold voltage through R_f, and the circuit is ready for the next sampling. In this type of sample-and-hold circuit great speed is gained at the price of a considerable increase in complexity and some loss in accuracy.

Equivalent Time Conversion of Repetitive Signals. After sampling, a sample-and-hold circuit must remain in the hold mode until the measurement system which follows has measured the held voltage. Thus the maximum sampling rate is limited by the response time of the measurement system. When the rate of change of the sampled signal exceeds the response time of the measurement system, the signal may vary considerably between sampled data points. Normally, the information about that variation would be lost. However, if the signal has a repetitive waveform, data points sampled from many waveform repetitions can be combined to produce a reasonable data point recording of the waveform. The relatively simple device which does this is shown in Fig. 3–104.

The repetitive input signal is applied to the input of a sample-and-hold circuit and to the trigger input of a fast sweep generator. The fast sweep generator is adjusted to be triggered by the feature of the signal which is to

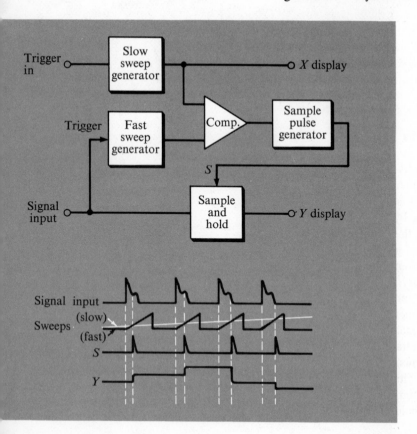

Fig. 3–104   Equivalent time converter.

be recorded and is set to complete its sweep in the period of interest after the trigger. A slow sweep generator which is slow enough for the *x-y* measurement system to follow is also started. The sampling instant occurs when the two sweep generators are of equal voltage as determined by a fast comparator circuit. There will thus be one sample taken each time the fast sweep generator is triggered. Since the sampling instant occurs a little later in the fast sweep each time, the sampled point is taken at increasing times in the signal waveform. The time between the fast sweep trigger and the sampling instant is called the **sampling delay time.** It is clear from the waveforms of Fig. 3–104 that if the sampled amplitude is plotted against the sampling delay time, points on the original waveform will result. The amplitude of the slow sweep is proportional to the sampling delay time and thus provides an equivalent time signal for the slow measurement system. The ratio of equivalent time to real time is equal to the ratio of the slow sweep rate to the fast sweep rate. The number of data points recorded in a single slow sweep is equal to the number of fast sweep triggers that occur in that time. But whether the triggers occur regularly or not, the point recorded will have the proper *x-y* value. Irregular triggering intervals only result in possible gaps in the resulting data point recording.

Equivalent time conversion is used with both oscilloscopes and *x-y* recorders. An equivalent time oscilloscope is called a **sampling oscilloscope.** Equivalent time sampling is used to extend oscilloscopic observations into the picosecond time range. Since the fastest sample-and-hold circuit is required, the error-sampled circuit discussed above is generally used. Much slower versions of the equivalent time converter are available for *x-y* recorders. The primary application for these devices is plotting the steady displays of repetitive signals from an oscilloscope for a permanent record. Such a device is very simple, since the fast sweep generator is the oscilloscope sweep generator and the slow sweep generator can be the *x-y* recorder time base.

Experiment 3–21 *Analog Multiplexing*

The general principles of analog multiplexing are illustrated in this experiment. A general purpose two-channel analog multiplexer is constructed and its operation is studied. Manual switch control is first used and any crosstalk between channels is noted. Then automatic control of the multiplexer is illustrated. The two-channel multiplexer in a dual-trace oscilloscope is next studied in order to illustrate its characteristics and limitations. The oscilloscope is used in the "chop" and in the "alternate" modes. Differences and limitations in each mode are observed.

Experiment 3–22 *Sample-and-Hold Measurements*

The sample-and-hold or analog memory circuit is used in this experiment to provide measurements of the instantaneous amplitudes of signals. A sample-and-hold circuit is first constructed from an *RC* network, a FET switch, and a voltage follower. The circuit is first characterized by measurements on a static signal, and then used to measure the amplitudes of varying signals.

PROBLEMS

1. In the OA current measurement circuit of Fig. 3–16, which gives a high equivalent value of R_f, $R_1 = 5$ MΩ, $R_2 = 1$ MΩ, and $R_3 = 5$ kΩ. What is the equivalent feedback resistance $R_{f(eq)}$?

2. The OA circuit of Fig. 3–20b is constructed. What is the gain if (a) $R_1 = 500$ kΩ and $R_2 = 10$ kΩ, (b) $R_1 = R_2 = 50$ kΩ, (c) $R_1 = 100$ kΩ and $R_2 = 5$ kΩ?

3. The instrumentation amplifier of Fig. 3–25 is constructed with $R_1 = 1$ kΩ, $aR_1 = 5$ kΩ, $bR_1 = 3$ kΩ, $R_2 = 100$ kΩ, and $KR_2 = 1$ MΩ. What is the amplifier gain?

4. A square wave signal has an amplitude of 10 V peak-to-peak and a frequency of 1 kHz. Plot the frequency spectrum of the signal out to the eleventh harmonic.

5. At what frequency will a 0.01 μF capacitor have a reactance of 2.5 kΩ?

6. What is the capacitive reactance of a 0.05 μF capacitor at frequencies of (a) 100 Hz, (b) 1 kHz, (c) 10 kHz?

7. At what frequency will a 200 mH inductor have an inductive reactance of 1 kΩ?

8. What is the inductive reactance of a 20 mH inductor at frequencies of (a) 100 Hz, (b) 1 kHz, (c) 10 kHz?

9. If the total impedance of a series *RC* circuit is 98 Ω, the phase angle is $-54.5°$, and the capacitive reactance is 80 Ω, what is the value of resistance in the circuit?

10. A 100 Hz source is connected to a series *RC* circuit in which $R = 3$ kΩ and $C = 0.5$ uF. Find the magnitude of the impedance and the phase angle.

11. A 10 kHz source of 150 V peak is connected to a series *RC* circuit in which $X_C = 400$ Ω and $R = 500$ Ω. Find the magnitude of the impedance, the phase angle, the peak voltage across *R*, the peak voltage across *C*, and the power dissipated.

12. A series RC circuit is to be used as a high pass filter. If $R = 5$ kΩ, what value of C is necessary to give a lower cutoff frequency of 200 Hz?

13. An RC circuit is to be used as a low pass filter. If $R = 2.5$ kΩ and $C = 0.05$ μF, what is the upper cutoff frequency?

14. A 500 Hz, 25 V peak source is connected to a series RL circuit in which $X_L = 50$ Ω and $R = 100$ Ω. Find the magnitude of the impedance, the phase angle, and the peak voltages across R and L.

15. The total impedance of a series RLC circuit is 150 Ω. If $X_L = 200$ Ω and $X_C = 100$ Ω, find the resistance of the circuit.

16. A 50 pF capacitor is in series with a 150 μH inductor. At what frequency will resonance occur?

17. An amplifier has a high pass input network in which $R_{in} = 10$ kΩ and $C_{in} = 1$ μF. The amplifier high frequency response is limited by a shunt capacitance of 100 pF and a series resistance of 100 kΩ. Find the upper and lower cutoff frequencies of the amplifier. Plot the relative gain of the amplifier vs. frequency.

18. A four-stage amplifier is connected in which the upper cutoff frequency f_2 of each stage is 100 kHz and the lower cutoff frequency f_1 of each stage is 50 Hz. Find the upper and lower cutoff frequencies of the four cascaded stages.

19. A sine wave voltage source has an average voltage of 25 V. Find (a) the peak voltage, (b) the peak-to-peak voltage, (c) the rms voltage.

20. The power line voltage is nominally 110 V rms, 60 Hz. Find (a) the average voltage, (b) the peak voltage, (c) the peak-to-peak voltage.

Time Domain Measurements and Conversions

One cannot distinguish a time domain signal from a varying analog domain signal by waveform. In both cases the signal amplitude varies with time. The difference is in what characteristic of the signal must be measured or quantified to obtain the desired information. In analog domain signals, the signal by waveform. In both cases the signal amplitude varies with time. data are in the amplitudes nevertheless. In time domain signals, it is the time relationship of certain amplitude variations that contains the critical information. The amplitude variations provide the timing cue, but the information is in the *time* of the variation, not the amplitude. Thus time domain measurements involve the measurement of the time intervals between signal features on an absolute or a relative basis.

In this section, precision time measurements are discussed, frequency ratio measurements by analog and digital techniques are explored, and the interconversion between analog and time domain signals is illustrated. Because of the relative freedom from noise, the ready availability of high precision time standards, and the close relation of time domain to digital domain signals, the encoding of measurement signals in the time domain is increasingly common.

3–3.1 AMPLITUDE-TIME FUNCTION MEASUREMENTS

In Section 3–2.5 it was stated that the recording of the complete amplitude-time function contains all the signal information and that it can be decoded in any way. From such a plot it is as easy to obtain, for instance, the time between two peaks as it is to obtain the peak amplitudes. In the first instance a time-to-number domain conversion is performed, and in the second an analog measurement is made. Actually, in both cases, the amplitude and the time are converted to the scale position domain by the recording process. The final answers are obtained by "reading" the scale. In recording signals with time-encoded information, one's attention is drawn to the characteristics of the time-to-scale position converter, i.e., the time base or sweep of the recording device. The following discussion will review the time-base characteristics of the strip chart recorder and the oscilloscope in this respect. The use of the comparator circuit to indicate the time of specified amplitude variations is also discussed.

Recorder Chart Drive

In order to record signal variations as an accurate function of time, the recorder chart paper must move under the recorder pen at a constant and accurate rate. This motion is accomplished by an electric motor which is coupled to the paper feed sprockets and writing platten by gears. The time accuracy of the drive depends on the speed accuracy of the motor, the quality of the gears and sprockets, and the accuracy of the chart paper dimensions. Chart drive motors for ac-line-powered recorders are generally synchronous motors for single speed models or stepper motors for multi-speed chart drives. In either case, the motor speed is often determined by the frequency of the ac power line. In most areas, the ac line frequency is accurate to 0.02% of its nominal value and is not the limiting factor. However, in many systems an internal crystal-controlled clock is used to control the step rate of a stepper motor for better accuracy. The gears are not likely to introduce error either, unless the timing measurement involves starting and stopping the chart motor or there is uneven tension on the recorder chart. Under these conditions, backlash in the gears can add an uncertainty of several hundredths of an inch. The paper dimension stability depends on the paper quality and the humidity. Variations of several tenths of a percent are possible. Precision time marks are sometimes added to the chart while the signal is being recorded. This technique provides a time calibration check and the possibility of a time comparison measurement. It is assumed, of course, that the recorded signal variations that are to be time-measured do not vary at rates exceeding the response speed of the recorder pen. Delay in pen response can cause time errors when the chart speed is such that there is discernible motion of the chart during a period equal to the full scale pen response time.

Oscilloscope Time Base

The accuracy of the oscilloscope time base depends on the sweep signal linearity, the sweep rate accuracy, and the stability and uniformity of the CRT deflection sensitivity. The CRT characteristics are generally limiting to ±3 to ±5% overall relative time-base accuracy. For this reason, the sweep circuits in an oscilloscope are often designed to a sweep rate tolerance of about ±2%. The oscilloscope time base can be used to provide a "ball-park" time measurement, but a comparison method using a standard time source with the oscilloscope as a difference detector is far more precise. Comparison techniques for frequency measurements are described in Section 3–3.2.

Experiment 3–23   *Measurements with the Recorder Time Base*

The time base of a servo recorder is studied in this experiment and used to make time domain measurements. First the chart drive of the recorder is studied and measurements are made of the chart drive rate and the precision with which time measurements can be made. Then the frequency of a sawtooth waveform is determined from the recorder time base and the limitations of the recorder for measuring rapidly changing signals are observed.

Experiment 3–24   *Measurements with the Oscilloscope Time Base*

This experiment illustrates the use of an oscilloscope for measurements involving rapidly changing signals. A neon bulb driven from a multitapped transformer provides a light signal which varies with time. A phototransistor transducer is used to convert the radiation into a time-varying electrical signal. Oscilloscope measurements of the turn-ON and turn-OFF time of the neon lamp provide interesting information about the lamp's operating characteristics. Then a neon bulb relaxation oscillator is constructed and its frequency measured on the oscilloscope as a function of the RC time constant of the oscillator. The maximum oscillation frequency is determined. If a multispeed recorder is available, the frequency of the signal which drives the stepper motor is determined with the oscilloscope.

The Comparator

In a time-encoded signal, the data are encoded as the relative times at which particular amplitude variations occur. The data in the signal are converted to numbers by detecting the specified amplitude variations and then measuring the appropriate time relationship between them. The comparator is the circuit used to provide an indication each time the specified signal amplitude variation occurs. The methods used to measure the time relationship of the resulting signals are the subject of the remainder of Section 3–3. In most time measurement techniques, the comparator circuit will play an important part.

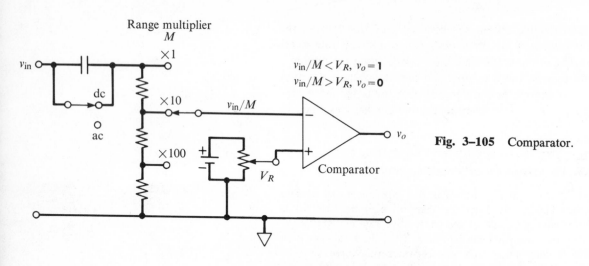

$$v_{in}/M < V_R, \; v_o = \mathbf{1}$$
$$v_{in}/M > V_R, \; v_o = \mathbf{0}$$

Fig. 3–105 Comparator.

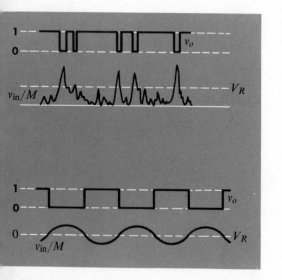

Fig. 3–106 Input and output waveshapes for comparator.

A comparator circuit is shown in Fig. 3–105. The characteristics of the ideal comparator are such that the output is always at one of its two output limits. The negative and positive limits are sometimes called LO and HI. Also, because there are just two states, they are sometimes called **0** (zero) and **1** (one) for the two numerals in the base-two (binary) number system. When the comparator input voltage v_{in}/M is less than the reference voltage V_R, the output is HI, but when v_{in}/M is greater than V_R, the output is LO. Thus whenever a signal at v_{in}/M exceeds the threshold level V_R, the output v_o makes a sudden transition from HI to LO and then goes back to HI again when v_{in}/M is less than V_R, as in Fig. 3–106a. The direction of the comparator output voltage transition relative to the direction of the input voltage change can be reversed by reversing the comparator input connections as described in the oscilloscope sweep generator trigger of Fig. 3–96. The sharp transitions at the comparator output contain a great deal of information concerning the time relationships of the V_R crossings of the input signal. An output pulse appears each time an input signal pulse exceeds V_R; this allows the counting of events which meet this criterion, the measurement of the rate of such events, or the measurement of the time intervals between such events. In addition, the output pulse duration is equal to the input pulse width at the threshold value. The comparator output voltage levels are matched to the input level requirements of the circuits which follow it. The attenuator and ac connection for the input signal are often included to bring the signal amplitude within the adjustment range of V_R and to avoid overdriving the comparator.

For signals such as those in Fig. 3–106a the setting of the threshold level can be very critical. A variation in the level of V_R will clearly have a dramatic effect on the number and duration of the output pulses. More easily discriminable signal variations in the threshold region are preferred. The time information in the sinusoidal signal of Fig. 3–106b is obviously the frequency or the period. The time between successive HI-LO (or LO-HI) comparator output transitions can be measured to decode this signal. The accuracy of such time measurements can be quite dependent on the accuracy and stability of the comparator threshold level. In addition to instability in the threshold level, noise in the input signal can cause an error in the time of the comparator transition, as shown in Fig. 3–107. The greater the rate of change of the signal voltage as it traverses the threshold region, the less will be the error in transition time. The variation in comparator output transition time for a repetitive input signal is called the **jitter.** The amount of the jitter depends on the characteristics of both the comparator and the signal. In the case of Fig. 3–106b the capacitor coupling (often labeled AC) was used and the threshold set at 0 V. The zero-crossing setting, sometimes called "AUTO," used with the "AC" input, automatically ensures that this type of signal will cross the threshold (trigger the transition of v_o from HI to LO) at the maximum rate of change of v_{in}.

Note that the transition of v_o from HI to LO in Fig. 3–106b occurs during the positive-going slope of the input signal. An additional **slope** switch is often used with the comparator circuit which inverts the input or output signal so that the HI-to-LO transition of v_o can occur on the negative-going slope of the input signal. This is particularly useful when the time-measuring circuit which follows is triggered only by (for example) HI-to-LO transitions. A variation of the comparator circuit called the **Schmitt trigger** is discussed in Section 3–4.3.

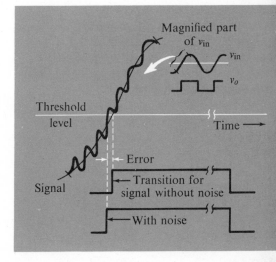

Fig. 3–107 Comparator time error.

Experiment 3–25 *The Comparator*

The input comparator of a counter-timer frequency meter is studied in this experiment in order to determine how the comparator operates, how the level control may be varied, and how the slope control, attenuator, and input coupling controls influence the output waveform. The output of the comparator or the Schmitt trigger in a counter-timer frequency meter is observed while various input controls and input signals are changed.

Separate instructions are given for the UDI comparator and the instrumentation ADD Schmitt trigger.

3–3.2 FREQUENCY MEASUREMENT AND CONVERSION

Of the several techniques for encoding data as the time of electrical signal variations, frequency encoding is the most often used. It combines the advantage of less noise susceptibility than analog signals with the availability of a large variety of simple and effective measurement and conversion devices. Also, there is a growing number of input transducers which produce frequency domain signals. These include quartz crystal oscillators for the measurement of temperature and mass, and a variety of radiation detectors when used for radiation intensity measurements.

This section begins with an introduction to oscillators, the most common sources of frequency domain signals and the reference standard in time domain measurements. Then circuits which convert a signal frequency to a proportional analog voltage are described. The most accurate frequency measurements are made by determining the frequency ratio between standard and unknown frequency sources. Both oscilloscopic and digital frequency ratio techniques are presented.

Variable Frequency Oscillators

A laboratory oscillator is a unit which contains a repetitive waveform generating circuit, an output circuit to provide the required voltage and current levels, controls for the circuit, connections to the output, a power supply for the whole unit, and a case. The range of output frequencies available from oscillators is from a fraction of one Hertz to many gigahertz (10^9 Hz), though no single oscillator circuit operates over more than a small portion of that range.

The devices and circuits used to generate sine wave signals differ greatly from one end of the scale to the other. The features, too, cover a wide range, from highly stable fixed frequency oscillators to variable frequency oscillators with simple or elaborate means of controlling the frequency and amplitude of the output signal. Some use the simple harmonic motion concept or its electronic analog described in Module 4, while others use the function generator wave-shaping approach described in Section 2–3 of Module 2. The frequency of some types of oscillators depends on the values of some of the resistors, capacitors, and/or inductors in the oscillator circuit. The frequency of such an oscillator can then be varied by adjusting one (or more) of the frequency-dependent components. When the adjustment is convenient and calibrated, the device is often called a **variable frequency oscillator** (VFO). Oscillators that depend on the stability of $R, L,$ and C components generally have frequency instability of several tenths of one percent.

Voltage-Controlled Oscillators.   The frequency of some oscillators, especially of the waveform generator type described in Module 2, is determined by the rate at which a capacitor is charged and discharged. This rate depends not only on the value of the capacitor and the resistor through which it charges, but also on the voltage source applied to the charging circuit. A voltage-controlled oscillator (VCO) circuit is designed to operate with a variable charging source voltage and has an input connection to supply that voltage from an external source. The base frequency of a VCO is switch-selected. The applied external voltage has a linear frequency control range of one-to-three orders of magnitude from the base value. The voltage-controlled oscillator is thus a kind of voltage-to-frequency converter. One application of a VCO is in recording slowly changing analog signals with a magnetic tape recorder. The rate of change of the analog signal is often well below the minimum frequency response of the tape recorder. When the analog signal is applied to a VCO, the resulting frequency-encoded output signal is readily recorded. When the recording is played back through a frequency-to-voltage converter, the original analog signal variation is recovered.

Voltage-to-Frequency Converter. A voltage-to-frequency (V-F) converter is a type of VCO in that its output frequency is directly proportional to the input voltage. Some V-F converters are linear over five or six orders of magnitude up to a maximum frequency of 10^3 to 10^7 Hz, depending on the design.

The general form of the V-F converter is illustrated in Fig. 3–108. The analog voltage v_{in} is the input to an operational amplifier integrator. The integrating capacitor C is charged at a rate determined by the time constant RC and the analog voltage v_{in}. The output of the integrator is one of the inputs to a comparator. When the capacitor has charged to the comparator reference level V_R, the comparator changes states and triggers a precision pulse generator. The pulse generator produces a pulse of precise charge content which rapidly discharges the integrating capacitor. The rate of charging and discharging the capacitor provides a signal frequency that is directly proportional to v_{in}.

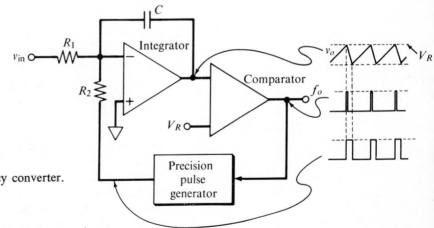

Fig. 3–108 Voltage-to-frequency converter.

The output frequency f_o is directly proportional to the input voltage, $f_o = Kv_{in}$. The proportionality constant K can be adjusted by changing the charge content of the precision pulse generator so that the frequency can be directly related to volts or other desired units.

Another type of V-F converter utilizes a unity-gain inverter in front of the integrator. The input voltage is switched alternately to the inverter-integrator combination and then directly to the integrator. The switching occurs at two preset voltage levels (e.g., 0 and 5 V). Therefore, the input signal alternately charges and discharges the integrator capacitor C. The slope of charge and discharge, and thus the output switching frequency from the comparator, are directly proportional to the input voltage.

Experiment 3–26   *Voltage-Controlled Oscillator*

This experiment demonstrates that an external dc voltage source can control the output frequency of a voltage-controlled oscillator (VCO). The output frequency of a voltage-controlled function generator is measured as a function of an external input voltage. The change in output frequency per unit change in input voltage is determined on an oscilloscope. Then a square wave input is applied to the VCO and the resulting waveforms are observed on the oscilloscope.

Experiment 3–27   *Voltage-to-Frequency Converter*

The characteristics of a voltage-to-frequency (V-F) converter are studied in this experiment. Various input voltages are applied to the converter and the output frequency is measured on an oscilloscope. The converter's transfer function is obtained from the measurements.

Frequency Standard Oscillators

One occasion when it is desirable to have a fixed frequency oscillator occurs when that oscillator is to provide a standard frequency signal. Such an oscillator is an integral part of most interdomain converters which convert from any time domain to any other domain. The most common standard frequency oscillator is the crystal-controlled oscillator. This type of oscillator, which is described in Module 4, employs a quartz piezoelectric crystal resonator. The crystal has a natural mechanical vibration frequency which depends on the material and its dimensions. The oscillator circuit provides impulses of electrical energy to the crystal at the oscillator frequency. This energy is converted by the piezoelectric effect into mechanical strain. Between pulses, the strain relaxes and most of the energy is returned to the oscillator circuit. The amount of energy required to keep the crystal vibrating falls very sharply to a minimum at the natural vibration frequency of the crystal. This effect is used in the circuit to keep the oscillator frequency exactly equal to the crystal vibration frequency.

The stability of crystal oscillators greatly exceeds that of oscillator circuits for which the frequency is determined by values of R, L, and C. Frequency stability and accuracy of 1 part in 10^5 are common, and stability of 1 part in 10^9 is attainable with care. Crystals are available over the frequency range from a few kilohertz to tens of megahertz. The higher frequency crystals (100 kHz and above) are preferred for their smaller size, lower cost, and better stability. Thus the crystal-controlled oscillator provides an easily obtainable reference frequency signal of extremely high accuracy. Through the use of electronic scaling (frequency division) circuits an almost limitless range of precision frequencies can be obtained from a single crystal oscillator.

Frequency Scalers. The frequency of the signal at the output of a frequency scaling circuit is a precise submultiple of the input signal frequency. Frequency scaling is now almost universally accomplished with digital counting circuits. The counting circuit is chosen to have a count capacity equal to the desired frequency division factor. The circuit then simply counts the cycles of the input signal. Each time the circuit's counting capacity is reached, the output signal completes one cycle and the counting cycle begins again.

The input and output signal waveforms are shown in Fig. 3–109 for the popular decade (divide-by-10) scaler. Logic level signals are required for digital counting circuits. The waveform with the arrow at the circuit input indicates that the circuit responds to (counts) HI-to-LO transitions of the input signal. The circuit counts from 0 to 9 and then returns to the "0" state on the tenth count. The indication that the counter capacity has been exceeded is the HI-to-LO transition of the output signal. The output of this particular circuit goes HI on the eighth count and returns to the LO state on the tenth count. This cycle is repeated every ten cycles of the input signal; hence the repetition rate of the output signal is precisely 1/10 that of the input signal.

As will be shown in Section 3–4, counting circuits with capacities of 2, 3, 5, 16, or any other whole number are readily obtained or constructed. Thus a wide variety of frequencies can be obtained from a single, fixed frequency source. Individual scaling circuits may be cascaded to obtain a scaling factor that is the product of the individual scaling circuits used. Clearly, if the output signal of the decade scaler of Fig. 3–109 were connected to the input of another decade scaling circuit, the frequency at the output of the second scaler would be 1/10 that at the first scaler output

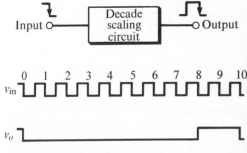

Fig. 3–109 Decade scaling circuit and waveforms.

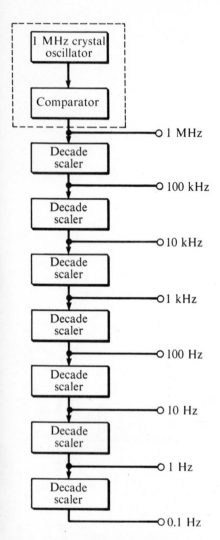

Fig. 3–110 Cascaded scalers to provide multiple standard clock frequencies.

and 1/100 that at the first scaler input. A popular clock oscillator circuit consisting of a 1 MHz crystal oscillator and seven decade scaling circuits is shown in Fig. 3–110. The comparator is used to convert the sinusoidal waveform produced by the crystal oscillator to oscillations between the two logic levels required by the digital counting circuits. A separate comparator circuit is not needed with many currently available crystal oscillator modules and integrated circuit oscillators that have incorporated the comparator circuit and thus provide an output signal suitable for direct connection to digital circuits. The seven outputs provide seven frequencies covering a span of seven decades. The nominal frequency at each output is as precise as that of the 1 MHz crystal oscillator. The complete oscillator and scaler assembly can be made so compactly and inexpensively from integrated circuits that it is increasingly common to find such sophisticated clock circuits built into modestly priced instruments which require an accurate time base. Micro-miniaturized versions of this same type of crystal oscillator and scaler circuit provide the accurate time base for the recently introduced electronic watches.

Experiment 3–28   *Frequency Standard Oscillators and Scalers*

In this experiment a high accuracy crystal oscillator is studied for use as a frequency standard. In conjunction with decade scalers, the oscillator is shown to be useful as a standard over a wide range of frequencies. The undivided output frequency of the oscillator is first measured on an oscilloscope, then the oscillator output is sent to decade scalers. The relationship between the input frequency to the scaler and the output frequency is determined for several decades of scaling.

Frequency-to-Voltage Converters

Several input transducers in scientific instruments produce a frequency domain signal directly. Such devices as photomultiplier tubes and radioactive particle detectors are direct radiation intensity-to-frequency domain transducers. Often it is desirable to continuously display or record the radiation intensity on a strip chart recorder or other analog measurement device. Devices which convert frequency domain signals to analog signals are known as **count rate meters.** Most count rate meters use the design shown in the block diagram of Fig. 3–111. A comparator input is used for pulse-amplitude discrimination of pulse signal sources or for squaring of repetitive waveform signal sources. Each transition (e.g., HI-to-LO) of the comparator output triggers the generation of a precisely reproducible current or voltage pulse. These pulses are then integrated for known periods to

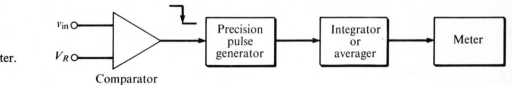

Fig. 3–111   Basic count rate meter.

yield the average rate for that period, or they are connected to an average-reading meter which reads the "average" rate continuously.

Diode Pump Pulse Generator.   A simple but extremely effective pulse generator called the **diode pump** is shown in Fig. 3–112. The two comparator output voltage levels differ by a *constant* ΔV volts. The time constant R_1C_1 is chosen so that the capacitor C_1 can completely charge to the full comparator output voltage after each transition. After positive transitions, the capacitor charging current goes through diode D_1. The charging current after negative transitions comes through diode D_2 from whatever source maintains its anode at a virtual ground. The charge q in coulombs after each transition is

$$q = C_1 \, \Delta V. \tag{3–144}$$

For an input signal of frequency f, the average charging current through D_2 is

$$-i_{\mathrm{av}} = f C_1 \, \Delta V, \tag{3–145}$$

where the minus sign indicates that the current must be supplied to the

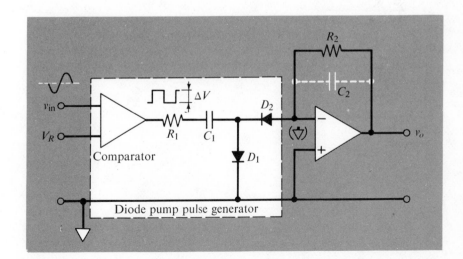

Fig. 3–112   Diode pump rate meter.

generator. The charging current is converted to a proportional voltage by the operational amplifier current-to-voltage converter circuit, which has an output voltage $v_o = -i_{in}R_2$. Thus

$$v_{o(\text{av})} = fR_2C_1 \, \Delta V. \qquad (3\text{--}146)$$

This voltage may be read by an average-reading voltmeter or the capacitor C_2 may be added to provide an averaging function in the current-to-voltage converter. The response time constant for the averaging function is R_2C_2. If $\Delta V = 5.0$ V, $C_1 = 0.02$ μF, and $R_2 = 10$ kΩ, the transfer function for the conversion is $v_{o(\text{av})} = 10^{-3}f$, a proportional relationship of 1 V per kilohertz of input frequency.

Integrating Rate Meter. An alternative to the average-reading meter or the averaging circuit is the integrator circuit shown in Fig. 3–113. The precision pulses from the diode pump pulse generator are integrated by the operational amplifier integrator for a precise period t_c. The voltage across the integrating capacitor C_2 increases by q/C_2 volts for each input cycle. The number of cycles occurring during each integration period is ft_c. Thus the peak value of v_o, the output voltage just before the switch closure, is

$$v_{o(\text{peak})} = \frac{qft_c}{C_2}. \qquad (3\text{--}147)$$

Repeated integrations of the pulses from the pulse generator will produce a repetitive sawtooth waveform with a peak amplitude of $v_{o(\text{peak})}$ volts, which can be measured with a peak-reading voltmeter. If $t_c = 0.1$ sec,

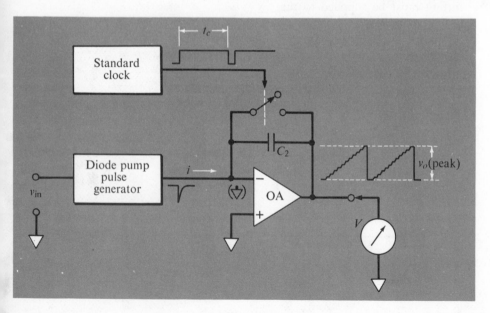

Fig. 3–113 Integrating rate meter.

$q = 10^{-7}$ C, and $C_2 = 1$ μF, the transfer function for the conversion is $v_{o(peak)} = 10^{-2}f$. The integration method is especially suitable for low frequency conversions, since its transfer function constant can readily be increased by increasing t_c or decreasing C_2. The average value of v_o is also proportional to the input frequency if the integrating periods come at regular intervals. If the capacitor discharging time is negligible compared to t_c, the average value of v_o will be equal to $v_{o(peak)}/2$.

Experiment 3–29 *Count-Rate Meter, Time-to-Analog Converter*

It is frequently necessary to convert time-encoded information into an analog signal. This experiment illustrates the use of a count-rate meter for producing an output voltage proportional to an input frequency. The operational amplifier circuit is constructed and output voltages are measured on a DVM as a function of input frequency. The transfer function of the circuit is then plotted.

Dual-Display Frequency Comparison

The relative frequencies of two signals can be compared by observing both signals on the same time scale, as on the dual-trace oscilloscope. Such a comparison is illustrated in Fig. 3–114. The standard reference frequency signal, or **clock,** has been "squared" with a comparator circuit for easier observation of the standard time period t. The observed number of input signal cycles per clock period cycle is the ratio of the input signal to clock signal frequencies. This type of direct observation of the ratio is obviously limited to relatively small ratios and is not often done. However, it is a useful technique for illustrating the principle of counting cycles per cycle for measuring frequency ratio. Figure 3–114 also illustrates the one-count uncertainty that exists in such count-ratio measurements. This minimum value of uncertainty results from the lack of synchronization between the input and clock signals. Of course, greater uncertainties result when one or both of the frequencies vary during or between measurements. Even though the two clock signals shown in Fig. 3–114 have identical periods, the number of input cycles counted changes by one as the relative timing of the two signals is shifted. It is easy to see that the maximum uncertainty due to lack of synchronization is one count from the true ratio.

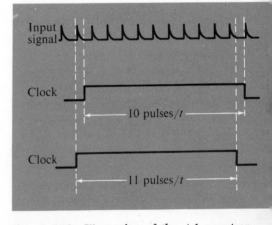

Fig. 3–114 Illustration of the ± 1 count error in nonsynchronized measurement modes.

Experiment 3–30 *Manual Counting Frequency Ratio Measurements*

In order to introduce the basic principles of frequency ratio measurements by counting methods, a manual technique is studied in this experiment. A frequency standard source and an unknown frequency source are applied to a light driver and indicator light. The number of cycles of the unknown source which occur during one cycle of the standard source is determined from observing the indicator lights. Counting errors due to nonsynchronized sources are observed.

Experiment 3–31   *Dual-Trace Oscilloscope Frequency Ratio Measurements*

For measurements of frequencies much higher than a few Hertz by manual counting (frequency ratio) methods, the oscilloscope can be used in the dual-trace mode. In this experiment the standard frequency signal is displayed on one channel of a dual-trace oscilloscope and the unknown frequency signal is displayed on the other channel. The time base of the oscilloscope is expanded to display one complete cycle of the standard frequency. The number of cycles of the unknown per cycle of the standard is then determined by manual counting from the oscilloscope display.

X-Y Display Comparison

When two signals are connected to the horizontal and vertical inputs of an oscilloscope, the oscilloscope becomes an *x-y* plotter displaying the functional relationship between the two signals. Instead of two signals vs. time, the values of one signal are plotted vs. the corresponding time values of the other, although neither axis is time. If two periodic signals are used and the time relationship of the two signals shifts, the pattern will change. Since the time-amplitude relationship of one signal is being plotted vs. that of the other, the resulting pattern must contain information about the time relationship between the two signals. These patterns, called **Lissajous patterns,** are useful for phase angle and frequency ratio measurements.

Phase Angle. Since the sine wave is such a simple waveform, only two pieces of information are obtainable from a sine wave signal (besides the fact that it is a sine wave): the frequency and the amplitude. If another signal of the same frequency is available for a time reference, the phase angle between the signals may contain information. Phase angle encoding of information is discussed in Section 3–3.4. The form of the Lissajous display as a function of phase difference between two signals is shown in Fig. 3–115. Note that the peak-to-peak amplitudes of the input signals can be obtained from the maximum horizontal and vertical excursions of the trace. For the equal-amplitude signals shown, the phase angle θ, the major measure of phase difference, can be determined by measuring the quantities a and b shown in Fig. 3–116 and applying Eq. (3–148):

$$\sin \theta = \frac{a}{b}.$$

(3–148)

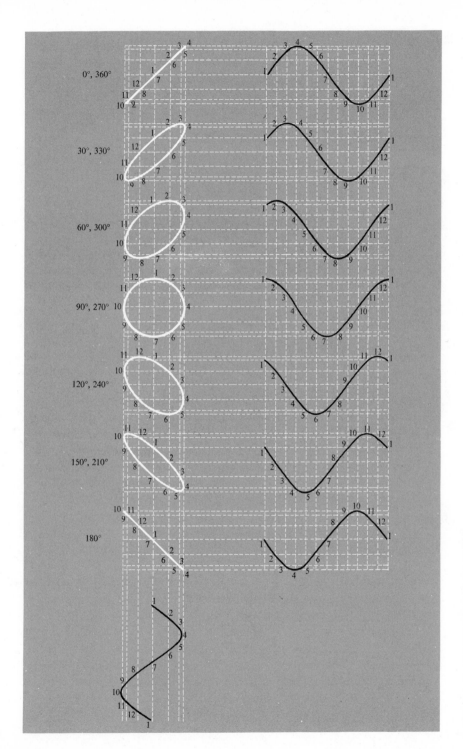

Fig. 3–115 Lissajous figures indicating phase difference.

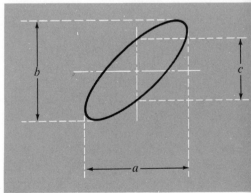

Fig. 3–116 Phase angle measurement.

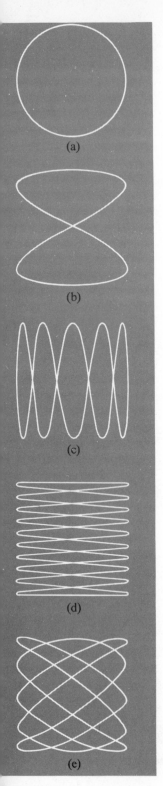

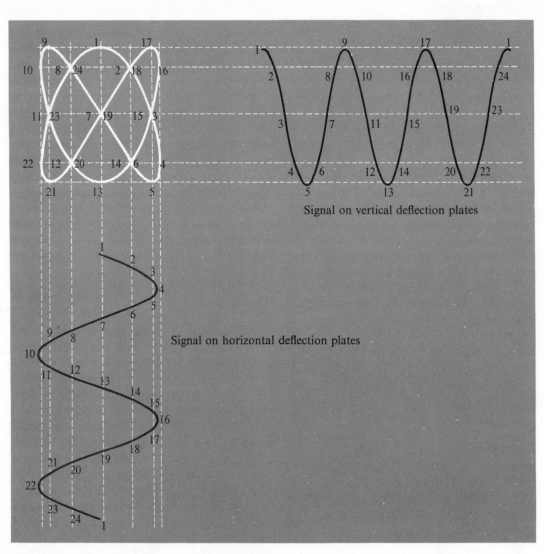

Fig. 3–117   Lissajous figure for 2:3 frequency ratio.

Fig. 3–118   Lissajous figures for various
horizontal-to-vertical ratios: (a) 1:1, (b) 2:1,
(c) 1:5, (d) 10:1, (e) 5:3.

Frequency Ratio. If two signals which have a frequency ratio that can be expressed in a small whole number or a simple fraction are applied to the horizontal and vertical inputs, patterns are obtained similar to those in Figs. 3–117 and 3–118. Patterns of this type can be used to determine the frequency ratio as shown. For example, in Fig. 3–117 each of the points (1, 2, 3, . . . , etc.) indicates an identical time for the signals applied to the horizontal and vertical deflection plates. There are two complete sine waves on the horizontal plates over the same time in which there are three complete sine waves on the vertical plates, to provide a pattern that indicates a 2:3 frequency ratio. Note from the examples given in Fig. 3–118 that the frequency ratio is the ratio of nodes along the horizontal and vertical extremities of the pattern. If the frequencies of the two signals are not locked together, a slight difference in the frequency ratio from the small whole number ratio will give a rotating appearance to the pattern because of the constantly changing phase angle.

Experiment 3–32 *Lissajous Figures*

Frequency ratio and phase angle measurements are frequently made by *x-y* plotting of two signals on an oscilloscope. The resulting patterns are called Lissajous figures. In this experiment *x-y* comparison of an unknown frequency signal with a standard frequency signal is first carried out to illustrate frequency ratio measurements with Lissajous figures. Then the *x-y* plotting method is used to determine the phase relationship between two signals.

Digital Frequency Ratio Measurement

A digital frequency ratio measurement is simply an automation of the dual-display frequency ratio method described above and illustrated in Fig. 3–114. In the digital method for frequency ratio measurement, a digital counting circuit is used to count the number of cycles of one signal that occur during a known number of cycles of the other signal. To perform this operation automatically, it is necessary to have a digital counting circuit which is incremented by one signal and is enabled (allowed to count) by the other, and to have comparator circuits that will condition the input signals into signal variations appropriate to increment and enable the counter.

A digital frequency ratio measurement system is shown in Fig. 3–119. The signal which is to increment the counter is applied to comparator 2. When one counts objects, events, pulses, cycles, or anything, it is necessary to decide whether or not each unit observed belongs in the set of units to be counted. A critical part of the counting process is the positive identification

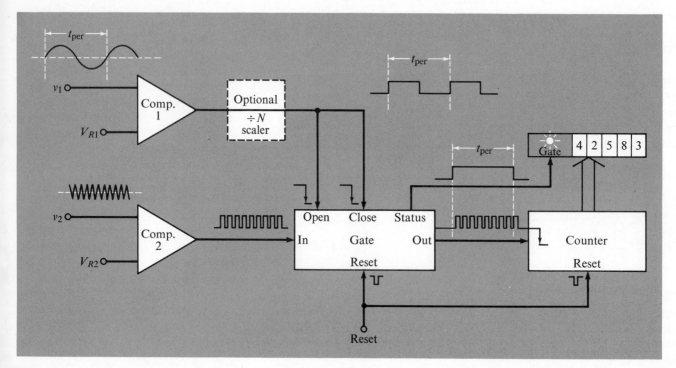

Fig. 3–119 Digital frequency ratio measurement.

of the desired units and the discrimination against units which are not in the desired set. When the counting process is automated, this critical decision-making part of the process must be automated as well. The device which performs this task in most digital counting applications is the comparator. The transitions from comparator 2 pass through the gate and are counted by the counter *when the gate is open.* The gate circuit is opened and closed in response to the open and close signals from comparator 1. In the system shown, the gate will open on the first HI-to-LO transition from comparator 1. The gate close input is not active until the gate is open. During the open period, the counter counts the HI-to-LO transitions from comparator 2. The gate status output is often used to light a "gate on" indicator. The gate closes on the second HI-to-LO transition from comparator 1, that is, after exactly one period of the v_1 signal. The counter has thus counted the cycles of v_2 per cycle of v_1 and the displayed count is the frequency ratio of the input signals. The gate will not respond to additional gate open signals until the gate and the counter have been reset. This allows time to read the result from the digital display before the next measurement begins.

The accuracy of the measurement is equal to the nearest whole number of the ratio plus or minus 1 plus any error due to the jitter of comparator

1. The number of significant figures for measurements of signals with small frequency ratios can be improved by dividing the output frequency of comparator 1 by N so that the gate is open for N periods of the v_1 signal. For each decade divider placed in the position marked "optional scaler," the decimal point in the resulting digital readout moves one place to the left. This allows the measurement of frequency ratio to virtually any number of significant figures. The introduction of the scaler after comparator 1 also reduces the relative error due to comparator jitter because the jitter becomes smaller compared to the total gate time.

Measurement of Frequency and Period.   The circuit of Fig. 3–119 can be used for precise frequency and period measurement when one of the input signals is obtained from a precision clock oscillator. To measure frequency, the precision oscillator is connected to the v_1 input to control the counter gate interval and the signal of unknown frequency is connected to the v_2 input. If the precision oscillator output is scaled down to 1 Hz exactly, the counter gate will be open for exactly 1 sec and the counter will display the input signal frequency directly in cycles per second. Other time bases may be used, depending on the input frequency range and the number of significant figures required. If the time base is 1 msec, the counter will read directly in kHz.

The period of an unknown signal is measured by interchanging the precision oscillator and unknown signals at the v_1 and v_2 inputs. The result is that the counter counts the number of clock oscillations at v_2 in one period of the unknown signal at v_1. If the precision oscillator is exactly 1 MHz, the period of the unknown signal will be read to the nearest microsecond. The optional scaler in Fig. 3–119 may be used with period measurements to increase the number of significant figures or reduce the error due to comparator jitter. When the optional scaler is used, the time of N periods is measured by the counter and clock. This type of measurement is called **multiple period average** measurement.

Experiment 3–33   *The Digital Counter and Input Gate*

In digital frequency ratio measurements, the frequency standard signal is used to open a gate to a digital counter for a preset precise length of time. The unknown frequency signal is allowed through to the counter during the gate open period, and the oscillations of the unknown frequency source are counted. In the experiment some characteristics of manual gating of the digital counter are studied along with methods for scaling down the unknown frequency signal when it is high enough to overrange the counter. In addition, several frequency ratio measurements are performed by using an external oscillator to open and close the counter input gate.

Experiment 3–34 *Frequency and Period Measurements*

In this experiment the internal, precision oscillator in a digital counter-timer is used as the counter gate control signal in order to perform accurate measurements of the frequencies of unknown signals. Counting errors are observed, and the influence of trigger level on frequency measurements is studied. Then, period measurements are made by counting the internal frequency standard oscillator for one cycle of the unknown frequency signal. The complementary use of frequency and period measurements for different signals is noted.

3–3.3 PULSE WIDTH MEASUREMENT AND CONVERSION

In the pulse width encoding of data, the duration of electrical pulses is related to the magnitude of the encoded data. Pulse width data encoding is used increasingly in data communications and some types of digital voltmeters. One advantage of pulse width encoding over frequency encoding is that the data quantity can be represented by a single pulse. Most often a series of pulses at regular intervals is used to represent the data quantity at successive times.

Examples of circuits which convert analog-encoded data to the pulse width domain and vice versa are described in this section. In addition, the basic technique of digital pulse width measurement is introduced.

Voltage-to-Pulse-Width Converters

There are at least two general ways of making voltage-to-pulse-width conversions that are widely used, the **single-ramp** and the **dual-slope** techniques. The basic ramp technique is illustrated in Fig. 3–120. A linearly increasing reference voltage, v_i, is generated which rests at 0 V and upon a trigger signal, v_t, begins to sweep to a preset full scale value. It is continuously compared with the input voltage, v_{in}. An OA integrator is used to produce the reference voltage, so that

$$v_i = -tV_R/RC. \qquad (3\text{–}149)$$

At the instant v_i slightly exceeds v_{in}, the comparator output voltage, v_c, changes logic level so as to cause the pulse generator to terminate the pulse. Since the pulse output controls the integrating switch, the integrator is reset, v_i drops, v_c rises, and the circuit is in its initial state, ready for the next trigger pulse. At the termination of the pulse, $v_i = v_{in}$ and $t = t_{pw}$, the pulse width. Substituting in Eq. (3–149) above and solving for t_{pw}, we obtain

$$t_{pw} = -v_{in}(RC/V_R). \qquad (3\text{–}150)$$

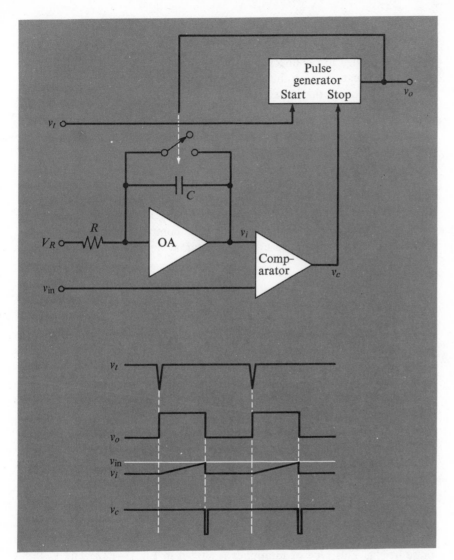

Fig. 3–120 Gated integrator voltage-to-pulse-width converter.

Because of the inverting nature of the integrator, V_R must be negative for positive values of v_{in} and vice versa. In the practical circuit, the values R, C, and V_R must be chosen so that v_i can reach the maximum value of v_{in} in a time shorter than the time t_{per} between trigger pulses. Equation (3–150) suggests that this circuit could also be used to produce a pulse width that is proportional to the ratio v_{in}/V_R. Since the average output voltage of the pulse generator is proportional to the pulse width (at constant trigger rate), the circuit of Fig. 3–120 can be used as an analog divider. A low pass filter is connected to the pulse generator output to provide an output voltage proportional to v_{in}/V_R.

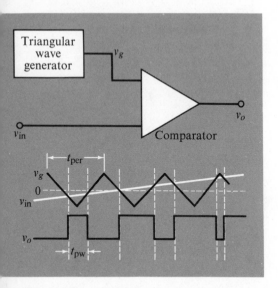

Fig. 3–121 Function generator voltage-to-pulse-width converter.

An extremely simple voltage-to-pulse-width converter of another type is shown in Fig. 3–121. A stable triangular wave generator and the unknown voltage are connected to the inputs of a comparator. From the waveforms it can be seen that as v_{in} varies from the negative to the positive extremes of v_g, the pulse width t_{pw} varies from 0 to 100% of t_{per}, the period of the triangular waveform, according to the relationship

$$\frac{t_{pw}}{t_{per}} = \frac{v_{in} + V_{p-}}{V_{p-p}}, \qquad (3-151)$$

where V_{p-} and V_{p-p} are the negative peak and peak-to-peak amplitudes of the triangle wave. This type of converter can be used for negative, positive, or bipolar values of v_{in} by simply adjusting the peak values of v_g to encompass the desired range (within the comparator input limits, of course). For the special case of a bipolar symmetrical triangular wave ($V_{p-} = V_{p-p}/2$),

$$\frac{t_{pw}}{t_{per}} = \frac{v_{in}}{V_{p-p}} + \frac{1}{2}. \qquad (3-152)$$

So long as the triangular waveform is linear, a linear relationship between pulse width and input voltage will result. A linear sawtooth waveform could also be used. Consideration of the relationship between the waveshape and the voltage–pulse-width transfer function suggests that non-linear waveshapes (e.g., sinusoidal, logarithmic) could be used to obtain specific nonlinear transfer functions if desired.

The dual-slope technique can be considered as a voltage-to-pulse-width ratio converter, but it is not discussed until Section 3–4 because its implementation generally involves the counter-timer circuits presented in that section.

Time Division Multiplier. The voltage-to-pulse-width generator of Fig. 3–121 is the basis of the **time division** type of analog multiplier, in which one input voltage controls the pulse width while the other controls the pulse height. This type of multiplier is shown in Fig. 3–122. The v_y input signal is inverted by OA1. The resulting $-v_y$ is summed with $+2v_y$ when the analog switch is ON. The analog switch is controlled by the comparator output to be OFF during t_{pw}. Thus, $v_o = +v_y$ during t_{pw} and $v_o = -v_y$ the rest of the time. The average output voltage, v_o, is

$$v_o = v_y \frac{t_{pw}}{t_{per}} - v_y \left(1 - \frac{t_{pw}}{t_{per}} \right). \qquad (3-153)$$

Substituting Eq. (3–152) for t_{pw}/t_{per} in Eq. (3–153) and simplifying,

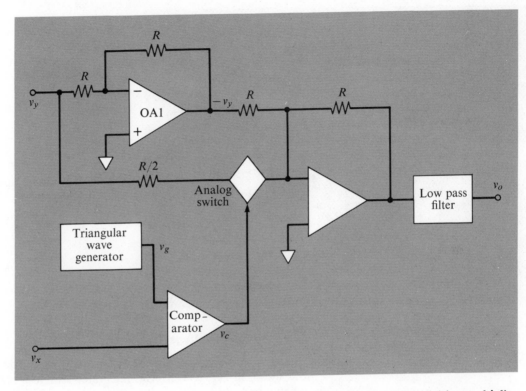

we obtain

$$v_o = 2\,\frac{v_x v_y}{V_{p\text{-}p}}. \tag{3-154}$$

Fig. 3–122   Time division multiplier.

Equation (3–154) demonstrates the validity of the multiplier function in all four quadrants. The accuracy of the time division multiplier is limited by the linearity of the triangular wave generator and by the requirement that the switching time of the analog switch be very much less than t_{per}. The frequency response of the circuit is necessarily less than $1/t_{per}$.

Experiment 3–35   *Voltage-to-Pulse-Width Converter*

A voltage-to-pulse-width converter is constructed in this experiment from a triangular wave generator or sawtooth generator and a comparator. Some characteristics of a comparator with added hysteresis are first studied before the complete converter is wired. Then the voltage-to-pulse-width converter is characterized by applying various input voltages and measuring the time required for the comparator to change states.

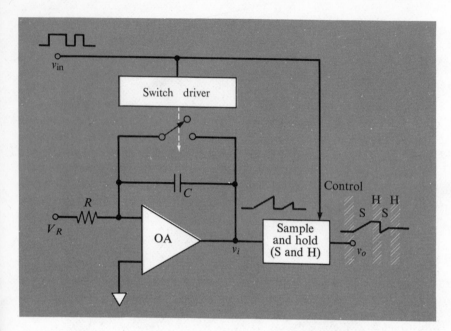

Fig. 3–123 Gated integrator pulse-width-to-voltage converter.

Pulse-Width-to-Voltage Converters

A pulse-width-to-voltage convertor that is suitable for individual pulses is shown in Fig. 3–123. The integrator switch is open for the pulse duration, producing a sawtooth voltage at the integrator output which has a peak value of

$$v_{i\text{(peak)}} = t_p V_R / RC. \tag{3–155}$$

This voltage can be measured directly by a peak-reading voltmeter, or the duration of the peak voltage can be extended to the beginning of the next pulse with the sample-and-hold circuit shown.

If the input pulses are of uniform amplitude and if they occur at a constant frequency, the pulse-width-to-voltage conversion can be accomplished by simply connecting the pulse string to an average-reading voltmeter. The voltmeter reading will vary from 0 to the full pulse amplitude in proportion to the fraction of the repetition period occupied by the pulse. If the pulse amplitude is not constant, the pulse can be used to control a switch placed between a constant voltage and an average-reading voltmeter, or an average-reading voltmeter can be connected to v_i in the circuit of Fig. 3–123.

Digital Pulse Width Measurement

A digital pulse width measurement is accomplished by simply counting the number of cycles from an accurate clock oscillator which occur during

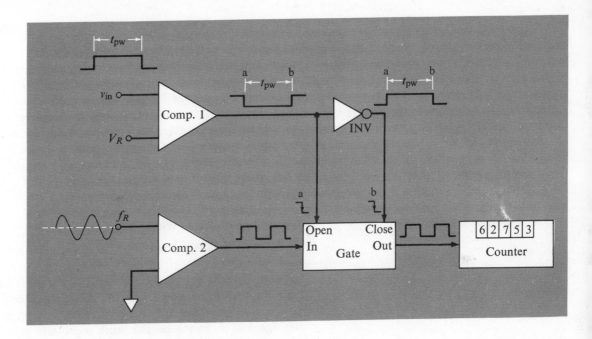

the pulse. The pulse to be measured is used to control the counter gate as shown in Fig. 3–124. The leading edge of the pulse (inverted by the comparator) causes the gate to open, enabling the counter to count the cycles of the standard frequency source, f_R. To obtain a falling transition at the end of the pulse to close the gate, the pulse is inverted again. (The circle on the amplifier symbol indicates an inverting amplifier.) If the standard frequency source is 1 MHz, the counter will display the pulse width in microseconds. The resolution of the measurement is equal to one period of the standard frequency source. The accuracy depends on the number of cycles counted and the jitter in comparator 1.

Fig. 3–124 Digital pulse width measurement.

Experiment 3–36 *Digital Pulse Width Measurement*

This experiment illustrates the use of a digital counter-timer to make time measurements. First a manual timing measurement is made to illustrate the general features of digital timing. Then a time *A-B* measurement is made to obtain pulse width information on an asymmetrical signal.

3–3.4 PHASE ANGLE MEASUREMENTS AND CONVERSION

Phase angle encoding of data requires two signals: one signal of reference phase, and the other whose phase difference from the first represents the encoded datum. When the phase angle between the two signals is constant, the frequency of the two signals must be identical. However, if the phase angle between the two signals is changing, the frequencies of the two signals must be different. Thus phase angle encoding is closely related to frequency encoding. Several methods of phase angle detection are described in this section, which concludes with an introduction to the phase-locked loop, a method of locking the frequency of one oscillator to another through the use of phase angle detection.

Phase-Angle-to-Pulse-Width Converters

One way to measure phase angle is to convert the time difference of the zero crossing of the two signals into pulse width. This is accomplished by the simple technique shown in Fig. 3–125. The waveform shown at v_1 leads the waveform shown at v_2 by the time t_θ. The positive zero crossings at v_1 and v_2 start and stop the pulse generator respectively. As the lag of the signal at v_2 increases, the pulse width will increase until it is almost equal to the signal period, t_{per}. At this point, the phase angle is almost equal to 360°. At the 360° point, the signals are again in phase. The

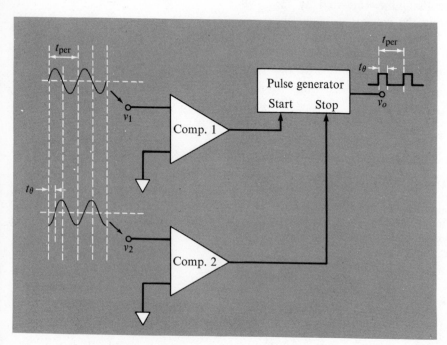

Fig. 3–125 Phase-angle-to-pulse-width converter.

phase angle $\theta_2 - \theta_1$ in degrees is obtained by multiplying the fraction of t_{per} represented by t_θ by 360:

$$\theta_2 - \theta_1 = 360 \frac{t_\theta}{t_{per}}. \qquad (3\text{--}156)$$

If the reference signal frequency is constant, it is generally unnecessary to measure t_{per}. The average output voltage, $v_{o(av)} = v_{o(peak)}(t_\theta/t_{per})$, so that if an average-reading voltmeter is connected to the output,

$$v_{o(av)} = (\theta_2 - \theta_1) \frac{v_{o(peak)}}{360}. \qquad (3\text{--}157)$$

If v_2 lags v_1 slightly, the output pulse will be very short, but if v_2 leads v_1 slightly, the pulse width will be almost equal to t_{per}. If the average phase difference between the input signals is 0 degrees, the sharp discontinuity of pulse width which occurs at zero phase difference makes great demands on the pulse generator capabilities and can make decoding the pulse width awkward. In such cases it is customary to shift the phase of one of the signals by some set amount. The signal at v_2 would be shifted exactly 180° if the common and signal connections to the comparator were reversed. In this case, t_θ would be $t_{per}/2$ when $\theta_2 - \theta_1$ is zero, and the range of pulse width from 0 to t_{per} would represent phase angles of $-180°$ to $+180°$. If the average output voltage is measured,

$$v_{o(av)} = (180 + \theta_2 - \theta_1) \frac{v_{o(peak)}}{360}, \qquad (3\text{--}158)$$

so that when $\theta_2 = \theta_1$, $v_{o(av)} = v_{o(peak)}/2$.

Phase Angle Measurement by Multiplication

It is well known that when two sinusoidal waveforms are multiplied, the resulting waveform contains frequency components equal to the sum and the difference of the input signal waveforms. If both waveforms have the same frequency, the output signal contains a sinusoidal signal at twice the input frequency and a dc component related to the phase angle between the input signals. Thus a multiplier circuit can be used as a practical phase-angle-to-voltage converter as shown in Fig. 3–126.
Consider two sinusoidal signals having the general form

$$v_1 = V_1 \sin(\omega_1 t + \theta_1),$$
$$v_2 = V_2 \sin(\omega_2 t + \theta_2). \qquad (3\text{--}159)$$

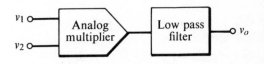

Fig. 3–126 Multiplier phase-angle-to-voltage converter.

Note 5.   Phase Shifter Derivation.

Taking C and R_c as a voltage divider,

$$v_+ = v_{in} \frac{R_c}{R_c + (1/j\omega C)}. \quad (1)$$

Also taking R and R as a voltage divider,

$$v_- = v_{in} + \frac{v_o - v_{in}}{2} = \frac{v_{in} + v_o}{2}, \quad (2)$$

from which

$$v_o = 2v_- - v_{in}. \quad (3)$$

For an OA within its input and output limits, $v_- \simeq v_+$. Substituting in Eq. (3) for $v_+ = v_-$ from Eq. (1), we obtain

$$v_o = v_{in} \left(\frac{2R_c}{R_c + 1/j\omega C} - 1 \right)$$

$$= -v_{in} \frac{1 - j\omega R_c C}{1 + j\omega R_c C} \quad (4)$$

or

$$\frac{v_o}{v_{in}} = -\frac{1 - j\omega R_c C}{1 + j\omega R_c C}. \quad (5)$$

The vector magnitudes of the numerator and denominator of the complex fraction on the right are both $\sqrt{1 + (\omega R_c C)^2}$, which shows that the amplitude ratio of v_o/v_{in} is -1, a unity-gain phase shift of $180°$. The phase angles of the vectors are $-\tan^{-1} \omega R_c C$ for the numerator and $+\tan^{-1} \omega R_c C$ for the denominator. These combine to provide a total phase shift of $2 \tan^{-1} \omega R_c C$ for the result:

$$v_o = v_{in} \quad \text{shifted by the angle}$$
$$180° - 2 \tan^{-1} \omega R_c C. \quad (6)$$

The product of v_1 and v_2 is v_o:

$$v_o = \frac{V_1 V_2}{2} [\cos (\omega_1 t + \omega_2 t + \theta_1 + \theta_2)$$

$$- \cos (\omega_1 t - \omega_2 t + \theta_1 - \theta_2)]. \quad (3\text{--}160)$$

The two terms in Eq. (3–160) are the frequency sum and the frequency difference terms respectively. Note that the time constant of the low pass filter can be chosen so that for two nearly equal signal frequencies, the sum term will be greatly attenuated compared to the difference term. In the special case where the frequencies of v_1 and v_2 are the same ($\omega_1 = \omega_2 = \omega$),

$$v_o = \frac{V_1 V_2}{2} [\cos (2\omega t + \theta_1 + \theta_2) - \cos (\theta_1 - \theta_2)]. \quad (3\text{--}161)$$

The first term in Eq. (3–161) represents a sinusoidal signal of twice the input frequency. The second term is a constant, since the phase angle $(\theta_1 - \theta_2)$ between the input signals must be constant if the frequencies of v_1 and v_2 are identical. If the time constant of the filter is chosen so that the $2\omega t$ sinusoidal signal is averaged to zero,

$$v_{o(av)} = \frac{-V_1 V_2}{2} \cos (\theta_1 - \theta_2) = \frac{V_1 V_2}{2} \sin (\theta_2 - \theta_1 - 90).$$
$$(3\text{--}162)$$

Thus the output has a maximum value of $+V_1 V_2/2$ when $\theta_2 - \theta_1 = 180°$ and a minimum of $-V_1 V_2/2$ when $\theta_2 - \theta_1 = 0°$. The output voltage is 0 when $\theta_2 - \theta_1 = 90°$ or $270°$. If v_1 is a signal of constant amplitude and a reference phase angle, $v_{o(av)}$ is proportional to the vector component of v_2 *in phase* with v_1. Then by shifting the phase of v_1 exactly $90°$, $v_{o(av)}$ is proportional to the component of v_2 that is $90°$ out of phase (the **orthogonal component**) with respect to the reference phase angle.

The multiplier can also be used as a frequency and phase angle null detector. In this application, it is necessary to have a signal of precise and variable phase angle with respect to the reference or the input signal. This is accomplished by a phase angle shifter as shown in Fig. 3–127a. The phase of one of the input signals is adjusted by the phase shifter until the null detector reads zero. This will occur when the phase of one of the signals has been shifted exactly $90°$ with respect to the other. A simple circuit which will allow the phase of a signal to be adjusted without affecting the amplitude is shown in Fig. 3–127b. The phase shift obtained[5] is $180° - 2 \tan^{-1} \omega R_c C$. As R_c is varied from $0 \, \Omega$ to ∞, the phase shift will vary from $180°$ to $0°$, with a $90°$ phase shift occurring for $\omega R_c C = 1$.

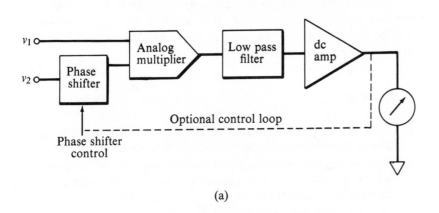

(a)

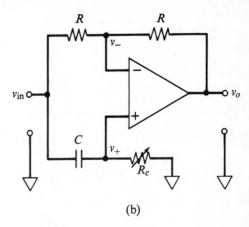

(b)

Fig. 3–127 Phase angle measurement by null detection. (a) Block diagram. (b) Phase shifter circuit.

In practice, it is only possible to approach the extremes within a few degrees with this circuit. When adjustments around 0° to 180° are required, an OA integrator circuit can be used to provide an additional constant 90° phase shift [$\int \sin (\omega t + \theta) \, dt = -\cos (\omega t + \theta)$]. If R_c were a voltage-controlled resistor, the output of a dc null detector could be used to control the phase shift to bring the circuit of Fig. 3–127a to an automatic null. The output voltage of the null detector amplifier would then be directly related to the phase angle between v_1 and v_2.

Phase-Locked Loop. Another important application for the multiplier phase angle detector is in the feedback control of the frequency of an oscillator. Such a control system is called a **phase-locked loop** and is shown in Fig. 3–128.

The multiplier output is filtered to allow only dc and low frequency signals through to the high gain dc amplifier. The amplifier output, v_o, will be zero if the oscillator output is of identical frequency and exactly 90° out of phase with the input signal, f_{in}. If the frequency of the two signals begins to differ, a change in phase angle will result and the amplifier output voltage will change. The amplifier output voltage is applied to the voltage control input of the oscillator with a sense such that the phase shift is decreased. Thus as the frequency of f_{in} changes, the frequency of the oscillator is controlled to track it precisely. Therefore $f_o = f_{in}$. The output voltage v_o is related to the frequency change through the transfer function of the VCO. The phase-locked loop is a widely used frequency-to-voltage converter. The finite values of v_o required as the frequency changes do require finite differences from the 90° phase angle between f_o and f_{in}, but the high gain amplifier ensures that the required difference is slight.

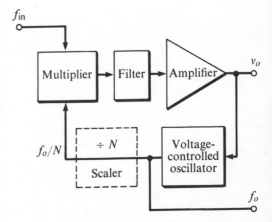

Fig. 3–128 Basic phase-locked loop.

Another application of the phase-locked loop is to produce an exact multiple of an input frequency. If a scaler is placed as shown in Fig. 3–128, the oscillator frequency will be controlled to be $f_o = Nf_{in}$, since the signal f_o/N is compared with f_{in}.

PROBLEMS

1. Over what range of differential input voltages will a comparator OA with a gain of 40,000 and an output limit of ± 13 V not be at a limit?

2. Show by drawing the sweep and signal waveforms that a triggered-sweep oscilloscope can be used to observe a fraction of the input cycle, while a recurrent-sweep scope cannot.

3. Sketch a block diagram of the scaling circuits required to produce pulses every 0.1 sec, every minute, and every hour for a digital clock. The standard frequency source is the 60 Hz power line frequency.

4. An x-y oscilloscope is used to make a phase angle measurement on signals of 40 Hz. One of the oscilloscope inputs was in the ac position and the other in the dc position. The input blocking capacitor is 1 μF, and the oscilloscope input resistance 1 MΩ. What phase angle measurement error is introduced by this oversight?

5. A digital frequency ratio measurement is to be made on two signals that are both about 100 Hz. It is desired to know the frequency ratio to within one part per thousand. What scaling factor should be used with the count gate control signal? Approximately how much time will be required for each frequency ratio measurement?

6. Calculate the relative error due to the ± 1 count uncertainty when the frequency ratio circuit of Fig. 3–119 is used to measure the frequency of 50 Hz, 3000 Hz, and 250 kHz signals. The standard reference frequency is exactly 1 Hz. Similarly calculate the relative error in the period measurements of the same signals with a standard reference frequency of exactly 1 MHz.

7. A multiplier is used as a frequency and phase angle null detector for signal frequencies in the 1 kHz range. Choose an RC time constant for the low pass filter of Fig. 3–126 such that the frequency sum term is effectively attenuated. However, the detector should operate at nearly full sensitivity for frequency differences up to 50 Hz. Calculate the resulting amplitudes of the sum and difference terms in units of $V_1 V_2/2$.

Digital Converters and Measurement Concepts

Section 3–4

Signals in the digital domain represent numbers, characters, or other specific information unambiguously. To achieve this, the signal must have as many clearly distinguishable states as the total number of numbers or characters which it might be required to represent. That is, a digital signal which might represent any of the numbers from 0 to 999 must have 1000 unambiguously identifiable states. A digital signal which might represent any of the characters and symbols on a standard 44 key typewriter (upper and lower case plus line feed and carriage return) would require 90 clearly distinguishable states. In order to represent all possible numerals in a decimal digit (0 through 9) a digital signal must have ten unambiguous states. In electronic circuits, signals, and data transmission, the fewer the required number of unambiguously distinguishable states, the greater the reliability and simplicity of the circuit.

The minimum number of distinguishable states a circuit can have and still be useful is two. It might seem unnecessary to go to this extreme, but the advantages of two-state circuits and signals over those with more than two states are that two states are most easily distinguished; arithmetic, logic, and counting operations are simple; and many components such as switches, relays, and diodes which have only two states (ON or OFF) can be used in digital circuits. Two-state signals can be used to represent the two numerals (0 and 1) in the base 2 (binary) number system or the two states (TRUE and FALSE) in logic operations. Two-state circuits are often called **binary circuits** and two-state signals—high (HI) or low (LO)—are often called **binary signals** or **logic level signals.**

In order to achieve the large number of distinguishable states required for practical data encoding, a number of binary signals or circuits are grouped together to represent the data. The data are thus encoded as a com-

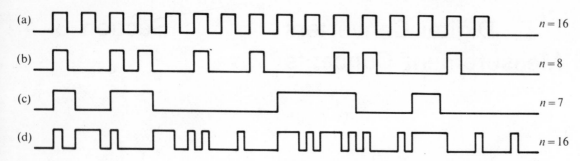

Fig. 3–129   Serial count digital signals.

bination of signals which are either HI or LO. The group of binary signals may be generated by a single binary circuit in succession (**serial digital signal**) or by a group of binary circuits simultaneously (**parallel digital signal**). If each HI signal represents one unit, the number represented is equal to the number of HI signals in the group. This equal-weight-for-all-signals code is called a **count digital signal** because it is decoded by simply counting the HI signals. Several serial count digital signals are shown in Fig. 3–129. In Fig. 3–129a a time is allotted for each binary signal and HI signal pulses are generated in each time space until the desired number of pulses has been generated. In Fig. 3–129b, the HI signal pulses are not successive, but may occur in any signal time space. If the signal time space is carefully defined, the signal need not be a separate pulse for each unit. Figure 3–129c shows a signal in which the data are encoded as the number of time spaces in which the signal is HI. In the signals of Fig. 3–129a, b, and c, the total number of different states available is equal to the number of signal time spaces. Serial count digital signals are sometimes aperiodic, as the randomly spaced pulses of Fig. 3–129d show.

In the count digital signal, it is not necessary to distinguish among count pulses (or HI level time spaces) because all have the same unit value in the number represented by the signal. If it is possible to individually identify each binary signal or time space in the group of signals or time spaces used to represent a number, much more efficient numerical encoding systems can be used. For example, if two distinguishable binary signals or time spaces are called A and B, there are four distinguishable combinations of binary signals for the combination BA (LO-LO, LO-HI, HI-LO, HI-HI, or using 0 and 1 for LO and HI, 00, 01, 10, and 11). A group of four binary signals $DCBA$ has 16 different states as shown in Table 3–1. If the A, B, C, and D signals are assigned the numerical values of $2^0 = 1$, $2^1 = 2$, $2^2 = 4$, and $2^3 = 8$ respectively, the group of signals is **binary coded.** Each binary signal is used to represent a single **binary digit,** or **bit** as it is often abbreviated. The word bit is also used for a single binary signal or time space. The decimal equivalents of the binary encoded numbers are obtained by adding the assigned binary values for all signals in the group

Table 3–1 Table of states for four binary signals

D $2^3 = 8$	C $2^2 = 4$	B $2^1 = 2$	A $2^0 = 1$	Decimal equivalent
0	0	0	0	0
0	0	0	1	1
0	0	1	0	2
0	0	1	1	3
0	1	0	0	4
0	1	0	1	5
0	1	1	0	6
0	1	1	1	7
1	0	0	0	8
1	0	0	1	9
1	0	1	0	10
1	0	1	1	11
1	1	0	0	12
1	1	0	1	13
1	1	1	0	14
1	1	1	1	15

that are 1. For example, the four-bit binary signal group or base 2 number $1011_2 = 1 \times 2^3 + 0 \times 2^2 + 1 \times 2^1 + 1 \times 2^0 = 8 + 2 + 1 = 11_{10}$, where the subscript after a number indicates the number base.

From Table 3–1 it can be seen that a three-bit signal (CBA) has eight different states, and that the addition of a fifth bit to the four shown would double the number of states to 32. In general, a group of n bits can have 2^n different combinations of binary signal levels. Thus any required number of states N can be attained by using a group of n individually identifiable binary signals or time spaces, with the condition that $2^n \geqslant N$. From the table of values of 2^n given in Table 3–2 it can be seen that a group of seven bits is required to represent numerical data to 1% (one part in 128), 10 bits for 0.1%, and 14 bits for 0.01%. The relatively small number of bits required to represent even large numbers in binary code makes parallel digital data transmission practical. A more extensive table of the powers of 2 is given in Appendix A. A parallel digital signal uses a separate wire for each bit, with appropriate binary signal levels appearing on all wires simultaneously. Parallel digital transmission can be faster than serial because the entire group of bits (often called a **word**) can be transmitted in a single time space. Parallel transmission also relieves the data receiver of the necessity of identifying the time spaces in order to decode the word, but parallel transmission obviously uses much more wire and more complex connectors

Table 3–2   Powers of two

n	2^n	n	2^n
0	1	11	2,048
1	2	12	4,096
2	4	13	8,192
3	8	14	16,384
4	16	15	32,768
5	32	16	65,536
6	64	17	131,072
7	128	18	262,144
8	256	19	524,288
9	512	20	1,048,576
10	1,024		

than serial. In practice, serial data transmission is almost always used for long distances ($>$100 meters) and parallel transmission for short distances ($<$10 meters).

Data other than numbers, such as letters or characters, and special instructions, indicators, or messages can also be transmitted in digital form. This is done by simply assigning a number equivalent to each different character or message that is in the transmitter's vocabulary. A common code for transmitting alpha-numeric (alphabetical, numerical, and common symbols) information is an ASCII (American Standard Code for Information Interchange) code that uses seven bits for the character code and one bit for parity check.

The 2^n states available from a word of n bits is a maximum. Certain of the possible states are unused in various data encoding schemes in order to increase convenience or reliability. A commonly used code that is very much more convenient than the binary code when decimal (base 10) numbers are to be represented is the **binary-coded-decimal** (BCD) code. In the most popular form of this code a group of four bits is allowed to have only the first nine states shown in Table 3–1, which then represent the numerals 0 through 9. This group of four bits can represent the numeral in the units place of a decimal number; a second group of four bits can represent the tens place numeral, a third group the hundreds place, and so on. In this code three groups of four, or 12 bits, are required to represent all the numbers from 0 to 999 (compared to 10 bits in binary code to represent 0 to 1023).

Codes which are used to increase the reliability of data transmission are called **error detecting** or **error correcting codes.** Additional bits are added to the number to impose some condition that can be readily detected. For instance, a **parity bit** can be added which makes an even number of 1's in the number. If an error is made in one bit, a test for an even number of 1's will fail and an error will be detected. When the error rate is very low, the probability of two errors in a single word is usually negligible and the simple parity check is sufficient. Other codes using additional redundant bits can detect two or more errors in a word and, in some cases, provide a "corrected" word that is probably right.

The encoding of numerical or other data as digital signals involves the logical translation of the data to the appropriate digital code. Similarly, the decoding of the digital signal into the number, character, or message it represents is a logical operation performed on the digital signal. The logical operations of digital encoding and decoding as well as those of computation with the digital data are performed by logic circuits or **gates** according to the simple theorems of **Boolean algebra.** The basic gate functions are introduced in Section 3–4.1.

In practice, digital circuits are quite forgiving about the exact logic voltage levels of the signals and the rates of change from one logic level to the other. However, there are limits to their tolerance, and signals which are not within those limits must be conditioned in some way. The techniques of digital signal conditioning, including signal squaring (sharpening of the transition times and levels), conversion of signals from one pair of logic levels to another, pulse generation and shaping, and digital data transmission, are discussed in Section 3–4.2.

Digital data are stored in binary circuits called a **memory.** A memory circuit in which the number represented by the stored logic level signals can be incremented on command is called a **counter.** The counter circuit is used in a wide variety of counting, time and frequency measurement, and sequencing devices. As shown in Section 3–4.3, the counter and other types of memory circuits are used for converting serial digital signals to parallel form and vice versa, as well as for digital data sampling and storage.

Conversion from the digital domain to an analog signal is accomplished by a digital-to-analog converter (DAC) which works by adding analog quantities proportional to the numerical value of each binary signal. Digital-to-analog conversion techniques and applications in display and digital control are described in Section 3–4.4.

Digital measurement systems for analog quantities involve some sort of analog-to-digital converter (ADC). The basis of the voltage comparison type of ADC introduced in Section 1–4.1 of Module 1 is the adjustment of a digital signal which is connected to a DAC until the output voltage of the DAC equals the input voltage. Section 3–4.5 describes the common methods of adjusting the digital signal in the voltage comparison type of ADC. Three types of integrating ADC's are also described. This section and module ends with a discussion of several examples of hybrid measurement systems (analog and digital), including sequencing systems, digital waveform generation, photon counting, and data acquisition systems.

3–4.1 LOGIC GATES

Two-level logic operations, however complex, are always combinations of three basic logic operations: AND, OR, and INVERT. The postulates of two-level logic operations, combinations, and equivalents are contained in Boolean algebra. Boolean algebra provides a means by which logic functions are easily expressed in algebraic equations which can be combined and simplified according to rules similar to ordinary algebra. The circuits which perform logic operations are logic gates. The output voltage level (HI or LO) of a gate is the logical result of the voltage levels at the inputs. The gate is named according to the logic function it performs; e.g., an AND gate performs the AND logic function.

The HI and LO voltage levels used in digital circuits are actually voltage ranges as shown in Fig. 3–130. Any voltage within either range is defined as HI or LO. There is generally a gap between the two ranges to reduce the probability of noise altering a signal voltage enough to take it into the other range. The voltage ranges for the frequently used TTL logic, for instance, are 0 to +0.8 V for LO and +2.4 to +4.0 V for HI. Many other combinations of ranges have been used and several different ranges are currently popular. Both ranges may be positive or negative, or zero may fall between the ranges. In any given logic circuit family, the logic levels for all gate circuits are the same. The two voltage levels (HI or LO) represent the two logic levels, **1** or TRUE and **0** or FALSE. If the more positive voltage level (HI) is assigned the **1** or TRUE logic level, the logic signal is said to be **positive TRUE.** If the less positive voltage level (LO) is assigned to the **1** or TRUE logic level, the signal is called **negative TRUE.**

Modern gate logic circuits are made of integrated circuits in which dozens of diodes, transistors, and resistors are fabricated on a tiny chip of silicon and encapsulated in a small metal, ceramic, or plastic case. Each integrated circuit package may contain from one to four separate gates or hundreds of gate circuits internally connected to perform a particular complex function.

The AND Logic Function

If two switches are wired in series as in Fig. 3–131, the circuit is completed only when switch A is CLOSED *and* switch B is CLOSED. These two switches are said to perform the AND operation with respect to transmitting current through the circuit by connecting the source to the load. The neces-

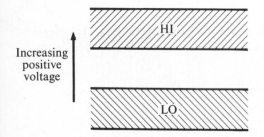

Fig. 3–130 Logic level signal ranges.

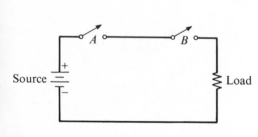

Fig. 3–131 AND operation implemented by manual switches.

sary condition for transmission can be expressed by the following equation in Boolean algebra:

$$A \cdot B = T \quad \text{or} \quad AB = T, \tag{3-163}$$

which reads, "If A is CLOSED AND B is CLOSED, then transmission will occur." The symbol A stands for the CLOSED state of switch A.

The symbols A, B, C, etc., could each represent the truth of one of the following specific statements: "the error in the resistance is less than 1%," "the straightness of the shaft is within specifications," "the temperature is above 24.6°C," "the weight of the sample is equal to 10.2 ± 0.1 g," "the radioactivity is at a dangerous level," etc.

In more general terms Eq. (3–163) could read, "If one specific statement (represented by the symbol A) is TRUE AND another specific statement (represented by the symbol B) is TRUE at the same time, then it can be logically concluded that a specific result (represented by the symbol T) is TRUE for that time."

Since the CLOSED or ON state of a switch can be used to represent the truth of a specific statement (A, B, C, etc.) and the OPEN or OFF state of the same switch to represent the falsity of the same specific statement, it is possible to use a switch to represent the state (TRUE or FALSE) of each logic statement, and a switching circuit to represent the AND function—whereby A AND B AND C AND . . . , etc., must all be TRUE statements simultaneously for a specific result T to be TRUE. As seen in Fig. 3–131, the AND function can be implemented by manual or relay switches operating in series.

It is customary to represent conditions or statements that are TRUE at a particular time by a **1**, and FALSE statements by a **0**. The equation $A = 1$ means that the A statement or condition exists (is TRUE) at a predetermined time, and $A = 0$ means that the A condition does not exist at the selected time.

It is also convenient to have a symbol for the opposite of a statement. If a condition such as "the line power is on" is given the symbol A, the opposite of that statement ("the line power is NOT on") is written $\overline{A}$ or A' and is read NOT A. Clearly, when A is TRUE, $\overline{A}$ is FALSE and vice versa.

Truth Table. A **truth table** or **table of combinations** provides a table of all possible combinations of states for all variables at a specific time and the results of each combination. A truth table is given in Table 3–3 for the AND operation with variables A and B and result T. At some selected time the variables A and B can both be TRUE (**1**), or FALSE (**0**), or A TRUE (**1**) and B FALSE (**0**), or vice versa—thereby making four possible combinations. For the AND operation all specified conditions must exist simultaneously for the result to be TRUE. Therefore, for the AND truth table

Table 3–3 Truth table for $A \cdot B = T$

Inputs		Output
A	B	T
0	0	0
0	1	0
1	0	0
1	1	1

Table 3–4   Truth table for $A \cdot B \cdot C = T$

Inputs			Output
A	B	C	T
0	0	0	0
0	0	1	0
0	1	0	0
0	1	1	0
1	0	0	0
1	0	1	0
1	1	0	0
1	1	1	1

(Table 3–3), the T column will have a **1** when, and only when, $A = 1$ AND $B = 1$. This table can be verified by considering a **0** as an OPEN switch and a **1** as a CLOSED switch in Fig. 3–131, or by wiring the circuit of Fig. 3–131 and trying all combinations of A and B, observing the conditions under which transmission occurs.

For the case of performing the AND operation with three variables (A, B, and C) the truth table would be as shown in Table 3–4. To be certain that all possible combinations of conditions are considered in an orderly fashion when constructing any truth table, it is simplest to count in binary numbers, starting with 0 for all variables in the top row and proceeding sequentially with the binary number 1 in the second row, 10 in the third row, 11 in the fourth row, etc., until in the final row each condition column (A, B, C, etc.) is filled with 1, as in Table 3–4. For three variables the number of combinations is $2^3 = 8$; for n variables the number of condition combinations is 2^n.

AND Theorems.   Boolean algebra provides a simple mathematical technique for the design, simplification, and analysis of logic circuits as well as a symbolism for the description of logic functions. Most of the basic theorems of Boolean algebra are readily understood in relation to their switching circuit equivalents. These theorems can then be used to aid the design or analysis of more complex logic circuits or functions. Table 3–5 gives the basic Boolean theorems concerning the AND function and the equivalent series switching circuit in which a **1** or TRUE logic function is represented by a CLOSED switch.

Table 3–5   Boolean AND theorems

AND Logic Gates

An AND gate is a circuit with input and output logic levels that correspond to the truth table for the AND function as shown in Table 3–4. The AND gate has two or more input terminals, A, B, C, etc., and one output terminal. Only when *all* the signals connected to the input terminals are simultaneously at the logic **1** level will the output be at logic **1**. Under all other conditions the output logic level will be **0**. One form of the AND gate which uses relay switches is shown in Fig. 3–132. A logic level **1** signal ($\sim$+5 V) at the A, B, or C inputs is sufficient to actuate the relay, but a **0** level signal ($\sim$0 V) will not. Only when all three relays are actuated ($A = B = C = \mathbf{1}$) will the signal from the +5 V source be connected to the output. At all other times the output is grounded. Frequently a transistor switch is used to drive the relay coil so that the entire coil current does not have to be supplied by the binary input signals.

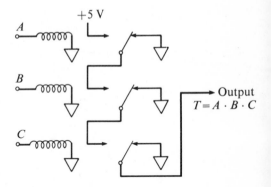

Fig. 3–132   Relay AND gate.

When drawing circuits using logic gates, it is awkward to draw the complete circuit for each gate. Thus symbols are used to represent the entire gate circuit and/or function. The symbol for an AND gate is shown in Fig. 3–133. This symbol is used regardless of the kind of switches used in the circuit.

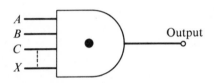

Fig. 3–133   AND gate symbol.

Solid state switches have obvious advantages over mechanical switches for logic gate applications. They are faster, smaller, and more economical and are thus universally used for modern logic circuits.

Diode AND Gate.   A diode circuit which performs the AND function is shown in Fig. 3–134. The logic level of the signal sources is +4.0 V and 0 V for **1** and **0**, respectively. (A 0 V signal source is a connection to the common through the output impedance of the signal source; it is *not* an open circuit.) Any one of the three diodes may conduct current from the +5 V source through R and its signal source to the common. The diode that will conduct is the one connected to the least positive signal source. The output voltage will be equal to the lowest input voltage level plus V_γ, the forward conduction voltage of the diode. If the lowest input voltage is 0 V, a logic **0**, the output voltage will be about 0.6 V (for silicon diodes), still within the **0** logic level range. Note that a **0** level source must be able to conduct (or "sink") 5 V/5 kΩ = 1 mA of current to ground without deviating significantly from 0 V output. In other words, a **0** level source must have a very low output resistance. The other diodes, connected to more positive input voltages, will not be biased to forward conduction. Their input sources are thus effectively isolated from the output and from each other. The logical function of this circuit can be described as "a **0** at any input results in a **0** at the output." Thus a **1** at all inputs is required to

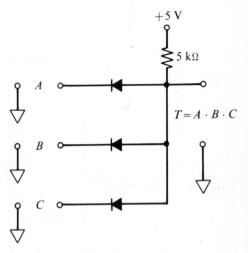

Fig. 3–134   Diode AND gate.

produce a **1** at the output. These conditions are seen to correspond exactly to the AND truth table of Table 3–4.

This simple diode resistor AND circuit is basic to a majority of the circuits in digital instruments and computers. The reason it is often called a "gate" is shown in the following example. Consider a two-input circuit. When one input is held at logic level **1**, the output will follow the level changes at the other input, i.e., the gate is OPEN. On the other hand, if one of the inputs is held at **0**, the output will be **0** regardless of the other input signal level, i.e., the gate is CLOSED.

TTL NAND Gate.   Logic gates are now available in integrated circuit forms which offer high performance and convenience at very low cost. One of the most popular integrated circuits is the TTL (Transistor-Transistor Logic) gate shown in Fig. 3–135. Transistor Q_1 is a multiple-emitter transistor. Grounding any one or more inputs forward biases transistor Q_1, which puts the collector of Q_1 at a low voltage and turns Q_2 OFF. This, in turn, turns Q_4 ON and Q_3 OFF and results in a logic **1** output. However, if all inputs are HI (or unused), the base-collector junction of Q_1 will conduct, forward biasing the base-emitter junction of Q_2. When Q_2 is ON, Q_3 is ON and Q_4 is OFF, resulting in a **0** at the output. In other words, a LO at any input produces a HI output and a LO output is obtained only when all inputs are HI. This is the exact inverse of the AND function and is called the NOT AND or NAND. The NAND and AND functions are both shown for comparison in Table 3–6. The Boolean algebra notation for the NAND function is $T = \overline{ABC}$ or $\overline{T} = ABC$. The NAND gate symbol shown in Fig. 3–135 is the AND gate symbol with a circle at the output terminal. The circle at a gate connection is used to indicate an inversion of the logic level at that point. Note that if only one of the NAND gate inputs is used,

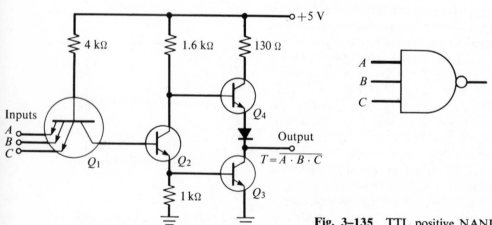

Fig. 3–135   TTL positive NAND gate and NAND symbol.

Table 3–6 Truth table for AND and NAND functions

Inputs			Outputs	
A	B	C	AND	NAND
0	0	0	0	1
0	0	1	0	1
0	1	0	0	1
0	1	1	0	1
1	0	0	0	1
1	0	1	0	1
1	1	0	0	1
1	1	1	1	0

the output is the inverse of the function at that input. That is, if the input function is A, the output will be $\overline{A}$.

Integrated Circuit Gates.   There are currently several types of integrated circuit gates in common use and a dozen or so more available for special situation application. These gates differ in the following important gate characteristics.

1.   Logic levels.   The voltage ranges corresponding to the two logic levels differ greatly among the various gate circuit types. Another type of logic level difference is whether a gate output is expected to have a very low output impedance at its 0 V logic level to sink current from the gate inputs connected to it **(current-sinking logic),** or whether the gate output impedance must be low at the nonzero output level to supply current to the gate inputs connected to it **(current source logic).** The TTL gate described above is of the current-sinking type with nominal logic levels of 0 V and $+4$ V.

2.   Noise immunity.   The noise immunity specification frequently given for logic gates is actually the **dc noise margin.** For the HI logic level it is the difference between the lowest voltage a gate output will produce in the HI state and the highest voltage a gate input will interpret as a LO level input. In principle, a HI level output could sustain a negative noise pulse of that amplitude without providing a false LO input to a gate. Similarly, for the LO logic level, the noise margin is the difference between the highest LO output voltage and the lowest voltage which will appear as a HI. These two margins are often similar and are quoted as noise immunity. The TTL gates described have a guaranteed noise margin of 400 mV and a typical value of 1.0 V.

3. Fan-out.   The output of a logic gate is frequently connected to several gate inputs. Because of the current-sinking or current source requirements each gate input makes on the gate output, there is a limit to the number of gate inputs that can be connected to a single gate output. The number of gate inputs that a gate type is guaranteed to drive properly is called the **fan-out.** For the TTL gates described, the fan-out is 10. Gate or circuit inputs which load an output more or less than the standard gate input are rated in terms of the number of standard gate inputs or **input load units** to which the input load is equivalent.

4. Propagation delay.   For all gates, there is a finite time required for the output logic level to respond to a logic level change at the input. While these times are generally very short (8–10 nsec for the TTL gates), when several gates are used in sequence the delay in producing the correct output level can be significant in very high-speed circuits.

The characteristics of the popular logic gate types are described and compared in Appendix C. The component-minded reader or anyone about to design a logic circuit is encouraged to read that appendix.

The OR Logic Function

If two switches are put in parallel as in Fig. 3–136, the circuit is completed when $A = 1$ OR $B = 1$ OR both $= 1$. This circuit is said to perform the OR operation. The Boolean algebra equation is written

$$A + B = T, \tag{3–164}$$

where the $+$ sign is a logic symbol that is read as OR. Table 3–7 gives the truth table for Eq. (3–164).

In general, the result T will exist only when either condition A OR B OR both conditions exist.

OR Theorems.   The theorems related to the OR function can also be most readily appreciated when seen with respect to the equivalent parallel switch function. The basic OR theorems are presented in this manner in Table 3–8.

AND and OR Theorems.   Finally, some very useful theorems involving both AND and OR functions are given in Table 3–9. The commutation theorems indicate that the order of the conditions in the logic statement is not significant. The association theorems show that the indicated AND or OR operations may be performed in any order. Redundant terms in an expression can be eliminated by the absorption theorems. The expression $A + AB$ depends only on the state of A. If $A = 0$, $0 + 0 \cdot B = 0$, and if A is 1, $1 + 1 \cdot B = 1$, so that $A + AB = A$. Similarly $A(A + B) = A$. Inspection of the equivalent switch network would also show this to be true. The distribution theorems indicate that combinations of variables with AND

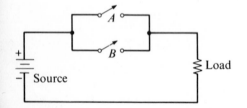

Fig. 3–136 OR operation implemented by manual switches.

Table 3–7 Truth table for $A + B = T$

Inputs		Output
A	B	T
0	0	0
0	1	1
1	0	1
1	1	1

Table 3–8   Boolean OR theorems

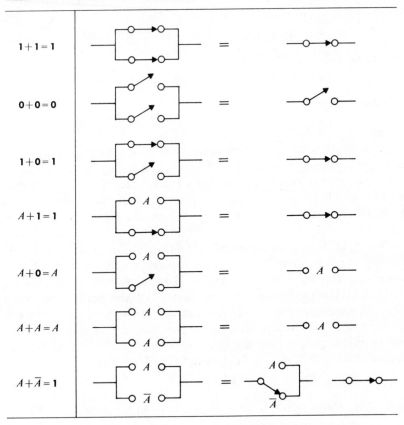

Table 3–9   Boolean AND and OR theorems

Commutation theorems:	$A + B = B + A$ $AB = BA$
Absorption theorems:	$A + AB = A$ $A(A + B) = A$
DeMorgan's theorems:	$\overline{A + B} = \overline{A} \cdot \overline{B}$ $\overline{AB} = \overline{A} + \overline{B}$
Association theorems:	$A + (B + C) = (A + B) + C$ $A(BC) = (AB)C$
Distribution theorems:	$A + BC = (A + B)(A + C)$ $A(B + C) = AB + AC$

and OR operations may be factored or distributed according to the rules of ordinary algebra. For instance, when $(A + B)(A + C)$ is "multiplied out,"

$$(A + B)(A + C) = AA + AC + AB + BC.$$

From Table 3–5, $AA = A$. Thus

$$(A + B)(A + C) = A + AC + AB + BC.$$

From the absorption theorem, $A + AB + AC = A$, so

$$(A + B)(A + C) = A + BC.$$

DeMorgan's theorems, dealing with the NOT function applied to entire terms, are especially interesting and useful. They can be proven by the truth table method, that is, by making a truth table for all four combinations of A and B for $\overline{A + B}$ and another for $\overline{A} \cdot \overline{B}$. If the T column is identical in the two tables, then the two functions are proven to be logically equivalent.

The application of the theorems of Boolean algebra is illustrated by the following example, in which equivalent expressions for the inverse of the statement $AB + BC + CA$ are sought. Each equivalent expression suggests a different logic gate implementation. All the implementations of equivalent statements are equally valid, of course, but some might be more convenient than others, depending on the gate functions available. To begin, draw an inversion bar over the entire expression and apply DeMorgan's theorem, treating AB, BC, and CA as individual quantities:

$$\overline{AB + BC + CA} = \overline{AB} \cdot \overline{BC} \cdot \overline{CA}. \tag{3–165}$$

Again, apply DeMorgan's theorem to each term on the right:

$$\overline{AB + BC + CA} = (\overline{A} + \overline{B})(\overline{B} + \overline{C})(\overline{C} + \overline{A}). \tag{3–166}$$

Apply the distribution theorem to the second two terms:

$$\overline{AB + BC + CA} = (\overline{A} + \overline{B})(\overline{B}\overline{C} + \overline{A}\overline{B} + \overline{C} + \overline{A}\overline{C}).$$

Since $\overline{C} + \overline{B}\overline{C} + \overline{A}\overline{C} = \overline{C}$ from the absorption theorem,

$$\overline{AB + BC + CA} = (\overline{A} + \overline{B})(\overline{A}\overline{B} + \overline{C}).$$

Apply the distribution theorem once more:

$$\overline{AB + BC + CA} = \overline{A}\overline{B} + \overline{B}\overline{C} + \overline{A}\overline{C}. \tag{3–167}$$

Then apply DeMorgan's theorems twice more, first to each term and then to the entire expression:

$$\overline{AB + BC + CA} = \overline{(A + B)} + \overline{(B + C)} + \overline{(A + C)} \tag{3–168}$$

$$= \overline{(A + B)(B + C)(A + C)}, \tag{3–169}$$

thus also proving that

$$AB + BC + CA = (A + B)(B + C)(A + C). \qquad (3\text{--}170)$$

A summary of Boolean theorems and postulates is given in Appendix B.

OR Logic Gates

The output of an OR gate circuit will have a logic **1** level when any one or more of the inputs is a logic **1**. A relay OR circuit is shown in Fig. 3–137. An examination of Fig. 3–137 will verify that the input and output levels of the circuit correspond to the OR truth table, Table 3–7. The general symbol for an OR gate is shown in Fig. 3–138.

Diode OR Gate.   If the diodes and the power supply polarity of the AND gate are reversed, the gate will perform the OR function. The circuit is shown in Fig. 3–139. Whichever diode has the highest input voltage will be forward biased, transmitting that input voltage to the output (reduced by V_γ). The other diodes will be reverse biased. Therefore, if any combination of the inputs have a logic level of **1**, the output will be **1**. Only when all inputs are **0** will the output be **0**. The output level thus corresponds to the OR function of the input levels. Additional inputs may be accommodated by adding more diodes.

The OR circuit also functions as a "gate." This action is accomplished by holding one of the inputs at logic **1**. During this period, variations in logic level at the other inputs cannot appear at the output.

Integrated Circuit OR Gates.   The OR gate function is available in several integrated circuit types. The inverted output OR gate (NOT OR or NOR gate) is more common. The availability of more than one logic function in a given integrated circuit type is for convenience only, since (as is shown in the next section) all logic functions can be obtained from either NAND gates or NOR gates alone. The primary gate function in the TTL gate family is the NAND gate. In some other gate families, the NOR gate is the basic gate function.

Logic Functions and Logic Levels

Four logic gate functions, AND, NAND, OR, and NOR, have been described. Table 3–10 is a truth table for all four functions. From this table, we see the expected inverted relationship between the AND and NAND and between the OR and NOR outputs. From the symmetry of the table, a relationship between the AND and NOR functions and the NAND and OR functions might also be suspected. These pairs of functions are shown below to be equivalent if the input quantities are inverted.

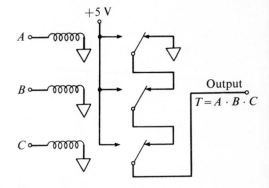

Fig. 3–137   Relay OR gate.

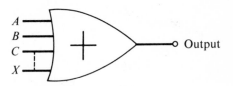

Fig. 3–138   OR gate symbol.

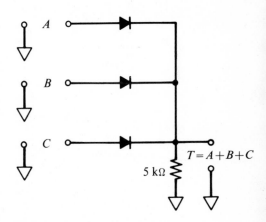

Fig. 3–139   Diode OR gate.

Table 3–10 Comparison of four basic logic functions

Inputs		Outputs			
		AND	NAND	OR	NOR
A	B	$A \cdot B$	$\overline{A \cdot B}$	$A + B$	$\overline{A + B}$
0	0	0	1	0	1
0	1	0	1	1	0
1	0	0	1	1	0
1	1	1	0	1	0

In the discussions of binary information signals and logic gates so far, the **1** or TRUE logic level has been taken as the HI logic level voltage and the **0** or FALSE logic level as the LO logic level voltage. This choice of logic level assignments is completely arbitrary. It happens to correspond to the current trend, but is by no means universal. It is interesting to consider the result of reversing this choice of logic levels, i.e., making LO the **1** state and HI the **0** state. This is called "negative logic" because the more negative level is the **1** state. The Boolean algebra does not change, of course, since its laws are entirely independent of the method of implementation. However, the gates do not perform the same operation for negative logic as they do for positive logic.

Consider the diode AND gate of Fig. 3–134. The output voltage will be 0 V or LO (logic **1**) when any one or more of the inputs is LO (logic **1**). The output can only be +5 V or HI (logic **0**) when all the inputs are HI (logic **0**) simultaneously. The truth table for this gate now corresponds exactly to the OR function. This is to say that a positive logic AND gate is an OR gate when negative logic is used. A similar consideration of the positive logic OR gate (Fig. 3–139) will show that it is a negative logic AND gate. An inverter can be thought of either as logic statement inverter (the NOT operation previously discussed) or as a logic level inverter, since a positive logic **1** at the input becomes a negative logic **1** at the output.

The above arguments suggest that the OR function could be performed on positive logic signals by inverting the logic level of each input to a positive AND gate (negative OR) and then returning to positive logic by using another inverter at the gate output. This is shown in Fig. 3–140a. The input inverters convert the positive logic representation of statements A and B to negative logic representations. The positive AND gate performs the OR function on negative logic signals to produce the quantity $A + B$ in negative logic level. The logic level of this signal is converted to positive by the output inverter. Boolean algebra can be applied to the positive logic quantities

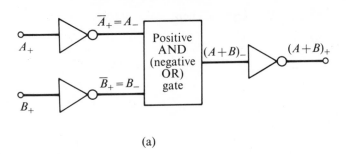

(a)

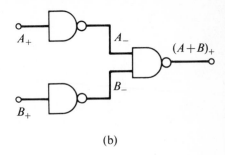

(b)

to demonstrate the validity of Fig. 3–140a. The input inverters change A to $\overline{A}$ and B to $\overline{B}$. After the $+$AND gate, we have $\overline{A} \cdot \overline{B}$. DeMorgan's theorem says that $\overline{A} \cdot \overline{B} = \overline{A + B}$, which, after inversion, becomes $A + B$. This would be implemented with positive NAND gates as shown in Fig. 3–140b. NAND gates with only one input used serve as the input inverters, while the two-input gate serves the $+$AND ($-$OR) function and provides the output inversion.

It is often helpful to draw logic diagrams with the equivalent gates as determined from DeMorgan's theorem. These equivalents are shown for the NAND and NOR gates in Fig. 3–141. Since $\overline{ABC} = \overline{A} + \overline{B} + \overline{C}$ according to DeMorgan's theorem, the NAND gate which produces $\overline{A \cdot B \cdot C}$ from $A, B,$ and C at the input is exactly equivalent to an OR gate with an inverter at each input. A similar argument holds for the NOR equivalent. Thus when an OR function is to be implemented with NAND gates, the equivalent OR gate symbol can be used and additional inverters inserted where necessary. The implementation of all four basic logic functions with only NAND or NOR gates is shown in Table 3–11. DeMorgan's theorem equivalent gate symbols have been used where appropriate.

Fig. 3–140 Positive OR function using (a) positive AND gate and inverters, (b) positive NAND gates.

Fig. 3–141 DeMorgan's equivalent gates.

Table 3–11   Logic functions using NAND and NOR gates

Function	NAND gate circuit	NOR gate circuit
AND		
OR		
NAND		
NOR		

Experiment 3–37   *Logic Levels*

Information is transmitted in digital systems in the form of logic level signals. In any binary digital system there are only two logic levels (HI or **1** and LO or **0**). In this experiment the generation of logic level signals, their detection, and the binary number system are studied in order to become familiar with binary information.

Experiment 3–38 *Basic Logic Gates*

Several basic logic functions are illustrated in this experiment. The logic function of the NAND gate is studied using logic level sources and detectors. The NAND function is shown to be the logic inverse of the AND function. When negative logic is used (less than 0.8 V represents a logic **1**), the positive logic NAND gate is seen to perform the NOR function. The logic level inverter is studied and used with a NAND gate to provide the AND logic function. It is shown how various combinations of NAND gates and inverters can achieve all four basic logic functions (NAND, NOR, AND, OR). Then an integrated circuit AND-OR-INVERT (AOI) gate is studied and used to perform NOR and AND-NOR functions.

Table 3–12 Exclusive-OR function

Inputs		Output
A	B	$A \oplus B$
0	0	0
0	1	1
1	0	1
1	1	0

Logic Gate Applications

Logic gates are used in digital instruments and digital information systems to control the flow of information, to produce, detect, and alter digital data encoding, and to perform logical operations upon digital data. The applications given in this section are only a few of the most basic and common of the endless variety of important and useful functions that can be performed by combinations of logic gates.

Exclusive-OR Function.   The exclusive-OR function of the variables A and B can be stated as A OR B but NOT A AND B. The truth table for this function is given in Table 3–12. The exclusive-OR is not obtainable from the AND or OR functions by simple inversion of inputs or outputs. It is in a different family from the AND/OR group, as its uniquely symmetrical truth table shows. Some people consider the exclusive-OR to be a third basic logic gate function (after AND/OR and INVERT). The exclusive-OR function is used so frequently that it has a separate symbol $\oplus$. From the truth table it can be seen that the output is true when $A = \mathbf{0}$ AND $B = \mathbf{1}$ OR when $A = \mathbf{1}$ AND $B = \mathbf{0}$. This statement is written in algebraic form as

$$A \oplus B = \overline{A}B + A\overline{B}. \tag{3–171}$$

This can be implemented by gates performing the AND operation on A and $\overline{B}$, and on B and $\overline{A}$, and then the OR operation on $A\overline{B}$ and $B\overline{A}$, as shown in Fig. 3–142.

For exclusive-OR operations with only one independent variable, the identities shown in Table 3–13 can be derived. Note that the first two entries in Table 3–13 give fixed level outputs regardless of the value of A. The last two entries suggest controlled inverter application of the exclusive-

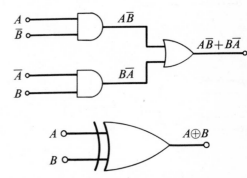

Fig. 3–142 Gate diagram for exclusive-OR function and exclusive-OR gate symbol.

Table 3–13 Exclusive-OR operations

$A \oplus A = A\overline{A} + \overline{A}A = \mathbf{0}$
$A \oplus \overline{A} = A \cdot A + \overline{A}\,\overline{A} = \mathbf{1}$
$A \oplus \mathbf{1} = A \cdot \mathbf{0} + \overline{A} \cdot \mathbf{1} = \overline{A}$
$A \oplus \mathbf{0} = A \cdot \mathbf{1} + \overline{A} \cdot \mathbf{0} = A$

OR gate. If variable A is connected to one input, the output will be A or $\bar{A}$ depending on whether a **1** or **0** logic level signal is connected to the other input.

Exclusive-OR gates may be cascaded to provide the exclusive-OR function for more than two variables, as shown in Fig. 3–143 for three, four, and eight variables. The exclusive-OR function applied to n variables provides a **1** output if an odd number of input variables are **1**. The n-input exclusive-OR function is thus used for the generation of the parity bit in a word coded for error detection by parity, and for the parity check error detector.

The exclusive-OR function can also be implemented with NAND gates as shown in Fig. 3–144a. Equivalent circuits could be obtained by deriving equivalent expressions. For instance, by adding the terms $B\bar{B}$ and $A\bar{A}$ to Eq. (3–171),

$$A \oplus B = B\bar{B} + \bar{A}B + A\bar{A} + A\bar{B},$$

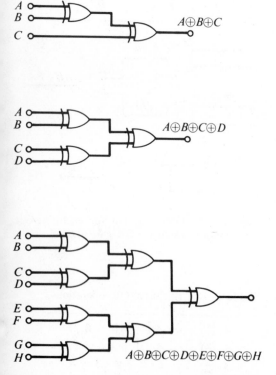

Fig. 3–143 Cascading exclusive-OR gates for three, four, and eight inputs.

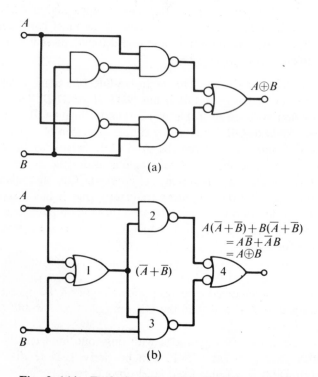

Fig. 3–144 Exclusive-OR using NAND gates: (a) from Eq. (3–171); (b) from Eq. (3–172).

and factoring B from the first two terms and A from the second two, we obtain

$$A \oplus B = B(\overline{A} + \overline{B}) + A(\overline{A} + \overline{B}). \qquad (3\text{–}172)$$

While Eq. (3–172) looks more complex than Eq. (3–171), it requires fewer NAND gates to implement it as shown in Fig. 3–144b. Equation (3–172) can be further altered by factoring $(\overline{A} + \overline{B})$ from both terms to obtain

$$A \oplus B = (B + A)(\overline{A} + \overline{B}), \qquad (3\text{–}173)$$

which suggests an implementation with NOR gates similar in form to that of Fig. 3–144a.

The manipulation of logic statements obtained from a truth table—such as Eq. (3–171), obtained from Table 3–12—is simplified significantly by the **P** and **S notation.** The P notation assigns a separate P term for every possible AND combination of the input variables. When there are two input variables, A and B, there are four AND combinations, $\overline{A}\overline{B}$, $\overline{A}B$, $A\overline{B}$, and AB. These are assigned the symbols P_0, P_1, P_2, and P_3 respectively. Thus Eq. (3–171) could be rewritten with P notation as

$$A \oplus B = P_1 + P_2. \qquad (3\text{–}174)$$

There are also four possible combinations of A and B with the OR function: $\overline{A} + \overline{B}$, $\overline{A} + B$, $A + \overline{B}$, and $A + B$. These are called S_3, S_2, S_1, and S_0 respectively. (Note the reverse order with respect to the P assignment.) From DeMorgan's theorem $(\overline{A}\overline{B} = \overline{A + B})$, it can be seen that $P_0 = \overline{S}_0$ and, in general,

$$P_i = \overline{S}_i. \qquad (3\text{–}175)$$

The P and S notation is often convenient in dealing with complex functions. A very helpful theorem for quantities that can be written with this notation states that if a function is in the form $P_i + P_j + \cdots$ (sums of products form), it can also be written in the form $S_a \cdot S_b \cdots$ (products of sums form), where the S terms are all those that did not appear in the equation with P's. Taking the exclusive-OR function for an example again, Eq. (3–174) could be written in S notation as

$$A \oplus B = P_1 + P_2 = S_0 S_3 = (A + B)(\overline{A} + \overline{B}), \qquad (3\text{–}176)$$

which is identical to Eq. (3–173). The P and S notation is used in several of the descriptions of gate applications below.

Equality Function. The **equality** circuit, also sometimes called the **comparator circuit** or **coincidence circuit** produces a **1** output whenever the two

Table 3–14 Equality function

Inputs		Output
A	B	$A \oplus B$
0	0	1
0	1	0
1	0	0
1	1	1

input signals have the same logic level. The truth table for the equality function is shown in Table 3–14. From the truth table it is clear that the output function is the exact inverse of the exclusive-OR function. For this reason it is also called the **exclusive-NOR function.** In algebraic form, the function is $AB + \bar{A}\bar{B}$, which, using the concepts given above, can be written in several equivalent ways:

$$AB + \bar{A}\bar{B} = P_0 + P_3 = S_1 S_2 = (A + \bar{B})(\bar{A} + B). \quad (3\text{--}177)$$

The forms of the equality function shown in Eq. (3–177) suggest two implementations of this function with NAND and NOR gates. A comparison of the equality function with the exclusive-OR results in these equivalent expressions:

$$AB + \bar{A}\bar{B} = \overline{\bar{A} \oplus B} = \bar{A} \oplus B = A \oplus \bar{B}. \quad (3\text{--}178)$$

The equality function can thus be implemented by an exclusive-OR circuit with an inverter added at the output or at either input. Another implementation can be made of NOR gates in a circuit analogous to Fig. 3–144b. This is shown in Fig. 3–145.

The equality function can be expanded to additional pairs of bits so that the output is **1** when all pairs of inputs are equal. This is accomplished by combining the outputs of the equality gates for each pair into an AND gate

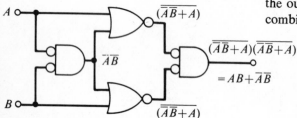

Fig. 3–145 Equality gate with NOR gates.

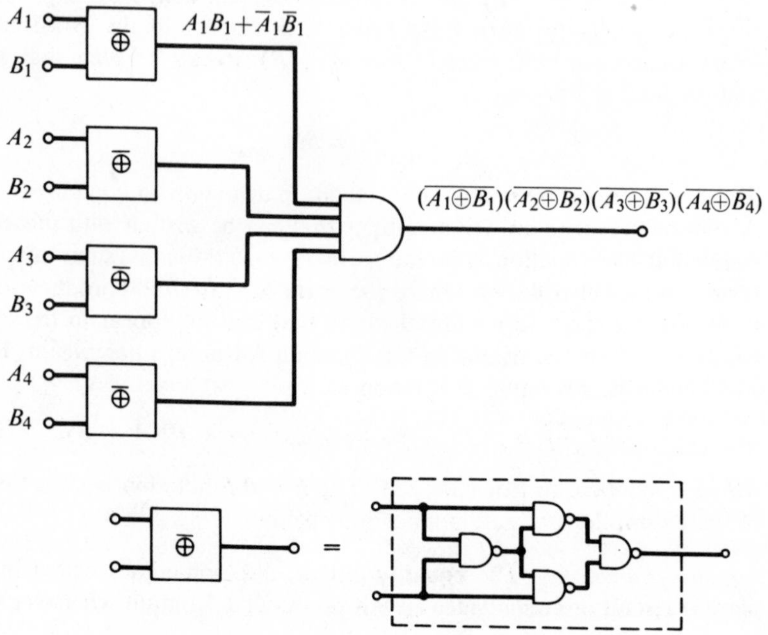

Fig. 3–146 Equality detector.

as shown in Fig. 3–146. Only when the word represented by bits $A_1 A_2 A_3 A_4$ is identical to the word represented by bits $B_1 B_2 B_3 B_4$ will the output be **1**.

Digital Multiplexer. It has been shown in this section that the logic gate acts as a logic-signal-actuated switch for logic level signals. The switching action is analogous to that of a single-pole, single-throw (SPST) switch as shown in Fig. 3–147.[6] If several logic signals are to be gated **ON** and **OFF** simultaneously (analogous to the multiple pole switch), several different gates are connected to the same control signal.

Note 6.   Gate and Switch Terminology.

A difference in terminology for switch and gate circuits should be noted. An **OPEN** *switch prevents the passage of the signal, whereas an* **OPEN** *gate allows the signal to pass. Similarly, the term* **CLOSED** *has opposite meanings for switch and gate circuits.*

Up, OPEN
Down, CLOSED

0, CLOSED
1, OPEN

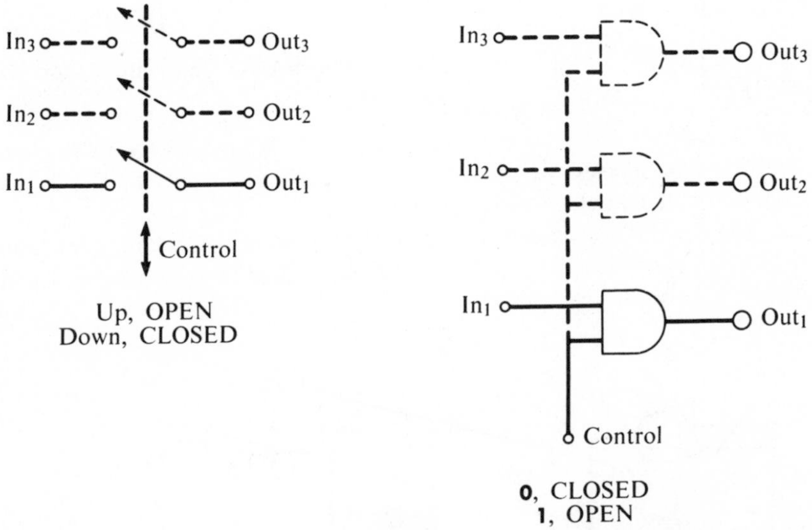

Fig. 3–147   Comparison of switch and gate.

The input of a logic circuit such as a counting register might have to be switched from one signal source to another to accommodate various desired applications. This requires the digital equivalent of a multiple-throw switch, as shown in Fig. 3–148. Such a gate circuit is called a **digital multiplexer** and is analogous to the analog signal multiplexer shown in Fig. 3–97. The outputs of the four signal-switching gates are combined in an OR operation to provide a single output. This is drawn using NAND gates in Fig. 3–148. According to the output function, the state of the control inputs A_c, B_c, C_c, and D_c determines which of the input signals A, B, C, and D will appear at the output. In practice, only one control input will be **1** at a time. For instance, if $A_c = $ **1** and B_c, C_c, and D_c are all **0**, the output function is A regardless of the states of the signal inputs. Thus the input signal that appears at the output depends on which of the control inputs is **1**.

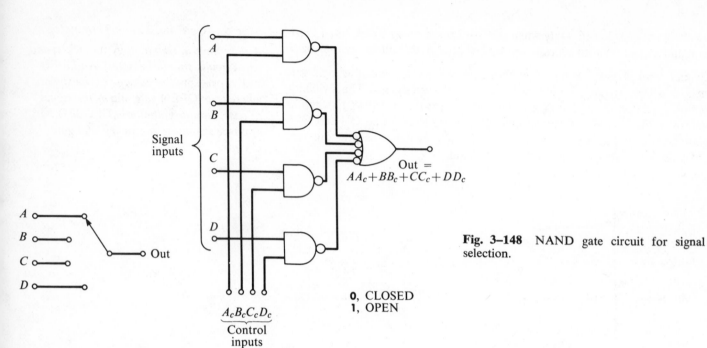

Signal inputs

A

B

C

D

Out = $AA_c + BB_c + CC_c + DD_c$

A o

B o

C o

D o

Out

Fig. 3–148 NAND gate circuit for signal selection.

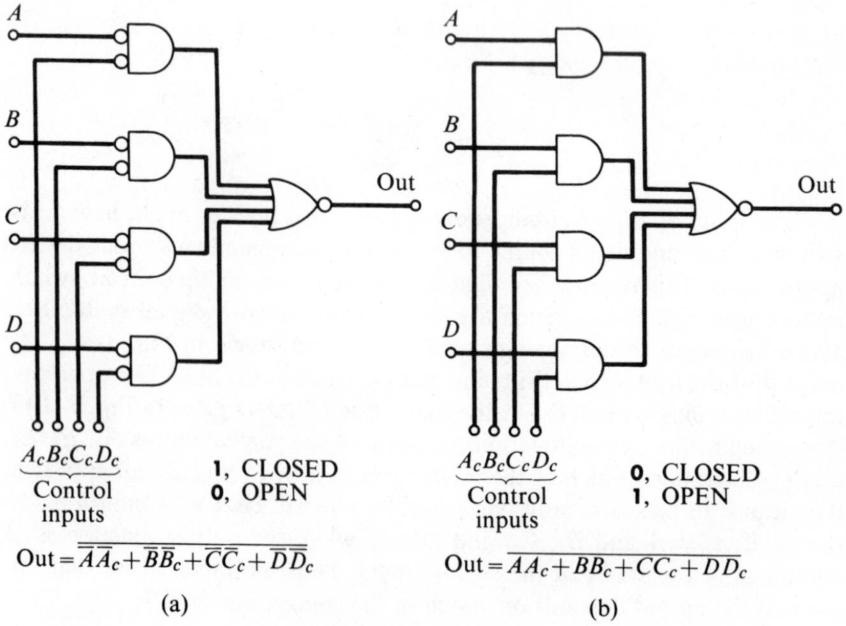

0, CLOSED
1, OPEN

$\underbrace{A_c B_c C_c D_c}_{\text{Control inputs}}$

Fig. 3–149 Signal selection gate circuits: (a) NOR gates; (b) AND-OR-INVERT gate.

A

B

C

D

Out

$\underbrace{A_c B_c C_c D_c}_{\substack{\text{Control} \\ \text{inputs}}}$

1, CLOSED
0, OPEN

Out = $\overline{\overline{A}\,\overline{A}_c + \overline{B}\,\overline{B}_c + \overline{C}\,\overline{C}_c + \overline{D}\,\overline{D}_c}$

(a)

A

B

C

D

Out

$\underbrace{A_c B_c C_c D_c}_{\substack{\text{Control} \\ \text{inputs}}}$

0, CLOSED
1, OPEN

Out = $\overline{AA_c + BB_c + CC_c + DD_c}$

(b)

This circuit can be extended to any number of inputs and can also be implemented with NOR gates or with the AND-OR-INVERT gate package as shown in Fig. 3–149. For the NOR gate circuit, a **0** at a control input is required to OPEN a gate. If A_c, B_c, and D_c are **1** and C_c is **0**, the output will be C. The AND-OR-INVERT gate circuit of Fig. 3–149b is exactly the same as the NAND gate circuit of Fig. 3–148 except for the inverted output logic level.

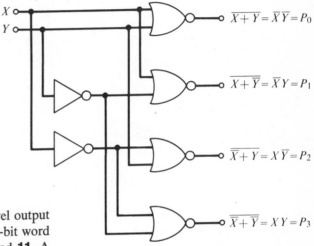

Fig. 3–150 Two-bit binary decoder.

$$\overline{X + Y} = \overline{X}\,\overline{Y} = P_0$$

$$\overline{X + \overline{Y}} = \overline{X}\,Y = P_1$$

$$\overline{\overline{X} + Y} = X\,\overline{Y} = P_2$$

$$\overline{\overline{X} + \overline{Y}} = X\,Y = P_3$$

Binary Decoders. A binary decoder produces a separate logic level output for each combination of input logic levels that is significant. A two-bit word can have four different combinations of logic levels, **00**, **01**, **10**, and **11**. A complete two-bit binary decoder has four outputs, one to indicate the occurrence of each of the four states. Such a circuit is shown in Fig. 3–150. The outputs could be connected to other circuits or indicators which would then respond only to the appropriate combination of logic levels at the inputs. For instance, the outputs could be connected to lights labeled 0, 1, 2, and 3 to indicate the number represented by the two-bit word XY. Or this circuit could be used to operate the control inputs of Figs. 3–148 and 3–149 to reduce the number of control connections and eliminate the possibility of more than one control input being 1 at the same time. If there were three inputs, there would be eight combinations. Thus eight three-input gates would be required for complete decoding, and eight outputs (P_0 through P_7) would be produced by a complete converter. Such a converter could decode a three-bit binary word into signals to turn on one of eight numerical indicators (0 through 7) corresponding to the decimal value of the three-bit binary word. The numerals 0 through 7 are all the numerals used in the octal (base 8) number system. Thus a longer binary word can be decoded into its octal equivalent by converting each group of three-bits of the binary number into a single numeral in the octal number. An octal number is written in about one-third the number of characters used by the equivalent

Table 3–15 Comparison of binary, decimal, octal, and hexadecimal number bases

Decimal	Binary	Octal	Hexadecimal
0	0	0	0
1	1	1	1
2	10	2	2
3	11	3	3
4	100	4	4
5	101	5	5
6	110	6	6
7	111	7	7
8	1000	10	8
9	1001	11	9
10	1010	12	*A*
11	1011	13	*B*
12	1100	14	*C*
13	1101	15	*D*
14	1110	16	*E*
15	1111	17	*F*
16	10000	20	10
17	10001	21	11
18	10010	22	12
19	10011	23	13
20	10100	24	14
32	100000	40	20
50	110010	62	32
60	111100	74	3*C*
64	1000000	100	40
100	1100100	144	64
255	11111111	377	*FF*

binary number. Thus octal numbers are more efficient than binary for written copy and human recognition.

A binary word can also be divided into four-bit sections which are then decoded into their base 16 (hexadecimal) equivalents. The octal, decimal, and hexadecimal equivalents of some binary numbers are given in Table 3–15. As expected, the decimal system is more compact than the octal and less than the hexadecimal. The right three digits of the binary number convert to the units place in the octal system according to the relationship established for 0 to 7. The next three binary digits convert to the second column in the octal number (the eights place) by the same code. Thus $110\ 101\ 011_2$ is 653_8 (the subscript refers to the number base). Using the

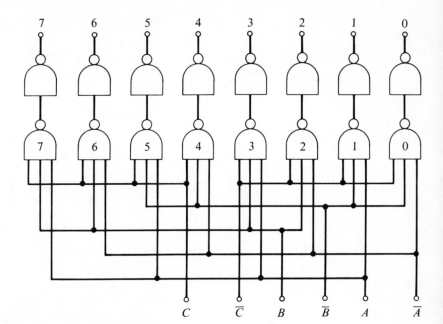

Fig. 3–151   Binary-to-octal decoder.

binary-hexadecimal conversion given in the numbers 0 to $(F)_{16}$, that same binary number ($1\ 1010\ 1011_2$) is equal to $(1AB)_{16}$. Thus decoding a large binary number to octal or hexadecimal is a simple repetition of the decoding of each group of three or four binary digits.

A binary-to-octal decoder using NAND gates is shown in Fig. 3–151. This decoder uses a three-input NAND gate and inverter to obtain the AND function. Each pair of gates decodes one line in the first eight lines of Table 3–15. That is, if $\overline{C}$ and $\overline{B}$ and $\overline{A}$ are **1**, then a **1** level appears at the zero output ($\mathbf{1 \cdot 1 \cdot 1 = 1}$). The 5 output is **1** when $C\overline{B}A = \mathbf{1}$ [$(101)_2 = (5)_8$]. The decoded octal output can be used to activate the appropriate elements of a neon numeral indicator lamp, a printer, or other readout device. Generally a decimal readout is employed with the 8 and 9 numerals unused.

Decoding a binary word into its decimal equivalent is much more complex, since a separate set of bits does not represent each decimal digit. A direct conversion can be performed by a logic gate circuit, but it becomes very complex for numbers larger than two decimal digits (99). When a decimal readout or conversion is required, the signal is generally encoded in the binary-coded decimal code.

BCD Decoding and Decimal Display.   In a binary-coded decimal signal, four bits are encoded to represent each decimal digit. The four-bit BCD group can be decoded by a ten-gate version of the octal decoder circuit

Table 3–16 Binary-to-decimal decoding table, natural code

Decimal	Flip-flop outputs				Logic statement				Minimized statement			
	D	C	B	A								
0	0	0	0	0	$\overline{A}$	$\overline{B}$	$\overline{C}$	$\overline{D}$	$\overline{A}$	$\overline{B}$	$\overline{C}$	$\overline{D}$
1	0	0	0	1	A	$\overline{B}$	$\overline{C}$	$\overline{D}$	A	$\overline{B}$	$\overline{C}$	$\overline{D}$
2	0	0	1	0	$\overline{A}$	B	$\overline{C}$	$\overline{D}$	$\overline{A}$	B	$\overline{C}$	
3	0	0	1	1	A	B	$\overline{C}$	$\overline{D}$	A	B	$\overline{C}$	
4	0	1	0	0	$\overline{A}$	$\overline{B}$	C	$\overline{D}$	$\overline{A}$	$\overline{B}$	C	
5	0	1	0	1	A	$\overline{B}$	C	$\overline{D}$	A	$\overline{B}$	C	
6	0	1	1	0	$\overline{A}$	B	C	$\overline{D}$	$\overline{A}$	B	C	
7	0	1	1	1	A	B	C	$\overline{D}$	A	B	C	
8	1	0	0	0	$\overline{A}$	$\overline{B}$	$\overline{C}$	D	$\overline{A}$			D
9	1	0	0	1	A	$\overline{B}$	$\overline{C}$	D	A			D

shown in Fig. 3–151. The 1 2 4 8 BCD decoding logic is given in Table 3–16. The logic statements are those for the first ten terms in a hexadecimal decoder. However, the elimination of the other six terms (P_{10} through P_{15}) allows some simplification of the logic statement, as shown in the right-hand column. The state of D needs to be specified only in order to distinguish between 0 and 8 and between 1 and 9. If D is **1**, the number must be 8 or 9 and the state of A is enough to distinguish between these. The minimized statement can be used only when the other six states (10 to 15) are rigorously excluded. If an undesired state, such as $\overline{A}\overline{B}CD$, should appear (at power turn-on, for instance), the 4 and 8 outputs would both be ON.

One of the most common kinds of decimal readouts is the NIXIE* tube, which is a neon lamp with one anode and ten separate cathodes. Each cathode is in the form of one of the decimal numerals. The anode is connected to a positive high–voltage (150–200 V) supply through a current-limiting resistor. Whichever cathode is grounded will light. To operate all the elements of the NIXIE tube, a driver (transistor switch) must be connected between each cathode and ground, i.e., ten drivers in all. The driver transistor inputs are each connected to the appropriate decoder output terminal.

Another frequently used decimal readout device is the seven-segment display shown in Fig. 3–152. Seven independent lamps are arranged in the array shown. Combinations of these lamps are lit to form the shapes of the numerals. Decoding for this readout requires an OR gate for each segment in addition to the BCD decoder. For instance, element A is lit for 0, 1, 4, 5, 6,

Fig. 3–152 Seven-segment decimal display.

* Trademark, Burroughs Co.

8, and 9. Therefore, the driver for element A is connected to the output of an OR gate which provides the function $A = 0 + 1 + 4 + 5 + 6 + 8 + 9$, or $A = P_0 + P_1 + P_4 + P_5 + P_6 + P_8 + P_9$ in the P notation. Some elements are used in so many of the numerals (element E appears in eight of them) that it is simpler to list the numerals they don't appear in: for instance, element C is OFF for numerals 1, 5, and 6. Therefore, $\overline{C} = P_1 + P_5 + P_6$ can be used instead of $C = P_0 + P_2 + P_3 + P_4 + P_7 + P_5 + P_9$.

Integrated circuit (IC) decoders and decoder-driver packages are now available for either the ten-element cold-cathode displays or the seven-segment displays. The BCD information is connected to four input terminals, and the indicator lamp and its supply are connected to the output terminals of the IC decoder-driver. This electronic convenience reduces the choice of readout devices to mechanical, esthetic, and economic considerations.

Other BCD Codes.   The 1 2 4 8 BCD code has been used exclusively thus far in the discussion of binary-coded decimal data. Though it has become an industry standard for most instrumentation, other codes have been used and have advantages in certain applications. Table 3–17 shows six useful BCD codes. The 8 4 2 1 (or 1 2 4 8), the 4 2 2 1, and the 5 1 1 1 1 codes are *weighted* codes, which means that a numerical weight can be assigned to each bit and the decimal number is equal to the sum of the weights of the **1** level bits.

Gray codes have the characteristic that only one bit changes state from one numeral to the next. These are used extensively in position encoders because the maximum reading error during the transition from one incremental position to the next is the adjacent numeral. If the 8 4 2 1 were used

Table 3–17   BCD codes

Deci-mal	Natural 8 4 2 1	4 2 2 1	Excess 3	Gray	2 out of 5	Biquinary 5 1 1 1 1
0	0000	0000	0011	0000	00011	00000
1	0001	0001	0100	0001	00101	00001
2	0010	0010	0101	0011	00110	00011
3	0011	0011	0110	0010	01001	00111
4	0100	1000	0111	0110	01010	01111
5	0101	0111	1000	1110	01100	10000
6	0110	1100	1001	1010	10001	10001
7	0111	1101	1010	1011	10010	10011
8	1000	1110	1011	1001	10100	10111
9	1001	1111	1100	1000	11000	11111

for position encoding, all bits to be changed must change at precisely the same position to avoid large error outputs during such transitions as 3 to 4 and 7 to 8. This precision usually imposes unreasonable mechanical tolerances. The Gray codes, however, are not weighted and thus generally require conversion to one of the other system codes.

There are literally thousands of useful codes for binary, octal, decimal, and hexadecimal numbers. The techniques of converting the various binary-coded numbers to decimal form or to some other binary-coded form generally follows the same approach as illustrated earlier in this section.

Binary Addition.   Any operation that can be stated in logic terms can be performed by an appropriate logic circuit. A good example is that of the binary addition of digits A and B. The binary addition table can be presented logically as follows:

$$0 + 0 = 0, \quad 0 + 1 = 1, \quad 1 + 0 = 1, \quad \text{and} \quad 1 + 1 = 0 \text{ and carry } 1.$$

The addition table is summarized in Table 3–18, where S is the sum and C is the carry. It can be seen from the table that

$$S = A \oplus B \quad \text{and} \quad C = AB. \tag{3–179}$$

The addition function can be thus implemented with an exclusive-OR gate and an AND gate. An implementation with NAND gates is shown in Fig. 3–153. The exclusive-OR circuit of Fig. 3–144b is used to provide S, with an inverter on the NAND function to provide C. This circuit is called a **half-adder** because it will not accept a carry signal from the addition of the less significant bits in the two binary numbers being added.

When two binary numbers are added, any carry resulting from adding the first column must be added to the two digits in the second column. This is illustrated by adding 1011 to 0101. Starting with the 2^0 column, $1 + 1 = 0$, carry 1. In the 2^1 column, $0 + 1 + \text{carry } 1 = 0$, carry 1. In the 2^2 column,

Table 3–18   Binary addition

Inputs		Outputs	
A	B	S	C
0	0	0	0
0	1	1	0
1	0	1	0
1	1	0	1

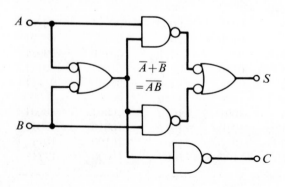

Fig. 3–153 Half-adder.

$1 + 1 +$ carry $1 = 1$, carry 1 and in the 2^3 column, $1 + 0 +$ carry $1 = 0$ carry 1.

	2^3	2^2	2^1	2^0	
	1	1	0	1	A
$+$	0	1	1	1	B
1	0	1	0	0	

The final carry is placed in the 2^4 column. This is the binary equivalent of the decimal addition $13 + 7 = 20$. A circuit which will perform the addition for columns 2^1, 2^2, etc., must add digits A, B, and C_i and produce outputs S and C_o, where C_i and C_o are the carry in and carry out, respectively.

Following the laws of binary addition, a truth table or table of combinations can be made up for all combinations of input states and for both S and C_o outputs. This table, Table 3–19, is made up by considering what S and C_o should be for each input combination. From this truth table we see that S is **1** whenever there are an odd number of **1** level inputs. Thus

$$S = A \oplus B \oplus C_i. \qquad (3\text{–}180)$$

The expression for C_o can be obtained by combining the terms which produce a **1** output:

$$C_o = \bar{C_i}AB + C_i\bar{A}B + C_iA\bar{B} + C_iAB$$

$$= AB + C_i(\bar{A}B + A\bar{B})$$

$$= AB + C_i(A \oplus B). \qquad (3\text{–}181)$$

The quantities S and C_o are usually generated by two half-adders as shown in Fig. 3–154. If a NAND gate is used in place of the OR gate shown, the inverters in the C output of the half-adders can be omitted. To simplify the construction of binary addition circuits, integrated circuit manufacturers have made exclusive-OR gates, half-adder, full-adder and four-bit full-adder circuits available in single packages.

Table 3–19 Truth table for full-adder

Inputs			Outputs	
C_i	A	B	S	C_o
0	0	0	0	0
0	0	1	1	0
0	1	0	1	0
0	1	1	0	1
1	0	0	1	0
1	0	1	0	1
1	1	0	0	1
1	1	1	1	1

Fig. 3–154 Full-adders using half-adder circuits.

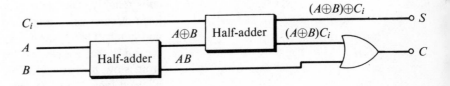

Experiment 3–39   *Exclusive-OR and Equality Functions*

The exclusive-OR logic function is sometimes called the third basic logic function. By inverting the inputs or output or both of AND, NAND, OR, or NOR gates, any of the other three functions can be performed. However, the exclusive-OR and equality functions are quite different from the rest and related to each other. The exclusive-OR and equality functions can be made of NAND gates or a combination of NAND and AOI gates as shown in this experiment. These logic functions are extremely useful, as will be seen in the multiplexer, and in other experiments in this module.

Experiment 3–40   *Binary Adders*

A binary half-adder performs all the logic required to add two binary digits. It has two inputs, one for each bit, and two outputs, one for the sum and one for the carry. It is called a half-adder because two half-adder circuits are required to add two bits *and* the carry from the two less significant bits.

   In this experiment, the half-adder is wired using NAND gates and/or a combination of NAND and AND-NOR gates. The sum function is shown to be the exclusive-OR function and the complete half-adder circuit is thus a modified exclusive-OR gate. A binary full-adder can also be wired.

Experiment 3–41   *Digital Multiplexers*

The digital multiplexer is a signal selector gate for digital signals. A number of signal sources are gated in such a way that any one of the signals can appear at the output. In this experiment digital multiplexer gates are made with NAND gates and AND-NOR gates. Control input decoding is used to reduce the number of control lines and ensure that only one input gate is open at a time.

3–4.2   DIGITAL SIGNAL CONDITIONING

Two basic conditions must be met by signals that are to operate logic circuits reliably. The logic levels must be well within the range for the specific logic circuits used, and the transitions between logic levels must be clean and quick. The reason for the first condition is obvious. The reason for the second is that for several logic families, a condition of instability can arise from slow movement of an input signal through the region between the allowed logic levels. In this section, we describe the Schmitt trigger circuit and its application for sharpening transitions, discriminating, and/or converting from one set of logic levels to another. The conversion between contact logic and voltage logic levels is also discussed. The monostable multivibrator is introduced as a pulse generator and pulse-shape restoring circuit, and several circuits and applications relevant to pulse signal gating are described. Finally, some of the principal requirements for the transmission of digital data without serious degradation of the logic levels or transition times are considered.

Level Conditioning

Ideally, the voltage level of a digital signal unambiguously identifies the signal as logic HI or LO. This is achieved by a substantial separation of the HI and LO logic level voltage ranges. In practice, many digital signal sources do not provide easily distinguishable signal logic levels. A particle or radiation detector with an electron multiplier produces a pulse for each particle or photon detected, i.e., a digital signal of the count digital form. However, the pulses produced are not of uniform amplitude and some pulses are produced by noise or spurious phenomena in the detector.

The signals from the magnetic and optical transducers that convert objects or phenomena to be counted or sorted into electrical pulses vary in amplitude as a result of variations in the physical parameters of the sensing process, variations in ambient light and magnetic fields, and electrical noise. Digital signals that have undergone a data transmission process through an imperfectly controlled medium (radio waves, telephone lines, etc.) are often of variable level and subject to spurious signals introduced in transmission. In each case, a circuit is needed which can discriminate against the pulses or signal variations that are not of interest and produce clean logic level transitions for those that are.

A natural choice for such a task is the comparator introduced in Section 3–3.1. The comparator threshold level is adjusted to the minimum level that is likely to represent a desired phenomenon, or to a level just above the maximum level signal likely to result from spurious effects. If these two levels overlap, perfect discrimination is impossible. Another problem can arise when the signal level traverses the threshold region at a rate much slower than the response time of the comparator. As the comparator input and output waveforms of Fig. 3–155 show, a small amount of noise on the signal can result in extra output logic level transitions. In many cases this is completely unacceptable. The system response time may be slowed to match that of the signal or, as we shall see later, the first transition can cause successive transitions occurring in a given time to be ignored. Neither of these techniques can be used in cases where the signal variation rate is unknown or unpredictable. Another method of avoiding multiple transitions is the Schmitt trigger circuit discussed below.

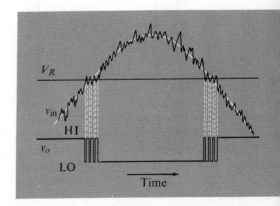

Fig. 3–155 Multiple transitions produced by a noisy signal undergoing a single threshold crossing.

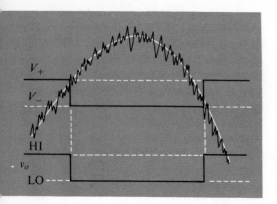

Fig. 3–156   Schmitt trigger waveforms.

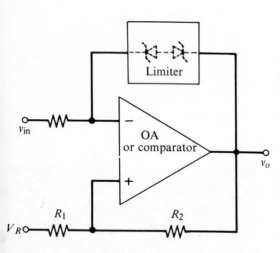

Fig. 3–157   Schmitt trigger circuit.

Schmitt Trigger.   The Schmitt trigger circuit is a modification of the comparator circuit in which there are two threshold voltages, V_+ and V_-. When a positive-going signal first crosses V_+, the output level changes state and the threshold level changes to the less positive value, V_-. The threshold is now, presumably, out of the range of the noise and no multiple transitions will occur. This situation is illustrated in Fig. 3–156. As the signal, now negative-going, again approaches the threshold region, the first crossing of V_- changes the output level and returns the threshold voltage to the V_+ value. So long as the peak noise voltage is less than the difference between V_+ and V_-, the signal can traverse the threshold region at any rate without multiple transitions. In most applications, the difference between positive-going and negative-going threshold levels creates no measurement error.

The simple modification of the comparator circuit that accomplishes the shift in threshold level is suggested by the correspondence between the v_o and threshold voltage waveshapes of Fig. 3–156. A portion of the v_o signal is summed with the threshold setting voltage V_R as shown in Fig. 3–157. The actual threshold voltages, V_+ and V_- are given by

$$V_{+,-} = V_R + (v_o - V_R)\frac{R_1}{R_1 + R_2}$$

$$= V_R\frac{R_2}{R_1 + R_2} + v_o\frac{R_1}{R_1 + R_2} \qquad (3\text{--}182)$$

where both values of v_o are used. The difference between the HI and LO levels of v_o is generally very much greater than the difference between V_+ and V_-. Therefore, R_2 in Fig. 3–157 is often much larger than R_1, so that Eq. (3–182) can be written

$$V_+ \approx V_R + \frac{R_1}{R_2}\,v_o \quad \text{for } R_2 \gg R_1. \qquad (3\text{--}183)$$

The limiter circuit provides well-defined HI and LO values of v_o and prevents amplifier limiting by providing negative feedback paths at the desired HI and LO values. Several limiting circuits suitable for this application are described in Section 2–3.6 of Module 2. Given well-defined HI and LO values for v_o, these can be used in Eq. (3–182) or Eq. (3–183) to obtain V_+ and V_- respectively.

If a logic level signal has been degraded by noise, logic level voltages, and transition sharpness, it should be "cleaned up" before it is applied to a gate input. The hysteresis (difference in response due to direction of input change) of the Schmitt trigger is an advantage in this application. If V_+ is set at the minimum voltage of the HI range and V_- is set at the maximum voltage of the LO range, the response shown in Fig. 3–158 results. Note that the noise rejection is much better than one would obtain with a comparator set at any level. A Schmitt trigger level-restoring circuit for TTL logic levels is available as an integrated circuit.

The Schmitt trigger circuit is also an ideal circuit for converting a digital signal from one set of logic levels to another. This function is needed if one uses circuits from different logic families in various parts of a system. In pulse-counting applications, for example, ECL logic may be used for high speed counting, TTL logic for medium speed counting, and MOS logic for memory or computation, in order to take advantage of the special characteristics of each type. In such an application, the V_+ and V_- levels of the Schmitt trigger are set as in Fig. 3–158 for the input logic levels, while the limiter is designed to provide the appropriate output logic levels. Interlevel converters for use between some of the most popular logic types are available in integrated circuit packages.

Contact Logic Interface. A frequently encountered digital data source is the switch contact. The state of the switch contact (**OPEN** or **CLOSED**) is a form of digital data that must be converted to a logic level signal in order to be used with gate logic circuits. The switch can be used to switch logic level voltage sources as in Fig. 3–159. The circuits of Fig. 3–159b and c take advantage of the fact that most types of logic gate inputs require much less current when the input is in one state than when the input is in the other state which allows the switch contact to the higher current source to override the resistive contact to the other.

A major difficulty with contact logic interfaces is contact bounce, which occurs when the switch state is changed. In each of the circuits of Fig. 3–159, contact bounce will cause multiple transitions. When the switch has a positive contact at each state, a logic gate circuit can eliminate the multiple transitions caused by contact bounce. Such a contact bounce eliminator circuit is shown in Fig. 3–160. When the moveable switch contact first touches the right contact, a **LO** signal is applied to NAND gate 1, forcing its output **HI**. Both inputs to gate 2 are now **HI**, and the result is a **LO** output which is applied to gate 1. The **LO** from gate 2 holds the gate 2 output **HI** even though the switch bounces off the contact and produces several **HI-LO** transitions. This state (gate 1 output **HI**, gate 2 output **LO**) will remain until the movable contact is changed and makes its first contact with the left terminal. Now the gate 2 output will be forced **HI** and the gate 1 output will go **LO** to hold it in that state. The gate 2 output is always opposite to that of gate 1. This circuit can also be implemented with NOR gates. Note that the quality of the output signal does not depend on maintenance of good switch contacts.

The transition from logic level signals to contact closures is also important in many applications. This is accomplished either by activating a relay

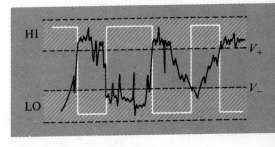

Fig. 3–158 Schmitt trigger used to restore logic levels.

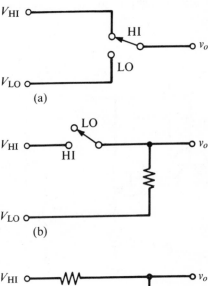

(a)

(b)

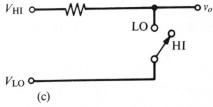

(c)

Fig. 3–159 Switch contact–to–logic level converters: (a) general; (b) for current-sourcing logic; (c) for current-sinking logic.

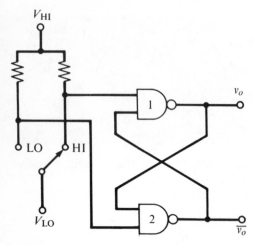

Fig. 3–160   Logic gate contact bounce eliminator.

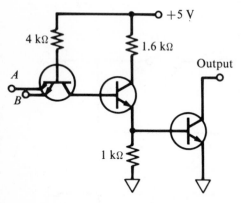

Fig. 3–161   Open collector NAND gate.

drive circuit with the logic level signal or by simulating a contact closure with a solid state switch. The reader is referred to Module 2, Section 2–3, for a discussion of relays and solid state switches. To be considered in this regard are logic gates that have an output transistor available for use as a gate-actuated switch. An example is the TTL open collector gate circuit shown in Fig. 3–161. A comparison of this circuit with that of Fig. 3–135 reveals that the portion of the normal gate circuit that provides the HI level output when the output transistor is OFF is missing. For this reason, such gates are called **open collector gates.** The output transistor is now a switch for which the logic function is $A \cdot B = $ CLOSED.

Pulse Generation and Shaping

Digital signals can have the data encoded as the logic level (HI, LO), as the direction of the logic level transitions (HI → LO, LO → HI), or as logic level pulses (⊓). We have seen in the above discussion how the Schmitt trigger circuit can be used to restore or convert the logic levels and transitions of digital signals. In this section, we shall see how logic level pulses are generated and some of the applications of pulse generating circuits.

The principal kind of pulse generator is a monostable version of a family of two-state circuits called, for historical reasons, **multivibrators.** As the name implies, the monostable multivibrator is stable in just one state. It can be triggered into its other "semistable" state, where it will remain for a time and then spontaneously return to its initial stable state. The time spent in the semistable state is not determined by the triggering signal but by the choice of component values in the monostable circuit itself. This circuit can be thought of as a triggered, adjustable-width, pulse generator. It is used as a pulse shaper to provide uniform-width pulses from a variable-width input pulse train. Its other principal application is as a delay element, since it provides a logic output transition at a fixed time after the trigger signal.

Logic Gate Monostables.   Simple monostable pulse generators can be made from integrated circuit logic gates. They are based on the use of a low pass RC circuit to delay the occurrence of a logic level transition at one gate relative to another. The simplest such circuit is shown in Fig. 3–162. Before the trigger pulse, the T input is LO and the output of gate 2 is HI. The capacitor C is charged through R to this level. When T goes HI, the gate 2 output goes LO and the capacitor discharges toward V_{LO} through R and the gate 2 output impedance. The voltage at $\overline{T}_d$ decreases and crosses the HI-LO logic threshold for the gate 1 input. At this time the LO at $\overline{T}_d$ brings the gate 1 output HI. The pulse duration t_{pw} thus depends on R, C, and the logic threshold level and is generally of the same magnitude as the RC time constant.

For most gate types, R in Fig. 3–162 is limited to small values. With current-sinking gates such as the TTL and DTL, the LO level input current must pass through R while at the same time the IR drop must not bring the gate 1 input voltage too near the HI-LO threshold. Approximately 220 Ω is considered a safe upper-limit input resistance for the standard TTL gate. This low R limits the practical pulse duration from such a circuit, for even if a 1 µF capacitor is used the pulse will be only about 250 µsec long. In any case, a Schmitt trigger gate should be used for gate 1 with time constants more than about 10 µsec because of the slow threshold transition occurring at its input.

The circuit of Fig. 3–162 requires that the T pulse be longer than the output pulse. Because of the direct connection between T and gate 1, whenever T is LO, the output is HI. Thus if the T signal returns LO before the T_d signal, the output pulse will be terminated prematurely. This problem is solved by gating the T signal and using the output pulse to control the gate as shown in Fig. 3–163. Because of the inverting gate 3, the circuit now triggers on the HI-LO transition of T. When T goes LO, the monostable is triggered and the LO at the output is applied to gate 3 to keep the gate 3 output HI as long as the monostable output is LO. The T pulse need only be long enough to change the states of gates 3 and 1. The **duty cycle** (the ratio of t_{pw} to the minimum pulse repetition period) for these circuits is relatively low, since the same RC time constant determines the charging and discharging times. Since the duration of the pulse is only about one discharge time constant, but the charging must go on for about three time constants to be complete, the maximum duty cycle is about $1/(1 + 3) = 25\%$.

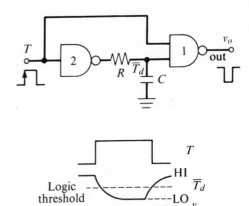

Fig. 3–162 *RC* delay monostable.

Fig. 3–163 Gated *RC* delay monostable.

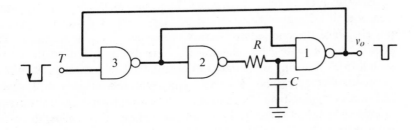

A circuit almost identical in its action to the circuit of Fig. 3–163, but which uses only two gates, is shown in Fig. 3–164. In this circuit, R is small enough to keep the input to gate 1 LO (for TTL, 220 Ω or less). The output and T input are normally HI, making the gate 2 output LO. When T becomes LO, the gate 2 output becomes HI. Since the voltage across C cannot change instantly, the gate 1 input also goes HI and the output goes

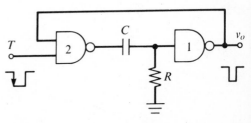

Fig. 3–164 Gated stored-charge monostable.

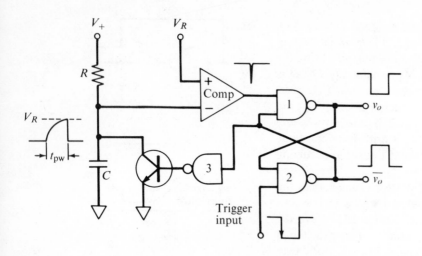

Fig. 3–165   Comparator monostable circuit.

LO. The positive charge on C at the gate 1 input discharges through R. When the gate 1 voltage falls below the logic threshold, gate 1 reverses state again and the pulse is over. The connection from the output to the input of gate 2 holds the gate 2 output HI for the duration of the output pulse.

Comparator Monostable and Timing Circuits.   The combination of the comparator, switching, and logic circuits provides a very versatile pulse generating and timing technique. The basis of this approach is llustrated in Fig. 3–165. The cross-connected NAND gates used in the switch contact bounce eliminator are used in this circuit to convert the pulse start and stop signals into a clean logic level pulse. The pulse is initiated by the application of a HI-LO transition at the trigger input, which drives the gate 2 output HI. Since the comparator output is normally HI, the output of gate 1 goes LO to hold gate 2 HI. The now LO output of gate 3 turns the transistor switch OFF, allowing capacitor C to charge from V_+ through R. When the capacitor voltage reaches V_R, the comparator output goes to **0**, reversing the states of gates 1 and 2, closing the transistor switch, and returning the circuit to its initial state. The comparison voltage V_R is generally obtained from the supply voltage V_+ with a 2/3 voltage divider. The pulse width is then about 1.1 RC, where RC can be any time from a few microseconds to many seconds.

The comparator type of monostable circuit is the basis for the wide variety of integrated circuit monostable and timing circuits avaliable today. Some have more versatile gate circuits which allow resetting and/or re-triggering during the pulse, triggering on positive or negative trigger signal slopes, and other features.

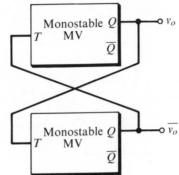

Fig. 3–166 Dual monostable square wave generator.

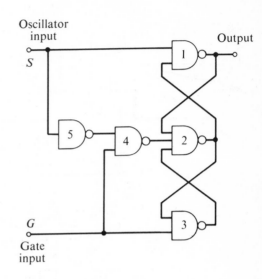

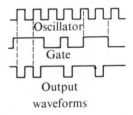

Fig. 3–167 Gating an oscillator.

Square Wave Generator. It is sometimes necessary to make a square wave generator when the logic system being used does not provide separate oscillator packages. A square wave generator can be made with two monostable multivibrators connected to trigger each other alternately, as in Fig. 3–166. The trailing edge of the signal at $\overline{v_o}$ triggers the upper monostable to generate a pulse. At the end of that pulse, the v_o signal triggers the lower monostable, and so on. Duty cycle restrictions on the monostable must be observed. Some integrated circuit monostables have connections available that allow a "free-running" or self-triggering mode suitable for generator applications.

Gating an Oscillator. The train of pulses generated by an oscillator must often be gated in such a way that no partial pulses are produced. An ordinary gate cannot be used, because if the gate ON or OFF signal occurs during the pulse, a shortened pulse will occur. A circuit to perform this function is shown in Fig. 3–167. If S is LO, the output is HI, gate 4 is OPEN, and cross-coupled gates 2 and 3 have G and $\overline{G}$ respectively at their outputs. If G is HI the output of gate 2 is HI, and gate 1 is OPEN and provides the inverted signal S at the output. However, if G goes LO before S goes LO and gate 3 changes state, gate 2 cannot change because the LO output holds gate 2 in the HI state until S returns to the LO state. Thus the gate cannot cut an S pulse short. If G now stays LO, all inputs to gate 2 are HI, and the LO output of gate 2 holds the output HI. The gate is CLOSED.

If the G input goes HI while S is LO, gate 2 goes HI opening gate 1 for the next S pulse. If G goes HI while S is HI, gate 4 is CLOSED and the coupled gates 2 and 3 cannot respond to the change in gate level until S is LO again. Thus a partial S pulse at the beginning of the gating signal is also blocked.

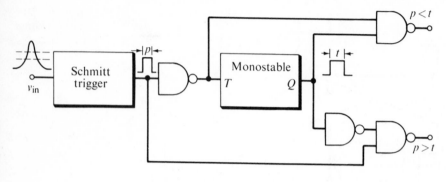

Fig. 3–168   Pulse width discriminator.

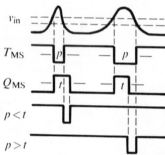

Pulse Width Discriminator.   Sometimes the fact that a given pulse represents the occurrence of an event of interest is evidenced by its pulse width rather than (or in addition to) its amplitude. In such cases, it is useful to be able to sort input pulses by their duration. Figure 3–168 shows a circuit that provides a pulse at one output or another, depending on whether the input pulse is wider or narrower than a reference pulse from a monostable circuit. The input pulse is shaped by a Schmitt trigger and inverted, so that the leading edge can trigger the monostable. The input and monostable pulses are thus inverted with respect to each other in each case and compared by NAND gates. If the LO-going pulse is the wider, it will hold the gate output HI through the comparison; if not, a pulse results.

Pulse Coincidence and Anti-Coincidence Gating.   Another criterion that can often be used to help sort pulses representing phenomena of interest from those that are not is whether their occurrence is coincident (or not coincident) with a pulse from a related source. An elegant illustration of coincidence gating is the laser beam reflection measurement of the distance to the moon (Fig. 3–169). A laser that emits a very short pulse of high intensity light is aimed at a spot on the moon. A telescope observing the same area collects an almost infinitesimal fraction of the reflected light along with all the other light observed from that area. The intention is to start a

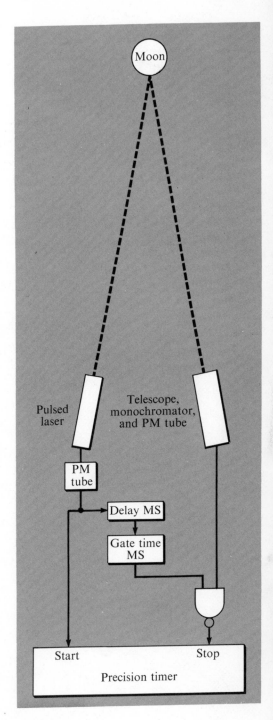

Fig. 3–169 Coincidence gating in measurement of distance to moon.

precision timer with light from the laser and stop it with the pulse of reflected light. From the time delay and a knowledge of the speed of light, the distance to the moon can be precisely calculated. The monochromator is used to select only those photons of the laser-emitted wavelength; the detector PM tube then emits a pulse for each photon detected. Even so, the reflected light is so weak that the pulses that are not due to the laser reflection outnumber those that are, by many orders of magnitude. Advantage is taken of the fact that the experimenter knows approximately when to expect the desired pulses. By means of monostable multivibrators, a pulse is generated which is coincident with the expected reflection delay time and wide enough to encompass the time uncertainty. The generated pulse and the detected pulses are connected to a coincidence gate (AND gate) so that only those pulses occurring during the expected time will stop the timer. The experiment was successful even though a reflected photon was detected only every hundredth flash, on the average. Clearly, the measurement was repeated thousands of times and the data were analyzed by a computer for coincidence of measured times. For the round trip time of about 5 sec, a timer resolution of 1 nsec will result in a distance measurement resolution of about 16 cm!

An example of anti-coincidence pulse gating occurs in the **pulse height amplitude window discriminator,** in which only pulses with amplitudes in a predetermined range are to be detected. To perform this function, two Schmitt trigger circuits are used with their respective V_+ values set at the lower and upper boundaries of the amplitude range or "window." Both Schmitt trigger circuits are connected to the pulse source as shown in Fig.

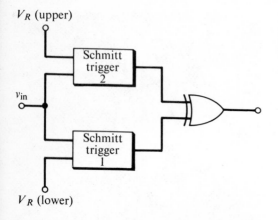

V_R (upper)

v_{in}

V_R (lower)

Fig. 3–170 Pulse height window discriminator.

3–170. Pulses below the window in amplitude will trigger neither Schmitt, and pulses too large will trigger both. Therefore, the desired function is a **1** from Schmitt trigger 1 coincident with a **0** from Schmitt 2. This function is satisfied by the exclusive-OR gate shown.

Experiment 3–42 *Monostable Multivibrators*

The monostable multivibrator has a variety of uses in digital circuits. It is used as a pulse shaper to provide uniform-width pulses and as a delay element. In this experiment monostable multivibrators are first made from logic gates by using RC circuits to delay the transmission of the logic level from one gate to another. Two logic gate monostable circuits are studied: an RC delay monostable and a stored-charge monostable. Then experience is gained with two integrated circuit monostable multivibrators.

Digital Data Transmission

Even though digital signals are much less susceptible to noise and distortion than analog signals, careless transmission techniques can lead to errors. Modern digital circuits produce signals with transition and response times of a few nanoseconds. Interconnection wires more than a few inches long can produce reflections of transient signals within the response times of the circuits. To avoid this problem, appropriate transmission line precautions should be taken. For short distances, the driving and receiving circuits are generally single-ended and the line is a wire, printed circuit board foil, coaxial cable, or ribbon cable with possible grounds at either end as shown in Fig. 3–171a. Driver type permitting, the cable is terminated at its receiving end in its characteristic impedance (for hookup wire or printed circuit board foil, about 200 Ω) to avoid reflections. The terminating resistor on the driver end can not always be used, though it is safest to do so. In the special case of open collector driver gates, the terminating resistor on the receiver end is connected to the gate supply voltage and serves the dual purpose of collector load and cable terminating resistor. Refer to Appendix C for some manufacturer's suggestions regarding data transmission in TTL logic systems.

For reliable data transmission over distances of many meters or more, special line drivers and receivers are available in most logic systems. The single-ended types are applied as shown in Fig. 3–171a. Some provide balanced output drivers and differential input receivers. These are for use with balanced lines such as the shielded twisted pair shown in Fig. 3–171b. The balanced line transmission is less susceptible to induced

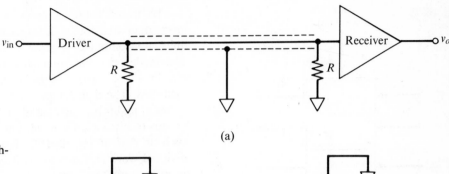

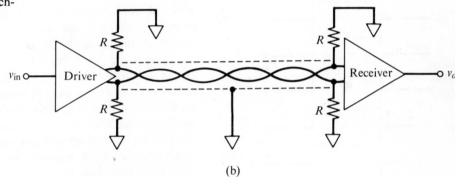

(a)

(b)

Fig. 3–171 Digital data transmission techniques: (a) single-ended; (b) balanced.

noise because both lines are driven identically, exposed to the same influences, and detected differentially. Note that both lines in a balanced line must be terminated.

3–4.3 DIGITAL DATA STORAGE REGISTERS

A digital data storage register is a circuit or device that stores one or more bits of digital data for subsequent use. Each memory unit stores a single bit of information and may be represented as shown in Fig. 3–172. The data to be stored (a **1** or **0** logic level) are connected to the data in terminal. At the appropriate time, a logic level transition, for instance, **1** → **0**, is applied to the load command and the data level at that instant is loaded into the memory. In one simple latch form of memory, the input information appears immediately at the output, which follows the input data as long as the load command is activated. This form of memory can serve as a digital "track-and-hold." In another form of memory unit, the acquired input data do not appear at the output until the load command signal returns to its nonload state. This type is used in situations where the data outputs must remain stable during loading. Activating the clear command input with a logic level transition will **clear** or **reset** the data stored in the memory to **0**.

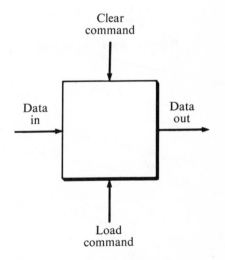

Fig. 3–172 Single memory unit.

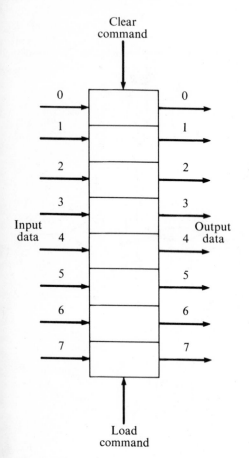

Clear command

Input data

Output data

Load command

Fig. 3–173 Parallel input/output register.

To be of value in most applications, a digital memory must store more than one bit of information. One of the ways in which memory units are combined to store more than one bit is shown in Fig. 3–173. The load and clear inputs for all bits are connected together so that all bits load and clear together. The input and output bits are individually identified. The digital input and output data are in the parallel form. Most systems requiring substantial data storage operate on the data in small groups of bits called a **word.** One word of data may be stored in a register such as that of Fig. 3–173. Where many words must be stored, the individual word registers are collected in a **stack.** The circuits used to make memory units and the ways they are interconnected into registers and stacks are discussed in this section.

There are two other forms of digital data, count and coded serial. There are ways of interconnecting memory units to store data in these forms as well. In the case of count data, it is impractical to store the data in the count form since this would require as many memory units as the maximum number of counts expected. For this reason, count information is stored in the more efficient coded form. The memory register should store a coded word representing the number of counts that occurred to that time and then increment (increase by one) the coded number as each new count bit is detected. A register that has an increment command input is called a **counter** and may be represented as shown in Fig. 3–174a. Activating the clear command input sets the register contents to the code that represents the number zero (generally all **0**'s in the memory units). The count signal is converted into an appropriate pulse or transition to activate the increment command and advance the coded number by one unit each time. The most commonly used codes are binary and binary-coded decimal. The encoded count data are stored in the memory units and available at their outputs in parallel form. Therefore, the counting register is also a count-to-coded-parallel converter in the digital domain. Some counting registers can perform additional functions such as counting up (incrementing) or down (decrementing). An **up/down counter** provides a subtraction function for count digital signals. Some counting registers provide for the parallel loading of the memory units. The counter represented in Fig. 3–174b is an up/down counter with parallel inputs. The memory units that are used for counting and a variety of counting circuits are described in this section.

The storage of coded serial data is performed by a combination of memory units called a **shift register.** The memory units are connected so that the contents of each one are transferred to the next one in line upon a shift command input. The shift register is shown in block form in Fig. 3–175a. When the shift command input is activated, the logic level at the serial input is shifted into memory unit 0, the level that was in unit 0

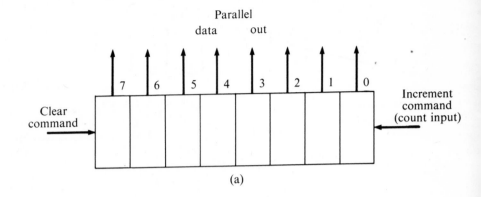

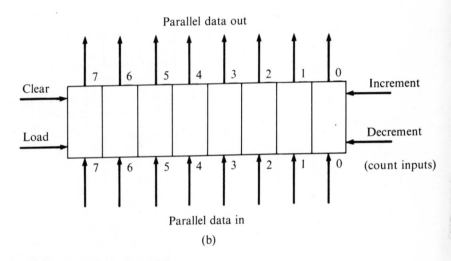

Fig. 3–174 Counting registers: (a) simple; (b) up/down counter with parallel input.

is shifted to unit 1, and so on. A shift command is then issued during each data space thereafter. After eight shifts, the register is full of new data, which are available in parallel form from the memory unit outputs. Thus the shift register is useful as a serial-to-parallel converter. If another shift command is applied, the logic level that entered unit 0 eight shifts before is shifted out of unit 7. The shift register is therefore also useful as a serial data delay circuit, in which the delay in data spaces is equal to the number of memory bits in the shift register. More complex shift register circuits can provide shifts to the left or the right and/or parallel data inputs. A shift right/left, parallel input/output shift register is represented by Fig. 3–175b. If a data word is loaded into the shift register through the parallel inputs and then a succession of shift commands is applied, the data word will appear in serial form at the serial output. Thus a parallel-to-serial conversion is performed. Several shift register circuits and applications are described in this section.

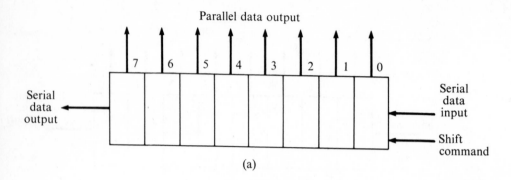

(a)

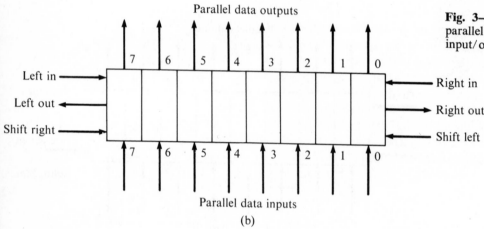

(b)

Fig. 3–175 Shift registers: (a) simple, with parallel output; (b) shift right/left, parallel input/output.

Latch Registers

The basic bistable memory unit is the simple cross-coupled gate circuit shown in Fig. 3–176. Recall that with the NAND gate, the output will be **1** if any input is **0**, and that the output can be **0** only when all inputs are at logic **1** or unused. Assume that the clear and set inputs are **1**. For either gate, then, a **1** at the cross-coupled input would produce a **0** at the output or vice versa. Since the output of one gate is the input of the other, the two outputs must be in opposite states, that is, $Q = 1$ while $\bar{Q} = 0$ or $Q = 0$ while $\bar{Q} = 1$. Follow the logic levels through on the diagram to show that either of the above is a stable state, and that Q and $\bar{Q}$ cannot both be **0** or **1** when clear and set are **1**.

If a **0** is applied to the clear input and a **1** to the set input, the output $\bar{Q}$ must become a **1**, and Q a **0**. If the set input is **0** and the clear input is **1**, then $Q = 1$ and $\bar{Q} = 0$. By applying a momentary **0** level to the clear

Truth table for Fig. 3–176

Inputs		Outputs	
Clear	Set	Q	$\overline{Q}$
0	1	0	1
1	0	1	0
0	0	1	1
1 (last **0**)	1	0	1
1	1 (last **0**)	1	0

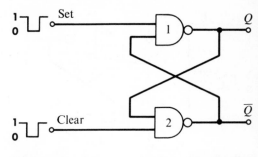

Fig. 3–176   Memory circuit using NAND gates.

or set input, the output will become **1** or **0** as desired. Since the output will remain in this state until a **0** is applied to the alternate input, the circuit "remembers" which input had the latest momentary **0** applied to it. In a typical application, the memory will be "cleared" by applying a momentary **0** to the clear input, producing a **0** at the Q output. Then some sequence of events will take place which might produce a **0** pulse at the set input to make $Q = $ **1**. Later the Q output will be "read" to see whether or not the events produced a set pulse. If **0** levels are applied to both set and clear inputs, both Q and $\overline{Q}$ outputs must be **1**. In a completely bistable circuit this condition would be avoided.

Truth table for Fig. 3–177

Inputs		Outputs	
Clear	Set	Q	$\overline{Q}$
0	1	1	0
1	0	0	1
1	1	0	0
0	0 (last **1**)	1	0
0 (last **1**)	0	0	1

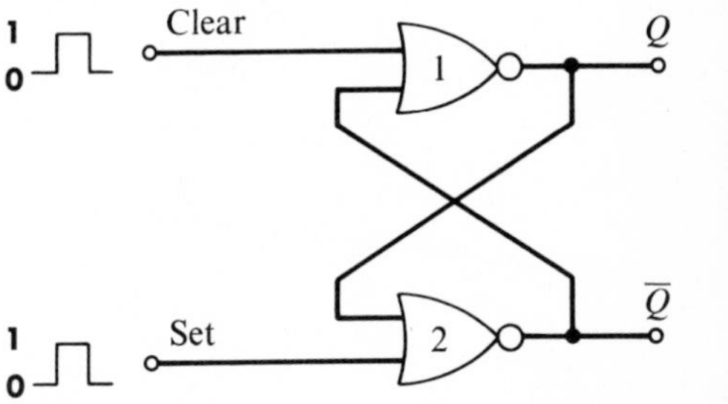

Fig. 3–177   Basic memory circuit using NOR gates.

Cross-coupled NOR gates can also be used to form the basic memory, as shown in Fig. 3–177. The NOR gate has a **0** output level if any input is at logic **1**. For this reason the truth table is very similar to the NAND gate memory except that the set and clear inputs are reversed and the Q and $\overline{Q}$ output levels are opposite. Note that the NOR gate inputs are normally **0** between set and clear pulses.

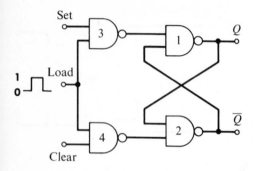

Fig. 3–178   Gated memory circuit.

Truth table for Fig. 3–178

| Inputs, t_n | | Output, t_{n+1} |
Clear	Set	Q
0	0	Q_n (Q at t_n)
0	1	1
1	0	0
1	1	Indeterminate

Gated Memory.   The output of the basic cross-coupled NAND gate memory circuit indicates which input had a momentary **0** level last. By the addition of gates to the inputs of this circuit, the memory can be made to respond to input levels only during a specified time interval, as shown in Fig. 3–178. So long as the load input is at **0**, gates 3 and 4 have **1** at their outputs regardless of the set and clear input levels. Recall that with a **1** level at both inputs to the basic NAND gate memory, the output level depends on which input was **0** last. To allow new information to reach the basic memory, the load input is changed to the **1** level. A **1** at the set input and a **0** at the clear input will now result in a **0** input to gate 1 and a **1** input to gate 2, and Q will be **1**. If clear were **1** and set were **0**, Q would be **0**. If both set and clear are **0**, the inputs to the basic memory remain **1**'s and the output does not change from its previous state. When the desired input level sampling interval is over, the load returns to **0** and further changes in set and clear levels will have no effect until the next load pulse.

This memory circuit is said to have a "clocked" input because the sampling interval can be "timed" to coincide with the appearance of the desired information at the set and clear inputs. For this reason, the load input is often called the **clock input.** The truth table for a clocked circuit is usually written as shown in Fig. 3–178 when the output state after the load or clock pulse depends on its previous state. The set and clear input signals to the gated memory do not have to be pulses or momentary level changes, as the basic memory required. In this case the input pulsing is accomplished by the clock signal. This is a very useful feature when set or clear inputs are obtained from other logic circuits.

A frequently used variation of the gated memory circuit is the **data latch** circuit shown in Fig. 3–179. This circuit has a single data input and an inverting gate, with the gated memory to apply the data level to the set input and its complement to the clear input. Thus the operation of the data latch is confined to the middle two entries in the truth table for Fig.

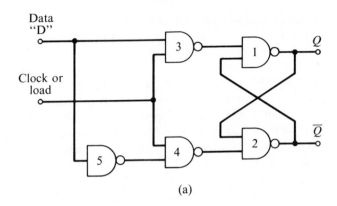

(a)

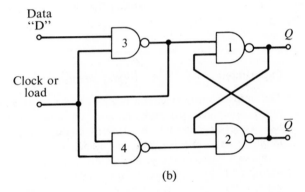

(b)

Input, t_n	Output, t_{n+1}
D	Q
0	**0**
1	**1**

Fig. 3–179 Data latch: (a) inverted input; (b) minimized.

3–178. The information present at the Q output becomes that of the D input when the clock becomes **1** and the Q output follows the D input as long as the clock is **1**. The latch stores the information at the D input when the clock signal returns to **0**. The data latch is frequently used for storing output information from counters and computers while the readout is taking place. The use of this circuit allows the counting or arithmetic circuits to begin their next cycle while the readout from the previous cycle is still taking place.

The data latch circuit can be reduced by one gate as shown in Fig. 3–179b. This is accomplished simply by obtaining the $\overline{D}$ information required at the gate 4 input from the gate 3 output instead of the added inverter, gate 5. Gate 3 acts as an inverter for D when the clock is HI, which is the only time the $\overline{D}$ is needed at gate 4. There are, of course, analogous data latch circuits that are made up of NOR gates. However, the construction of latches from gates is now unnecessary, since the complete latches are available in integrated circuit form. They come with four or eight latches in a package and two clock connections, each of

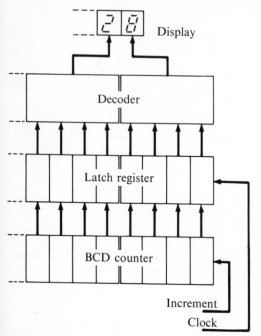

Fig. 3–180 Application of latch register in counting measurement.

which is connected to half the latches on the chip. Some packages provide both Q and $\overline{Q}$ outputs from each bit, others only Q.

A common application of the latch memory is the memory register used in counting measurement instruments. This application is shown in Fig. 3–180. At the end of the counting measurement, the latch register clock input is pulsed, loading the count information into the latch register. The counter may now be cleared and the next measurement begun while the latch register holds the data for decoding and display, as shown, or for printing, transfer to a computer, or whatever. If the clock input is held at **1**, the counter contents appear continuously at the latch outputs and the counting operation can be observed in the display. The quadruple latch packages are very convenient for use with the binary-coded decimal counter, since each package stores the binary information for one digit of display.

Experiment 3–43 *Basic Digital Memory*

The simplest memory circuit is a pair of inverting gates which have the output of one connected to the input of the other, and vice versa. This circuit has two stable states which are easily distinguished and it can be readily set in either state. In this experiment, a pair of NAND gates are cross-coupled to make a basic memory. Signals are applied to each input to observe the setting and storage characteristics and to verify the truth table. The same experiment is performed using AOI gates as NOR gates. The similarities and differences between the NAND and NOR memories are noted. A NAND memory circuit is used as a contact bounce eliminator for a switch.

Experiment 3–44 *Gated Memory and Data Latch*

By the addition of gates to the inputs of the basic memory, the memory circuit can be made to respond to input levels only during a specified time interval. In this experiment gated NAND and NOR memories are wired and tested. A data latch circuit that stores the input data level when the input gates are opened is constructed and tested. Then a 12-bit data latch for storing parallel digital data is investigated.

Memory Stacks

Where many words of data are to be stored, individual word latch registers are combined in "stacks." A 64-bit stack arranged as 16 words of four bits each is shown in Fig. 3–181. One set of parallel inputs and outputs

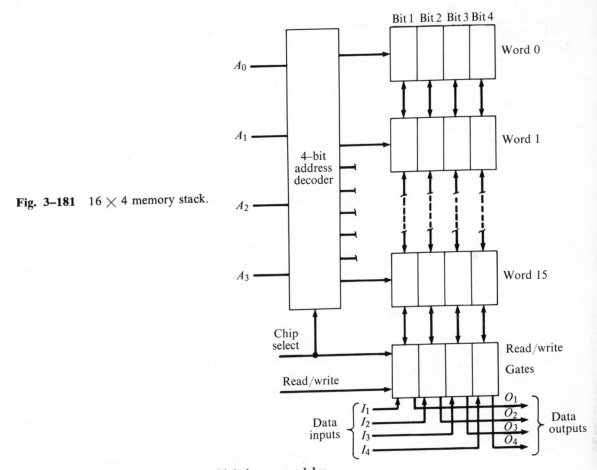

Fig. 3–181   16 × 4 memory stack.

serves the entire stack. The particular word register which is accessed by (made accessible to) the input and output connections is determined by coded word-select information called the **address.** Since there are 16 words in the memory, a four-bit binary-coded address word is used. This word is decoded into 16 separate word enable lines. When a word register is enabled (addressed), the content of that register will appear at the data outputs if the read/write control line is in the read state, or the information present at the data inputs will be stored in that word register if the read/write line is in the write mode. Such a memory stack is called a **random access memory (RAM)** because any word in the memory can be accessed directly in any order.

Solid State RAM.   The entire memory stack shown in Fig. 3–181 is available in a single integrated circuit package from several sources. Bipolar transistors are used in the memory and gate circuits. The input logic levels

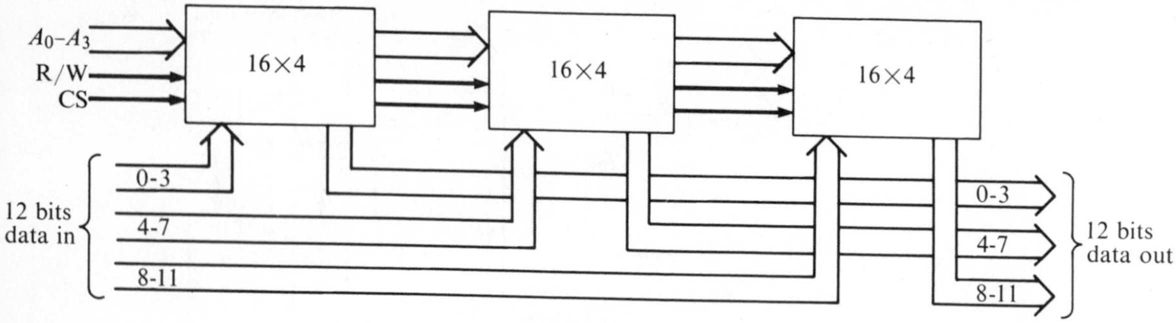

Fig. 3–182 16 word $\times$ 12 bit memory stack.

are TTL and the data outputs are open-collector for compatability with current sinking logic or data bus connections.

Memory stacks with more and/or longer words can be easily made by combining a number of the 16×4 memory chips shown in Fig. 3–181. The combination of three chips to store 16 12-bit words is shown in Fig. 3–182. The address, read/write, and chip select lines of all three chips are connected together. Each chip then provides four bits of the combined 12-bit word. The combination of four chips to provide a 64×4 memory stack is shown in Fig. 3–183. The address, read/write, and data lines of all four chips are connected together. The four chips are addressed by the two address bits A_4 and A_5, which are decoded into four chip select lines. The chip select lines are used to disable all but the chip in which the desired word exists. The complete address word is now six bits long, the minimum number of bits required for 64 addresses. The output connections of the disabled chips are all at logic **1**, which, for an open-collector output, means an open circuit. Thus the data outputs of all the chips can be connected together without interference (so long as only one chip is selected at a time). Obviously, the techniques of Figs. 3–182 and 3–183 could be combined to provide random access memory stacks of virtually any size. Large stacks of bipolar memory units are generally used only when extremely high speed access is important.

MOS Random Access Memory. For large semiconductor memory stacks, MOS integrated circuits are generally used. The lower power consumption per gate of MOS circuits allows the fabrication of larger stacks on one chip for lower cost. Response speeds of 1 μsec or faster, which are adequate for many purposes, are available. The MOS integrated memory circuits usually have TTL logic level compatible inputs and outputs. The basic storage elements in a popular 1024-bit MOS memory chip are arranged in a 32×32 bit array as shown in Fig. 3–184. The five-bit row address (A_0 through A_4) activates a 32-bit register (one row). The 32-bit word is not used as such because it is a longer word than needed

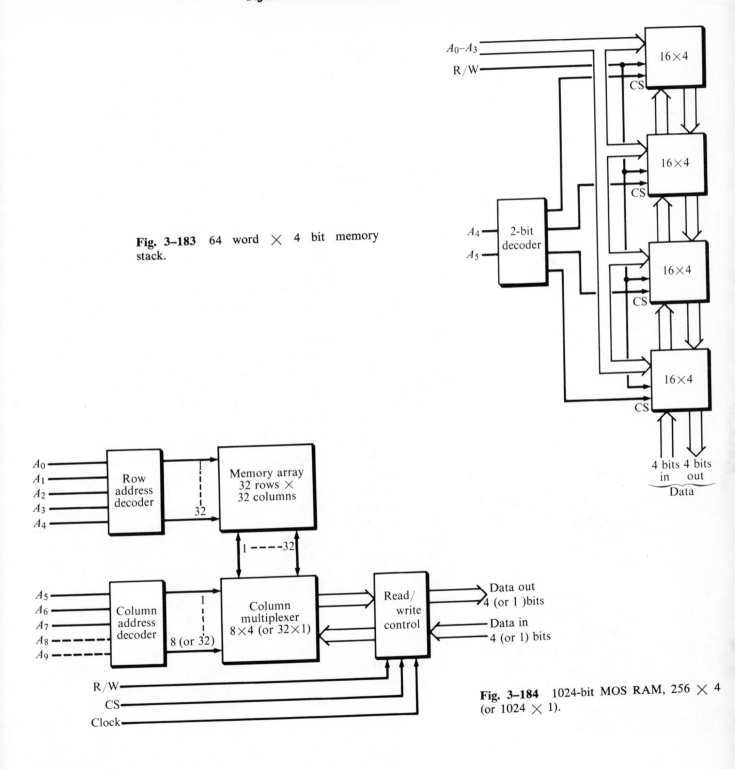

Fig. 3–183 64 word × 4 bit memory stack.

Fig. 3–184 1024-bit MOS RAM, 256 × 4 (or 1024 × 1).

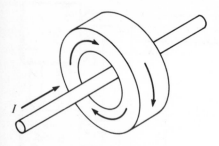

Clockwise magnetic flux

(a)

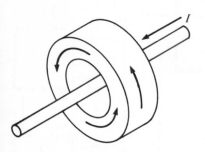

Counterclockwise magnetic flux

(b)

Fig. 3–185   Magnetizing the core.

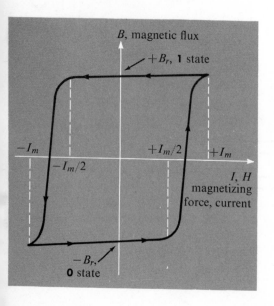

Fig. 3–186   Hysteresis loop for a ferrite magnetic core.

for many applications, but even more because it would require too many connections to the chip. The 32-bit word is divided into 8 four-bit words which are selected by 4 eight-input multiplexers. The multiplexer select commands are decoded from three additional address bits A_5, A_6, and A_7. In some models, a single 32-bit multiplexer is used to access one bit at a time in the 32-bit row. The chip then has a single input and output bit and a 10-bit address word. A number of such chips equal to the desired number of bits per word are then wired as shown in Fig. 3–182 to make a 1024-word memory.

MOS memory circuits are of two types, **static** and **dynamic.** In a static register, the information is held in bistable circuits that hold the stored information indefinitely (as long as the power to the circuit is maintained). In a dynamic register, the information is held as charge stored on a very small capacitance. Since this charge will eventually leak off, the capacitance must be recharged every few milliseconds or less. This is accomplished by reading the stored information and rewriting it on a regular basis. A dynamic MOS RAM is in a continuous **refresh** mode, reading and re-writing the memory in a row-by-row sequence, interrupted only when the chip is selected for reading or storing data. The advantage of the dynamic MOS RAM over the static is that the individual memory unit circuits are smaller and require less power. This makes it easier to fabricate greater storage capacity in a single integrated circuit. The additional complexity of the refresh circuitry, which is included in the chip, is more than offset by the simpler bit-storage circuits.

Magnetic Core RAM.   A magnetic core is a very small doughnut-shaped piece of ferromagnetic material of high retentivity. When a current is passed through a wire threading the hole in the core, a magnetization is induced in the core as shown in Fig. 3–185. Because of the high retentivity, the magnetism remains when the current is turned off. The core is now set in the **1** state. If the current is now reversed, it is found that the magnetism of the core does not change appreciably until the magnitude of the current is great enough to counteract the stored magnetic flux in the core. When the reverse current exceeds this minimum value, the direction of the magnetic flux is reversed, as shown in Fig. 3–185b. This behavior is illustrated in the *B-H* curve of Fig. 3–186. A magnetizing

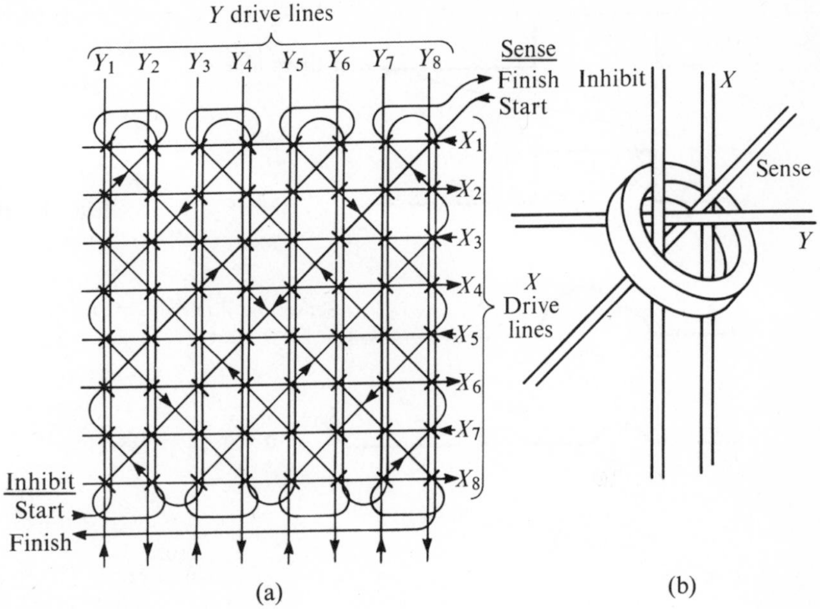

Fig. 3–187 A core memory bit plane.

force or current $+I_m$ is applied to put the core in the $+B_r$ or **1** state, and a current $-I_m$ is applied to put the core in the $-B_r$ or **0** state.

To write information into and read information out of the core, the cores are wired in a matrix called a **bit plane,** shown in Fig. 3–187a. This arrangement takes advantage of the hysteresis of the core to accomplish some addressing logic. If the core at the intersection of X_3 and Y_5 is to be set, half of the required set current or $+I_m/2$ is applied to the X_3 and Y_5 drive lines. Only the core X_3Y_5 receives the required setting current. The other cores on the X_3 and Y_5 drive lines receive only $I_m/2$, which, as Fig. 3–186 shows, is not sufficient to set the core. The part-current drive line matrix reduces the number of current drivers for N cores from N to $2N^{1/2}$. Each core has a unique matrix intersection or address.

The state of a particular core is read or "sensed" by applying a clear pulse to that core. If the core was in the **1** state, the resulting change in magnetic flux induces a current pulse in a **sense line** that threads through all the cores in the plane. If the core was in the **0** state, a much smaller pulse, induced by the clear pulse, appears in the sense line as shown in Fig. 3–188. Even though the sense line threads through all the cores, the induced current pulse can only come from the core address that received the clear pulse. The reading operation is "destructive", hence, if it is desired to continue to store the information, some provision must be made to write the bit back into the addressed core.

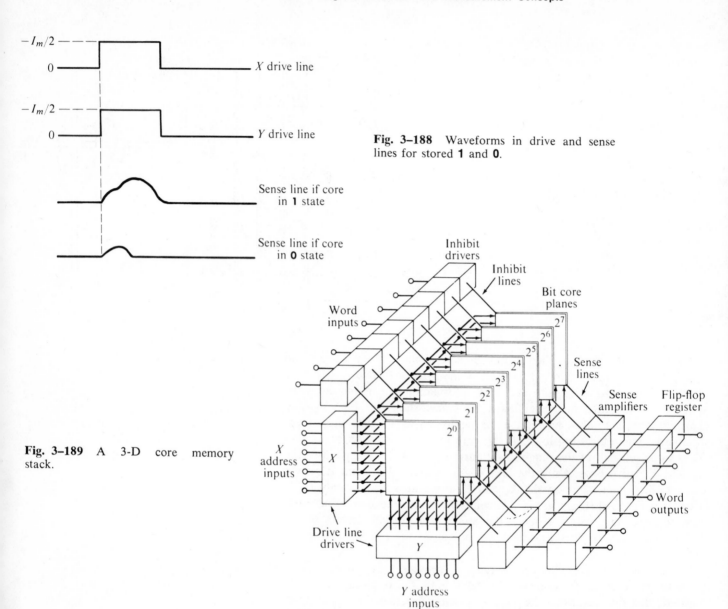

Fig. 3–188 Waveforms in drive and sense lines for stored **1** and **0**.

Fig. 3–189 A 3-D core memory stack.

Core memories store words by using as many bit planes as there are bits in the word. One method of organizing the bit planes into words is the "3-D" technique shown in Fig. 3–189. Corresponding X and Y drive lines for each bit plane are wired together so that all bit planes are addressed together. Each bit plane stores one bit in the addressed word. The memory will store as many words as there are cores in each plane.

A separate sense amplifier is used for the sense line in each bit plane, so that when an address is read (when a $-I_m/2$ pulse is applied to one X and one Y drive line) the outputs of the sense amplifiers present the word in parallel digital form. This information is read into a register that provides a stable output for transmission of the word to the desired location and stores the word until it can be applied to the word inputs and written back into the core address.

When a write pulse is applied to an address in a 3-D core stack, a **1** will be written into every bit of the word. To enable a **0** to be written, an **inhibit line** is threaded through every core in a plane. When a $-I_m/2$ current is applied to the inhibit line, the $+I_m$ setting current has a net effect of $I_m/2$, which is insufficient to set the core. Since the $-I_m/2$ inhibit current is opposite to the two $+I_m/2$ drive line currents, no core receives sufficient magnetizing force to change state. Since the addressed bit was just read, it must be in the **0** state. A word is then written into that address by applying set currents to the appropriate X and Y drive lines and an inhibit current to the inhibit line of each plane where a **0** bit is to be stored. Figure 3–187b shows the X and Y drive lines and the sense and inhibit lines threading through a typical core.

Another form of memory organization that is increasingly popular is the "$2\frac{1}{2}$-D" memory. In this memory, the corresponding X drive lines of each plane are wired in series or parallel, but the Y drive lines of each plane are driven separately. The read operation is the same as for the 3-D memory, but the write is accomplished simply by not applying a set pulse to the Y driver line of the planes that are to remain **0**. The $2\frac{1}{2}$-D requires more drivers than the 3-D, but the need for the long inhibit line is eliminated. As attempts are made to design smaller (and consequently faster) core memories, the elimination of the fourth wire can be an important performance and economy factor.

Current core memories are made in packages of 512 words by eight bits per word up to almost any conceivable size. The switching speed of standard current designs allows a complete read-write cycle in about 1 μsec, and faster designs are available. The size of the core is of the order of 1 mm or less outside diameter, with a complete 32×32 bit plane taking less than 6 cm^2 of area. One of the advantages of the core RAM over the solid state RAM is that it is "nonvolatile." The data are held in the magnetic memory even when the power is shut off.

Read-Only Memory

In many applications, digital data which do not change must be accessed. An example of such permanent data is a coded instruction list for a fixed sequence system. The coded instructions are placed in a memory stack in order at successive addresses. The address is set at that of the first instruction, which is then read out, decoded, and executed. When that step is complete, the address is set to the next location and the next instruction is read and performed. Another example is a look-up table of data such as log x vs. x. Each value of x is assigned a memory address and the value of log x for that value of x is stored at that location. Then applying any value of x to the address input will cause the value of log x to appear at the memory output. In such cases, a memory with the desired data permanently stored in it can be a great advantage. It cannot be lost through a power-down or accidental erasure. A memory in which the stored data cannot be altered is called a **read-only memory (ROM).** The memory units in a ROM are much simpler than in a RAM, in one case consisting of a single diode that is connected or not, depending on the datum stored at that bit location.

For most ROM's, the data content is determined by the interconnection pattern applied during the device fabrication. Some look-up table ROM's such as log x, sin x, and dot pattern character generators are of sufficiently general interest to be standard catalog items available from several sources. Special purpose ROM's can be ordered by providing a complete listing of the desired memory contents. This is sometimes justified where a substantial number of ROM's with the same pattern is needed. Many manufacturers now make ROM's that can be programmed by the user. These **field programmable ROM's** (PROM's) have **1**'s (or **0**'s) in every location. During the programming, a word is addressed and a special signal is applied to each bit in that word that is to be changed. The applied signal generally destroys by heating an active element or connection in that memory unit so that it permanently changes state. The entire PROM is programmed in this way.

In some PROM's, the programming can be reversed by ultraviolet radiation or by another special signal, so that the device can be reprogrammed. One such memory, made of amorphous semiconductor material, is called an RMM (read mostly memory). Much less seriously, another manufacturer has attempted to promote a WOM (write only memory), apparently without much success.

Counting Registers

As described in the introduction to this section, a counting register is one in which the count information is stored in a coded digital form such as binary or BCD and which will increment the stored number on command. After each increment command, the code that represents the number of counts received is revised to represent the next higher number. To perform the increment operation, the output of each memory unit is applied through logic circuits to the inputs of other memory units in such a way that the stored number plus 1 will be loaded into the register upon each increment command. This requires memory units which never allow the output state to track changes in the input state during loading. If the output state tracked the input state during loading, as in the data latch circuits described above, the following situation would occur. The increment command would cause the memory units to load the stored number N (from their outputs) plus 1. While loading, the outputs would become $N + 1$, which, through the logic circuits, would be converted to $N + 2$ and appear at the inputs, and thus the outputs, and so on. The only way to limit the cycle to a single increment is not to allow the memory output levels to change until the input load cycle is completed. Consequently, memory unit circuits have been developed that do not allow the output to track the input. The two most widely used memory circuits for counting registers are the **master-slave flip-flop** and the **edge-triggered flip-flop**. Both are described below.

Master-Slave Flip-Flops.   The master-slave flip-flop is composed of two gated memory circuits: one to hold the output state and the other to hold the information present at the input during the clock pulse for later transfer to the output memory. One such combination of memories is called an RS master-slave flip-flop and is shown in Fig. 3–190. The operation of the RS flip-flop through one clock pulse is described below. The clock input is normally **0**, which keeps the outputs of gates 7 and 8 at **1** and prevents the R and S inputs from having any effect on the circuit. With a **1** at each input, the memory bistable made up of gates 5 and 6 could be in either state. The slave circuit will be recognized as the gated memory circuit or latch, with the master gated memory circuit supplying the input signals on the inverted clock signal. When the clock input is **0**, the output of gate 9 is **1**, and hence gates 3 and 4 to the slave flip-flop are open. Therefore, the gates 1–2 latch will be in the same state as the gates 5–6 latch.

The **0** level current at the gate 9 input is sunk to the clock source through a 220 Ω resistor. This makes gate 9 a little closer to the **1** input state than gates 7 and 8. When the clock pulse is applied, a four-step sequence occurs as outlined in Fig. 3–190. First, as the clock goes positive, because of the 220 Ω resistor mentioned above, gate 9 reaches the **1** input state before gate 7 or 8. A **1** at the gate 9 input results in a **0** at the output, which closes gates 3 and 4 and isolates the slave from the master. This isolation occurs before any change in the state of the master could take place. Thus the state of the master is stored at the slave outputs Q and $\overline{Q}$.

Second, gates 7 and 8 are opened by the clock signal and the information at the S and R inputs determines the state of the master according

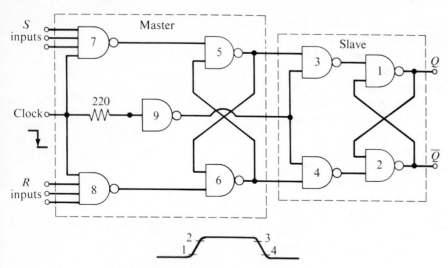

Fig. 3–190 *RS* master-slave flip-flop from NAND gates.

1. Gates 3 and 4 close, isolating slave from master
2. Gates 7 and 8 open, connecting master to inputs
3. Gates 7 and 8 close, isolating master from inputs
4. Gates 3 and 4 open, connecting master to slave

Truth table for Fig. 3–190

	Inputs, t_n		Output, t_{n+1}
	$S =$ $(S_1 \cdot S_2 \cdot S_3)$	$R =$ $(R_1 \cdot R_2 \cdot R_3)$	Q
	0	0	Q at t_n
	0	1	0
	1	0	1
	1	1	Indeterminate

to the gated memory truth table. Third, the clock pulse begins to fall, closing gates 7 and 8 and isolating the master from the *S* and *R* inputs. Fourth, gate 9 achieves a **0** input, which opens gates 3 and 4. At this time, the master outputs are transferred to the slave and appear at the circuit output terminals. Thus, the output change does not occur until the clock pulse is over and, therefore, the effects of output changes cannot appear at the input terminals during the clock pulse. The trailing (**1 → 0**) edge of the clock pulse is shown with an arrow in the circuit diagram to indicate the portion of the clock signal that can result in a change in state of the output.

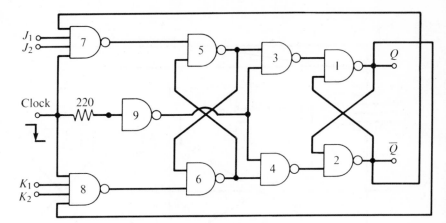

Fig. 3–191 *JK* master-slave flip-flop from NAND gates.

Truth table for Fig. 3–191

Inputs, t_n		Output, t_{n+1}
$J_1 \cdot J_2$	$K_1 \cdot K_2$	Q
0	0	Q at t_n
0	1	0
1	0	1
1	1	$\bar{Q}$ at t_n

There are several types of master-slave flip-flops designed on this basic circuit. They are generally named for the type of inputs provided. The circuit of Fig. 3–190 is called an *RS* master-slave flip-flop because the inputs simply reset (clear) or set the output levels as shown in the truth table. Additional inputs on the master latch input gates are sometimes useful. As seen from the circuit and the truth table, a **0** at any *S* input results in $Q = $ **0**, and a **0** at any *R* input results in $Q = $ **1**. A **0** at both *S* and *R* results in a **1** output level for gates 7 and 8, which leaves the master flip-flop unchanged from its previous state. The action of the *R* and *S* inputs in this circuit is very similar to that of the set and clear inputs of the gated memory (Fig. 3–178).

JK flip-flop. One of the most popular forms of the master-slave flip-flop is the *JK* flip-flop. As Fig. 3–191 shows, the *JK* is identical to the *RS* except that the outputs are connected to the inputs to obtain a **toggle** operation in which the state of the flip-flop outputs is inverted on each clock pulse. According to the truth table of Fig. 3–190, if *Q* is **1** and should

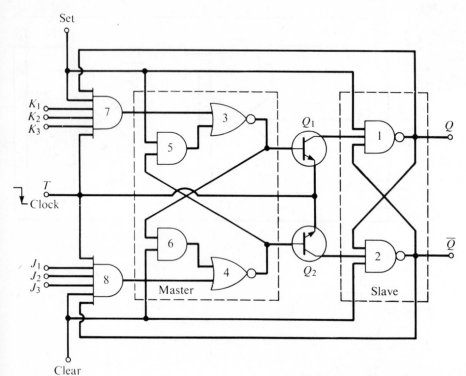

Fig. 3–192 Functional block diagram of an integrated circuit *JK* master-slave flip-flop.

go to **0** on the next clock pulse, one of the *S* input must be connected to **0** ($\overline{Q}$). Similarly, to return *Q* to **1** on the pulse after that, one of the *R* inputs should be connected to *Q*.

The remaining inputs are called *J* and *K* instead of *R* and *S*, and the truth table is changed to show the alternation of output state when *J* and *K* inputs are both **1** or unused. Thus the *JK* flip-flop can be used as a **toggle flip-flop** or the *J* and *K* inputs can be used to establish particular output states as required.

When a master-slave flip-flop is fabricated in an integrated circuit, several special design techniques are used which result in an improved circuit. Figure 3–192 shows the functional block diagram and the pin connections for a TTL *JK* master-slave flip-flop. The master latch is a pair of cross-coupled AND-OR-INVERT gates and the slave uses NAND gates, cross-coupled. Multiple-input AND gates provide the inputs to the master while transistor switches are used to connect the information between the master and the slave.

The action of the clock pulse in loading data into the master and transferring it to the slave is exactly the same as in the master-slave flip-flops of Figs. 3–190 and 3–191. The function of the *J* and *K* inputs is the same as in the NAND gate *JK* flip-flop of Fig. 3–191. The action of the direct set

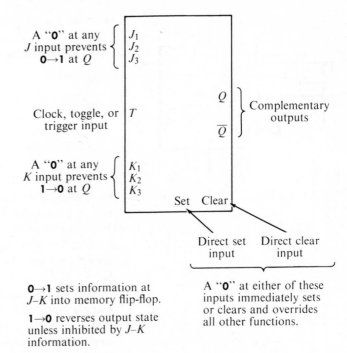

Fig. 3–193 Summary of relationships between input and output functions.

A "**0**" at any
J input prevents
0→1 at Q
$\left.\begin{array}{l} J_1 \\ J_2 \\ J_3 \end{array}\right.$

Clock, toggle, or
trigger input $\quad T$

A "**0**" at any
K input prevents
1→0 at Q
$\left.\begin{array}{l} K_1 \\ K_2 \\ K_3 \end{array}\right.$

Q
$\overline{Q}$
$\left.\begin{array}{l} \end{array}\right\}$ Complementary
outputs

Set   Clear

Direct set   Direct clear
input   input

0→1 sets information at
J–K into memory flip-flop.

1→0 reverses output state
unless inhibited by J–K
information.

A "**0**" at either of these
inputs immediately sets
or clears and overrides
all other functions.

and clear inputs is readily seen in Fig. 3–192. A **0** at the set input forces an immediate **1** at the master gate 3 output through both input AND gates 5 and 7. The **0** applied to the slave gate 1 also forces a **1** at Q. Thus the K gate 7 is disabled and both flip-flops are set. The clear input operates the same way on the lower half of the circuit. The direct set and clear inputs act immediately regardless of the state of the clock input, and they override all other input functions. Figure 3–193 and the accompanying truth table summarize the relationships among the input and output functions.

Truth table for Fig. 3–193

Inputs, t_n		Output, t_{n+1}
J	K	Q
0	**0**	Q at t_n
0	**1**	**0**
1	**0**	**1**
1	**1**	$\overline{Q}$ at t_n

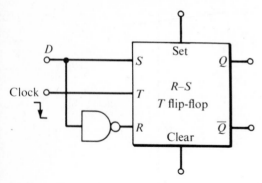

Fig. 3–194   *D* flip-flop.

Truth table for Fig. 3–194

Inputs, t_n	Output, t_{n+1}
D	*Q*
0	0
1	1

D flip-flop.   The third type of master-slave flip-flop is the "*D*" flip-flop, which is a "data" input modification of the *RS* flip-flop of the same type used to obtain the data latch from the gated memory circuit. The resulting circuit is shown in Fig. 3–194. The data level is applied to the *S* input and its complement is applied through an inverter to the *R* input. When the clock signal goes HI, the data level appears at the master flip-flop output, and when the clock pulse is over, the data level appears at *Q*.

The *D* flip-flop is very useful when the flip-flop output must respond to a single input level. When the *D* input is connected to $\overline{Q}$, the input information is always opposite to the state of the output, so that the flip-flop will toggle on successive clock pulses.

Edge-Triggered Flip-Flops.   Another type of flip-flop circuit which overcomes the timing problem is the edge-triggered flip-flop. This type reads the input information and makes the output transition in response to the leading or trailing edge of the clock pulse. There is an advantage—in system design, speed of operation, and clock width control—in not having the inputs active during the clock pulse.

An example of an edge-triggered flip-flop design is the *D* flip-flop of Fig. 3–195. With this circuit the data input and transfer occur on the leading edge of the clock pulse as shown by the arrow. Gates 1 and 2 comprise the bistable circuit with the set and clear functions. The output levels of gates 3 and 4 determine the state of the cross-coupled output gates. The remaining gates 5 and 6 determine whether the output of gate 3 or 4 (but not both) will become **0** in response to a trigger signal applied to the clock input.

A description of the operation of this circuit follows. A positive transition (from **0** to **1**) triggers the circuit. Therefore, prior to the trigger the clock input is at **0**. Assume also that the set and clear are both at **1**. The level at the *D* input determines which transition of the bistable circuit will be allowed. To begin, assume that *D* is **1**. Since the clock input is **0**, the outputs of gates 3 and 4 must be **1**. Now, since all inputs to gate 6 are **1**, its output is **0**. This forces the output of gate 5 to **1**. With all **1**'s at the inputs

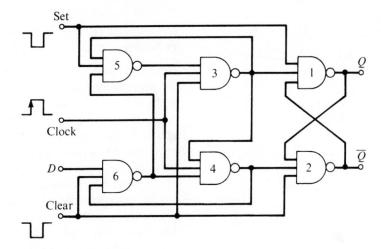

Fig. 3–195 *D*-type edge-triggered flip-flop.

Truth table for Fig. 3–195

Input, t_n	Output, t_{n+1}
D	*Q*
0	**0**
1	**1**

to gates 1 and 2, the bistable circuit could be in either state. Now when the clock input goes to **1**, all inputs to gate 3 are **1**, causing its output to become **0**. However, the output of gate 4 stays at **1** because of the **0** output of gate 6 and the new **0** level from gate 3. The **0** at the gate 1 input will cause *Q* to become **1**. If *Q* had been **1** prior to the trigger, no output transition would have occurred.

Now consider the circuit again, but with a **0** level applied to the *D* input. Prior to the **0** → **1** transition at the clock input the outputs of gates 3, 4, and 6 are **1**; gate 5 is **0**, and the bistable may be in either state. When the clock input goes from **0** to **1**, all inputs to gate 4 are **1**; thus its output becomes **0** and the output of gate 3 cannot change. In this case, the $\bar{Q}$ output will become **1** or remain **1** if it was in that state prior to the trigger. Therefore, the action of the triggering circuit is to cause *Q* to become whatever level was applied to the *D* input just prior to the trigger signal. Note that any change in the level at *D* which occurs after the trigger transition has set the outputs of gates 3 and 4 cannot affect the outputs of gates 1, 2, 3, or 4 until the next triggering edge.

The versatile *JK* flip-flop can also be made as an edge-triggered circuit. Both *D* and *JK* edge-triggered flip-flops are available in integrated circuit packages in several logic types.

Binary Counter.   The register of a binary counter stores information about the accumulated number of input pulses as a binary coded number. That is, if the state of each memory unit in the register is sequentially written as a **1** or a **0**, the resulting binary number is equal to the number of input pulses counted. It is a popular counter because it is very simple, stores the maximum possible amount of information per memory unit, and is related to a basic number system.

A basic binary counter is shown in Fig. 3–196. Toggle or complementing flip-flops are used. The outputs of each flip-flop change state when the *T* or clock input signal goes from **1** to **0**. Assume that the four flip-flops have been cleared ($Q = $ **0**) by an appropriate pulse at the clear input. When the first pulse appears at the count input, flip-flop *A* toggles and *A* becomes

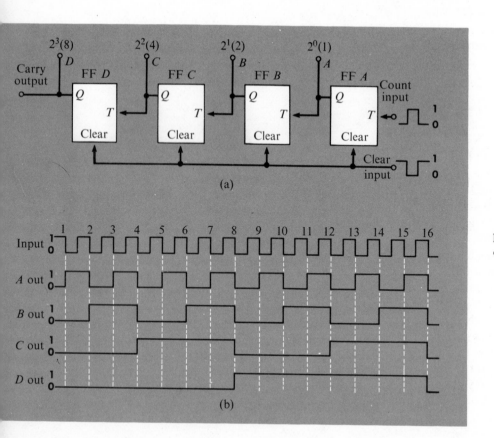

(a)

(b)

Fig. 3–196 Basic binary counter: (a) circuit; (b) waveforms.

logic **1** ($2^0 = 1$). On the next pulse, *A* returns to logic **0**, which toggles flip-flop *B*. Now *B* is **1** and *A* is **0** ($2^1 = 2$). After the next pulse, *A* is **1** again and *B* is unchanged ($2^1 + 2^0 = 3$), and so on. The table of waveforms in Fig. 3–196b shows how the *A* flip-flop is set after every odd input pulse to represent the number 1. The *B* flip-flop output represents the number 2, and the *C* and *D* outputs represent 4 and 8, respectively. The sum of the values of the set flip-flops represents the cumulative count at any instant. The outputs could be connected to indicators, and the instantaneous count read out in binary form. The counter of Fig. 3–196 can be extended to any desired number of flip-flops by adding flip-flops *E*, *F*, *G*, and so on, in like manner. Since each successive flip-flop output represents another binary digit (power of 2), *n* flip-flops will have 2^n states and can count from zero to $2^n - 1$. In Fig. 3–196b the waveforms of the four-flip-flop (four-bit) counter show the progression from 0 to 15. Then, on the sixteenth pulse the register returns to the zero state.

Binary Down-Counter.   The counter of Fig. 3–196 is called an **up-counter** because the register counts up to the next higher number for each input pulse. A counter can also be made which counts down or subtracts one from the count for each input pulse. Such a counter, a **down-counter,** and its output waveforms are shown in Fig. 3–197. The circuit is identical to the up-

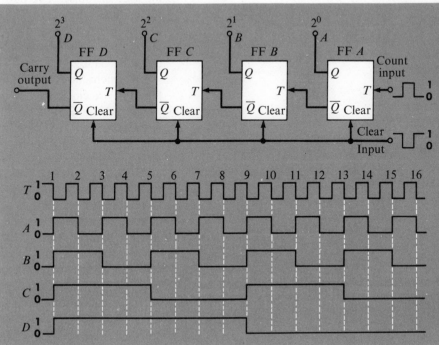

Fig. 3–197   Binary down-counter and wave-forms.

counter except that $\overline{Q}$ instead of Q is connected to the following T input. When the first $\mathbf{1} \rightarrow \mathbf{0}$ transition appears at the count input, flip-flop A toggles, changing Q of A from $\mathbf{1}$ to $\mathbf{0}$. The A signal toggles flip-flop B, and so on, resulting in a full count of 15 in the register. The next pulse brings flip-flop A back to $\mathbf{0}$ for a 14, and so on, until the register is cleared.

Synchronous Binary Counter.   The counting circuits of Figs. 3–196 and 3–197 are not synchronous because the flip-flops are not all clocked at the same time. In each flip-flop, there is a delay between the triggering edge of the clock pulse and the logic level change at the output due to the propagation delay through the flip-flop gates. If the output of either of these asynchronous counters is sampled while the new count information is "rippling through" the counter chain, serious errors in the apparent count could result. For instance, with the up-counter, as the eighth count pulse is applied, the waveforms of Fig. 3–196b show that the counter output will be $\mathbf{0111}$ (7), then $\mathbf{0110}$ (6), then $\mathbf{0100}$ (4), and finally $\mathbf{1000}$ (8). A serious error exists at the output for two of the four propagations required in the example and in general could occur for up to $N - 2$ propagations, where N is the number of counter bits. If the propagation delay is 50 nsec through each flip-flop, a 10-bit counter will have error states for as long as 400 nsec. Two solutions to this problem are used. The first is to trigger a monostable multivibrator, of pulse duration equal to the maximum error period, with each input count pulse. The output of the monostable is then used to hold off the counter output sampling circuit during the ripple-through period. The second technique is to use synchronous counters in which all the flip-flops are clocked simultaneously and all therefore change output state at essentially the same time.

A synchronous four-bit binary counter is shown in Fig. 3–198. All four flip-flops are clocked together and the J and K inputs are used to determine which transitions will be allowed. Note that the waveform chart in Fig. 3–198 should be identical to that of Fig. 3–196. Flip-flop A is to change state on every other count, and flip-flop B is to alternate only when A is $\mathbf{1}$ after the previous count. Therefore, the output Q of A is connected to J_1 and K_1 of B to prevent B from changing state when A is $\mathbf{0}$. Similarly C should change only when A and B are $\mathbf{1}$. Connecting A and B outputs to J and K inputs of C allows C to alternate only on every fourth count. Connections to D are obtained in the same way. Note the convenience of multiple J and K inputs in this application.

The carry out connection is $\mathbf{1}$ only after the fifteenth count. When this is connected to the carry in of another four-bit synchronous counter, the input and clear lines of both circuits can be connected together to make an eight-bit synchronous counter. In this way, synchronous counting registers of any bit size can be made. The availability of four-bit synchronous

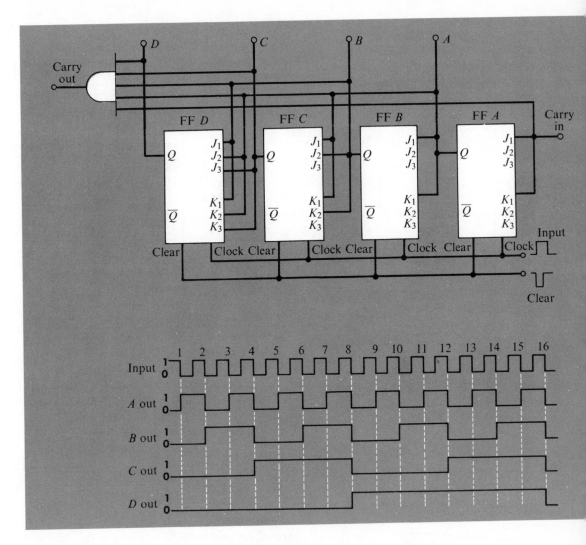

Fig. 3–198   Synchronous binary counter.

counters in an integrated circuit package makes such counters very easy to build.

Up-Down Binary Counter.   In the synchronous binary up-counter, each flip-flop is allowed to toggle, or complement, if all the less significant bits are **1**'s. In the synchronous down-counter, the J and K inputs of each flip-flop are connected to the $\overline{Q}$ outputs of all previous flip-flops so that each flip-flop complements only if all less significant bits are **0**'s. To make an up-down counter, selector gates are connected at the JK inputs so as to determine whether the Q's or $\overline{Q}$'s of the previous flip-flops affect the JK

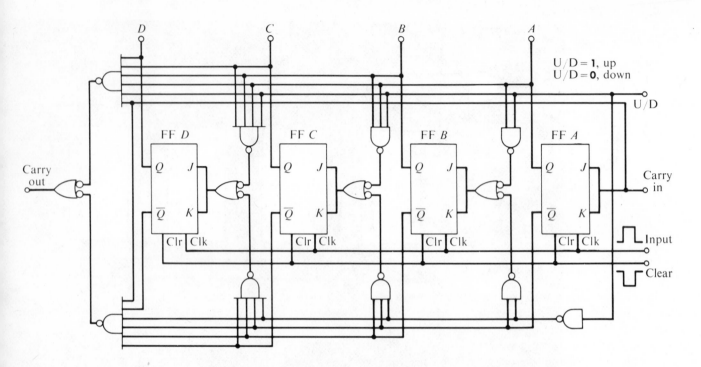

Fig. 3–199   Synchronous up/down counter.

level. Figure 3–199 gives the circuit diagram for the synchronous binary up-down counter. A **1** at the U/D input commands an up count and a **0** commands a down count. The selector gates in this illustration are made with NAND gates. The carry in and carry out connections allow several such four-bit units to be connected to make a larger counting register. Fortunately, it is no longer necessary to build such circuits from gates and flip-flops, since the four-bit unit shown in Fig. 3–199 and other similar units are available in IC form.

BCD Counter.   In instruments or systems in which the readout convenience of decimal numbers outweighs the ease of computation and economy of binary numbers, the registers are arranged to store each decimal digit in a group of four flip-flops. Up to 16 combinations of output states are available from four binary circuits, but, when they are used to store a decimal number, only 10 of these states are used. Each combination is assigned to one of the decimal numerals 0 to 9. A group of four flip-flops connected as a counting register with 10 states is called a **decade counting unit (DCU).** For instance, to store numbers up to 9999, four DCU's are required. Since the decoding for each DCU is identical, more DCU's and decoding circuits can readily be added to store and read out as large a decimal number as necessary.

Table 3–20 Binary-coded decimal storage

Decimal	Flip-flops			
	D	C	B	A
0	0	0	0	0
1	0	0	0	1
2	0	0	1	0
3	0	0	1	1
4	0	1	0	0
5	0	1	0	1
6	0	1	1	0
7	0	1	1	1
8	1	0	0	0
9	1	0	0	1

The most common storage code for decimal numbers follows the first 10 states of the binary counter as shown in Table 3–20. It is called the 1 2 4 8 or natural code because the values of 1, 2, 4, and 8 can be assigned to flip flops A, B, C, and D respectively, and the stored decimal number can be obtained by adding the values of the set (**1** level) flip-flops. There are thousands of other possible BCD codes, some of which have advantages in various applications. However, the 1 2 4 8 code has become the industry standard because of its widespread adoption in integrated circuits.

The synchronous decade counting unit of Fig. 3–200 is similar to the binary counter of Fig. 3–198 except that it has a 10-count cycle and a specially coded output. The desired output levels for each input pulse in the cycle are shown for the 1 2 4 8 code. The J and K inputs are used to make the counter conform to this specific pattern. Flip-flop A alternates with every input pulse, so that no J or K connections are required. Flip-flop B alternates whenever A is **1**, but should not go to **1** on the tenth pulse when D is **1**. Connecting J_1 and K_1 of B to Q of A satisfies the first condition, and connecting J_1 of B to $\bar{Q}$ of D prevents Q of B from becoming **1** when $\bar{Q}$ of D is **0**, to satisfy the second condition. The transitions of C are affected only by A and B outputs as in the binary counter. The conditions for the $\mathbf{0} \rightarrow \mathbf{1}$ transition of D require A, B, *and* C to be **1**. However, since D should return to **0** the next time A is **1**, only the $A = \mathbf{1}$ condition is put on the K input of flip-flop D. This method, i.e., plotting the desired output waveforms and using the J and K inputs to allow transitions only at the appropriate counts, allows counters with any special output sequence to be easily designed and constructed. If multiple J and K inputs are not available or are insufficient, other gates can be used at the J and K inputs.

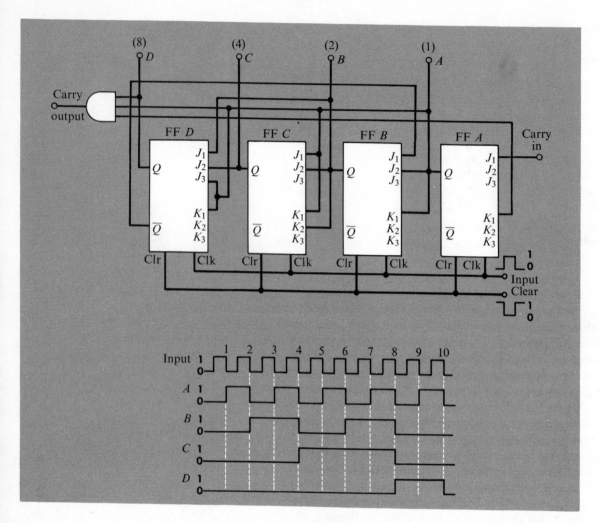

Fig. 3–200   Synchronous BCD counter.

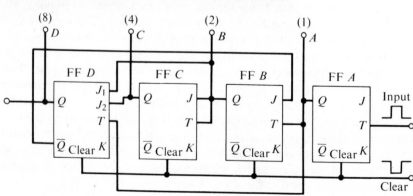

Fig. 3–201   Asynchronous BCD counter.

The 1 2 4 8 BCD counter is considerably simplified if asynchronous counting is permitted, as shown in Fig. 3–201. The output waveforms of this counter are identical to those shown in Fig. 3–200. Flip-flop A is toggled by the input signal. The output of A toggles B on input pulses 2, 4, 6, and 8, but the connection of $\overline{Q}$ of D to J of B keeps B from setting on the tenth pulse. Flip-flop C is toggled by the B output. The output of C cannot be used to toggle D because D must clear on the tenth pulse when only the A output changes. Thus D is toggled by the A output but is prevented from setting until B *and* C are **1**.

The simpler asynchronous circuit of Fig. 3–201 has been used to make the most popular integrated circuit BCD counter. In the integrated circuit form, this counter can be obtained with maximum count rates of over 50 MHz. Generally the connection between the Q output of A and the T input of B (and D) is not completed inside the integrated circuit package. Bringing these connections out instead allows the A circuit with two stable states to be used independently of the circuit composed of flip-flop B, C, and D which has five stable states. Because this BCD counter is made up of a two-counter followed by a five-counter, it is sometimes referred to as a **biquinary counting circuit.**

Decimal counters are frequently used for decade scaling, that is, to divide an input frequency or pulse rate by an exact factor of 10. As the waveform of Fig. 3–200 shows, the D output of the 1 2 4 8 BCD counter goes to **1** on the eighth pulse and back to **0** on the tenth pulse. Thus one output pulse is produced for every 10 input pulses. If the input frequency is exactly 1 MHz, the frequency at the D output will be exactly 100 kHz. The 100 kHz signal can be further divided by additional decade counting units to obtain precise 10 kHz, 1 kHz, 100 Hz, etc., signals. With six decades of division a precise 1.000000 Hz signal can be obtained from a 1.000000 MHz crystal oscillator. This application is described in Section 3–3.2.

A "preset to 9" input available on some integrated circuit BCD counters is particularly convenient in this application. If, prior to a timing operation, the preset inputs of all the DCU scalers are activated, the counting register contents will be all 9's. This means that on the first cycle of the crystal oscillator after the preset control is released, a **1** → **0** transition will appear at every DCU output. This provides control over the time of initiation of timing cycles.

Scaling circuits are sometimes used at the inputs to counters to reduce the input events to a number which will not overrange the counter. Extremely high speed scaling circuits are also used at frequency meter inputs to reduce the signal frequency to a value that is within the measurement capability of the basic frequency meter.

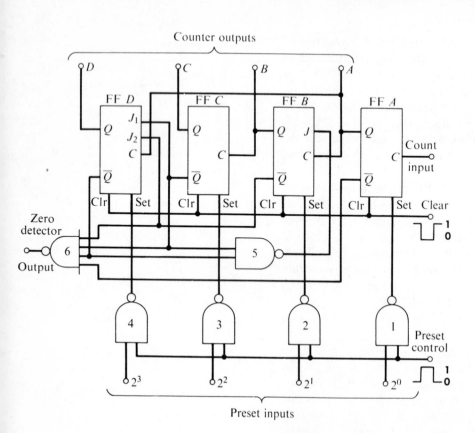

Fig. 3–202   Preset BCD down-counter.

Presettable Counters.   A presettable counter is a counting register that can be loaded by a parallel digital signal. A BCD down-counter which can be present to any count and which indicates when it has counted down to zero is shown in Fig. 3–202. The counter is first cleared by applying a **0** to the clear input. Then a **1** is applied to the preset control input, causing those flip-flops to set for which a **1** level exists at the preset inputs. Thus the preset number is transferred into the counting register. Now the counter is allowed to count. When the counter has counted down to zero, the output of gate 6 becomes **0**. The zero detector output signal can be used to stop further counting by a connection to J and K of flip-flop A or to indicate to another circuit that the countdown has been completed.

Additional decades can be ganged to achieve any required count capacity. Ten-position BCD-coded panel switches are frequently used to provide the logic levels to the preset inputs for each decade. This kind of preset counter is used, for instance, to measure how long it takes to accumulate the preset number of counts. Another application is as a parallel-to-count converter for digital signals, as shown in Fig. 3–203. The parallel digital data

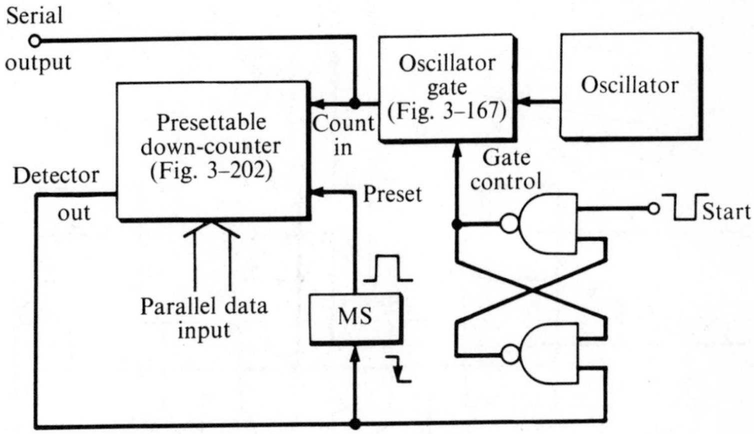

Fig. 3–203 Parallel-to-count converter.

are entered into the counter through the preset inputs. A momentary **0** at the start input sets the bistable gate circuit so that the oscillator gate circuit is **OPEN**. The oscillator pulses now appear at the serial output and are counted by the down-counter. When the down-counter reaches zero, the zero detector output clears the bistable gate circuit, turning the oscillator gate **OFF**; this terminates the pulse train to the serial output and the counter. A monostable generates a pulse to preset the parallel data once more.

The counter of Fig. 3–202 must be cleared before the preset control pulse is applied to ensure that only the desired flip-flops will be set. A positive preset circuit that uses both direct set and clear inputs is illustrated in Fig. 3–204. When the data strobe input goes to **0**, the logic levels at the parallel data inputs are transferred to the corresponding flip-flops. Note also that so long as the data strobe is **0**, the counting register is disabled. The circuit of Fig. 3–204 is available in integrated circuit form as the asynchronous four-bit counter shown and the corresponding asynchronous BCD counter. Many presettable counting registers are available as IC's, including the elegant yet inexpensive synchronous up/down, presettable BCD or four-bit counters.

Odd or Variable Modulus Counters. The modulus of a counter is the number of count input pulses for a complete count cycle. For instance, the four-flip-flop binary counter of Fig. 3–196 has a modulus of 16. If it starts at **0000**, in 16 pulses it will be at **0000** again. Similarly, a three-flip-flop binary counter is a modulo-8 counter and the 1 2 4 8 BCD counter of Fig. 3–200 is a modulo-10 counter. Many counting and scaling applications require counters with a modulus other than 2^n or 10^n. Furthermore, it is sometimes desirable to be able to select or vary the modulus of a counter as needed.

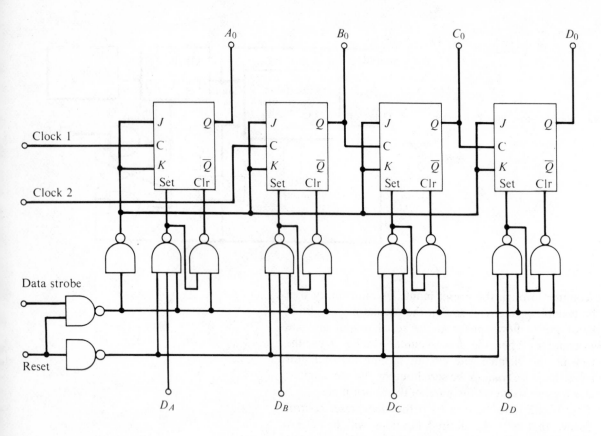

Fig. 3–204   Presettable four-bit counter.

A counter can be made to have a particular modulus in several ways: the flip-flops can be wired to repeat the output cycle every M counts (fixed-modulus counter), or a decoding circuit can be used to detect the $M - 1$ count and stop or reset the counter, or the counter may be preset to a value from which it will proceed until it is full or clear. Examples of each of these techniques are illustrated below.

The modulus of a binary counter with n flip-flops is 2^n. Therefore, simple binary counting circuits provide modulo-2, -4, -8, -16, -32, etc., counters. In each case the binary counter provides the largest modulus attainable with a given number of flip-flops. The number of flip-flops required to count to any given modulus will be at least equal to the number required to count to the next larger power of 2. For instance, a modulo-5 counter will require three flip-flops $(2^2 < M < 2^3)$, a modulo-28 counter will require five flip-flops, and so on.

Two flip-flops are required for a modulo-3 counter. One possible circuit is shown in Fig. 3–205. As shown by the count sequence table, the counter outputs follow the natural binary sequence. When both flip-flops are clear, there is a **1** at J of flip-flop A (J_A) and a **0** at J_B. On the first input pulse, FF A can set, but FF B cannot. Now there is a **1** at J_B. On the second pulse, FF B sets and FF A clears. The resulting **0** at J_A prevents FF A from setting on the third pulse, but FF B clears, leaving both flip-flops cleared to begin again. The modulo-3 counter can be combined with additional flip-flops (modulo-2 counters) to make modulo-($3 \times 2^n = 6$, 12, 24, etc.) counters.

Another example illustrates fixed-modulus counter design. The modulo-7 counter shown in Fig. 3–206 requires some gating in order to achieve the

Count sequence table for Fig. 3–205

	Outputs	
Count n	B	A
0	**0**	**0**
1	**0**	**1**
2	**1**	**0**
3, 0	**0**	**0**

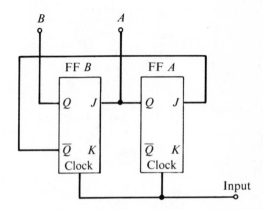

Fig. 3–205   Modulo-3 counter.

Count sequence table for Fig. 3–206

	Outputs		
Count n	C	B	A
0	**0**	**0**	**0**
1	**0**	**0**	**1**
2	**0**	**1**	**0**
3	**0**	**1**	**1**
4	**1**	**0**	**0**
5	**1**	**0**	**1**
6	**1**	**1**	**0**
7, 0	**0**	**0**	**0**

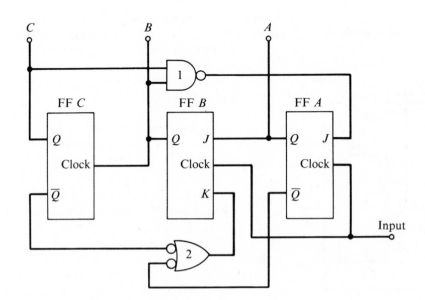

Fig. 3–206   Modulo-7 counter.

desired count sequence and modulus. The desired count sequence table shows that flip-flop A is to alternate states except on the sixth pulse, when it is to stay clear. Thus J_A should be **0** when B AND C are **1**. This can be written as $\overline{J_A} = B \cdot C$ or $J_A = \overline{B \cdot C}$ and implemented by gate 1 as shown. Flip-flop C changes state on each $\mathbf{1} \rightarrow \mathbf{0}$ transition of FF B, requiring only a simple Q-to-clock connection between FF B and FF C. Flip-flop B sets on the first pulse after FF A is **1**, which is done by clocking FF B with the signal and connecting Q of FF A to J of FF B. Flip-flop B is to clear on the fourth pulse (after A is **1**) and on the sixth pulse (after C is **1**). Therefore, K_B should be **1** when A OR C is **1**. This is written $K_B = A + C$ and implemented by the gate 2 circuit. Of course, this circuit can be used with binary circuits to provide modulo-7, -14, -28, etc., counters and with modulo-3 counters to provide modulo-21, -42, etc., counters, and so on.

In the above examples of fixed-modulus counters, the counter circuit uses logic gates and/or J and K flip-flop inputs to control the counting sequence and bring the counter to zero after the maximum count has been reached. Another approach is to apply a pulse to the direct clear inputs of the flip-flops when the desired maximum count is reached. A counter of this type with a modulus of 11 is shown in Fig. 3–207. On the count of 11 the output will be **1011** and all inputs to the NAND gate will be **1**. The resulting **0** output from the gate clears all the flip-flops. The clear pulse is very short, since the clearing of the flip-flops returns the gate output to a **1**. The counter is then free to count again to 11. Note that reset occurs on the eleventh count; hence 10 is the maximum count and 11 is the modulus of this counter.

A counter of any modulus can be quickly and easily made in this way. The flip-flop outputs which are **1** when the desired modulus is reached are

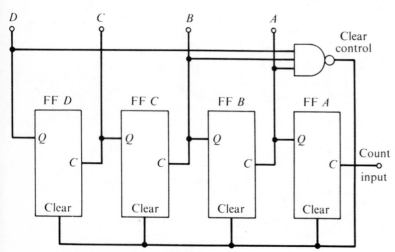

Fig. 3–207 Modulo-11 counter with direct clearing.

connected to the clear control gate. It is not necessary to ensure that the flip-flops which are to have a **0** output actually have a **0**. For instance, the gate output in Fig. 3–207 will be **0** for a count of **1011** or **1111** since the level of flip-flop C is not detected. However, since the counter resets at **1011**, the count of **1111** can never be attained.

For some counting and scaling applications, it is desirable to be able to change the modulus of the counter quickly and easily. This is generally done by allowing an ordinary binary or BCD counter to advance until the desired maximum count is detected, and then stopping or clearing the counter. A general block diagram of such a variable modulus counter is shown in Fig. 3–208. A circuit is used to compare the outputs of the counting register with inputs representing the desired maximum count. When the desired count is reached, the digital comparison circuit output changes logic level. The comparator output can be used to close the counting gate, thus stopping the counter, or to reset the counter, or to cause the counter register to clear on the next input pulse. Another rather similar approach, commonly used, is to invert the zero detector output of the presettable down-counter of Fig. 3–202 and connect it to the preset control input. The counter will then preset to the parallel data input automatically upon reaching zero and begin to count down again. The modulus of this counter is equal to the preset binary-coded number.

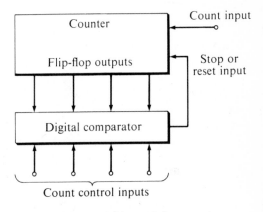

Fig. 3–208 Variable modulus counter.

Experiment 3–45   RS, JK, and D Master-Slave Flip-Flops

A basic master-slave flip-flop is wired using NAND gates. The characteristics of the RS flip-flop are studied. It is noted that the input gates can accept information from the R and S inputs at any time during the clock pulse, the final state of the master being determined by whichever was **1** last. The RS flip-flop is modified to make a JK flip-flop. The JK flip-flop truth table is verified and its toggling and frequency division are observed. A D flip-flop is made from the basic RS flip-flop circuit. It is noted that the D input remains active during the clock pulse. When the $\overline{Q}$ output is connected to the D input, the toggling action is observed. The master-slave type of flip-flop will be used in almost all remaining flip-flop and register experiments.

Experiment 3–46   Integrated Circuit JK Flip-Flop

The JK flip-flop is the most versatile flip-flope type. In its integrated circuit form, it is also very convenient to use. An integrated circuit JK flip-flop is studied and its characteristics are verified experimentally. The action of each of the inputs and outputs is observed and compared with the expected response. The JK flip-flop is wired to perform the function of a D flip-flop and a latching circuit. The integrated circuit *JK* flip-flop will be used extensively as the basic circuit element in counters, shift registers, and other memory circuits.

Experiment 3–47   *Binary Counters*

The binary counter is the simplest and most efficient form of the counting register. The output of each flip-flop in the register represents one binary digit, so that an *n*-flip-flop register can store any binary number from 0 to $2^n - 1$. In this experiment a four-bit binary up-counter is wired using *JK* flip-flops. Binary counting is observed by connecting the flip-flop outputs to indicator lamps. The 16 states of the counter are verified. Then the binary up-counter is modified to make a down-counter. Two down-counting circuits are studied and the response of counting and clearing pulses is observed and analyzed.

Experiment 3–48   *Binary-Coded Decimal Counters*

A synchronous BCD decade counting unit is wired and shown to follow the binary equivalent of counting from 0 to 9. The output signal from the fourth flip-flop is observed to be exactly one-tenth the frequency of the input signal. Precise frequency scaling (dividing) is another very common application of this circuit. Then the asynchronous BCD counter is wired. Where propagation delay is not critical, the asynchronous BCD counter is often used because of its simplicity. The count sequences of the asynchronous and synchronous BCD decade counting units are compared.

Experiment 3–49   *Variable Modulus Counter by Direct Clearing*

One way to control the modulus of a counter is to use a gate circuit to clear the flip-flops when the desired count has been reached. A direct-clearing modulo-11 counter is wired and the count sequence is observed. If possible, the maximum input frequency for reliable counting is determined. Counters with several moduli between 1 and 16 are wired by connecting other combinations of flip-flop outputs to the gate input.

Counting Measurements

Earlier portions of this section have shown how flip-flops are used to count and store information. Such circuits have virtually endless applications in

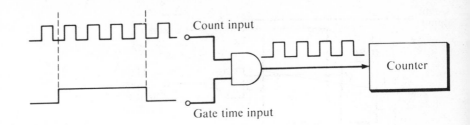

Fig. 3–209   Counter input gating.

digital instrumentation and computation. However, in every case, some auxiliary circuits have to be used to control what is counted or stored and when. In this discussion, typical counting measurements are used to illustrate the counter and register control for particular applications. First the problem of gating the count signal is discussed, and then an automatic measurement cycling circuit is developed.

The simplest counter input gate is shown in Fig. 3–209. When the gate time input is **0**, the AND gate is closed. Pulses at the count input will be counted only when the gate time input is **1**. This simple gate is satisfactory where the gate time signal is **1** only when the count measurement is to be made and will not become **1** again until the count reading is taken, the counter reset, and the next count measurement is desired. To ensure that this is the case usually requires that the gate time signal itself be gated or controlled. Control is certainly required when the gate time signal is a continuous square wave from a standard clock oscillator. In other words, the problem is not that of gating, but of gating just once per count measurement.

There are numerous circuits for solving this problem. One of the most direct is to use two flip-flops, one to start the count and the other to stop it. Each flip-flop is triggered by the appropriate part of the gate time input signal. Such a circuit is shown in Fig. 3–210. The start and stop flip-flops are normally cleared so that $Q_{FF A}$ is **0** and $\overline{Q}_{FF B}$ is **1**. Gate 1 is closed by the **0** input. The **0** at the K inputs of the flip-flops causes them to operate in a "latching" mode. They can be set by a clock input signal but not cleared. The clear must be performed through the clear input. Since Q from FF A is connected to J of FF B, flip-flop B cannot respond to its clock input until FF A is set. On the first **1** $\rightarrow$ **0** transition after reset at the start input, FF A will be set, gate 1 will open, and the count will begin. The counting gate will continue to be open, as indicated by a **1** at the output of gate 2, until a **1** $\rightarrow$ **0** transition occurs at the stop input. At this time, gate 1 closes and the counting stops. The closing of the gate is indicated by a **1** $\rightarrow$ **0** transition at the output of gate 2. The gate can be reopened only by the application of a reset pulse to the flip-flops and a start signal to FF A.

This single counting gate circuit is useful for a variety of counting measurements. For ordinary counting measurements, the start and stop

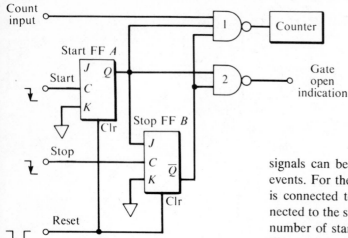

Count
input

Start FF *A*

Start

Stop FF *B*

Stop

Reset

Counter

Gate
open
indication

Fig. 3–210 Counter input gating.

signals can be derived from manual switches or appropriate experimental events. For the measurement of time between two events, a standard clock is connected to the count input and the first and second events are connected to the start and stop inputs respectively. The counter then counts the number of standard time units between events. The gate control circuit of Fig. 3–210 also satisfies the counter gate requirements for frequency and period measurement as shown in Fig. 3–119. For frequency measurement, the input signal is connected to the count input and the clock signal to both start and stop inputs. The first **1 → 0** transition of the clock will start the count and the next one will stop it. The counter will thus have counted the number of input cycles for one cycle of the standard clock oscillator. The period of a signal is measured by counting time units per signal period. In this case the clock output is counted and the input signal is connected to the start and stop inputs. To measure the duration of a pulse, apply the pulse to the start and stop inputs with an inverter interposed at one of the inputs so that the leading edge of the pulse produces a **1 → 0** transition at the start input and the trailing edge of the pulse produces a **1 → 0** transition at the stop input.

Most modern digital counting instruments provide for automatic repetitive measurements. This involves controlling a sequence of events within the instrument generally as follows: (a) the count gate control is allowed to be triggered; (b) the count gate is opened and closed for the measurement interval; (c) the counter information is transferred to the memory; (d) an adjustable display time (or repetition delay time) passes; (e) the counter is cleared; (f) the count gate control is allowed to be triggered again; and so on. A typical circuit that provides such a sequence for frequency or period measurement is developed below.

The circuit of Fig. 3–211 contains a series of BCD counters such as in Fig. 3–201, memory registers such as in Fig. 3–179, and decoder and readout circuits such as discussed in Section 3–4.1. The counter is gated by the count control circuit of Fig. 3–210. When the count interval is over, $\bar{Q}_{FF\,B}$ goes from **1** to **0**, triggering the delay monostable. The output of the delay monostable is used to activate the transfer of information from the counting

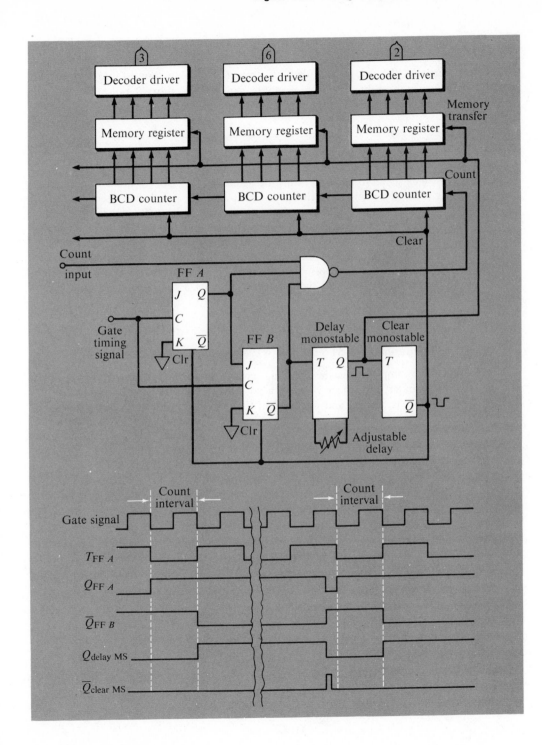

Fig. 3–211 Automatic cycling frequency or period meter.

registers to the memory registers. The duration of the delay monostable output pulse is adjusted for the desired display time or measurement repetition rate. At the end of the display time, the display monostable triggers the clear monostable. This monostable generates a short **0** pulse to clear the counting registers and the flip-flops in the gate control circuit.

Experiment 3–50   *Counter Input Gating*

The signal to a counter must be gated **ON** and **OFF** in order to control the counting interval. In this experiment a simple gate is first used to illustrate counting and frequency measurements with a manually controlled counting interval. When a repetitive oscillator is used to provide a more precise counting interval, a circuit is needed which prevents the gate from opening until the count has been read out and the next count measurement is to begin. The single pulse gate is wired to perform this function. The gate is then combined with the counter to provide a complete count gate control circuit.

Experiment 3–51   *Complete Frequency/Period Meter*

A complete decade counting circuit including an asynchronous BCD counter, a memory register, a decoder-driver, and a neon decimal display tube is studied and operated in this experiment. Several such circuits are used together to make a three- or four-digit decimal counter. A precise time base is added from a crystal oscillator and scaler to open and close the counting gate for a precise time interval. Then a complete period or frequency meter with automatic cycling is wired and tested.

Shift Registers

It is frequently convenient to be able to shift the information from one flip-flop to the next in the register. Registers in which this is possible are called **shift registers.** Shift registers are used for accepting binary data in serial form, for aligning decimal places for binary addition, and for shifting the multiplicand in binary multiplication.

Serial Entry.   A basic shift register circuit is shown in Fig. 3–212. Data are entered in the form of logic levels at input I. If the complement of I is available, it is connected to $\bar{I}$; if not, the inverting gate shown is used. When the clock input goes to **1**, the level at I is entered into the master section of FF A to appear at Q of FF A when the clock signal returns to **0**. Thus the information at I appears at A one clock pulse later. On the next clock pulse,

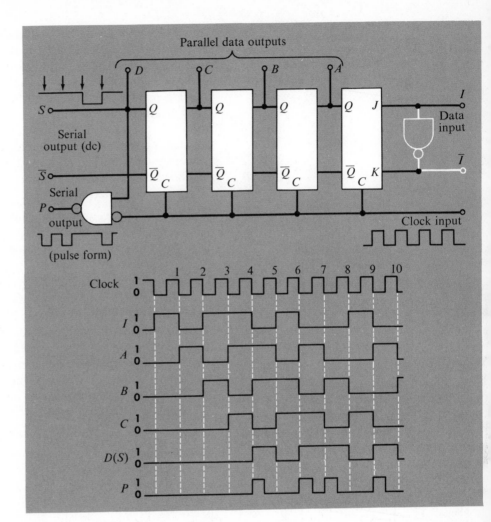

Fig. 3–212 Basic four-bit shift register.

the information at A will appear at B while the new information at I will be transferred to A. The waveform chart of Fig. 3–212 shows the response of the shift register to a hypothetical signal at I. Note how the information I is shifted to the next flip-flop output on each clock pulse, appearing at A one pulse later, at B two pulses later, and so on. The output at S has exactly the same form and sequence as that at I except that it appears four clock pulses later. A logic level train (serial-form binary signal), such as I, can be delayed by any desired number of clock pulses by using that number of flip-flops in the shift register circuit. If the clock pulses are stopped after the serial word has been loaded into the shift register, the serial data will remain stored. They may later be read out serially by simply resuming the clock pulses. If a pulse-form output is desired, it may be obtained by gating S with a signal derived from the clock source as shown.

Another application of the shift register is the conversion of data from serial to parallel form. Note that the first four levels at I in Fig. 3–212 are **1011** and that after four clock pulses the outputs $DCBA$ are **1011** in parallel form. If eight flip-flops were used, an eight-bit word could be read into the register in eight clock pulses and appear at the eight flip-flop outputs simultaneously. This application is most useful for converting serial digital information telemetered from a remote source to a parallel output suitable for printing or computer input.

Parallel Entry.   As in the case of the counting register, the shift register can be modified for parallel data entry (preset). A four-bit shift register with parallel data entry is shown in Fig. 3–213. When the enter input is pulsed with a **1** pulse, the logic levels at the data inputs are set into the corresponding flip-flops by the direct set and clear inputs. If the clock input is now pulsed, the data will be shifted one flip-flop to the left for each clock pulse. This circuit is used to align the decimal places of two binary numbers prior to addition or subtraction and to shift the multiplicand for each step in a binary multiplication. Another obvious application of Fig. 3–213 is as a parallel-to-serial information converter. Data entered at the parallel inputs will appear at S in serial form as the clock input is pulsed.

Integrated circuit versions of the direct-set parallel-input shift registers are available with up to eight bits or with two data inputs for each bit with

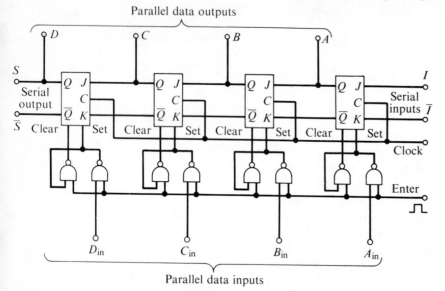

Fig. 3–213 Parallel input, parallel output shift register.

separate enter control lines. A particularly versatile parallel entry shift register available in IC form is shown in Fig. 3–214. All data enter the flip-flops through the *RS* inputs by clocking. The *RS* inputs have two-input multiplexers to determine whether the data entered are the shift data from the adjacent flip-flop or the parallel data input connections. This provides master-slave time isolation between the parallel inputs and outputs. The mode control input determines the state of the data input and clock input multiplexers. When the mode control is **1**, a **1 → 0** transition at the clock 2 input will cause a shift transfer. When it is **0**, a similar transition at the

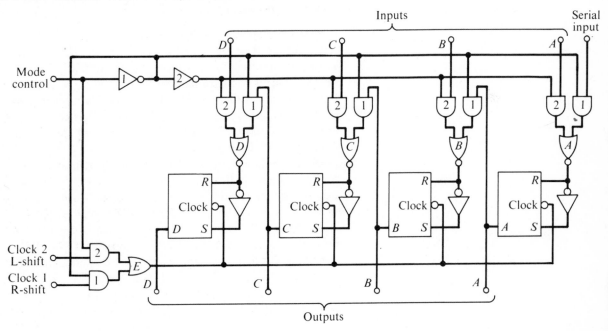

Fig. 3–214 Parallel transfer, shift transfer register.

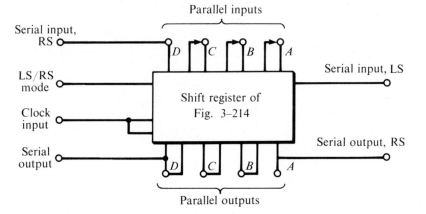

Fig. 3–215 Left shift/right shift register.

clock 1 input will cause a parallel input transfer. One application of this device is the **left shift/right shift** (LS/RS) **register** shown in Fig. 3–215. The left shift operation is performed by the internal shift connections when the mode control is **1**. The right shift is made by connecting each data output to the input to its right so that the parallel transfer operation is, in effect, a shift to the right. This type of application would not be possible with a direct set type of parallel entry. Many other ingenious applications in counting, generating, and data manipulation have been found for the versatile shift register. For information on these, the reader is referred to the bibliography at the end of this module.

Circulating Register.   When the shift register can be used as a serial in/ serial out memory register, it is sometimes desirable to be able to read the word out in serial form without losing it, that is, without destroying its storage. This is accomplished by entering each bit back into the data input I as it appears at the output S. If the clock input were pulsed continuously, the word would then circulate through the register continuously and appear at the serial output S every n clock pulses, where n is the number of flip-flops in the register.

Parallel data outputs are often omitted from the shift register designed for circulating applications. This means that only five or six external connections to the register are required (input, output, clock, and power) no matter how many flip-flops are used. For this reason, serial input/output shift registers for circulating storage applications were one of the first circuits developed for large-scale integration (LSI). LSI shift registers of over 1000-bit lengths are made by several manufacturers and are becoming an increasingly popular form of data storage.

A circuit for controlling a circulating register is shown in Fig. 3–216. A two-input multiplexer gate circuit is used. When the circulation control is

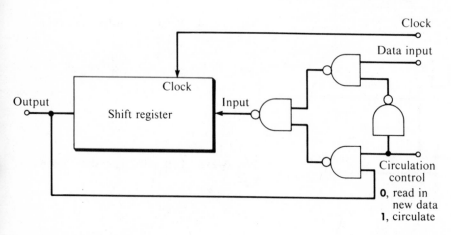

Fig. 3–216   Circulating register.

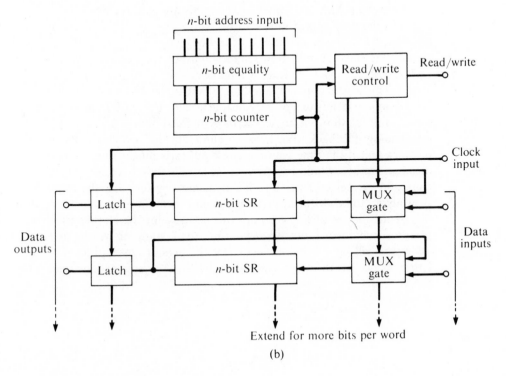

Clock

Write/recirculate
control

n = bit SR

MUX
gate

n = bit SR

MUX
gate

n = bit SR

MUX
gate

Parallel
data
outputs

Parallel
data
inputs

Extend for more bits per word

(a)

n-bit address input

n-bit equality

Read/write
control

Read/write

n-bit counter

Clock
input

Latch

n-bit SR

MUX
gate

Latch

n-bit SR

MUX
gate

Data
outputs

Data
inputs

Extend for more bits per word

(b)

Fig. 3–217 Circulating registers for parallel words: (a) automatic recycling; (b) addressable.

1, the register output is connected to the register input and the data input gate is CLOSED. When the circulation control goes to **0**, the data at the data input will be entered into the register. Additional circulating registers can be combined to provide a memory that is parallel by word and serial by address, as shown in Fig. 3–217a. An n-bit sequence of parallel data words can be written into this memory and then that same sequence can be reproduced indefinitely at the parallel data outputs. Such a memory is not a random access memory since the stored words appear in sequence. The time required for one complete cycle of the entire memory is n/f_s, where f_s is the data shift rate determined by the clock frequency. For example, a circulating memory made of 1024-bit shift registers and using a 1 MHz shift rate will produce all the stored words at its output once every 1.024 msec.

If it is desired to write or read specific words in a circulating register rather than the entire sequence, it is necessary to be able to identify each shift position or word address. This is done by adding a shift counter and address comparator as shown in Fig. 3–217b. When the counter and address input are equal, the desired word is at the SR output and is transferred into the output latches in the read mode. Since this same bit is the input bit on the next clock cycle, the data inputs will be active during that cycle in the write mode.

As in the RAM, MOS integrated circuits are used for the large capacity registers and they appear in two forms, static and dynamic. In the dynamic registers, the data are stored as a capacitor charge which must be refreshed to hold the data. A shift transfer reestablishes the normal data charge in each memory unit. Therefore, a minimum shift clock rate must be maintained. This is generally not less than 10 kHz. The maximum data shift rate for dynamic registers can be very fast, generally 1 to 5 MHz or more. In the static shift registers, at some sacrifice in maximum data rate, a charge-maintaining feedback transistor is automatically switched ON in each memory unit if the time between clock pulses is longer than is reliable for dynamic storage. Static registers are generally slower, of less capacity per chip, and greater cost per bit than dynamic registers. On the other hand, their ability to be clocked at any rate makes them more versatile in application. Thus the circulating memory can be used for the storage or retrieval of any specific word, but the delay time for reading or writing at random addresses will be a variable fraction of the memory cycle time. The circulating memory is therefore much slower for RAM applications than a true RAM.

Experiment 3–52   *Shift Registers*

A shift register is a chain of flip-flops connected so that the state of each flip-flop will be transferred to the next flip-flop in the chain when clocked. A four-bit serial input shift register is wired in the first part of this experiment and the shifting action of the circuit is observed as the input data are applied and the register is clocked. A four-bit parallel input shift register is next wired and tested. The data are entered into each flip-flop in the register simultaneously. Once they are entered, shifting occurs on each clock pulse. A four-bit circulating register is then wired and studied in the final part of the experiment.

3–4.4   DIGITAL-TO-ANALOG CONVERSION

A digital-to-analog converter (DAC), as the name suggests, is a circuit that converts a digital signal into an electrical analog quantity directly related to the digitally encoded number. The digital domain input signal may be in the count, serial, or parallel form, though the parallel form is the most common. The analog output signal is generally in the current or voltage domain. Since the input quantity is a number, the basis of all the conversion techniques is to convert the number to a corresponding number of units of current, voltage, or charge, and then sum these units in an analog summing circuit. Two analog summing circuits come to mind, both of which are used in DAC's: the OA integrator that sums charge, and the OA current follower that sums current. When the digital signal is in the count form, where each pulse is of equal weight, each pulse in the pulse train can be converted to a unit of charge that is accumulated in an integrator. A monostable multivibrator or the diode pump pulse generator is used to generate the charge units, which are integrated as shown in the integrating rate meter of Fig. 3–113. The integration period would, of course, be made to correspond to the transmission time of the count-form signal. The converted data are now in the voltage domain. Held as the charge on a capacitor, it will not remain stable indefinitely.

The summing of current, on the other hand, produces a stable output so long as the current inputs are stable. The standard approach for a current-summing converter is to convert the digital data to parallel form, generate a current of appropriate magnitude for each bit in the word that is a **1**, and sum the currents. The basic DAC then takes the form shown in Fig. 3–218. There is a current generator for, and controlled by, each bit of the input data word. When a current generator is turned **ON** by a **1** level input bit, it generates a current proportional to the weight value of its input bit. Com-

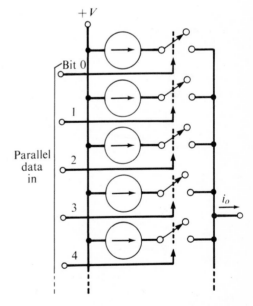

Fig. 3–218   Basic   digital-to-analog   converter.

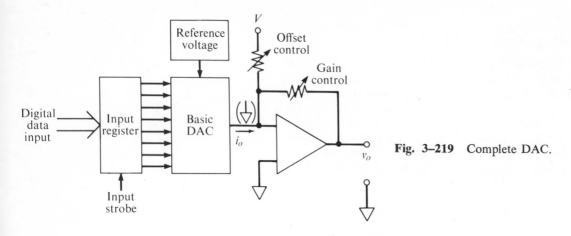

Fig. 3–219   Complete DAC.

bining the generated currents then produces a total current that is proportional to the digitally encoded number. This is by far the most popular form of the DAC. Two limitations, however, are immediately clear: (1) the digital data must be encoded in one of the weighted codes (where constant weight values can be assigned to all the bits and the encoded number is equal to the sum of the weights of the **1** level bits), and (2) a DAC designed for use with one code will not work with another.

The basic DAC shown in Fig. 3–218 is generally used in conjunction with one or more of the additional circuit functions shown in Fig. 3–219. The input register samples the digital data source at the appropriate times and holds the data in parallel digital form for a steady input to the DAC. The input register will take a different form, depending on the form of the input data. For count digital, it will be a counter; for serial digital, a shift register; and for parallel digital, a latch. If the input register is a counter or shift register, the DAC will follow the changes in the register contents as the digital data are being loaded into the register and will thus be in error during this time. Where this error is not acceptable, a latch register must be used between the conversion register and the DAC. The latch register is reloaded when the conversion is completed and no erroneous data are applied to the DAC.

The reference voltage is a voltage source used to supply power to the current generators. In many designs, the output current per bit weight is proportional to the reference voltage. For current summing a current follower circuit is almost always used to provide a virtual common summing point for the current generators and to convert the current signal to a proportional voltage. Offset and gain controls can be added to adjust the overall number-to-voltage transfer function as shown.

The following discussion of DAC devices and their operation begins with a description of the most commonly used digital input codes. Then several of the most popular methods for current generation and switching are introduced. A discussion of the limitations of DAC's in the light of manufacturers' specified characteristics serves to summarize this section.

Digital Input Codes

Commercial DAC's are made which accommodate the two weighted codes encountered in earlier discussions, the binary and the 1 2 4 8 BCD. The relations between the input word and the output signal for an eight-bit binary DAC and an eight-bit BCD DAC are given in Tables 3–21 and 3–22. The weighting factor for each bit is clearly seen in these tables.

Table 3–21 DAC output for eight-bit binary-coded input

Binary	Fraction full scale	v_o for 10 V f.s.
11111111	255/256	9.961 V
10000000	1/2	5.0
01000000	1/4	2.5
00100000	1/8	1.25
00010000	1/16	0.625
00001000	1/32	0.312
00000100	1/64	0.156
00000010	1/128	0.078
00000001	1/256	0.039
00000000	0	0

Table 3–22 DAC output for eight-bit BCD input

BCD	Percent full scale	v_o for 10 V f.s.
10011001	99	9.90 V
01010000	50	5.0
01000000	40	4.0
00100000	20	2.0
00010000	10	1.0
00001001	9	0.9
00000100	4	0.4
00000010	2	0.2
00000001	1	0.1
00000000	0	0

Looking at the binary table, we see that the most significant bit (MSB) has a weight of 1/2 full scale; the next most significant bit, 1/4 full scale; and so on. With an eight-bit word, the least significant bit (LSB) has a weight of 1/256 full scale, or 39 mV for 10 V full scale. This value corresponds to the smallest increment, or the **resolution,** of the DAC. Note that the maximum output voltage is the full scale voltage less 1 LSB. For a 10-bit binary converter, the LSB has a weight of 1/1024 or 9.77 mV/10 V. The BCD code uses the bit positions less efficiently, as we have seen. For an eight-bit BCD converter, the LSB (and thus the resolution) is 0.1 V/10 V, about 2.5 times larger than that of the binary. Comparing the resolutions of 12-bit binary and BCD converters, we have 2.44 mV/10 V for the binary vs. 10 mV/10 V for the BCD.

Bipolar Codes.   The binary and BCD codes described above are **unipolar,** in that the sign of the input number is assumed to be positive and the entire range of output voltage or current is of a single polarity. Digital control and computation systems use both positive and negative numbers by including the sign information in a **bipolar code.** When a signed number is converted to the analog domain, it is generally desired to have the signal polarity

Table 3–23   Bipolar binary codes

Number	Sign + magnitude	Offset binary	Two's complement	One's complement
+7	0111	1111	0111	0111
+6	0110	1110	0110	0110
+5	0101	1101	0101	0101
+4	0100	1100	0100	0100
+3	0011	1011	0011	0011
+2	0010	1010	0010	0010
+1	0001	1001	0001	0001
+0	0000	1000	0000	0000
−0	1000	1000	0000	1111
−1	1001	0111	1111	1110
−2	1010	0110	1110	1101
−3	1011	0101	1101	1100
−4	1000	0100	1100	1011
−5	1101	0011	1011	1010
−6	1110	0010	1010	1001
−7	1111	0001	1001	1000
−8		0000	1000	

indicate the sign of the number. A **bipolar DAC** will accept a signed or bipolar coded input and produce a bipolar output. There are several bipolar codes in current use. All of them use the MSB to indicate the sign. Four of the most common bipolar binary codes are illustrated in Table 3–23 for four-bit (three bits for magnitude plus one bit for sign) numbers. The **sign-and-magnitude code** is a direct solution to the problem. The MSB gives the sign (**0** = +, **1** = −) and the remaining bits the absolute magnitude in straight binary code. The most direct implementation of this code in a DAC is to use the MSB to control the direction of the currents generated by all the less significant bits. This is done fairly easily when the generation and switching circuits allow the reference voltage to have either polarity. The fact that there are two conditions that result in a zero output can be awkward in some applications. Note that no significant bits are lost by using the MSB as a sign bit. The sign bit doubles the range of the magnitude bits, which is just what the MSB does for the remaining bits in a straight binary number. Though the sign-and-magnitude code is not very popular for binary-coded signals, it is almost universally adopted for bipolar BCD signals.

Another simple bipolar code comes from the concept of dividing the succession of states of a binary number into two halves with the higher valued half (MSB = **1**) positive and the lower valued half (MSB = **0**) negative. The number at 1/2 full scale (1000 . . . 0) is arbitrarily assigned the zero value. The result is the **offset binary code.** It is very easy to implement with an ordinary DAC by adding a 1/2 full scale current offset to the current summing point so that when the DAC output is 0, the net output is −1/2 full scale; when the DAC output is +1/2 full scale, the net output is 0; and so on.

The bipolar code most often used in small digital computers is the **two's complement code** because of its convenience for computation. Positive numbers use a **0** MSB to indicate the sign and the normal binary code for the magnitude. The negative number code is formed by complementing (exchanging **1**'s for **0**'s and vice versa) the corresponding positive number and then adding **1**. Thus −3 is $\overline{0011} + 1 = 1100 + 1 = 1101$. (An additional sign change to return to +3 must follow the same process. Thus $\overline{1101} + 1 = 0010 + 1 = 0011$.) A comparison of the two's complement and offset binary codes reveals that they are the same except for the state of the MSB, which is exactly the opposite for the two cases. Thus a DAC designed for offset binary can be converted to two's complement by inverting the MSB between the data source and the DAC input.

In the **one's complement code,** the negative number code is obtained by just complementing the code of the corresponding positive number. This results in a rather awkward to implement code in which both extremes represent zero. Fortunately, this code is falling from popularity, but where encountered, one can often make do with a two's complement converter by applying a 1/2 LSB offset. Thus **0000** is $+1/2$, **1111** is $-1/2$, **0111** is $7\frac{1}{2}$, **1000** is $-7\frac{1}{2}$, etc. The 1/2 LSB error added to the $\pm 1/2$ LSB maximum error usually specified by the manufacturer results in a maximum 1 LSB error that is often tolerable.

From this discussion, we see that the unipolar binary DAC can be adapted to accommodate the offset binary code by a 1/2 scale output offset, the two's complement code by inverting the MSB and using the 1/2 scale offset, and, with some error, the one's complement code by an additional 1/2 LSB offset.

DAC Circuits

The conceptually simplest form of the DAC is the weighted resistor DAC shown in Fig. 3–220. The current generators are precision resistors in series with the reference voltage, V_R. The current generator controls are solid state analog switches driven in response to the logic levels at the digital inputs. Each current generator produces an appropriate fraction of the full scale current, I. The generated currents are summed by the OA current follower. Resistance values for a binary-coded converter are shown. The current generated by each bit is proportional to V_R and inversely proportional to the resistance. We shall call R the resistance necessary to generate the full scale current I. Therefore,

$$I = V_R/R. \tag{3-184}$$

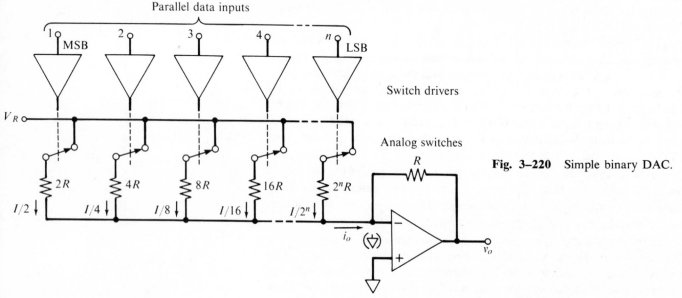

Fig. 3–220 Simple binary DAC.

Since the MSB is to generate a current of $I/2$, its resistance should be $2R$. The resistance for bit 2 is $4R$ for a current of $I/4$, and in general, the resistance is $2^n R$, where n is the number of the bit with number 1 as the MSB. The sum of the generated currents, i_o, is

$$i_o = I \left(\frac{a_1}{2} + \frac{a_2}{4} + \frac{a_3}{8} + \frac{a_4}{16} + \cdots + \frac{a_n}{2^n} \right), \qquad (3\text{--}185)$$

where the a's are the logic levels (**0** or **1**) of the binary bits. Equation (3–185) can also be written as

$$i_o = I \sum_{n=1}^{n} \frac{a_n}{2^n} = \frac{V_R}{R} \sum_{n=1}^{n} \frac{a_n}{2^n}. \qquad (3\text{--}186)$$

For the current follower, $v_o = -i_o R$, and so

$$v_o = -V_R \sum_{n=1}^{n} \frac{a_n}{2^n}. \qquad (3\text{--}187)$$

A V_R of 10 V and an I of 2 mA are common values. For these values, R is 5 kΩ. The resistor in the MSB generator is 10 kΩ, and in the LSB

Fig. 3–221 DAC with series-shunt voltage switches and attenuators for quad groups, binary or BCD.

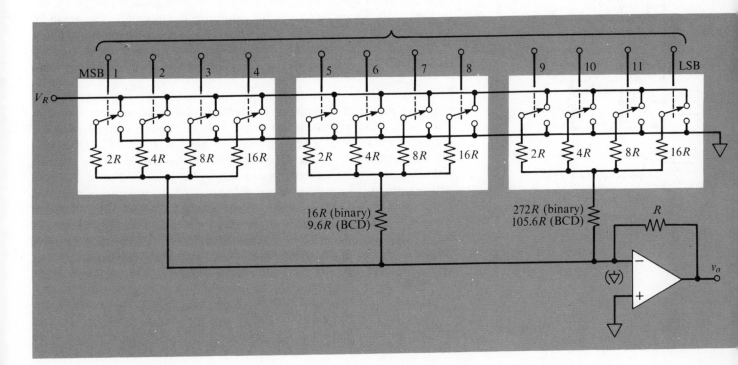

generator, $2^n \times 5$ kΩ. For an eight-bit converter, the LSB resistor is 1.28 MΩ, and for a 12-bit converter, 20.48 MΩ. The large range of resistance values required seriously limits the usefulness of this simple circuit. Not only are resistance tolerances hard to hold over this range, but the requirements for the analog switch ON and OFF resistances ($R_{ON} \ll 10$ kΩ and $R_{OFF} \geqslant 20$ MΩ) are extreme.

The switch and resistance requirements are eased greatly by the circuit of Fig. 3–221. The series-shunt switch reduces the resistance range requirement of the switch as explained in Section 2–3 of Module 2. In addition, each group of four generators is the same as the most significant four bits, with the output current from the less significant groups appropriately attenuated by a series resistor. As the diagram shows, the range of resistances required is greatly reduced. To calculate the series attenuator resistor required, consider the group of four generators to be a voltage generator with an equivalent source resistance of $(16/15)R$. (Note that the constant source resistance would not exist if the resistors were not always connected to common or V_R.) For binary coding, the current from bits 5–8 should be $1/16$ the current from bits 1–4. The total resistance should then be 16 times greater, requiring 15 times the generator source resistance to be added. Since $15 \times (16/15)R = 16R$, that is the appropriate value. If bits 5–8 represent the next most significant decimal digit in a BCD coded input, the series attenuator resistor can be chosen to give $1/10$ the current from the most significant group. The result of that calculation, $2 \times (16/15)R$, is $9.6R$. Similarly, another group of four bits can be added to make a 12-bit binary or three-digit BCD DAC as shown.

Ladder Circuits.   Another technique for reducing the required range of resistance values still farther is the **ladder network** of resistors shown in Fig. 3–222. The expression for i_o can best be found by considering the current sources one at a time. If only the MSB is **1**, $i_1 = V_R/2R = I/2$. All this current goes to the summing point of the OA, since the effective resistance to ground at that point is negligible compared to that of the network upwards from point $N1$. If only bit 2 is **1**, a current i_2 is generated which splits at $N2$, part going to $N3$ and part going to $N1$ to be summed. The net resistance to common of the R-$2R$ network upwards of point $N2$ is $2R$. This is a characteristic of the network. The resistance to common downwards from $N2$ is R. The combined resistance to common at $N2$ is thus $2R/3$. The current i_2 is then

$$i_2 = V_R/(2R + 2R/3) = 3V_R/8R. \tag{3–188}$$

The current i_2 is split, $2/3$ going to $N1$ and $1/3$ to $N2$. Thus the current

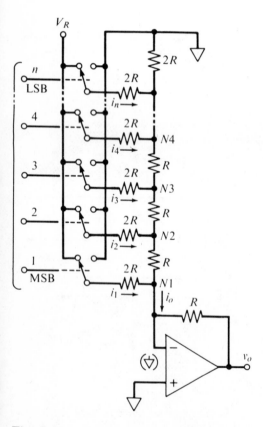

Fig. 3–222   Ladder network binary DAC.

from bit 2 to be summed is

$$\text{bit 2 current} = \frac{2}{3}\left(\frac{3V_R}{8R}\right) = \frac{V_R}{4R} = \frac{I}{4}. \qquad (3\text{--}189)$$

If only bit 3 is **1**, the resistance upwards from $N3$ is $2R$ and downwards is $5R/3$. The current $i_3 = 11V_R/32R$. This current splits at $N3$ with $3V_R/16R$ going toward $N2$. At $N2$, this current splits again with $2/3$ or $V_R/8R$ or $I/8$ going to $N1$ to be summed. The remaining bits may be solved similarly to demonstrate that i_o and v_o for this DAC follow Eqs. (3–185) through (3–187) exactly. The ladder network is very popular for binary DAC's, particularly hybrids, and IC's because of the use of only two resistance values despite the larger number of resistors required. Note that the ladder network will not work without the series-shunt form of switching. For a noninverting output, point $N1$ could instead be connected to a voltage follower input to provide a full scale output of V_R.

Current Switching. The switches in the DAC's discussed above are applied in the **voltage-switching mode.** That is, they are used to determine what voltage will be applied to the current-generating resistance network. The voltage changes in the switched circuits and the voltage across the open switch contacts approach V_R V. This requires large switch drive voltages and reduces switching speed. Since currents are being summed, the series-shunt **current mode switch** can be used to advantage. (Analog switches and switching circuits are described in detail in Section 2–3 of Module 2.)

A current switching ladder network DAC is shown in Fig. 3–223. The switches direct the generated currents to the summing point or to the common. All three contacts of each DPST switch are at the common voltage. Since there is no voltage change and no contact voltage, the switch drive signal can be smaller and the switching faster. Very often FET switches are used in this application to allow V_R to have either polarity. The currents i_1–i_n are constant. The value of i_1 is clearly $V_R/2R = I/2$. The resistance to common at $N2$ is the $2R$ that generates i_2 in parallel with the $2R$ of the network above $N2$. The voltage at $N2$ is thus $V_R/2$, resulting in a current $i_2 = I/4$. Similarly, the voltage at $N3$ is $V_R/4$ and the current $i_3 = I/8$, and so on up the ladder. The current-switching technique is most often found in high speed and miniaturized designs.

Summing and Output Amplifiers. The DAC output amplifier serves two purposes: summing the currents and converting the i_o current to a voltage signal. As the switching speed of DAC's has increased, the output OA response is often the limiting factor in DAC response speed. For this reason, many DAC's come with terminals to connect the i_o current output

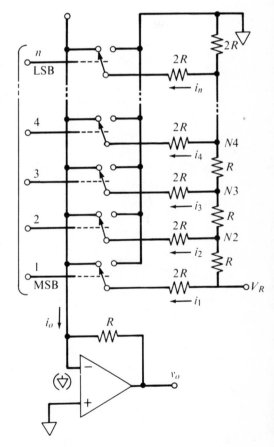

Fig. 3–223 Ladder network DAC with current switching.

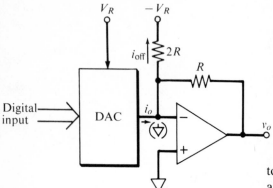

Fig. 3–224   DAC with output amplifier.

to the summing amplifier externally. This allows the substitution of a faster amplifier if desired. To obtain the fastest response, i_o is connected directly to a load resistor. The load resistor is chosen to obtain a full scale voltage of 0.1 V or less. For $I = 2$ mA, a load resistor of 50 Ω is ideal. If the DAC is to drive a high speed scope display, the i_o output is connected to the scope input with a terminated 50 Ω coaxial cable, which serves as both connection and load.

The output amplifier can serve several other useful functions, including voltage output to 10 V full scale, offsetting for bipolar codes using unipolar DAC's, and isolation between the DAC current generators and the load circuit. These functions of the summing OA are illustrated in Fig. 3–224. The virtual common of the current follower summing point provides a near-perfect load for the DAC current source. Assume first that there is no offset. Since the full scale current range from the DAC is 0 to V_R/R, the current follower output v_o range will be 0 to $-V_R$. Clearly, other feedback resistors can be chosen to change the range. Now, if offset is added to accommodate the offset binary or two's complement codes, the offset current should be exactly half the range. For the circuit shown, the range of v_o will be $+V_R/2$ to $-V_R/2$. Since the OA will generally go to V_R in both directions, the range can now be increased to $+V_R$ to $-V_R$ by making the feedback resistor $2R$. Obviously, a number of useful combinations are possible.

Multiplying DAC's.   A multiplying DAC is one for which the reference voltage is supplied externally and for which the output current is accurately proportional to the value of V_R supplied. Frequently, a follower amplifier is included in the package so that the V_R source does not have to supply the ladder current. As seen from Eq. (3–187), the output voltage is equal to the product of V_R and the input digital number. The multiplying DAC is thus a multiplier with one digital and one analog domain input and an analog domain output. It is very convenient for the digital control of analog signals, controlled-gain amplifiers, and so on. A full four-quadrant multiplier accepts bipolar coded digital data and a value of V_R of either polarity and produces an output signal of appropriate polarity.

DAC Characteristics

As a summary to this discussion of DAC's, some of the critical characteristics and limitations of DAC devices will be described in the terms used by manufacturers in their specifications.

Resolution.   The resolution of a DAC is the fraction of its full scale range represented by the smallest possible change of the input number. A 10-bit DAC should have 2^{10} or 1024 distinct output states and thus it resolves its range to 1 part in 1024. The resolution is equal to the weight of the LSB.

Accuracy.   The accuracy of a converter is measured by the difference between the expected and measured output voltage (or current) in terms of the change caused by changing the LSB. Most converters are specified to be accurate to at least $\pm 1/2$ LSB.

Linearity.   If the analog output is plotted vs. increasing digital code, the result should be a straight line. Linearity does not imply absolute accuracy, since the line may not go through the zero axis or the full scale value. Nonlinearity should also be no worse than $\pm 1/2$ LSB even if absolute accuracy is unimportant in a particular application.

Monotonicity.   A particularly aggravating form of nonlinearity in some applications is a lack of monotonicity, that is, a momentary reversal in the expected direction of change. For instance, if on increasing the code 1 LSB, the output decreases 1/4 LSB instead and then increases $2\frac{1}{4}$ LSB on the next increment, the DAC is nonmonotonic. If the DAC is part of a control loop (as in an A-to-D converter), the reversal of transfer function slope can cause a control system hangup at that point.

Zero Offset.   The output of a DAC should be zero for an input code of zero. This is generally true within 1/2 LSB for unipolar DAC's, but for bipolar DAC's an error in the offset circuit can cause considerable zero offset.

Stability.   The stability of a DAC is the constancy of its full scale output with age and with temperature and power supply variations. Each cause of instability is usually specified separately. Temperature is generally the most important since it affects the reference source, the resistance values, and the switch resistances and offsets.

Output Impedance.   This is normally very low because of the OA output amplifier. However, the amplifier often has little current to spare after satisfying its full scale feedback requirements.

Settling Time.   The response speed of a DAC is the time required for the output signal to settle within 1/2 LSB of its final value after a given change in input code (usually full scale). As mentioned above, this is generally limited by the output amplifier response.

Glitches. When new digital data are applied to a DAC, not all the switches change simultaneously. During the time from the first change to the last, erroneous output signals can be generated. These appear as transients or spikes, called glitches, in the output quantity. Where glitches cause trouble in high speed systems, they can be removed by using a fast sample-and-hold circuit at the output, putting it in the hold mode only during the glitch-producing transition period.

Experiment 3–53 *Digital-to-Analog Converters*

This experiment demonstrates several types of digital-to-analog conversion techniques. A three-decade weighted-resistor DAC is first constructed. The digital output from a switch register is converted to an analog voltage and recorded on a strip chart recorder. An OA staircase integrator is then constructed and used as a serial DAC. In the final part of the experiment a commercial 10-bit DAC is tested by applying digital signals from a switch register and measuring the output voltage.

3–4.5   ANALOG-TO-DIGITAL CONVERSION

An analog-to-digital converter (ADC) is sometimes called a quantizer because it determines the number of units, quanta, or increments that comprise the measured quantity. There are two general methods by which the determination of the number of quanta can be accomplished: counting the quanta, and matching the unknown quantity with a quantity of a known and variable number of quanta. We shall call analog-to-digital converters that employ the first method **quantum counting systems;** and those using the second method, **digital servo systems.** Several examples of each technique are described in this section.

Quantum Counting Systems

Of the three basic electrical analog domains—current, voltage, and charge —the easiest to store, count, and quantize is charge. For this reason all the currently favored quantum counting systems count units of charge. The measurement of current then becomes the measurement of the number of charge units per time period. This can be done by accumulating the charge (integrating the current) for a given time and then counting the charge units accumulated (dual-slope method). Or it can be done by keeping the rate of a standard charge unit generator at a charge generation rate equal to the measured current, while counting the number of charge units generated in a given time (current-to-frequency-to-digital and charge balance methods).

Voltage is often measured by using a resistor to convert it to a current and then applying one of the above techniques. In fact, most of the above current-to-digital converters come with an input resistor built in and are sold as voltage-to-digital converters. Another voltage measurement technique involves counting the number of standard charge units that must be applied to a capacitor to make its voltage equal to the unknown (ramp method). There are two techniques used for the standard charge unit generating circuit. One is a precision pulse generator that produces a reproducible quantity of charge in response to a trigger input signal. The other is a timed constant current source in which a charge unit is produced during each increment of time. The charge units are then counted by counting time increments. All the above-mentioned quantum counting and charge unit generating techniques are illustrated in the quantum counting ADC systems described below.

Ramp ADC.   In the ramp ADC, charge units from a standard charge unit generator are integrated on a capacitor until the capacitor voltage equals the unknown voltage. The number of standard charge units required is proportional to the unknown voltage. The block diagram of a ramp ADC is shown in Fig. 3–225. A clock oscillator is connected directly to a counter that has the desired output code. An additional flip-flop is triggered by the most significant bit on the counter so that it will produce **1** and **0** on alternate cycles of the quantum counter. The cycle is shown beginning with this flip-flop in the reset condition. Switch S of the integrator is closed and the integrator output, v_C, is 0 V. When the counter completes the count through its modulus, it starts again at zero and the reset line goes to **0**, opening the integrating switch. The units of charge q_u now accumulating on the capacitor are equal to the charging current times the time of one count, or

$$q_u = \frac{V_R}{R}\, t_c = \frac{V_R}{Rf_c}, \qquad (3\text{–}190)$$

where t_c and f_c are the clock oscillator period and frequency respectively. The voltage across the capacitor after n counts is

$$v_C = \frac{nq_u}{C} = \frac{nV_R}{RCf_c}. \qquad (3\text{–}191)$$

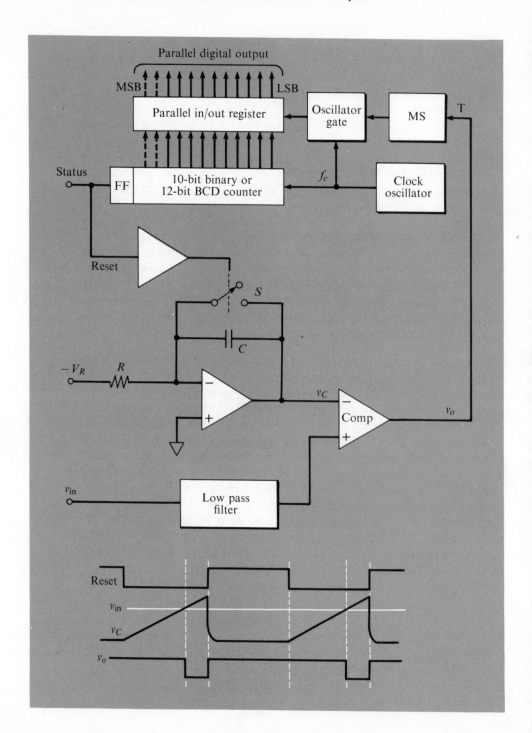

Fig. 3–225 Ramp-type ADC.

When v_C first exceeds v_{in}, the comparator output, v_o, goes to **0**, triggering the monostable, which will hold the oscillator gate open long enough for one clock pulse. This pulse transfers the data in the counter at count n_r to the latch register for readout. Substituting n_r and v_{in} for n and v_C in Eq. (3–191) and solving for n_r, we obtain

$$n_r = v_{in}RCf_c/V_R. \qquad (3\text{–}192)$$

Thus the number readout is proportional to the input voltage. The full scale input voltage is equal to the maximum value of v_C, or

$$V_{fs} = \frac{n_m V_R}{RCf_c}, \qquad (3\text{–}193)$$

where n_m is the counter modulus.

When the quantum counter reaches its modulus, the reset flip-flop is set and switch S is closed, resetting the integrator.

With the fast counters and comparators currently available, the ramp ADC can be fairly fast. A three-digit BCD conversion will be made every 2000 clock pulses, which, at $f_c = 10$ MHz, is 200 μsec per conversion or 5000 conversions per second. The main weakness of the ramp converter is its dependence on the stability of R, C, f_c, and V_R. Because of its great simplicity, it should be an attractive technique for conversion accuracies of the order of 0.1 to 1%.

Dual-Slope ADC.   In the dual-slope technique, the input quantity is converted to a charge by integration over a set period and then the quantity of charge is determined by counting the number of charge units required to discharge the integrator. A practical dual-slope ADC is shown in Fig. 3–226. Before the conversion cycle begins, the counter is at the count of the previous conversion, and the S signal is **1**, holding the integrator input switch in the V_R position. The positive reference voltage will drive the integrator output, v_C, negative, but this is clamped at -0.6 V by the diode. The comparator output, v_o, is **0**, which holds the gate between the oscillator and counter **CLOSED**. The conversion cycle is begun by triggering the monostable clear-pulse generator. The clearing of the counter makes S a **0**, causing the integrator input to be connected to the input voltage inverting amplifier. The positive input voltage, inverted, drives the integrator output in the positive direction. As v_C crosses zero, v_o becomes **1** and the counter input gate **OPENS**.

The oscillator and counter determine the input signal integrating time. In the simple converters usually depicted in textbooks, the integration period is one complete cycle of the counter, so that it is at zero when the counting of the accumulated charge units begins. In the waveform chart of Fig. 3–226, this is shown by the dashed line for v_C. To count the charge

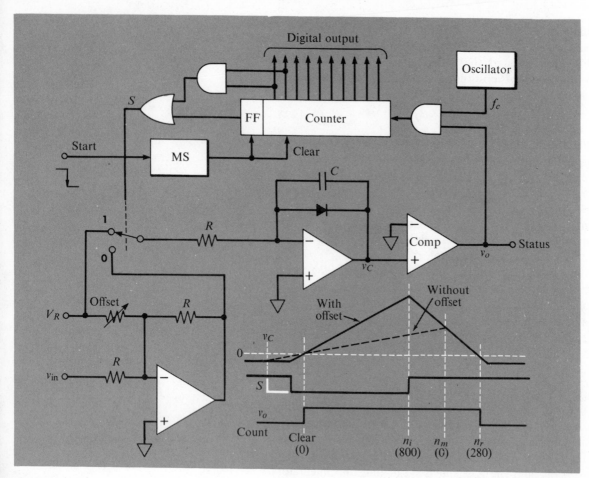

Fig. 3–226   Dual-slope ADC.

units, the integrator input switch is put in the V_R position and the number of clock cycles required to discharge the integrator capacitor with the constant input current is counted. When the integrator output again crosses the comparator threshold, the comparator output becomes **0** and the counting gate CLOSES, completing the measurement. The converted data remain at the parallel output of the counting register as long as v_o is at the **0** level. The dual-slope technique gets its name from the shape of the v_C waveform.

The converter without offset as described will not work with a zero input signal because the integrator would never go positive through the comparator threshold. Thus it is necessary to add a positive offset to the input signal to ensure a threshold crossing even when v_{in} is zero. The charge units accumulated due to the offset signal need to be subtracted from the final count. This is done simply by starting the charge unit count before the end of the first counter cycle as shown in the solid line of the v_c waveform, so that at the time the counter clears, the charge due to the input offset has been subtracted. Now the counter is at zero when the charge remaining on the integrator is due to v_{in} alone. The inversion of S prior to the end of the first counter cycle is accomplished by the S gating circuit and one or two of the most significant bits of the counter. If a BCD counter is used, the MSB goes to **1** on 8 (80% of the counter cycle) to make S a **1**. At the end of the first counter cycle, the extra flip-flop becomes **1**, holding S at **1** through the remainder of the measurement. If a binary counter is used, the two MSB's are used to bring S to **1** at 75% of the counter cycle.

We will now obtain an equation for n_r so that the critical components and quantities in the converter can be identified. At the end of the integration period, the charge on capacitor C is Cv_C, which is

$$Cv_C = \frac{v_{in} + kV_R}{R} \cdot \frac{n_i}{f_c}, \qquad (3\text{-}194)$$

where kV_R is the offset voltage gain. Each charge unit q_u counted is V_R/Rf_c as given by Eq. (3-190). The number of charge units counted is $n_m - n_i$ for the offset and n_r for v_{in}; hence the total charge unit count equal to Cv_C is

$$Cv_C = (n_m - n_i + n_r)q_u = (n_m - n_i + n_r)\frac{V_R}{Rf_c}. \qquad (3\text{-}195)$$

Equations (3-194) and (3-195) can be combined to eliminate Cv_C:

$$\frac{(v_{in} + kV_R)n_i}{Rf_c} = \frac{(n_m - n_i + n_r)V_R}{Rf_c}. \qquad (3\text{-}196)$$

Thus we see that C, R, and f_c, which affected the conversion relationship for the ramp ADC, cancel out in the relationship for the dual-slope ADC. Slow drift in the comparator threshold voltage also has no effect. The reason is, of course, that the reference and input signals are integrated by the same integrators, timed by the same clock oscillator, and referenced to the same comparator threshold voltage. When Eq. (3-196) is solved for n_r,

$$n_r = n_i\frac{v_{in}}{V_R} + n_i(k + 1) - n_m. \qquad (3\text{-}197)$$

When the offset factor k is adjusted so that $n_i(k + 1) - n_m = 0$,

$$n_r = n_i\frac{v_{in}}{V_R} \quad \text{when} \quad k = \frac{n_m - n_i}{n_i}, \qquad (3\text{-}198)$$

showing the result to be a simple ratio of v_{in} to V_R. If $n_m = 1000$ and $n_i = 800$, $k = 1/4$ and the offset resistor in Fig. 3–226 is $4R$. An offset binary conversion could also be performed if k were adjusted so that $n_r = n_m/2$ when $v_{in} = 0$. Imposing this condition on Eq. (3–197), we obtain $k = 3n_m/(2n_i) - 1$. For a binary counter, $n_i = 3n_m/4$; hence $k = 1$ and the offset resistor $= R$.

The dual-slope converter is widely used in digital multimeters and digital panel meters because of its accuracy and simplicity. It offers the further advantage of integrating the input signal over the period n_i/f_c. This integrating action is sometimes very helpful in reducing output errors due to input signal noise. (See Module 4 for a discussion of signal integration.) The dual-slope ADC is thus one of a class of ADC's referred to as the **integrating-type ADC.**

Voltage-to-Frequency-to-Digital ADC. The concept of converting an analog domain signal to the frequency domain was introduced in Section 3–3. Obviously one could combine a voltage-to-frequency converter with a digital frequency meter (discussed in Sections 3–3 and 3–4.3) to obtain a voltage-to-digital conversion. This combination is discussed briefly here to place it in the perspective of the general ADC concepts and to prepare for the discussion of the charge-balancing ADC.

A block diagram of a voltage-to-frequency converter is shown in Fig. 3–227. The input voltage is converted to a current (charge units per time) and integrated on the capacitor. When v_C goes negative, the comparator output v_o becomes **1**, causing the charge unit generator to add a unit of charge, opposite in sign to i_{in}, to the summing point of the integrator. This drives v_C positive across the comparator threshold, making v_o a **0** again. The integrator output v_C is thus kept very near 0 V as charge units are generated at a rate that exactly compensates for the input current. The input current is v_{in}/R and the average current from the charge

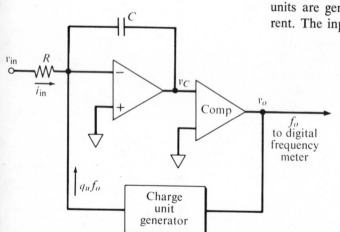

Fig. 3–227   Voltage-to-frequency-to-digital ADC.

unit generator is $q_u f_o$, where q_u is the charge unit generated and f_o is the frequency of alternation of v_o. Thus

$$v_{\text{in}}/R = q_u f_o$$

and

$$f_o = v_{\text{in}}/q_u R. \tag{3–199}$$

The accuracy of the voltage-to-frequency conversion is clearly dependent on the constancy of q_u. Because of the difficulty of designing a charge generator for which the output charge is truly independent of frequency, temperature, aging, etc., the voltage-to-frequency technique fell from favor as other methods were developed. However, recent improvements in integrated circuit switching and timing circuits have made the voltage-to-frequency conversion an attractive approach once again. Several voltage-controlled oscillators available in IC form are linear enough to form the basis of ADC's in the 0.1 to 1% accuracy category. If a second oscillator, with matching temperature coefficient, is used to time the frequency meter counting gate, the temperature drift of frequency is compensated.

The voltage-to-frequency-to-digital technique offers several advantages. It is an integrating technique in which the integration period is equal to the frequency meter counting gate period. The resolution of the conversion can be varied by simply varying the gate time of the frequency meter. For the conversion of voltages generated at a point remote from the display, the location of the voltage-to-frequency converter at the source and the frequency meter at the display offers convenient single-line frequency domain transmission.

Charge-Balance ADC. The charge-balance ADC is conceptually similar to the voltage-to-frequency-to-digital ADC, except that the charge generator is of the timed constant current variety. The block diagram is shown in Fig. 3–228. The input voltage is converted to a current and integrated on C. As v_C goes negative the comparator output v_o goes to logic **1**. A **1** at J and **0** at K of the flip-flop allows Q to go to **1** on the next clock pulse, which **CLOSES** switch S. While Q is **1**, a current $-V_R/R_R$ equal to the full scale input current is applied to the integrator to reverse the direction of charge. When v_C crosses the comparator threshold going positive, v_o goes to **0**, making J a **0** and K a **1**, so that Q will go to **0** on the next clock pulse. In this way, charge is added for an exact number of clock pulses in an amount to exactly offset (within one charge unit) the input current. Two counters are used: one to count the total clock pulses n_m in a measurement, and the other to count the number of clock pulses during which the reference charge unit generator is on. In the circuit of Fig. 3–228, the oscillator and 10-bit binary counter determine the total clock pulses per

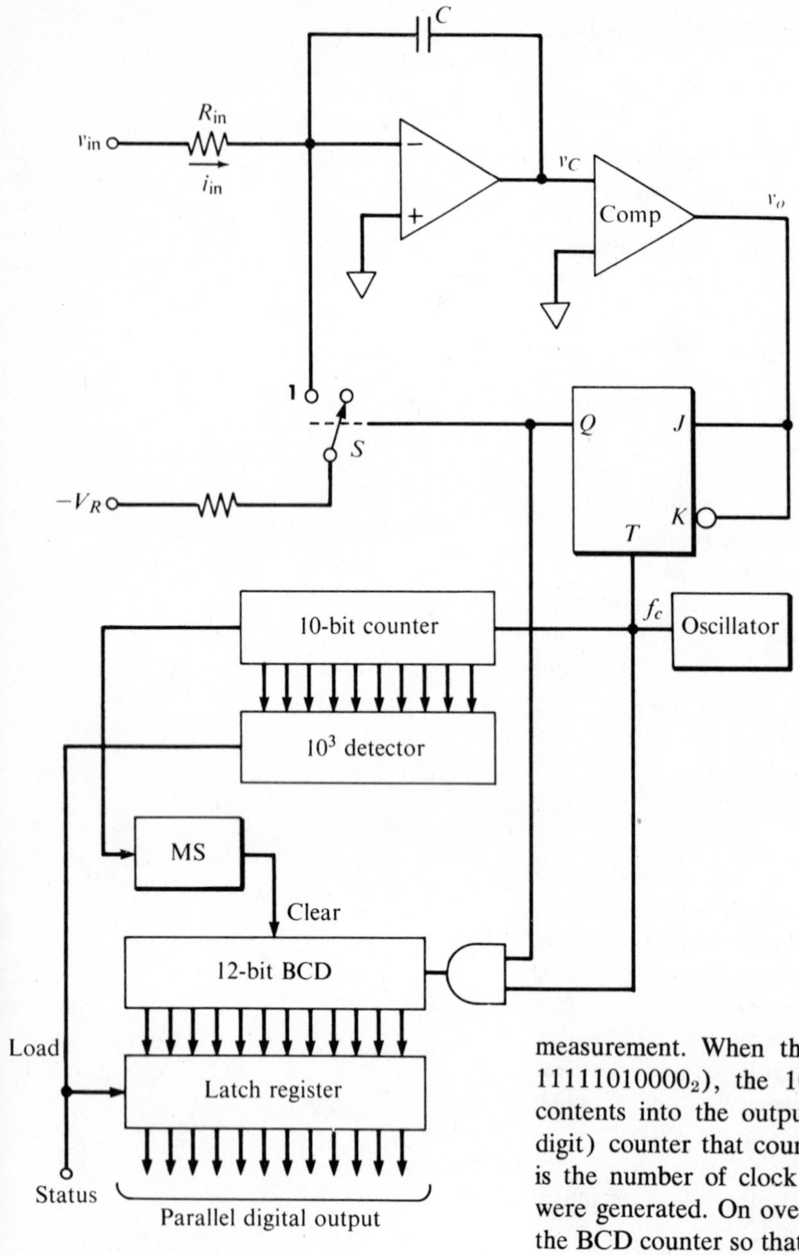

Fig. 3–228 Charge-balance ADC.

measurement. When the binary counter reaches a value of 1000_{10} (or 11111010000_2), the 10^3 detector causes the other counter to load its contents into the output latch register. This counter is a 12-bit (three-digit) counter that counts the clock pulses when Q is **1**. Thus the output is the number of clock pulses n_r in 1000 during which the charge units were generated. On overflow, the 10-bit counter generates a reset pulse to the BCD counter so that both are cleared for the next cycle.

The total charge generated by the input signal q_{in} is

$$q_{in} = \frac{v_{in}}{R_{in}} \cdot \frac{n_m}{f_c}. \qquad (3\text{--}200)$$

Since the circuit is set up to generate charge units as needed to balance the input signal charge,

$$n_r q_u = q_{in}. \tag{3-201}$$

Substituting for $q_u = V_R/(R_R f_c)$ from Eq. (3-190) and for q_{in} from Eq. (3-200) in Eq. (3-201), we obtain

$$\frac{n_r V_R}{R_R f_c} = \frac{v_{in}}{R_{in}} \cdot \frac{n_m}{f_c},$$

which can be solved for n_r to give

$$n_r = n_m \cdot \frac{v_{in}}{V_R} \cdot \frac{R_R}{R_{in}}. \tag{3-202}$$

This shows the output reading to be dependent on the stability of V_R and the ratio R_R/R_{in}. If $R_R = R_{in}$, the full scale input voltage is V_R.

The charge-balancing technique integrates over the entire measurement period. With the current-source switching shown, it can be quite fast for a counting type ADC.

Digital Servo ADC's

Many null balance conversion systems operate by placing the inverse of the desired function in the feedback loop of a control system. In the digital servo ADC, a DAC is used to provide an analog signal to be compared with the signal that is to be measured. The digital input to the DAC is changed in a direction determined by the results of the comparison until the input signal and the DAC output are equal. The digital servo ADC is shown in block form in Fig. 3–229. Upon the application of a start signal, the register sequence controller alters the register contents until the DAC output voltage is within 1 LSB of the input voltage. The direction of change is determined by the comparator output. The rate of change is determined by the clock. This ADC is called a digital servo because it is a servo system in which the feedback information is in the digital domain. There are several types of digital servo ADC's based on different digital registers and methods for adjusting their contents. Three types of digital ADC's predominate: the staircase, the tracking, and the successive approximations. These three ADC types are described below.

Staircase ADC. This is the simplest servo digital ADC. The register is a counter that counts clock pulses, increasing the DAC output by 1-LSB increments. It is the "staircase" appearance of the DAC output waveform that gives this converter its name. When the DAC output is equal to v_{in}, the counter stops counting. The sequence controller is simply the counting gate of Fig. 3–210, in which the start signal clears the counter and opens the gate and the comparator output closes it. The gate open signal indicates that

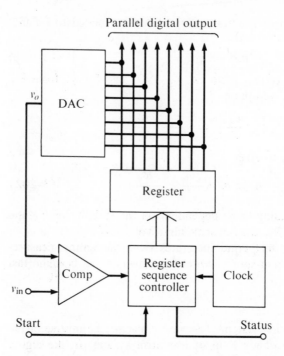

Parallel digital output

Fig. 3–229   Digital servo ADC.

the converter is busy and the output is not correct. In another form, the clock is connected directly to the counter and a data latch is used at the digital output. The edge of the comparator output is used to trigger the latch as in the ramp converter of Fig. 3–225. Such a converter recycles automatically at a regular interval and has a continuous readout. With the advent of DAC's in IC form, this is an extremely simple type of converter to build. It is capable of accuracy to about 0.1% with a 10-bit DAC. Using a 10 MHz clock, a conversion can be made every 100 μsec or 10,000 conversions per second. It is not an integrating type of converter and it is the slowest of the three types of servo digital ADC's. The range of input voltage and the output code are obviously determined by the DAC.

Tracking ADC.   The true digital servo should be able to adjust the output in increments in either direction in order to follow or "track" the input quantity changes. This is achieved within the block diagram of Fig. 3–229 by using an up-down counter for the register. The clock is connected directly to the counter input and the comparator output simply controls the count direction. This too, is a very simple ADC to build, with the advantages of a continuous output and updating at the clock rate. If a 10-bit DAC with a 10 V output range and a 10 MHz clock is used, the converter will follow input voltage changes as fast as 10^5 V per sec.

Successive Approximations ADC. The major variation in types of digital servo ADC's is the method by which the number in the register is adjusted to give a DAC output equal to the input. In the staircase type, each numerical code is tried in order until the correct one appears. This is a simple but not very efficient way of going about a search for the right number. A much more efficient way is to divide the range into a small number of fields (usually two, but sometimes four) to identify the field with the desired number in it. Then that field is divided into smaller segments, and so on until the final result is determined. This procedure is illustrated in Fig. 3–230. The register is first cleared, giving a 0 V output for v_o. Then the MSB only is made a **1** for a test period (marked T). This puts v_o at 1/2 scale and the comparator output indicates whether v_o is too high or too low. In effect, this tests whether v_{in} is in the upper or lower half of the range. Since the test shows v_{in} to be in the upper half, the MSB is a **1** and that is kept in the register during the posting interval (marked P). Next the upper half of the range is divided in two and tested by making the next most significant bit a **1**. This test shows v_{in} to be in the lower quarter and the **1** is replaced by a **0** during the posting period. Next the quarter between 1/2 and 3/4 is tested and v_{in} found to be in the upper eighth; therefore, the **1** is kept. Now the eighth between 5/8 and 3/4 is tested and v_{in} is found in the lower sixteenth; consequently the test **1** level is removed. This process is continued until the LSB of the converter has been posted.

This procedure requires only one clock cycle per bit of conversion. A 10-bit converter with a 10 MHz clock can complete a conversion every microsecond. However, this requires a DAC and comparator that can settle to within 1/2 LSB from a half-scale step input in 0.5 μsec—in other words, an extraordinary combination. A more common conversion time for 10 bits would be 10–20 μsec—still very fast. The register sequence controller is often a shift register with the test bit circulating and the decision on whether to load the **1** or clear the bit made by the comparator output. A requirement of this converter is that the input voltage remain absolutely constant during the conversion time. If it does not, errors in the more significant bit tests can be made. An analog sample-and-hold circuit is often used to acquire the voltage to be converted and hold it constant for the successive approximations conversion.

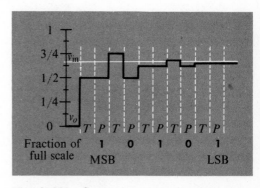

Fig. 3–230 Successive approximations search.

Bipolar ADC's

In the discussion of ADC types above, some could be made to be bipolar and some are definitely limited to unipolar input signals. The servo digital ADC's can be bipolar if bipolar DAC's are used. Two general techniques that can be used to adapt unipolar ADC's to bipolar input signals are shown in Fig. 3–231. The first (Fig. 2–231a) is best suited to the sign-and-

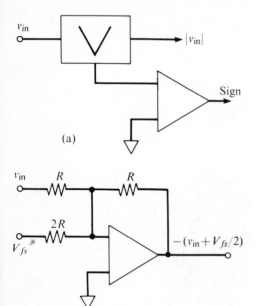

v_{in} → |v_{in}|

Sign

(a)

v_{in} R R

$2R$

V_{fs}

$-(v_{in} + V_{fs}/2)$

(b)

Fig. 3–231 Input circuits for unipolar ADC's to accept bipolar signals: (a) sign and magnitude; (b) offset binary.

magnitude form of bipolar encoding. An absolute value circuit such as that of Fig. 3–31 is used to provide a unipolar signal regardless of the input polarity. A comparator is connected to the inverting OA output to provide information on the sign of v_{in}.

The other technique, shown in Fig. 3–231b, is simply a 1/2 full scale offset so that v_{in} plus the offset will always be a unipolar signal. When used with a straight binary ADC, the output will be in offset binary form.

Experiment 3–54   *Analog-to-Digital Converters*

The conversion of an analog voltage to a digital word is performed in this experiment by several techniques. First the staircase ramp generator of Experiment 3–53 is used in a voltage comparison method, and then a linear ramp is used with a digital timer in an A-T-D method. In the final part of this experiment, experience is gained with a commercial 10-bit ADC. The converter is tested for various analog input voltages.

3–4.6 EXAMPLES OF HYBRID SYSTEMS

The old controversy whether analog or digital electronic systems are better has been laid to rest. Both are essential. Operations in both domains are frequently found in the most elegant system designs. It is also true that advances in the technology in each area have been both the reason for and the effect of advances in the other, so that the development of digital and analog circuits and devices has gone hand in hand.

In this last section of this module, we have chosen several examples of hybrid (analog and digital) systems which illustrate some important points in system design and which are systems or system elements of general interest and utility.

Sequencing System Operations

Automatic sequencing of the operations required for measurement or control is a necessity for total automation but may also provide increased convenience and speed and less chance of errors for many applications. Consider the photographic observation of the impact of a bullet hitting a glass bottle. The sequence of operations is (a) turn out the room light, (b) open the camera shutter, (c) fire the rifle, (d) trigger the strobe light flash to illuminate the bullet and target at the desired instant, (e) close the camera shutter, and (f) turn on the room lights. If any step occurs out of sequence, the experiment will be spoiled. In addition, there is one critical time in the sequence—the time between the firing of the rifle (c) and the triggering of the strobe light (d)—since this determines what part of the impact phenomenon will be observed. An automatic precision electronic sequencer is clearly called for in this experiment.

In this section it is assumed that all required operations are ON-OFF operations (not continuous adjustments) and that each operation to be performed can be controlled by a logic level signal. Therefore, a sequencer is a circuit or device that provides a logic level signal output for each operation to be controlled and that determines the time relationship of the logic level changes appearing at the outputs. An electromechanical sequencer can be made by operating switch contacts by cams attached to a motor shaft. One cam and switch is required for each operation. Commercial electromechanical sequencers are available with adjustable cams for varying the sequence as required.

Monostable Multivibrator Sequencers. An electronic sequencer can be made with monostable (MS) multivibrators as shown in Fig. 3–232. The output pulse of each MS triggers the input of the next. Thus pulses appear at the A, B, C, D, and E outputs in succession which can be used to actuate the required operations. Note that the time at the start of each pulse is the sum of the durations of all previous pulses. If the pulse width of MS A is increased, the time of the beginning of all successive pulses (relative to the start pulse) is increased. This is a convenient feature when it is important to maintain or control the time between successive operations.

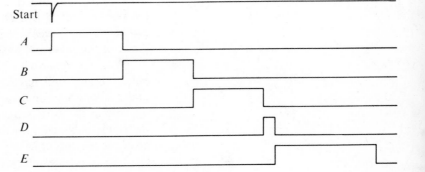

Fig. 3–232 Monostable sequencer.

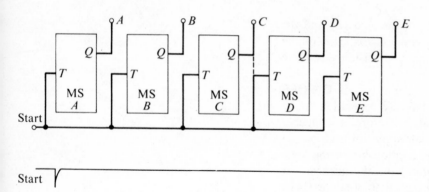

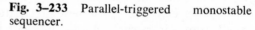

Fig. 3–233 Parallel-triggered monostable sequencer.

The serial-triggered sequencer can be inconvenient if operation C, for instance, is to have a fixed time relationship with the start pulse while the time of operation B is to be varied. In this case, a parallel-triggered circuit such as Fig. 3–233 might be more desirable. The time of the termination of each pulse is related to the start pulse only by its own duration, so that the "absolute time" of each operation can be varied independently. Note that the D pulse is to terminate shortly after C terminates. If the time difference between C and D must be known and constant (as is the case when C triggers the oscilloscope sweep and D initiates the event to be observed), extremely stable multivibrators will be required. A better solution would probably be a combination of the serial and parallel trigger approaches, as indicated by the dashed lines in Fig. 3–233. If MS D is triggered by MS C, the sequence of C and D and the time interval between C and D can be controlled by conventional MS circuits without affecting the time independence of the other outputs.

When working with MS sequencers, one must keep in mind that the better circuits will have a relative pulse duration stability of about 1% under laboratory conditions. In the serial-triggered case, the time errors for the start of the pulse are cumulative but the duration of each pulse is within the single pulse width error limits. In the parallel-triggered case, the time error for the end of each pulse is the error for a single pulse width, but the error for the time interval between two pulse terminations is as much as

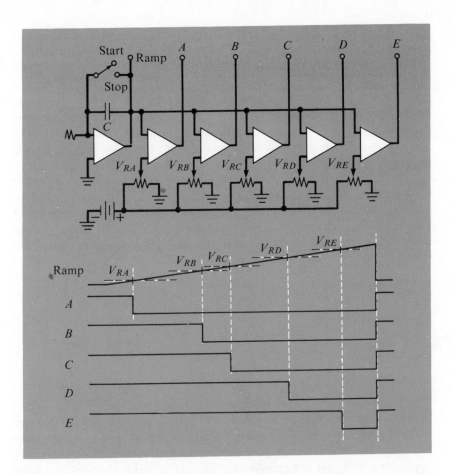

Fig. 3–234 Ramp and comparator sequencer.

two times the total pulse duration error, or about 2% of the time from the start pulse to the time interval in question. Careful thought must be given to the critical time invervals to evaluate the relative merits of the serial and parallel approaches for each operation.

Ramp and Comparator Sequencer. An analog approach to the design of a sequencer is shown in Fig. 3–234. A ramp signal which has a duration longer than the complete sequence is generated by an OA integrator. When the shorting switch across the integrating capacitor is opened, the sequence starts. As the ramp voltage reaches the comparison voltage of each comparator, the comparator output changes logic level as shown in the waveform chart. This circuit is analogous to the parallel-triggered MS sequencer in that each comparison voltage is set independently of the others. The time error for each comparator output transition depends on the stability, linearity, and magnitude of the ramp signal, the stability and sensitivity of the comparators, and the stability and resolution of the comparison sources. A good situation would be the following: a 5 V ramp amplitude with $\pm 0.1\%$ stability and linearity; comparators with ± 1 mV sensitivity and ± 3 mV stability; comparison sources of ± 5 mV accuracy and ± 1 mV stability and noise. In this case the stability of the sweep source (± 5 mV) limits the reproducibility to $\pm 0.1\%$ of the ramp duration. The accuracy of the comparison sources and the linearity of the sweep make the absolute time error as much as $\pm 0.2\%$ of the ramp duration. Note that errors for this kind of sequencer are relative to the duration of the complete sequence.

A kind of "serial" sequencer can be made by putting the comparison voltages V_{RA}, V_{RB}, etc., in series. Although this is not a true serial sequencer, the possibility of getting close events out of order is reduced. A one-step sequencer of the ramp-comparator type is used to trigger the delayed sweep generator in oscilloscopes that have the delayed sweep feature.

Fig. 3–235 Counter and decoder sequencer.

Counting Sequencer. Each of the two previous sequencers is quite limited in relative and absolute timing accuracy. When the time relation between operations in a measurement affects the accuracy of the measurement, more accurate time intervals will generally be required. An extremely accurate approach is to count out precise time increments from a crystal-controlled clock oscillator. A simple but very effective sequencer is the oscillator, counter, and decoder shown in Fig. 3–235. When the counter gate is open, the counter will advance at the clock rate, producing a **1** at each decoder output in sequence. Up to 16 equally spaced tasks could be sequenced by this simple system.

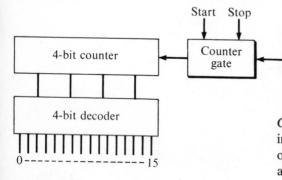

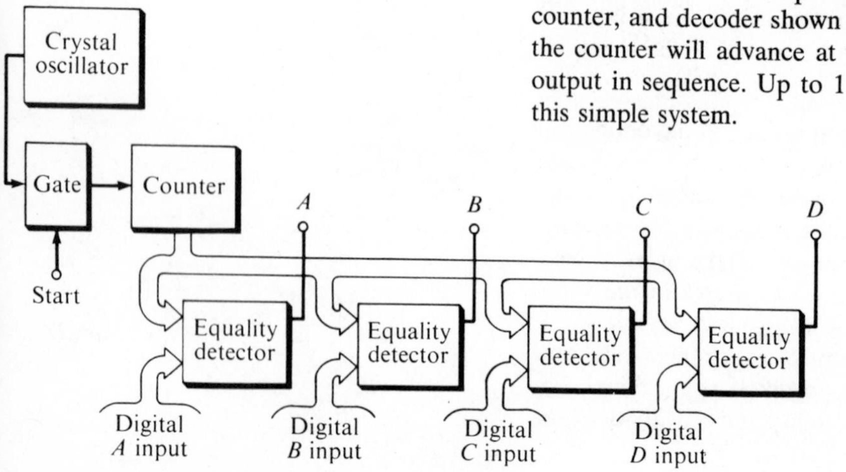

Fig. 3–236 Counting sequencer.

A more sophisticated counting sequencer is shown in Fig. 3–236. Instead of decoding all possible output states of the counter, equality detectors are used to detect when the desired states occur. For instance, if the clock oscillator has a frequency of 1.00000 MHz and if the digital A input signal has the same logic levels as the counter output when the count is 1467, then exactly 1467 μsec after the gate opens, a 1 μsec pulse will appear at output A. If the digital B input is set at 1567, a pulse will appear at B exactly 100 μsec after the pulse at A.

The counting sequencer combines the advantages of independent adjustment with high accuracy. The timing accuracy depends on the clock oscillator. The resolution (smallest difference between the time interval that can be set) is one clock period. The maximum duration of the sequence is n_m/f_c, where n_m is the modulus of the counter used. The digital inputs are often derived from switches. Ten-position rotary or thumbwheel switches with BCD-coded contact arrangements are very convenient for this application.

Digital Waveform Generator

A waveform generator is a generator that produces a particular waveform at regular intervals. A waveform is digitally encoded by converting the waveform to a digital word at regular intervals. Conversely, if a succession of digital words representing the waveform were applied to a DAC, the waveform would be reproduced at the DAC output. A digital waveform generator uses digital circuits to store and/or generate the succession of digital words representing the waveform. The concept is most simply illustrated in Fig. 3–237a. A ramp function is desired which, digitally encoded, is a sequence of digital words, each word one increment larger than the previous one. Such a sequence is generated by a counter. When the counter is incremented by a clock oscillator and the parallel output is connected to a DAC, a ramp (actually, if you look closely, a staircase) is generated. The waveform is repeated automatically as the counter cycles through its modulus. If a triangular waveform were desired instead, an up/down counter could be used with a logic circuit that reverses the direction of the count when the counter reaches all **1**'s or all **0**'s.

This technique is particularly well suited to slow ramp generation because the output is essentially drift free. The analog integrator type of ramp generator generally has drift errors for sweep periods of 100 sec or more. The output waveform frequency is f_c/n_m, which for a 10-bit counter is $f_c/1024$. A variable frequency oscillator (VFO) can be used to provide a variable output waveform frequency, though obviously very high frequencies are achieved with difficulty. Reducing n_m and the DAC resolution increases the increment step and makes the output a rougher approximation of a linearly increasing signal.

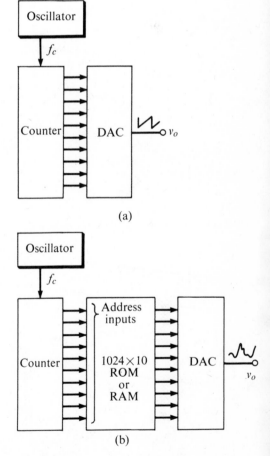

Fig. 3–237 Digital waveform generators: (a) ramp or triangle; (b) any function.

A ROM or RAM can be used between the counter and the DAC to provide virtually any waveform. The counter is used to sequence the memory stack address, and the memory contents at each address are read out to the DAC in sequence. The waveform produced is whatever sequence was stored in the memory. If a sine function ROM is used, a sinusoidal output waveform will result. If a RAM is used, any encoded waveform read into the RAM can be repeated indefinitely at the DAC output. This system gives independent control over the time and amplitude resolution. The amplitude resolution is determined by the word length, which will be matched by the DAC resolution. The time resolution (fraction of a waveform cycle per increment) is determined by the word capacity of the memory, which is matched by the counter modulus.

Photon Counting

The technique known as photon counting provides an interesting example of a hybrid measurement system whose components must have rather special characteristics. The photon counting approach to light measurement involves digital processing of the discrete electron pulses from the anode of a photomultiplier (PM) tube so that the number of counted pulses is directly related to the number of photons incident on the PM photocathode. The basic principles underlying the use of the PM tube as a pulse output transducer were developed in Section 1–2.5 of Module 1, where it was shown that if all anode pulses are resolved by the counting circuitry, the number of anode pulses, N_a, over certain boundary conditions is related to the number of photons incident on the photocathode, P_k, by

$$N_a = bP_k, \tag{3–203}$$

where b is a factor made up of the quantum efficiency of the photocathode, the collection efficiency of the first dynode in the PM tube, and the transfer efficiency of electron bursts between dynodes. If the light source is a continuous source, then P_k is the arrival rate of photons at the photocathode and N_a is the number of anode pulses per unit time.

The apparent advantages of the photon counting technique over other light-measurement systems include direct digital processing of discrete spectral information, discrimination against dark current, sensitivity to very low light levels, high system stability against drift, and improved signal-to-noise ratio under certain circumstances. Because of these advantages the photon counting approach is attractive for a variety of light-measurement applications.

In order to make use of the potential advantages of the photon counting approach, the measurement system must be carefully designed and its influence on the final readout must be well understood. The discussion here includes some of the considerations in photon counting system design, a discussion of some of the errors which can result from pulse pileup effects, and some of the applications of photon counting in laboratory measurements.

Photon Counting Systems.   The usual system necessary to implement photon counting for measurements on continuous light sources is shown in block diagram form in Fig. 3–238. The anodic current pulses are usually coupled through an *RC* load into a pulse amplifier. The pulse height discriminator (voltage comparator or tunnel diode discriminator) allows all pulses above a preset threshold level to enter the counter and shapes the pulses. The preset timer opens and closes the counting gate for an accurate time interval. The total count per integration period is read from the digital counter or printer. A count rate meter is often employed to produce an analog output for graphical display of data, often on a strip chart recorder.

To determine the final number read in the counter and its relationship to the incident light level, the influence of each of the different counting

Fig. 3–238   Block diagram of photon counting system.

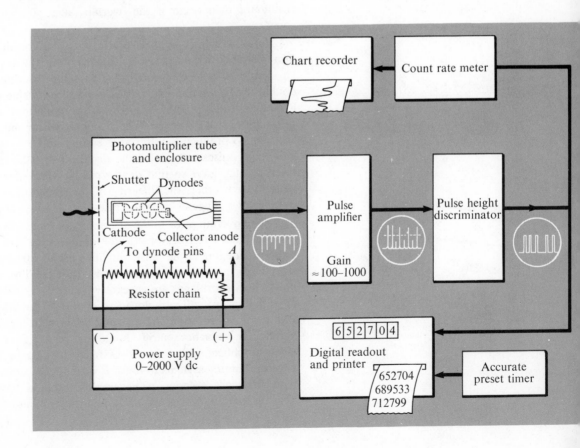

circuit components must be considered. The finite time duration of the anode pulses from the PM or of the voltage pulses in the counting circuitry results in a finite probability of pulses overlapping and not being resolved. The resolving time or dead time of the counting system is defined as the minimum time interval between two anode pulses from the PM that can be distinguished. Pileup of pulses in the PM is often negligible or, in other words, the average anode pulse width is much less than the dead time of the counting system. Hence the design of a photon counting system which is linear over a wide dynamic range of incident light intensities often centers on minimizing the system dead time. Often one component of the system limits the frequency response and hence determines the dead time and the relationship between the true pulse rate and the observed pulse rate.

In the majority of photon counting systems the counting circuit is said to be "paralyzable" in that each pulse that occurs extends the dead time whether or not the pulse is counted. The effect of pulse pileup on the observed readout depends on where in the counting system the overlap occurs. Different effects are found if pileup occurs prior to the pulse height discriminator than occur if the overlap takes place in the discriminator or counter.

Pulse overlap before the discriminator can result in a net loss of counts or a net gain of counts. At very low discriminator level settings, pulse overlap results in a loss of counts because the discriminator sees overlapped double or higher multiple pulses as one resolvable pulse. This dead time loss effect is also possible when the discriminator is set above the 0 V level. In addition, dead time gain can occur if two or more small pulses pile up and sum to produce a pulse large enough to be counted when none of the individual pulses was above the discriminator threshold. By adjusting the discriminator level so that dead time loss and gain effects partially compensate each other, the linearity of the counting system can be extended over a wider range of counting rates.

Pulse overlap prior to the discriminator occurs because of the frequency response characteristics of the *RC* load and the pulse amplifier. To reduce pulse broadening at the pulse amplifier input, care must be taken to reduce stray capacitance in the connections from the PM anode to the amplifier input. Very short leads are necessary for low pulse degradation. AC-coupled amplifiers are usually employed to eliminate the effects of dc drifts in the PM. For low distortion and for good linearity at high pulse rates, the amplifier upper frequency cutoff should be very high (>100 MHz). Special video amplifiers with bandwidths from 10 kHz to 100 MHz are used in photon counting systems to reject low frequency noise but still provide good high frequency response. The amplifier should have good gain stability. Sharing the gain between the PM and the amplifier requires that the amplifier voltage gain be about 100–1000 for use with most PM tubes. With some high gain PM tubes the pulse amplifier is unnecessary.

If the frequency response of the pulse amplifier is higher than that of the discriminator-counter system, pulse pileup can still occur at high pulse rates during or after pulse height discrimination. In this case, only a loss in counts is possible. Fortunately pulse height discriminators, such as tunnel diode discriminators, are available with high sensitivities (50 mV) and good high frequency characteristics (>100 MHz). High speed discriminators have a minimum output pulse width which can lead to dead time loss. The counter for photon counting at moderate light levels can also be the cause of pulse pileup because of propagation delays in the counting circuits. Fortunately many modern counters can operate at rates greater than 100 MHz.

Experimentally it is found that, to keep the dead time error less than 1% at the maximum rate, the combined frequency response of the measurement system should be at least 25 times greater than the maximum average pulse rate measured. With present photon counting systems it is necessary to shift from photon counting to conventional analog current measurements at high count rates. The exact crossover point depends on the frequency response (dead time) of the particular counting system, but is at present in the range of 100 kHz to 1 MHz maximum average count rate. With new high speed amplifiers and counters it should be feasible to extend this to a counting rate of 10 MHz or greater.

Applications.   The photon counting approach has been applied to a variety of light-measurement applications. Since it is particularly well suited for low light level measurements and long-term averaging, photon counting is used extensively in astronomy. The inherent stability of photon counting systems allows the accumulation of pulses over long periods of time for high precision without the troublesome drifts which occur with analog techniques.

In Raman and luminescence spectrometry (molecular and atomic fluorescence), the photon counting technique has found widespread application for the measurement of the low level light signals which are characteristic of these techniques.

Photon counting has also been advantageously used in high precision molecular absorption spectrometry. Even though molecular absorption spectrophotometry involves rather high light levels, photon counting makes it feasible to use very small slit widths for achieving high resolution and avoiding Beer's law deviations. In addition, the counting technique makes it very convenient to preset the desired measurement precision and vary the counting time to accumulate the appropriate number of counts.

Many additional applications of this versatile technique have been reported, and as photon counting technology improves, it is expected to become even more widely employed.

No-Droop Sample-and-Hold Circuit

All the analog sample-and-hold circuits described in Section 3–2.5 hold the analog data as a charge stored on a capacitor. As a result of leakage in the capacitor, stray currents to surrounding circuit points, and voltage amplifier input current, it is unavoidable that the capacitor will loose its charge over time and the held voltage will "droop." Data stored in a digital register, on the other hand, can be held indefinitely without loss or change of information. Thus a "no-droop" or "infinite hold" sample-and-hold circuit can be achieved by the conversion of the analog quantity to a digital signal for storage in a register with the continuous conversion of the register contents to an analog output signal. Such a circuit is shown in Fig. 3–239. At the desired sample time, a transition is applied to the analog sample-and-hold control input that shifts the sample-and-hold circuit from the track mode to the hold mode. This same transition triggers the ADC to begin a conversion of the held signal after a short delay to accommodate the settling time of the sample-and-hold circuit. At the completion of the conversion, the transition of the ADC status output causes the register to load the digital information. The DAC, connected to the register, provides a continuous analog output of the register contents. If track-and-hold operation is required, the conversion cycle can be made automatically continuous by using the ADC status output to trigger the sample-and-hold command input. The register contents and DAC output will then be updated at the maximum acquisition-conversion rate. To go into the hold mode, the gate in the automatic cycling connection is CLOSED.

The accuracy of the circuit is determined by the gain accuracy of the analog sample-and-hold circuit and by the conversion accuracies of the ADC and DAC. The aperture time is determined by the analog sample-and-hold circuit alone, while the output settling time is determined by the total time for both conversion processes. With fast converters, this can be as short as a few microseconds. In the track-and-hold mode, the aperture time

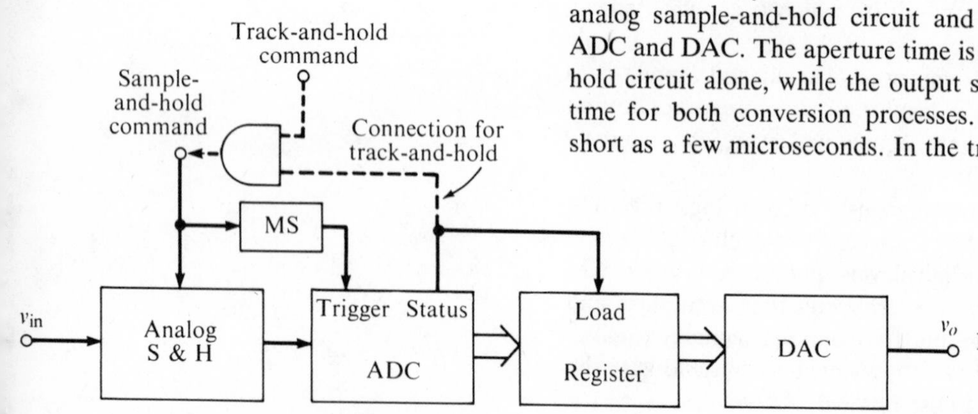

Fig. 3–239 No-droop sample-and-hold.

uncertainty is equal to the conversion cycle time. The availability of the major components of this circuit in convenient and relatively inexpensive IC and/or hybrid circuit modules makes this circuit practical where the no-droop requirement is desirable. The brief conversion of the signal into the digital domain is made solely for the purpose of taking advantage of the nonvolatile storage available in that domain.

Data Acquisition Systems

A data acquisition system is one which acquires and stores input data for later use. The recent trend in data acquisition systems is to use digital storage because it is nonvolatile (as explained above) and because the data are in a form compatible with digital data processors. A data acquisition system with digital storage thus consists of an ADC and a digital memory. In general, other elements are needed for a practical system as shown in Fig. 3–240. An instrumentation amplifier is used to provide a differential input and to adjust the input voltage level to the range of the ADC. A sample-and-hold circuit is used to provide precise acquisition timing and to hold the analog voltage constant during the conversion process. The memory address register specifies the memory location for recording the data. The sequencer times the acquisition, conversion, and storage of the data and increments the memory address register. This system could be considered to be a multiple memory no-droop sample-and-hold circuit, and the same considerations discussed above for that circuit apply to this one as well.

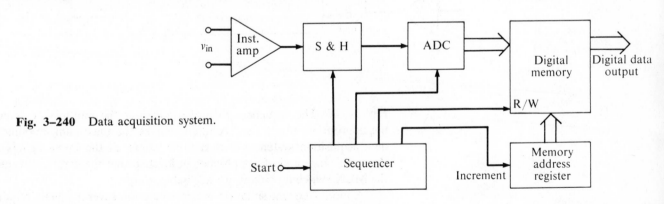

Fig. 3–240 Data acquisition system.

The measurement system producing the data that are to be sampled and stored often has more than one data source. In such cases, multiplexing is used so that the memory at least can be shared among several data sources. Three forms of multiplexing are shown in Fig. 3–241. In Fig. 3–241a an nalog multiplexer (MUX) is used at the input of the system shown in

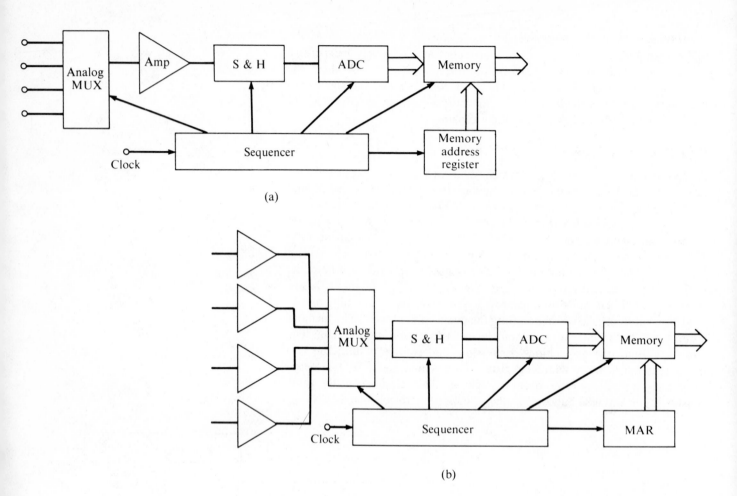

(a)

(b)

Fig. 3–240. The sequencer then also advances the multiplexer channel at the appropriate time. This technique shares the maximum amount of the data acquisition system and is suitable where all the input signals are in the same range and of a high enough level so that the errors introduced in the MUX switching circuit are negligible.

A more frequent situation is one where the several signals require different amplification and offset to fall within the optimum ADC range. In this case, each input channel can have its own instrumentation amplifier with the analog-conditioned signals multiplexed to the remainder of the converter system, as shown in Fig. 3–241b. This system provides much greater flexibility in accepting a variety of input sources and greatly reduces the demands on the multiplexer quality.

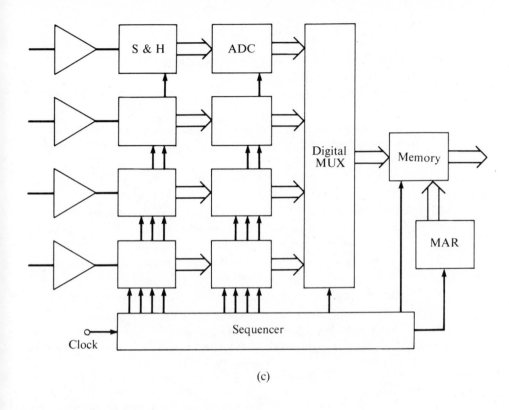

(c)

Fig. 3–241 Multiplexed data acquisition systems: (a) common system; (b) individual amplifiers; (c) individual conversion.

Where the data sources are scattered over several locations, such as testing stations in process monitoring, the data must be transmitted to the central data acquisition system. If the systems of Fig. 3–241a or b were used, the data would have to be transmitted in analog form. Digitally encoded data are far less susceptible than analog signals to errors due to induced noise. Therefore, it is preferable to perform the digital conversion at each site and transmit the digital data to the central control and storage unit. This form of multiplexed system is shown in Fig. 3–241c. As the price of good ADC's continues to drop, the sharing of the ADC in multiplexed systems is less and less important. The alternative of transmitting frequency-encoded data as mentioned in the description of voltage-to-frequency-to-digital converters should also be considered for this application.

PROBLEMS

1. Write the decimal equivalent of the following binary numbers: (a) 0110101, (b) 111010011, (c) 100001101.

2. Express in binary the following decimal numbers: (a) 132, (b) 57, (c) 28, (d) 103.

3. Express the decimal numbers of problem 2 in binary-coded decimal (BCD) code, and note the difference between the binary and the BCD representation.

4. Prove by means of truth tables the Boolean AND and OR theorems of Table 3–9.

5. Use the basic theorems of Boolean algebra to prove the following identities: (a) $A(\bar{A} + B) = AB$, (b) $AB + \bar{A}C = (A + C)(\bar{A} + B)$, (c) $(A + B)(\bar{A} + C) = AC + \bar{A}B$.

6. Verify the following identities using Boolean algebra: (a) $\overline{\overline{A + B} + \overline{A + \bar{B}}} = A$, (b) $AB + AC + B\bar{C} = AC + B\bar{C}$, (c) $AB + \bar{B}C + A\bar{C} = AB + \bar{B}C$.

7. Use Boolean algebra to show that the AND, OR, and NAND functions can all be implemented with NOR gates as shown in Table 3–11.

8. A lamp is to be controlled by logical combinations of three switches A, B, and C. Whenever switches B and C are in the same position, the lamp is to be ON. When B and C are in opposite states, the lamp is to be ON when switch A is CLOSED. (a) Draw the truth table for the above operation. (b) Write the Boolean expression for the lamp T in terms of A, B, and C. (c) Simplify the resulting expression and suggest a logic circuit to accomplish the desired function.

9. Write the decimal numbers 38 and 22 in binary and perform a binary addition. Construct a truth table showing the sum and carry outputs for each input combination.

10. It is desired to construct a synchronous BCD down-counter from JK flip-flops and NAND gates. Draw the desired output waveforms which show the countdown according to the 1 2 4 8 BCD code. Use four JK flip-flops and one NAND gate and draw a circuit which will perform the BCD down-counting operation.

11. Design a modulo-13 counter using JK flip-flops and gates to control the counting sequence and bring the counter to zero after the maximum count has been reached.

12. The four-bit binary DAC of Fig. 3–220 is used with $V_R = 5$ V and $R = 5$ kΩ. What are the values of v_o for the following digital inputs? (a) 1111, (b) 1001, (c) 0101, (d) 0001.

13. The table below gives some of the relations between bipolar binary codes in terms of instructions for the interconversions. Complete the table.

To convert From → To ↓	Sign magnitude	2's comple- ment	Offset binary	1's comple- ment
Sign magnitude	No change			If MSB = **1**, complement other bits
2's complement		No change	Complement MSB	
Offset binary	Complement MSB If new MSB = **0**, complement other bits, add **00** . . . **01**		No change	
1's complement				No change

14. For a dual-slope ADC, it is often desirable to make the integration period exactly equal to one period of the 60 Hz ac power line to reduce the effects of power line frequency noise on the conversion accuracy. If a $3\frac{1}{2}$-digit BCD counter (full scale count is 1999) is used and the signal is integrated until the two most significant bits of the counter are 1, what value of f_c must be used?

2^n	n	2^{-n}
1	0	1.0
2	1	0.5
4	2	0.25
8	3	0.125
16	4	0.062 5
32	5	0.031 25
64	6	0.015 625
128	7	0.007 812 5
256	8	0.003 906 25
512	9	0.001 953 125
1 024	10	0.000 976 562 5
2 048	11	0.000 488 281 25
4 096	12	0.000 244 140 625
8 192	13	0.000 122 070 312 5
16 384	14	0.000 061 035 156 25
32 768	15	0.000 030 517 578 125
65 536	16	0.000 015 258 789 062 5
131 072	17	0.000 007 629 394 531 25
262 144	18	0.000 003 814 697 265 625
524 288	19	0.000 001 907 348 632 812 5
1 048 576	20	0.000 000 953 674 316 406 25
2 097 152	21	0.000 000 476 837 158 203 125
4 194 304	22	0.000 000 238 418 579 101 562 5
8 388 608	23	0.000 000 119 209 289 550 781 25
16 777 216	24	0.000 000 059 604 644 775 390 625
33 554 432	25	0.000 000 029 802 322 387 695 312 5
67 108 864	26	0.000 000 014 901 161 193 847 656 25
134 217 728	27	0.000 000 007 450 580 596 923 828 125
268 435 456	28	0.000 000 003 725 290 298 461 914 062 5
536 870 912	29	0.000 000 001 862 645 149 230 957 031 25
1 073 741 824	30	0.000 000 000 931 322 574 615 478 515 625
2 147 483 648	31	0.000 000 000 465 661 287 307 739 257 812 5
4 294 967 296	32	0.000 000 000 232 830 643 653 869 628 906 25
8 589 934 592	33	0.000 000 000 116 415 321 826 934 814 453 125
17 179 869 184	34	0.000 000 000 058 207 660 913 467 407 226 562 5
34 359 738 368	35	0.000 000 000 029 103 830 456 733 703 613 281 25
68 719 476 736	36	0.000 000 000 014 551 915 228 366 851 806 640 625
137 438 953 472	37	0.000 000 000 007 275 957 614 183 425 903 320 312 5
274 877 906 944	38	0.000 000 000 003 637 978 807 091 712 951 660 156 25
549 755 813 888	39	0.000 000 000 001 818 989 403 545 856 475 830 078 125
1 099 511 627 776	40	0.000 000 000 000 909 494 701 772 928 237 915 039 062 5
2 199 023 255 552	41	0.000 000 000 000 454 747 350 886 464 118 957 519 531 25
4 398 046 511 104	42	0.000 000 000 000 227 373 675 443 232 059 478 759 765 625
8 796 093 022 208	43	0.000 000 000 000 113 686 837 721 616 029 739 379 882 812 5
17 592 186 044 416	44	0.000 000 000 000 056 843 418 860 808 014 869 689 941 406 25
35 184 372 088 832	45	0.000 000 000 000 028 421 709 430 404 007 434 844 970 703 125
70 368 744 177 664	46	0.000 000 000 000 014 210 854 715 202 003 717 422 485 351 562 5
140 737 488 355 328	47	0.000 000 000 000 007 105 427 357 601 001 858 711 242 675 781 25
281 474 976 710 656	48	0.000 000 000 000 003 552 713 678 800 500 929 355 621 337 890 625
562 949 953 421 312	49	0.000 000 000 000 001 776 356 839 400 250 464 677 810 668 945 312 5
1 125 899 906 842 624	50	0.000 000 000 000 000 888 178 419 700 125 232 338 905 334 472 656 25
2 251 799 813 685 248	51	0.000 000 000 000 000 444 089 209 850 062 616 169 452 667 236 328 125
4 503 599 627 370 496	52	0.000 000 000 000 000 222 044 604 925 031 308 084 726 333 618 164 062 5
9 007 199 254 740 992	53	0.000 000 000 000 000 111 022 302 462 515 654 042 363 166 809 082 031 25
18 014 398 509 481 984	54	0.000 000 000 000 000 055 511 151 231 257 827 021 181 583 404 541 015 625
36 028 797 018 963 968	55	0.000 000 000 000 000 027 755 575 615 628 913 510 590 791 702 270 507 812 5
72 057 594 037 927 936	56	0.000 000 000 000 000 013 877 787 807 814 456 755 295 395 851 135 253 906 25
144 115 188 075 855 872	57	0.000 000 000 000 000 006 938 893 903 907 228 377 647 697 925 567 626 953 125
288 230 376 151 711 744	58	0.000 000 000 000 000 003 469 446 951 953 614 188 823 848 962 783 813 476 562 5
576 460 752 303 423 488	59	0.000 000 000 000 000 001 734 723 475 976 807 094 411 924 481 391 906 738 281 25
1 152 921 504 606 846 976	60	0.000 000 000 000 000 000 867 361 737 988 403 547 205 962 240 695 953 369 140 625
2 305 843 009 213 693 952	61	0.000 000 000 000 000 000 433 680 868 994 201 773 602 981 120 347 976 684 570 312 5
4 611 686 018 427 387 904	62	0.000 000 000 000 000 000 216 840 434 497 100 886 801 490 560 173 988 342 285 156 25
9 223 372 036 854 775 808	63	0.000 000 000 000 000 000 108 420 217 248 550 433 400 745 280 086 994 171 142 578 125
18 446 744 073 709 551 616	64	0.000 000 000 000 000 000 054 210 108 624 275 221 700 372 640 043 497 085 571 289 062 5
36 893 488 147 419 103 232	65	0.000 000 000 000 000 000 027 105 054 312 137 610 850 186 320 021 748 542 785 644 531 25
73 786 976 294 838 206 464	66	0.000 000 000 000 000 000 013 552 527 156 068 805 425 093 160 010 874 271 392 822 265 625
147 573 952 589 676 412 928	67	0.000 000 000 000 000 000 006 776 263 578 034 402 712 546 580 005 437 135 696 411 132 812 5
295 147 905 179 352 825 856	68	0.000 000 000 000 000 000 003 388 131 789 017 201 356 273 290 002 718 567 848 205 566 406 25
590 295 810 358 705 651 712	69	0.000 000 000 000 000 000 001 694 065 894 508 600 678 136 645 001 359 283 924 102 783 203 125
1 180 591 620 717 411 303 424	70	0.000 000 000 000 000 000 000 847 032 947 254 300 339 068 322 500 679 641 962 051 391 601 562 5
2 361 183 241 434 822 606 848	71	0.000 000 000 000 000 000 000 423 516 473 627 150 169 534 161 250 339 820 981 025 695 800 781 25
4 722 366 482 869 645 213 696	72	0.000 000 000 000 000 000 000 211 758 236 813 575 084 767 080 625 169 910 490 512 847 900 390 625

Summary of Boolean Algebra Postulates and Theorems

Appendix B

Postulates

$X = 1$ or else $X = 0$

$1 \cdot 1 = 1$ 　　　　　　　$0 + 0 = 0$

$1 \cdot 0 = 0 \cdot 1 = 0$ 　　$0 + 1 = 1 + 0 = 1$

$0 \cdot 0 = 0$ 　　　　　　　$1 + 1 = 1$

$\bar{1} = 0$ 　　　　　　　　$\bar{0} = 1$

Theorems

1a. $0 \cdot X = 0$ 　　　　　　**1b.** $1 + X = 1$ 　　　　**2a.** $1 \cdot X = X$ 　　　　**2b.** $0 + X = X$

3a. $XX = X$ 　　　　　　　**3b.** $X + X = X$ 　　　　　**4a.** $X\bar{X} = 0$ 　　　　　**4b.** $X + \bar{X} = 1$

5a. $XY = YX$ 　　　　　　**5b.** $X + Y = Y + X$

6a. $XYZ = (XY)Z = X(YZ)$ 　　　　　　　　　　　　　**6b.** $X + Y + Z = (X + Y) + Z = X + (Y + Z)$

7a. $\overline{XY \cdots Z} = \bar{X} + \bar{Y} + \cdots + \bar{Z}$ 　　　　　　　**7b.** $\overline{X + Y + \cdots + Z} = \bar{X}\bar{Y} \cdots \bar{Z}$

8. $\bar{f}(X, Y, \ldots, Z, \cdot, +) = f(\bar{X}, \bar{Y}, \ldots, \bar{Z}, +, \cdot)$

9a. $XY + XZ = X(Y + Z)$ 　　　　　　　　　　　　　　**9b.** $(X + Y)(X + Z) = X + YZ$

10a. $XY + X\bar{Y} = X$ 　　　　　　　　　　　　　　　　**10b.** $(X + Y)(X + \bar{Y}) = X$

11a. $X + XY = X$ 　　　　　　　**11b.** $X(X + Y) = X$ 　　**12a.** $X + \bar{X}Y = X + Y$ 　　**12b.** $X(\bar{X} + Y) = XY$

12a′. $ZX + Z\bar{X}Y = ZX + ZY$ 　　　　　　　　　　　　**12b′.** $(Z + X)(Z + \bar{X} + Y) = (Z + X)(Z + Y)$

13a. $XY + \bar{X}Z + YZ = XY + \bar{X}Z$ 　　　　　　　　**13b.** $(X + Y)(\bar{X} + Z)(Y + Z) = (X + Y)(\bar{X} + Z)$

14a. $XY + \bar{X}Z = (X + Z)(\bar{X} + Y)$ 　　　　　　　　**14b.** $(X + Y)(\bar{X} + Z) = XZ + \bar{X}Y$

15a. $X \cdot f(X, \bar{X}, Y, \ldots, Z) = X \cdot f(1, 0, Y, \ldots, Z)$ 　　**15b.** $X + f(X, \bar{X}, Y, \ldots, Z) = X + f(0, 1, Y, \ldots, Z)$

16a. $f(X, \bar{X}, Y, \ldots, Z) = X \cdot f(1, 0, Y, \ldots, Z) + \bar{X} \cdot f(0, 1, Y, \ldots, Z)$

16b. $f(X, \bar{X}, Y, \ldots, Z) = [X + f(0, 1, Y, \ldots, Z)] \ [\bar{X} + f(1, 0, Y, \ldots, Z)]$

Logic Gate Types and Characteristics

The logic gate is the basic circuit element in all modern digital circuits. The characteristics of a computer or instrument are determined, in large part, by the speed, versatility, noise immunity, durability, etc., of the gate circuit, which is used hundreds of times over in the simplest digital instruments. Improvements in gate design have occurred frequently in recent years, resulting in currently available gates which are very nearly the ideal universal gate. The gate types described below are all available in integrated circuit form. They are presented in the approximate chronological order of their development.

RTL Gates

An RTL gate is simply a number of transistor switches in parallel, as shown in Fig. C–1. When any one or more of the transistors is ON (a positive logic **1** input), the output will be **0**. The output will be **1** only if all the inputs are **0**. This is the inverse of the positive OR function; thus the gate is a positive NOR or negative NAND gate. The RTL gate requires forward biased base current from a **1** source and virtually no current from a **0** source. Since the output impedance of the gate shown for a **1** level is 640 Ω, the fan-out may only be three or four gate inputs. With respect to noise immunity and speed also, the RTL gate is inferior to other logic gates developed later.

Gates such as the RTL which require current from the signal source to activate the gate input device are in a class of circuits called **current-sourcing logic.** This term is appropriate because the gate output acts as a current source for all the gate inputs connected to it.

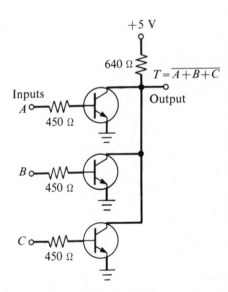

Fig. C–1 RTL logic gates.

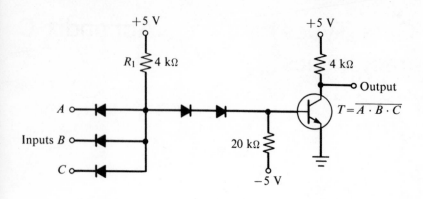

Fig. C–2   DTL NAND gate.

DTL Gates

A natural development from the diode gates discussed in Section 3–4.1 is the addition of an amplifier to the gate output to reduce the output impedance, as shown in Fig. C–2. This is a diode-transistor logic or DTL gate. The input diodes and R_1 are the AND circuit. When the AND output is near 0 V, the transistor is cut off and the output is +5 V, a logic **1**. A **1** output from the AND gate is sufficient to saturate the transistor, making the output very near 0 V. The gate function is now $T = \overline{A \cdot B \cdot C}$ or NAND. For negative logic levels, this is a NOR gate. When a **0** level signal source is connected to one of the inputs, a current of $5/R_1$ A must pass to ground through the signal source without raising the input voltage beyond the upper limit of the **0** level. Therefore, a signal source for the DTL gate is required to absorb, or "sink," current from a gate input in order to hold that input in the **0** state. Gates of this type are called **current-sinking logic.** Because of the different output and input requirements, current-sinking and current-sourcing logic circuits are not mixed in a single system without careful consideration or special interfacing circuits.

The number of gate inputs which can be connected to a gate output without jeopardizing the logic level is called the **fan-out,** an important gate characteristic. Very little current is required of a signal in the **1** state, and the amplifier is easily designed to supply enough current at the **1** logic level to equal the fan-out limitation set by the **0** state conditions. The noise immunity of a gate is the smallest **1** level output voltage minus the minimum effective **1** level input voltage, or the minimum effective **1** level input voltage minus the maximum **0** level output, whichever is smaller under the worst-case conditions. A noise pulse exceeding this value has a chance of affecting the state of the gate. The DTL gate shown has a noise immunity of about 1.0 V and a propagation delay of 50–100 nsec.

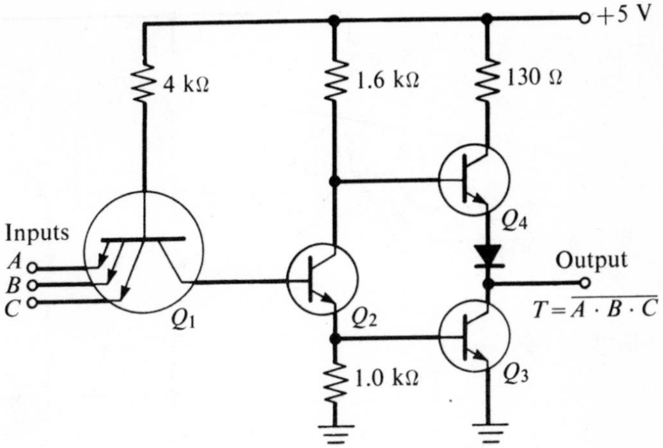

Fig. C–3   TTL positive NAND gate.

TTL Gates

Along with the development of RTL and DTL circuits described earlier came the realization that some circuit improvements were possible with IC's which were not practical with discrete components. One such example is the TTL gate shown in Fig. C–3. Here Q_1 is a multiple-emitter transistor. Grounding any one or more inputs forward biases transistor Q_1, which puts the collector of Q_1 at a low voltage and turns Q_2 OFF. This, in turn, turns Q_4 ON and Q_3 OFF and results in a logic **1** output. However, if all inputs are HI (or unused), the base-collector junction of Q_1 will conduct, forward biasing the base-emitter junction of Q_2. When Q_2 is ON, Q_3 is ON and Q_4 is OFF, resulting in a **0** at the output. This corresponds to the inverted AND or NAND function.

The circuit is actually very similar to the DTL gate with Q_1 replacing the gate and coupling diodes. Transistor Q_4, a dynamic collector resistor for Q_3, permits a larger current capability in the **1** state, and Q_2 provides gain for increased speed and noise immunity. Since the connecting leads inside the integrated circuit (IC) gate are extremely short, lead capacitance and inductance are minimized and very high speed is possible. The standard TTL gate has a propagation delay of only 11 nsec. Since the fan-out is about 10 and the noise immunity is at least 1.0 V, this is a very convenient general purpose gate.

TTL Gate Variations.   As the TTL gate applications increased, special versions of the basic TTL function were developed to meet the requirements of low power dissipation or high speed applications. For lower power dissipation, the resistances in the standard TTL gate are increased approximately tenfold. The result is a power dissipation of only 1 mW per gate at a sacrifice in propagation delay (typically 33 nsec). By comparison, the

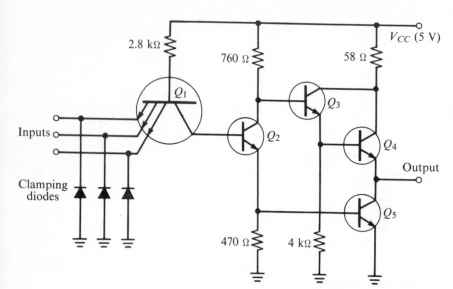

Fig. C-4 High speed TTL gate.

standard TTL circuit has a power dissipation of 10 mW per gate. Low power TTL gates are abbreviated LPTTL.

Conversely, for higher speed, the resistances are reduced to reduce the RC switching time constants, and the inputs are diode-clamped to eliminate negative transients on the input signals. The resulting gate circuit is shown in Fig. C-4. The propagation delay is reduced to 6 nsec and the power dissipation increased to 20 mW per gate. High speed TTL (HSTTL) flip-flops can be clocked at an average rate of 50 MHz.

For still greater speed, transistor saturation in the HSTTL gate is eliminated by using Schottky-diode-clamped transistors throughout. The resulting circuit is otherwise very similar to Fig. C-4. The gate delay is reduced to only 3 nsec, with no increase in power dissipation.

Many TTL gates and gate circuits are available with open collector outputs as shown in Fig. 3–161. Since the output levels of these gates are not driven, they are very convenient when the outputs of several gates are to be connected. When this is done, and an external collector load resistor is supplied, the circuit of Fig. C-5 results. The output will be LO if any of the gate outputs is LO; this corresponds to the negative logic OR or the positive logic AND function. A value of 1 kΩ for R_L will allow nine gate outputs to be connected and will provide a fan-out capability for the output of seven gate inputs.

The fan-out of all TTL gates is 10 when they are driving other TTL gates of the same type. However, the output capacity and input load of the several TTL types are different. When they are mixed in the same system,

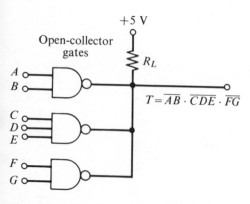

Fig. C-5 Wired-AND connection of open collector gates.

Table C–1   Loading rules for TTL families

Logic type	Fan-out to standard TTL load unit	Input load normalized to standard TTL gate
LPTTL	1.25*	0.25
TTL	10	1
HSTTL (and Schottky)	12.5	1.25

* Fan-out of LPTTL is 2.5 when only LPTTL gate inputs are being driven.

the fan-out capability for driving other types of inputs can be obtained from Table C–1. For example, a LPTTL gate can drive one standard TTL input *and* one LPTTL input. An HSTTL output can drive 12 standard TTL inputs or 50 LPTTL inputs.

TTL System Considerations.   The following rules have been established for the minimization of transmission line effects in TTL systems†:

1. Use direct wire interconnections that have no specific ground return for lengths up to about 10 in. only. A ground plane is always desirable.

2. Direct wire interconnections must be routed close to a ground plane if longer than 10 in. and should never be longer than 20 in.

3. When using coaxial or twisted-pair cables, design around approximately 100 Ω characteristic impedance. Coaxial cable of 93 Ω impedance (such as Microdot 293-3913) is recommended. For twisted pair, No. 26 or No. 28 wire with thin insulation twisted about 30 turns per foot works well. Higher impedances increase crosstalk, and lower impedances are difficult to drive.

4. Ensure that transmission line ground returns are carried through at both transmitting and receiving ends.

5. Connect reverse termination at driver output to prevent negative overshoot.

6. Decouple line driving and line receiving gates as close to the package V_{cc} and ground pins as practical, with a 0.1 µF capacitor.

7. Gates used as line drivers should be used for that purpose only. Gate inputs connected directly to a line driving output could receive erroneous

---

† From R. L. Morris, and J. R. Miller, *Designing with TTL Integrated Circuits,* McGraw-Hill, New York, N.Y., 1971.

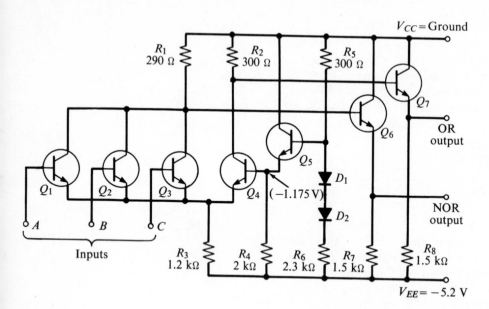

Fig. C–6   ECL gate circuit.

inputs due to line reflections, long delay times introduced, or excessive loading on the driving gate.

8. Gates used as line receivers should have all inputs tied together to the line. Other logic inputs to the receiving gate should be avoided, and a single gate should be used as the termination of a line.

9. Flip-flops are generally unsatisfactory line drivers because of the possibility of collector commutation from reflected signals.

ECL Gates

One of the limitations on the maximum switching speed of all the transistor gates described thus far except the Schottky TTL is the storage time of the saturated transistor. Several gate designs have been made which eliminate this delay by not allowing the transistor to saturate. One of the most successful of these circuits is the emitter-coupled logic or ECL. The circuit of a representative ECL gate is shown in Fig. C–6.

Transistors Q_3 and Q_4 are in a nonsaturating current-steering switch circuit. The voltage at the base of Q_4 is -1.175 V. If inputs A, B, and C are all logic **0** (-1.5 V), Q_1, Q_2, and Q_3 will be cut off; Q_4 then acts as an emitter follower, producing $-1.175\ V_{BE} \approx -1.9$ V at the emitter. The current through Q_4 and R_2 is nearly equal to the current through R_3. Therefore, the voltage across R_2 is $(300/1.2\ \text{k}\Omega)\,(5.2 - 1.9) = 0.82$ V. Thus the voltage at the base of emitter follower Q_7 is -0.82 V and the OR output voltage is -0.82 V $- V_{BE} \approx -1.5$ V (a logic **0**). Since Q_1, Q_2, and Q_3 are cut off, there is almost no current through R_1 and the voltage on the base of emitter follower Q_6 is nearly 0 V. Thus the NOR output voltage is $0 - V_{BE} \approx -0.75$ V (a logic **1**). The -1.175 V fixed bias voltage at the base of Q_4 is obtained from the voltage divider consisting of R_5 and R_6 and from the emitter-follower amplifier Q_5. Diodes D_1 and D_2 are to compensate for the temperature dependence of the base-emitter junction voltages of Q_5 and Q_4.

If one or more of the inputs A, B, or C is raised to a logic **1** (-0.75 V), the resulting increase in current through R_3 raises the emitter potential of Q_4, cutting that transistor off. The emitters are now at $-0.75 - V_{BE} \approx 1.5$ V. The forward bias on Q_4 is not sufficient for it to conduct appreciably, resulting in zero voltage drop across R_2 and ≈ -0.75 V (logic **1**) at the OR output. The *IR* drop across R_1 is $(290/1.2 \text{ k}\Omega)(5.2 - 1.5) = 0.77$ V, resulting in ≈ -1.5 V (logic **0**) at the NOR output. Thus the outputs are the stated logic function of the input levels. The great speed of the ECL gate is due to the elimination of the saturated state for the transistors and to the relatively small logic level swing required to go between **1** and **0** (≈ 1.5 V).

The current-steering switch naturally provides differential or "complementary" outputs. The gate designers have taken advantage of this to provide both OR and NOR outputs. Gates of this type are also sometimes called **current-steering logic** (CSL) and **current-mode logic** (CML). The standard ECL propagation delay is about 5 nsec, with flip-flop clocking rates of over 100 MHz and power dissipation of 25–50 mW per gate.

ECL Translators.   In digital systems where various sections operate at different speeds, it is common to use high speed logic where necessary, and to use lower speed logic for economy and lower power dissipation where possible. Systems employing both TTL and ECL have become sufficiently common that manufacturers provide translators in IC form to convert from ECL logic levels to TTL logic levels and vice versa. These translators have input and output circuits that are compatible with the logic systems to which they connect.

C-MOS Gates

Gates made with field-effect transistors offer the advantages of high input impedance, low power consumption, greater circuit simplicity, and further reduced size. With these advantages goes a disadvantage—significantly slower speed due to high input capacitance. A number of FET gate types have been manufactured, including designs based on p-channel FET's (P-MOS) and n-channel FET's (N-MOS). While successful for many large scale integrated circuits (LSI) such as large memories, calculators, long

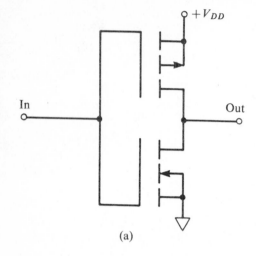

(a)

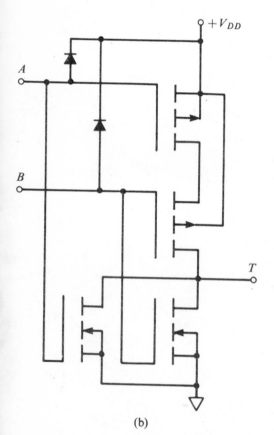

(b)

shift registers, etc., neither of these types has caught on as a substitute for TTL for basic gate and flip-flop applications. As a result, almost all P-MOS or N-MOS LSI circuits provide TTL compatible input and output logic levels for interfacing to other digital circuits.

A logic family made of complementary p-channel and n-channel pairs of MOSFET transistors (C-MOS) appears likely to dominate the MOS IC field soon.

The C-MOS family is based on the inverter shown in Fig. C–7a. The simplicity of the inverter is apparent from the diagram. Only the complementary transistors are required. If the input voltage is high, the n-channel FET turns ON, providing a low resistance path from the output to ground. The p-channel FET is OFF when the input voltage is high, providing a very high resistance path to the positive supply. With a low input voltage, the upper p-channel FET is ON, while the n-channel FET turns OFF. This provides a low resistance path from the output to the positive supply and a high resistance path to ground. Hence a high input voltage provides a low output voltage, and vice versa. C-MOS logic functions are designed around the basic inverter as illustrated by the two-input NOR gate circuit of Fig. C–7b. The n-channel FET's driven by the two inputs are in parallel across the output, so that if either A OR B is HI, the output is LO. Thus the equation $\overline{T} = A + B$ applies, which is the NOR function. The p-channel FET's are in series between the positive supply voltage and the output, so that the output is HI only when A AND B are both LO. Thus the equation $T = A \cdot B$, applies, which by DeMorgan's theorem is the NOR function.

The C-MOS logic family has a variety of significant characteristics which make it attractive for many applications. One of the most significant parameters is that C-MOS logic can operate over a wide range of power supply voltages. Typical supply voltages can range from 3 to 18 V, while TTL logic must operate in a narrow range of supply voltages (generally 4.75 to 5.25 V). Low power dissipation is also a significant advantage of C-MOS. Power dissipation is often two to three orders of magnitude lower than that of TTL logic. The exact power dissipation depends on the operating frequency, but is usually in the µW range.

Fig. C–7 C-MOS gate circuits: (a) inverter; (b) two-input NOR gate.

Since the gate output is driven to both HI and LO levels and since the gate input current is very low, the fan-out of this series is very high (>50 to other C-MOS gate inputs). The propagation delay depends on the supply voltage, but is about 60 nsec for 5 V. The flip-flop clock rate for a 5 V supply is about 4 MHz. The noise immunity is claimed to be 30–40% of the supply voltage.

Interfacing C-MOS to TTL.   When C-MOS circuits are operated from a +5 V power supply, their outputs are directly compatible with TTL input logic levels. The fan-out of a standard C-MOS gate is about 0.25 standard TTL load units, making it directly compatible with a single LPTTL input. Driver gates are available with a fan-out of 2 standard TTL load units. The TTL gate outputs can drive C-MOS inputs if a 2 kΩ resistor is connected from the gate output to the +5 V supply. This is required to ensure a positive HI level signal under the low current drain of the C-MOS input. The dc current available would provide a very large fan-out. However, the C-MOS input capacitance causes a propagation delay of 3 nsec per input connected to the TTL output.

Glossary of Digital Terms

Appendix D

Many of these definitions were obtained from the *Pocket Dictionary of Integrated Circuit Terminology* (Sylvania Electric Products, Inc., Semiconductor Division, Woburn, Mass.).

Access time   (1) The time interval required to communicate with a memory or storage unit. (2) The time interval between the instant at which a circuit calls for information from the memory and the instant at which the information is delivered.

Adder   Switching units that combine binary bits to generate the sum and carry of these bits. Takes the bits from the two binary numbers to be added (*addend* and *augend*) plus the carry from the preceding less significant bit and generates the sum and the carry.

Address   Noun: a location, either name or number, where information is stored in a computer. Verb: to select or pick out the location of a stored information set for access.

AND   A Boolean logic expression used to identify the logic operation wherein given two or more variables, all must be logic **1** for the result to be logic **1**. The AND function is graphically represented by the dot ($\cdot$) symbol.

Anticipated carry adder   A parallel *adder* in which each stage is capable of looking back at all *addend* and *augend* bits of less significant stages and deciding whether the less significant bits provide a **0** or a **1** *carry in*. Having determined the *carry in,* it combines it with its own *addend* and *augend* to give the *sum* for that bit or stage. Also called *fast adder* or look ahead *carry adder*.

Asynchronous inputs   Those terminals in a flip-flop that can affect the output state of the flip-flop independent of the clock. Called set, preset, reset or dc set and reset, or clear.

Binary-coded decimal (BCD)   A binary numbering system for coding decimal numbers in groups of four bits. The binary value of these four-bit groups ranges from 0000 to 1001 and codes the decimal digits 0 through 9. To count to 9 takes four bits; to count to 99 takes two groups of four bits; to count to 999 takes three groups of four bits.

Binary logic   Digital logic elements which operate with two distinct states. The two states are variously called TRUE and FALSE, HI and LO, ON and OFF, or **1** and **0**. In signals they are represented by two different voltage levels. The level which is more positive (or less negative) than the other is called the HI level, the other the LO level. If the true (**1**) level is the more positive voltage, such logic is referred to as positive true or positive logic.

Bit   A synonym for binary numeral. Also refers to a single binary numeral in binary word.

Boolean algebra   The mathematics of logic which uses alphabetic symbols to represent logical variables and **1** and **0** to represent states. There are three basic logic operations in this algebra: AND, OR, and NOT. (See also NAND, NOR, INVERT, which are combinations of the three basic operations.)

Buffer   A circuit element which is used to isolate between stages or to handle a large fan-out or to convert input and output circuits for signal level compatibility.

Chip (die)   A single piece of silicon which has been cut from a slice by scribing and breaking. It can contain one or more circuits but is packaged as a unit.

Clear   An asynchronous input. Also called "reset." It is used to restore a memory element or flip-flop to a "standard" state, forcing the Q terminal to logic **0**.

Clock   A pulse generator that controls the timing of switching circuits and memory states. It serves to synchronize all operations in a digital system.

Clock input   That terminal on a flip-flop whose condition or change of condition controls the admission of data into a flip-flop through the synchronous inputs and thereby controls the output states of the flip-flop. The clock signal performs two functions: (1) it permits data signals to enter the flip-flop; (2) after entry, it directs the flip-flop to change state accordingly.

CML (current mode logic)   Logic in which transistors operate in the unsaturated mode, as distinguished from most other logic types, which operate in the saturation region. This logic has very fast switching speeds and low logic swings. Also called ECL or MECL.

Complement   Verb: to reverse the state of a storage device or of a control level (e.g., changing a flip-flop from the **1** to the **0** state or vice versa). Noun: the complement of a number (base or base minus 1).

Counter A register capable of changing states in a specified sequence upon receiving appropriate input signals. The output of the counter indicates the number of pulses which have been applied. (See also Divider.) A counter is made from flip-flops and some gates. The output of all flip-flops is accessible to indicate the exact count at all times.

Counter, binary An interconnection of flip-flops having a single input so arranged as to enable binary counting. Each time a pulse appears at the input, the counter changes state and tabulates the number of input pulses for readout in binary form. It has 2^n possible counts, where n is the number of flip-flops.

Counter, ring A special form of counter sometimes called a Johnson or shift counter which has very simple wiring and is fast. It forms a loop or circuits of interconnected flip-flops so arranged that only one is **0** and that as input signals are received, the position of the **0** state moves in sequence from one flip-flop to another around the loop until they are all **0**, then the first one goes to **1** and this moves in sequence from one flip-flop to another until all are **1**. It has $2 \times n$ possible counts, where n is the number of flip-flops.

Data Term used to denote facts, numbers, letters, symbols, binary bits presented as logic levels. In a binary system data can only be **0** or **1**.

Decimal A system of numerical representation which uses ten numbers 0, 1, 2, 3, . . . , 9. Each numeral is called a digit. A numbering system to the radix 10.

Delay The slowing up of the propagation of a pulse either intentionally, such as to prevent inputs from changing while clock pulses are present, or unintentionally, as caused by transistor rise and fall time pulse response effects.

Discrete circuits Electronic circuits built of separate, individually manufactured, tested and assembled diodes, resistors, transistors, capacitors, and other specific electronic components.

Divider, frequency A counter that has a gating structure added which provides an output pulse after receiving a specified number of input pulses. The outputs of all flip-flops are not accessible.

Driver An element that is coupled to the output stage of a circuit in order to increase its power or current handling capability or fan-out; for example, a clock driver is used to supply the current necessary for a clock line.

DTL (diode-transistor logic) Logic employing diodes with transistors used only as inverting amplifiers.

Dynamic register A MOS memory or shift register which must be refreshed, clocked, or cycled periodically to maintain the stored data.

Enable   To permit an action or the acceptance or recognition of data by applying appropriate signals (generally a logic **1** in a positive logic) to the appropriate input. (See Inhibit.)

Exclusive-OR   A logic function whose output is **1** if either of the two variables is **1** but whose output is **0** if both inputs are **1** or both are **0**.

Fan-out   The number of input load units that can be driven by a circuit output.

Fast adder   See Anticipated carry adder and Adder.

Flip-flops (storage elements)   A circuit having two stable states and the capability of changing from one state to another with the application of a control signal and remaining in that state after removal of signals.

Flip-flop, D   *D* stands for delay. A flip-flop whose output is a function of the input that appeared one pulse earlier; for example, if a **1** appeared at the input, the output after the next clock pulse will be a **1**.

Flip-flop, JK   A flip-flop having two inputs designated *J* and *K*. At the application of a clock pulse, a **1** on the *J* input and a **0** on the *K* input will set the flip-flop to the **1** state; a **1** on the *K* input and a **0** on the *J* input will reset it to the **0** state; and **1**'s simultaneously on both inputs will cause it to change state regardless of the previous state. $J = $ **0** and $K = $ **0** will prevent change.

Flip-flop, RS   A flip-flop consisting of two cross-coupled NAND gates having two inputs designated *R* and *S*. A **1** on the *S* input and a **0** on the *R* input will reset (clear) the flip-flop to the **0** state, and **1** on the *R* input and **0** on the *S* input will set it to the **1**. It is assumed that **0**'s will never appear simultaneously at both inputs. If both inputs have **1**'s, it will stay as it was; **1** is considered nonactivating. A similar circuit can be formed with NOR gates.

Flip-flop, RST   A flip-flop having three inputs, *R, S,* and *T*. This unit works like the *RS* flip-flop except that the *T* input is used to cause the flip-flop to change states.

Flip-flop, T   A flip-flop having only one input, *T*. A pulse appearing on the input will cause the flip-flop to change states. Used in ripple counters.

Gate, AND   All inputs must have **1** level signals at the input to produce a **1** level output.

Gate, NAND   All inputs must have **1** level signals at the input to produce a **0** level output.

Gate, NOR   Any one input or more than one input having a **1** level signal will produce a **0** level output.

Gate, OR   Any one input or more than one input having a **1** level signal will produce a **1** level output.

Gates (decision elements)   A circuit having two or more inputs and one output. The output depends upon the combination of logic signals at the input.

Half-adder   A switching circuit that combines binary bits to generate the sum and the carry. It can only take in the two binary bits to be added and generate the sum and carry. (See also Adder.)

Hybrid   A method of manufacturing integrated circuits by using a combination of monolithic and thick film techniques.

Inhibit   To prevent an action or acceptance of data by applying an appropriate signal to the appropriate input (generally a logic **0** in positive logic). (See Enable.)

Input load unit   The load provided by a single gate input of the standard gate in a given logic family.

Integrated circuit (EIA definition)   "The physical realization of a number of electrical elements inseparably associated on or within a continuous body of semiconductor material to perform the functions of a circuit." (See Slice and Chip.)

Inverter   A circuit whose output is always in the opposite state from the input. This is also called a NOT circuit. (A teeter-totter is a mechanical inverter.)

Logic swing   The voltage difference between the two logic levels **1** and **0**.

Modulus   The number of states through which a counting register sequences during each complete cycle.

Monolithic   Refers to the single silicon substrate in which an integrated circuit is constructed. (See Integrated circuit.)

MOS   Abbreviation for metal-oxide semiconductor and for MOS integrated circuits incorporating MOS field-effect transistors.

NAND   A Boolean logic operation that yields a logic **0** output when all logic input signals are logic **1**.

Negative logic   Logic in which the more negative voltage represents the **1** state; the less negative voltage represents the **0** state. (See Binary logic.)

Noise immunity   A measure of the insensitivity of a logic circuit to triggering or reaction to spurious or undesirable electrical signals or noise, largely determined by the signal swing of the logic. Noise can be in either of two directions, positive or negative.

NOR   A Boolean logic operation that yields a logic **0** output with one or more true **1** input signals.

NOT   A Boolean logic operation indicating negation, not **1**. Actually an inverter. If input is **1**, output is NOT **1** but **0**. If the input is **0**, output is NOT **0** but **1**. Graphically represented by a bar over a Boolean symbol such as A. $\overline{A}$ means "when A is not **1**. . . ."

OR   A Boolean logic operation used to identify the logic operation wherein two or more TRUE **1** inputs only add to one TRUE **1** output. Only one input needs to be TRUE to produce a TRUE output. The graphical symbol for OR is a plus sign $(+)$.

Overflow   In a counter or register, the production of a number that is beyond the storage capacity of the counter or register. The extra number may be held in an "overflow element."

Parallel   This refers to the technique for handling a binary data word that has more than one bit. All bits are transmitted and/or acted upon simultaneously. (See also Serial.)

Parallel adder   A conventional technique for adding where the two multibit numbers are presented and added simultaneously (parallel). A ripple adder is still a parallel adder; the carry is rippled from the least significant to the most significant bit. Another type of parallel adder is the "look ahead," or "anticipated carry" adder. (See Ripple adder, Anticipated carry adder, and Adder.)

Parallel operation   The organization of data manipulation where all the digits of a word are transmitted simultaneously on separate lines in order to speed up operation, as opposed to serial operation.

Parity check   Use of a digit, called the "parity digit," carried along as a check which is **1** if the total number of ones in the machine word is odd, and **0** if the total number of ones in the machine word is even (odd parity). Even parity uses the reverse conditions.

Positive logic   Logic in which the more positive voltage represents the **1** state; e.g., **1** $= +3.45$ V, logic **0** $= +0.45$ V. (See Binary logic.)

Preset   An input like the "set" input and which works in parallel with the set.

PROM   Abbreviation for programmable read-only memory. A ROM that can be programmed (usually irreversibly) by the user.

Propagation delay   A measure of the time required for a change in logic level to be transmitted through an element or a chain of elements.

Propagation time   The time necessary for a unit of binary information (high voltage or low) to be transmitted or passed from one physical point in a system or subsystem to another, for example, from input of a device to output.

Q **output**   The reference output of a flip-flop. When this output is **1**, the flip-flop is said to be in the **1** state; when it is **0**, the output is said to be in the **0** state. (See also State and Set.)

Random access memory (RAM)   A memory in which every stored word is as easily and quickly accessed as any other.

Read   To acquire information, usually from some form of storage.

Read-only memory (ROM)   A memory in which the stored words are permanent, that is, they cannot be altered or lost. It serves as a look-up table.

Real time   In solving a problem, a speed sufficient to give an answer in the actual time during which the problem must be solved.

Real-time operations   Processing data in the time scale of a physical process so that the results are useful in guiding the physical process. Also, solving problems in real time.

Register   A number of storage devices (usually flip-flops) used to store a certain number of bits. For example, a four-bit register requires four flip-flops.

Reset   Also called "clear." Similar to set except that it is the input through which the *Q* output can be made to go to **0**.

Ripple   The transmission of data serially. It is a serial reaction analogous to a bucket brigade or a row of falling dominoes.

Ripple adder   A binary adding system similar to the system most people use to add decimal numbers, that is, add the units column, get the carry, add it to the 10's column, get the carry, add it to the 100's column, and so on. Again it is necessary to wait for the signal to propagate through all columns even though all columns are present at once (parallel). Note that the carry is rippled.

Ripple counter   A binary counting system in which flip-flops are connected in series. When the first flip-flop changes, it affects the second, which affects the third, and so on. If there are ten in a row, the signal must go sequentially from the first flip-flop to the tenth.

RTL (resistor-transistor logic)   Logic is performed by resistors; transistors are used to produce an inverted output.

Serial   This refers to the technique for handling a binary data word which has more than one bit. The bits are acted upon one at a time.

Serial operation   The organization of data manipulation within computer circuitry where the digits of a word are transmitted one at a time along a single line. The serial mode of operation is slower than parallel operation, but utilizes less complex circuitry.

Set   An input on a flip-flop not controlled by the clock (see Asynchronous inputs) and used to affect the Q output. It is this input through which signals can be entered to get the Q output to go to **1**. Note that it cannot get Q to go to **0**.

Shift   The process of moving data from one place to another. Generally many bits are moved at once. Shifting is done synchronously and by command of the clock. An eight-bit word can be shifted sequentially (serially); that is, the first bit goes out, second bit takes the first bit's place, the third bit takes the second bit's place, and so on, in the manner of a bucket brigade. Generally referred to as shifting left or right. It takes eight clock pulses to shift an eight-bit word or all bits of a word can be shifted simultaneously. This is called parallel load or parallel shift.

Shift register   An arrangement of flip-flops which is used to shift serially or in parallel. Binary words are generally parallel loaded and then held temporarily or serially shifted out.

Skewing   Refers to time delay or offset between any two signals in relation to each other.

Slice   A single wafer cut from a silicon ingot forming a thin substrate on which all active and passive elements for multiple integrated circuits have been fabricated, utilizing semiconductor epitaxial growth, diffusion, passivation, masking, photo resist, and metallization technologies. A completed slice generally contains hundreds of individual circuits. (See Chip.)

State   This refers to the condition of an input or output of a circuit as to whether it is a logic **1** or a logic **0**. The state of a circuit (gate or flip-flop) refers to its output. The flip-flop is said to be in the **1** state when its Q output is **1**. A gate is in the **1** state when its output is **1**.

Static register   A memory or shift register which will maintain the stored data without periodic refreshing. The opposite of the dynamic register.

Synchronous   Operation of a switching network by a clock pulse generator. All circuits in the network switch simultaneously. All actions take place synchronously with the clock.

Synchronous inputs   Those terminals on a flip-flop through which data can be entered, but only upon command of the clock. Unlike those of a gate, these inputs do not have direct control of the output, but only when the clock permits and commands. Called *JK* inputs or ac set and reset inputs.

Thick film   A method of manufacturing integrated circuits by depositing thin layers of materials on an insulated substrate (often ceramic) to perform electrical functions; usually only passive elements are made this way.

Toggle   To switch between two states, as in a flip-flop.

Transfer   To transfer information from one register to another without modifying it, i.e., to copy, exchange, read, record, store, transmit, transport, or write data.

Trigger   A timing pulse used to initiate the transmission of logic signals through the appropriate circuit signal paths.

Truth table   A chart that tabulates and summarizes all the combinations of possible states of the inputs and outputs of a circuit. It tabulates what will happen at the output for a given input combination.

TTL, T²L (transistor-transistor logic)   A logic system which evolved from diode-transistor logic wherein the multiple-diode cluster is replaced by a multiple-emitter transistor but is commonly applied to a circuit which has a multiple-emitter input and an active pull-up network.

Turn-on time   The time required for an output to turn on (to sink current, to ground output, to go to 0 V). It is the propagation time of an appropriate input signal to cause the output to go to 0 V.

Turn-off time   Same as turn-on time except that the output stops sinking current, goes off, and/or goes to a high voltage level.

Wired OR   Externally connected separate circuits or functions arranged so that the combination of their outputs results in an AND function. However, common usage is that the point at which the separate circuits are wired together will be **0** if any one of the separate outputs is a **0**. The same as a dot AND.

Word   An ordered set of characters which has at least one meaning and is stored, transferred, or operated upon as a unit. Also called "machine word" or "information word."

Bibliography

Eimbinder, Jerry (Ed.), *Application Considerations for Linear Integrated Circuits,* Wiley-Interscience, New York, N.Y., 1970.

A collection of papers, largely from industry sources, which are very useful in the selection and application of several types of linear integrated circuits including operational amplifiers, transconductance analog multipliers, C-MOS analog switches, voltage regulators, transmission line drivers and receivers, and comparators.

Enke, C. G., "Data Domains—An Analysis of Digital and Analog Instrumentation Systems and Components," *Analytical Chemistry,* **43,** Jan., 1971, p. 69A.

A formal development of the data domain concepts as they are related to the basic measurement process and applied in scientific instrumentation.

Gillie, Angelo C., *Pulse and Logic Circuits,* McGraw-Hill, New York, N.Y., 1968.

Basically a solid and readable book on *RC* wave-shaping circuits, this book contains an unusually helpful section on coaxial cable transmission and delay lines.

Ingle, J. D., Jr., and S. R. Crouch, "Pulse Overlap Effects on Linearity and Signal-to-Noise Ratio in Photon Counting Systems," *Analytical Chemistry,* **44,** April, 1972, p. 777.

A mathematical development of the influence of pulse overlap on photon counting systems. The treatment reveals that the relationship between the observed and true pulse rate depends on the type of counting system, the system dead time, and the discriminator level.

Kime, R. C. Jr., "The Charge-Balancing A-D Converter: An Alternative to Dual Slope Integration," *Electronics,* **46,** No. 11, p. 97, May 24, 1973.

A clear and well-illustrated presentation of the principle of the charge-balance ADC.

Kintner, Paul M., *Electronic Digital Techniques,* McGraw-Hill, New York, N.Y., 1968.

A digital circuits book which, though not centered around IC logic, contains many clever solutions to a variety of practical problems.

Malmstadt, H. V., and C. G. Enke, *Digital Electronics for Scientists,* W. A. Benjamin, Menlo Park, Calif., 1969.

A treatment of digital techniques directed toward instrumentation applications.

Malmstadt, H. V., C. G. Enke, and E. C. Toren, Jr., *Electronics for Scientists,* W. A. Benjamin, Menlo Park, Calif., 1972.

An earlier work containing several still relevant sections on analog measurement techniques and devices.

Malmstadt, H. V., M. L. Franklin, and G. Horlick, "Photon Counting for Spectrophotometry," *Analytical Chemistry,* **44,** July, 1972, p. 63A.

A general treatment of photon counting systems especially as applied to spectrophotometry. The discussion includes instrumentation, practical considerations in system design, and a variety of spectrometric applications.

Millman, J., and H. Taub, *Pulse, Digital and Switching Waveforms,* McGraw-Hill, New York, N.Y., 1965.

The standard sourcebook for switching and wave-shaping and component gate and flip-flop circuits.

Morris, R. L., and J. R. Miller (Eds.), *Designing with TTL Integrated Circuits,* McGraw-Hill, New York, N.Y., 1971.

A very helpful guide to the application of TTL integrated circuit gates, flip-flops, and some MSI circuits.

Oliver, B. M., and J. M. Cage (Ed.), *Electronic Measurements and Instrumentation,* McGraw-Hill, New York, N.Y., 1971.

A collection of scholarly and comprehensive articles on a variety of measurement systems and problems, including frequency and time measurement, impedance measurement, instrumentation amplifiers, recorders and oscilloscopes, and voltage and current measurement.

Schmid, H., *Electronic Analog/Digital Conversions.* Van Nostrand Reinhold Co., New York, N.Y., 1970.

A modern and comprehensive treatment of DAC and ADC techniques and circuits.

Sheingold, D. H. (Ed.), *Analog-Digital Conversion Handbook.* Analog Devices, Inc., Norwood, Mass., 1972.

A manufacturer's book which contains a good introduction to basic DAC and ADC devices and techniques. The treatment is practical and oriented toward data acquisition systems. Descriptions of multiplexers, signal amplifiers, and sample-and-hold circuits are also included.

Smith, J. I., *Modern Operational Circuit Design,* Wiley-Interscience, New York, N.Y., 1971.

An applications-oriented treatment of operational amplifiers, which contains a great number and variety of useful and practical circuits.

Smith, R. J., *Circuits, Devices, and Systems,* 2nd ed., Wiley and Sons, New York, N.Y., 1971.

A basic book in electrical engineering, which includes relevant sections on ac circuit theory, digital devices, and logic circuits. There are many relevant problems and illustrations.

Tobey, G. E., J. G. Graeme, and L. P. Huelsman (Eds.), *Operational Amplifiers,* McGraw-Hill, New York, N.Y., 1971.

A comprehensive treatment of the design and applications of operational amplifiers. The applications discussion includes linear and nonlinear circuits, waveform generators, ADC's, DAC's, and sampling networks.

Solutions to Problems

SOLUTIONS TO PROBLEMS IN SECTION 3–2

1. Substitute into Eq. (3–9) $R_1 = 5 \times 10^6 \ \Omega$, $R_2 = 10^6 \ \Omega$, and $R_3 = 5 \times 10^3 \ \Omega$.

$$R_{f(eq)} = \frac{5 \times 10^6 \times 10^6 + 10^6 \times 5 \times 10^3 + 5 \times 10^6 \times 5 \times 10^3}{5 \times 10^3},$$

$$R_{f(eq)} = 10^9 \ \Omega.$$

2. From Eq. (3–18), $\beta = \dfrac{R_2}{R_1 + R_2}$, $\dfrac{v_o}{v_{in}} = \dfrac{1}{\beta}$.

a) $\beta = \dfrac{10 \ k\Omega}{500 \ k\Omega + 10 \ k\Omega} = \dfrac{1}{51}$, gain $= 51$.

b) $\beta = \dfrac{50 \ k\Omega}{50 \ k\Omega + 50 \ k\Omega} = \dfrac{1}{2}$, gain $= 2$.

c) $\beta = \dfrac{5 \ k\Omega}{100 \ k\Omega + 5 \ k\Omega} = \dfrac{5}{105}$, gain $= \dfrac{105}{5} = 21$.

3. From Fig. 3–25, $v_o = K(a + b + 1)(v_2 - v_1)$ and the difference gain $G_d = K(a + b + 1)$. Since

$$a = \frac{aR_1}{R_1} = \frac{5 \ k\Omega}{1 \ k\Omega} = 5, \quad b = \frac{bR_1}{R_1} = \frac{3 \ k\Omega}{1 \ k\Omega} = 3,$$

and $K = \dfrac{KR_2}{R_2} = \dfrac{10^6 \ \Omega}{10^5 \ \Omega} = 10$, $G_d = 10(5 + 3 + 1) = 90$.

4. The Fourier series of a square wave, Eq. (3–60), can be generalized as

$$v(t) = \frac{4V}{\pi} \sum_{n=0}^{\infty} \frac{\sin(2n+1)\omega t}{2n+1},$$

where $2n+1$ is the harmonic. For the problem, n runs from 0 to 5 to generate the fundamental through eleventh harmonic. The amplitude coefficient A of a given harmonic is

$$A = \frac{4V}{\pi}\left(\frac{1}{2n+1}\right),$$

where $V = 5$ V peak. Calculating the amplitude at each odd harmonic, we find

$$A\,(1\text{ kHz}) = \frac{4 \times 5\text{ V}}{3.14} \times 1 = 6.38\text{ V},$$

$$A\,(3\text{ kHz}) = \frac{6.38\text{ V}}{3} = 2.12\text{ V},$$

$$A\,(5\text{ kHz}) = \frac{6.38\text{ V}}{5} = 1.28\text{ V},$$

$$A\,(7\text{ kHz}) = \frac{6.38\text{ V}}{7} = 0.91\text{ V},$$

$$A\,(9\text{ kHz}) = \frac{6.38\text{ V}}{9} = 0.71\text{ V},$$

$$A\,(11\text{ kHz}) = \frac{6.38\text{ V}}{11} = 0.58\text{ V}.$$

The frequency spectrum would be similar to Fig. 3–25 with the above values used for A and ω.

5. Rearranging Eq. (3–64), we have

$$f = \frac{1}{2\pi X_C C} = \frac{1}{2 \times 3.14 \times 2.5 \times 10^3\ \Omega \times 10^{-8}\text{ F}},$$

$$f = 6.4\text{ kHz}.$$

6. Substituting into Eq. (3–64), we find

$$X_C = \frac{1}{2 \times 3.14 \times 5 \times 10^{-8}\,\mathrm{F} \times f} = \frac{3.2 \times 10^6}{f}\,\Omega.$$

Thus

a) $X_C = \dfrac{3.2 \times 10^6}{10^2} = 32\ \mathrm{k}\Omega,$

b) $X_C = \dfrac{3.2 \times 10^6}{10^3} = 3.2\ \mathrm{k}\Omega,$

c) $X_C = \dfrac{3.2 \times 10^6}{10^4} = 320\ \Omega.$

7. Rearranging Eq. (3–71), we have

$$f = \frac{X_L}{2\pi L} = \frac{1\ \mathrm{k}\Omega}{2 \times 3.14 \times 0.2\ \mathrm{H}} = 80\ \mathrm{Hz}.$$

8. Substituting into Eq. (3–71), we find

$$X_C = 2 \times 3.14 \times 0.02f = 0.125f \text{ in } \Omega.$$

Thus

a) $X_C = 0.125 \times 100 = 12.5\ \Omega,$

b) $X_C = 0.125 \times 10^3 = 125\ \Omega,$

c) $X_C = 0.125 \times 10^4 = 1.25\ \mathrm{k}\Omega.$

9. Rearrange Eq. (3–80), and solve for R:

$$R^2 = Z^2 - X_C{}^2 = (98)^2 - (80)^2 = 3204,$$

$$R = 56.5\ \Omega.$$

Or use Eq. (3–81):

$$R = \frac{X_C}{\tan \theta} = \frac{80}{1.41} \approx 56.5\ \Omega.$$

10. Use Eq. (3–80) to find Z and Eq. (3–81) to find θ. Find X_C from Eq. (3–64):

$$X_C = \frac{1}{2 \times 3.14 \times 100 \times 5 \times 10^{-7}} = 3.18 \text{ k}\Omega,$$

$$Z = \sqrt{(3 \text{ k}\Omega)^2 + (3.18 \text{ k}\Omega)^2} = 4.35 \text{ k}\Omega,$$

$$\theta = \arctan \left(\frac{3.18 \text{ k}\Omega}{3 \text{ k}\Omega} \right) = 46.7°$$

with the current leading the voltage.

11. Use Eqs. (3–80) and (3–81) to find Z and θ:

$$Z = \sqrt{(500 \ \Omega)^2 + (400 \ \Omega)^2} = 640 \ \Omega,$$

$$\theta = \arctan \frac{400}{500} = 38.6°$$

with the current leading the voltage.

From Eq. (3–80), $I_p = \dfrac{V_p}{Z} = \dfrac{150 \text{ V}}{640 \ \Omega} = 234 \text{ mA}.$

$$V_{pR} = I_p R = 0.234 \times 500 = 117 \text{ V},$$

$$V_{pC} = I_p X_C = 0.234 \times 400 = 93.6 \text{ V},$$

$$P_R = I_{\text{rms}}^2 R = (0.707 \times 0.234)^2 \times 500 = 13.75 \text{ W}.$$

12. Rearrange Eq. (3–87) to find C:

$$C = \frac{1}{2\pi f_1 R} = \frac{1}{2 \times 3.14 \times 200 \times 5 \times 10^3} = 0.16 \ \mu\text{F}.$$

13. Substitute into $f_2 = \dfrac{1}{2\pi R C}$ to give

$$f_2 = \frac{1}{2 \times 3.14 \times 2.5 \times 10^3 \times 0.05 \times 10^{-6}} = 1.27 \text{ kHz}.$$

14. Use Eqs. (3–99) and (3–100) to find Z and θ:

$$Z = \sqrt{(100)^2 + (50)^2} = 112 \ \Omega,$$

$$\theta = \arctan \frac{50}{100} = 26.5°$$

with the current lagging the voltage.

$$I_p = \frac{V_p}{Z} = \frac{25 \text{ V}}{112 \ \Omega} = 223 \text{ mA},$$

$$V_{pR} = I_p R = 0.223 \text{ A} \times 100 \ \Omega = 22.3 \text{ V},$$

$$V_{pL} = I_p X_L = 0.223 \text{ A} \times 50 \ \Omega = 11.15 \text{ V}.$$

15. Substitute into Eq. (3–105) and solve for R:

$$Z = \sqrt{R^2 + (X_L - X_C)^2},$$

$$Z^2 = R^2 + (X_L - X_C)^2,$$

$$R^2 = Z^2 - (X_L - X_C)^2 = (150)^2 - (200 - 100)^2,$$

$$R^2 = 12500 \ \Omega^2,$$

$$R = 112 \ \Omega.$$

16. Substituting into Eq. (3–107) $L = 150 \ \mu\text{H}$ and $C = 50 \text{ pF}$, we have

$$f_o = \frac{1}{2 \times 3.14 \times \sqrt{150 \times 10^{-6} \times 50 \times 10^{-12}}} = 1.84 \text{ MHz}.$$

17. Assuming that the source resistance is negligible, Eq. (3–130) can be used for f_1. To find f_2 use Eq. (3–132):

$$f_1 = \frac{1}{2\pi C_{\text{in}} R_{\text{in}}} = \frac{1}{2 \times 3.14 \times 1 \times 10^{-6} \times 10 \times 10^3} = 16 \text{ Hz},$$

$$f_2 = \frac{1}{2\pi C_s R_s} = \frac{1}{2 \times 3.14 \times 100 \times 10^{-12} \times 100 \times 10^3} = 16 \text{ kHz}.$$

Draw a Bode diagram as in Fig. 3–79 with slopes of ± 20 dB/decade outside the bandpass and 3 dB points at 16 Hz and 16 kHz.

18. Find f_{24} from Eq. (3–137) and f_{14} from Eq. (3–138):

$$f_{14} = (2^{1/4} - 1)^{-1/2} \, 50 \text{ Hz} = 115 \text{ Hz},$$

$$f_{24} = (2^{1/4} - 1)^{1/2} \, 10^5 \text{ Hz} = 43.5 \text{ kHz}.$$

19. a) Rearranging Eq. (3–139), we have

$$V_p = \frac{V_{\text{av}}}{0.637} = \frac{25 \text{ V}}{0.637} = 39.3 \text{ V}.$$

b) $V_{\text{p-p}} = 2V_p = 2 \times 39.3 \text{ V} = 78.6 \text{ V}.$

c) Substituting $V_p = 39.3$ V into Eq. (3–141) and rearranging, we have

$$V_{rms} = \frac{39.3 \text{ V}}{1.414} = 27.8 \text{ V}.$$

20. Do part (b) first to obtain V_p.

b) From Eq. (3–141),

$$V_p = 1.414 V_{rms} = 1.414 \times 110 \text{ V} = 155.5 \text{ V}.$$

a) From Eq. (3–139), $V_{av} = 0.637 \times 155.5$ V $= 99.5$ V.

c) $V_{p\text{-}p} = 2V_p = 2 \times 155.5$ V $= 311$ V.

SOLUTIONS TO PROBLEMS IN SECTION 3–3

1. The comparator will be at limit if v_{in} exceeds

$$\frac{\pm \text{limit}}{\text{open-loop gain}}.$$

So if $v_{in} > \pm 13$ V/$(4 \times 10^4) = \pm 325$ μV, the comparator is at limit. The comparator is not at limit only for voltages within 325 μV of common.

2. In a triggered sweep oscilloscope, the input signal starts the sweep when it crosses a preset threshold level. The calibrated time base makes it possible to observe fractions of cycles or even single events which are not repetitive. The synchronized sweep generator, however, must oscillate at the same frequency as the input signal or some lower multiple. This makes it impossible to observe fractions of cycles. For a diagram, see Fig. 1–66 of Module 1.

3. The diagram would consist of the 60 Hz source, a comparator for shaping purposes, and three scalers in series. The first scaler would divide by 6 to produce a 10 Hz output; the second scaler would divide by 600 to produce an output period of $3600/60 = 60$ sec/cycle. The third scaler would divide by 60 to produce an output period of 3600 sec/cycle.

4. The phase shift error θ_e is the phase shift contributed by the oscilloscope. From Eqs. (3–81) and (3–64),

$$\theta_e = \arctan \left(\frac{X_C}{R} \right) \quad \text{and} \quad X_C = \frac{1}{2\pi f C}.$$

Thus

$$\theta_e = \arctan\left(\frac{1}{2\pi fRC}\right) = \arctan\left(\frac{1}{2 \times 3.14 \times 40 \times 10^6 \times 10^{-6}}\right)$$

or

$$\theta_e = 0.234°.$$

5. Since there is a ± 1 count uncertainty due to the fact that the two signals are not synchronized, a scaling factor of 1000 is necessary for the gate control signal to obtain the ratio within 0.1%, or one part per thousand. Since the measurement requires 1000 cycles and $f = 100$ Hz, the time required is $1000/100 = 10$ sec.

6. The relative error RE in percent can be found from

$$RE = \frac{\text{count uncertainty}}{\text{total counts}} \times 100.$$

For example, for the 50 Hz signal in the frequency mode,

$$RE = \frac{1}{50} \times 100 = 2\%.$$

For the 250 kHz signal in the period mode,

$$RE = \frac{1}{4} \times 100 = 25\%.$$

The other values can be found similarly and the table below constructed.

f of signal	RE in frequency mode	RE in period mode
50 Hz	2%	0.005%
3 kHz	0.033%	0.33%
250 kHz	0.004%	25%

7. We want signals of frequency $\omega_1 + \omega_2$ effectively attenuated by the low pass filter, but signals of frequency $\omega_1 - \omega_2$ unattenuated. For $f_1 \approx f_2 \approx 1$ kHz, this can be accomplished if the upper cutoff frequency of the filter is set to 1 kHz. Thus

$$1 \text{ kHz} = \frac{1}{2\pi RC}$$

and

$$RC = \frac{1}{2 \times 3.14 \times 10^3} = 1.59 \times 10^{-4} \text{ sec.}$$

From Eq. (3–91) the attenuation for the frequency-doubled term $f = 2f_2$ can be found:

$$|H(j\omega)| = \frac{1}{\sqrt{1 + (2)^2}} = 0.45.$$

Thus $v_o = 0.45(v_1v_2/2)$, so the high frequency term is attenuated approximately in half. For the low frequency term $f/f_2 = 50/1000 = 0.05$, and

$$|H(j\omega)| = \frac{1}{\sqrt{1 + (0.05)^2}} = 0.999.$$

Hence, $v_o = 0.999(v_1v_2/2)$. If more attenuation of the sum term is desired, the difference term will also be more attenuated. One can set the permissible attenuation for the low frequency term, and solve for f_2 from Eq. (3–91). The new RC time constant can then be found.

SOLUTION TO PROBLEMS IN SECTION 3–4

1. The decimal equivalents are found by summing the assigned binary values for every signal in the group that is 1:

    a) $1 \times 2^0 + 1 \times 2^2 + 1 \times 2^4 + 1 \times 2^6 =$
    $$1 + 4 + 16 + 32 = 53.$$

    b) $1 \times 2^0 + 1 \times 2^1 + 1 \times 2^4 + 1 \times 2^6 + 1 \times 2^7 + 1 \times 2^8 =$
    $$1 + 2 + 16 + 64 + 128 + 256 = 467.$$

    c) $1 \times 2^0 + 1 \times 2^2 + 1 \times 2^3 + 1 \times 2^8 =$
    $$1 + 4 + 8 + 256 = 269.$$

2. To convert to binary from decimal, find the largest power of 2 which the number contains; then take the remainder and do likewise. Repeat until the binary equivalent is found.

    a) For example, 132 contains 2^7 (128), so a 1 is placed in the 2^7 column. The remainder (4) contains only 2^2, so a 1 is placed in the 2^2 column. All other powers of 2 contain zeros. So the binary equivalent of 132 is

    $$1\ 0\ 0\ 0\ 0\ 1\ 0\ 0.$$

b)

Remainder	2^6 (64)	2^5 (32)	2^4 (16)	2^3 (8)	2^2 (4)	2^1 (2)	2^0 (1)
57		1					
25			1				
9				1			
1					0	0	1
0							

So $57_{10} = 111001_2$.

c)

Remainder	2^6	2^5	2^4	2^3	2^2	2^1	2^0
28			1				
12				1			
4					1	0	0
0							

So $28_{10} = 11100_2$.

d)

Remainder	2^6	2^5	2^4	2^3	2^2	2^1	2^0
103	1						
39		1					
7			0	0	1		
3						1	
1							1
0							

So $103_{10} = 1100111_2$.

3. In BCD code, each set of four binary bits represents one decimal digit.

a) Thus 132 is represented by

0001　0011　0010

1　　3　　2

b) Decimal 57 is

0101　0111

5　　7

c) Decimal 28 is

$$0010 \quad 1000$$

$$2 \qquad 8$$

d) Decimal 103 is

$$0001 \quad 0000 \quad 0011$$

Note the increased number of bits required in BCD as opposed to binary coding (problem 2).

4. Only a few examples are given here. DeMorgan's Theorems $\overline{A + B} = \overline{A} \cdot \overline{B}$ and $\overline{AB} = \overline{A} + \overline{B}$ can be readily proven as follows:

A	B	$A + B$	$\overline{A + B}$	A	B	$\overline{A}$	$\overline{B}$	$\overline{A} \cdot \overline{B}$
0	0	0	1	0	0	1	1	1
0	1	1	0	0	1	1	0	0
1	0	1	0	1	0	0	1	0
1	1	1	0	1	1	0	0	0

Noting the equality of the right-hand columns of each proves $\overline{A + B} = \overline{A} \cdot \overline{B}$.

A	B	AB	$\overline{AB}$	A	B	$\overline{A}$	$\overline{B}$	$\overline{A} + \overline{B}$
0	0	0	1	0	0	1	1	1
0	1	0	1	0	1	1	0	1
1	0	0	1	1	0	0	1	1
1	1	1	0	1	1	0	0	0

Again the equivalence of $\overline{AB}$ and $\overline{A} + \overline{B}$ proves this form of DeMorgan's Theorem.

5. a) Starting with the left-hand term, we find

$$A(\overline{A} + B) = A\overline{A} + AB.$$

But $A\overline{A} = 0$, so $A(\overline{A} + B) = AB$.

b) Starting with the right-hand term, we find

$$(A + C)(\overline{A} + B) = A\overline{A} + \overline{A}C + AB + BC = \overline{A}C + AB + BC.$$

Multiplying the last term by $(A + \bar{A}) = 1$, we have

$$(A + C)(\bar{A} + B) = \bar{A}C + AB + BC(A + \bar{A})$$
$$= \bar{A}C + AB + ABC + \bar{A}BC$$
$$= \bar{A}C(1 + B) + AB(1 + C) = \bar{A}C + AB.$$

c) Expanding the left-hand term, we find

$$(A + B)(\bar{A} + C) = A\bar{A} + \bar{A}B + AC + BC = \bar{A}B + AC + BC$$
$$= \bar{A}B + AC + BC(A + \bar{A})$$
$$= \bar{A}B + AC + ABC + \bar{A}BC$$
$$= \bar{A}B(1 + C) + AC(1 + B) = \bar{A}B + AC.$$

6. a) By DeMorgan's Theorem,

$$\overline{\bar{A} + B} + \overline{\bar{A} + \bar{B}} = A\bar{B} + AB = A(\bar{B} + B) = A.$$

b) $AB + AC + B\bar{C} = AB(C + \bar{C}) + AC + B\bar{C}$

$$= ABC + AB\bar{C} + AC + B\bar{C}$$
$$= AC(B + 1) + B\bar{C}(A + 1) = AC + B\bar{C}.$$

c) $AB + \bar{B}\bar{C} + A\bar{C} = AB + \bar{B}\bar{C} + A\bar{C}(B + \bar{B})$

$$= AB + \bar{B}\bar{C} + AB\bar{C} + A\bar{B}\bar{C}$$
$$= AB(1 + \bar{C}) + \bar{B}\bar{C}(1 + A) = AB + \bar{B}\bar{C}.$$

7. For the AND function, we want to prove that $\overline{\bar{A} + \bar{B} + \bar{C}} = ABC$. This can be done directly using DeMorgan's Theorem, or the truth table method can be used.

A	B	C	$\bar{A}$	$\bar{B}$	$\bar{C}$	OR	NOR	
0	0	0	1	1	1	1	0	
0	0	1	1	1	0	1	0	
0	1	0	1	0	1	1	0	Note that this is
0	1	1	1	0	0	1	0	the AND function.
1	0	0	0	1	1	1	0	
1	0	1	0	1	0	1	0	
1	1	0	0	0	1	1	0	
1	1	1	0	0	0	0	1	

For the OR function the statement from Table 3–11 is

$$\overline{\overline{A + B + C}} = A + B + C$$

and is directly proven. For the NAND function, the Boolean logic from Table 3–11 is

$$\overline{\overline{A} + \overline{B} + \overline{C}} = \overline{ABC}$$

and this is directly proven by DeMorgan's Theorem.

8. Let $T = 1$ be the lamp ON and $T = 0$ be the lamp OFF and 1's signify closed states of switches A, B, or C.

a)

A	B	C	T
0	0	0	1
0	0	1	0
0	1	0	0
0	1	1	1
1	0	0	1
1	0	1	1
1	1	0	1
1	1	1	1

b) From the truth table,

$$T = \overline{A}\overline{B}\overline{C} + \overline{A}BC + A\overline{B}\overline{C} + A\overline{B}C + AB\overline{C} + ABC.$$

c) This can be simplified as follows:

$$T = \overline{B}\overline{C}(\overline{A} + A) + BC(\overline{A} + A) + A\overline{B}C + AB\overline{C},$$

$$T = \underbrace{\overline{B}\overline{C} + BC}_{1} + \underbrace{A(\overline{B}C + B\overline{C})}_{2} = \underbrace{\overline{B}\overline{C} + BC + A}_{1}$$

Term 1 is obtained by putting logic level signals representing B and C into an equality gate. Term 2 is simplified by recognizing the inequality function. The final result is obtained from an OR gate with terms 1 and A as the inputs.

9. The decimal number 38 is 1 0 0 1 1 0 in binary, while the decimal number 22 is 1 0 1 1 0 in binary. The binary addition is

$$
\begin{array}{cccccc}
1 & 0 & 0 & 1 & 1 & 0 \\
0 & 1 & 0 & 1 & 1 & 0 \\
\hline
1 & 1 & 1 & 1 & 0 & 0
\end{array}
$$

The result $1\ 1\ 1\ 1\ 0\ 0_2$ is 60 in decimal. The truth table for binary addition of the least significant binary bit (2^0) can be written

A_0	B_0	S_0	C_0
0	0	0	0
0	1	1	0
1	0	1	0
1	1	0	1

For every other binary bit the truth table may be written

A_n	B_n	C_{n-1}	S_n	C_n
0	0	0	0	0
0	0	1	1	0
0	1	0	1	0
0	1	1	0	1
1	0	0	1	0
1	0	1	0	1
1	1	0	0	1
1	1	1	1	1

10. The desired count sequence table for four flip-flops A, B, C, and D is

Count	D	C	B	A
0	0	0	0	0
1	1	0	0	1
2	1	0	0	0
3	0	1	1	1
4	0	1	1	0
5	0	1	0	1
6	0	1	0	0
7	0	0	1	1
8	0	0	1	0
9	0	0	0	1
10,0	0	0	0	0

It can be seen that flip-flop A should alternate on every count. Hence no J or K connections to A are required. Looking at the count table

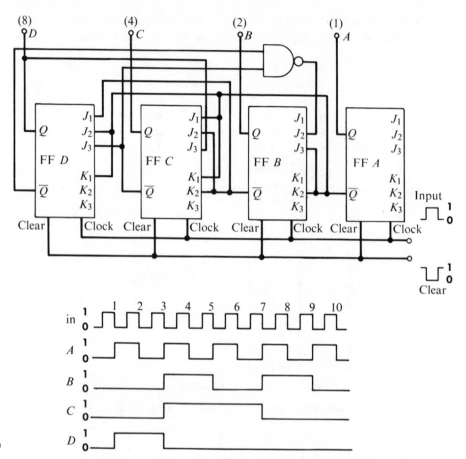

Fig. S–10

for flip-flop B, we can see that the $\mathbf{0} \rightarrow \mathbf{1}$ transition of B should be inhibited whenever $A = \mathbf{1}$. This condition is not sufficient, however, since the $\mathbf{0} \rightarrow \mathbf{1}$ transition must also be inhibited on the first count. Hence the $\mathbf{0} \rightarrow \mathbf{1}$ transition of B must also be inhibited if both C and D outputs are zero. These two conditions can be satisfied by connecting $\bar{Q}_A$ to J_B and $\bar{Q}_C$ and $\bar{Q}_D$ through a NAND gate to J_B. Connecting these $\bar{Q}$ outputs to a NAND gate performs the OR operation, which allows the $\mathbf{0} \rightarrow \mathbf{1}$ transition of B to occur if C OR D is $\mathbf{1}$ (inhibits if C AND D are $\mathbf{0}$). The $\mathbf{1} \rightarrow \mathbf{0}$ transition of flip-flop B should be inhibited if $A = \mathbf{1}$. Hence $\bar{Q}_A$ is also connected to K_B (see Fig. S–10).

Flip-flop C should be inhibited from undergoing a $\mathbf{0} \rightarrow \mathbf{1}$ transition whenever $A = \mathbf{1}$ or $B = \mathbf{1}$ or $D = \mathbf{0}$. This is accomplished by connecting $\bar{Q}_A$ to a J_C input, $\bar{Q}_B$ to a J_C input, and Q_D to a J_C input as shown in Fig. S–10. The **1 → 0** transition of C should be inhibited whenever $A = $ **1** or $B = $ **1**. This is accomplished by connecting $\bar{Q}_A$ and $\bar{Q}_B$ to K inputs of C.

Flip-flop C should be inhibited from the **0 → 1** transition if A or B or C is **1**; i.e., it should undergo this transition only when all other outputs are **0** on the first count. Hence the $\bar{Q}$ outputs of A, B, and C are connected to J inputs of D. The **1 → 0** transition of D should be inhibited if $Q_A = $ **1**. Hence $\bar{Q}_A$ is connected to K_D.

The resulting circuit diagram is shown in Fig. S–10.

11. The count sequence table for a Modulo-13 counter with four flip-flops A, B, C, and D is

Count	D	C	B	A
0	0	0	0	0
1	0	0	0	1
2	0	0	1	0
3	0	0	1	1
4	0	1	0	0
5	0	1	0	1
6	0	1	1	0
7	0	1	1	1
8	1	0	0	0
9	1	0	0	1
10	1	0	1	0
11	1	0	1	1
12	1	1	0	0
13,0	0	0	0	0

The count sequence table shows that flip-flop A is to alternate on each input count except when both D and C are **1**. Thus J_A should be **0** when D and C are **1**. This can be written $J_A = \overline{DC}$ and implemented with gate 1 as shown in Fig. S–11. Flip-flop D changes state on each **1 → 0** transition of flip-flop C, which requires only a connection from Q_C to the clock input of D.

For flip-flop C, the **0 → 1** transition should be inhibited if A or $B = $ **0**. This can be accomplished by clocking FF C with the input signal and connecting Q_A and Q_B to J inputs of C. The **1 → 0** transition of C should be inhibited unless Q_A and Q_B are **1** or Q_D is **1**. Thus $K_C = AB + D = \overline{AB} \cdot \bar{D}$, which can be implemented with

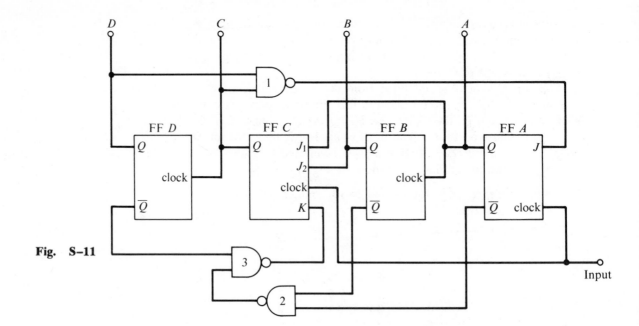

Fig. S–11

NAND gate 2 and 3 as shown in Fig. S–11. Flip-flop B should alternate on each $\mathbf{1} \rightarrow \mathbf{0}$ transition of A. Hence Q_A is connected to the clock input of B. The resulting circuit is shown in Fig. S–11.

12. Use Eq. 3–187, $v_o = -V_R \sum_{n=1}^{4} (a_n/2^n)$.

a) $v_o = -5 \text{ V}\left(\dfrac{1}{2} + \dfrac{1}{4} + \dfrac{1}{8} + \dfrac{1}{16}\right) = \left(\dfrac{15}{16}\right)(-5 \text{ V}) = 4.69 \text{ V.}$

b) $v_o = -5 \text{ V}\left(\dfrac{1}{2} + \dfrac{1}{16}\right) = \dfrac{9}{16}(-5 \text{ V}) = -2.81 \text{ V.}$

c) $v_o = -5 \text{ V}\left(\dfrac{1}{4} + \dfrac{1}{16}\right) = \dfrac{5}{16}(-5 \text{ V}) = -1.56 \text{ V.}$

d) $v_o = -5 \text{ V}\left(\dfrac{1}{16}\right) = -0.312 \text{ V.}$

13. The table is completed below:

To convert From / To	Sign magnitude	2's complement	Offset binary	1's complement
Sign magnitude	No change	If MSB = 1, complement other bits, add 00 . . . 01	Complement MSB If new MSB = 1, complement other bits, add 00 . . . 01	If MSB = 1, complement other bits
2's complement	If MSB = 1, complement other bits, add 00 . . . 01	No change	Complement MSB	If MSB = 1, add 00 . . . 01
Offset binary	Complement MSB If new MSB = 0, complement other bits, add 00 . . . 01	Complement MSB	No change	Complement MSB If new MSB = 0, add 00 . . . 01
1's complement	If MSB = 1, complement other bits	If MSB = 1, add 11 . . . 11	Complement MSB If new MSB = 1, add 11 . . . 11	No change

14. When the two MSB's of the counter are first **1**, the count is 1800. Thus $n_i = 1800$. The integration period is n_i/f_c, and it is desired that this period be $1/60$ sec. Solving for f_c yields

$$\frac{n_i}{f_c} = \frac{1}{60},$$

$$f_c = 60n_i = 60 \times 1800 = 108 \text{ kHz}.$$

OPTIMIZATION OF ELECTRONIC MEASUREMENTS

Introduction

No measurement is perfect. There are always uncertainties in measurements which may arise from several different sources or from one predominant source. Thus an experimenter, in carrying out a particular electronic measurement, is usually involved with optimizing the measurement in order to minimize the uncertainties.

Basic to any systematic approach to optimization is a consideration of the signal-to-noise ratio of a measurement. The general properties of signals and noise are treated in the first section of this module, with particular emphasis on the frequency composition of signals and noise. Then the basic approaches to the improvement of the signal-to-noise ratio are discussed.

The importance of controlling the frequency response of an electronic measurement system is clear after the discussion of the frequency dependent properties of signals and noise. Signal-to-noise ratios can often be significantly improved with such control. The properties of analog filters which combine *RC* circuits with operational amplifiers to form active filters are discussed in section two. These circuits can be used to construct high pass, low pass, narrow bandpass, and rejection filters which play important roles in controlling measurement system frequency response.

The improvement in signal-to-noise ratios which can be obtained by simple analog filtering is often not enough. For many signals it is necessary to utilize more sophisticated techniques. Two important operations in many of these more sophisticated techniques are **modulation** and **demodulation.** The characteristics and implementation of modulation and demodulation techniques are discussed in the third section, along with their relation to modern signal processing techniques and data conversion operations.

Many of the modern signal processing techniques require as an inherent step in their implementation the sampling of a time varying continuous analog signal. Thus an important aspect in the overall optimization of many measurements involves the correct choice of sampling parameters with respect to the specific characteristics of the signal. The choice of these sampling parameters is discussed in section four.

In the final section the increasingly important more sophisticated signal-to-noise ratio enhancement techniques are presented. The techniques covered include low pass filtering, integration, lock-in amplification, boxcar integration, multichannel averaging, and correlation techniques. This section concludes with a brief discussion of some computer-based techniques, such as digital filtering, which are closely related in conceptual approach to the hardware techniques emphasized in this module.

Signals and Noise

Electrical signals in scientific measurements and control applications are carriers of encoded information about some chemical or physical quantity. Electrical signals consist of a desirable signal component, which is related to the physical or chemical quantity of interest, and an undesirable component, which is termed **noise.** Electrical noise may thus be defined as any part of the observed electrical signal which is unwanted. Implicit in this definition is the concept that what is considered noise in one experiment may be useful information in a different experiment, and vice versa. For example, consider the flame emission spectrometer shown in the pictorial diagram of Fig. 4–1.

A solution containing calcium, barium, sodium, and potassium is sprayed into a flame where thermal energy excites each of the atoms and causes them to emit radiation of a specific wavelength. The optical signal from the flame contains information concerning the concentration of the four elements in solution, as well as emitted radiation from the flame itself. If the only information desired is the calcium concentration, all the other radiation contributes to unwanted fluctuations in the optical signal and thus to unwanted electrical fluctuations, once the optical signal is converted into an electrical signal. These fluctuations are considered noise. To reduce the unwanted components of the signal, a wavelength isolation device, such as a monochromator, is used before the transducer, as shown in Fig. 4–1. Thus the optical signal reaching the transducer consists only of the calcium emission and any flame background at the same wavelength.

In addition to the inherent fluctuations in the calcium emissions, this flame background is also a source of noise. A photomultiplier tube is used to transduce the optical signal into an electrical current. The output current from the photomultiplier tube consists of components due to calcium emission, flame background emission, and the dark current of the photomultiplier tube, which represents an additional source of noise. Finally, in this measurement example, the current signal is converted to a voltage; and

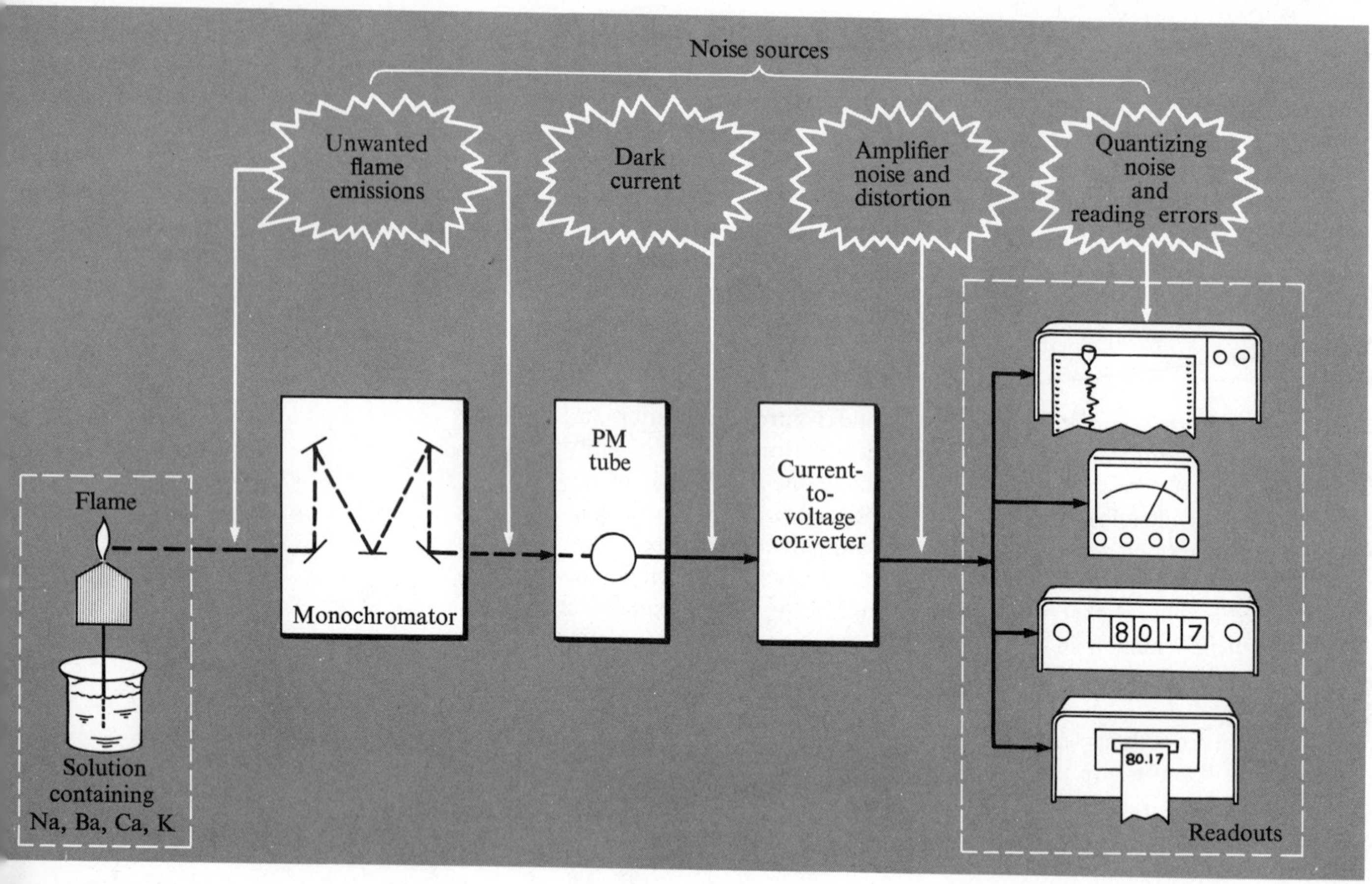

Fig. 4-1 Flame emission spectrometer.

the resulting voltage may be read out, using a variety of devices including strip chart recorders, meters, or digital displays. Again, additional noise may be introduced at either or both of these stages.

In order to optimize a measurement such as that described above, a knowledge of the general properties of electrical signals and noise is required. In the next section, the frequency composition of typical signal sources and noise sources is stressed, and the basic principles which underlie optimization methods are discussed. Then the specific types and characteristics of typical noise sources are discussed, and the signal-to-noise ratio as a measure of precision is introduced.

4-1.1 FREQUENCY COMPONENTS OF SIGNALS AND NOISE

Electrical signals almost always vary with time. A plot of signal amplitude as a function of time results in a signal waveform which can be characterized by its **frequency spectrum.** A knowledge of this frequency composition is an important consideration if the electrical signal is to be processed without loss of information. Also, there are often differences in the frequency distribution of signal and noise information that facilitate selective measurement of the signal information.

The analysis of the frequency composition of a waveform is called **Fourier analysis.** The basic concepts of Fourier analysis were discussed in Section 3–2.2 of Module 3. In Fourier analysis an amplitude-vs.-time waveform is transformed into its spectrum, which is the amplitude of each frequency component plotted vs. frequency. These two representations of a waveform are often called the **time domain** and **frequency domain** representations.[1] The amplitude axis in the frequency domain is often plotted as an amplitude density, in order to avoid mathematical problems which occur with integrals going to infinity. If the amplitude-vs.-frequency plot has units of volts/Hz vs. Hz, the plot is referred to as the **amplitude spectrum;** and if the units are watts/Hz vs. Hz, the plot is referred to as the **power density spectrum** or simply the **power spectrum.**

Fourier analysis can be used to characterize the distribution of signal and noise information as a function of frequency. This information can be used as a guide in selecting the best method of signal processing for a specific case. In addition, the analysis and understanding of many signal processing techniques, particularly systems involving modulation, are greatly facilitated by considering the frequency domain representation of the signal. Thus this method of representing the signal information will be used extensively throughout the rest of the module.

Direct Current Signals

A direct current signal is one in which the current is always in the same direction and the magnitude of the current is constant. Many signals which are derived from physical or chemical quantities are often considered to be dc signals. However, no electrical signal can be totally dc, that is, constant indefinitely. Consider, for example, the system for measuring solution pH shown in Fig. 4–2. The voltage output of the glass-reference electrode combination is related by a transfer function to the solution pH. A typical plot of the transducer output voltage vs. time is shown in Fig. 4–3a. The signal power density spectrum which results from a Fourier analysis is shown in Fig. 4–3b. It is clear from the power spectrum that most of the signal power is at or very near 0 Hz (dc). However, there is some information in the signal at higher frequencies. In the specific example discussed here, these frequency components may arise from changes in the transfer characteristics of the transducer or pH changes in solution as a result of chemical reactions, and these changes may be of direct interest to the investigator. In that case such variations would not be considered noise.

Note 1. Fourier Domains and Data Domains.

The data domain refers to the manner in which a single data point is encoded. The Fourier domain (time or frequency) refers to the way in which the information in an entire waveform is represented.

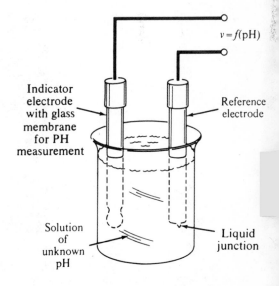

Fig. 4–2 pH/reference electrode pair.

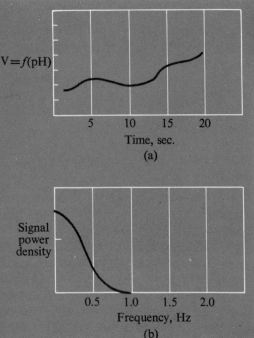

$V = f(\text{pH})$

5   10   15   20

Time, sec.

(a)

Signal power density

0.5   1.0   1.5   2.0

Frequency, Hz

(b)

Fig. 4–3 (a) Output voltage vs. time for a pH measurement. (b) Power spectrum for a pH measurement.

Fig. 4–4 Signals with alternating current components.

Similar spectra result from other transducers that are usually thought of as producing dc outputs. A general definition of a dc signal would be a signal whose power is concentrated in a narrow band of frequencies near 0 Hz like the signal shown in Fig. 4–3. For such signals the power is proportional to the square of the dc current or voltage. An electronic instrument for amplifying, modifying, and displaying such dc signals must have a low frequency response that extends to 0 Hz.

Alternating Current Signals

An alternating current signal is one in which the direction of the current periodically reverses. The simplest example of an alternating current is a sinusoidal signal, as shown in Fig. 4–4a. Many electrical signals from transducers have both ac and dc components. Two such signals are shown in Figs. 4–4b and 4–4c. In a strict sense neither of these signals can be classified as ac because the direction of the current does not reverse. However, they can be broken down into a dc component, which is the long term average value of the signal, and an ac component, which is the chang-

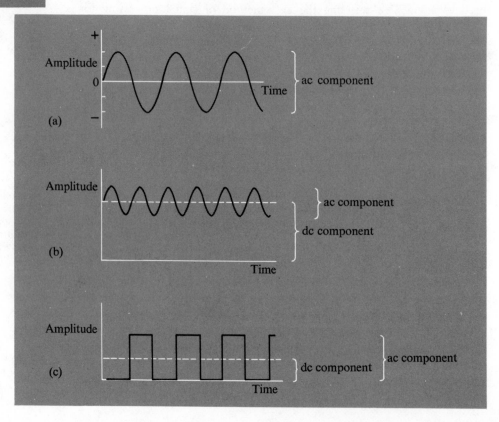

+

Amplitude

0

−

Time

} ac component

(a)

Amplitude

Time

} ac component

} dc component

(b)

Amplitude

Time

} dc component

} ac component

(c)

ing part of the signal. A series capacitor will block the dc component of such a signal and convert the changing part of the signal into a true ac signal. This is the action of the ac–dc coupling switch on most laboratory oscilloscopes.

Alternating electrical signals can be usefully characterized by their power spectra. In contrast to dc signals, the power in an ac signal occurs at higher frequencies. The output of a microphone, for example, is an alternating electrical voltage whose frequency composition is related to that of the audio input signal. Such a signal and its power spectrum are shown in Fig. 4–5a. The audio signal is an example of a broadband ac signal in that it generally contains frequencies anywhere in the range of 20 Hz to 20 kHz. Electronic instruments for amplifying and measuring such signals must be capable of responding to a wide range of frequencies.

In laboratory measurements, a dc signal is often converted to an ac signal before it is electronically processed. This is carried out by some type of mechanical, electromechanical, or electronic chopper. For example, a radiation source such as a tungsten lamp can be made to produce an alternating light output by powering it from the 60 Hz power line, by powering it from a pulsating power supply, or by passing the radiation through a rotating disk (mechanical chopper) which is half transparent and half opaque. The chopper input of a servo recorder discussed in Module 1 is another example of a device which converts a dc voltage into an ac voltage.

Fig. 4–5 Waveforms and power spectra of some ac signals.

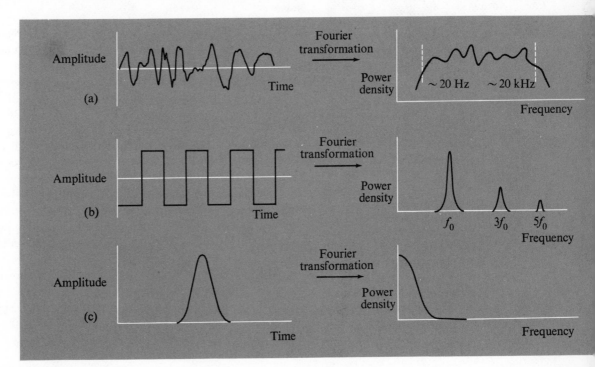

Such a chopped electrical signal and its power spectrum are shown in Fig. 4–5b. The chopped signal looks much like a square wave. The signal power appears in a narrow band of frequencies surrounding the frequency at which the signal is chopped and the odd harmonics of this fundamental. Since most of the signal power appears near the fundamental, the harmonics are often neglected when measuring such a signal. Thus, an electronic instrument for amplifying the band of frequencies around the fundamental frequency is not required to have anywhere near the wide frequency response required for the audio signal from the microphone.

A frequently encountered signal in scientific measurements that has both an ac component and a dc component is a peak signal. A peak signal and its power spectrum are shown in Fig. 4–5c. The power spectrum is similar, in a way, to that of the slowly changing dc signal. However, while the signal power tends to be concentrated near 0 Hz, appreciable signal power can extend out to very high frequencies. The power spectra of most common peak shape functional forms such as Gaussian, Lorentzian, and exponential, theoretically extend to infinite frequencies. Thus, in order to accurately amplify and measure these signals the frequency response may have to extend from dc up to the point where little signal power is left. For narrow peak signals this can be a surprisingly high frequency.

Noise

In order to carry out measurements which are relatively free from the effects of electrical noise, an analysis of the frequency content of noise is quite important. A noise power density spectrum, analogous to a signal power density spectrum, is a plot of noise power density in watts/Hz versus frequency. Noise sources give rise to three distinct types of power spectra, which are illustrated in Fig. 4–6. In Fig. 4–6a, the noise power density is of nearly equal amplitude at all frequencies. Such noise is known as **white noise,** since the power spectrum is essentially flat. White noise can be considered to be a mixture of signals of all frequencies with random amplitudes and phase angles.

The noise in Fig. 4–6b is known as **flicker noise,** or **one-over-f ($1/f$) noise,** and has a spectrum in which the power density increases approximately with the reciprocal of frequency at low frequencies. Such a spectrum is typical of low frequency drifts, which are common in transducers, amplifiers, and measurement systems.

A third common type of noise spectrum is shown in Fig. 4–6c. Here noise components peak at certain discrete frequencies rather than over a broad range of frequencies. This type of power spectrum is typical of **interference noise** such as noise originating from the 60 Hz power lines.

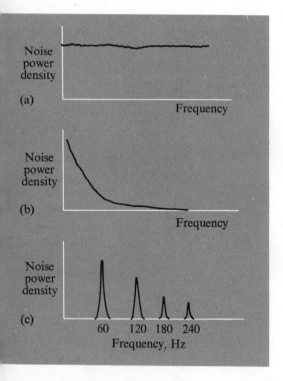

Fig. 4–6 Noise power spectra: (a) white noise; (b) flicker or $1/f$ noise; (c) interference noise from power lines.

Discrete frequency noise sources will cause the noise power spectrum to have peaks at the fundamental and harmonics of this frequency. For 60 Hz interference, noise power will be found at 60 Hz, 120 Hz, 180 Hz, etc. For noise, as for ac signals, the power is proportional to the mean square value of the waveform.

In a real measurement system, all three types of noise are likely to be encountered, so that the overall noise is a mixture of flicker noise, white noise, and interference noise. The power density spectrum which results from combining these three types of noise is shown in Fig. 4–7. Note that at low frequencies the $1/f$ noise tends to predominate, while at high frequencies white noise becomes the major contributor.

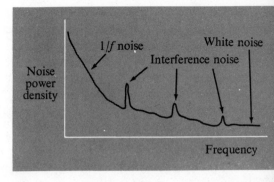

Fig. 4–7   Combined noise power spectrum.

Amplitude and Phase Spectra of a Noisy Signal

It is important to appreciate the basic distinguishing characteristics of signals and noise that allow measurement systems to be set up that can discriminate against noise components with respect to signal components. From the above discussion of power density spectra one important distinction is clear. The distributions of signal and noise information as a function of frequency may be quite different.

Many signals have or can be made to have relatively narrow and well-defined frequency regions where most of the signal power occurs. A dc signal has most of its power concentrated in a narrow band of frequencies near 0 Hz. A chopped signal normally has its power concentrated around the fundamental frequency and the odd harmonics. However, such is not always the case. Some signals, such as the audio signal mentioned above, and most pulse-like signals have relatively broad power density spectra.

In general, noise tends to have a rather broad and featureless power density spectrum. The power spectrum, as mentioned above, is flat at high frequencies but does tend to increase sharply at low frequencies. Some noise sources, such as 60 Hz, do have narrow power spectra, but the location is often predictable and the location of the signal information in the same region can usually be avoided. The general differences in the distribution of signal and noise information as a function of frequency is an important distinguishing property and forms the basis for frequency selective measurement systems.

A second important difference between the frequency components of noise and signals is their phase relation. The frequency components that make up a signal are, in general, phase related. Noise frequencies, on the other hand, are typically not related in phase to the signal frequencies or, for that matter, to other noise frequencies. This important difference forms the basis of all integration type measurement systems.

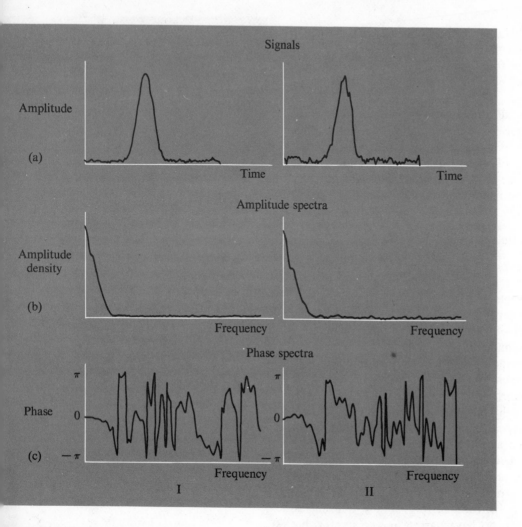

Signals

Amplitude

(a)

Time   Time

Amplitude spectra

Amplitude
density

(b)

Frequency   Frequency

Phase spectra

Phase   0

(c) $-\pi$

Frequency   Frequency

I   II

Fig. 4–8 (a) Noisy peak signals. (b) Amplitude spectra. (c) Phase spectra.

These two distinguishing properties of noise and signal frequencies (relative distribution and phase relation) are illustrated in Fig. 4–8. The two signals shown in Fig. 4–8a are two typical noisy peak-like signals that might be measured scanning an emission line in an optical spectrum. The signal shown in column II has a higher noise level than that in column I. The amplitude spectra (amplitude of the frequencies that make up the signal plotted vs. frequency) are shown in Fig. 4–8b. The main frequency components of this peak signal are located in a narrow band near dc (0 Hz). The higher frequencies present in this noisy signal are primarily due to noise. Note that the amplitude of these higher frequencies is greater in the amplitude spectrum of the noisier signal. Attenuation of these

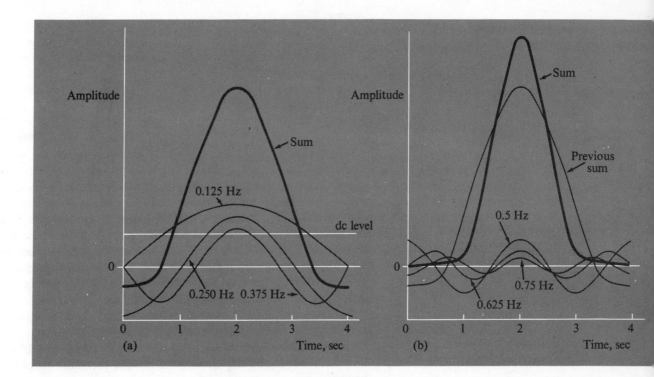

higher frequencies will significantly reduce the noise level in the original signal. This type of difference between the frequency distribution of signal and noise components forms the basis for improving the signal-to-noise ratio by techniques such as *RC* low pass filtering.

The phase spectra of these two signals are shown in Fig. 4–8c. These spectra are plots of the actual phases of the frequency components of the noisy signals at the point where the peak maximum occurs. The phase values are normalized to $\pm\pi$. At the location of the peak maximum, all the frequency components that make up the peak should be in phase, and the value of the phase should be zero. This, indeed, is the reason for the existence of the peak when, in the Fourier sense, it is interpreted as a summation of various sinusoidal frequencies. This is shown in Fig. 4–9 for a Gaussian peak that has a full width at half-height of 1 sec. The Fourier analysis of this peak was carried out from 0 to 1 Hz at intervals of 0.125 Hz. The results are summarized in Table 4–1. The summation of the first four frequency components of this signal (0 to 0.375 Hz) is shown in Fig. 4–9a. Summing in the next three components (Fig. 4–9b) results in a peak which is beginning to resemble a Gaussian and also has a full width at half-height of approximately 1 sec. It is also

Fig. 4–9 Generation of a Gaussian peak from its frequency components.

Table 4–1 Fourier components of a Gaussian peak

Frequency, Hz	Relative amplitude
0	0.52
0.125	1.0
0.250	0.84
0.375	0.63
0.500	0.43
0.625	0.26
0.750	0.14
0.875	0.07
1.000	0.03

obvious from Fig. 4–9 that the frequency components which make up the peak are in phase at the peak maximum and the value of the phase is zero considering the frequency components as cosine waves.

Now referring to the phase spectra of the noisy peaks (Fig. 4–8c), it is seen that the lower frequency components of *both* noisy signals are in phase, as indicated by the relatively flat phase spectrum in this region; and the value of the phase in both cases is approximately zero. From the discussion of Fig. 4–9, this would be expected for those frequency components making up the peak signal. The similarity of the phase spectra in this region indicates that the phases of the frequency components responsible for the peak are essentially the same in these two noisy signals. The phases of the higher frequency components fluctuate at random, as would be expected if these frequencies are noise. It is really only by chance that a noise frequency would have the same phase as the signal frequencies at the point in time at which the peak maximum occurs. The very abrupt changes in phase are a result of the normalization to $\pm\pi$. It is also important to note that the two phase spectra are quite dissimilar in the higher frequency region. This indicates quite clearly why integration techniques are effective in enhancing the signal-to-noise ratio of a measurement. If the two signals in Fig. 4–8a were added together, the frequency components making up the signal would add in phase since they have essentially the same phase, while the noise frequencies would add randomly or noncoherently, tending to cancel out.

These figures (Fig. 4–8b and 4–8c) are not schematic, but are the actual amplitude and phase spectra of the signals shown in Fig. 4–8a, as calculated by Fourier transformation. They illustrate the importance of knowing the properties of signals and noise in both the conventional amplitude-time domain and the frequency domain. In the following section, specific noise sources in experimental systems are presented and classified.

4–1.2   NOISE SOURCES

The total noise in a measurement system can be considered to result from two distinct types of noise: fundamental noise and nonfundamental (or excess) noise. Fundamental noise is generated by the motion of discrete charges in circuits and cannot be completely eliminated. **Johnson** or **thermal noise,** and **shot noise** are the two most important types of fundamental noise. **Excess noise,** such as $1/f$ noise, interference noise, power supply and amplifier drifts, arises from imperfect instrumentation or nonideal component behavior, and can in principle be reduced to insignificant levels by careful practice and instrument design. Noise is also introduced in the process of converting an analog signal into a digital representation, a step common to all scientific measurements. This type of noise is called **quantizing noise.**

Johnson Noise

Johnson noise or thermal noise is produced by the random motion of electrons in resistive elements because of thermal agitation. It is a basic thermodynamic phenomenon which can only be eliminated at a temperature of absolute zero. The random motion of charges in resistors gives rise to a fluctuation in voltage, called the noise voltage, across the resistor even though the average voltage, in the absence of an external source, is zero. Johnson noise has a white power density spectrum; i.e., the noise power density is independent of frequency.

Because Johnson noise is due to thermal motion, the magnitude of the noise voltage increases with temperature. Thermal noise also increases as the resistance of the component increases. Since the noise power density is equal at all frequencies, the total noise voltage observed across a resistor depends upon the range of frequencies which the measurement system allows to pass, i.e., the system **bandwidth** Δf. Bandwidths of measurement systems are discussed in Section 4–2.

The quantitative relationship expressing Johnson noise voltage in terms of resistance, temperature, and bandwidth is known as the Nyquist relationship and is given by Eq. 4–1,

$$\bar{v}_{\text{rms}} = (4kTR\,\Delta f)^{1/2}, \tag{4--1}$$

where $\bar{v}_{\text{rms}}$ is the rms noise voltage contained in bandwidth Δf, k is the Boltzmann constant, T is the temperature of the resistor in °K, and R is the resistance of the element. For a 1 MΩ resistor at 25°C and a measurement bandwidth of 1 MHz, the rms noise voltage is predicted by Eq. (4–1) to be

$$\bar{v}_{\text{rms}} = \left(4 \times 1.38 \times 10^{-23}\, \frac{\text{J}}{°\text{K}} \times 298°\text{K} \times 10^6\, \Omega \times 10^6\, \text{sec}^{-1} \right)^{1/2}$$

$$= 1.3 \times 10^{-4}\, \text{V}.$$

Note that the Johnson noise power:

$$\frac{(\bar{v}_{\text{rms}})^2}{R} = 4kT\,\Delta f \tag{4--2}$$

is independent of the size of the resistor.

Shot Noise

Shot noise results from the random movement of discrete charge carriers, usually electrons, across a junction. Examples of such conduction mechanisms include the flow of charges between the cathode and anode in a vacuum tube or phototube or across junctions in a semiconductor. In a vacuum tube, electrons are released from the heated cathode at random times and thus arrive at the anode with a random time distribution. Likewise, in a phototube, the random arrival time of photoelectrons at the anode results from the random arrival of photons at the photocathode and the subsequent random emission of photoelectrons. Electrons flowing in a metallic conductor, such as a wire, do not exhibit the shot effect because there is not a discrete, random emission process but rather a continuous conduction mechanism.

If the actual arrival of single photoelectrons at the anode of a vacuum phototube could be observed, the current signal would look like that shown in Fig. 4–10. Statistically, if the arrival of electrons is truly random, the number N arriving in an observation time t fluctuates with a standard deviation of $N^{1/2}$. Since the current i is proportional to the number of photoelectrons arriving at the anode in a unit time, the current must also fluctuate on account of the random arrival of electrons. It is this fluctuation in current that is called shot noise.

A quantitative expression for the rms shot noise current can be readily derived. It will be shown in Section 4–1.3 that the standard deviation of a signal parameter is equivalent to its rms value. Thus the relative standard deviation of the current can be expressed as:

$$\frac{\bar{i}_{\mathrm{rms}}}{i} = \frac{N^{1/2}}{N} = \frac{1}{N^{1/2}}. \tag{4–3}$$

If the charge carriers giving rise to the current are electrons, the average current i observed during the time interval t is NQ_e/t, where Q_e is the charge on a single electron. Substituting into Eq. (4–3) for N and solving for the rms shot noise current in terms of the average current i, the result is

$$\bar{i}_{\mathrm{rms}} = \left(\frac{Q_e i}{t}\right)^{1/2}. \tag{4–4}$$

Equation (4–4) is the basic shot effect equation and reveals that the rms shot noise increases with the square root of the average current and decreases with increasing observation time. For an average current of 1.6×10^{-8} A and an effective observation or integrating time of 0.1 sec, the rms shot noise current is

$$i_{\mathrm{rms}} = (1.6 \times 10^{-19}\,\mathrm{C} \times 1.16 \times 10^{-8}\,\mathrm{A} \times 10\,\mathrm{sec}^{-1})^{1/2}$$
$$= 1.6 \times 10^{-13}\,\mathrm{A}.$$

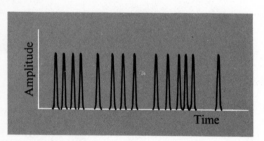

Fig. 4–10   Photoelectron pulses at the anode of a phototube.

The shot noise equation or Schottky equation (Eq. 4–4) may also be expressed in terms of bandwidth, as was the Nyquist equation. In order to express Eq. (4–4) in terms of bandwidth, it is necessary to develop a relation between the integration time t and bandwidth Δf. All amplitude–time waveforms have a bandwidth. In order to facilitate comparisons of bandwidth it is useful to define an equivalent bandwidth. This is usually defined as the width of a rectangular band of frequencies whose amplitudes are all equal to the amplitude of the most intense frequency of the waveform, and whose total area is equivalent to the total area of the actual power spectrum of the waveform. For example, for the power spectrum shown in Fig. 4–5c, the equivalent bandwidth spectrum would be a rectangle with the same height as the 0 Hz component and width Δf extending from 0 Hz to a frequency f_0 such that the area of the rectangular bandwidth was equal to the area of the actual power spectrum. The calculation of the equivalent bandwidth for a particular waveform can be quite complex, and involves the use of Fourier transforms. For a rectangular gating pulse t sec in duration, the equivalent bandwidth Δf is $1/(2t)$. Thus, rms shot noise current can be written as $(2Q_e i\, \Delta f)^{1/2}$. Shot noise, like Johnson noise, has a white power density spectrum.

In photomultiplier tubes (Fig. 4–11) the photoelectrons released from the cathode by photon bombardment are amplified by the process of secondary electron emission down a dynode chain. As a result of this process, each photoelectron gives rise finally to a relatively large pulse of charge at the anode, consisting of 10^5 to 10^6 electrons. The random nature of secondary electron emission at each dynode gives these pulses an amplitude distribution. Unlike single photoelectrons in a phototube, the current pulses at the output of a photomultiplier tube can be directly observed. A typical output signal is shown in Fig. 4–12. Thus, the amplification in a photomultiplier adds a small amount of additional shot noise. The amount of noise added by secondary emission, however, increases i_{rms} by a factor of only 1.1 to 1.4, and is often neglected. Since internal amplification may increase i by a factor of 10^5 to 10^6, this small amount of additional noise is a small price to pay.

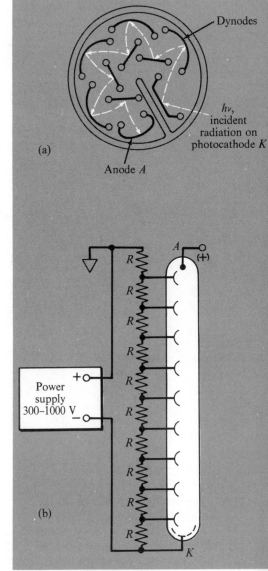

Fig. 4–11 Photomultiplier transducer. (a) Cross-sectional view of side-illuminated PM tube. (b) Schematic showing connection to power supply.

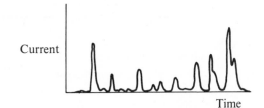

Fig. 4–12 Photomultiplier tube current pulse output.

Excess Noise

Excess noise may be thought of as any noise in a system which is above and beyond that due to Johnson and shot noise. In contrast to fundamental noise, excess noise is almost always frequency dependent. Interference noise and various sources of flicker noise are classified as excess noise sources.

Interference Noise.   Any conductor can act as an antenna for the pickup of electromagnetic interference, which is ever present. Much of the interference which occurs in measurement systems is of discrete frequencies, for example, the pickup of 60 Hz signals from the ac power lines. Other sources of interference may produce a wider noise power spectrum and thus be more difficult to eliminate. Examples of this type of interference include electrical discharges (lightning, ignition systems, relay arcing, electrical motorswitch brushes, etc.) and ionospheric phenomena. The sensitivity of a wire or component to electromagnetic interference is roughly proportional to its length and cross section, the resistance to ground at that point in the circuit, and the voltage amplification which follows.

In principle, interference noise can be greatly reduced by proper circuit design and shielding. Shielding consists of enclosing the circuit (or critical wires) with a conducting material which is usually connected to earth ground, power supply ground, or the system common. Ideally the space enclosed by an electrical shield is free from the effects of external electrical or electromagnetic fields. Metal chassis and cabinets are used for enclosing electrical instruments, and the metal parts are connected to the desired circuit ground. Shielded wires are quite useful for external signal connections and for some critical internal connections. A typical shielded wire and its schematic are shown in Fig. 4–13.

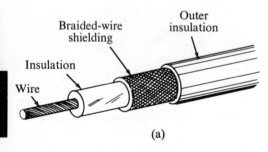

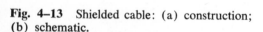

(a)

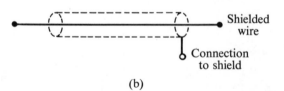

(b)

Fig. 4–13   Shielded cable: (a) construction; (b) schematic.

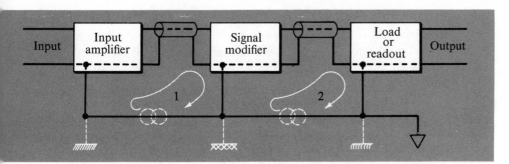

Fig. 4–14 Ground loops when multiple grounds are used.

In high gain systems, care must be taken to attach all shields to one common point without introducing "ground loops," which are closed paths in the system that may induce noise signals in nearby components. Such ground loops are illustrated in Fig. 4–14, where three functional modules of a measurement system are shown interconnected with shielded cable. If the shields are connected to separate grounding points, it is possible to develop loop currents primarily because of power line interference pickup in the grounding system. If the entire system of electronic modules is grounded at one point, such as the input terminal, it is possible to reduce ground-loop currents to low levels.[2]

Note 2.   Grounding and Shielding.

It is wise to follow certain basic time-tested guidelines for effective shielding and grounding. Some of the practical guidelines and rules are presented in Appendix A.

Flicker Noise.   In vacuum tube and transistor amplifiers, there is excess noise with a $1/f$ power spectrum whose origin is not well understood. Flicker noise is often considered to be synonymous with drift. In amplification elements it is assumed to be due to regions of enhanced electron emission. It can also be introduced by long term power supply fluctuations, changes in component values, temperature drifts, etc.

Flicker noise can also occur in transducers and in some physical signals. Many transducers have power supplies associated with them which can drift with time and temperature, and hence cause low frequency fluctuations in the transfer function. Likewise, physical signals, such as radiation sources, may fluctuate because of line voltage changes, lamp aging, arc wander, convection, etc. Many of the fluctuations in transducers and signals have a strong $1/f$ component in the power density spectrum.

Quantizing Noise

The final result of any measurement is a number. The number obtained in a measurement is an integer value composed of a finite number of digits. In most measurements the sought-for information is encoded as an analog electrical signal which is continuously variable and hence, in principle, of infinite resolution. Noise can be introduced in the process of converting this analog value to a number. This type of noise is called **quantizing noise.** Quantizing noise is basically a result of the finite resolution of an A-to-D conversion. If the phenomenon of interest is converted directly to a digital form by an input transducer, quantizing noise is related to the transducer resolution.

Quantization noise is often thought of only in terms of A-to-D conversions. However, quantizing noise will be present in any process which converts a continuous, infinite resolution signal to a finite number of digits, regardless of whether the technique involves reading a scale position to generate a number or converting to a digital signal by an electronic analog-to-digital converter.

The two main variables in quantization are the quantization level and the dynamic range. The dynamic range of a quantizer is the range of signal amplitude that can be accurately quantized. The quantization level is the smallest increment of amplitude that can be measured within the dynamic range. For any input within the dynamic range of a digitizer, the output will be a digital representation of the input, to the *nearest integral multiple* of the quantization level. Typically, a quantizer will round off to the nearest quantization level or to the next lower quantization level.

It is important to note that the noise or error introduced by quantization depends on the relative intensity of the noise in the original signal with respect to the size of the quantization level. If, for example, the noise level on a signal is considerably less than the magnitude of the quantization level, the quantization noise or error will be determinate. If the quantizer rounds off to the next lower quantization level, the quantized value of the signal could be in error by as much as the value of the quantization level. The precision and accuracy of the measurement would be limited by the resolving power of the digitizer, and the maximum achievable precision and accuracy would not be obtained. In most cases the sensitivity of the quantizer can be increased by amplifying the signal and using offset methods, so that the intensity of the noise is large relative to the size of the quantization level. This prevents sequences of determinate errors of the same sign and greatly reduces the overall effects of quantization, allowing the accurate determination of a signal to less than the quantization level. If the standard deviation of the signal noise is at least as large as the magnitude of the quantization level, the effect of quantization is to add white random noise to the signal noise. The standard deviation of the quantization noise is $v_l/\sqrt{12}$, where v_l is the magnitude of the quantization level in volts. The derivation of this result is quite complex. An outline of it may be found in a paper by Kelly and Horlick which is listed in the bibliography. Some specific considerations with respect to common digitization operations will now be discussed.

Analog Readouts. Whenever an analog meter, recorder, or oscilloscope is read by the human eye and a numerical result is recorded, a quantizing error is introduced in converting from the scale position domain to the number domain. Consider the strip chart recorder tracing shown in Fig. 4–15. The full scale range (dynamic range) is 100 mV and there are graduations every 1 mV. A person reading such a recorder tracing makes an estimate of the smallest voltage difference that can be detected in order to record the numerical result with an appropriate number of significant figures. For the recorder tracing shown in Fig. 4–15, it might be estimated that 0.1 mV can just be detected. Hence 0.1 mV represents the quantization level and the numerical result might be in error by as much as ± 0.1 mV. Such an error is often called a resolution error or reading error, and is present whether or not the signal contains a detectable amount of electrical noise.

As mentioned above, in the absence of a significant amount of electrical noise, this quantization error can limit the obtainable precision and accuracy in a given measurement. The exact effect of this error is difficult to assess. It will likely add a small nonrandom determinate error to the result, as human scale readers are very seldom unbiased. Typical biases occur for 0's and 5's, or for even numbers over odd numbers, in the last digit of a reading. To decrease the effect of such errors, the sensitivity of the meter or recorder can be increased and offset methods can be used if necessary. Once the sensitivity is high enough to observe the signal noise level on the recorder, the effects of quantization become small; and the experimenter knows that measurement precision is not being limited by instrument readout, but by signal noise. Hence it is always sound practice to expand the scale to the point at which electrical noise can be observed.

Time-to-Digital Conversion. A quantizing error is also present whenever a time encoded signal is converted to a digital signal. For example, in a digital frequency ratio measurement, the standard clock oscillator opens a gate for a precise time interval to allow the time encoded signal to enter a counter. Because there is usually no synchronization of the signal with the clock oscillator, a quantizing error of ± 1 count can occur in the conversion. Note that this is a worst-case error. Such a quantizing error can again limit the measurement precision if only a small number of counts are accumulated. Again, scale expansion by opening the counter gate for a longer period of time can be used to reduce the relative quantizing error to a small amount.

Analog-to-Digital Converters. An electronic analog-to-digital converter will also introduce quantization noise to a digitized signal. A-to-D converters (ADC's) have a specific dynamic range, and the number of quanti-

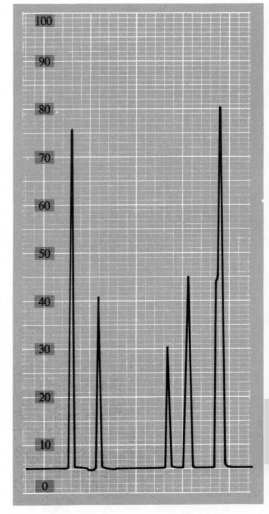

Fig. 4–15 Strip chart recorder tracing.

zation levels that can fit within the dynamic range is often expressed as a number of bits. The number of bits is the $\log_2$ of the dynamic range divided by the size of the quantization level. For example, a typical ADC might have a dynamic range of 10.00 V and a quantization level of 0.01 V. Thus, it would be a 10-bit converter with a least significant bit value of 0.01 V. If the digitizer rounds off to the next lower quantization level, the digitized value of a noise-free analog signal could be in error by as much as the value of the least significant bit (0.01 V).

Again, this is a worst-case error. If the analog input signal contains electrical noise that is large in magnitude compared to the value of the least significant bit, the relative effect of quantization noise is greatly reduced, as discussed above. Scale expansion techniques are also useful in A-to-D conversion to reduce the effect of quantization noise.

Addition of Noise Sources

In most measurement applications, it is the total noise present with the desired signal that is of interest. It is often impossible to determine the separate sources of noise unless one type is totally predominant. If noise sources are completely independent, the total mean-square noise current or voltage is the simple sum of the individual mean-square noise components (variances). Equation (4–5) expresses the total mean-square noise voltage $\bar{v}_t{}^2$ in terms of individual components $\bar{v}_{n1}{}^2$, $\bar{v}_{n2}{}^2$, etc.,

$$\bar{v}_t{}^2 = \bar{v}_{n1}{}^2 + \bar{v}_{n2}{}^2 + \bar{v}_{n3}{}^2 \ldots \qquad (4\text{–}5)$$

Thus, shot noise and Johnson noise, which are random and independent, will add quadratically, as shown in Eq. (4–5).

In some cases, dependent noise sources exist; that is, the noise from one source is dependent on the noise from a second. In this case, the mean-square total noise includes a cross term, whose magnitude depends on the degree of coherence or correlation of the two sources.

4–1.3   THE SIGNAL-TO-NOISE RATIO

The preceding discussion has centered upon the general characteristics of signals and noise. In all electronics-aided measurements, it is desirable to extract the signal from the noise with as much selectivity as is necessary for the desired **accuracy** and **precision**.[3] Because some of the noise is of a fundamental nature, it must be realized that complete freedom from noise can never be realized in a real measurement. The figure of merit used to describe the quality of a measurement is the signal-to-noise ratio (S/N) rather than the absolute signal strength or noise strength. This section examines the S/N, its relationship to measurement precision, and its mea-

Note 3.   Precision and Accuracy.

The **precision** *of a measurement is related to its degree of repeatability and resolution. The* **accuracy** *of a measurement is the degree of agreement between the true and measured values.*

surement. The general principles of S/N enhancement, which will be expanded upon throughout this module, are briefly summarized.

Expression of Signal-to-Noise Ratios

One of the most common ways of expressing the signal-to-noise ratio is as an amplitude ratio. In general, the final output signal from a measurement system is a dc signal with a fluctuating ac noise component. In this case the signal-to-noise ratio can be expressed as:

$$\frac{S}{N} = \frac{\text{average signal}}{\text{rms noise}}. \tag{4-6}$$

A noisy dc current signal is shown in Fig. 4–16. The average current and rms noise current are marked on the figure. The S/N in this case is the average current divided by the rms noise current = 10.3. Large values of S/N imply noise free and precise signal levels. At signal-to-noise ratios of about 3 or less the signal information is usually difficult to discern visually. A noisy base line with an rms noise level of 0.265 V is shown in Fig.4–17a. The low signal-to-noise ratio signal resulting when a peak type signal with a peak height of 0.6 V (Fig. 4–17b) is added into this base line is shown in Fig. 4–17c. In this case the S/N, calculated as peak height/rms noise, is 2.26.

The signal-to-noise ratio may also be expressed as a power ratio in terms of decibels. The expression for S/N becomes:

$$\frac{S}{N} \text{ (dB power)} = 10 \log \frac{\text{signal power}}{\text{noise power}}. \tag{4-7}$$

Fig. 4–16 Noisy signal.

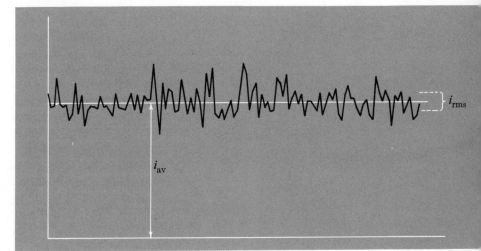

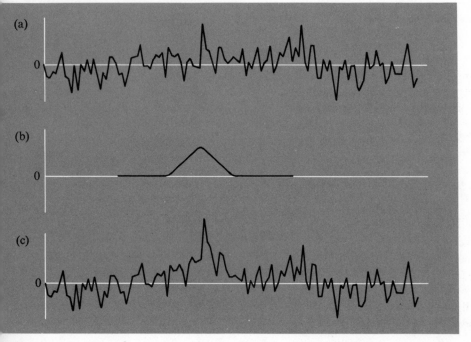

Fig. 4–17 Composition of a noisy signal: (a) noisy base line; (b) peak type signal; (c) noisy signal.

The decibel format is also used for voltage and current ratios where

$$\frac{S}{N} \text{ (dB voltage)} = 20 \log \frac{\text{signal voltage}}{\text{rms noise voltage}}$$

and

$$\frac{S}{N} \text{ (dB current)} = 20 \log \frac{\text{signal current}}{\text{rms noise current}}.$$

Another important relationship for a specific unit in a measurement system, such as an amplifier, readout device, or signal modifier, is the noise factor NF, which is defined as

$$NF = \frac{S/N \text{ (input)}}{S/N \text{ (output)}}. \tag{4–8}$$

The noise factor reveals how much noise was contributed by the part of the system under consideration. For an ideal electronic device, the noise factor would be unity, whereas real components all have NF's greater than unity.

In many fields, specific figures of merit are often defined which are related to the signal-to-noise ratio. For example, with infrared detectors the term **spectral noise equivalent power NEP$_\lambda$,** is used to indicate the rms signal power of a monochromatic wavelength which is required to just produce a S/N of unity at the output of the detector. The reciprocal of NEP_λ when it is normalized to take into account the area of the detector and the bandwidth of the measurement system, is called the **spectral D-star** (D*).

Relationship of S/N to Measurement Precision

There are many factors which can limit the overall precision of the measurement of a physical or chemical quantity. For example, consider the measurement of the pH of a solution. Among the factors which can influence the overall reproducibility of the measurement are the reproducibility of preparing the sample, the ability of the pH meter to detect small changes in pH, and electrical noise. In many cases, when measurement sensitivity is made high enough and sample preparation and other external factors are extremely reproducible, it is random electrical noise that limits measurement precision. In such cases the signal-to-noise ratio is directly related to the overall measurement precision.

For dc signals, the rms noise can be defined as the square root of the average squared deviation of the signal from its mean value, as shown in Eq. (4–9):

$$\text{rms noise} = \left[\frac{(S - S_1)^2 + (S - S_2)^2 + \ldots (S - S_n)^2}{n} \right]^{1/2}, \quad (4\text{–}9)$$

where S is the mean value of the signal, $S_1, S_2, \ldots, S_n$ are the instantaneous values at times $t_1, t_2, \ldots, t_n$, and n is the total number of values. Note that the rms noise as defined by Eq. (4–9) is the standard deviation of the signal σ_s. Hence, the signal-to-noise ratio is merely S/σ_s, or the reciprocal of the relative standard deviation. If a measurement gives an S/N of 1,000 and the overall precision is limited by electrical noise, we would expect the measurement to be precise to one part per thousand, or 0.1 percent. The relative standard deviation of a measurement is often determined by making repetitive discrete measurements, while the S/N is usually determined from a single result. However, if electrical noise is the limiting factor, the S/N is equivalent to the reciprocal of the relative standard deviation.

For ac signals, the relationship between S/N and measurement precision is less straightforward. However, in many cases ac waveforms are converted to dc by rectification techniques before they are displayed on a readout device. In these situations the S/N and relative standard deviation are related, as for dc signals.

Measurement of Signal-to-Noise Ratios

For dc signals, the S/N can be measured in several ways. One of the methods, suggested by Eq. (4–9), is to repeat the measurement several times. The average value of the signal is obtained, and the noise is computed by Eq. (4–9). Enough measurements must be made to obtain a reliable average and a statistically good estimate of the standard deviation. For a small number of measurements, n should be replaced by $(n - 1)$ in Eq. (4–9).

A second way of estimating the S/N is to record the dc signal on an oscilloscope or recorder. If the noise present with the signal is white noise, it is 99 percent probable that individual excursions from the mean value are within 2.5 standard deviations. Since random variations occur in both positive and negative directions around the mean value, it is 99 percent probable that the total peak-to-peak fluctuations of the signal lie within 5 standard deviations. Hence, the standard deviation can be estimated to be $\frac{1}{5}$ the peak-to-peak fluctuations in the signal. Again the signal should be recorded for a time which is long enough to obtain a reliable estimate of the standard deviation.

With ac signals, signal-to-noise ratios can be measured in several ways. If the signal is converted to dc before the measurement is completed, the S/N is estimated in exactly the same manner as for a dc signal. If the signal remains ac, an rms voltmeter can sometimes be used for S/N measurements.

Principles of S/N Enhancement

Once a signal has been transduced to an electrical quantity, no amount of amplification will increase the signal power with respect to the noise power unless specific measures are taken to discriminate against amplifying noise. Thus enhancement of S/N is always achieved by discrimination against noise. As seen from the above discussion, signals and noise have several different characteristics. A signal may contain information in a narrow band of frequencies or it may be of a specific waveshape. In addition, properties of a signal waveform such as its phase, ON-OFF time, and repeatability may be unique. Any of these features which characterize a signal may be used to extract the signal from the noise.

One of the most common ways of improving the signal-to-noise ratio is to control the frequency response of the measurement. Specific approaches for controlling the frequency response of electronic systems are discussed in the next section. Because most system noise exhibits a strong $1/f$ component, it is often advantageous to move the signal information away from dc before amplifying and processing the signal. This is carried out utilizing modulation and demodulation techniques which are discussed in Section 4–3. A number of modern techniques for enhancing the S/N involve a combination of several basic approaches which rely on different characteristics of signals and noise. A discussion of these methods is found in Section 4–5.

PROBLEMS

1. Several interesting properties of waveforms and their spectra can be illustrated by calculating the Fourier transform of a finite length cosine wave. If a function is even $[f(-x) = f(x)]$, its Fourier transform can be calculated using the following equation:

$$F(f) = 2 \int_0^t f(t) \cos 2\pi ft \, dt,$$

where $F(f)$ is the Fourier transform (amplitude spectrum) of waveform $f(t)$. (a) Using this equation (and standard integral tables), calculate the amplitude spectrum of a cosine waveform $f(t) = \cos 2\pi f't$. (b) Plot the spectrum for $f' = 100$ Hz and $t = 0.1$ and 1 sec. (c) Plot the square of the amplitude spectrum (power spectrum) for $f' = 100$ Hz and $t = 0.1$ sec. (d) What happens to the power spectrum when (i) t gets large? (ii) t gets small? Explain.

2. Table 4–1 contains the cosine frequency components of a Gaussian peak which has a full width at half-height of 1 sec. The unit height equation for such a Gaussian peak is $\exp[-4(\ln 2)t^2]$, where t is time in sec. For each frequency component f with relative amplitude A, calculate $A \cos 2\pi ft$ for $t = 0, \pm0.5, \pm1.0, \pm1.5$, and ±2.0 sec. Tabulate the values and sum them to get the resulting peak similar to Fig. 4–9. Compare the resulting peak to the actual Gaussian as described by the equation.

3. Calculate the rms Johnson noise at 300°K for a 100 MΩ resistor and a bandwidth of 10 Hz.

4. Eleven measurements were made of a fluctuating dc voltage. The measurements were 50.44, 45.28, 46.40, 60.41, 65.92, 44.46, 67.13, 63.87, 61.50, 61.69, 59.88. (a) Calculate the S/N. (b) Plot the values and estimate the S/N with the approximate method discussed in Section 4–1.3.

Control of Frequency Response

The control of the frequency response in a measurement system is an important aspect of the optimization of an electronics-aided measurement. As seen in Section 4–1, the specific frequency composition of a waveform is often unique and thus the frequency response should be tailored to the frequency characteristics of the signal. Where differences in frequency composition of signal and noise information exist, the response should discriminate against the noise as much as possible with minimal distortion of the signal. Essentially, all electronic systems designed to enhance the S/N of a measurement rely in whole or in part on this type of frequency response control or filtering.

The frequency response characteristics of a system can also impose limitations on the measurement if some subsystem, component, or transducer is controlling the frequency response in an undesirable manner. Such a limitation is frequently encountered in attempting to achieve high frequency response with reasonable gains.

In assessing the overall frequency response of a measurement system, it is quite important to look at all components and subsystems from the input transducer through to the output transducer. Some general conclusions in this area are discussed in the next section. Then the basic concepts and characteristics of amplifiers with feedback are discussed. Feedback is an important factor in the control of amplifier frequency response.

Analog filters are perhaps the most important circuit networks utilized to control frequency response. Simple passive low and high pass filters were introduced in Section 3–3 of Module 3. Here, these approaches are extended to active filters, which are being increasingly used in measurement systems. They can be used to control the upper and lower frequency response of a system as well as to pass or reject narrow frequency bands.

An electronic oscillator is a device that produces an output signal which continuously repeats the same time varying pattern. It can also be inter-

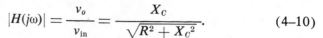

Fig. 4–18 Block diagram of a generalized measurement system.

preted as a circuit in which the frequency response is controlled so that a specific frequency or group of frequencies is continuously available at the output terminals. Many types of oscillators have been presented throughout the previous modules. These are summarized, and then oscillators based on frequency selective positive feedback are discussed.

4–2.1 SYSTEM FREQUENCY RESPONSE

In a real measurement system a variety of components, networks, and systems are present along with input and output transducers. A generalized measurement system in which the major components are grouped together is illustrated in Fig. 4–18. The overall system frequency response will be determined by the combination of the individual response characteristics of each subsystem. A common way of representing the frequency response of a system is a Bode plot, which is a plot of the system gain in decibels as a function of the log of the frequency (dB vs. log f). The Bode plot was introduced in Section 3–2.3 of Module 3. In a surprisingly large majority of cases, the Bode plot of a system or subsystem has the basic characteristics of that for the simple low pass filter, since it is often this circuit or its equivalent that limits the upper frequency response of a system. The Bode plot for a typical system component (an operational amplifier) is shown in Fig. 4–19. At low frequencies the gain of the system component is flat; and at higher frequencies, above the 3 dB point, the gain drops off at a rate of 20 dB/decade (6 dB/octave). The limiting slope is easy to derive, starting from the equation for the amplitude transfer function of the low pass filter:

$$|H(j\omega)| = \frac{v_o}{v_{\text{in}}} = \frac{X_C}{\sqrt{R^2 + X_C{}^2}}. \qquad (4\text{--}10)$$

In the high frequency region of the Bode plot, where the gain is decreasing linearly with log f, X_C is much smaller than R. Remember $X_C = 1/(2\pi f C)$. Therefore, in this region Eq. (4–10) simplifies to:

$$\frac{v_o}{v_{\text{in}}} = \frac{X_C}{R}. \qquad (4\text{--}11)$$

Converting to dB, we get:

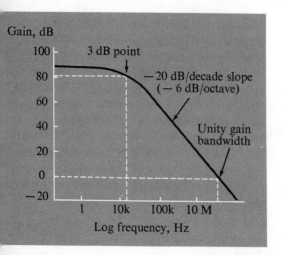

Fig. 4–19 Bode plot for a system component.

$$\text{Gain (in dB)} = 20 \log \frac{v_o}{v_{\text{in}}} = 20 \log X_C - 20 \log R. \qquad (4\text{--}12)$$

Substituting for X_C and rearranging yields:

$$\text{Gain (in dB)} = -20 \log f - 20 \log 2\pi RC. \qquad (4\text{--}13)$$

The result is a linear equation in dB of gain vs. log f with a negative slope of 20 dB per decade change in frequency. This simply means that, in this region, if the signal frequency is increased by a factor of 10, the gain goes down by a factor of 10. The slope is frequently stated as -6 dB per octave. A change of one octave is equivalent to a factor of two change in frequency. A change of 6 dB in gain is also equivalent to a change in gain by a factor of two. Thus a 6 dB/octave rolloff simply means that if the frequency is doubled, the gain is cut in half.

The frequency response of a system can be, at best, only as good as the poorest component in the signal path. The component with the lowest 3 dB point will be the main limiting factor with respect to high frequency gain. In the block diagram of Fig. 4–18 there are two transducers to consider in the basic measurement process. The input transducer converts the quantity of interest into an electrical quantity in one of the electrical domains, whereas the output transducer provides the opposite function. Since the overall system response characteristics are often limited by these devices, a brief review of their characteristics is presented here.

Input Transducers

The input transducer can be described by its input-output relationship, which may or may not be linear. The input transducer converts the input signal quantity Q_i into an electrical output quantity E_o and its transfer characteristic may be written in general terms as

$$E_o = f(Q_i). \qquad (4\text{--}14)$$

The sensitivity S of the input transducer is defined as the slope of the transfer characteristic

$$S = \frac{dE_o}{dQ_i}. \qquad (4\text{--}15)$$

The sensitivity S is constant if the relationship between input and output is linear, while S varies with Q_i for a nonlinear transducer.

The transfer function of the input transducer is always frequency dependent. For example, radiation transducers in the infrared region of the spectrum, such as thermocouples and bolometers, have relatively long response times and are incapable of following rapid variations in the input optical signal. The frequency response of a transducer may be expressed by a Bode diagram where the magnitude of the ratio E_o/Q_i in dB is plotted vs. log f. A limited transducer frequency response can cause distortion, just as was the case with amplifiers. Amplitude distortion may also be introduced if the transducer transfer function is nonlinear over the range that the signal varies.

There are many other characteristics of input transducers which are important in overall system response. The usable input and output ranges are often restricted. Likewise the output impedance is an important characteristic, for it determines whether power can be supplied by the transducer and what the input impedance of the remainder of the measurement system must be to avoid loading errors.

Output Transducers

Output transducers can be characterized in ways similar to those of input transducers. The frequency response of output transducers varies widely. Some oscilloscopes, for example, have essentially flat response characteristics from dc to several MHz. Others have more limited frequency response. A servo recorder may have excellent response from dc to 0.1 Hz, but its response may fall off rapidly above this frequency. An analog-to-digital converter in conjunction with a computer storage system may have a very wide frequency response, which enables μsec variations to be recorded.

The choice of output transducer is often dictated by the frequency components of the signal. Again a Bode diagram can be used to express the dependence of the output transducer transfer function on frequency.

System Response

The overall system response $H(j\omega)$ of a system of n components can be found from the transfer function of each of the components, as shown in Eq. (4–16):

$$H(j\omega) = H_1(j\omega)H_2(j\omega) \ldots H_n(j\omega). \tag{4–16}$$

For a simple system consisting of an input transducer, an amplifier, and an output transducer, the overall response is just the product of the response of each component as long as there is no loading of one component by another. The magnitude of the overall response function $|H(j\omega)|$ can similarly be written as the product of the magnitudes of each of the three transfer functions.

The Bode diagram is a simple method for obtaining the overall system response because it is a logarithmic plot. To obtain a Bode diagram for an entire system, it is necessary only to add the response characteristics of each component. A Bode diagram is shown in Fig. 4–20 for a simple three-component system, where the output transducer has the worst frequency response characteristics. Note that the system response follows that of the worst component at low frequencies. Thus, the system response has a corner

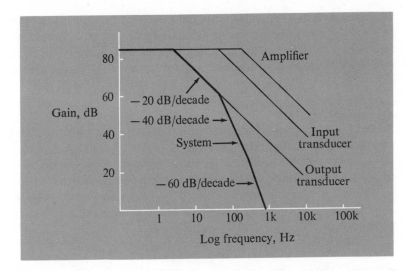

Fig. 4–20 Overall Bode plot for a measurement system.

frequency equal to that of the output transducer, for the example in Fig. 4–20, and begins to roll off with the same slope. At the corner frequency of the input transducer, the system response begins to roll off with a slope equal to the sum of the rolloff slopes of the two transducers. Finally at the corner frequency of the amplifier, the rolloff slope of the system equals the sum of the slopes of all three components. If each component rolls off with a —20 dB/decade slope, the system would show 4 different slopes of 0 dB/decade, —20 dB/decade, —40 dB/decade, and —60 dB/decade, as shown in Fig. 4–20.

4–2.2 AMPLIFIERS WITH FEEDBACK

An amplifier is said to have feedback when part of its output signal is returned to its input. **Negative feedback** occurs when the magnitude of the input signal is decreased and **positive feedback** when the magnitude is increased. In general, negative feedback has a stabilizing influence on a system while positive feedback has a destabilizing influence. The presence of feedback affects almost every electrical characteristic of an amplifier. Feedback can be used to control the gain of an amplifier, improve its stability and impedance characteristics, and reduce noise and distortion. Also, feedback is very useful in controlling and modifying the frequency response characteristics of an amplifier. Often it is simpler, and in many cases more satisfactory, to tailor an amplifier's characteristics by adding feedback to a standard amplifier circuit than to design an unfedback amplifier with the required characteristics. On the other hand, feedback can be undesirable in that distortions and spurious oscillations can be introduced to the signal by improper design. This widespread influence of feedback on an amplifier's characteristics means that an understanding of it is essential to an understanding of modern amplifiers. In the next sections some of the basic characteristics of amplifiers with feedback are discussed.

Gain

A quantitative relation for the effect of feedback on the gain of an amplifier can be derived through reference to Fig. 4–21. A fraction β of the output signal is fed back and added to the input signal, as shown in Fig. 4–21. The fraction β is determined solely by the simple voltage divider circuit in the feedback loop, and is equal to $R_1/(R_1 + R_2)$. The input voltage to the amplifier v_{in} is the sum of the signal input voltage v_{sig} and the feedback voltage βv_o:

$$v_{in} = v_{sig} + \beta v_o. \qquad (4\text{--}17)$$

From the amplification A of the amplifier,

$$v_o = A v_{in}. \qquad (4\text{--}18)$$

The gain of the amplifier with feedback A_f is the ratio of v_o to v_{sig}, and from Eqs. (4–17) and (4–18) is expressed by:

$$A_f = \frac{v_o}{v_{sig}} = \frac{A}{1 - \beta A}. \qquad (4\text{--}19)$$

The gain of the fedback amplifier A_f is often called the closed-loop gain, while A is called the open-loop gain.

When $1 - \beta A$ is greater than 1, the gain of the amplifier is less than A and the feedback is said to be negative. An important limiting case of Eq. (4–19) occurs when $\beta A \gg 1$ or, essentially, when the open-loop gain is very large. In this case, the gain of the fedback amplifier A_f reduces to $-1/\beta$. This result means that the closed-loop gain of the fedback amplifier can be made virtually independent of the amplifier's open-loop characteristics. Note, from Fig. 4–21, that the feedback fraction β can be selected at will, using high precision resistors, making the gain $-1/\beta$ very accurate. Also relatively large changes in A influence A_f only slightly. For example, if A is -10^5 (100 dB) and $\beta = 0.01$, the closed-loop gain A_f is -99.9. Note that this is essentially equal to $-1/\beta$, which equals -100. If the amplifier characteristics were to change (because of component aging, replacement of the amplifier, power supply variations, etc.) such that A became -1.1×10^5, the new closed-loop gain A_f is -99.91. Hence a 10% change in A results in less than 0.1% change in A_f.

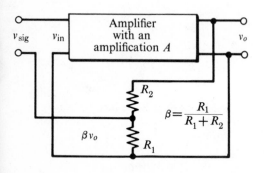

Fig. 4–21   Amplifier with feedback.

Frequency Response and Distortion

It was shown above that, as the amount of negative feedback is increased, the gain of the amplifier becomes less dependent on the amplifier and more a function of the feedback network. If the impedance of the feedback network is nonreactive, so that the fraction β is independent of frequency, the gain A_f of the fedback amplifier will be virtually independent of frequency and the bandpass of the amplifier will be extended. Positive feedback, on the other hand, tends to decrease the bandpass of the amplifier.

The exact relationship between the increase in frequency response and gain is easily visualized on a gain vs. frequency plot, as shown in Fig. 4–22. In Fig. 4–22a the frequency response of an ac amplifier is plotted. The dotted lines show the open-loop gain vs. log f. The midfrequency gain is A and the amplification rolls off at 20 dB/decade. The original bandwidth is $f_2 - f_1$, as shown. If a fraction β is now fed back in such a way that closed-loop gain A_f results, it is apparent that the bandwidth has been extended to $f_2' - f_1'$. The same improvement in bandwidth for a dc amplifier such as an operational amplifier is shown in Fig. 4–22b. The bandwidth has been extended from f_2 for the unfedback amplifier to f_2' for the fedback amplifier. The open-loop unity-gain bandwidth f_t is related to the closed-loop bandwidth f_2' by

$$f_t = \frac{f_2'}{\beta} = A_f f_2' \cong \text{constant.} \qquad (4\text{--}20)$$

Thus the bandwidth of the fedback amplifier f_2' is proportional to β. The constant $A_f f_2'$ is often called the **gain-bandwidth product**.

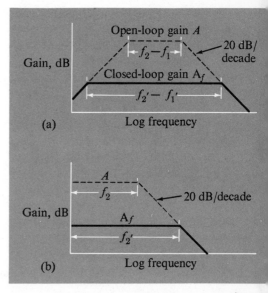

Fig. 4–22 Frequency response of amplifiers with feedback: (a) ac amplifier; (b) dc amplifier.

Experiment 4–1   *Bandwidth of Amplifiers with Feedback*

The bandwidth of an amplifier with feedback depends on the gain of the amplifier. In this experiment the upper 3 dB point of a dc amplifier is determined for three different gains. The data obtained are used to determine the slope of the open-loop rolloff and to determine the unity-gain bandwidth of the amplifier.

The phase relationship between a sine-wave signal and the feedback signal is shown for negative feedback in an ac amplifier in Fig. 4–23a. Note that the voltage at the amplifier input is the same phase as the signal voltage, but reduced in magnitude. Suppose that the signal is delayed in the amplifier high pass RC networks by 45°. The feedback signal would thus have a

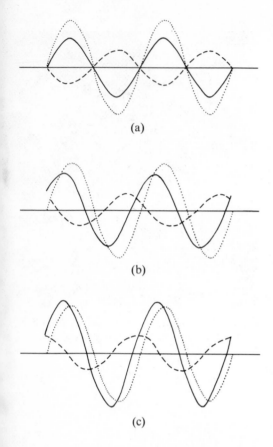

(a)

(b)

(c)

Fig. 4–23 Sum of signals of different phase angle: (a) $\theta = 180°$; (b) $\theta = 225°$; (c) $\theta = 270°$. (Dotted line, input signal; dashed line, feedback signal; solid line, resulting input signal.)

phase angle of $180° + 45°$ with respect to the input signal. The resultant amplifier input signal for this case is shown in Fig. 4–23b. Note that the resultant input signal is advanced in phase, which tends to compensate for the delay in the amplifier. Thus negative feedback decreases phase distortion. The same corrective action could be shown had the phase angle of the feedback signal been less than $180°$ instead of greater. The effect of a phase shift in the amplifier of greater than $45°$ is shown in Fig. 4–23c. Here it can be seen that a phase angle of $180° + 90°$ corrects the phase less than does the $45°$ phase shift. Note too that the amplitude of the input signal is actually augmented. In this case the combined phase shift of the amplifier ($90°$) and the feedback network ($180°$) results in positive feedback. Thus negative feedback can reduce moderate amounts of phase distortion, but a larger phase distortion can actually result in positive feedback and an enhancement of the distortion. In this way signals well outside the normal bandpass of the amplifier, which are attenuated and shifted in phase in the amplifier, can be regenerated by the feedback loop and appear in the output signal. In general, the presence of these distorted and augmented signals is undesirable. Increasing the negative feedback only increases the regeneration of these unwanted signal components. They can be eliminated only by a better rejection of the unwanted frequencies either in the amplifier or in the feedback network.

Noise

Nonlinearity of amplification and noise from the amplifier are also reduced by negative feedback. A noise voltage at the amplifier output which is fedback (attenuated) to the input out of phase causes the amplifier to counteract the spurious signal at the output. Suppose that a noise voltage N is generated *in* the amplifier. Because of the feedback, the actual noise voltage at the output will be different from N. The output noise will be designated N'. A voltage $\beta N'$ will be fedback to the amplifier input. The actual output noise N' will then be the sum of N, the noise generated in the amplifier, and $A\beta N'$, the amplified noise fed back to the input. Solving for N',

$$N' = N + A\beta N', \tag{4–21}$$

$$N' = \frac{N}{1 - \beta A}. \tag{4–22}$$

It is seen that noise generated in the amplifier is reduced by the same factor as the gain. Just how negative feedback might be used to increase the signal-to-noise ratio depends on the source of the noise. Three cases will be considered:

1. Suppose that the noise generated in the amplifier N is independent of the size of the signal. Examples of this would be thermal noise in the components and 60 Hz interference noise in the amplifier. In this case, negative feedback will reduce the output signal and the noise by the same amount, and the signal-to-noise ratio will not be improved. Improvements in the signal-to-noise ratio could be gained only by increasing the size of the input signal (output signal increases proportionately, but noise remains constant), using less noisy components, or increasing the ratio of the signal gain-to-noise value of the amplifier.

2. Suppose that the noise is actually distortion due to the nonlinearity of the amplifier. Negative feedback reduces this distortion by two effects. The input signal is reduced, which helps to keep the signal within more linear regions of amplification. The remaining distortion component in the output is further reduced by the factor $1 - \beta A$. If the size of the input signal had been increased to compensate for the reduction in gain, the output signal would be the same magnitude as with the unfedback amplifier but the distortion component would be reduced by the factor $1 - \beta A$. Thus increasing the input signal magnitude and introducing negative feedback will increase the signal-to-noise ratio.

3. If the noise is present in the signal to be amplified, there must be some means for distinguishing between the signal and the noise if the signal-to-noise ratio is to be improved. Some sort of circuit which rejects the non-signal frequencies is ordinarily used. There is no way in which just amplification, with or without feedback, can improve the signal-to-noise ratio of a noisy input signal. Negative feedback reduces only that noise which is generated in the amplifier or the feedback circuit.

The Possibility of Oscillation

The gain of the voltage feedback amplifier was given in Eq. (4–19):

$$A_f = \frac{A}{1 - \beta A}.$$

When A is positive and the quantity $1 - \beta A$ is less than 1, A_f is greater than A. In other words there is positive, or regenerative, feedback. As βA approaches 1, the gain A_f approaches infinity. Another way of saying this is that the signal voltage v_{sig} required for a given output voltage decreases to 0 as βA approaches 1. When βA equals 1, the feedback voltage is sufficient

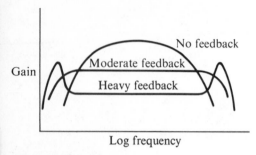

Fig. 4–24 Effect of negative feedback on frequency response.

to maintain an output signal even though the input signal voltage is 0. Such an amplifier is called an **oscillator.** All frequencies for which the condition $\beta A = 1$ is fulfilled will be present in the output. When a feedback path exists in an amplifier, it is essential that the conditions necessary for oscillation do not exist.

Even when the feedback is negative at normal frequencies, the possibility of positive feedback at frequencies on the fringe of the bandpass exists. A set of curves showing the effect of negative feedback on frequency response is shown in Fig. 4–24. A moderate amount of feedback flattens the response curve a great deal. Frequencies at the extremes of the response curve are shifted in phase through the amplifier so that the feedback for them is actually positive. This is shown by the fact that the gain at these frequencies is greater than for an unfedback amplifier. When a larger amount of feedback is applied, a greater regeneration of the extreme frequencies occurs, as shown. The response shown is sufficient to cause definite distortion of the input waveform. Were the feedback to be increased still further, the condition for oscillation might well be fulfilled for either the very low or the very high frequencies.

If heavy negative feedback is required, the possibility of oscillation can be eliminated by putting filters in the feedback loop to eliminate regeneration of the offending frequency, or by deliberately limiting the bandwidth of one of the amplifier stages.

Positive feedback is useful if it is desired to make an oscillator. Feedback oscillators are discussed in Section 4–2.4.

4–2.3 ANALOG FILTERS

Simple analog filtering is often all that is required to provide high measurement precision for many signals. The section begins with a discussion of the characteristics of low pass filters, including the many types of active filters which have relatively recently been developed with operational amplifiers. Second- and higher-order low pass filters with Butterworth, Bessel, or Tchebyscheff responses are difficult to design from first principles. However, many manufacturers of operational amplifiers market a wide range of these types of filters. Thus the discussion of these filters will be directed primarily towards their basic characteristics rather than their design. Active high pass filters are briefly considered, and then narrow bandpass filters and their use in tuned and band rejection amplifiers are discussed.

Single Time-Constant Low Pass Filters

In a large number of measurement situations, simple RC low pass circuits are completely adequate for controlling the upper frequency response of a system.

The series RC circuit followed by a voltage follower to reduce loading is often used and is illustrated in Fig. 4–25. Recall that the simple low pass filter has an upper 3 dB point of $f_2 = 1/(2\pi RC)$. At higher frequencies the filter attenuation increases to a limiting slope of -6 dB/octave (-20 dB/decade).

The noise bandwidth of a simple RC low pass filter may be readily found from its complex transfer function $H(j\omega)$. For the simple RC low pass circuit $H(j\omega)$ is the ratio of the complex capacitive reactance to the total complex impedance as shown in Eq. (4–23),

$$H(j\omega) = \frac{v_o}{v_{in}} = \frac{-jX_C}{R - jX_C} = \frac{1}{1 + j\omega RC}, \qquad (4\text{–}23)$$

where v_o and v_{in} are instantaneous output and input voltages. Note that the amplitude of the transfer function, $|H(j\omega)|$, is merely the magnitude of the divider fraction without regard to phase, as expressed previously in Eq. (4–10). For noise which has a constant spectral density, the equivalent noise bandwidth Δf is related to the power transfer function $|H(j\omega)|^2$, and is found by integration of $|H(j\omega)|^2$ with respect to frequency, as shown in Eq. (4–24),

$$\Delta f = \int_0^\infty |H(j\omega)|^2 \, d\omega = \frac{1}{4RC}. \qquad (4\text{–}24)$$

Note that the noise equivalent bandwidth is slightly larger than the upper cutoff frequency of the network f_2. If $RC = 1$ sec, for example, the noise bandwidth is 0.25 Hz, while the upper cutoff frequency is 0.16 Hz.

In order to decrease the noise bandwidth of a measurement system by filtering, the value of RC must be increased, which leads to decreased response time of the circuit. Since the voltage at the output of the filter approaches the signal voltage exponentially, an exact value of the response time is impossible to give. However, after 5 time constants, the output voltage is greater than 99 percent of the input voltage. Hence, for a noise bandwidth of 0.25 Hz, five seconds are required for the measurement error to be less than 1 percent. Higher accuracies require more time.

A second arrangement of a simple RC filter is to incorporate it into the feedback network of an operational amplifier, as shown in Fig. 4–26. The circuit shown is the simplest example of what is known as an **active filter.** An active filter results when RC networks are used in conjunction with active elements such as transistors or operational amplifiers. There are several advantages in using the operational amplifier to perform filtering. The OA has a very low output impedance, in contrast to the **passive** RC filter of Fig. 4–25. Thus, several filtering stages can be designed independently and used together with minimal interaction. The operational

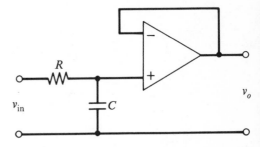

Fig. 4–25 RC low pass filter with buffer amplifier.

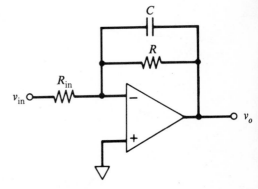

Fig. 4–26 First-order active low pass filter.

amplifier also permits tuned circuits to be realized with only resistors and capacitors, as will be discussed later in this section. There can also be voltage gain associated with active filters. The circuit of Fig. 4–26, for example, has a dc voltage gain of $-R/R_{in}$.

The operational amplifier active filter of Fig. 4–26 is known as a **multiple-feedback filter** since there is a feedback path through R and a second path through C. It has a time constant of RC and a noise bandwidth of $\Delta f = 1/(4RC)$. The response time is exactly equivalent to that of the simple low pass filter.

Second-Order Low Pass Filters

In order to have a better compromise between noise bandwidth and response time, second-order filters can be used. These filters have sharper rolloff characteristics than do single time-constant filters. Such second-order filters can be achieved with RLC circuits or with multiple time-constant active RC circuits. Since inductors are bulky and expensive, the latter approach is normally preferred in modern circuits. In order to understand the behavior of such second-order filters, it is instructive to review the frequency response behavior of the RLC circuit.

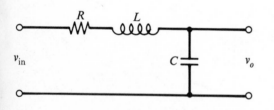

Fig. 4–27   *RLC* low pass filter.

Series RLC Filter.   The series RLC circuit of Fig. 4–27 was previously considered as a resonant circuit in Section 3–2.4 of Module 3. The circuit can also be used as a low pass filter with certain advantages over single time-constant RC circuits.

The transfer function of the network can be expressed by Eq. (4–25),

$$H(j\omega) = \frac{v_o}{v_{in}} = \frac{-jX_C}{R + jX_L - jX_C} = \frac{1}{1 + (j\omega)^2 LC + j\omega RC}.$$
(4–25)

In Module 3, the resonant frequency was defined as

$$f_0 = 1/(2\pi\sqrt{LC}) \quad \text{or} \quad \omega_0 = 1/\sqrt{LC}.$$

If $LC = 1/\omega_0^2$ is substituted into Eq. (4–25), and the numerator and denominator multiplied by ω_0^2, Eq. (4–26) results:

$$H(j\omega) = \frac{\omega_0^2}{\omega_0^2 + (j\omega)^2 + j\omega RC\omega_0^2}.$$
(4–26)

Equation (4–27) results when $\omega_0 = 1/\sqrt{LC}$ is substituted into the right-hand term in the denominator of Eq. (4–26):

$$H(j\omega) = \frac{\omega_0^2}{\omega_0^2 + (j\omega)^2 + \dfrac{j\omega RC\omega_0}{\sqrt{LC}}} = \frac{\omega_0^2}{\omega_0^2 + (j\omega)^2 + j\omega R\sqrt{\dfrac{C}{L}}\,\omega_0}.$$

(4–27)

A second parameter called the damping factor,

$$d = \frac{R}{2}\sqrt{\frac{C}{L}},$$

can be substituted into Eq. (4–27) to yield Eq. (4–28):

$$H(j\omega) = \frac{\omega_0^2}{\omega_0^2 + (j\omega)^2 + j\omega d\omega_0} = \frac{1}{[1 - (\omega/\omega_0)^2] + j(2d\omega/\omega_0)}.$$

(4–28)

The amplitude of the transfer function $|H(j\omega)|$ can be expressed as:

$$|H(j\omega)| = \frac{1}{\sqrt{[1 - (\omega/\omega_0)^2]^2 + (2d\omega/\omega_0)^2}}.$$ (4–29)

Several important consequences of the second-order RLC response can be obtained from Eq. (4–29). First, at frequencies much higher than the natural or resonance frequency, $\omega/\omega_0 \gg 1$, Eq. (4–29) reduces to

$$|H(j\omega)| \approx \frac{1}{\sqrt{(\omega/\omega_0)^4 + 4d^2(\omega/\omega_0)^2}} \quad \text{for } \omega \gg \omega_0.$$ (4–30)

At high frequencies and at damping factors $\leqslant 1$, the left-hand term in the denominator of Eq. (4–30) is much larger than the right-hand term, and the transfer function reduces to

$$|H(j\omega)| \cong \frac{1}{(\omega/\omega_0)^2} \quad \text{for } \omega \gg \omega_0.$$ (4–31)

Hence, the high frequency attenuation in dB can be expressed by

$$\text{Attenuation (in dB)} = 20 \log |H(j\omega)| = -20 \log \left(\frac{\omega}{\omega_0}\right)^2$$

$$= -40 \log \frac{\omega}{\omega_0}.$$ (4–32)

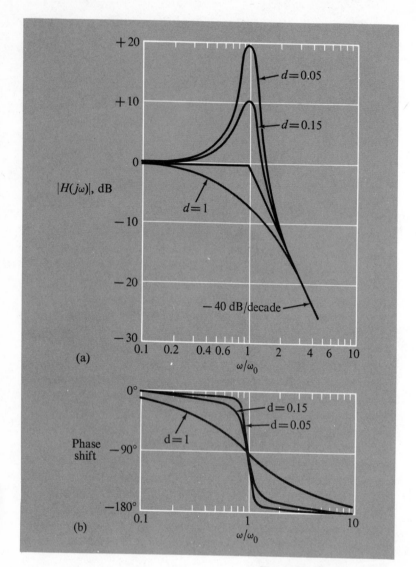

(a)

(b)

Fig. 4–28 Bode plot for an *RLC* circuit. (a) Transfer function magnitude (dB) vs. log relative frequency. (b) Phase shift vs. log relative frequency.

For each decade change in frequency at high frequencies there is a -40 dB/decade rolloff characteristic for the second-order filter, as compared with -20 dB/decade for a simple first-order low pass filter.

A Bode diagram for the *RLC* circuit is shown in Fig. 4–28, where the damping factor d is used as a parameter. The damping factor is inversely related to the quality factor Q of the circuit. Since $Q = 2\pi f_0 L/R = \omega_0 L/R$, the exact relationship is

$$Q = \frac{\omega_0 L}{R} = \frac{L}{\sqrt{LC}\,R} = \frac{1}{R}\sqrt{\frac{L}{C}} = \frac{1}{2d}. \qquad (4\text{–}33)$$

A circuit with a value of $d > 1$, and hence $Q < 0.5$, is said to be **over-damped** and no oscillations appear in the output. If $d < 1$, the circuit is **underdamped** and the output to a step input will be a damped sine wave whose amplitude decays with time. For a sinusoidal input, the RLC network with a high Q factor can be a frequency selective filter for the resonant frequency, as was shown previously in Module 3. At $d = 1$ the circuit is said to be critically damped and acts much like a "low pass filter" with a 40 dB/decade rolloff.

Cascaded RC Networks. A second-order response can be obtained by cascading two RC networks, as shown in Fig. 4–29. If the two RC circuits have equivalent time constants ($R_1C_1 = R_2C_2 = RC$), and the loading of the first network by the second network is neglected, the transfer function has the form

$$H(j\omega) = \frac{1}{(1 + j\omega RC)^2}. \tag{4-34}$$

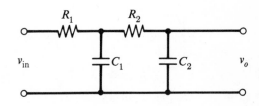

Fig. 4–29 Cascaded RC filter.

This equation is similar to that of the series RLC circuit at high frequencies if RC is equated to the reciprocal of the natural frequency $1/\omega_0$. The cascaded RC circuit can thus simulate a critically damped RLC circuit. Again, a rolloff characteristic of -40 dB/decade can be achieved with this second-order filter.

The noise bandwidth of the cascaded RC network can be found as before by integration of the power transfer function,

$$\Delta f = \int_0^\infty |H(j\omega)|^2 \, d\omega = \frac{1}{8RC}. \tag{4-35}$$

Thus the cascaded RC network has a bandwidth of $\frac{1}{2}$ the value for the single RC filter of the same time constant. The response time of a critically damped filter is also improved over a single RC filter. In Fig. 4–30, the response of a single RC network and a cascaded RC circuit are compared. Both circuits have the same bandwidth. Although the single RC filter output rises more rapidly at the start, the critically damped filter output quickly overtakes it.

Second-order filters from cascaded RC networks are not extremely practical because the stages interact, unless the impedance of the second stage is much greater than that of the first. Because of this restriction, large ratios of resistance values and capacitance values must be used. Also, the equivalent damping factor of the cascaded RC filter is such that perfect critical damping cannot be obtained. Hence, the second-order active filters described below are most commonly used.

Second-Order Active Low Pass Filters. A second-order response can be quite readily obtained with an operational amplifier active filter. A simple

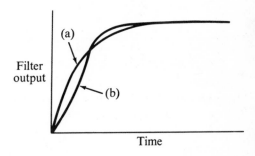

Fig. 4–30 Step response of RC filters: (a) first-order RC filter; (b) second-order critically damped filter.

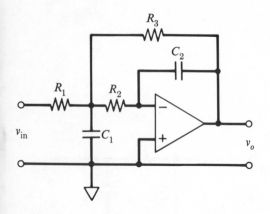

Fig. 4–31 Multiple-feedback second-order active low pass filter.

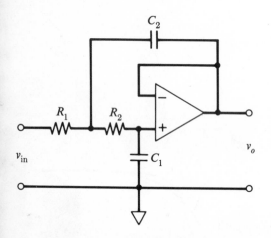

Fig. 4–32 Second-order active low pass filter using noninverting configuration.

multiple-feedback low pass filter is shown in Fig. 4–31. Note that the circuit is a combination of an input low pass filter and a feedback filter.

The active filter can be characterized by a quality factor Q or damping factor d which describes the shape of its frequency response curve. The cutoff frequency or resonant frequency ω_0 is given by Eq. (4–36):

$$\omega_0 = \sqrt{\frac{1}{R_2 R_3 C_1 C_3}} \qquad (4\text{–}36)$$

and the magnitude of H_0, the transfer function at ω_0, is $-R_3/R_1$.

The Bode diagram for the second-order active filter is exactly analogous to that of the RLC circuit shown in Fig. 4–28, and the filter attenuation rolls off at -40 dB/decade at high frequencies. The damping factor d is a complicated function of all R and C values. For the method of choosing component values to obtain the desired d, ω_0, and H_0, refer to the book by Tobey *et al.* in the Bibliography. For a critically damped filter, the value of H_0 can be as high as 100, so that the filter can provide gain. Filters are also useful with values of $d < 1$, which often provide other desirable characteristics, as described below.

A second type of low pass active filter can be realized using the noninverting follower configuration of the operational amplifier, as shown in Fig. 4–32. For the unity-gain follower of Fig. 4–32, $v_o/v_{in} = 1$ at low frequencies. The upper cutoff frequency is

$$\omega_0 = \sqrt{\frac{1}{R_1 R_2 C_1 C_2}} \qquad (4\text{–}37)$$

and the damping factor is again a complicated function of all R's and C's. The follower with gain can also be used in the circuit of Fig. 4–32 to provide a low pass filter with a preselected gain.

Although the critically damped filter provides an excellent compromise between noise bandwidth and response time, other characteristics of the filter are often equally important. An excellent filter which has the characteristic of essentially flat amplitude-vs.-frequency response up to the cutoff frequency is known as the **Butterworth** filter, and is achieved when the damping ratio is 0.707. The sharpness of the cutoff of this filter can be seen in Fig. 4–33 where it is compared with the critically damped filter ($d = 1$) and another filter known as the **Bessel** filter. Note that all three filters reach a rolloff slope of -40 dB/decade at frequencies much greater than the cutoff frequency. Although the Butterworth filter achieves the sharpest attenuation, the phase shift as a function of frequency is nonlinear. The Bessel filter provides the ideal phase characteristic with an approximately linear phase response up to nearly the cutoff frequency. The Bessel filter attenuation sharpness is in between that of the Butterworth

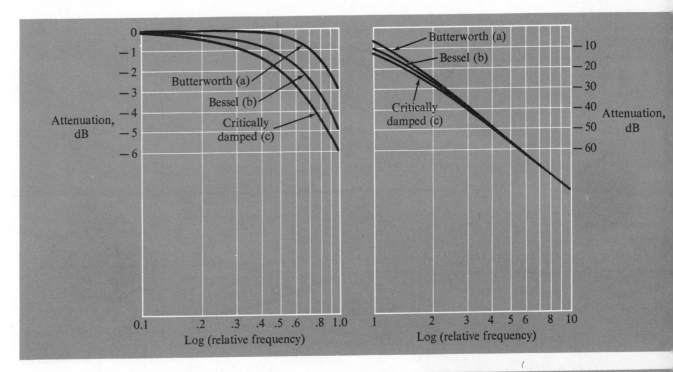

Fig. 4–33 Bode plots for second-order active filters: (a) Butterworth; (b) Bessel; (c) critically damped.

and the critically damped filter, as can be seen from Fig. 4–33. The Bessel filter response is achieved when the damping ratio $d = 0.866$.

In addition to attenuation sharpness and phase response, the step response of the filter is an important characteristic. In Fig. 4–34, the relative output response time and overshoot are indicated for the three second-order filters. The Butterworth filter displays the largest overshoot of 4.3 percent at the peak. The critically damped filter has no overshoot, but a relatively long response time. The Bessel filter offers an excellent compromise between response time and overshoot, as can be seen in Fig. 4–34.

The choice of filter to use, among the second-order types, is usually dictated by which characteristics are most important: phase, attenuation sharpness, or response time. All three types and combinations may be implemented with the circuits of Figs. 4–31 and 4–32 by simple changes in component values to achieve the desired damping factors. In addition, active low pass filters with preselected characteristics are available from a variety of commercial sources.

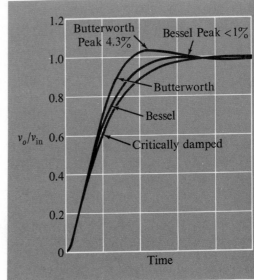

Fig. 4–34 Response time and overshoot characteristics of second-order active filters.

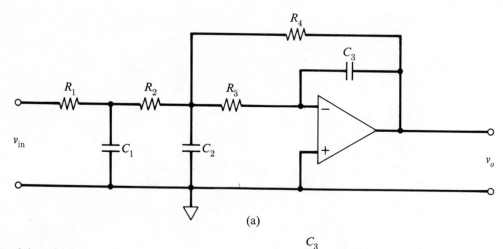

(a)

Fig. 4–35   Third-order active filters: (a) multiple-feedback implementation; (b) noninverting amplifier configuration.

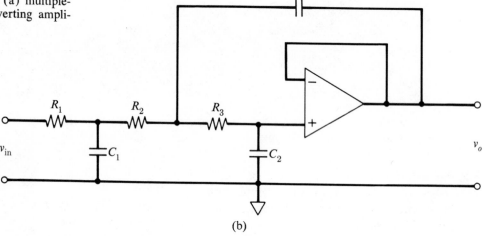

(b)

Higher-Order Active Low Pass Filters

Filters with even sharper rolloff characteristics can be designed by cascading active filters. In general the rolloff beyond the cutoff frequency is $-6n$ dB/octave ($-20n$ dB/decade), where n is the order of the filter. An active third-order filter is simply achieved by cascading an active second-order filter with a passive first-order (single time constant) RC network. The multiple-feedback implementation of a third-order filter is shown in Fig. 4–35a, where the passive network of R_1C_1 has been added to the second-order filter of Fig. 4–31. The noninverting amplifier implementation of a third-order filter is shown in Fig. 4–35b.

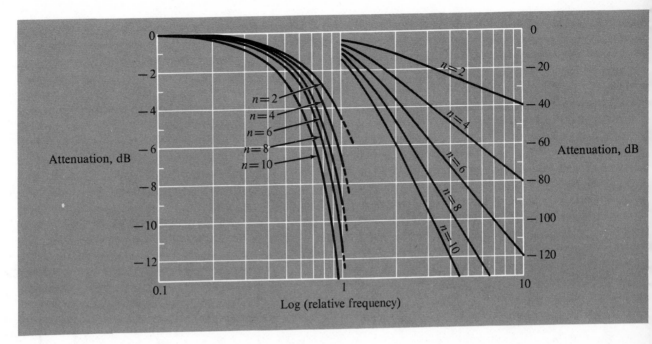

Attenuation, dB

$n=2$
$n=4$
$n=6$
$n=8$
$n=10$

$n=2$
$n=4$
$n=6$
$n=8$
$n=10$

Attenuation, dB

0.1   1   10

Log (relative frequency)

To obtain even higher-order filters, second-order and/or third-order filters are cascaded together. Filters of even order need only cascaded second-order filters, whereas odd-order filters are achieved with second-order sections combined with one third-order section. A sixth-order filter, for example, is achieved by three cascaded second-order active filters, whereas a fifth-order filter is achieved with one second-order section and one third-order section. Filter-orders up to 10 are readily achieved with operational amplifier active filters. Again, component values can be chosen to achieve the desired damping factor for Butterworth, Bessel, or characteristics between the two. The Bode diagrams for Bessel filters of orders from 2 to 10 are shown in Fig. 4–36. Note that there are two different attenuation scales for the low frequency and high frequency characteristics.

Higher-order filters are often used because of their sharper rolloffs. If very sharp rolloff is desired a **Tchebyscheff** filter can be used. The Bode plot for this response is shown in Fig. 4–37. Note, however, that the sharp rolloff of the Tchebyscheff filter is obtained at the expense of having gain ripple in the low frequency bandpass.

High Pass Active Filters

High pass active filters are also easily implemented with operational amplifier circuits. Both multiple-feedback and noninverting configurations

Fig. 4–36 Bode plots for Bessel filters of orders 2 to 10.

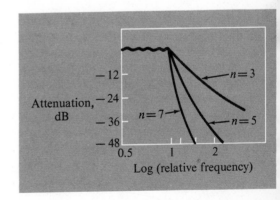

Fig. 4–37 Bode plots for Tchebyscheff filters of orders 3, 5, and 7.

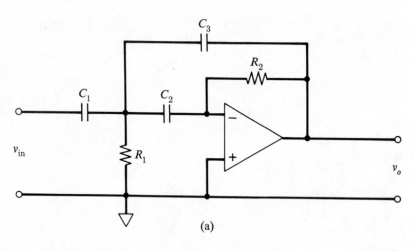

(a)

Fig. 4–38 Active high pass filters: (a) multiple-feedback configuration; (b) noninverting configuration.

(b)

are used, as shown in Fig. 4–38. For the multiple-feedback arrangement of Fig. 4–38a, an input high pass network and a high pass feedback network are combined to give a second-order response. The lower cutoff frequency or natural frequency ω_0 is given by:

$$\omega_0 = \left(\frac{1}{R_1 R_2 C_2 C_3}\right)^{1/2}, \tag{4-38}$$

and H_0, the magnitude of the transfer function at ω_0, is C_1/C_3. The damping factor is again a complex function of all the circuit components. The noninverting configuration of Fig. 4–38b has a lower cutoff frequency of

$$\omega_0 = \left(\frac{1}{R_1 R_2 C_1 C_2}\right)^{1/2}, \tag{4-39}$$

and a unity gain at low frequencies for the unity-gain follower shown. A follower with gain can also be used in the circuit of Fig. 4–38b to provide filter gain at low frequencies.

The Bode diagrams for the high pass active filters are mirror images of those shown in Fig. 4–33 for low pass filters. Butterworth, Bessel, and critically damped filters can all be achieved with the circuits of Fig. 4–38. Higher-order high pass active filters can be implemented analogously to higher-order low pass filters by cascading second- and third-order sections.

Experiment 4–2 *Active Low and High Pass Filters*

Active filters are used extensively to control the bandpass of electronic circuits. In this experiment the relative responses of some first- and second-order active filters are determined as a function of frequency. The four active filters studied include a single stage first-order low pass filter, a two stage second-order low pass filter, a single stage second-order low pass filter, and a single stage second-order high pass filter.

Narrow Bandpass Filters

For many electronic measurements it is necessary to have circuits that respond to or reject a narrow band of frequencies, or, as much as possible, a single frequency. Such tuned amplifiers and notch or band rejection amplifiers can be realized by combining both low and high pass networks with amplifiers to form active filters. Simple first-order tuned systems based primarily on the twin T network will be emphasized in this section, as practical filters are relatively easy to design. Higher-order narrow bandpass active filters can also be realized using the approaches discussed in the last section. The resonant characteristics of the second-order circuits such as the *RLC* circuit are obvious in Fig. 4–28. However, as for the higher-order active low pass filters, these circuits are difficult to design from first principles but are readily available from commercial sources.

Fixed Frequency Tuned Amplifiers. A common frequency selective circuit is the twin T network. It is shown in Fig. 4–39 along with its response function. Note from its response function that it is a rejection filter. The circuit is a combination of a low pass filter and a high pass filter arranged in a "twin T" manner which gives the name to the network. The equation for the rejection frequency is:

$$f_0 = \frac{1}{2\pi RC}. \tag{4-40}$$

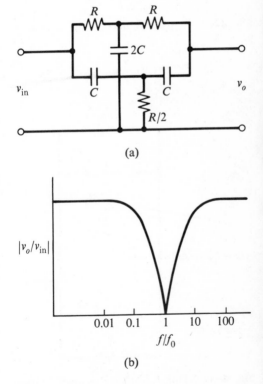

Fig. 4–39 (a) Twin T network. (b) Frequency response of twin T network.

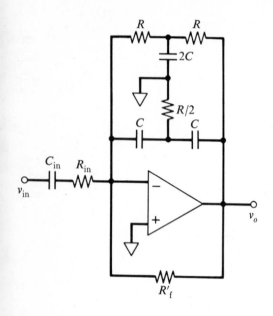

Fig. 4–40 Tuned amplifier using a twin T network.

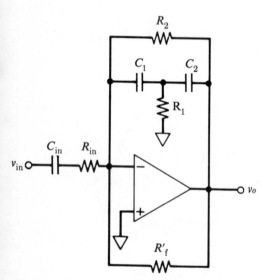

Fig. 4–41 Tuned amplifier using a bridged T network.

For an input of this frequency the impedance is very large and may be as high as 2000 MΩ if the components are well matched.

A tuned amplifier can be constructed by inserting a twin T filter in the negative feedback loop of an operational amplifier, as shown in Fig. 4–40. The voltage gain of an operational amplifier is:

$$\frac{v_o}{v_{in}} = -\frac{\mathbf{Z}_f}{\mathbf{Z}_{in}}. \tag{4–41}$$

The feedback impedance $\mathbf{Z}_f$ for the amplifier shown in Fig. 4–40 is determined by the parallel combination of R'_f and the impedance of the twin T network. If $\mathbf{Z}_f$ were determined solely by the twin T network, the gain of the amplifier would be too large at the f_0 of the network, and the circuit would be very unstable and subject to oscillation. This problem is averted by addition of a conventional negative resistive feedback loop R'_f to control the gain at the tuned frequency to approximately R'_f/R_{in}. For other than the resonant frequency, the impedance of the twin T network is low, and thus the gain of the amplifier is also low.

A simpler tuned network is the bridged T filter. It is shown in Fig. 4–41 in the feedback loop of an operational amplifier. The filter itself is a rejection filter like the twin T filter, and thus the amplifier shown in Fig. 4–41 is a tuned amplifier. The resonant frequency is:

$$f_0 = \frac{1}{2\pi(R_1C_1R_2C_2)^{1/2}}. \tag{4–42}$$

The rejection is not as sharp as that of the twin T and it is used in less critical applications. Also the precision of the component values is much less critical than in the case of the twin T filter.

Variable Frequency Tuned Amplifiers. Many times it is useful to be able to tune or vary the resonant frequency of a tuned amplifier without redesigning the complete network. A simple network in which tuning can be accomplished by adjustment of a single component is shown in Fig. 4–42a. The rejection frequency of this circuit is:

$$f_0 = \frac{1}{2\pi RC(1 - a^2)^{1/2}}, \tag{4–43}$$

where a is the fractional setting of the potentiometer $R/2$. The tuning range is not large. With a basic frequency ($1/2\pi RC$) of about 800 Hz, the filter can be tuned up to about 4000 Hz. In order to be able to tune over a much larger range, it is necessary to use a filter in which the value of two components are varied. Such a filter is shown in Fig. 4–42b. It

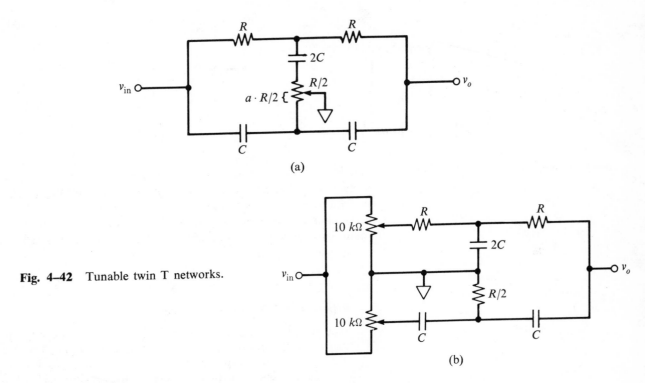

(a)

Fig. 4–42   Tunable twin T networks.

(b)

is a tunable twin T filter. The two 10 kΩ pots must be ganged together. The base frequency is the same as for the twin T $[1/(2\pi RC)]$. With an f_0 of about 2000 Hz, this filter can be tuned from about 200 Hz to 20 kHz, using the ganged 10 kΩ pots.

Notch Amplifiers.   A notch or narrowband rejection amplifier can be implemented by inserting the twin T network in the input circuitry of an operational amplifier, as shown in Fig. 4–43. A normal input resistor R_{in} is required in series with the twin T network to maintain reasonable gains for those frequencies outside the rejection band. Any of the tuned and

Fig. 4–43   Notch amplifier using a twin T network.

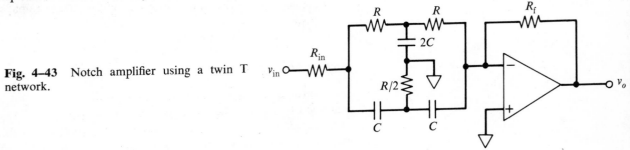

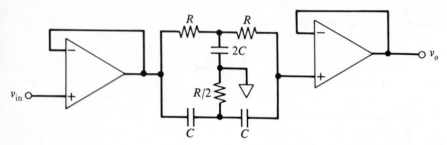

Fig. 4–44   Buffered twin T rejection filter.

tunable networks mentioned in the last section could be used in place of the twin T. A band rejection filter could also be implemented by buffering the twin T with followers, as shown in Fig. 4–44. This avoids disturbance of the characteristics of the twin T filter by loading. Notch filters are very useful for rejecting specific interference noise frequencies that may occur in a measurement system.

Experiment 4–3   *Tuned Amplifiers*

Both fixed and variable frequency tuned amplifiers are often required in many measurement situations. Some practical tuned amplifiers are built and characterized in this experiment. They are all based on the insertion of fixed or variable frequency twin T networks in the negative feedback loop of an operational amplifier.

Experiment 4–4   *Notch Filters*

Often in an experimental system it is necessary to reject specific interference frequencies. In this experiment a notch filter and a narrow band rejection amplifier are built and characterized. The notch filter is simply a standard twin T circuit with buffered input and output. The narrowband rejection amplifier is constructed by putting a twin T network in series with the input resistance of an operational amplifier voltage amplification circuit.

4–2.4   OSCILLATORS

An electronic oscillator is a device that produces an output signal which continuously repeats the same pattern of current and voltage variations with respect to time. Oscillator circuits can produce sine waves, square waves, sawtooth waves, pulse trains, repetitive ramps, or essentially any other desired ac repetitive waveform.

There are a large number of applications for oscillators in modern scientific instrumentation and measurements. One of the important applications is as a clock or time base for computers or digital instruments. An oscillator provides, in a sense, the heart of many digital and analog–digital instruments as it may control the timing, triggering, and gating of events and signals throughout the instrument and the measurement system. Oscillators also provide many of the control waveforms for the oscilloscope and test waveforms for evaluating a circuit's performance. Oscillators are basic to modern communication and signal processing systems because they provide the carriers for modulation of audio, video, or any other specific type of signal information. Modulation and demodulation will be discussed in Section 4–3.

A large number of approaches to the design of oscillator circuits is possible. Many have already been discussed in previous modules. The unijunction and neon bulb relaxation oscillators were discussed in Sections 2–3.3 and 2–3.6 of Module 2. Repetitive sweep generators, triangular and square wave generators, and the use of diode wave shaping to generate sine waves were discussed in Section 2–3.6 of Module 2. The instrument which generates a variety of repetitive waveforms by these techniques is generally referred to as a function generator rather than an oscillator. Other important oscillator-based circuits such as the voltage controlled oscillator, voltage-to-frequency converter, and crystal oscillator controlled time bases were discussed in Section 3–3.2 of Module 3.

Oscillators may also be designed using frequency selective circuits in the feedback loop of an amplifier to restrict the regenerative feedback to a single frequency. A block diagram of such an oscillator is shown in Fig. 4–45. With the proper choice of the frequency selective circuit the output signal is essentially harmonic (a pure sinusoidal waveform). The principles of feedback amplifiers and frequency selective components, such as the *RLC* and twin T circuits, have already been considered in this module. Specific examples of feedback oscillators will be discussed in the next section.

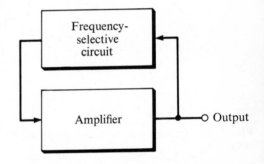

Fig. 4–45 Block diagram of harmonic oscillator.

Feedback Oscillators

The basic form of the feedback oscillator is shown in Fig. 4–45. The frequency selective circuit is in the feedback loop of an amplifier. To sustain oscillation the amplifier must counteract the losses in the frequency selective circuit. The fraction of the amplifier output signal which is transferred in the feedback loop is β. The amplifier gain A must then be at least $1/\beta$; in other words, $\beta A \geqslant 1$. The formula for the gain of the feedback amplifier (Eq. 4–19) is $A/(1 - \beta A)$. Notice that, when $\beta A = 1$, the resulting gain is infinite. In other words, the feedback is positive. As the amount of positive feedback is increased in an amplifier, the gain approaches infinity. When the gain of the amplifier is infinite, it no longer needs an external signal source to amplify. Thermal noise within the amplifier is sufficient to start the regeneration process at all frequencies. If it were not for the frequency selective circuit, which restricts the signals which can be regenerated, the feedback oscillator would oscillate at all frequencies for which $\beta A \geqslant 1$. It is now apparent that two conditions must be met for this system to operate properly as a harmonic oscillator. One is that βA must be equal to or greater than 1. The other is that the first condition be satisfied only by the particular frequency which is to be generated. Specific examples of harmonic oscillators are discussed in the remainder of this section. It should be pointed out that the particular arrangement of blocks shown in Fig. 4–45 is not sacred. The frequency selective circuit may be in the amplifier, the feedback loop, or both, and the output may be taken wherever it is convenient.

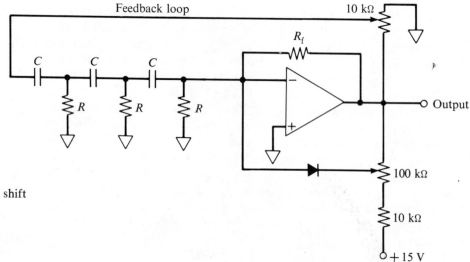

Fig. 4–46 Operational amplifier phase shift oscillator.

Phase Shift Oscillator. A simple feedback oscillator is shown in Fig. 4–46. The amplifier is an operational amplifier. Since the inverting input is used, the signal at the output is 180° out of phase with the signal at the input. The feedback network that couples the output to the input must shift the phase an additional 180° in order to have regenerative feedback, and it must do this for only one frequency. The network shown is a series of high pass filters each of which shifts the phase of the feedback signal approximately an equal amount. The degree of phase shift depends on the frequency, and the total phase shift will be the required 180°. The phase shift network thus fulfills the oscillation condition of regenerative feedback for only one frequency. For a 60° phase shift, each high pass filter attenuates the signal to about one-third. The output of the feedback network of Fig. 4–46 is 1/29 of the input. The amplifier must have a voltage gain of at least 29 to fulfill the other condition of oscillation. When R and C are the same for all three high pass filters, the frequency of oscillation f_0 is:

$$f_0 = \frac{1}{2\pi\sqrt{6}RC}. \qquad (4\text{–}44)$$

This type of oscillator is called the phase shift oscillator. Other networks which provide a 180° phase shift for only one frequency could be used equally well.

Twin T Oscillator. A twin T network in the negative feedback loop of an operational amplifier was used to make a tuned amplifier. Because the gain of the operational amplifier becomes very large at the resonant frequency of the twin T, a normal resistive feedback is necessary to prevent oscillation. Without this feedback resistor, oscillation would occur because the gain of the amplifier is very large for the rejection frequency, and the small fraction of the rejection frequency that is passed by the twin T is actually positively fed back because of the 180° phase shift of the amplifier and the 180° phase shift which the twin T gives to the rejection frequency. A more controlled oscillation can be provided by the circuit modifications which are shown in Fig. 4–47. The biased diode limits the output so that the amplifier does not saturate during oscillation.

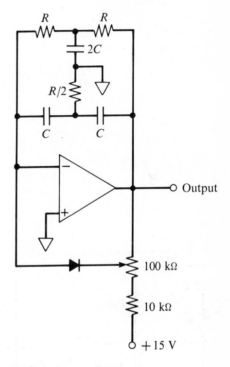

Fig. 4–47 Twin T oscillator.

Experiment 4–5 *Feedback Oscillators*

Oscillators may be designed using frequency selective circuits in the feedback loop of an amplifier. In this experiment oscillators are built using an *RC* phase shift network and a twin T filter as the frequency selective networks and an operational amplifier as the amplification element.

Wien Bridge Oscillator. Because of their frequency and phase sensitivity, certain bridges have been used in the feedback loops of feedback oscillators. By far the most common of these is the Wien bridge. A diagram showing the Wien bridge in an oscillator feedback loop is shown in Fig. 4–48. In order to have regenerative feedback, there should be no phase shift in the feedback network. The phase angle of the series *RC* network decreases with increasing frequency, while that of the parallel *RC* network increases. At the frequency for which the series and parallel impedances are equal, the phase angle will be unchanged in the feedback loop. This condition would exist at the bridge balance point. The magnitude of the amplifier input voltage depends on the fraction of the bridge voltage across the parallel impedance. This voltage is a maximum when the series and parallel impedances are equal. When the frequency increases, the parallel impedance decreases; when the frequency decreases, the series impedance increases. The ratio R_A/R_B is then adjusted so that the circuit just oscillates. This creates a very stable low-distortion oscillator because any frequency other than the balance frequency not only is attenuated in the feedback loop but also is of the wrong phase to sustain oscillation.

The Wien bridge oscillator is usually constructed with the bridge resistors and capacitors equal in value, as shown. Under these circumstances the frequency of oscillation is $1/(2\pi RC)$.

Crystal Oscillators.   By far the most stable oscillators are those which use a quartz crystal as the frequency-determining tuned circuit. A piece of quartz crystal sandwiched between metallic plates deforms mechanically when a voltage is applied between the plates. When the external voltage is removed, the relaxation of the crystal will induce a voltage between the metal plates. There will be a natural vibration frequency of the crystal which depends on its cut and size. An ac voltage of the resonant frequency of the crystal applied across it will maintain the resonant vibration in much

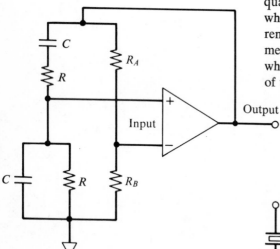

Fig. 4–48   Wien bridge oscillator.

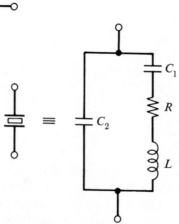

Fig. 4–49   Equivalent circuit of a quartz crystal.

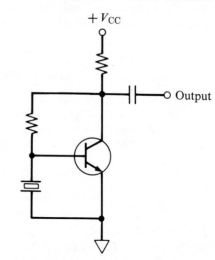

Fig. 4–50   Basic Pierce crystal oscillator.

the same way as oscillation is maintained in a tank circuit. The crystal is superior to the *LC* resonant circuit because the frequency is virtually independent of external circuit parameters and because the resonance peak is very much sharper (*Q* much higher). An equivalent circuit for a quartz crystal is shown in Fig. 4–49 and the circuit of a simple Pierce crystal oscillator is shown in Fig. 4–50.

The crystal oscillator has a fixed frequency. To change the frequency it is necessary to substitute another crystal of appropriate dimensions. However, most crystal oscillators in laboratory measurements are used to generate time bases and sequence events. In this case all the necessary frequencies and time intervals are generated using digital techniques as discussed in Sections 3–3 and 3–4 of Module 3.

PROBLEMS

1. Feedback also affects both the input and output impedances of an amplifier. (a) If the *IR* drop across the output impedance R_o of an amplifier is considered, Eq. (4–18) is modified to

$$v_o = Av_{in} - IR_o.$$

   Starting with this equation, calculate the effect of voltage feedback on the output impedance of the amplifier. (b) Starting with Eq. (4–17) and the assumption that the output load is small (no *IR* drop across R_o), calculate the effect of voltage feedback on the input impedance of the amplifier.

2. Starting with Eq. (4–13) show that a 6 dB change in gain results in a factor of two change in frequency.

3. Show that a 6 dB change in gain is a factor of two change in linear gain.

Modulation and Demodulation

In the optimization of an electronic measurement it is frequently necessary to employ modulation. **Modulation** is the alteration, in a specific way, of some property of a carrier wave by a signal. Two basic types of carrier waves can be distinguished. A carrier wave may be sinusoidal, and modulation schemes have been developed that employ alteration of the amplitude or the frequency of a sinusoidal carrier wave by a signal. The carrier wave may also be a pulsed waveform. The amplitude, width, and position of pulses have been used to carry signal information. In addition, a specific coded sequence of pulses can convey information about a signal.

Modulation may be necessary in measurement systems for a variety of reasons. Three general reasons are common. It is frequently necessary to transmit a signal, the basic properties of which may not allow for efficient direct transmission. This is the common reason for the extensive utilization of modulation in communication systems. Audio and video signals are used to modulate relatively high frequency carrier waves that can be efficiently transmitted through the earth's atmosphere.

In most laboratory scientific measurements, this is not of primary importance, as signals can be transmitted in cables. However, two important reasons do exist for modulation in such systems. For the most efficient processing of a signal, modulation is frequently necessary. This is particularly true in systems designed to enhance the signal-to-noise ratio of a measurement. Utilizing modulation, the signal can be processed in a region of minimal noise and can also be distinguished from the noise because of its unique modulation pattern. This is important for measurement systems such as lock-in amplifiers and boxcar integrators.

A second important use of modulation in scientific measurements is in the general area of conversions between the various data domains. Many ADC's, for example, use a modulation scheme in order to generate the digital value. Voltage-to-frequency converters employ frequency modulation; those that convert a voltage level to a time interval utilize pulse width modulation, and a successive approximation ADC is an example of pulse code modulation. Thus modulation is not only used to facilitate data transmission but also is a very important part of data processing and data conversion systems.

Modulation must be reversible. The process of recovering the signal from the modulated carrier wave is called **demodulation.** In the following sections, amplitude modulation, frequency modulation, and pulse modulation are discussed, as well as a number of demodulation techniques, including synchronous demodulators and phase-locked loop demodulators. The emphasis is on the signal processing and data conversion aspects of modulation, as these tend to be the significant reasons for the utilization of modulation in laboratory scientific measurements.

4–3.1 AMPLITUDE MODULATION

Amplitude modulation involves the alteration of the amplitude of a sinusoidal carrier wave by a signal. In general, the frequency of the carrier wave is much greater than the frequencies that make up the signal. Amplitude modulation results in an upward frequency translation of the signal frequencies. This is one of the more useful characteristics of amplitude modulation as applied to signal processing, since it provides a means of moving dc and low frequency signal information out of the $1/f$ noise region and into a region of minimal noise. The two amplitude modulation methods most often encountered in laboratory scientific measurements are double sideband modulation and AM modulation. These are described in the next section along with demodulation techniques.

Double Sideband Modulation

Double sideband modulation is relatively easy to describe in mathematical terms. The carrier wave $A_c \cos (2\pi f_c t)$ is a sinusoidal wave; for simplicity, assume that the signal $A_s \cos (2\pi f_s t)$ consists of a single relatively low frequency sinusoidal oscillation. Double sideband modulation of the carrier wave by the signal is carried out simply by multiplying these two together:

$$M_c(t) = A_c \cos (2\pi f_c t) \times A_s \cos (2\pi f_s t), \qquad (4\text{–}45)$$

where $M_c(t)$ is the modulated carrier wave, A_c and A_s are the amplitudes of the carrier and the signal, and f_c and f_s are the frequencies of the carrier and signal waves. In general, $f_c \gg f_s$.

This operation can be carried out using a four quadrant multiplier, and is depicted in Fig. 4–51.

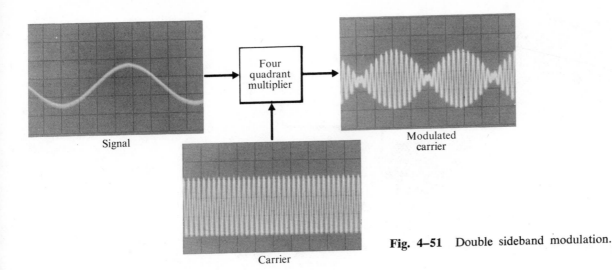

Signal

Four
quadrant
multiplier

Modulated
carrier

Carrier

Fig. 4–51 Double sideband modulation.

Using the trigonometric identity for the product of two cosines, Eq. (4–45) becomes:

$$M_c(t) = \frac{A_c A_s}{2} [\cos \{2\pi(f_c - f_s)t\} + \cos \{2\pi(f_c + f_s)t\}].$$

$$(4\text{–}46)$$

This equation indicates that the modulated waveform consists of two frequencies centered on both sides of the carrier wave. The separation of these frequencies from the carrier wave is equal to the frequency of the signal. The spectrum of this simple double sideband modulated waveform is shown in Fig. 4–52a.

If the signal contains a continuous range of frequencies, then a sum-and-difference frequency is generated for each signal frequency and the spectrum of the modulated carrier is as shown in Fig. 4–52b. The group of frequencies greater than the carrier frequency is called the **upper sideband** and that group below the carrier frequency, the **lower sideband.** Note that the bandwidth has been doubled from that of the signal alone. The complete spectrum is called a double sideband, and thus this particular type of modulation is called **double sideband modulation** and is abbreviated DSB. It is also referred to as double sideband suppressed-carrier modulation, because the carrier wave is not present in the spectrum of the modulated waveform. This aspect will be clarified shortly.

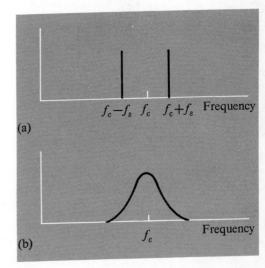

$f_c - f_s$ f_c $f_c + f_s$ Frequency

(a)

f_c Frequency

(b)

Fig. 4–52 Spectra of double sideband modulation: (a) single frequency signal; (b) broadband signal.

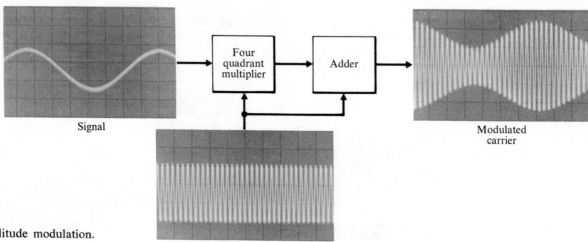

Signal

Carrier

Modulated carrier

Fig. 4–53 Amplitude modulation.

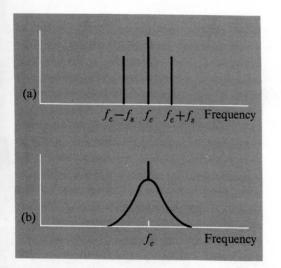

Fig. 4–54 Spectra of amplitude modulation: (a) single frequency signal; (b) broadband signal.

Amplitude Modulation

The traditional amplitude modulation utilized for AM radio differs from DSB in that the carrier wave is specifically added to the modulated carrier wave. This modulation scheme (abbreviated AM) is depicted in Fig. 4–53. AM is also easy to describe mathematically and the basic equation analogous to Eq. (4–45) is:

$$M_c(t) = [1 + A_s \cos (2\pi f_s t)] A_c \cos (2\pi f_c t). \qquad (4\text{–}47)$$

The modification of Eq. (4–45) for DSB to Eq. (4–47) for AM involves simply the addition of an $A_c \cos (2\pi f_c t)$ term. The spectrum for the AM waveform is shown in Fig. 4–54a for a single frequency signal and in Fig. 4–54b for a continuous range of frequencies. Thus DSB is the same as AM except for the suppression of the carrier wave.

This may appear to be somewhat of a trivial difference but it has very important consequences with respect to the demodulation of these carriers. It is important to appreciate the differences between the resulting DSB and AM carrier waveforms. These are shown in Fig. 4–55 for an ac ramp signal. Note that the envelope of the AM carrier is the same as the signal, while this is not true, in general, for the DSB carrier. If the signal goes negative, that information is encoded in the DSB modulated carrier as a phase reversal. In this case, recovery of only the envelope would not faithfully reproduce the original signal. Circuits for envelope recovery can be quite simple and thus an AM carrier is considerably easier to demodulate than a DSB carrier.

Demodulation

The general method for demodulation of amplitude modulated carriers (both DSB and AM) is synchronous multiplication by a sinusoidal signal exactly phase-locked to the carrier, followed by low pass filtering. This results in downward frequency translation of the signal information to its original location, plus the generation of some higher frequency components. Demodulation is depicted in Fig. 4–56. The mathematical description of this process is quite simple. It is necessary only to multiply Eq. (4–46) by $A_c \cos (2\pi f_c t)$. The result is

$$D_c(t) = \frac{A_c^2 A_s}{2} \cos (2\pi f_s t) + \frac{A_c^2 A_s}{2} \cos (2\pi f_s t) \cos (4\pi f_c t),$$

$$(4\text{–}48)$$

where $D_c(t)$ is the demodulated carrier. The first term is the original signal and the second term is an amplitude modulated carrier wave at twice the frequency of the original carrier. It is this term that is removed by low pass filtering.

The synchronization of this multiplication with the carrier wave is not always easy. Often a small amount of the original carrier is added to the DSB signal so that it can be used to synchronize the local oscillator used for demodulation. Alternatively, two extremely stable oscillators can be used to generate the carrier and demodulating waveforms. (Periodic synchronization of the oscillators is necessary to ensure adequate stability.) Finally, as is common in most laboratory measurements using DSB, the carrier wave or a waveform phase-locked to it can be transmitted, via a separate cable connection, directly from the modulator to the demodulator. This is the approach used, for example, in the lock-in amplifier measurement system. This technique is discussed in detail in Section 4–5.3 of this module.

Synchronous demodulation works for AM as well as DSB. However, as mentioned earlier, the envelope of the AM carrier is identical to the original signal. Thus, a considerably simpler demodulation method can

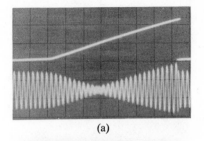

(a)

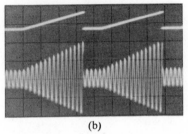

(b)

Fig. 4–55 Comparison of DSB (a) and AM (b) modulations.

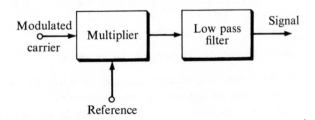

Fig. 4–56 Synchronous demodulation.

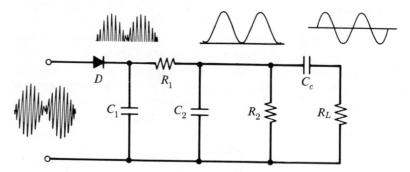

Fig. 4–57 Diode envelope demodulator.

be used. A simple envelope detector is shown in Fig. 4–57. The basic steps are rectification and low pass filtering.

Experiment 4–6   *Double Sideband Modulation (DSB)*

In this experiment, double sideband suppressed-carrier modulation and demodulation are studied. Double sideband modulation is carried out by multiplying a relatively high frequency carrier wave by a signal waveform using a four quadrant multiplier. Demodulation is more complex in that a synchronous demodulation step is necessary. This amounts to synchronous full wave rectification and is carried out by multiplying the modulated carrier wave by a signal phase-locked to, and of the same frequency as, the original carrier wave. A final low pass filtering step is used to recover the original signal waveform.

Experiment 4–7   *Amplitude Modulation (AM)*

Amplitude modulation involves multiplying a carrier wave and a signal waveform together as did DSB but, in addition, the carrier wave is added to the DSB type carrier to produce the final AM waveform. In this experiment, an AM modulator is constructed using a four quadrant multiplier and an operational amplifier summing circuit. A diode envelope detector is wired and used as the demodulator.

Other Amplitude Modulation Techniques

It was mentioned earlier that the DSB waveform has a bandwidth that is twice that of the original signal. In situations where bandwidth conservation is important, two other amplitude modulation techniques may be used. These are **single sideband modulation** (SSB) and **vestigial sideband** modulation (VSB). SSB involves suppression of both the carrier and one of the sidebands. This is often difficult to do in practice, and part of the carrier and other sideband may still be transmitted. In this case, the modulation technique is called VSB since a vestige of the second sideband is transmitted. In a sense, it is practical SSB.

Another important amplitude modulation technique is **pulse amplitude modulation** or PAM, where the carrier wave is a pulse sequence or a square wave. This modulation technique is discussed in conjunction with other pulse modulation techniques in Section 4–3.3 of this module.

4–3.2 FREQUENCY MODULATION

Frequency modulation is the alteration of the instantaneous frequency of a carrier wave by a signal. The amplitude of the signal determines the extent of the frequency change, and the frequency of the signal determines the rate at which the change occurs. A simple method of carrying out frequency modulation is to connect the signal to a voltage controlled oscillator (VCO). The oscilloscope traces of a signal and the FM carrier output from a VCO are shown in Fig. 4–58. Note that the frequency of the FM carrier is greatest when the signal amplitude is at its maximum value and decreases to its minimum frequency at the signal minimum. Thus FM is conceptually quite simple. However, unlike amplitude modulation, the mathematics of FM is very complex and will not be dealt with in this discussion.

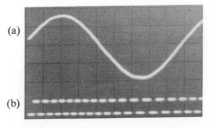

Fig. 4–58 Signal (a) and FM carrier (b) waveforms.

Despite the importance of FM to the communications industry it is not frequently encountered as a modulation technique *per se* in many laboratory measurements. The most common example is FM data recording systems where low frequency and dc signal information can be recorded on magnetic tape, using frequency modulation. In recent years direct digitization of data has replaced many FM recording systems.

An important application of FM to scientific measurements is for the conversion of analog data to digital data. The voltage-to-frequency converter is one of the common approaches to A-to-D conversion. This technique was discussed in Module 3.

Of more practical importance in many scientific measurements are the techniques for demodulation of an FM carrier. In many situations, the information of interest is fundamentally encoded as a frequency modulation of a carrier wave. The measurement of the rate of radioactive decay and photon counting are two important examples. In the following sections, several FM demodulation techniques are discussed, among them the increasingly important phase-locked-loop approach.

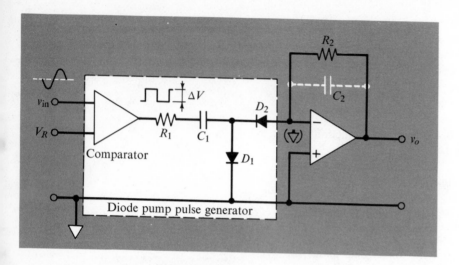

Fig. 4–59   Diode pump rate meter.

Rate Meter FM Demodulator

A very simple approach to the demodulation of an FM carrier is the analog rate meter. A classic analog rate meter is the so-called diode pump rate meter shown in Fig. 4–59. This circuit operates by first differentiating the input frequency modulated carrier, which is squared up by a comparator before applying it to the circuit. The negative spikes resulting from the differentiation are clipped off by diode D_1 and the positive spikes are passed by diode D_2 to the low pass filter, where they are averaged to provide the demodulated signal.

Another type of analog rate meter is shown in Fig. 4–60. It consists of a monostable multivibrator and a low pass filter. Again, a comparator should be used to square up the FM carrier before applying it to this circuit. In this circuit, a uniform pulse is generated by the monostable for each negative-going edge of the FM carrier, and the pulses are averaged by the low pass filter to give an analog level representative of the frequency of the FM carrier.

Fig. 4–60   Monostable–low pass filter rate meter.

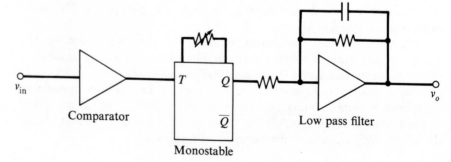

Both these averaging demodulators are useful only when the carrier frequency is changing at a relatively slow rate, or if an average rate over several hundred cycles is the desired signal information. Such circuits can provide a simple analog output for photon counting measurements where, in general, the average frequency is measured over a few seconds or tenths of seconds.

Experiment 4–8   *Rate Meter FM Demodulator*

In this experiment a monostable–low pass filter rate meter is built and used to demodulate an FM carrier. This type of averager demodulator is useful when the changes in the frequency of the carrier are quite slow compared to the carrier wave frequency. A voltage-to-frequency converter is used as the source of the FM carrier.

FM Demodulation by Period Measurement

If it is necessary to observe relatively rapid changes in the frequency of a carrier wave, demodulation techniques other than the rate meter methods must be used. One approach is to measure the period of every cycle. A circuit for period measurement is presented in Fig. 4–61. The FM carrier is squared up with a comparator and then fed to a sequencer. The sequencer gates a linear integrator for the duration of one period of the FM carrier. At the end of the period the output of the integrator is

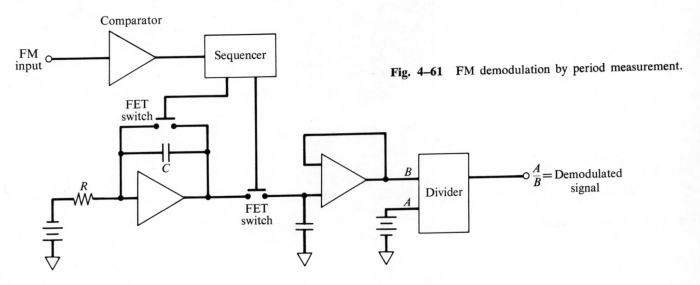

Fig. 4–61 FM demodulation by period measurement.

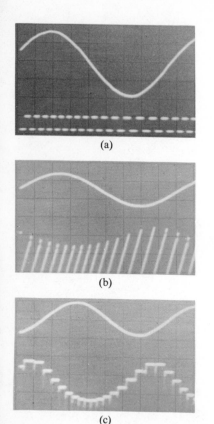

(a)

(b)

(c)

Fig. 4–62   Waveforms from period FM de-modulator.

sampled and held, the integrator is rapidly reset, and the same operation is repeated for the next period of the FM carrier. The output of the sample-and-hold amplifier is a series of analog levels representative of the instantaneous period of the FM carrier. The reciprocal of this output provides a signal representative of the frequency; the analog divider carries out this operation.

Several waveforms throughout this circuit are shown in Figs. 4–62 and 4–63. The modulating signal is the upper trace in Figs. 4–62a through c. The frequency-modulated carrier wave which would be observed at the output of the comparator is shown in the lower trace of Fig. 4–62a. The output of the integrator is shown in the lower trace of Fig. 4–62b, and the output of the sample-and-hold amplifier is shown in the lower trace of Fig. 4–62c. The stepped amplitudes represent the period of the FM carrier at each cycle. Note that this is not a sinusoidal waveform. This is shown more clearly in Fig. 4–63. The lower trace in both cases is the output of the sample-and-hold follower. The upper trace in Fig. 4–63a is simply this same signal low pass filtered. The upper trace in Fig. 4–63b, however, is the low pass filtered output of the divider and hence is the fully demodulated signal. Note that it does have the proper sinusoidal shape, as compared to the upper trace in Fig. 4–63a.

Period may also be measured rapidly using digital counting techniques, as discussed in Section 3–3 of Module 3. These techniques can be used for period FM demodulation provided the successive period measurements can be stored or read out rapidly.

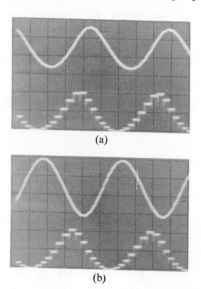

(a)

(b)

Fig. 4–63   Waveforms from period FM de-modulator.

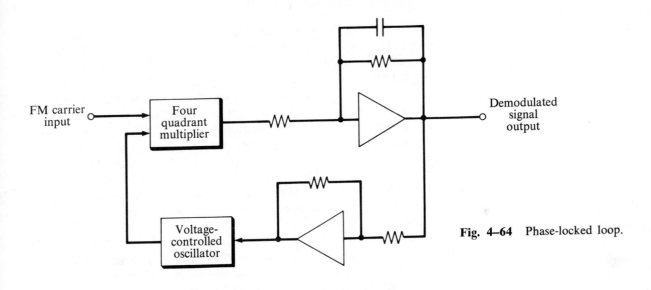

Fig. 4–64 Phase-locked loop.

Experiment 4–9 *Period Measurement FM Demodulator*

Demodulation of an FM carrier can be accomplished by measuring the duration of each period of the FM carrier. The basic idea is to gate a linear integrator for the duration of one period of the FM carrier, and then sample and hold the output value of the integrator at the end of the period. The integrator is reset, and the operation is repeated for the next period of the FM carrier. The sample-and-hold output represents the period of the FM carrier and its reciprocal must be taken to yield the fully demodulated signal. This final step is carried out using an analog divider.

Phase-Locked-Loop Techniques and Applications

The phase-locked loop is a very versatile circuit and among its many applications is FM demodulation. This circuit was briefly discussed in Section 3–3.4 of Module 3. A circuit for a phase-locked loop is shown in Fig. 4–64. The FM carrier and the output of a voltage controlled oscillator are multiplied together using a four quadrant multiplier. The VCO is controlled by the low pass filtered output of the multiplier. The output of the low pass filter is also the demodulated signal. When the loop is locked, the input FM carrier and the VCO output are 90° out of phase. Waveforms from an operating phase-locked loop are shown in Fig. 4–65. The output of the VCO is shown in Fig. 4–65a. Note that it is a unipolar square wave. The input FM carrier is shown in Fig. 4–65b.

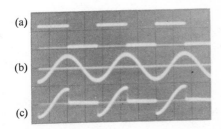

Fig. 4–65 Waveforms from the phase-locked loop circuit: (a) VCO output; (b) FM carrier input; (c) multiplier output.

It is locked 90° out of phase with respect to the VCO. The direction of the phase lock depends on the sign of the loop feedback to the VCO. The output of the multiplier is shown in Fig. 4–65c. The 90° phase shift and the necessity for it are obvious from this figure. Phase lock occurs so as to make the output of the multiplier a minimum. However, the phase shift is not exactly 90°, because some voltage is required to drive the VCO to the required frequency. High gain in the loop keeps this required offset to a minimum. Note that, for the lock condition shown in Fig. 4–65, an increase in the frequency of the FM carrier (Fig. 4–65b) will positively increase the average output of the multiplier. This is fed back through two inverter stages (Fig. 4–64), so that this increase will increase the output frequency of the VCO, which maintains the lock condition.

There are a large number of applications for phase-locked loops other than FM demodulation. Also the availability of phase-locked loops in 16 pin integrated circuit packages should greatly expand their utilization in scientific instrumentation. The phase-locked loop can also be used for frequency multiplication and division. If a frequency divider is inserted in the loop between the VCO and the multiplier, the loop can be made to lock onto a submultiple of the VCO output. The VCO output can be considered as being multiplied by the frequency division factor of the loop. A versatile frequency multiplier can be constructed by putting a preset counter in the loop between the VCO and the multiplier, as shown in Fig. 4–66. A wide range of multiplication factors can be obtained, depending on the complexity of the preset counter. Also, the desired multiplication factor could be automatically controlled with this approach.

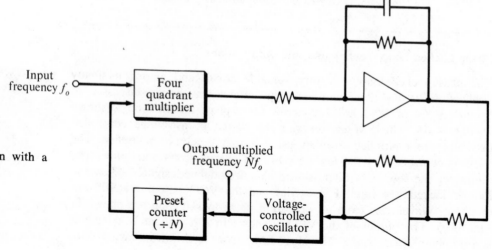

Fig. 4–66 Frequency multiplication with a phase-locked loop.

Input frequency f_o

Four quadrant multiplier

Output multiplied frequency $N f_o$

Preset counter ($\div N$)

Voltage-controlled oscillator

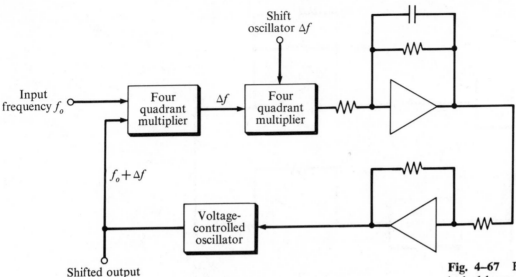

Fig. 4–67 Frequency shifting with a phase-locked loop.

Frequency division can be accomplished by inserting a frequency multiplier in the loop instead of a frequency divider; however, frequency division can also be accomplished easily by conventional digital techniques.

It is also possible to shift an input signal frequency f_0 by a specific amount, using a phase-locked loop. A circuit to accomplish this is shown in Fig. 4–67. The frequency to be shifted, f_0, is multiplied by the VCO output, which is initially set to the approximate shifted frequency $(f_0 + \Delta f)$. The output of the first multiplier is fed to a second multiplier, where it is multiplied by the exact offset or shift frequency which is desired. The output of this second multiplier provides the error signal to close the loop. The output of the VCO will be locked to $f_0 + \Delta f$. The two multipliers are simply carrying out double sideband modulation. In fact, single sideband modulation could be used to accomplish this same task, but carrier and sideband suppression are more difficult to achieve than with the above approach.

The phase-locked loop can also be used as a synchronous AM demodulator as shown in Fig. 4–68. If the input to the multiplier of the phase-locked loop is an AM modulated waveform, the VCO output provides a phase-locked reference for synchronous demodulation. It is simply necessary to shift the phase of the VCO output by 90° and then use a conventional multiplier–low pass filter combination for synchronous demodulation. This complete circuit, including the AM multiplier and low pass filter, is also available in a single 16 pin integrated circuit, making the implementation of this type of AM detection quite easy. Note also that this

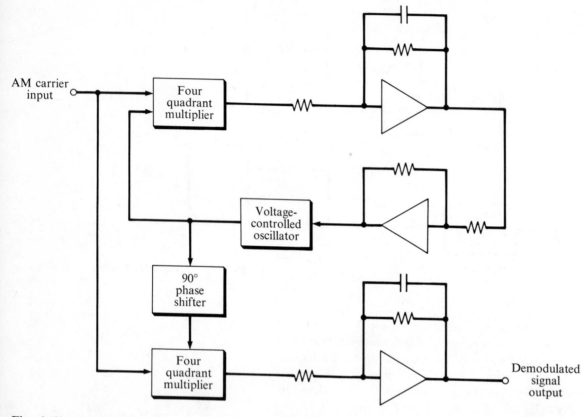

Fig. 4–68 Synchronous AM demodulation with a phase-locked loop.

circuit is really a lock-in amplifier for AM signals. Lock-in amplifiers are discussed in detail in Section 4–5.3.

The measurement of the frequency of a noisy signal is difficult, particularly if the frequency of the signal is drifting. A narrowband tuned amplifier will isolate the desired frequency, but frequency drift in the bandpass of the tuned amplifier creates problems. The standard phase-locked loop (Fig. 4–64) is an excellent system for this measurement, as it acts as a narrowband tracking filter. The desired information, of course, must be encoded in the frequency of the input signal and not in its amplitude. These few examples attest to the versatility of the phase-locked loop beyond the primary example of FM demodulation. It is a powerful circuit element for a wide variety of applications.

Experiment 4–10 *Phase-Locked Loop*

In this experiment a phase-locked loop circuit is constructed, using a four quadrant multiplier as the phase comparator and a voltage-to-frequency converter as the VCO. The phase-locked loop can be used to perform a wide variety of tasks. The circuit is used as an FM demodulator and for frequency multiplication.

Slope Demodulator

A simple FM demodulator can be built using a filter to convert the FM carrier to an AM carrier. This approach is shown in Fig. 4–69.

Fig. 4–69   FM slope demodulator.

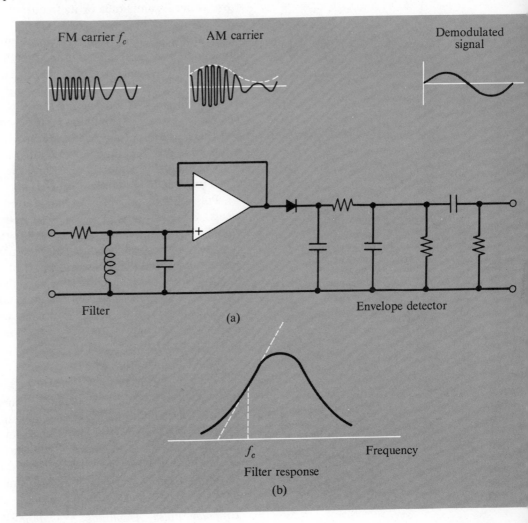

The frequency f_c of the carrier wave is centered on the slope of a band-pass filter. Frequency variations of the carrier are thus converted to amplitude variations and these are then demodulated, using envelope detection, as for an AM carrier.

4–3.3 PULSE MODULATION

In the two modulation techniques discussed so far, the signal information produces a continuous variation of a particular parameter of a carrier wave, either its amplitude or its frequency. With pulse modulation systems, some parameter of a pulse or a set of pulses represents the value of the signal *at a particular point in time*. Thus, the sampling operation is an inherent step in pulse modulation. Sampling is discussed in detail in Section 4–4. The parameters of a pulse that can encode information about a signal are its amplitude, its width or duration, and its position. Modulation techniques based on all these parameters have been developed. They are called **pulse amplitude modulation (PAM), pulse duration modulation (PDM),** and **pulse position modulation (PPM).** A fourth pulse modulation technique is completely digital in approach. The value of the signal is represented by a specific coded sequence of pulses. This technique is called **pulse code modulation (PCM).**

Pulse modulation methods are found throughout electronic measurements. Chopper amplifiers are an excellent example of pulse amplitude modulation. Sampling an analog waveform, an important step in A-to-D conversion, is also pulse amplitude modulation. Also, many digitizers convert a voltage to a time interval as part of their conversion process. This is, in effect, a form of pulse duration modulation. And the complete digitization and storage of a waveform in a computer, on paper tape, or on magnetic tape, can be considered pulse code modulation. Thus the pulse modulation techniques are very important in signal processing and data conversion techniques.

Pulse Amplitude Modulation

Pulse amplitude modulation (PAM) occurs frequently in measurement systems. Whenever a signal is chopped or sampled, the operation can be interpreted as PAM. Thus the chopper input amplifier discussed in detail in Module 1 is a PAM technique. PAM may be usefully thought of as a type of amplitude modulation and in particular it has many of the characteristics of DSB. Two basic types of PAM can be distinguished, unipolar and bipolar.

Unipolar Pulse Amplitude Modulation. Unipolar PAM (or unipolar chopping) is simply ordinary sampling. It is depicted in Fig. 4–70. PAM

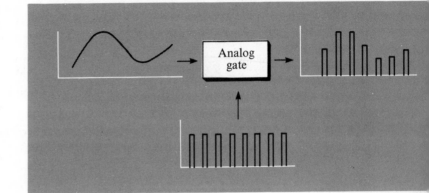

Fig. 4–70 Unipolar pulse amplitude modulation.

can be implemented by using an analog gate that is actuated by a pulse train. The result is a set of unipolar pulses whose amplitudes represent those of the original continuous waveform; i.e., the waveform has been sampled. The pulse train must have a repetition rate that is consistent with the criterion for accurate sampling. (This is discussed in detail in Section 4–4 of this module.) As with conventional amplitude modulation, the spectrum of the PAM waveform is particularly informative. It is shown in Fig. 4–71 and consists of the spectrum of the original signal and a DSB type spectrum centered at the frequency of the sampling rate and at each harmonic. Bandpass filtering can be used to isolate the first double sideband, producing a conventional DSB signal which can be amplified and then synchronously demodulated. This constitutes the chopper amplifier.

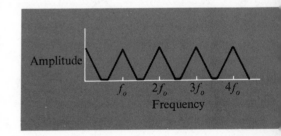

Fig. 4–71 Spectrum of a unipolar PAM waveform.

This PAM waveform is also prominent in A-to-D and D-to-A data handling systems. Generation of a PAM waveform is conceptually the first step in A-to-D conversion, the second being quantization of the pulse amplitude to generate the digital value. The complete process could be called pulse code modulation (PCM). During D-to-A conversion, an analog pulse sequence of the digitized waveform is generated, and then low pass filtered to regenerate the original analog waveform. This is a common operation on small computer systems, for displaying digital data on analog devices such as recorders and oscilloscopes.

Demodulation of a PAM waveform is carried out simply by low pass filtering. This isolates the signal frequencies from the upper double sidebands (see Fig. 4–71).

Bipolar Pulse Amplitude Modulation.   Many chopper systems are bipolar, and this type of pulse modulation is called bipolar PAM. It is best described as a multiplication operation between the signal and a square wave, and is depicted in Fig. 4–72. (Note that this is equivalent to DSB except that the carrier wave is square rather than sinusoidal.) As discussed in

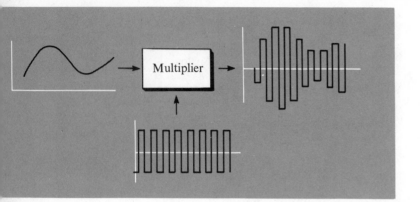

Fig. 4–72 Bipolar pulse amplitude modulation.

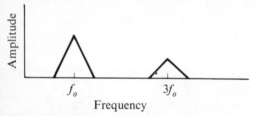

Fig. 4–73 Spectrum of bipolar pulse amplitude modulation.

Section 3–2.2 of Module 3, a square wave is composed of a sine wave of the same frequency as the square wave, plus all the odd harmonics. Thus, the spectrum of the bipolar PAM waveform has a DSB type spectrum centered at the frequency of the square wave and at each of the odd harmonics, as is shown in Fig. 4–73. It is easy to isolate the first double sideband by bandpass filtering, to generate a conventional DSB waveform, if desired. Demodulation of bipolar PAM must be carried out by using synchronous techniques analogous to demodulation of DSB.

Pulse Duration Modulation

The amplitude information at a particular point on a signal can be converted to a specific pulse width. This type of modulation is referred to as pulse duration modulation (PDM). PDM is used as an intermediate step in many A-to-D conversion techniques. The ramp type of ADC and the dual-slope ADC both convert the input voltage to a time interval and measure the time interval by gating an oscillator to a counter. The conversion of the input voltage to a time interval is PDM.

These ADC's use an integration step to generate a pulse duration equivalent to a voltage amplitude (see Module 3). Another approach to PDM is shown in Fig. 4–74. This circuit will convert a continuous analog input voltage into a pulse duration modulated waveform. It relies on summing the input waveform with a sawtooth waveform and then feeding the composite waveform to a comparator.

Demodulation of a PDM waveform is accomplished by a simple averager. Thus, a low pass filter will demodulate a PDM waveform.

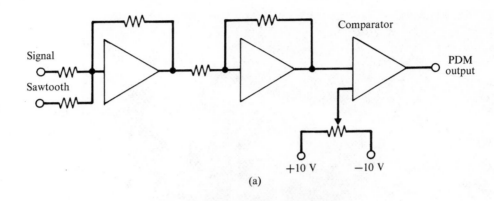

(a)

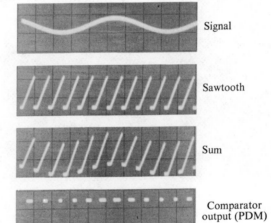

Fig. 4–74 Sawtooth based pulse duration modulation: (a) circuit; (b) waveforms.

(b)

Experiment 4–11 *Pulse Duration Modulation*

Pulse modulated signals are common in many scientific measurements. In this experiment a simple circuit is built to demonstrate pulse duration modulation. The signal is summed with a repetitive ramp waveform, and the result is connected to a comparator to produce the PDM carrier.

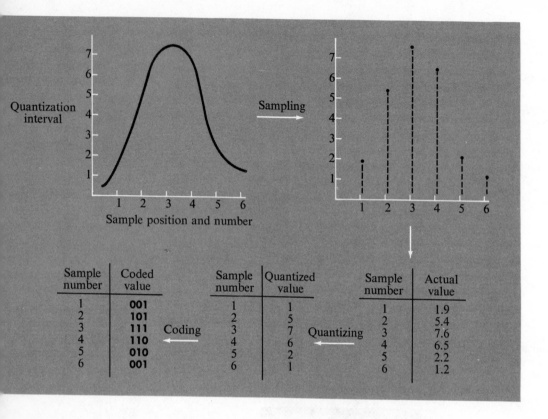

Fig. 4–75 Pulse code modulation as performed by a digitizer.

Pulse Code Modulation

The pulse modulation techniques discussed above all use some continuously variable parameter of a pulse to represent the signal information. Pulse code modulation (PCM) is quite different. In this case, a specific coded sequence of pulses is used to represent the signal amplitude at a particular point in time. Digitization of a continuous waveform may be thought of as a form of PCM. A digitizer such as a 10-bit successive approximation ADC, with a sample-and-hold input amplifier, first samples the analog signal and then quantizes the sample and represents the magnitude as a sequence of 10 bits, typically in a natural binary code. These same steps, sampling, quantization, and encoding, are common to all PCM systems. The sampling step is not unique to PCM but is present in all pulse modulation techniques. The PCM operation of a digitizer is depicted in Fig. 4–75.

PROBLEMS

1. Rearrange Eq. (4–47):

$$M_c(t) = [1 + A_s \cos (2\pi f_s t)] A_c \cos (2\pi f_c t),$$

using trigonometric identities, into three terms that show the frequency composition of the AM waveform.

2. Demodulation of the above waveform can be carried out by multiplication with $A_c \cos (2\pi f_c t)$:

$$D_c(t) = [1 + A_s \cos (2\pi f_s t)]A_c^2 \cos^2 (2\pi f_c t).$$

Rearrange this equation to show the frequency composition of the synchronously demodulated AM waveform before it is low pass filtered.

Relation of Sampling Parameters to Signal Characteristics

The implementation and optimization of many electronic measurement systems involves sampling a time-varying continuous analog signal. These include gated dc integration, boxcar integration, multichannel averaging, correlation techniques, and all computer-based techniques. Sampling is also necessary when converting analog signals to digital signals with ADC's and when manually digitizing data from strip chart recorder tracings and meters. Such digitization is frequently necessary in a number of electronic measurement systems. The sampling of analog signals is also an integral operation in the recording of very rapid, repetitive events by a sampling oscilloscope. This widespread occurrence of the sampling operation in electronic measurement systems means that it is necessary for the experimenter to have a knowledge of the parameters which affect the validity and accuracy of sampling.

The sampling operation is not, of course, limited to electronic systems and waveforms, but is quite ubiquitous. It is important to have an intuitive feel for the parameters and characteristics of a system so that a finite set of samples can accurately represent a continuous flow of information. For example, you are presently carrying out a sampling operation as you read this book. It is not necessary for you to read every letter, word, or even every complete sentence in order to understand all the information that is presented. Several readily definable constraints control this sampling operation. First, the text is written in the English language and the reader knows that it is very unlikely that the next word or set of words would be a mixture of German, Spanish, or French. If the text language was likely to change in such a random fashion, the reader would have to carry out a much finer sampling of the text. Secondly, the English language has a reasonable amount of redundancy. It is often not necessary to read a complete sentence in order to grasp its complete meaning. Also, the constraints of grammar do not give the authors complete freedom of choice in spellings, and in many cases, in the next word, or in sentence construction. Thus the reader can anticipate certain portions of text. The third constraint is that of content. From sentence to sentence, the reader is reasonably assured that the content will be about electronic measurements and will not suddenly switch to history or geography.

The above three constraints all impose a finite limit to the rate of change that is possible for the text information. Since this rate of change is finite, it is reasonable to assume that a finite number of samples would be sufficient to represent the information contained in the text. This is a general characteristic of any system and, stated in electronic terminology, the text is band limited. As seen in Section 4–2 of this module, the same is true of all electronic circuits. There always exists some finite maximum rate of change for an analog waveform in a specific circuit, as determined by the overall electronic bandwidth. Thus a finite number of samples should be able to accurately represent the information contained in a continuous analog waveform, and this number should be related to the bandwidth of the circuit or waveform.

Another important factor that affects the number of samples that must be taken is prior knowledge. Various readers will sample the text of this book differently, depending on their present level of electronics knowledge and, to some extent, on how rapidly the new knowledge is assimilated. A reader with greater prior knowledge samples the text less frequently. The same holds true for sampling analog waveforms. The sampling rate for a signal can be relaxed if there is some prior knowledge of the signal shape. A signal that is known to be a straight line can be characterized by only two points, perhaps a few more if noise is present. However, a greater number of samples would be necessary to prove that a signal *is* a straight line. In the same way, prior knowledge that a peak shape is Gaussian reduces the number of samples necessary to characterize the peak, because the sampled set of data can be fitted to a Gaussian function.

Analog waveforms are sampled in two basic ways. In general, the waveform is simply sampled at fixed time intervals in real time. For some signals that are very rapid and repetitive, an equivalent time sampling may be used. In this method, at some point in time after a trigger pulse has occurred, a sample of the signal is rapidly acquired. Additional samples are then acquired at different delay times on each repetition of the signal, and a plot of sample amplitude vs. sampling delay time is used to reproduce the original curve. The sampling delay time becomes equivalent to real time. This technique can be used to sample extremely rapid signals, and is the principle used in some boxcar integrators and sampling oscilloscopes.

Several distinct steps which require finite amounts of time can be identified in the process of sampling a continuous analog signal in real time. The **sampling interval** (Δt) is the time between samples. The reciprocal of the sampling interval ($1/\Delta t$) is the **sampling rate** and is the rate or frequency at which samples are taken. The **sampling duration** (t_d) is the total time over which samples are taken. The **aperture time** (t_a) is the actual time required for the acquisition of each sample. In a sampling operation associated with A-to-D conversion, an additional process is present as the sample must also be quantized. The time required for quantizing the sample is called **quantizing time.** The total time for A-to-D conversion of a sample is the combination of the aperture time and the quantizing time, and this is called the **digitization time.**

Each of these times encountered in the sampling operation can influence the accuracy with which the resulting set of analog or digital samples represents the original continuous analog signal.

4–4.1 SAMPLING RATE

The rate at which samples are taken is an important factor in determining how well a set of samples represents the original continuous analog signal. As discussed in the introduction, the bandwidth of a signal is important in determining how often the signal should be sampled. It would seem that, if the sampling rate is high compared to the highest frequency components in the signal, then accurate sampling would result. This means that it is quite important, when sampling signals, to have a knowledge of the spectrum of the signal.

The Nyquist sampling theorem provides a quantitative basis for the rate at which samples must be taken, based on the bandwidth of the signal. The Nyquist sampling theorem states that a signal must be sampled at a rate that is *twice the highest frequency component* in the signal. In other words, if the sampling rate is $1/\Delta t$ (sampling interval $= \Delta t$), the Fourier transform of the signal must be zero at all frequencies greater than $1/(2\Delta t)$. The critical frequency, $1/(2\Delta t)$, is called the **Nyquist frequency.** Thus, a signal with Fourier components extending from 0 to 300 Hz should be sampled at a rate of at least 600 samples per second, or every 1.7 msec.

This Nyquist limit can be appreciated by thinking of the sampling operation in terms of pulse amplitude modulation. (PAM was discussed in Section 4–3.3 of this module.) An analog waveform and two different PAM representations of it are shown in Fig. 4–76. The spectra of the two PAM waveforms are also shown in Fig. 4–76. Remember that the spectrum of a PAM waveform consists of the spectrum of the original signal and a double sideband (DSB) type spectrum centered at the frequency of the

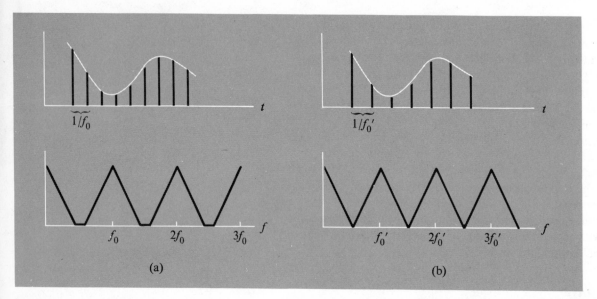

(a)

(b)

Fig. 4–76 PAM waveforms and their corresponding spectra. Note that the smaller sampling interval in case (a) results in less overlap of the double sidebands.

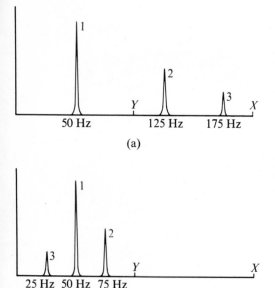

(a)

(b)

sampling rate and at each harmonic. It is clear from these spectra that the sampling frequency must be at least twice the highest frequency component in the signal; otherwise the double sidebands would begin to overlap. If they do overlap, accurate recovery of the original signal is not possible from the sampled waveform (PAM waveform).

These errors are manifested in two ways. The frequency information in the signal above the Nyquist frequency is lost, and the undersampled high frequencies show up as spurious low frequencies. This second error is referred to as **aliasing** and is depicted in Fig. 4–77. The spectrum of an analog waveform composed of three sinusoidal frequencies (50 Hz, 125 Hz, and 175 Hz) is shown in Fig. 4–77a. A sampling rate of 400 Hz, providing a Nyquist frequency of 200 Hz (X in Fig. 4–77), would be adequate for such a signal. However, if a sampling rate of 200 Hz (Y in Fig. 4–77) had been chosen, the spectrum of the undersampled waveform would be as shown in Fig. 4–77b. The high frequency information has been lost and spurious low frequency components have been added to the signal. In this specific example of exactly halving the sampling rate, the frequency position of the aliased high frequencies can be predicted as a folding-over about the central position of the original frequency axis. The 50 Hz component of the waveform is still sampled adequately, but the 125 and 175 Hz components

Fig. 4–77 Aliasing.

have aliases of 75 and 25 Hz. The manner of this aliasing can be appreciated by means of a simple example. The 175 Hz signal component is shown in Fig. 4–78, with sampling points indicated at the original 400 Hz sampling rate. If every other point is dropped out, simulating the 200 Hz sampling rate and the remaining points connected, a 25 Hz sine wave results. Actual sampled and aliased signals are shown in Fig. 4–79. A continuous sine wave of about 150 Hz is shown in Fig. 4–79a, the sine wave sampled at a rate of 400 Hz in Fig. 4–79b, and the *same* sine wave sampled at 200 Hz in Fig. 4–79c. Note that it has an alias of about 50 Hz. Low pass filtering was used to reconstruct the sine wave from the samples for Figs. 4–79b and 4–79c. Finally a familiar example of aliasing is the backward, and at times, motionless appearance of stagecoach wheels in western movies as the frequency of spoke rotation is undersampled by the framing of the movie camera.

It is interesting to point out, with respect to Fig. 4–77b, that if no frequency components below 100 Hz had been present in the original signal, then the fold-over or undersampling would not have been serious. The aliased high frequencies would not have overlapped with any other signal information, and the position of fold-over is accurately predictable. This points out the more general statement of the Nyquist sampling theorem, that a waveform or signal must be sampled at twice its bandwidth and not necessarily twice its maximum frequency. Thus if all the frequency components of a signal are contained in a 100 Hz bandwidth, a 200 Hz sampling rate will be adequate, no matter where the 100 Hz bandwidth is located along the frequency axis. This can result in a considerable reduction in the sampling rate for narrow bandwidth, high frequency signals.

With many signals some compromises must be made in choosing the sampling interval because they contain Fourier components from 0 Hz out to very high frequencies. The Fourier transform of a Gaussian peak, for example, approaches zero only at infinitely high frequencies, and it is impractical to rigorously apply the Nyquist sampling theorem to such signals. The Nyquist sampling theorem gives the rate at which samples must be acquired to preserve all the information in the original waveform. In many cases, the experimenter can tolerate errors of 0.1% or 0.01%, and the Nyquist rule can be relaxed. Also, as mentioned earlier, some prior knowledge of the signal shape often allows for a further relaxation of the Nyquist rule.

The numbers of samples necessary in order to ensure a given accuracy are presented in Table 4–2 for triangular, exponential, Lorentzian, and Gaussian peaks. The accuracy criterion is based on the maximum difference that could be expected between the original analog waveform and an analog waveform regenerated from the sampled set of data using a minimum of prior information, i.e., no assumptions about peak shape. This difference

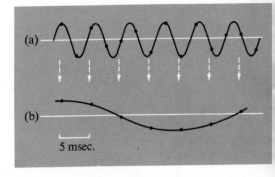

Fig. 4–78   Aliasing of 175 Hz to 25 Hz.

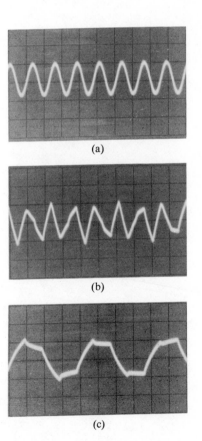

Fig. 4–79   Oscilloscope traces of aliasing: (a) input ∼150 Hz sine wave; (b) 400 Hz sampling rate; (c) 200 Hz sampling rate.

Table 4–2   Number of samples for common peak shapes

Maximum error, % of peak height	Samples per full width at half-height			
	Triangle	Exponential	Lorentz	Gaussian
10	5.5	5.9	1.8	1.5
1	40	50	3.6	2.2
0.1	357	454	4.8	2.6
0.01	—	—	6.3	3.0
0.001	—	—	8.3	3.3

is expressed as a percentage of the peak maximum. For peaklike signals, the sampling rate error tends to occur near the peak maximum and thus is approximately the error to be expected in measuring the peak height, using the sampled set of data. Note that this does not mean that these are the percentage errors to be expected in measuring other peak parameters, such as position, half-width, or area. In many cases, the percentage errors for these parameters will be lower. However, the values presented in Table 4–2 can serve as a relative guide to the magnitude of errors to be expected when the Nyquist sampling theorem is relaxed for peaklike signals.

Another factor that may influence the choice of the sampling interval is the presence of noise at a single frequency. For example, if 60 Hz pickup is a problem in a measurement system, taking samples at 16.6 msec intervals, or some multiple thereof, with a constant phase relation to the line frequency, is an excellent way of discriminating against the 60 Hz interference. It is best to phase the sampling to the maximum or the minimum of the 60 Hz waveform, as this is where its rate of change is least. The VCO output of a phase-locked loop (see Section 4–3.2 of this module) can be used to provide the appropriate sampling signal, as it is locked 90° out of phase with respect to the input frequency.

Experiment 4–12   *Aliasing*

This experiment is designed to demonstrate aliasing. Aliasing is the generation of spurious low frequencies in a sampled waveform by the undersampling of high frequencies. Various sinusoidal frequencies between 0 and 200 Hz are sampled at 400 and 200 Hz sampling rates, and the frequencies of the waveforms regenerated from the sampled data are measured.

4–4.2 SAMPLING DURATION

The total time over which samples are taken, the sampling duration, can also cause errors in the sampled signal. Consider the sampling of the peak-shaped signal shown in Fig. 4–80. Sampling is begun at time t_1 and ended at time t_2, resulting in a total sampling duration $t_d = t_2 - t_1$. Even if the sampling is done at the appropriate rate, an error will still occur because of the loss of signal information beyond the truncation points. The error is often called a truncation error, and it is concentrated at the ends of the signal. Normally the truncation error is trivial, because the information loss is quite low, or can be made so simply by taking a few more samples, although long trailing signals such as Lorentzian peaks may give problems in some cases.

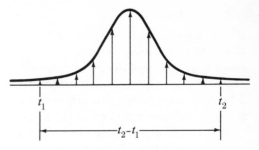

Fig. 4–80 Sampling duration.

The value of the signal at the truncation point is a good approximation of the magnitude of the truncation error. An idea of the magnitude of this error can be obtained from Fig. 4–81, which is a log-log plot of the amplitude of some common peaks vs. time in units of peak width at half-height. For a maximum truncation error of less than 10^{-3} of the peak height, sampling would have to be continued for 1.6, 5, and 16 peak widths at half-height on each side of the peak maximum for Gaussian, exponential, and Lorentzian peaks respectively.

The importance of truncation error may depend on the final use to be made of the sampled data. If the peak height is to be measured, it is likely quite unimportant. If an attempt is being made to fit the sampled data to a specific functional form, the truncation error may be quite serious, particularly for exponential and Lorentzian shapes. The Fourier transform of

Fig. 4–81 Comparative shapes of common peaks.

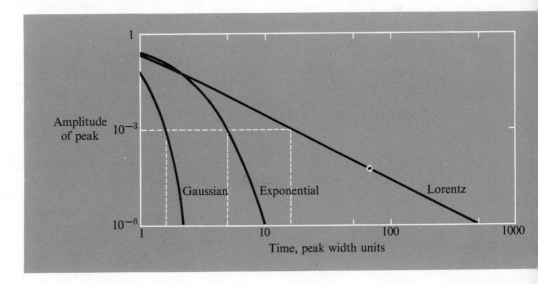

Amplitude of peak

Gaussian   Exponential   Lorentz

Time, peak width units

Note 4.   Apodization.

Apodization involves multiplication of a signal by a weighting function such as a linear or exponential decay, in order to ensure that the signal goes smoothly to zero at its truncation points. It is primarily used when the Fourier transform of the signal is the desired information, and controls spurious oscillations in the transformed signal.

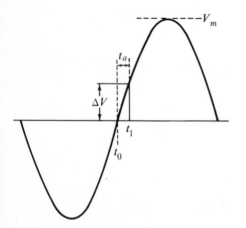

Fig. 4–82   Aperture time error.

the sampled signal may be the final desired information, as in the case of Fourier transform spectroscopy. The truncation error can affect the line shape and resolution throughout the final spectrum, although apodization[4] can be used to some extent to control the effects.

4–4.3   APERTURE TIME

A sample cannot be acquired instantaneously. In order to obtain a sample of a continuous analog signal, it must be observed for a finite length of time. This time is called the **aperture time.** The effect on a signal is analogous to the effect of a finite slit width on a spectrum, in that the signal is smoothed or averaged over the aperture time.

In general, the aperture time should be a small fraction of the sampling interval when sampling a continuous analog waveform. The maximum error due to a finite aperture time can be estimated for a signal that is a sine wave. Consider the signal shown in Fig. 4–82. The maximum error for this sine wave signal is calculated at the zero crossing of the signal where its rate of change is a maximum. An upper limit estimate of the error (ΔV) is given by:

$$\Delta V = V_m \sin (2\pi f t_a), \qquad (4\text{–}49)$$

where V_m is the amplitude of the sine wave, f is its frequency, and t_a is the aperture time.

In order to ensure accurate sampling, the ratio of the sampling interval to aperture time should be kept large. For a signal with a maximum frequency of 200 Hz, a sampling rate of at least 400 Hz would be chosen. This sets the sampling interval at 2.5 msec. If a sampling system is used that has an aperture time of 25 μsec, thereby providing a sampling interval to aperture time ratio of 100, the maximum error for a signal frequency of 200 Hz is 3% of its amplitude. This error will be progressively less for the lower frequency Fourier components of the signal.

The above discussion of aperture time concerned the sampling of continuous analog signals. Several experimental systems are such that the parameter of interest is stepped in time. For example, many commercial monochromators use stepping motor scanning systems. *In a stepped system, the aperture time with respect to the parameter being stepped is zero.* This allows the sampled signal to be integrated in order to improve the S/N without distorting the signal information represented with respect to the stepped parameter, provided, of course, that the sampling interval is correctly chosen.

4–4.4   DIGITIZATION TIME

A very important electronic measurement situation where it is necessary to sample a continuous analog signal is during A-to-D conversion. The above discussions of sampling rate and sampling duration are directly applicable to A-to-D conversion of a continuous waveform. However, the discussion of aperture time must be expanded.

An ADC cannot digitize a signal instantaneously. Digitization time is the time from the start of the sampling operation to the appearance of the digitized signal at the output terminals of the digitizer. The digitization time can be conceptually broken down into an *aperture* time and a *quantization* time. The aperture time is the time to acquire the analog sample and is essentially identical to that discussed in the last section. The quantization time is the time necessary to quantize the analog sample, i.e., generate a digital electronic signal that represents the amplitude of the analog sample. In ADC's which use input sample-and-hold amplifiers, these two times are quite separate. A sample-and-hold input successive approximation ADC is a good example of this type of system. Another digitizer is a dual-slope integrating ADC where the signal is integrated for a fixed period of time and then quantized, although the quantization time does depend on the amplitude of the sample. However, with many ADC's, it is difficult to distinguish the aperture and digitization times. With a voltage-to-frequency ADC, the aperture and quantization times can be considered to be equal, as sampling and quantization occur simultaneously. For a stand-alone successive approximation ADC, the aperture time is difficult to define, but the quantization time is readily definable and normally a constant. With these last two ADC's, the total digitization time can be thought of as equal to the quantization time.

The basic types of errors that were discussed in connection with aperture time are also associated with digitization time. In general, when digitizing continuous analog signals, it is necessary that the digitization time, or a distinct aperture time associated with the digitization, be kept small with respect to the sampling interval if errors are to be avoided. A specific example was discussed in the previous section. This requirement is particularly difficult to meet with voltage-to-frequency ADC's because of the relatively long times required to measure the frequency. However, these ADC's can be particularly efficient when used with systems in which the parameter of interest is stepped in time and an integrated measurement is desired.

One of the most versatile digitizing systems is a sample-and-hold input amplifier coupled to a successive approximation ADC. This combination provides a short aperture time that minimizes signal distortion with a distinct and relatively short quantization time, enabling rapid sampling and digitization. Presently available sample-and-hold amplifiers have aperture times as low as 100 nsec, and many successive approximation ADC's can make a 10-bit conversion in about 10 μsec. This type of digitizer can also be used for much longer sampling intervals. The effective aperture time can be set by analog low pass filtering; an excellent alternative is to average a specific number of fast acquisitions to define the aperture time. This also has the desirable effect of increasing the effective number of bits, as averaging several 10-bit conversions gives a result which has greater than 10-bit resolution. For example, adding only four full scale 10-bit values results in a 12-bit answer in certain circumstances as described in Section 4–5.5.

PROBLEMS

1. The mathematical forms of some common peak shapes are given below. The heights are normalized to unity and the widths at half-height (in seconds) are normalized to unity.

Gaussian	Lorentz	Exponential
$\exp\left[-4\,(\ln 2)t^2\right]$	$(1 + 4t^2)^{-1}$	$\exp\left[-2\,(\ln 2)|t|\right]$

Calculate the total number of samples that must be taken for each of these peak forms, if the value of the peak at the truncation points is to be less than 1% of the peak maximum. Assume that the number of samples taken per full width at half-height is such as to give a 1% maximum error in the peak height (see Table 4–2).

Signal-to-Noise Ratio Enhancement

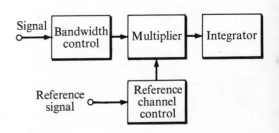

Section 4–5

Scientists are finding it increasingly necessary to measure small electrical signals in the presence of noise. As limits of sensitivity are lowered and weaker physical effects are utilized to provide scientific information, the problem of discriminating between an information-conveying signal and extraneous unwanted signals becomes increasingly difficult. Several electronic measurement techniques have been developed to assist the scientist in carrying out measurements in situations where the signal-to-noise ratio is very small.

In general there are two distinguishing characteristics between signals and noise that make possible the design of measurement systems that improve or enhance the signal-to-noise ratio. Signals often have, or can be made to have, unique and distinguishing spectra; and, secondly, the time of occurrence or phase coherence of the signal frequency components can be controlled in a predictable manner. Fundamental noises, on the other hand, have rather broad spectra; and their time of occurrence or phase coherence is predictable only in a statistical sense. Essentially all signal-to-noise ratio (S/N) enhancement techniques depend on the above two properties to discriminate between signals and noise. An important point to keep in mind is that the S/N can be improved only by noise reduction, for once the signal indicative of the parameter of interest has been transduced into an electrical signal, no amount of amplification will enhance the signal information with respect to the noise.

A generalized block diagram of an S/N enhancement technique is shown in Fig. 4–83; most S/N enhancement techniques can be reduced to this block diagram. Typically, the noisy signal is amplified and the amplifier bandwidth is controlled to minimize the passage of noise frequencies with respect to signal frequencies. The next step, in most systems, is a multiplication step, in which the amplified signal is multiplied by a reference signal. While the reference channel control is shown only as a single block, several

Fig. 4–83 Generalized block diagram of an S/N enhancement technique.

complex control and sequencing operations throughout the instrument may be required at this step. In general, it is this multiplication step that discriminates against noise on the basis of the predictable time behavior or phase coherence of the signal information. Finally the output from the multiplier is integrated or time-averaged. This last step can often be carried out using a simple low pass filter.

The following treatment of S/N enhancement begins with a brief discussion of correlation. A feel for the basic concepts involved in the correlation of two signals is important, because this forms the basis of the multiplication–integration step mentioned above, as well as providing an accurate description of electronic filtering in the time domain. Then several S/N enhancement techniques are discussed. These include low pass filtering, integration, lock-in amplification, boxcar integration, multichannel averaging, correlation techniques, and computer-based techniques.

4–5.1 INTRODUCTION TO CORRELATION

Most of the S/N enhancement techniques discussed in this section involve a multiplication–integration operation that is best described in terms of correlation. Correlation involves multiplying one signal by a delayed version of a second signal and integrating or time-averaging the product. When this time-averaged product is evaluated over a range of relative displacements, a correlation pattern or function is generated that is a function of the relative displacement. If correlation is carried out with continuous functions, it can be mathematically described by the following integral:

$$cc_{ab}(\pm \tau) = \lim_{x \to \infty} \frac{1}{2x} \int_{-x}^{+x} a(x)b(x \pm \tau)\, dx, \qquad (4\text{–}50)$$

where $cc_{ab}(\pm \tau)$ is the correlation pattern of the two signals $a(x)$ and $b(x)$, and τ is their relative displacement. With most practical correlators, the correlation operation is carried out on sampled waveforms. Correlation is then best described by the following summation:

$$cc_{ab}(\pm n\, \Delta x) = \sum_{x} a(x)b(x \pm n\, \Delta x), \quad n = 0,1,2,\ldots, \qquad (4\text{–}51)$$

where Δx is the sampling interval. The relative displacement is $n\, \Delta x$ and is identical to τ in Eq. (4–50).

Two general types of correlation are commonly distinguished. If the two signals $a(x)$ and $b(x)$ are different, the process is called **cross-correlation,** and if they are the same signal, it is called **auto-correlation.**

A simple pictorial example of correlation is shown in Fig. 4–84a, where the auto-correlation of a rectangular pulse signal is depicted. At each relative displacement, the ordinate of the auto-correlation function (Fig. 4–

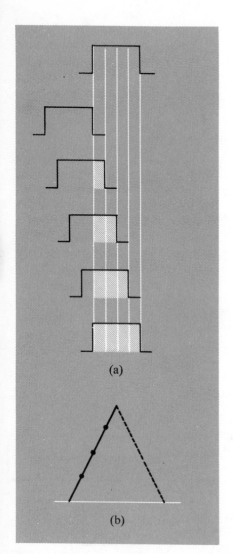

(a)

(b)

Fig. 4–84 Auto-correlation of a rectangular pulse.

84b) is simply the mutual area of the signal and the displaced version of itself. Note that the auto-correlation of a pulse signal results in a triangular waveform whose full width at half-height is equal to the width of the original pulse.

An important theorem dealing with the correlation operation states that correlation of two signals is equivalent to multiplication of their Fourier transforms. Thus Eq. (4–50) may be restated:

$$CC_{AB}(f) = A(f) \cdot B(f), \qquad (4\text{–}52)$$

where $A(f)$ and $B(f)$ are the Fourier transforms of $a(x)$ and $b(x)$, and $CC_{AB}(f)$ is the Fourier transform of $cc_{ab}(\tau)$. It was seen in Section 4–1 that the amplitude–time waveform and the spectrum of a signal were related by the Fourier transform; therefore, correlating two waveforms is equivalent to simply multiplying their spectra together. This is depicted in Fig. 4–85 for the auto-correlation of the rectangular pulse. The Fourier transform of a rectangular pulse is the $\sin x/x$ function, as mentioned in Section 3–2.2 of Module 3. The product of these two functions is $\sin^2 x/x^2$. This is the Fourier transform of the auto-correlation function. Since it is the product of two spectra, it is sometimes referred to as the **cross spectrum.** Inverse Fourier transformation of this product yields the correct triangular auto-correlation function. The effects of a particular correlation operation are often clarified by thinking in terms of multiplying spectra rather than correlating waveforms.

As mentioned earlier, correlation is involved in most S/N enhancement techniques as a multiplication–integration operation. It is an operation that provides discrimination between phase-related signal components and randomly phased noise components. In some cases, only a simplified correla-

Fig. 4–85 Fourier transform representation of correlation (* indicates correlation and × multiplication).

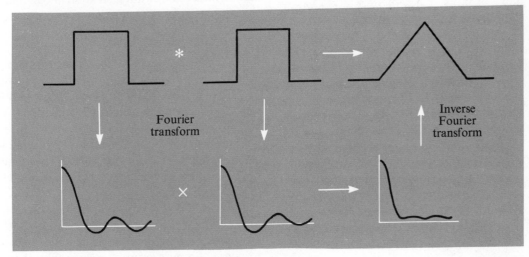

Note 5.   Correlation and Convolution.

The multiplication-integration operation given by Eq. (4–50) occurs widely in scientific measurements. For example, it accurately describes the effect of an instrument on a waveform as depicted in Fig. 4–86 for an RC low pass filter. Two of the more common names associated with this operation are correlation and convolution, although terms such as weighted running average, smoothing, and scanning are also frequently encountered. However, in a strict sense the correlation of two functions is not equivalent to the convolution of the two functions. Typically the response function of an instrument is defined in such a way that to describe the output waveform in terms of Eq. (4–50) the response function must be reversed from left to right before carrying out the multiplication-integration operation. When one of the starting functions must be reversed, the operation is referred to as **convolution** *rather than correlation, and the shifting function, $b(x \pm \tau)$, of Eq. (4–50) is replaced with $b(-x \pm \tau)$. For the correlation depicted in Fig. 4–84, the scanning function is symmetrical and thus correlation and convolution give equivalent results.*

tion operation is carried out, in the sense that the correlation function is evaluated at only a single relative displacement; this is true of the lock-in amplifier. With the boxcar integrator, the cross-correlation function between a square gating pulse and a repetitive waveform is evaluated, often over a range of relative displacements. In general, a measurement technique is referred to specifically as a correlation technique when the complete correlation function of any two waveforms is evaluated. A number of examples are presented in Section 4–5.6, along with a discussion of the S/N enhancement, signal processing, and signal detection capabilities of correlation techniques.

Equation (4–52) also indicates that correlation may be used to describe electronic filtering.[5] The effect of a particular filter on a waveform can be described by the multiplication of the waveform spectrum by the transfer function of the filter. This means that the effect of a filter on a waveform can also be described as a cross-correlation between the waveform and the Fourier transform of the transfer function. For most practical analog filters, the mathematics involved in the Fourier transformations is quite complex, involving both real and imaginary Fourier transforms. However, the effect of an *RC* filter on a signal can be presented pictorially, and this approach is followed in the next section.

4–5.2   LOW PASS FILTERING AND INTEGRATION

Perhaps the most common methods of enhancing the S/N of a measurement are low pass filtering and dc integration. Many signals of interest have their main Fourier components at dc (0 Hz) with bandwidths extending only a few hertz from dc. In these cases, simple *RC* low pass filtering can be effectively used to limit the measurement system bandpass to that necessary to pass the signal frequencies. DC integration over precise time periods is also a powerful technique for reducing random noise on a signal. Typically the signal adds directly with respect to integration time, while white noise adds as the square root of the integration time; thus the S/N improves as the square root of the integration time if the noise is white noise.

There are two basic problems or limitations associated with these techniques. The measurement system bandpass is centered at dc, and it has already been shown that this is the region of maximum $1/f$ noise; therefore, both these methods may be seriously affected by long-term drifts. Secondly, severe distortions can be introduced when applying these techniques to continuous signals where the parameter of interest is being scanned as a function of time.

Low Pass Filtering

Low pass filtering is a simple method of controlling the upper frequency response of a measurement system. The characteristics of first- and higher-order active low pass filters have been discussed in detail in Section 4–2. It is simply necessary to choose the *RC* time constant, and hence the bandwidth and phase shift characteristics, in such a way that the signal frequencies of interest are affected as little as desired. However, some trade-offs must be made at this point. Some distortion of the signal must be tolerated if the maximum S/N is desired. If it is also desirable to preserve, as well as possible, the observed signal shape, then a less-than-optimum S/N must be accepted. This simply reflects the definition typically chosen for the S/N, average peak signal divided by the rms noise level. For peak type signals, the definition does not include any criteria for preservation of peak shape or position.

The distortions in the shape of the signal occur primarily in systems where the parameter of interest is being continuously scanned as a function of time. The type of distortion which occurs for a peaklike signal is shown in Fig. 4–86. The distortion can be accurately described in terms of the cross-correlation of the impulse response function of the *RC* low pass filter with the signal. This response function has the form of a trailing exponential function, shown in Fig. 4–86b. The response function simply reflects the fact that previous signal values contribute to the current output value with exponentially diminishing weight. The distortion results in the alteration of the peak height, a shift of the peak maximum, and skewing of the peak with the generation of a trailing edge. The effects can be minimized by ensuring that the time constant is short relative to the time required to scan across the peak. The situation depicted in Fig. 4–86 is for a Gaussian peak with a full width at half-height of 1 sec being cross-correlated with a first-order low pass filter whose time constant is 0.25 sec.

When low pass filtering is applied to signals where the parameter of interest is stepped in time, the output signal level at a new step position will not reach its final value until about five time constants have passed. In this case, if a time equal to five time constants is allowed between steps, little or no distortion of the signal information occurs if the sampling rate criterion discussed in Section 4–4 is satisfied.

Integration

Integration is a widely used technique for improving the S/N of a measurement. With an integration technique, the S/N improves approximately as the square root of the measurement time. The improvement results because the frequency components that make up a signal add in phase, and the noise frequencies add randomly as they are not, in general, phase related. These points were illustrated in Fig. 4–8 of Section 4–1.

In this section, integration techniques which are primarily applicable to dc signals will be discussed. Integration of dc signals is accomplished in two basic ways. Active and passive low pass filters can be used where the *RC*

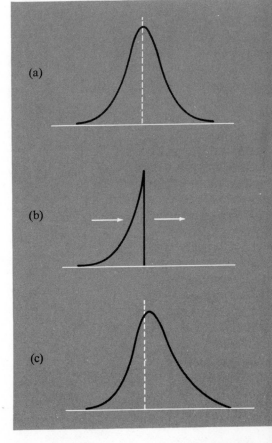

Fig. 4–86 Distortion of a peak signal by *RC* low pass filtering. (a) Raw peak. (b) Shape of convolving *RC* filter. (c) Filtered peak.

time constant is typically much larger than the integration time. The other main approach is to use an integrating digital voltmeter such as those employing voltage-to-frequency converters. These two approaches are somewhat different, and will be discussed separately.

In some measurement situations, the magnitude of the parameter of interest is encoded in the rate of a series of pulses, and integration can be implemented simply by counting the pulses for an accurate period of time. Photon counting is an example of such a measurement situation. Thus, some signals can be integrated simply by counting. This type of integration is discussed at the end of this section.

RC Integration.   The response of a first-order low pass *RC* filter to a step signal is shown in Fig. 4–87a. The equation of the curve is:

$$v_o = v_{in} [1 - e^{-t/RC}]. \tag{4–53}$$

If a low pass filter is to be used as a linear integrator, integration times must be restricted to the initial, short, linear portion of this curve. Using the series expansion $e^x = 1 + x + x^2/x! \ldots$, we may reduce Eq. (4–53) to:

$$v_o = \frac{v_{in}}{RC} \cdot t \cdot \left[1 - \frac{t}{2RC} \ldots \right]. \tag{4–54}$$

In order for the output voltage to increase at a rate linearly related to the input voltage (i.e. be a linear integrator), the integration time t must be small relative to the *RC* time constant. In this case, Eq. (4–54) becomes:

$$v_o = \frac{v_{in}}{RC} \cdot t, \tag{4–55}$$

which is the desired linear relation between the input voltage v_{in}, the output voltage v_o, and time t. To make this equation valid within 1 percent, the integration time must be less than 0.02*RC*; or, in other words, the *RC* time constant must be 50 times the longest integration time that is to be used. When used as a linear integrator in this manner, the upper 3 dB point f_2 of the low pass filter is very low. For a 10 sec integration time, linear to within 1 percent, the time constant would have to be at least 500 sec, resulting in an f_2 point of 0.003 Hz. This tends to make this integrator very susceptible to $1/f$ noise. The response of a "linear" integrator to a step input (Eq. 4–55) is shown in Fig. 4–87b for comparison.

In general, a simple passive *RC* filter is not used to perform linear integrations. An operational amplifier with a capacitor in the negative feedback loop is an excellent linear current amplifier. OA integrators were discussed in Section 3–2.1 of Module 3.

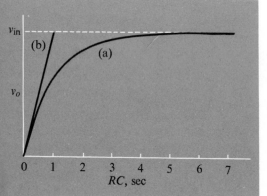

Fig. 4–87   (a) Response of first-order low pass filter to a step input. (b) Response of an "ideal" integrator to a step input.

The low pass filter does not have to be used in this linear region to be effective as an integrator. It is perfectly acceptable to integrate for two or three time constants, as long as the integration time is constant for each measurement that is being carried out. In fact, integration may be allowed to continue for several time constants, but little increase in signal results after five time constants. Therefore, five time constants represents a practical upper limit for the integration time. However, it is important to note that this does *not* represent the actual effective integration time. The effective integration time of an *RC* low pass filter is compared to that of a gated linear integrator in the next section.

Integrating DVM.   An integrating DVM is an excellent measurement system for dc signals. Typical integrating DVM's use the dual-slope approach or a voltage-to-frequency converter. In general, with integrating digital voltmeters it is easy to trade off integration time vs. signal level as they are often linear integrators over several orders of signal magnitude.

The frequency response of such an integrator is a little more difficult to assess than that of the *RC* integrator. Physically, the input signal is being multiplied by a rectangular gating function and the product integrated. This, then, is the familiar cross-correlation operation which was discussed in Section 4–5.1. Thus the input signal is being cross-correlated with a rectangular pulse whose width equals the integration time. This is equivalent, in the frequency domain, to multiplying the spectrum of the signal by the spectrum of the rectangular pulse. The power spectrum (PSF) of a rectangular gate pulse *t* sec in duration has the form:

$$\text{PSF} = \frac{\sin^2{(\pi f t)}}{(\pi f t)^2}. \tag{4–56}$$

This function is plotted in Fig. 4–88a for a gate time of 1 sec.

Note that this function has nodes at 1 Hz, 2 Hz, 3 Hz, . . . Thus, when using a 1 sec integration time, noise signals at these frequencies will be completely attenuated. This makes intuitive sense since, irrespective of the phasing of these frequencies to the 1 sec integrating gate, their average values will always be zero. However, a frequency of 1.5 Hz will, on the average, have a finite power transfer through the 1 sec gate. In a certain sense, a 1 sec integrating gate can be thought of as a filter with the frequency response shown in Fig. 4–88a. Many dual-slope integrating DVM's use the node characteristic of this frequency response to attenuate 60 Hz noise. If a gate time or integration time of 16.667 msec is used, the first node falls at 60 Hz. This is the shortest time that can be used and still provide nodal rejection of 60 Hz. The filtering characteristic of the integrating DVM is sometimes used to con-

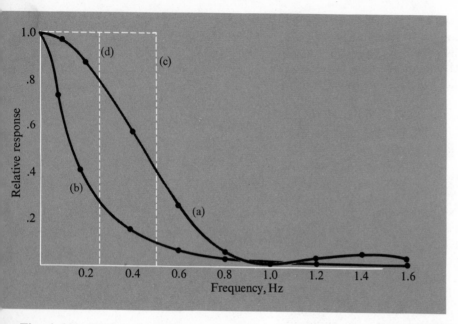

Fig. 4–88 Comparative frequency responses of a 1 sec rectangular integrating gate of a DVM (a) and an *RC* low pass filter with a 1 sec time constant (b). Equivalent bandwidths are shown by curves (c) and (d).

trol aliasing in sampled data systems, and is also an important characteristic to remember in connection with boxcar integrators.

As a comparison the power spectrum of an *RC* integrator with a 1 sec time constant is shown in Fig. 4–88b. These responses can be compared on the basis of equivalent bandwidth. The idea of an equivalent bandwidth was introduced in Section 4–1.2 during the discussion of shot noise. The equivalent bandwidth is simply a rectangular bandwidth whose area is the same as the actual power spectrum under consideration. The equivalent bandwidth Δf for a rectangular gating pulse t sec in duration is $1/2t$ and is shown in Fig. 4–88c. The equivalent bandwidth Δf for an *RC* low pass filter was given in Eq. (4–24) of Section 4–2.3. It is $1/4RC$ and is also plotted in Fig. 4–88 (curve d). This means that for a gated linear integrator to have a bandwidth equivalent to an *RC* low pass filter, the duration of the gate must be twice the time constant of the filter ($t = 2RC$).

The contributions of signals such as 1.5 Hz to the output integral of the 1 sec gate time can be greatly reduced by averaging or summing ten 1 sec integrations. The 1.5 Hz signal will contribute at random, both positively and negatively, to the output integral because it is not phase locked to the integrating gate and its period is not an integer multiple of 1 sec. This is essentially equivalent to using a 10 sec integration time which would have its first node at 0.1 Hz.

Integration using low pass filters or an integrating DVM has two main limitations as an S/N enhancement technique. The measurement bandwidth is centered at dc, which is the region of maximal $1/f$ noise, and the range of signals to which integration may be applied is essentially limited to dc or relatively slowly changing signals. With measurement systems such as the lock-in amplifier and the boxcar integrator, these limitations can be overcome.

Experiment 4–13   *Integrating DVM*

The filtering and averaging characteristics of a gated integrating DVM are studied in this experiment. In particular the specific frequency response of the gate is measured and the existence of rejection nodes is verified.

Counting.   In some measurement systems, the value of the parameter of interest is encoded in the rate of a series of pulses or in the frequency of a signal. In these cases, the signal can be integrated simply by accurately counting the number of pulses that occur in a definite interval of time. An integrating DVM that uses a voltage-to-frequency converter is really integrating by counting. An excellent example of a signal that can be directly integrated by counting is the current pulse output of a photomultiplier tube; this output signal was shown in Fig. 4–12. When the output signal from the photomultiplier is measured by counting these current pulses, the measurement is referred to as photon counting. A block diagram of a complete photon counting system is shown in Fig. 4–89. The output signal from the photomultiplier tube is amplified with a pulse amplifier, and then all pulses greater in amplitude than a preset discrimination level are counted for an accurate preset time interval. Thus if the counting time is 10 sec, the integration time is 10 sec. The precision of the integrated count value in this measurement is related to the total number of counts acquired during the integration time. For a random phenomenon such as photoelectron emission, the standard deviation of the total integrated count is the square root of the total count; thus, the precision will improve as the square root of the integration time, as with the other integration techniques.

An important point to note is that a counting type measurement, as illustrated by photon counting, generally allows long integration times with little or no interference from $1/f$ noise. The pulse amplifier is normally ac coupled since pulse fidelity can be maintained with a bandwidth of about 10 kHz to 100 MHz. Thus, very long integration or counting times, in the range of hours if desired, can be effectively used in counting measurements such as photon counting. This could not be done using conventional analog amplification and integration of the photomultiplier tube output signal.

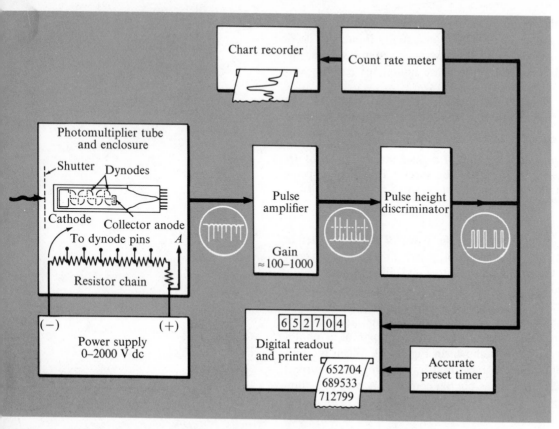

Fig. 4–89   Block diagram of a photon counting system.

4–5.3   LOCK-IN AMPLIFICATION

The lock-in amplifier is a complete signal measurement and processing system that is very efficient in discriminating against noise components in a signal. A complete lock-in amplifier measurement system consists of four main operations: modulation, selective amplification, synchronous demodulation, and low pass filtering. The lock-in amplifier itself normally carries out only the latter three operations.

The typical signal measured with a lock-in amplifier measurement system is a relatively slowly varying signal with its main Fourier component at dc. The modulation step puts this information on a carrier wave whose frequency is chosen to be well removed from $1/f$ noise, environmental noises such as 60 Hz, and other interferences. Thus the signal information can be amplified in a frequency region of minimal noise. The modulated carrier wave is selectively amplified often with a tuned amplifier. This step serves to limit the measurement system bandpass to that necessary to pass the modulated carrier wave. Any noise components outside of the

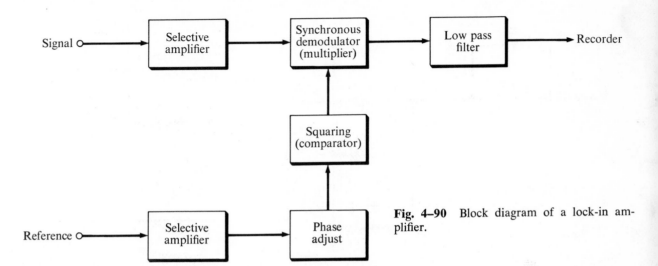

Fig. 4–90 Block diagram of a lock-in amplifier.

signal bandwidth are strongly attenuated at this step. Then the amplified carrier wave is synchronously demodulated, by multiplying the modulated carrier wave by a bipolar reference square wave signal that is equal in frequency and phase-locked to the carrier wave. Synchronous demodulation is very powerful in its ability to discriminate against random noise components, because, on the average, only the phase-locked carrier wave is demodulated by this multiplication operation. It is this step that has given the name lock-in amplifier to this signal processing technique. Finally, the output from the synchronous demodulator is low pass filtered, to regenerate the original signal. Note that the basic steps in the lock-in amplifier are directly analogous to those involved with double sideband (DSB) modulation and demodulation, which were discussed in Section 4–3.1.

A block diagram of the basic components of a lock-in amplifier is shown in Fig. 4–90. Normally the lock-in amplifier has two input channels, one for the modulated carrier (signal) and one for the reference signal which is used to control the synchronous demodulation. The signal is selectively amplified before being demodulated. The reference channel is more complex. Some form of selective amplification is often applied to the reference signal and then its phase must be adjusted relative to that of the modulated carrier. The reference and modulated carrier must be exactly in phase for the synchronous demodulation step. Finally the reference is converted to a bipolar square wave before multiplication. Note that the cross-correlation operation is present in the synchronous demodulation–low pass filtering step. In this case, the cross-correlation is carried out at only one relative displacement, namely at $\tau = 0$.

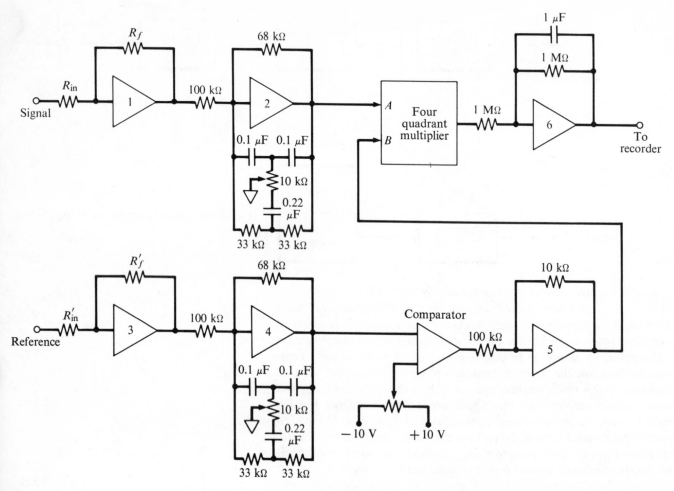

Fig. 4–91 Circuit for a simple lock-in amplifier.

The complete circuit of a simple lock-in amplifier is shown in Fig. 4–91. In this circuit, the selective amplification in both the signal and reference channels is provided by variable frequency tuned amplifiers (amplifiers 2 and 4). (These were discussed in Section 4–2.3.) Amplifiers 1 and 3 act simply as input amplifiers. The tuned amplifier in the reference channel is also used to provide phase adjustment for the reference signal, by fine tuning the resonant frequency of the tunable twin-T network. This provides adequate phase adjustment for this simple lock-in amplifier. The reference signal is converted to a bipolar square wave with a comparator, and then applied to a four quadrant multiplier for the synchronous demodulation step. Low pass filtering completes the signal processing. The main steps in the lock-in amplifier measurement system are discussed in more detail below.

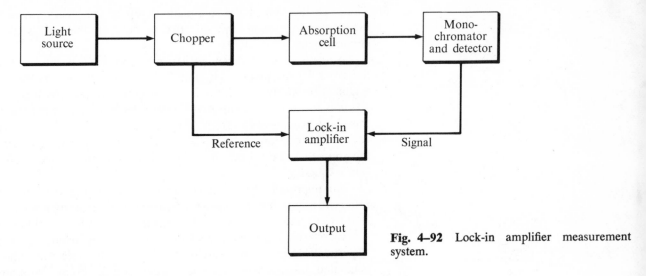

Fig. 4–92 Lock-in amplifier measurement system.

Modulation

In general, the basic signal sources to which lock-in amplifier techniques are applicable are those where the signal frequencies are at or very near dc. The general area of spectrophotometric measurements provides many examples. A general absorption spectrophotometric measurement system is depicted in Fig. 4–92. The first step in a lock-in amplifier measurement is modulation. Fundamentally, the modulation may take place anywhere before the lock-in amplifier. However, in a practical sense, the point in the experiment at which modulation takes place is important because often the carrier can be selectively modulated by the signal with respect to various noise sources in the experimental system.

Typically it is best to carry out the modulation step as soon as possible. For the absorption experiment shown in Fig. 4–92, the modulation step is carried out by inserting a mechanical chopper between the light source and the absorption cell. Modulation may also be effectively carried out by electronic modulation of the light source power supply. This is an excellent way of modulating hollow cathode lamps as used in atomic absorption and fluorescence spectroscopy. If modulation is carried out between the absorption cell and the detector, any interfering signals originating at the absorption cell will modulate the carrier. For example, if the absorption cell is a flame, as in atomic absorption spectroscopy, any radiation emitted by the flame would constitute an interfering signal.

Likewise, modulation between the detector and the lock-in amplifier results in modulation of the carrier by the detector noise as well as by all other noise sources and interferences in the complete experimental system.

Thus, the location of the modulation step in the measurement system is far from being trivial, and selective modulation can greatly aid the S/N enhancement.

A modulation frequency must also be chosen. Typically the best regions are from 10 through 35 Hz and from about 250 to 10^5 Hz. The lower region is primarily used for detectors which have poor frequency response. (This is true for many infrared thermal detectors.) However, even getting this short distance away from dc is very beneficial, as these detectors have severe $1/f$ noise problems. In source-modulated systems, the rate at which the source can be reliably modulated may limit the choice of the modulation frequency.

An important aspect of the modulation step is the generation of a reference signal that is the same frequency as the carrier wave (chopping frequency) and is phase-locked to it. (It need not be exactly in phase with the carrier but only phase-locked to it, as their relative phases can be adjusted later in the measurement system.) This reference signal can be generated, for example, at a rotating mechanical chopper with an auxiliary light source-detector combination. If the source is electronically modulated, the waveform used to modulate the source power supply can also be used as the reference signal. Many lock-in amplifiers have an internally generated reference signal that can be used to drive an external modulator. The resulting modulated carrier wave is demodulated by using this internal reference.

Selective Amplification

The next three steps, amplification, demodulation, and filtering, are normally carried out in the lock-in amplifier. First, the modulated carrier wave is selectively amplified. The primary purpose of this step is to amplify only the modulated carrier wave and to reject $1/f$ noise, interference noise (60 Hz), and as much of the white random noise as possible. Traditionally, this has been done by using a tuned amplifier with a bandpass sufficient to pass the carrier wave and its signal sidebands. It should be noted that use of a tuned amplifier necessitates a very stable modulation frequency. If the modulation frequency drifts within the bandpass of the tuned amplifier, its amplitude will fluctuate and it is impossible to distinguish these amplitude fluctuations from those caused by the modulation step.

In some measurement situations it is sufficient to ensure that only the $1/f$ and specific interference noises are rejected by this selective amplification step. Thus, the amplifier can be rather broadly tuned to the carrier frequency, making it less sensitive to frequency drifts. Any random noise is effectively discriminated against at the synchronous demodulation–low pass filtering step. Also, using a broader amplifier response at this stage facilitates the use of more complex and unique modulation methods than the typical symmetrical periodic waveforms. Pulselike modulation schemes may be desirable; or specific nonperiodic pulse sequences may be used as carriers for the signal information. At this point, the lock-in amplifier may be more like a boxcar integrator, depending on your point of view. Thus with a first amplification stage that is selective rather than tuned, the lock-in amplifier is a more flexible measurement instrument.

Synchronous Demodulation and Low Pass Filtering

The demodulation step of the lock-in amplifier provides the lock-in aspect of the measurement. Synchronous demodulation can be carried out using a four quadrant multiplier. Actual signal, reference, and demodulated waveforms are shown in Fig. 4–93. The signal is shown in Fig. 4–93a as a sine wave, and the reference as a bipolar square wave in Fig. 4–93b. When these two are in phase, the output of the four quadrant multiplier is as shown in Fig. 4–93c. Synchronous demodulation amounts to no more than synchronous full-wave rectification. The phase of the reference wave will, in general, have to be adjusted so that it is exactly in phase with the carrier wave.

The waveforms for the phase adjustment are shown in Fig. 4–94. The signal waveform is shown in the upper trace and the bipolar square wave reference in the middle trace of both figures. The two waveforms are in phase in Fig. 4–94a, and the output of the multiplier shown in the lower trace of Fig. 4–94a is the correct full-wave rectified signal. The

Fig. 4–93 Synchronous demodulation: (a) signal; (b) reference; (c) multiplier output.

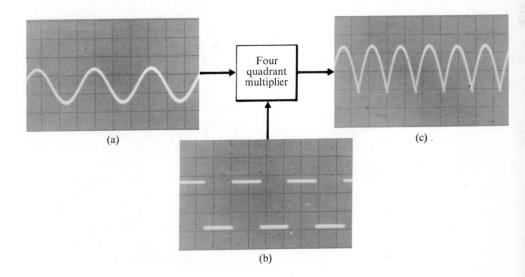

(a)

Four quadrant multiplier

(c)

(b)

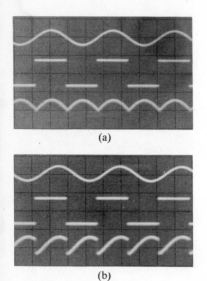

Fig. 4–94 Lock-in amplifier waveforms: (a) in-phase condition; (b) out-of-phase condition.

signal and carrier waveforms are shown out of phase in Fig. 4–94b. Now the output of the multiplier, as shown in the lower trace of Fig. 4–94b, is asymmetric.

The final step in the recovery of the signal information is low pass filtering. This step simply converts the synchronously full-wave rectified carrier into a dc level, the magnitude of which is representative of the amplitude of the carrier wave. The characteristics of the low pass filter should be chosen so as not to attenuate or phase-shift the recovered signal frequencies.

The performance of a lock-in amplifier is shown in Fig. 4–95. A noise-free input is shown in Fig. 4–95a. The resulting dc output, as recorded on a strip chart recorder, is shown in Fig. 4–95b. A random noise component was added to the signal to generate the noisy input shown in Fig. 4–95c. Now the signal is essentially completely obscured by noise. The oscilloscope horizontal and vertical sensitivities are identical for Figs. 4–95a and 4–95c. The resulting dc output of the lock-in amplifier for this

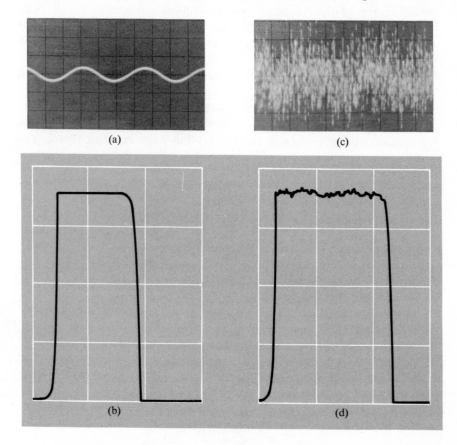

Fig. 4–95 Lock-in amplifier performance.

noisy input signal is shown in Fig. 4–95d. Despite the complete "burial" of the signal in noise, the output dc level has maintained a relatively high S/N and the amplitude of the obscured sine wave can easily be measured.

In summary, the lock-in amplifier measurement system discriminates against noise in several ways. Wherever possible, selective modulation is used, the modulation step moves the signal information to a region where it can be amplified with minimal amplification of noise frequencies, and demodulation is carried out in a synchronous manner to discriminate against any noise or interference frequencies that are not phase-locked to the modulated signal.

However, the lock-in measurement system does have some limitations. The signal must be capable of modulating a carrier wave. Many signals cannot do this very effectively; transient signals, signals with a high repetition rate, low duty cycle signals, and fast pulse signals are all examples of signals for which it is essentially impossible to use lock-in amplifiers. For any of the signals which can be made repetitive, the boxcar integrator and the multichannel averager may be more effective measurement systems. Also, both ends of the lock-in amplifier measurement are centered at dc. The original signal is typically a dc signal and may already be corrupted by $1/f$ noise before the modulation step. The final low pass filtering step is also centered at dc. Thus $1/f$ noise may still be a problem in lock-in amplifier systems.

Experiment 4–14 *Lock-In Amplifier*

A lock-in amplifier is constructed and characterized in this experiment. Separate amplification channels are provided for the signal and reference waveforms. Each channel consists of an input amplifier and a tuned amplifier, which has a tuning range of approximately 100 through 2000 Hz. After tuned amplification, the reference signal is converted to a bipolar square wave with a comparator. The signal and reference waveforms are multiplied together using a four quadrant multiplier. Phase adjustment of the reference waveform is easily accomplished by fine tuning the tuned amplifier. Finally, the output signal from the multiplier is low pass filtered to complete the demodulation. The output signal is recorded on a servo recorder.

4–5.4 BOXCAR INTEGRATION

The boxcar integrator is a versatile gated integrator used for the measurement of repetitive signals. Basically, the technique involves repetitively

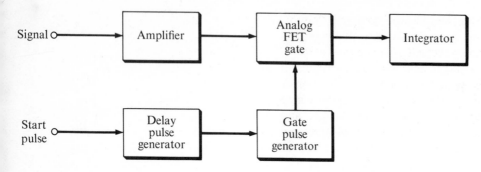

Fig. 4–96   Block diagram of a boxcar integrator.

gating out a particular section of a waveform and integrating the gated signals to improve the S/N. It is particularly useful for measuring repetitive short pulse signals and signals that have a slow repetition rate or low duty cycle. A block diagram of a boxcar integrator is shown in Fig. 4–96. It consists of an input amplifier, an analog FET gate, and an integrator. The analog FET gate is controlled by a reference channel gate pulse generator that opens the gate for a short period of time at a particular time with respect to the start of the repetitive signal.

The analog gating operation can be thought of as a multiplication operation in which the input waveform is multiplied by a normalized rectangular pulse (amplitude $= 1$). Hence, this analog gating–integration step is again the familiar cross-correlation operation. Thus, the boxcar integrator really involves cross-correlating a signal waveform with a rectangular gating pulse at a relative displacement, τ, set by the delay pulse. Some boxcar integrators are designed so that the delay can be slowly scanned, thus carrying out this cross-correlation across the complete waveform. This is a particularly effective measurement system for measuring very fast repetitive pulses. With a sufficiently narrow gate pulse and a relatively slow sweep on the delay time, a 10 nsec pulse can be recorded on a conventional strip chart recorder. Gate pulse widths as small as 100 psec are available on commercial boxcar integrators.

The key step in the boxcar integrator measurement system is the analog gating. This step is really a sampling operation, and therefore all the criteria discussed in Section 4–4 with respect to accurate sampling must be satisfied. The gate pulse width is the aperture time of the sampling operation. It must be narrow enough so as not to distort or average the sought-for structure in the particular waveform being investigated. If the delay pulse is moved incrementally across the waveform, the Nyquist sampling theorem must be satisfied. In addition, many of the points discussed in Section 4–5.2 with respect to low pass filtering and gated dc integration are applicable to the boxcar integrator.

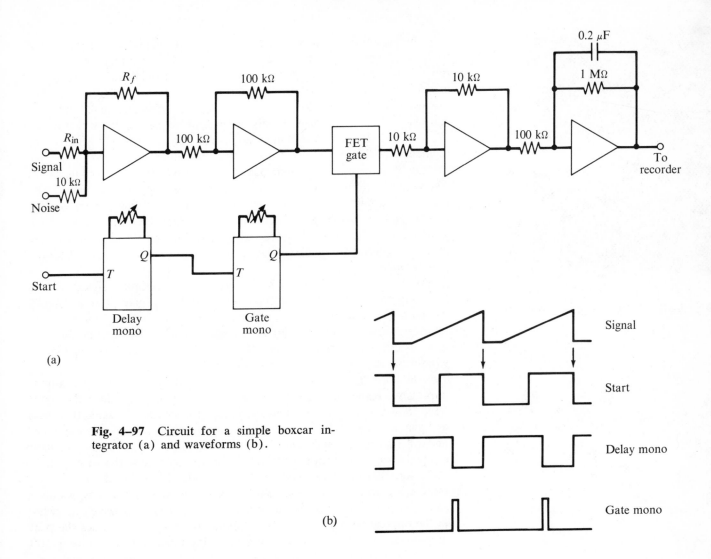

Fig. 4–97 Circuit for a simple boxcar integrator (a) and waveforms (b).

Simple Boxcar Integrator

The complete circuit for a simple boxcar integrator is shown in Fig. 4–97a, along with typical waveforms in Fig. 4–97b. The signal channel consists of an input amplifier, an analog FET gate, and a simple active low pass filter to average the output of the gate. The reference channel consists of two monostable multivibrators, one to generate a delay which can be manually adjusted, and a second to generate the gate pulse.

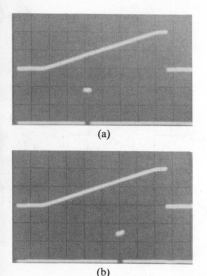

(a)

(b)

Fig. 4–98   Boxcar integrator waveforms.

The signal shown in Fig. 4–97b is a repetitive ramp. It is necessary to have available a start pulse indicative of the beginning of a signal repetition in order to trigger the boxcar integrator. This start pulse signal is also shown in Fig. 4–97b. The start pulse is used to trigger the delay monostable which is set so that the desired signal segment is gated out. The end of the delay pulse then triggers the gate pulse monostable which opens the FET gate for the desired integration time. The repetitively gated signal segments are then integrated with the low pass filter.

Oscilloscope traces of some actual waveforms are shown in Fig. 4–98. The repetitive ramp signal is shown in the upper trace of each figure. The gate pulse applied to the FET gate is shown in the lower trace of Fig. 4–98a; and the actual gated signal as observed at the output of the FET gate is shown in the lower trace of Fig. 4–98b. The position of the gate pulse, as controlled by the delay monostable, has been varied between photographs.

With a manually adjusted gate pulse delay, a complete signal such as the ramp can be sampled by successively setting the gate pulse at equally spaced intervals along the ramp and recording the output of the low pass filter on a recorder. The major division lines on the oscilloscope face can serve as convenient sampling points. Such a recording is shown in Fig. 4–99b for the ramp signal trace shown in Fig. 4–99a.

The noise discrimination capabilities of the boxcar integrator are quite impressive. Random noise was added to the signal to produce the signal trace shown in Fig. 4–99c. This oscilloscope trace has the same time base and gain settings as that in Fig. 4–99a. The particular random noise added had a flat frequency spectrum from about 20 Hz to 20 kHz. The manually stepped boxcar integrator output of the servo recorder is shown in Fig. 4–99d. Note that the noise level has been considerably reduced and the sawtooth waveform is readily discernible. The rms noise level is now only a small fraction of the maximum amplitude of the sawtooth waveform, while in the original waveform its rms amplitude is about 3 to 4 times the peak amplitude. The plot of these values vs. delay time, shown in Fig. 4–100, results in an excellent facsimile of the original waveform that was completely obscured by noise.

The noise rejection properties of the boxcar integrator can be strongly dependent on the specific frequency of the noise. The filtering characteristics are similar to those of the gated integrating DVM discussed in Section 4–5.2.

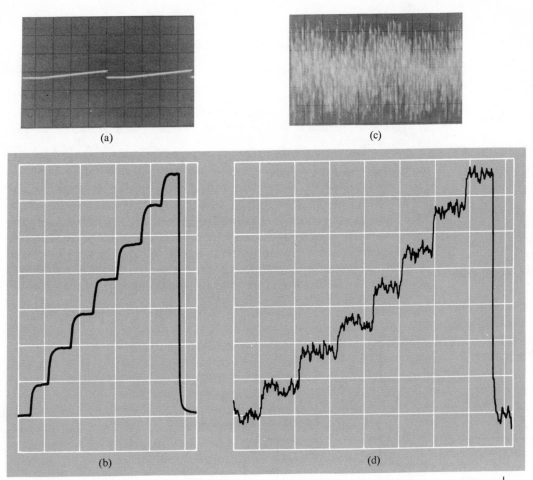

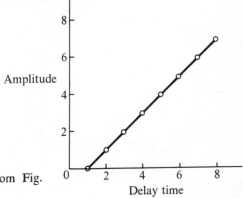

Fig. 4–99 Boxcar integrator performance.

Fig. 4–100 Reconstructed signal from Fig. 4–99d.

Experiment 4–15   *Simple Boxcar Integrator*

A simple boxcar integrator, in which the position of the gating pulse on the signal waveform is manually adjusted, is constructed in this experiment. Simple monostables control the delay and gating times, and the signal is gated by using a FET switch. A sine wave generator and a random noise generator are used to test the S/N enhancement characteristics of the boxcar integrator.

Digital Scanning Boxcar Integrator

Many boxcar integrators are designed to automatically scan the gate pulse across a specific waveform or section of a waveform. A block diagram of a self-scanning boxcar integrator is shown in Fig. 4–101. The key block is the *N*-bit circulating shift register. At all times when the circuit is operating it contains only one bit. This bit is circulated and shifted under control of the

Fig. 4–101   Block diagram of a digital scanning boxcar integrator.

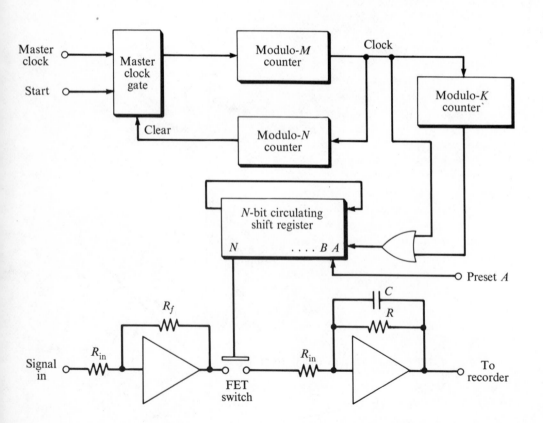

clock and the various counters in synchronism with the repetition of the signal. A start pulse indicative of the beginning of a signal repetition opens the master clock gate, gating the master clock to the modulo-M counter. A modulus of at least 20 is necessary for this counter, the output of which provides the clock pulses that cycle and control the integrator.

This modulo-M counter is necessary to ensure synchronization of the boxcar gate pulse to a specific point on the signal. The master clock frequency and the signal repetition rates are asynchronous. Thus, the master clock gate has a jitter of as much as one clock period. However, the clock signal from the modulo-M counter has a jitter, with respect to the signal, that is only one part in M of its period. The master clock must have a frequency that is M times greater than the clocking rate desired for the shift register. An M value of 20 or greater is quite adequate in many cases.

The clock pulses from the modulo-M counter are fed to the circulating register and also to the modulo-N and modulo-K counters. The modulo-N counter (same value of N as the length of the register) closes the master clock gate after N clock pulses. Assume N is 10, for the purposes of this discussion. Thus, if bit A of the circulating register is set (HI), it circulates *once* after application of a start pulse and stops in position A ready for the next signal repetition. Note that bit N of the circulating register is connected to the FET gate control. Thus the segment of the signal that occurs $(N - 1)$ clock periods after the start pulse will be gated out and integrated on successive signal repetitions.

The modulo-K counter provides the self-scanning action of the integrator. The modulus of this counter is normally large, 10^3 to 10^5. Assume that all counters are cleared and bit A of the register is set. When a repetitive start pulse and signal are applied to the circuit, the register will circulate once, each time a start pulse occurs, gating out the $(N - 1)$ segment of the signal as mentioned above. However, assuming $K = 10^4$, on the 10^4 clock pulse (1000 cycles of the 10 bit register), the register will receive one extra clock pulse through the OR gate. Thus the steady-state position of the circulating bit is now location B. Now the segment of the signal that occurs $(N - 2)$ clock periods after the start pulse will be gated out and integrated. The circuit cycles for 10^4 clock pulses before the circulating bit is again jogged one position by the modulo-K counter. In this manner, the complete signal is integrated in $(N - 1)$ steps, starting from the end of the signal and working back towards the start of the signal. The output signal can easily be displayed on a strip chart recorder.

While the self-scanning boxcar integrator shown in Fig. 4–101 may seem complex, it can easily be implemented using the basic circuit units discussed in Module 3. The complete circuit of a self-scanning boxcar integrator based on the block diagram of Fig. 4–101 is shown in Fig. 4–102.

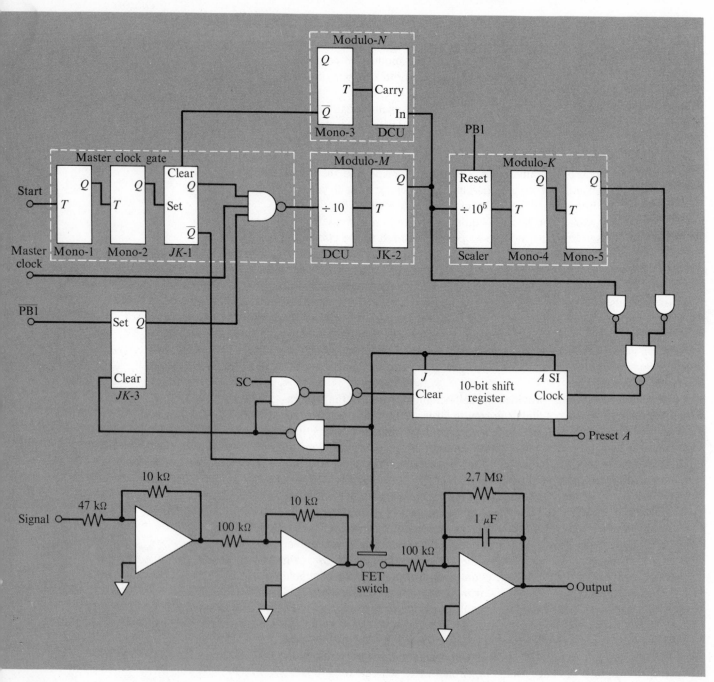

Fig. 4–102 Complete circuit of a digital scanning boxcar integrator.

The master clock gate consists of two monostables, *JK*-1, and a three-input NAND gate. The start pulse triggers monostable 1 which delays the opening of the master clock gate. The sampling pulses can be set at a specific section of the signal by adjusting this delay. After the delay, monostable 2 fires, setting *JK*-1 and gating the master clock to the modulo-20 counter. The other controlling input on the NAND gate is connected to *Q* of *JK*-3, which is set before a measurement begins.

The modulo-20 counter consists of a divide-by-ten counter (DCU) and a *JK* flip-flop (*JK*-2). The *Q* output of this flip-flop is the clock signal. The clock signal is connected to the modulo-10 counter, modulo-*K* counter, and the 10-bit circulating shift register. The modulo-10 counter is another DCU. Its output triggers a monostable (monostable 3) that closes the master clock gate (clears *JK*-1) after 10 clock pulses.

The modulo-*K* counter consists of seven DCU's cascaded to give a total modulus of 10^7. This type of unit is frequently referred to as a scaler. It can easily be set up as a counter with a modulus from 1 to 10^7 in decades of ten. The combination of monostables 4 and 5 provides the extra clock pulse that jogs the bit in the register one position forward. The waveforms from these monostables and the output of the OR gate when the *K*th clock pulse is applied to the modulo-*K* counter are shown in Fig. 4–103.

The *N*-bit circulating shift register can be constructed from a variety of integrated circuit shift registers that are available. Values of *N* from 4 to 1024 are available in single 16 pin dual in-line packages. For example a simple 10-bit register can be made by cascading two SN 7496N five-bit shift registers. Provision must be made for externally presetting the first bit of the shift register, and the *N*th bit is used to control the FET gate.

An end-of-scan detector is necessary to eliminate the *N*th bit position as a steady-state position for the circulating bit. If the bit resided in the *N*th location which also controls the FET gate, the FET gate would be closed between scans and open during a scan, exactly the opposite of the desired behavior. When the bit is in the *N*th location *and* the master clock gate is closed, it means that the steady-state location of the circulating bit has finally been jogged all the way to the end of the register. When this condition occurs, *JK*-3 is cleared, locking the master clock gate closed. The occurrence of this condition also clears the circulating register. To start another measurement cycle, bit *A* (the first bit) must be present, *JK*-3 set, and the modulo-*K* counter cleared.

The analog signal channel is quite simple. The signal is connected to the FET gate, using an OA as the input amplifier, and the gated signal segments are integrated with a low pass filter. The output signal can be displayed directly on a servo recorder. An example of the performance of the self-scanning boxcar integrator is shown in Fig. 4–104. The signal is a repetitive exponential decay and is shown in Fig. 4–104a. The output of a ten-bit self-

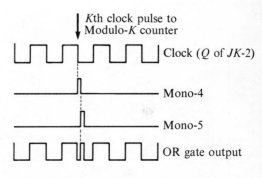

Fig. 4–103 Waveforms from modulo-*K* counter and OR gate.

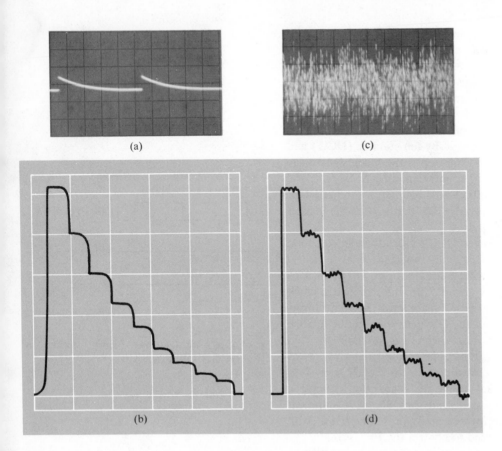

Fig. 4–104 Performance of scanning box-car integrator.

scanning boxcar integrator, as plotted on a recorder, is shown in Fig. 4–104b. The same signal obscured by random noise is shown in Fig. 4–104c, and the output as plotted on the recorder is shown in Fig. 4–104d. The S/N enhancement is quite dramatic. Again, full recovery of the signal is accomplished by plotting amplitude vs. delay time. This is shown in Fig. 4–105 for the output data (Fig. 4–104d) from the noisy signal of Fig. 4–104c.

Experiment 4–16 *Digital Scanning Boxcar Integrator*

With the boxcar integrator of Experiment 4–15 it was necessary to manually adjust the delay time in order to measure the complete shape of a waveform. The boxcar integrator to be constructed and studied in this experiment is capable of automatically scanning the gate pulse of the integrator across a waveform. The self-scanning operation is based on a

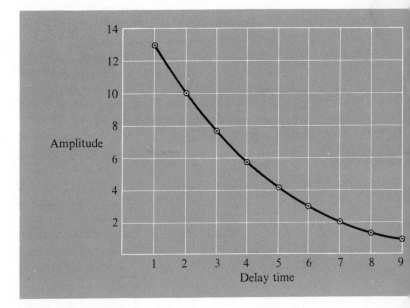

Fig. 4–105 Reconstructed signal from Fig. 4–104d.

digital clocking and sequencing system, constructed from a ten-bit circulating shift register. This register is circulated and shifted under control of a master clock and various counters, in synchronism with the repetition of the signal.

Analog Scanning Boxcar Integrator

A scanning boxcar integrator can also be designed based on analog timing and sequencing. The circuit of an analog-based scanning boxcar integrator is shown in Fig. 4–106. The timing and sequencing are controlled by two integrators whose outputs are summed and fed to a comparator. Integrator 1 (OA1) has a long integration time, and it controls the total measurement time for the scanning operation. Its output serves the purpose of slowly increasing the voltage at which the comparator will fire. To begin the measurement the shorting switch on this integrator is opened.

The second integrator (OA2) is a triggered integrator. The FET switch on OA2 is momentarily closed by the start pulse that also initiates the signal repetition. It has an integration time comparable to the period of the signal. The two integrator outputs are subtracted; thus the comparator fires at a slightly longer time after the start pulse on each successive signal repetition because of the slowly increasing output of integrator 1. The output of the comparator is used to trigger a gate pulse monostable which controls the FET gate in the signal channel.

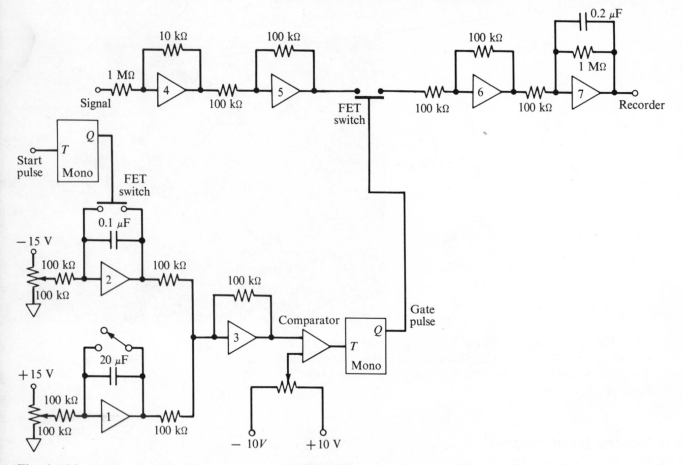

Fig. 4–106 Analog scanning boxcar integrator.

The waveforms for time t_1 after the beginning of a measurement are shown in Fig. 4–107a. The output of OA3 (input to the comparator) will have reached a value $+V_1$ as determined by the first integrator. Since its output is increasing (in a negative direction) slowly with respect to the signal repetition rate, it is shown as a dc level. After a time t_1 the second integrator will have decreased the output of OA3 to zero, which fires the comparator, generating a gate pulse at the monostable. Analogous waveforms are shown for a later time t_2 in Fig. 4–107b. Thus, in this manner, the gate pulse is slowly moved along the waveform.

Experiment 4–17 *Analog Scanning Boxcar Integrator*

An analog scanning boxcar integrator is constructed in this experiment. The self-scanning action is obtained by summing the outputs of two OA integrators and applying the sum to a comparator. One integrator has a

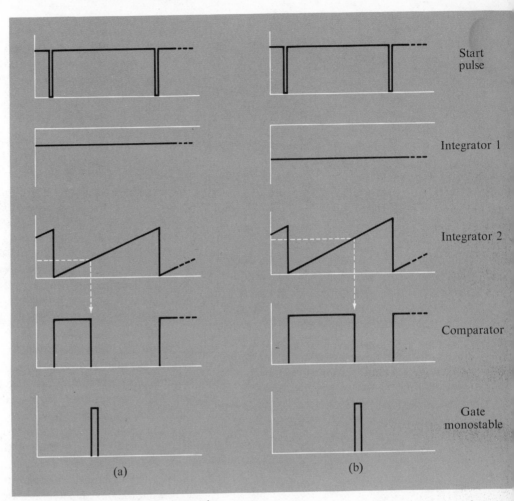

Start
pulse

Integrator 1

Integrator 2

Comparator

Gate
monostable

(a)   (b)

very slow integration time, which serves the purpose of slowly increasing the voltage at which the comparator will fire. The second integrator is triggered by the start pulse indicative of the beginning of the signal repetition, and it has an integration time comparable to the period of the signal. The integrator outputs are subtracted; thus the comparator fires at a slightly longer time on each signal repetition. The output of the comparator triggers a monostable which gates a signal segment to a low pass filter with a FET switch.

Fig. 4–107 Waveforms for analog scanning boxcar integrator.

Boxcar Integrator Applications

The boxcar integrator is an excellent measurement system for a wide variety of signals. The basic requirement is that the signal be repetitive. As shown

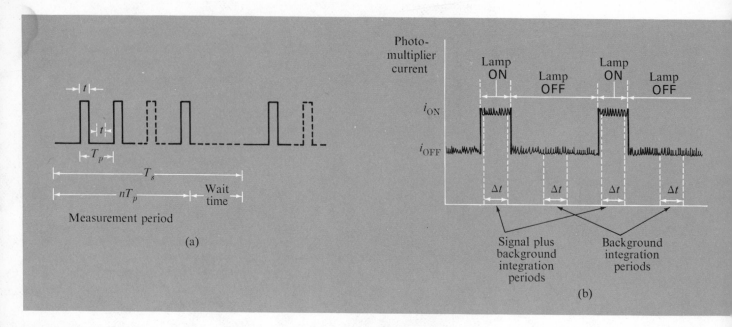

Fig. 4–108 (a) Pulse sequence applied to hollow cathode lamp. (b) Integration times Δt for the signal-plus-background and background information channels.

in some of the examples, high-speed, repetitive, noisy signals can easily be displayed on normal laboratory recorders with significant increases in S/N, using scanning boxcar integrators. The boxcar integrator is also an excellent measurement system for pulsed signals that have low duty cycles, i.e., a short ON time compared to OFF time. For example, assume that the entire area of a 1 μsec wide repetitive pulse signal with 100 μsec between pulses is to be integrated. For 99 percent of the time, only noise is present in the signal channel. The best way to discriminate against this noise is to simply shut off the signal channel and open it only when the signal is present. This is exactly what the boxcar integrator does. In a sense it is a lock-in amplifier for pulsed signals.

The lock-in nature and versatility of the boxcar integrator is illustrated by a dual channel system developed for the measurement of signals generated from a pulsed, hollow cathode tube light source. The pulse sequence applied to the lamp is shown in Fig. 4–108a. The lamp is pulsed ON for a time t and this is repeated n times with a period of T_p sec. For example, the ON time might be 10 msec and the OFF time 30 msec, giving a pulse period T_p of 40 msec. A typical measurement might involve repeating this 20 times and then waiting several seconds before repeating the complete sequence. The output signal from the experiment, as detected by a photomultiplier tube, is shown in Fig. 4–108b. The desired measurement is the total integral (over the 20 pulses) of *only the signal information*. This can be accomplished using the dual channel boxcar integrator circuit shown in Fig. 4–

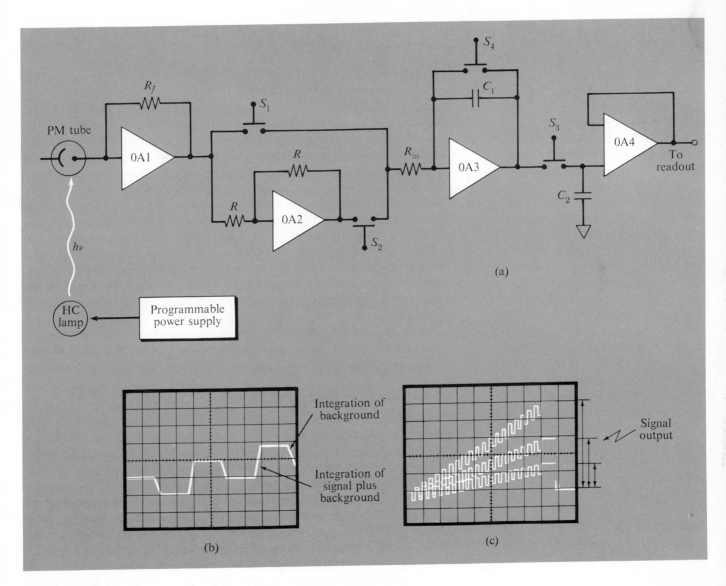

(a)

(b)

Integration of background

Integration of signal plus background

(c)

Signal output

Fig. 4–109 (a) Dual-channel boxcar integrator. (b) Subtraction of integrated background from integrated signal plus background to provide a net signal for each pulse (time scale = 20 msec/div). (c) Integration of 20 pulses in each of several series at different signal levels (time scale = 200 msec/div).

109a. The integrator (OA3) is gated with FET switch S1 to integrate the signal plus background (see Fig. 4–108b), and with FET switch S2 to subtract the background integral from the previous signal-plus-background result. This sequence is repeated for all 20 pulses, and then the final value of the signal integral is sampled and held, using FET switch S3 to control the sample-and-hold follower. Actual waveforms from the output of the integrator during a measurement cycle are shown in Figs. 4–109b and 4–109c.

Thus the boxcar integrator is a rather versatile measurement instrument. Also, it occupies somewhat of a pivotal position among the various S/N enhancement techniques, as it embodies characteristics of essentially all the other techniques. The most general description is simply that it is a triggerable gated integrator. As seen above it can also be thought of as a lock-in amplifier for pulsed signals. The boxcar integrator can, if desired, actually be used directly as a lock-in amplifier to perform synchronous demodulation of sinusoidal signals. However, it is capable of providing only half-wave synchronous rectification. As a gated integrator, the boxcar integrator is a one-channel "multichannel" averager for repetitive waveforms; and in its scanning mode, where the gate pulse delay is slowly scanned across a waveform, it is an analog cross-correlation computer.

Some of the advantages of the boxcar integrator are that very complex modulation techniques can be used as compared to conventional lock-in amplifiers. Much narrower gate pulses can be attained than are possible with most multichannel averagers, with the state of the art being about 100 psec. Hence, much faster waveforms and narrower pulses can be measured and recorded. However, for many signals, multichannel averagers offer significantly faster measurement times, simply because several hundred channels of data can be averaged simultaneously. Also, more general cross-correlation analysis requires more complex cross-correlations than those between a simple rectangular pulse and a signal waveform.

4–5.5 MULTICHANNEL AVERAGING

In many experimental measurements it is necessary to recover a complete repetitive signal from a noisy waveform. A multichannel averager can be used to carry out such an operation. Unlike the boxcar integrator which can acquire only one sample per signal repetition, the multichannel averager acquires a large number of evenly spaced samples across the complete waveform on *each* repetition. The samples from successive repetitions are averaged or integrated, as in the boxcar integrator. However, since the multichannel averager takes N samples across a waveform, the complete waveform can be recovered N times faster. As with any averaging or integrating technique, the signal builds up directly as the number of scans, and the noise only as the square root of the number of scans. Thus the S/N improves as $N^{1/2}$.

Most multichannel averagers are digital instruments. Each repetition of the waveform is digitized at a desired number of points, and these values are added to the previous values stored in a digital memory. Typical upper limits on the number of points range from 1024 to 4096 for most systems. This limit, of course, depends on the available digital memory.

Digital multichannel averagers, in general, are not extremely fast. Sampling rates in excess of 100 kHz would be considered high. Typical input circuitry might include a fast sample-and-hold amplifier with an acquisition time of 0.1 μsec and a 10-bit successive approximation ADC with a quantization time of 10 μsec. After sampling and quantization, some time is required to add the digitized sample to the previous value and store the sum, although this may be carried out while the next sample is being digitized.

The general structure of the digital multichannel averager may be based on a stand-alone hardware system, or often it is just a program in a small computer that is interfaced to an ADC. The specific hardware systems are faster than the software based averagers, usually because of the time it takes to run a small program that controls the sampling, quantizing, and averaging. A fast sampling rate for a small computer-based multichannel averager would be in the range of 50 kHz, assuming that at least ten-bit precision is required. Precision can be traded off for higher sampling rates if desired.

It should be noted that if averaging is not required, considerably faster sampling rates can be achieved. Recently, instruments called transient recorders have been developed that can digitize and store a complete single trace of a waveform. Sampling rates for a six-bit transient recorder can be as high as 10 MHz with a total of about 100 samples being stored.

Digital Multichannel Averager

A general block diagram of a digital multichannel averager is shown in Fig. 4–110. The start pulse that initiates the repetition of the signal waveform

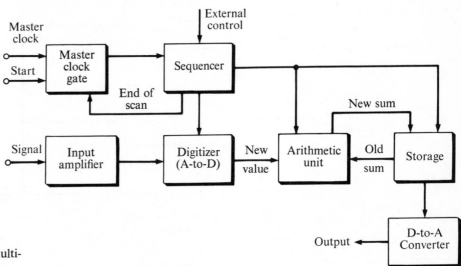

Fig. 4–110 Block diagram of a digital multichannel averager.

gates a master clock to the sequencer, which controls the sampling, digitization, addition, and storage of the new values. This sequencer may be a specific hardwired controller or a minicomputer sequencer under software control. The input signal is successively sampled and digitized, and each new value is added to the sum of the previous values and stored. As mentioned before, a typical digital system would be capable of storing up to 1024 samples of a signal waveform. For some measurements, it is necessary to be able to externally control the sampling sequence if the parameter of interest is not a linear or constant function of time. Thus, most systems have some provision for externally clocking the sequencer. A digital-to-analog converter (DAC) is usually part of the instrument, so that an analog output can be generated for display on an oscilloscope or a recorder.

The specific blocks in this instrument could be implemented in a variety of ways. A simple approach is shown in Fig. 4–111. The signal channel consists of an input amplifier, a 10-bit successive approximation analog-to-digital converter (ADC), a 20-bit parallel binary adder, and an N-word 20-bit shift register circulating memory. The clocking channel circuits are very similar to those used to control the digitally based scanning boxcar integrator. The master clock gate consists of a JK flip-flop and a NAND gate, a modulo-M counter for synchronizing, and a modulo-N counter to control the cycle time. The 20-bit parallel adder could be constructed by cascading integrated circuits such as SN 7483N's which are four-bit binary full-adders. The shift register memory could be made from

Fig. 4–111   Circuit for a digital multichannel averager.

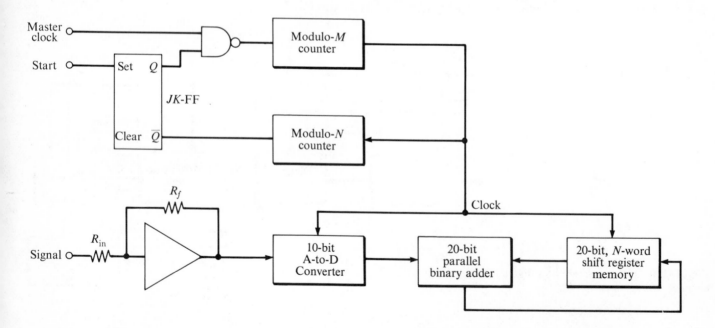

1024-bit static shift registers which are now available in 8 pin IC packages. (This type of memory system was discussed in Section 3–4.3 of Module 3.) Shift registers are also available in shorter lengths. Another excellent alternative for the digital storage are random access memories (RAM's). LSI RAM's are available in up to 4096 bit sizes. They require more complex control circuitry than the shift registers but offer more flexibility. These memory systems were also discussed in Section 3–4.3 of Module 3.

The transient recorder mentioned above is very similar to the circuit shown in Fig. 4–111, except that the binary adder is unnecessary. The digitized values are simply clocked directly into the shift register memory. This is quite a useful circuit and can be used to make a storage oscilloscope or for recording relatively high speed waveforms on conventional laboratory recorders. In the latter application, the data are clocked out of the memory to a DAC at a rate compatible with normal recorder speeds.

Analog Multichannel Averager

Multichannel averaging may also be implemented using analog averaging and storage. An advantage of the analog approach is that reasonably fast sampling rates can be achieved, in the range of 1 MHz or greater. Normally a relatively small number of samples is taken; 100 is a typical upper limit. The storage and averaging is accomplished using FET switched capacitors that are sequenced by a circulating shift register clocked in synchronism with the repetitive signal. A block diagram of an analog multichannel averager is

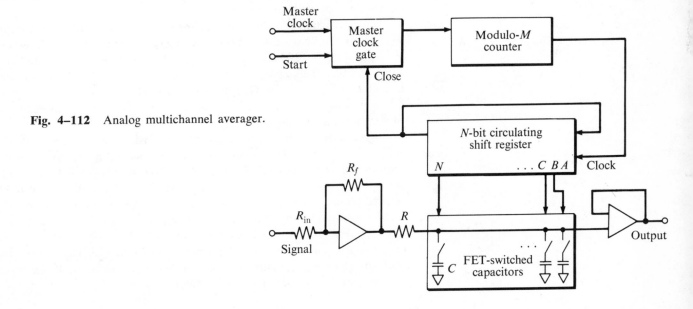

Fig. 4–112   Analog multichannel averager.

shown in Fig. 4–112. The start pulse that initiates the signal repetition gates a master clock to the modulo-M counter. This circuit generates the clock pulses that sequence a circulating bit in the shift register through one cycle. Between scans the one bit in the register normally resides in the first location, which is not connected to a FET switch control. As the bit circulates, the FET switched capacitors successively sample segments of the signal waveform and store or average the segments on capacitors.

Experiment 4–18   *An Analog Multichannel Averager*

A complete four channel analog averager for repetitive signals is constructed and characterized. The circuit averages four successive segments of a signal on four FET switched capacitors that are sequenced by a circulating shift register clocked in synchronism with the signal.

Implementation of Multichannel Averaging

In all types of multichannel averagers, the sampling operation as discussed in Section 4–4 is a very important step. All the criteria for accurate sampling with respect to sampling rate, sampling duration, aperture time, and quantizing time must be satisfied if meaningful results are to be obtained. For this reason, reasonably versatile bandwidth control should be available on the input amplifier of the signal channel, in order to minimize aliasing for a wide variety of waveforms.

For the signals to which multichannel averaging is applicable, impressive results can be obtained in the way of S/N enhancement. An example is shown in Fig. 4–113. A single scan of a noisy signal is shown in Fig. 4–113a, and the resulting signal when 100 repetitive scans have been averaged, is shown in Fig. 4–113b. One hundred scans should reduce the noise level by about a factor of 10. Thus, random noise can be significantly reduced by multichannel averaging.

Since most multichannel averagers are digital instruments, quantizing noise can also present a serious limit to measurement precision in some cases. Quantizing noise can be a problem if the ADC does not have enough bit resolution or, more often, when two large signals are subtracted to yield a small result. The latter situation occurs frequently when recovering a small signal superimposed on a relatively large background. The signal shown in Fig. 4–114a is just such a signal, which was obtained by subtracting a single scan of background from a single scan of signal plus background. The severe quantizing noise problems are obvious. If the noise in the system (ADC and signal) is much less than the magnitude of the

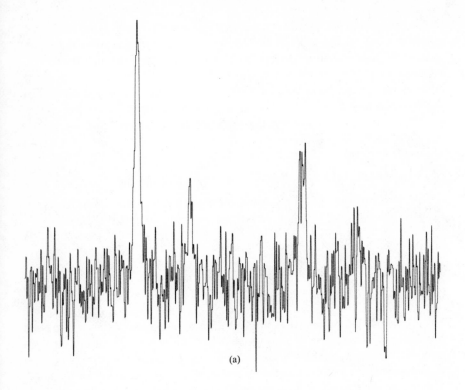

(a)

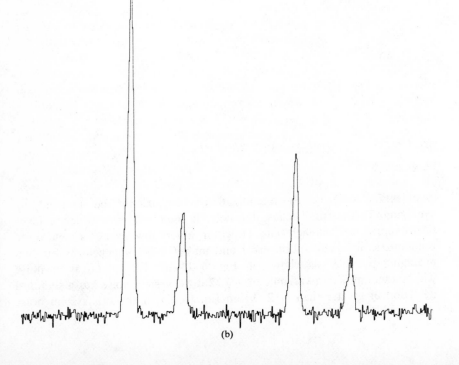

(b)

Fig. 4–113 S/N enhancement by multichannel averaging: (a) single scan; (b) 100 scans.

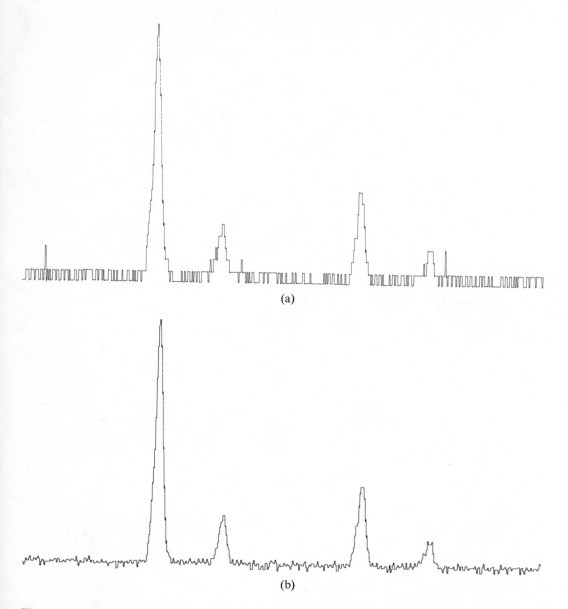

(a)

(b)

Fig. 4–114 Effect of multichannel averaging on quantization noise: (a) single scan; (b) 100 scans.

quantization level v_l, the quantization noise cannot be reduced by averaging. This is true because the noise is round-off error which is identically reproduced on each scan. However, a very low level of system noise randomizes the quantization error and makes S/N enhancement by data averaging possible, as shown in Fig. 4–114b. The quantization noise will be random with a variance of $v_l^2/12$ if the system noise has a standard deviation of greater than $v_l/2$. From Eq. (4–5), if the rms system noise

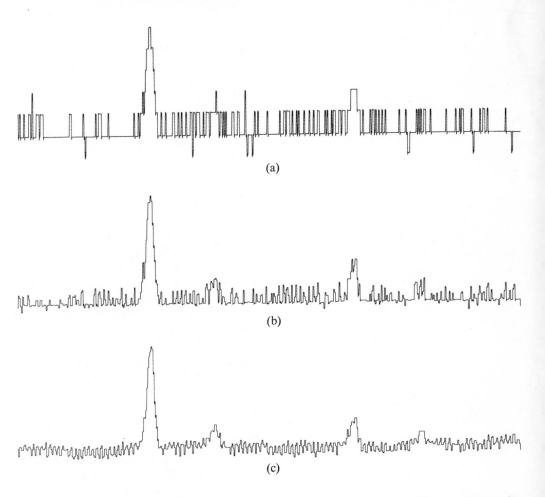

(a)

(b)

(c)

is $v_l/2$, the total mean-square measurement noise is $v_l^2/4 + v_l^2/12 = v_l^2/3$, resulting in an rms noise of about $0.6v_l$. If the average of N scans (computed without round-off) is taken, the S/N will be improved by a factor of $\sqrt{N}$. Hence, for 100 scans the standard deviation will be about $0.06v_l$. Starting with a 10-bit ADC and a system rms noise of $v_l/2$, the measurement resolution will be equivalent to about 14 bits after 100 scans. Note that the actual average, without round-off, will be a 17-bit number of which three bits are not significant. System noise of more or less than the optimum value will result in a greater number of insignificant bits.

A single scan of a very low level signal is shown in Fig. 4–115a, resulting again from the subtraction of two large signals. The quantization noise is so severe that it obscures all but the strongest peak in the signal. The signal resulting from the subtraction of 100 scans is shown in Fig. 4–115b and is improved because a small amount of system noise is present.

Fig. 4–115 Effect of adding random noise in order to recover low level signals obscured by quantization noise: (a) single scan; (b) 100 scans; (c) 100 scans with added random noise.

An even better improvement can be realized by *adding random noise to the original signal*. As long as the added random noise is not too intense, its final level can be reduced to an acceptable level by averaging. The signal resulting from the averaging of 100 scans in which random noise was added to the signal is shown in Fig. 4–115c. This signal, while not of high S/N, is still the best of the three signals shown in Fig. 4–115, since all four peaks are evident. Thus this technique of adding random noise to very low level signal variations may aid the detection of signal information that would otherwise be completely obscured by quantization noise.

4–5.6   CORRELATION TECHNIQUES

The basic correlation operation has been present to some extent in all of the S/N enhancement techniques discussed in this section. The lock-in amplifier and the boxcar integrator could easily be designated as correlation techniques. However, except for the self-scanning boxcar integrator, correlation was evaluated in these techniques at only one relative displacement and even in this case one of the functions could be only a unipolar binary pulse. When correlation is carried out between any two waveforms over a reasonably wide range of relative displacements, the signal processing is usually referred to specifically as a correlation technique.

The basic correlation operation was discussed in Section 4–5.1 of this module. Remember that the correlation pattern is generated by evaluating the time-averaged or integrated product of two complete signals as their relative displacement is shifted. The resulting cross-correlation pattern is a function of the shift, which is normally given the symbol τ. It is useful to distinguish two general correlation techniques. When two different signals are correlated, the operation is referred to as cross-correlation, and when a signal is correlated with itself, as auto-correlation. These two correlation techniques will be discussed first, and then specific approaches to correlation instrumentation will be covered, along with some further examples of correlation as implemented with relatively simple instrumentation.

Cross-Correlation

In many cases where cross-correlation is used, it is useful to distinguish a signal waveform and a reference waveform, and to think in terms of the reference waveform moving with respect to the signal waveform. A conceptually simple but excellent example of cross-correlation is the operation carried out by the self-scanning boxcar integrator which was discussed in Section 4–5.4. The scanning gate pulse of this boxcar integrator is the reference waveform for the cross-correlation operation. As was seen in Section

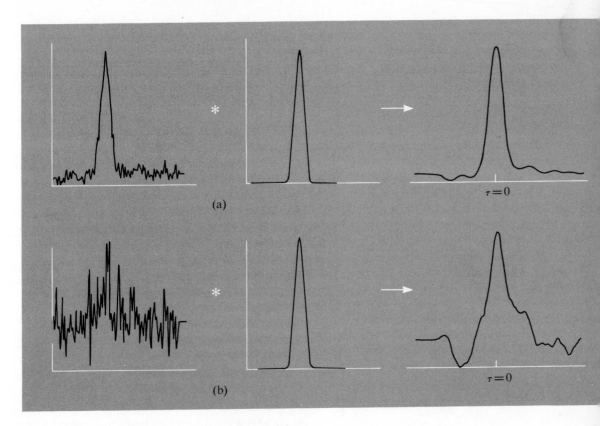

Fig. 4–116 S/N enhancement by cross-correlation (* indicates correlation).

4–5.4, cross-correlation of a noisy signal with a narrow rectangular reference pulse is an excellent way of improving the S/N of a measurement. Also the reference cross-correlating pulse does not have to be rectangular but could be triangular or any other desired shape. The optimal shape for the reference cross-correlation waveform, in order to achieve the maximum S/N ratio for a given measurement, has been thoroughly investigated. In the case of enhancing the S/N of repetitive single pulse signals by a cross-correlation operation, a noise-free version of the pulse itself provides a close-to-optimal reference pulse shape. Two examples of this cross-correlation operation are shown in Fig. 4–116. In both cases, the reference waveform is a noise-free version of the sought-for signal. Note that considerable smoothing and reduction of the noise level has occurred. However, the signal shape has also been slightly distorted, as it is broadened and rounded by the cross-correlation operation.

The goal of cross-correlation with simple pulse waveforms need not simply be S/N enhancement. Useful modifications of the signal waveform can be carried out by cross-correlation with certain types of bipolar pulses.

(a)

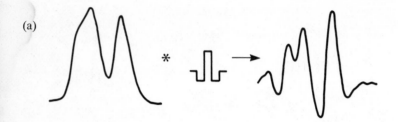

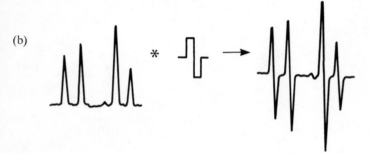

(b)

Fig. 4–117 Resolution enhancement (a) and differentiation (b) by cross-correlation.

Two examples are shown in Fig. 4–117. The sequence shown in Fig. 4–117a illustrates a cross-correlation that carries out a resolution enhancement operation on the signal waveform. The cross-correlation illustrated in Fig. 4–117b results in approximate differentiation of the signal waveform. These simple examples point out that cross-correlation can be much more than just an S/N enhancement technique. A wide range of powerful signal processing operations may also be effectively implemented.

Cross-correlations carried out with more complex reference waveforms can be usefully thought of as a measure of the similarities between the signal waveform and the reference waveform. The value of the cross-correlation function at zero displacement ($\tau = 0$) is the sum or time average of the product of the two waveforms; and its magnitude is a measure of all identical features shared by the two waveforms. If they do not have any common features, the value of the cross-correlation function will be zero at $\tau = 0$. All the other values of the cross-correlation function contain information about similarities between the two waveforms that occur at different relative points in time or position.

A simple example of the use of cross-correlation to aid the detection of specific information in a signal waveform is shown in Fig. 4–118. In this case, the signal waveform is a series of binary pulses and the reference waveform is a triplet sequence. The cross-correlation pattern is shown in Fig. 4–118c. The two relatively prominent maxima indicate the relative τ positions at which information in the signal that is most similar to the reference

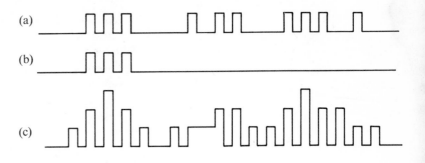

Fig. 4–118 Simple pattern detection by cross-correlation. (a) Signal waveform. (b) Sought-for pattern (reference waveform). (c) Cross-correlation pattern.

waveform can most likely be found. For simple binary signals such as these, the cross-correlation operation is equivalent to evaluating the logical AND operation between the two signals and summing the resultant 1's.

A more complex example of the signal detection capabilities of the cross-correlation operation is presented in Fig. 4–119. The two signals in this case are elemental emission spectra for cobalt (Fig. 4–119a) and nickel (Fig. 4–119b), which cover the region from 3430 to 3500 Å.

The correlation pattern between the cobalt spectrum and another identical cobalt spectrum is shown in Fig. 4–119c. In this case, since the two signals are identical, a large value at $\tau = 0$ would be expected and is ob-

Fig. 4–119 Pattern detection by cross-correlation. (a) Cobalt spectrum. (b) Nickel spectrum. (c) Cobalt-cobalt cross-correlation pattern. (d) Cobalt-nickel cross-correlation pattern.

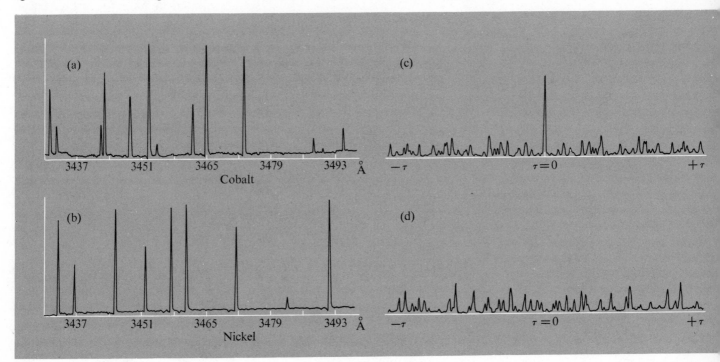

served. The rest of the correlation pattern contains information about relative peak separations and overlaps. The cross-correlation pattern between the cobalt spectrum and the nickel spectrum is shown in Fig. 4–119d. No prominent peaks are present, indicating that there are no significant similarities between these two signals other than two or three peaks that may overlap at certain τ values. Thus, these two signals could easily be distinguished on the basis of the behavior of the correlation function at $\tau = 0$, using the cobalt spectrum as the reference waveform. This is another example of the signal processing capabilities of correlation, in that similarities in signal waveforms as compared to a reference waveform can be easily evaluated. This is a first step in many automatic signal or pattern recognition systems. As with the simple pulse cross-correlations, these may also be effectively carried out under conditions of poor S/N.

In cases where the signal waveform and reference waveform are periodic, the similarities measured between the two waveforms by cross-correlation amount to what might be called mutual periodicities. For example, the cross-correlation pattern resulting from the correlation of a sine and a square wave of the same frequency results in a sine wave of the same frequency. This is the first harmonic of the square wave and hence the common periodic function between the two waveforms.

Auto-Correlation

When correlation is carried out between a signal waveform and a delayed version of the same signal waveform, it is called auto-correlation. This type of correlation can be useful in recovering a periodic signal from noise when *no* reference waveform is available. The signal becomes its own reference. However, any phase information in the original periodic signal is lost.

The auto-correlation patterns for several periodic waveforms are shown in Fig. 4–120. The auto-correlation pattern in each case is periodic but its form may be quite unlike the original waveform. The auto-correlation pattern of noise is quite different. It simply is an exponential decay (see Fig. 4–120d). The rate of the decay depends directly on the bandwidth of the noise. Thus if a periodic waveform is present in a noisy signal, its auto-correlation pattern will persist after that for the noise has decayed away. Thus, auto-correlation can be used to recover a periodic waveform from random noise.

The auto-correlation function, particularly its value at $\tau = 0$, is also useful in setting an upper limit of similarity that can occur between the reference waveform and a signal waveform. This is necessary when attempting to identify patterns by a cross-correlation operation, as illustrated in Figs. 4–118 and 4–119.

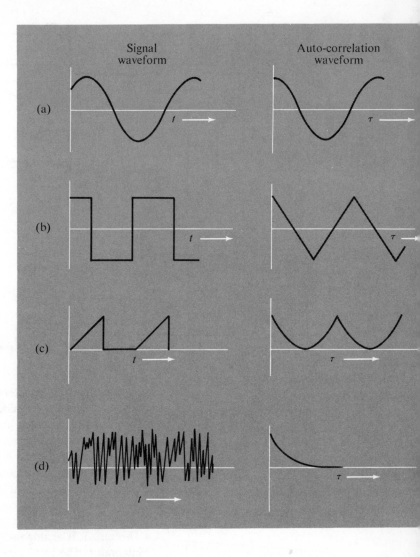

Fig. 4–120 Signals and auto-correlation waveforms.

Correlation Instrumentation

The instrumentation available specifically for general purpose correlation analysis is rather limited at this time. Perhaps the best approach is to use a minicomputer-based system. This provides the most flexibility in pretreatment of the data, in carrying out the correlation operation, and in interpreting the results as all these may be performed under program control. All that is required is that the computer be interfaced to an ADC capable of acquiring the signals of interest.

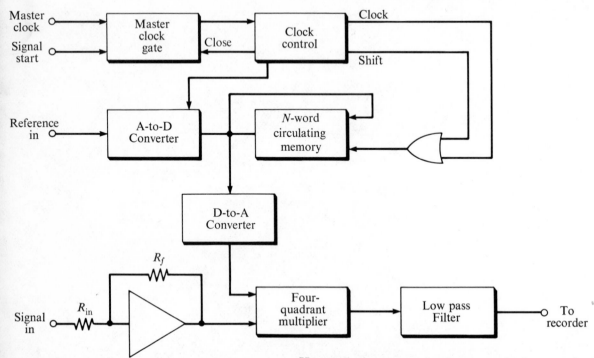

Fig. 4–121   Hardware cross-correlator.

However, the recent developments in large scale integrated circuits,
particularly shift registers and random access memories, have allowed the
development of stand-alone hardware correlators that are attractive alterna-
tives to minicomputer-based systems. One of the main advantages of the
hardware instruments, as was the case with multichannel averaging, is speed.
Sampling intervals and hence τ intervals of 0.2 µsec are possible. The main
limit is the number of points that may be calculated. At present this number
is less than 500, but this should rapidly improve with the availability of
1024-bit shift registers and RAM's.

A simple approach to a hardware correlator is depicted in Fig. 4–121.
This circuit is analogous, to some extent, to that of the self-scanning boxcar
integrator. The reference waveform is first read into the N-word circulating
memory via the ADC. Now a complete digitized waveform is circulated,
rather than just a single binary bit. Each start pulse initiates one complete
cycle of the circulating memory. The output of the memory is converted to
an analog signal by the DAC and multiplied by the input signal. The multiplier output is integrated with a low pass filter. After a sufficient number of
products have been integrated at one particular τ position, the clock control
sends an extra clock pulse to the memory register via the shift line to shift
the steady-state location of the waveform to the next τ location. This
continues until the complete cross-correlation pattern has been evaluated.

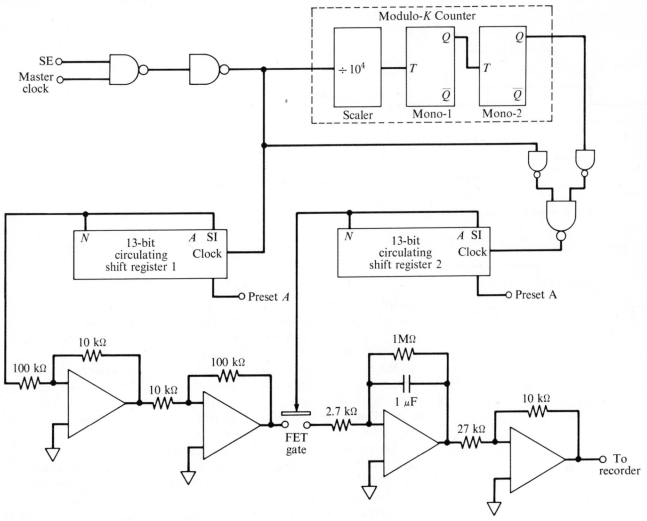

Note that for this hardware implementation the signal waveform must be repetitive.

This limitation is easily overcome by adding a second N-word circulating memory to store the signal waveform, which is then circulated in synchronism with the reference waveform memory. This second memory would also have to have its own DAC.

A correlator that is capable of evaluating the cross-correlation and auto-correlation functions for some simple waveforms is shown in Fig. 4–122. The waveforms are simple sequences of binary ones and zeros that are stored in two circulating shift registers. The signal waveform is stored in register 1 and the reference waveform in register 2. There are no external

Fig. 4–122 Simple correlator for binary waveforms.

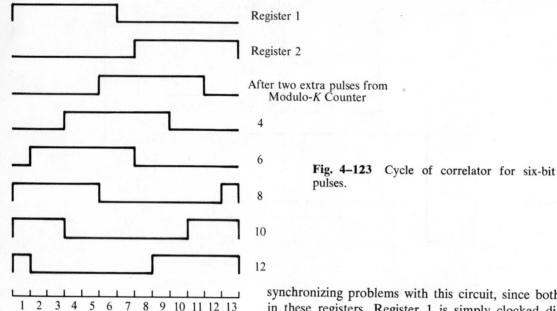

Register 1

Register 2

After two extra pulses from
    Modulo-K Counter

4

6

8

10

12

1 2 3 4 5 6 7 8 9 10 11 12 13

Fig. 4–123 Cycle of correlator for six-bit pulses.

synchronizing problems with this circuit, since both waveforms are stored in these registers. Register 1 is simply clocked directly from the master clock. Register 2's operation is very similar to that of the register in the digitally-based scanning boxcar integrator.

The relative position of the two signals must be sequenced in order to carry out the correlation over a range of τ values. This is accomplished by providing an extra clock pulse to register 2 after a specific number of clock pulses. This extra clock pulse is generated by the modulo-K counter and monostables 1 and 2. Typical values for K are 10^3 or 10^4.

The output of the signal register (register 1) is applied to an OA and then to a FET gate. The output of the FET gate is averaged, by a low pass filter, and the final output is recorded directly on a recorder. The output of the reference register (register 2) controls the FET gate which performs the multiplication step of the cross-correlation operation. A four quadrant multiplier is not necessary in this case because the reference waveform is always a unipolar binary signal.

This circuit is quite useful for studying some of the basic properties and results of simple correlations. A classic example of correlation is the auto-correlation of a rectangular pulse, as shown in Fig. 4–84 of Section 4–5.1. This auto-correlation can be carried out using this circuit. If a six-bit pulse is loaded into each 13-bit register, their relative positions at some time during the clocking and shifting of the register will be as shown in traces 1 and 2 of Fig. 4–123. The initial relative positions are not important, as the circuit will keep recycling the auto-correlation pattern on the recorder. The integrated product for traces 1 and 2 of Fig. 4–123 would be zero. After

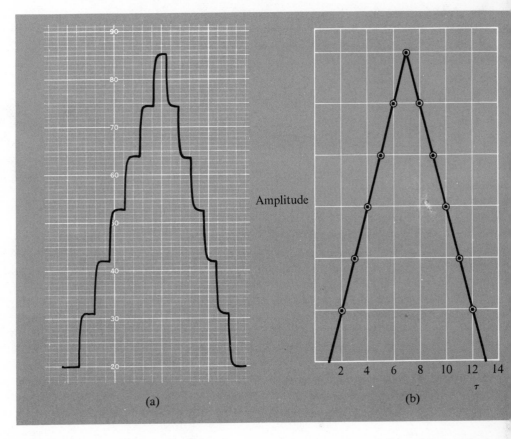

Fig. 4–124   Auto-correlation pattern output for the six-bit pulses: (a) recorder output; (b) plot of data.

two extra clock pulses from the modulo-K counter, the relative position of the pulse in register 2 would be as shown in trace 3. Additional traces are shown for 4, 6, 8, 10, and 12 modulo-K counter pulses. The integrated or averaged product for these and the other overlaps not illustrated yields the well known triangular auto-correlation function. The actual output of this correlator for these particular waveforms is shown in Fig. 4–124a, as recorded on a servo recorder. A plot of the value of the auto-correlation as a function of τ yields the expected result as is shown in Fig. 4–124b.

The cross-correlation operation, as mentioned earlier, is effective in aiding signal detection. The correlation pattern for two triplet pulse sequences (Fig. 4–125a) is shown in Fig. 4–125b. This correlation is similar to that discussed in connection with Fig. 4–118. It is important to note that this signal detection cross-correlation is also very effective in low S/N measurement situations. An oscilloscope trace of the triplet pulse sequence from register 1 (Fig. 4–122) is shown in Fig. 4–126a. Random noise was added to this signal at OA1 (see Fig. 4–122), and the resulting signal is shown in

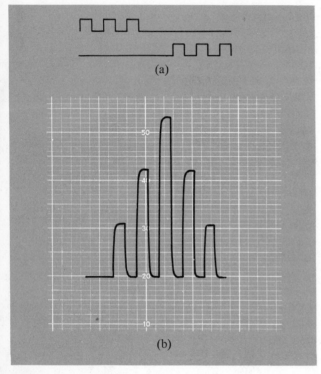

Fig. 4–125 Correlator output (b) for the auto-correlation of a triplet pulse sequence (a).

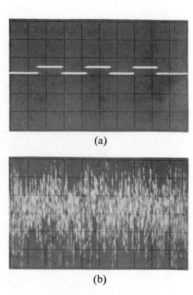

Fig. 4–126 Oscilloscope trace of triplet pulse sequence (a) and triplet pulse sequence in noise (b).

Fig. 4–126b. The cross-correlation pattern for this signal and the triplet pulse sequence in register 2 is shown in Fig. 4–127a. The characteristic pattern for the triplet sequence is easily identified.

It is interesting to note that if only one bit is set in register 2, this circuit simulates the action of the scanning boxcar integrator. The result of cross-correlating the noisy signal of Fig. 4–126b with a single pulse circulating in register 2 is shown in Fig. 4–127b. Note that, if the goal of the measurement was to detect the triplet pulse sequence in the noisy signal, the triplet–triplet cross-correlation pattern would be more effective than the boxcar approach because the central peak of the triplet–triplet pattern is the summation of all the information in the signal about the presence of the triplet.

It should be noted that the correlation of binary waveforms (no noise added) can be evaluated with a circuit that is a bit simpler. The input amplifiers and the FET gate can be replaced by an AND gate. This circuit modification is shown in Fig. 4–128.

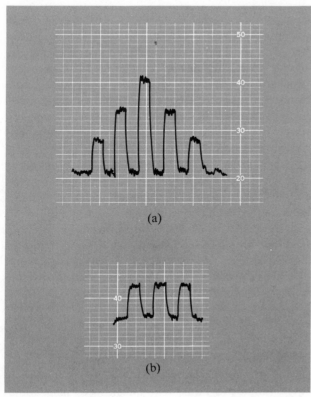

(a)

(b)

Fig. 4–127 Correlator output for noisy triplet pulse signal shown in Fig. 4–126b. (a) triplet pulse correlation; (b) single pulse correlation.

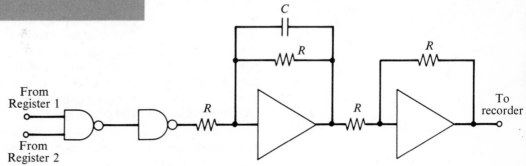

Fig. 4–128 AND sum correlator circuit for binary waveforms.

Experiment 4–19 *Correlator*

A correlator is constructed that is capable of evaluating the cross-correlation and auto-correlation functions for some simple waveforms. The waveforms are simple sequences of binary ones and zeros that are stored in circulating shift registers.

4–5.7 COMPUTER-BASED TECHNIQUES

The modern minicomputer is rapidly becoming a common laboratory instrument. Coupled to devices such as ADC's and DAC's the minicomputer can be used to perform a rather large number of data processing operations since data can be manipulated and acquired under software control. Essentially all of the hardware-based S/N enhancement techniques discussed in this section can be effectively implemented on a computer-based system.

A primary advantage of a computer-based system is flexibility. In many cases an S/N enhancement operation on a waveform is just a pretreatment step to subsequent data handling. If the S/N enhancement is computer based, this subsequent data handling is greatly facilitated. The main tradeoff that may be encountered is data acquisition speed, which, in many cases, is not a problem, particularly when compared with the added flexibility.

An inherent step in any computer-based S/N enhancement technique or signal processing technique is digitization. The signals of interest must be sampled and quantized. No amount of subsequent sophisticated data processing can make up for mistakes at this point. Thus the points discussed in Section 4–4 dealing with the accurate sampling of analog signals are pertinent to all computer-based data processing techniques.

The various computer-based approaches to S/N enhancement are too numerous to discuss in detail. It has already been mentioned that correlation techniques and multichannel averaging can be effectively implemented on computer-based systems. The boxcar integrator is also easy to implement on a computer, using quite simple programs. With a high speed sample-and-hold amplifier coupled to an ADC, the boxcar integrator approach allows the acquisition of higher speed repetitive waveforms than computer-based multichannel averaging. Also, when acquiring a relatively slow waveform with a high speed digitizer, often several quick samples are digitized and averaged to generate a single sample. This method of acquiring data simulates very closely the boxcar integrator operation.

It is also possible to simulate a lock-in amplifier in *real time* with an interfaced computer. The reference waveform generated in a lock-in amplifier measurement system can be used to trigger data acquisition of the sinusoidal carrier. If the frequency of the carrier is not too high, say less than 1000 Hz, several samples can be acquired across each half cycle of the carrier, with alternate half cycles added together with the correct polarity so as to carry out synchronous demodulation. The values for several hundred cycles would be averaged to provide a "low pass" filtered output via a DAC. Thus, an important point to appreciate in this brief discussion is that the minicomputer not only can be used to carry out S/N enhancement on stored waveforms and signals, but it can also be used to simulate, under real time operating conditions, the actual performance of techniques such as lock-in amplification and boxcar integration. This simulation can be such that, as far as the input and output terminals are concerned, the operator would find it impossible to tell whether the "lock-in amplifier" was a conventional hardware unit or a computer.

Simple analog filtering has been prevalent to some extent in all of the S/N enhancement techniques. Filtering operations may also be effectively implemented with a computer on a stored waveform. This example will now be discussed in detail, as it illustrates the versatility of a computer-based technique and it can be easily compared to the analog approach.

Digital Filtering

It has been seen, throughout this module, that control of a circuit's or signal's bandwidth is of paramount importance in optimizing the measurement of a signal. Bandwidth control was implemented using frequency selective active or passive networks of resistors, capacitors, and inductors. However, results were often somewhat less than desirable because of undesirable attenuation and phase shift of signal frequencies, which results in distortion of signal information or less than optimal enhancement of the S/N. Digital filtering can overcome many of these problems.

Digital filtering simply involves bandwidth control using software. A particularly powerful and informative route for the implementation of a digital filter is through the use of Fourier transforms. It has been stressed in this module that it is important to have a knowledge of the frequency composition of a signal. The frequency composition can be determined from the amplitude–time waveform by carrying out a Fourier transformation. The action of an analog filter on an amplitude–time waveform can be accurately described by multiplying the frequency spectrum of the waveform by the frequency response curve for the filter. This same process can be carried out on the computer after calculation of the Fourier transform of the signal waveform. However, essentially any desired frequency response curve can be set up, including many that simply would be impossible to design with hardware. Filters with no phase shift, filters with square cutoffs, high pass filters, differentiating filters, and unique discrete frequency filters are all easy to implement.

A simple example is presented in Fig. 4–129. A noisy signal is shown in Fig. 4–129a. The real output of the Fourier transform is shown in Fig. 4–129b. This plot is analogous to the amplitude spectra presented in Fig. 4–8b and can be thought of as one component of the amplitude spectrum. The important point is that the distribution of signal and noise information in this plot is the same as it was in the amplitude spectra with the signal information concentrated near the origin and the noise information spread throughout. A digital low pass filter is shown in Fig. 4–129c. This

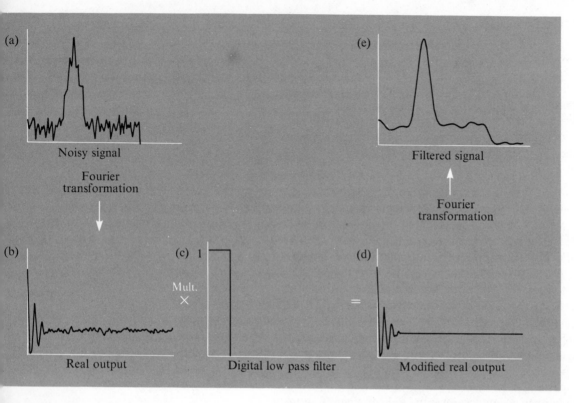

(a) Noisy signal

Fourier transformation

(b) Real output

(c) Digital low pass filter

Mult. ×

=

(d) Modified real output

(e) Filtered signal

Fourier transformation

Fig. 4–129 Digital filtering.

filter has the characteristics of no phase shift and an abrupt cutoff, which are characteristics impossible to achieve with analog filters. Filtering is implemented simply by multiplying the real output by this filter response function resulting in the modified real output shown in Fig. 4–129d. The filtered signal can be regenerated by inverse Fourier transformation. The result is shown in Fig. 4–129e. Note that the noise level has been significantly reduced. Analogous reduction of the noise level with analog techniques would have been difficult to achieve without distorting the signal information. Distortions can result from the digital filter if signal information is also attenuated by the filter. Simple precautions can make errors from this negligible.

A pictorial comparison of an *RC* low pass filter response to that of the simple digital low pass filter is shown in Fig. 4–130. Note that this is a linear plot as compared to the Bode plots presented earlier. The 3 dB point was set at the cutoff of the digital filter.

Although this example is quite simple, the basic power and flexibility of this approach to filtering should be obvious. Many unique and useful filters can be designed and implemented with this approach. It is also important

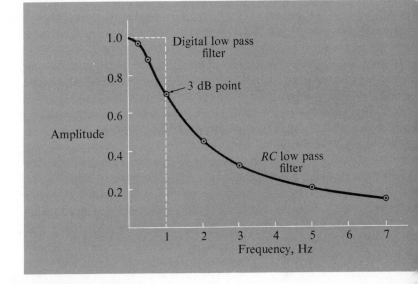

Fig. 4–130 Comparison of the frequency response of a digital software filter to a low pass filter.

to note that digital filtering as discussed in this section is exactly analogous to the moving average smoothing operations which are frequently utilized to enhance the S/N of signals. In this approach the signal is cross-correlated with a smoothing function. The smoothing function may not be any more complicated than a rectangular gate and the cross-correlation operation would essentially simulate a scanning boxcar integrator. A large number of more complex smoothing functions have been developed, among the most popular being those of Savitzky and Golay (see the Bibliography). In any case, no matter what shape of smoothing function is utilized, an equivalent digital filter can always be found simply by taking the Fourier transform of the smoothing function.

Grounding and Shielding   Appendix A

When measuring low level signals, problems are frequently encountered which can be traced to improper grounding, poor choice of input amplifier, and improper or inadequate shielding. Such problems are often difficult to assess and their elimination remains somewhat of an art. However, some basic system interconnection guidelines are developed in this appendix which, when followed, will minimize the occurrence of grounding and shielding problems.

Grounding

Voltage is not an absolute quantity but is the potential difference between two points. In order to establish and maintain reproducible and safe voltages in a circuit, a stable reference point from which all voltages are measured must be established. This single stable reference point is called the circuit common. When a circuit is linked to other circuits in a measurement system, the commons of the circuits are often connected together to provide the same common for the complete system. The circuit or system common may also be connected to the universal common, earth ground, by connection to a ground rod, water pipe, or power-line common (see Note 16, Module 1). The terms ground and common are often used interchangeably but are usefully distinguished as described above.

Most of the problems with grounds or commons arise because *two separate commons or earth grounds are seldom, if ever, at the same voltage.* It is quite possible for two commons in the same rack of electronics to be at different voltages; and any time a signal source is somewhat remote from the input amplifier, as is often the case, it can almost be guaranteed that the signal source common will be at a different voltage than the measurement system common.

A simple voltage signal source is shown in Fig. A–1a connected to an OA voltage measurement circuit. The signal common C1 is at a different

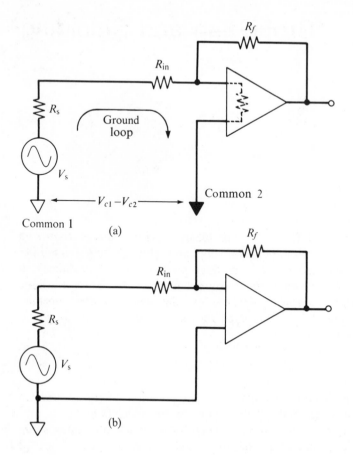

Fig. A–1 (a) Ground loop resulting from different commons. (b) Elimination of the ground loop by establishing a single common.

voltage than the amplifier common $C2$, and thus a ground loop is present which can give rise to an erroneous signal. This can be particularly troublesome if the two commons are unstable with respect to each other.

Ground loops can be eliminated by connecting all commons to a single point as shown in Fig. A–1b. However, it is important that the connections to common have very low resistance and high current carrying capacity so that ohmic drops along the connections are minimized. Typically a large copper wire or foil is used. This is particularly important if a number of connections to a single common point are made and if some of the connections are long, as they would be when the signal source must be remote from the measurement circuits. Even so, at RF frequencies the resistance is increased by the skin effect, and inductive reactance can be very large.

While only two components of a measurement system (signal source and input amplifier) are shown in Fig. A–1, a single common should be established for all circuits in the measurement system in order to eliminate

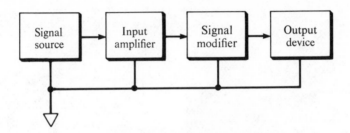

Fig. A–2 Single common for a measurement system.

and minimize ground loops. This is shown in Fig. A–2. In complex systems the necessity of low resistance to ground or common is very important, since ultimately a single connection must carry the sum of all the currents from every component in the system. It may in fact become impractical to have a single common point because the current carrying capacity cannot be provided. In this case it may be safer to have several stable grounds and tolerate some ground loops. This sort of compromise is often necessary in solving the ground problems associated with large installations and buildings, such as a computer center, or when the circuitry is subjected to interference which may cause large currents, such as interference from electrical storms.

However, even in laboratory measurement situations it may not be possible or practical to have a single common point, particularly if the signal source is remote from the measurement system. In these cases it is advantageous to use a differential input instrumentation amplifier as discussed in Section 3–2.1 of Module 3. The simple signal source of Fig. A–1 can be measured using an instrumentation amplifier, as shown in Fig. A–3.

Fig. A–3 Instrumentation amplifier used to cancel out the effect of ground loops.

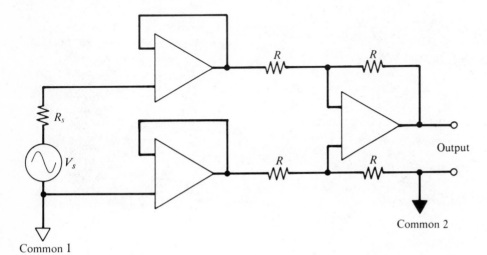

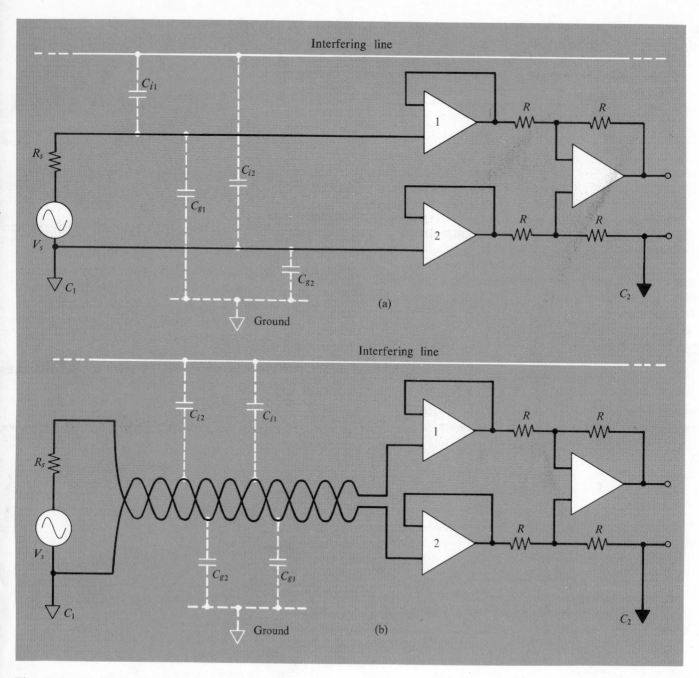

Fig. A–4   (a) Capacitive pickup on input signal lines. (b)
Equalization of pickup using a twisted wire signal pair.

Now, even though there is a difference in potential between the signal common and the amplifier common, the erroneous signals generated by the ground loops are common mode (common to both inputs) and as such are rejected by the differential amplifier. Therefore, it is unnecessary for the two common voltages to be stable with respect to each other. Note also that the input impedance of the instrumentation amplifier must be large with respect to the source resistance; otherwise, the common mode rejection ratio (CMRR) will be degraded, since the two lines are not identical. Better common mode rejection (CMR) is achieved when the source is balanced, as is the case with a Wheatstone bridge circuit.

Shielding

High CMR depends on the equality of the two input lines to the instrumentation amplifier. As mentioned above, a finite source resistance can create an imbalance. It is also possible to pick up interfering signals on the input lines as a result of capacitive coupling to a disturbing line. Ground loops may also be established by capacitive coupling to ground. These problems are illustrated in Fig. A–4a. Differences in the amount of pickup can be significantly minimized by using a twisted wire signal pair as shown in Fig. A–4b. Now the capacitive coupling to the disturbing line and ground is approximately equal in both lines and high CMR is maintained.

In addition to equalizing the pickup as above, the amount of pickup can be reduced by shielding the input lines. Shielding involves surrounding the input lines with a conductor. A high quality signal cable consists of a twisted wire signal pair, a foil shield, and a copper drain wire (see Fig. A–5). The shield should be connected to the signal ground (Fig. A–6a) so that the capacitance between the shield and the signal pair does not shunt the input impedance of the differential amplifier as it does in Fig. A–6b. In addition, the shield should not be connected to both the signal and measurement system commons, since this can establish a ground loop through

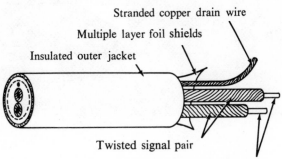

Stranded copper drain wire

Multiple layer foil shields

Insulated outer jacket

Twisted signal pair

Low resistance stranded copper conductors

Fig. A–5 High quality signal cable.

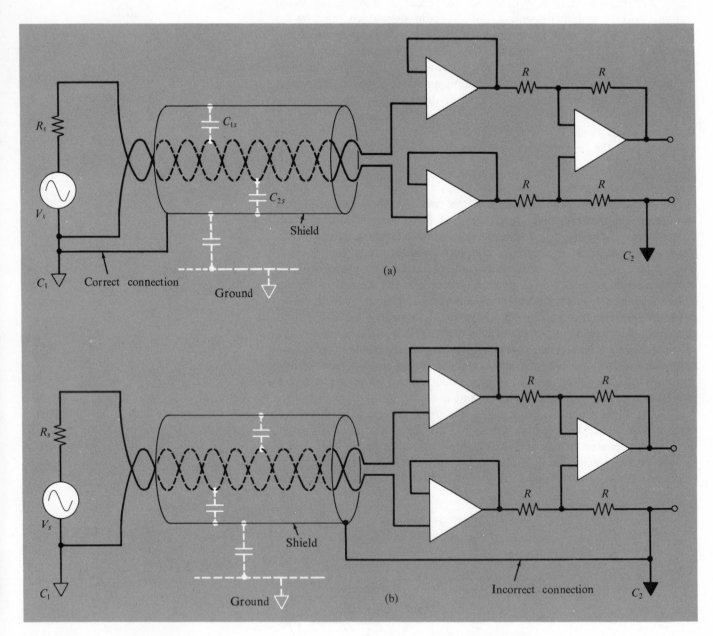

Fig. A–6 (a) Correct shield connection.
(b) Incorrect shield connection.

the shield and currents in the shield can induce currents in the signal pair via capacitive and inductive coupling. It is also possible to have a capacitively coupled ground loop to the shield which can in some cases result in induced currents in the signal pair.

Some instrumentation amplifiers are equipped with an internal floating shield which surrounds the input section. This floating shield should be connected to the shield on the input twisted wire signal pair, which in turn is connected to the signal common.

While signal lines are shielded as shown in Fig. A–5, instruments are shielded by their metal enclosures and chassis. In general, it is best if all shields are connected to the signal common, and no measurement system common is connected to the shield system except at the signal common. The arrangement is shown in Fig. A–7. Incorrect arrangements are shown in Fig. A–8. Since all the shields are capacitively coupled to ground, connecting a system component common to its shield can result in a ground loop as shown in Fig. A–8a. Also a system component common can be capacitively coupled to its shield, and connecting its shield to ground can result in ground loops both through the shield and the signal cables.

Most of the considerations in this appendix concern analog signals with frequencies that are not very large, certainly less than 1 MHz. At higher frequencies and with most digital signals, coaxial cable is often used. Considerations with respect to digital signal transmission are discussed in Appendix C and Section 3–4.2 of Module 3.

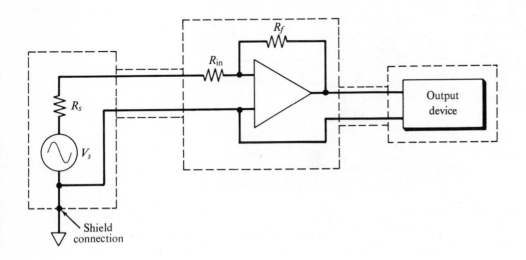

Fig. A–7 Shield around a measurement system.

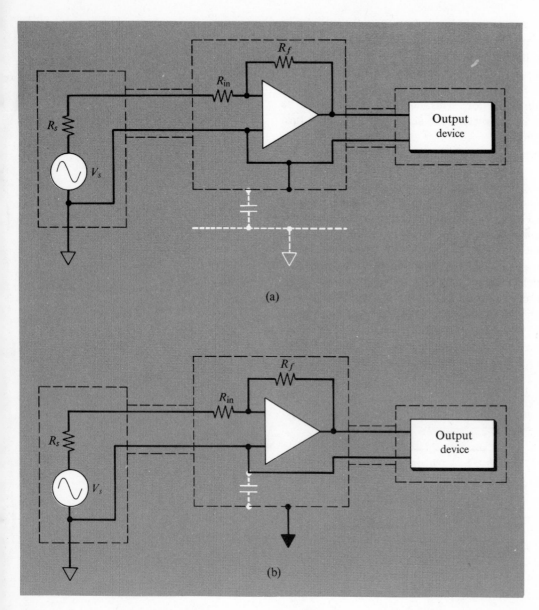

(a)

(b)

Fig. A–8 Two incorrect shield connections.

Isolation

Occasionally it is desirable to isolate one circuit from another. For ac signals below about 5 MHz an isolation transformer can be used, as shown in Fig. A–9. At higher frequencies stray capacitance in the transformer makes the isolation ineffective.

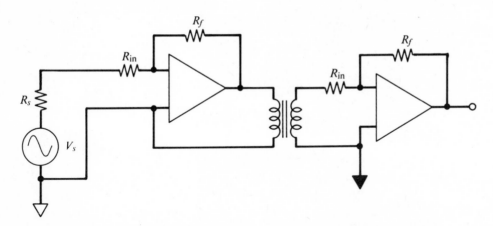

Fig. A–9 Isolation transformer.

For digital signals excellent isolation can be achieved using opto-isolators. These consist of a light source–detector pair which can couple binary signals. Typical light sources are tungsten bulbs, neon bulbs, and LED's. Photoconductors, phototransistors, or photodiodes are used as detectors.

RF Shielding

High frequency interference in circuits is frequently referred to as RF (radio frequency) interference. Numerous sources of RF interference can be found in laboratory environments. Spark sources, flash lamps, and gaseous discharges for lasers are but a few. RF interference can be quite serious, rendering many digital circuits completely inoperable.

Enclosing the circuit in a metal shield and using shielded cable can provide RF shielding. A conductor that has a high surface area (mesh or braid) makes an excellent RF ground. However, one main requirement of an RF shield clashes with that of the shield depicted in Fig. A–6a. The shield should be terminated at both ends, like the termination of a signal cable for high frequency signals (see Appendix C and Section 3–4.2 of Module 3). Thus for best shielding two separate shields should be used, since the desired features for RF and low frequency shields are incompatible. One shield should be as shown in Fig. A–6a and a second RF shield terminated at both ends can be used around this first shield. The termination prevents reflections of RF in the shield.

Bibliography

Bracewell, Ron, *The Fourier Transform and Its Application,* McGraw-Hill, New York, N.Y., 1965.

A classic reference work on the Fourier transform.

Buus, R. G., "Electrical Interference," in *Design Technology,* Vol. I, Prentice-Hall, Englewood Cliffs, N.J., 1970, p. 381.

An excellent discussion covering topics such as electromagnetic shielding, component interference reduction, interference reduction in cables and interconnections, and grounding techniques.

Carlson, A. Bruce, *Communication Systems: An Introduction to Signals and Noise in Electrical Communication,* McGraw-Hill, New York, N.Y., 1968.

A modern text on communication systems with excellent coverage of Fourier transform concepts as applied to signals, modulation, demodulation, and sampling.

Cordos, E., and Howard V. Malmstadt, "Dual Channel Synchronous Integration Measurement System for Atomic Fluorescence Spectrometry," *Analytical Chemistry,* **44**, 2277 (1972).

The synchronous integration measurement system can accurately subtract background and also average noise over a wide frequency spectrum.

Hieftje, G. M., "Signal-to-Noise Enhancement Through Instrumental Techniques. Part I. Signals, Noise, and S/N Enhancement in the Frequency Domain," *Analytical Chemistry,* **44**, No. 6, May, 1972, p. 81A; "Part II. Signal Averaging, Boxcar Integration, and Correlation Techniques," *Analytical Chemistry,* **44**, No. 7, June, 1972, p. 69A.

These two articles provide a good introduction to modern hardware-based signal processing techniques.

Horlick, Gary, "Digital Data Handling of Spectra Utilizing Fourier Transformations," *Analytical Chemistry,* **44**, 943 (1972).

Smoothing, differentiation, and resolution enhancement of spectra using Fourier transforms are described. A discussion of the distribution of spectral information in the Fourier domain is included.

Horlick, Gary, "Detection of Spectral Information Utilizing Cross-Correlation Techniques," *Analytical Chemistry,* **45,** 319 (1973).

The application of cross-correlation techniques to the detection of a single spectral peak in a noisy base line and the detection of complex spectral features is discussed and illustrated.

Kelly, P. C., and Gary Horlick, "Practical Considerations for Digitizing Analog Signals," *Analytical Chemistry,* **45,** 518 (1973).

The effects of sampling interval, sampling duration, quantization, digitization time, aperture time, and jitter are examined. Quantitative error criteria for sampling common peaklike signals are given.

Malmstadt, H. V., and C. G. Enke, *Digital Electronics for Scientists,* W. A. Benjamin, Menlo Park, Calif., 1969.

A treatment of digital techniques directed toward instrumentation applications.

Malmstadt, H. V., C. G. Enke, and E. C. Toren, Jr., *Electronics for Scientists,* W. A. Benjamin, Menlo Park, Calif., 1962.

An earlier work containing several still-relevant sections on analog measurement techniques and devices.

Morrison, Ralph, *Grounding and Shielding Techniques in Instrumentation,* Wiley, New York, N.Y., 1967.

A comprehensive treatment of grounding and shielding problems. Electrostatics, shielding, differential amplifiers, bridge systems, and magnetic and RF processes in instrumentation are among the topics discussed.

Savitzky, A., and Marcel J. E. Golay, "Smoothing and Differentiation of Data by Simplified Least Squares Procedures," *Analytical Chemistry,* **36,** 1627 (1964).

A classic paper on smoothing using weighted moving averages.

Schuartz, Mischa, *Information Transmission, Modulation, and Noise,* McGraw-Hill, New York, N.Y., 1970.

A general text on modern communication systems. Contains a brief but excellent discussion of equivalent bandwidth and the bandwidth–time inverse relationship.

Tobey, G. E., J. G. Graeme, and L. P. Huelsman (Eds.), *Operational Amplifiers,* McGraw-Hill, New York, N.Y., 1971.

A comprehensive treatment of the design and applications of operational amplifiers. The applications discussion includes active filters, modulation, and demodulation.

Solutions to Problems

SOLUTIONS TO PROBLEMS IN SECTION 4–1

1. a)
$$F(f) = 2\int_0^t f(t) \cos 2\pi f t \, dt$$

$$f(t) = \cos 2\pi f' t$$

$$\therefore F(f) = 2\int_0^t \cos 2\pi f' t \cos 2\pi f t \, dt$$

From integral tables,

$$\int \cos(mx)\cos(nx)\,dx = \frac{\sin(m-n)x}{2(m-n)} + \frac{\sin(m+n)x}{2(m+n)}.$$

In our specific problem,

$$x = t, \qquad m = 2\pi f', \qquad n = 2\pi f.$$

$$\therefore F(f) = \frac{\sin 2\pi(f'-f)t}{2\pi(f'-f)} + \frac{\sin 2\pi(f'+f)t}{2\pi(f'+f)}$$

For positive frequencies only the first term is significant.

$$\therefore F(f) = \frac{\sin 2\pi(f'-f)t}{2\pi(f'-f)} \Bigg|_0^t$$

b) The above function is plotted in Fig. S–1 for $f' = 100$ Hz and $t = 1$ sec, and in Fig. S–2 for $f' = 100$ Hz and $t = 0.1$ sec. Note that only the axes are different.

c) The power spectrum $[F^2(f)]$ is plotted in Fig. S–3 for $f' = 100$ Hz and $t = 1$ sec.

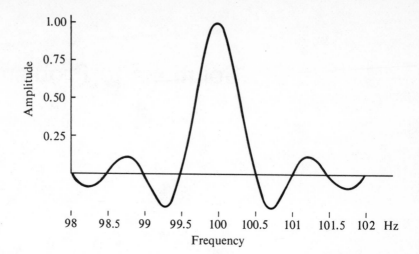

Fig. S–1 Fourier transform of cos $2\pi f't$ for $f' = 100$ Hz and $t = 1$ sec.

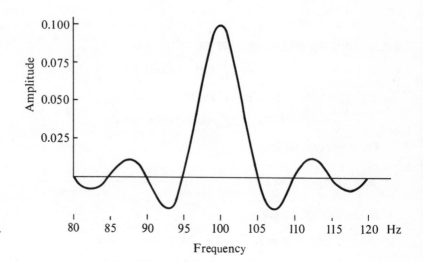

Fig. S–2 Fourier transform of cos $2\pi f't$ for $f' = 100$ Hz and $t = 0.1$ sec.

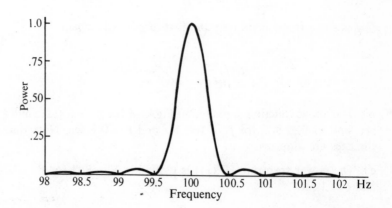

Fig. S–3 The power spectrum $[F^2(f)]$ for $f' = 100$ Hz and $t = 1$ sec.

d) When t gets large the spectrum is narrow, indicating that the longer cosine wave is composed of essentially a single frequency. When t is small the spectrum is quite broad.

2. Below is the complete table of values. The values are calculated using the equation $A \cos 2\pi f t$, where A is the relative amplitude, f the frequency, and t the time.

Frequency, Hz	Time, sec								
	−2.0	−1.5	−1.0	−0.5	0	0.5	1.0	1.5	2.0
0	0.52	0.52	0.52	0.52	0.52	0.52	0.52	0.52	0.52
0.125	0.00	0.38	0.71	0.92	1.00	0.92	0.71	0.38	0.00
0.250	−0.84	−0.59	0.00	0.59	0.84	0.59	0.00	−0.59	−0.84
0.375	0.00	−0.58	−0.44	0.24	0.63	0.24	−0.44	−0.58	0.00
0.500	0.44	0.00	−0.43	0.00	0.43	0.00	−0.43	0.00	0.44
0.625	0.00	0.24	−0.18	−0.10	0.26	−0.10	−0.18	0.24	0.00
0.750	−0.14	0.00	0.00	−0.10	0.14	−0.10	0.00	0.10	−0.14
0.875	0.00	−0.03	0.05	−0.06	0.07	−0.06	0.05	−0.03	0.00
1.000	0.03	−0.03	0.03	−0.03	0.03	−0.03	0.03	−0.03	0.03
Σ	0.01	0.01	0.26	1.98	3.92	1.98	0.26	0.01	0.01
$3.92 \exp[-4(\ln 2)t^2]$	0.00	0.00	0.007	1.96		1.96	0.007	0.00	0.00

3. $\bar{v}_{\text{rms}} = (4kTR\,\Delta f)^{1/2}$

$\qquad = (4 \times 1.38 \times 10^{-23} \times 300 \times 100 \times 10^6 \times 10)^{1/2}$ V

$\qquad = 4.069 \times 10^{-6}$ V

4. Ave = 57.00, $\sigma = 8.6$,   S/N = 6.6.

From Fig. S–4, max $\simeq$ 67, min $\simeq$ 45.

$\therefore \dfrac{\text{max} - \text{min}}{5} = \dfrac{67 - 45}{5} = 4.4 \simeq \bar{v}_{\text{rms}}$

$\therefore$ S/N = 57/4.4 = 13

Note: $\sigma^2 = \dfrac{n\,\Sigma x_i^2 - (\Sigma x_i)^2}{n(n-1)}$

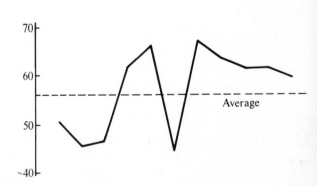

Fig. S–4 Plot of values for Problem 4 in Section 4–1.

SOLUTIONS TO PROBLEMS IN SECTION 4–2

1. a)
$$v_o = Av_{in} - IR_o$$

According to Eq. (4–17), $v_{in} = v_{sig} + \beta v_o$.

$$\therefore v_o = A\ (v_{sig} + \beta v_o) - IR_o$$

$$v_o = Av_{sig} + \beta Av_o - IR_o$$

$$v_o\ [1 - \beta A] = Av_{sig} - IR_o$$

$$v_o = \frac{Av_{sig}}{1 - \beta A} - \frac{IR_o}{1 - \beta A}$$

This final equation describes the output v_o of the amplifier when feedback is present, and is analogous to the starting equation for the output when no feedback was present. Note that the gain of the amplifier is $1/(1 - \beta A)$ of what it was without feedback (same results as before) and that the effective output impedance R_o is also $1/(1 - \beta A)$ of what it was without feedback. This reduction of the output impedance is another desirable characteristic of negative feedback.

b) Equation (4–17) is $v_{in} = v_{sig} + \beta v_o$. Rearranging yields

$$v_{sig} = v_{in} - \beta v_o.$$

Assuming no load is connected or that the IR drop across the output is very small:

$$v_o = Av_{in}$$

$$\therefore v_{sig} = v_{in} - \beta Av_{in}$$

$$v_{sig} = v_{in}\ [1 - \beta A]$$

Dividing by the input current i_{in} yields

$$\frac{v_{sig}}{i_{in}} = \frac{v_{in}}{i_{in}}\ [1 - \beta A]$$

or

$$R_{sig} = R_{in}\ [1 - \beta A].$$

Thus the input impedance of the amplifier with feedback R_{sig} is increased over that of the amplifier without feedback R_{in} by the factor $[1 - \beta A]$.

Therefore, in addition to achieving gain stabilization with negative feedback, the input impedance is increased and the output impedance is decreased, both of which are desirable in a voltage amplifier.

2. Equation (4–13) can be written

$$20 \log (\text{gain}) = -20 \log f - 20 \log 2\pi RC.$$

If the gain decreases 6 dB, the frequency must increase to a new value f':

$$20 \log (\text{gain}') = -20 \log f' - 20 \log 2\pi RC.$$

Subtracting these two equations must give a 6 dB difference.

Thus
$$6 = -20 \log f + 20 \log f'$$

$$20 \log \frac{f'}{f} = 6$$

$$\log \frac{f'}{f} = 0.3$$

$$\frac{f'}{f} = 1.995.$$

Thus decreasing the gain 6 dB results in a factor-of-two increase in the frequency on the Bode plot.

3. Voltage gain $v_{\text{out}}/v_{\text{in}}$ expressed in dB is

$$20 \log \frac{v_{\text{out}}}{v_{\text{in}}} = 20 \log (\text{gain}) \text{ in dB.}$$

If the gain decreases 6 dB, the new gain becomes

$$20 \log (\text{gain}') \text{ in dB.}$$

Subtracting these two equations must give a 6 dB difference.

Thus
$$6 = 20 \log (\text{gain}) - 20 \log (\text{gain}')$$

$$0.3 = \log \frac{(\text{gain})}{(\text{gain}')}$$

$$\frac{\text{gain}}{\text{gain}'} = 2.$$

Therefore, a 6 dB decrease in gain is equivalent to reducing the linear gain by a factor of two.

Another way to see this is to construct a table as follows:

Gain, dB	Gain (v_{out}/v_{in})
0	1.00
6	1.995
12	3.98
18	7.94

SOLUTIONS TO PROBLEMS IN SECTION 4–3

1. $M_c(t) = [1 + A_s \cos (2\pi f_s t)] A_c \cos (2\pi f_c t)$

    $M_c(t) = A_c \cos (2\pi f_c t) + A_s A_c \cos (2\pi f_s t) \cos (2\pi f_c t)$

    $M_c(t) = A_c \cos (2\pi f_c t) + \dfrac{A_c A_s}{2} \cos 2\pi (f_c - f_s)t$

    $\qquad + \dfrac{A_c A_s}{2} \cos 2\pi (f_c + f_s)t$

    The last step requires the trigonometric identity

    $$\cos \alpha \cos \beta = \tfrac{1}{2}[\cos (\alpha + \beta) + \cos (\alpha - \beta)].$$

    Negative frequency terms are avoided.

2. $D_c(t) = [1 + A_s \cos (2\pi f_s t)] A_c^2 \cos^2 (2\pi f_c t)$

    $D_c(t) = A_c^2 \cos^2 (2\pi f_c t) + A_s A_c^2 \cos^2 (2\pi f_c t) \cos (2\pi f_s t)$

    With the trigonometric identity $\cos^2 \alpha = (\cos 2\alpha + 1)/2$,

    $$D_c(t) = \frac{A_c^2}{2} \cos (4\pi f_c t) + \frac{A_c^2}{2} + \frac{A_s A_c^2}{2} \cos (2\pi f_s t)$$

    $$+ \frac{A_s A_c^2}{2} \cos (4\pi f_c t) \cos (2\pi f_s t).$$

    With the trigonometric identity $\cos \alpha \cos \beta = \tfrac{1}{2}[\cos (\alpha + \beta) + \cos (\alpha - \beta)]$,

    $$D_c(t) = \frac{A_c^2}{2} + \frac{A_s A_c^2}{2} \cos (2\pi f_s t) + \frac{A_c^2}{2} \cos (4\pi f_c t)$$

    $$+ \frac{A_s A_c^2}{4} \cos 2\pi (2f_c - f_s)t +$$

    $$\frac{A_s A_c^2}{2} \cos 2\pi (2f_c + f_s)t.$$

SOLUTION TO PROBLEM IN SECTION 4–4

1. The sampling rates for 1% maximum error in the peak height as taken from Table 4–2 are:

Gaussian	2.2 samples/sec
Lorentz	3.6 samples/sec
Exponential	50 samples/sec

The Gaussian peak will fall to 0.01 of its maximum value in a time t, which can be calculated by solving the following equation:

$$0.01 = \exp\left[-4(\ln 2)t^2\right]$$
$$-4.605 = -2.773t^2$$
$$t^2 = 1.661$$
$$t = 1.289$$

Therefore, the total time over which samples must be taken is 2.578 sec (both sides of the peak) and the total number of samples should be 5.67.

For the Lorentz peak,

$$0.01 = (1 + 4t^2)^{-1}$$
$$0.01 = \frac{1}{1 + 4t^2}$$
$$0.01 + 0.04t^2 = 1$$
$$0.04t^2 = 0.99$$
$$t^2 = 24.75$$
$$t = 4.97.$$

Therefore, the total time for the Lorentz peak is 9.94 sec and a total of 35.8 or 36 samples should be taken.

For the exponential peak,

$$0.01 = \exp\left[-2(\ln 2)|t|\right]$$
$$-4.605 = -1.386t$$
$$t = 3.323.$$

Therefore, the total time for the exponential peak is 6.645 sec and 332 samples should be taken.

Index

For definitions of digital terms, see also pp. 693–701.

SELECTED PHYSICAL CONSTANTS

Quantity	Symbol	Value
Electron charge	Q_e	1.603×10^{-19} C
Faraday's constant	$\mathcal{F}$	96487.0 C equiv^{-1}
Gas constant	R	$\begin{cases} 8.32 \text{ J mole}^{-1} \,^{\circ}\text{K}^{-1} \\ 0.0821 \text{ liter atm mole}^{-1} \,^{\circ}\text{K}^{-1} \end{cases}$
Planck's constant	h	6.63×10^{-34} J-sec
Boltzmann's constant	k	1.38×10^{-23} J $^{\circ}$K^{-1}
Ice point	T_0	$273.15\,^{\circ}$K

TABLE OF PREFIXES FOR UNITS

Order	Prefix	Symbol
10^{12}	tera	T
10^{9}	giga	G
10^{6}	mega	M
10^{3}	kilo	k
10^{2}	hecto	h
10	deka	da
10^{-1}	deci	d
10^{-2}	centi	c
10^{-3}	milli	m
10^{-6}	micro	μ
10^{-9}	nano	n
10^{-12}	pico	p
10^{-15}	femto	f
10^{-18}	atto	a